A Natural Calling

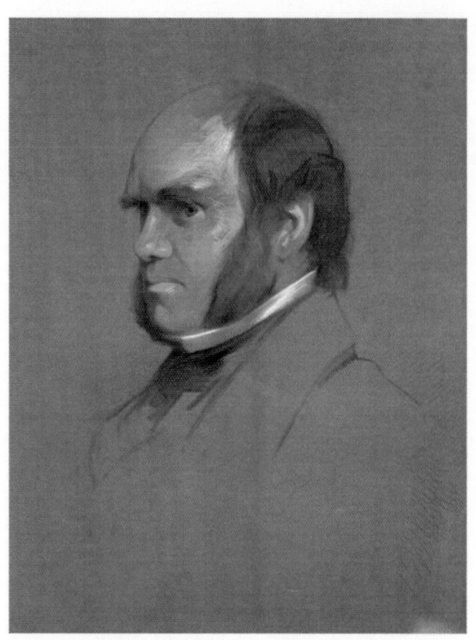

Charles Darwin in 1849. Pastel Drawing by Samuel Laurence. Copyright: English Heritage Photo Library, with permission.

Daguerrotype of William Darwin Fox. Taken in 1848 by William Kilburn, London. Copyright: Gerard Crombie and the Sydics of the University of Cambridge, with permission.

Anthony W. D. Larkum

A Natural Calling

Life, Letters and Diaries of Charles Darwin and William Darwin Fox

Prof. Anthony W. D. Larkum
University of Sydney
School of Biological Sciences
Sydney NSW 2006
Australia
alark@mail.usyd.edu.au

For permission to reproduce transcripts of the letters and diaries, the following are acknowledged and thanked: the Master and Fellows of Christ's College, Cambridge; the Syndics of Cambridge University Library for material from the University Library (Darwin Archive – DAR); the Editors of the Darwin Correspondence Project for access to their transcripts of Darwin/Fox letters; the Woodward Library, University of British Columbia; the American Philosophical Society, Philadelphia; the Dibner Library of the History of Science and Technology, Smithsonian Institution, Washington; The Sandown Museum, Isle of Wight, UK; and Mr Gerard Crombie, Norwich, UK.

For permission to reproduce drawings, paintings and photographs, the following are acknowledged and thanked: the Master and Fellows of Christ's College, Cambridge; the Syndics of Cambridge University Library for a drawing and a daguerrotype in the University Library (Darwin archive – DAR 204 and 250); the Woodward Library, University of British Columbia; The University Museum of Zoology, Cambridge; the Property of the Darwin Heirloom Trust, by permission of the Master and Fellows of Darwin College, Cambridge (Erasmus Darwin and Robert Darwin Portraits); English Heritage; and Mr Gerard Crombie, Norwich.

ISBN 978-1-4020-9232-9 e-ISBN 978-1-4020-9233-6

DOI 10.1007/978-1-4020-9233-6

Library of Congress Control Number: 2009920102

Printed on acid-free paper

9 8 7 6 5 4 3 2 1

springer.com

Preface

From 1965–1968, I held an Agricultural Research Council Research Fellowship at Christ's College, Cambridge. Later in 1981, when I was a Visiting Fellow in the Department of Biochemistry, University of Cambridge and renewed my contacts with Christ's College, my friend and colleague David Coombe, a Fellow of Christ's College, informed me that a collection of letters of Charles Darwin had just been uncovered in the Library storeroom, underneath the College. I had always maintained an interest in Charles Darwin, from the early age of thirteen, when I had first read his books, with I might say some difficulty! This collection was the 155 letters of Charles Darwin to his second cousin William Darwin Fox, which had been given in trust to the College, in 1909, by members of the Fox family at the time of the Darwin Centenary celebrations.

I was allowed access to these 155 letters and at that time made my own transcriptions. It seemed to me that this was a magnificent account of the lives of two naturalists of the nineteenth century, starting at the time that they were at Christ's together, in 1828, and going to 1880 when W D Fox died – just two years short of the death of Charles Darwin in 1882. Of course this valuable resource had not gone unnoticed before. Darwin's son, Francis Darwin had been given the letters in the 1880s, when he was preparing his *Life and Letters of Charles Darwin* in 3 volumes. This work therefore included a few of the letters, verbatim, some in an edited form and many with just a few highlights. Francis Darwin had been overwhelmed with material and so had no opportunity to go into these letters in depth and so had missed many interesting features. As far as he was concerned, and it is doubtful whether he ever met Fox or indeed knew of him before he saw the letters, Fox had been a close friend of Darwin at Christ's College, but had drifted apart thereafter, as their lives changed: Darwin to become the iconic figure of nineteenth century biology and Fox to become the country vicar that Darwin had once aspired to be. There is a grain of truth in this, they did become distanced by illness and large families, but this does not explain the extraordinary fidelity of their letter writing, with almost one letter a year between 1828 and 1878, and usually many more. Neither does it explain why biographers of Darwin have turned to these letters so frequently in their works. The answer seems to be that Darwin was able to relate much more freely to Fox, since he was a close friend from college days, a close relation, who knew all of Darwin's family, and someone who was a thoroughly competent natural historian and indeed had introduced Darwin to beetles at Cambridge. Only perhaps in his later correspondence with Joseph Dalton Hooker was Darwin able to speak so freely in writing to another soul.

However, there was an immediate problem before giving any thought to editing the work for publication. Very few letters of W D Fox to Darwin had survived; whereas Fox had been scrupulous in keeping Darwin's letters, Darwin, uncharacteristically, preserved very few of Fox's. Thus their relationship, as revealed by the letters was very much one-sided and for the project to be viable Fox needed some flesh on his bones. Unfortunately, at that time very little was known about William Darwin Fox and his family. From the *Darwin Pedigrees* (Freeman 1977), I was able to get Fox's family tree and to learn that he had sixteen children by two wives. He had been a clergyman of the Church of England, for most of his time the Rector of Delamere Parish, in Delamere Forest, Cheshire. Surely, however, more information than this remained. I then began a historical search for material related to Fox. His family had been active citizens of Derby, involved in the thriving silk trade of that city in the late eighteenth and early nineteenth century. Unfortunately, research in the Derby Local Studies Library and the Derby County Archives at Matlock, Derbyshire, revealed rather little information. So I then began a search for living family relatives, especially any who might have inherited Fox's library and letters.

Very quickly it became obvious that a search of births, marriages and deaths was unwieldy because there are just too many Foxes to search through. The search then turned to wills, first at Somerset House and then at the Principle Registry of the Family Division, High Holborn, London. The wills were very informative indeed as very often they indicated to whom family papers and books had been bequeathed. So it was not long before I had a fairly good indication of where any important papers might be located. The search narrowed to Ann (Nancy) Darwin Fox at Southwold in Suffolk. Upon 'phoning this lady of 89 years, I was instantly informed that there were indeed family papers to be had! And I arranged a visit at the first opportunity. She had a large black box, later bequeathed to her niece, Susan Crombie and Susan's husband, Col Richard Crombie, which contained a large store of William Darwin Fox's personal papers including some of his diaries. This was of course an untold horde of riches and I set to work immediately to assess the discovery. The diaries unfortunately did not cover the whole period of Fox's life; apparently the diaries had been divided up amongst his children on his death and only batches of years had survived. Nevertheless, there was enough there for a good reconstruction of his life (see Appendix 1). Eventually, I arranged with Susan and Richard Crombie and Peter Gautrey, Keeper of the Darwin Archives at the University Library, for the major part of the material to be deposited at the University Library Cambridge, where it is today.

Only a small part of the remainder of the missing material has turned up so far. This is mainly in the Woodward Library of the University of Vancouver and the strange history of this material is set own in Appendix 1. This collection includes 5 letters of the Darwin-Fox series. It also includes some fragments from diaries of years missing from the Crombie collection. And in addition, in this collection, there are a number of letters of W D Fox to other naturalists of his time, such as William

Hewitson, who wrote standard works on birds and insects in the nineteenth century. One further letter, for which facsimiles have survived, was held at the Sandown Museum, Isle of Wight. Probably one of Fox's children loaned this to the Museum, after his death at Sandown in 1880. And the letters of Darwin to Fox in the possession of the Smithsonian Museum and American Philosophical Society were no doubt passed on by Francis Darwin. I am grateful to these institutions for providing facsimiles of these letters.

During my early studies I came into contact with Fred Burckhardt, who gave me much-valued support and advice on a number of issues. Through Fred I was introduced to a constellation of Darwin scholars, including Sydney Smith, Nora Barlow, David Kohn, and many others. All of these have been vital to my development. Later I made contact with the Correspondence of Charles Darwin Project and have kept in close contact ever since. Alison Pearn gave me permission to view Darwin-Fox material, which has not yet been published in the Darwin correspondence; and since the Correspondence is at the year 1868, this means a total of 30 letters. All the rest of the letters that are published here have already been published with footnotes in the Correspondence of Charles Darwin (CCD) series. I have carefully compared my version with the CCD version. The two versions are closely comparable and any differences are set out in Appendix 6. I have formatted the letters in a slightly different way to the CCD version: ends of pages are shown in the text, as are insertions and deletions. In preparing footnotes I have followed CCD in many instances. Sometimes I followed a similar course of my own and came to similar footnotes; where I have relied on research of the CCD Project, I have always tried to point this out; another 40-50% of the footnotes, are original to my own research. The facility which was opened to me to refer to the Fox Diaries, to Fox family history and to other material, has been a great help in resolving obscure points. I also felt qualified to go a little bit further into biological points on a number of occasions. Moreover, the ability here to focus only on the Darwin-Fox letters, has provided the opportunity for a more comprehensive comparison and cross-referencing of these letters, resulting in a number of additional footnotes.

These then are the body of materials that have formed the backbone of this book. I have divided them into an introductory chapter detailing the history on the one side back to Charles Darwin's grandfather, Erasmus Darwin, and on the Fox side to William Darwin Fox's grandfather, Samuel Fox II, and beyond to his great great grandfather, Timothy Fox. The second chapter presents the first year of Fox's diary, when he was a student at Christ's College, Cambridge, studying to be a clergyman. The next six chapters present the extant letters between Darwin and Fox with detailed footnotes. The division of these letters is somewhat arbitrary. There are many ways to divide a cake! And I tried several ways to divide the letters! I have also given an introduction and conclusion to each chapter. These are to summarise the first impressions that can be gained by a close comparison of the letters against the known background of the two correspondents. These were aimed at the general reader. I avoided delving too deeply into the vast Darwin literature, as I think that the materials presented here will take some time and a number of different scholarly approaches to properly assimilate.

My work could not have gone ahead without the support of a large number of people whose help I now acknowledge. The following are thanked for specific help at various stages of the work: Prof David Sedgley, Christ's College, for help with Greek signs, which W D Fox used in his diaries; Nick Lane, University Library, Cambridge, for help in deciphering difficult sections of handwriting (while Charles Darwin's handwriting is notoriously bad in his letters, Fox's handwriting can be frustratingly difficult when he is writing in the small diaries of the times in a very diminutive hand); Rosy Clarkson, Sam Kuper, and Samantha Evans of the Correspondence of Charles Darwin Project for detailed discussions on letters in the Darwin-Fox correspondence; Richard Crombie, Susan Crombie and Gerard Crombie for allowing me access to the Fox materials in their possession, and Gerard Crombie for permission to publish Fox family material; the Syndics of the University of Cambridge to publish Darwin Fox material in their possession; the Master and Fellows of Christ's College for permission to publish Fox material in their possession and to Geoffrey Martin for showing me the Christ's College Admissions Book; members of the Manuscript Room at the University Library, Cambridge for expert support, especially Adam Perkins, Curator of Scientific Manuscripts and Dr Patrick Zutschi, Keeper of Manuscripts, and of course, Peter Gautry, former librarian at the University Library, Cambridge, who was in charge of of the Darwin Archive; staff of the University Library, Sydney and especially to Peter Stanbury and the Rare Book Department (Peter Stanbury was instrumental in acquiring a set of all Darwin's first or second editions in that Library; an unknown, earlier scholar who filled shelves with a wide range of early evolutionary books, such as Romanes, 1896; Candace Guite, Librarian at Christ's College and her staff, Colin Higgins and Carolyn Keim, for help over the Fox materials in their possession; Dr John van Wyhe for friendly help over the *Darwin Online* website and for discovering the rooms that W D Fox lived in while at Christ's College; Lee Perry, Woodward Library, University of British Columbia for help over their Fox manuscripts; Randal Keynes for early help over Emma Darwin's Diary (now on the *Darwin Online* website); Desmond King-Hele for help over Erasmus Darwin; Maxwell Craven, Derby, for information on William Darwin Fox, his father, his grandfather and for the image of Osmaston Hall; staff of the Local Studies Centre, Derby for locating obscure documents.

I wish to thank Jacco Flipsen and Noeline Gibson at Springer and Lakshmi Praba at Integra-India for their great help in expediting this work.

Research of this kind does not get done without a cost. I devoted much time to these studies over the last 20 years to the exclusion of my studies in plant physiology and biochemistry. To my colleagues in photosynthesis research, who must have wondered at times why I was not getting on with joint work, I offer my apologies and thanks for their forbearance; foremost amongst these is Professor Christopher Howe of the Biochemistry Department, Cambridge, whose continuing support of all my studies, I gratefully acknowledge. In earlier days, my friends Drs Derek and Fay Bendall, also gave me much help and encouragement. My father and mother, Frederick and Vera Larkum, my brother, Charles Larkum, and family (including

Penny Price-Larkum), and David and Jane Hindmarsh, provided accommodation and a welcoming environment during my many visits to Cambridge.

Finally, I wish thank my wife, Hilary, and children, Lesley and Matthew, for allowing me to escape on a number of occasions into a world where they could not reach me, and for their loving support.

Contents

List of Letters

No:	CCD #	Chapter	Christ's/other #	Address	Date	Description
1	42	Chapt 3	Christ's 1	Shrewsbury	12 June 1828	I am dying by inches : Figs
2	43	Chapt 3	Christ's 2	Shrewsbury	29 june 1828	Diagram of 3 insects
3	45	Chapt 3	Christ's 3	Barmouth	29 July 1828	To follow a Machaon on a windy day
4	46	Chapt 3	Christ's 4	Barmouth	19 Aug 1828	And if there is any bliss on earth that is it.
5	48	Chapt 3	Christ's 5	Shrewsbury	1 or 7 Oct 1828	Mahomet's noddle: Maer or Woodhouse
6	52	Chapt 3	Christ's 6	Shrewsbury	29 Oct 1858	Hope-perfect specimen of an Entomologist
7	54	Chapt 3	Christ's 7	Shrewsbury	24 Dec 1829	Father pleased with the Death Heads
8	55	Chapt 3	Christ's 8a	Shrewsbury	07 Jan 1829	Fox's dismal state in Cambridge
9	56	Chapt 3	Christ's 8	Shrewsbury	25-29 Jan 1829	Sappho, best of bitches and Dash of dogs
10	57	Chapt 3	Christ's 9	Christ's College	26 Feb 1829	London trip (Hope); no "little Go"
11	59	Chapt 3	Christ's 10	Christ's College	18 Mar 1829	A quiet life/ good deal of Van John
12	60	Chapt 3	Christ's 11	Christ's College	01 April 1829	That of all blackguards you are the greatest
13	61	Chapt 3	Christ's 12	Christ's College	11/12 April 1829	Perfect & absolute state of idleness
14	62	Chapt 3	Christ's 13	Christ's College	12 April 1829	Illness of Fox's sister, M.A.
15	63	Chapt 3	Christ's 14	Christ's College	23 April 1829	Death of Fox's sister, M.A.
16	64	Chapt 3	Christ's 15	Christ's College	18 May 1829	Invitation to Osmaston
17	66	Chapt 3	Christ's 16	Christ's College	07 June 1829	Settling accounts
18	67	Chapt 3	Christ's 17	Shrewsbury	03 July 1829	Welsh expedition with Hope
19	68	Chapt 3	Christ's 18	Shrewsbury	15 July 1829	Graham smiled and bowed so very civily.
20	69	Chapt 3	Christ's 19	Shrewsbury	29 July 1829	Possible trip to Osmaston
21	70	Chapt 3	Christ's 20	Shrewsbury	26 Aug 1829	Woodhouse & la belle Fanny
22	71	Chapt 3	Christ's 21	Maer Hall	04 Sept 1829	Just about to visit Osmaston
23	72	Chapt 3	Christ's 22	Shrewsbury	22 Sept 1829	Erasmus to give up doctoring
24	73	Chapt 3	Christ's 23	Christ's College	15 Oct 1829	Report of Music Meeting/ Christ's empty
25	74	Chapt 3	Christ's 24	Christ's College	03 Nov 1829	CD's father dangerously ill/ Mr Jenyns
26	75	Chapt 3	Christ's 25	Christ's College	03 Jan 1830	Cambridge is very empty
27	76	Chapt 3	Christ's 26	Christ's College	13 Jan 1830	My Dytici & Colym astonished his weak mind
28	78	Chapt 3	Christ's 27	Christ's College	25 Mar 1830	Through my little Go!!!
29	79	Chapt 3	Christ's 28	Christ's College	01 April 1830	Staying in Cambridge for vacation

No:	CCD #	Chapter	Christ's/other #	Address	Date	Description
30	80	Chapt 3	Christ's 29	Christ's College	09 May 1830	Missed invitations
31	81	Chapt 3	Christ's 30	Christ's College	31 May 1830	Yarrell - Bewick Swan
32	81	Chapt 3	Christ's 31	Barmouth	25 Aug 1830	CD at Maer, Fox in Cambridge with Henslow
33	82	Chapt 3	Christ's 32	Maer Hall	8 Sept 1830	Bad shooting this season, Charlotte agreeable
34	86	Chapt 3	Christ's 33	Christ's College	8 Oct 1830	CD rides his horse part of way to Cambridge
35	87	Chapt 3	Christ's 34	Christ's College	5 Nov 1830	Henslow is my tutor
36	88	Chapt 3	Christ's 35	Christ's College	27 Nov 1830	Items of Fox and accounts
37	89	Chapt 3	Christ's 36	Christ's College	23 Jan 1831	What a good place I got in the polls
38	92	Chapt 3	Christ's 38	Christ's College	9 Feb 1831	Many friends (Henslow among the foremost)
39	94	Chapt 3	Christ's 37	Christ's College	15 Feb 1831	Paying nat. hist. collectors
40	96	Chapt 3	Christ's 39	Shrewsbury	7 April 1831	A scheme . . . of going to the Canary Islands
41	100	Chapt 3	Christ's 41	Christ's College	11 May 1831	Genl Election in Cam, Microscope, Canary Scheme
42	101	Chapt 3	Christ's 40	Shrewsbury	9 July 1831	Arrived at this stupid place. . . .in the Kings bench
43	103	Chapt 3	Christ's 42	Shrewsbury	1 Aug 1831	Canary scheme postponed/ deliberate Falsehoods
44	120	4	Christ's 43	London	6 Sept 1831	Offering me the place of Naturalist in a vessel
45	132	4	Christ's 44	London	19 Sept 1831	Details on forthcoming Beagle voyage
46	149	4	Christ's 45	Devonport	17 Nov 1831	We have had delay after delay
47	168	4	Christ's 46	Botofogo Bay	? May 1832	But geology carries the day
48	175	4	Fox DAR 204 6.2	Epperstone	30 June 1832	Inflammation of the Lungs/Regular Journal
49	184	4	Fox DAR 204 6.2	Ryde, I.O.W	28 Sept 1832	Reform Bill/Cholera
50	189	4	Christ's 46a	Rio Plata	12-13 Nov 1832	I see no end to the voyage
51	197	4	Fox DAR 204.7	Ryde, I.O.W	23 Jan 1833	Erasmus Galtons ship came here
52	207	4	Christ's 46b	Rio Plata	23 May 1833	A wild man is indeed a miserable animal
53	223	4	Christ's 46	Buenos Ayres	25 Oct 1833	Rio Negra to Bahia Blanca to Buenos Ayres
54	261	4	Fox DAR 108.8	Osmaston	Nov 1834	expects to be confined at the end of December
55	270	4	Christ's 47	Valparaiso	Mar 1835	Passage from Conception/dreadful earthquake
56	282	4	Christ's 47a	Lima	July 1835	Women in these countries
57	279	4	Christ's 48	Hobart Town	15 Feb 1836	Visiting Australia. . .Empress of the South
58	319	4	Christ's 49	Gt Marlborough St	6 Nov 1836	No one so friendly and kind as Lyell
59	327	4	Christ's 50	Cambridge	15 Dec 1836	staying in Henslow's house

No:	CCD #	Chapter	Christ's/other #	Address	Date	Description
60	348	4	Christ's 51	Gt Marlboro St	12 Mar 1837	The new ostrich should be called "darwinii"
61	364	4	Christ's 52	Gt Marlboro St	7 July 1837	Caroline to marry Jos Wedgewood
62	374	4	Brit Columbia	Gt Marlboro St	28 Aug 1837	Proof sheets beginning to tumble in
63	393	4	Christ's 53	Gt Marlboro St	11 Dec 1837	Hensleigh. . .resigned his place
64	419	5	Christ's 54	Gt Marlboro St	15 June 1838	Crossing of animals. It is my prime hobby
64A	418	5	Fox DAR 164:173	Delamere	? Nov 1838	Dogs, bloodhounds, geese
65	541	5	Christ's 58	12 Up Gower St	24 Oct 1839	We give up/all parties
66	572	5	Christ's 60	Maer	7 June 1840	periodical vomiting to which I was subject
67	586	5	Christ's 59	12 Up Gower St	24 Jan 1841	All kinds of facts, "Varieties and Species"
68	606	5	Christ's 61	12 Up Gower St	23 Aug 1841	Coral volume
69	609	5	Christ's 62	12 Up Gower St	28 Sept 1841	The title of compadre
70	624	5	Christ's 57	12 Up Gower St	23 Mar 1842	Loss of your poor wife
71	625	5	Christ's 56	12 Up Gower St	31 mar 1842	What a comfort - the having children
72	654	5	Christ's 64	Down	9 Dec 1842	Our house lies high on the chalk
73	665	5	Christ's 66	Down	25 Mar 1843	An addition to our house
74	715	5	Christ's 68	Down	20 Nov 1843	Profoundly tranquil state
75	801	5	Christ's 70	Down	20 Dec 1844	With respect to mesmerism
76	827	5	Christ's 69a	Down	13 Feb 1845	News of your Noah's Ark
77	859	5	Christ's 69	Down	24 April 1845	Capitally-written book, Vestiges
77A	558	5	Fox DAR 205.7(2)	Delamere	>1 June 1846	Swan Goose Gander
78	13809	5	Christ's 107	Down	3 Oct 1846	Potato seeds
79	1222	5	Christ's 71	Down	6 Feb 1849	Watercure. Dr Gully's book
80	1233	5	Christ's 72	Down	6 Mar 1849	Emma writes "Bag and Baggage"
81	1235	5	Christ's 73	Malvern	24 Mar 1849	Cold scrubbing in morning
82	1240	5	Christ's 73A	Malvern	18 April 1849	Water cure has the most extraordinary effect
83	1249	5	Christ's 74	Down	7 July 1849	Birmingham for British Association
84	1292	5	Christ's 75	Down	17 Jan 1850	Chloroform used in birth of Leonard
85	1323	5	Christ's 76	Down	May 1850	I am having douche daily
86	1352	5	Christ's 77	Down	4 Sept 1850	You speak about homoeopathy
87	1362	5	Christ's 78	Down	10 oct 1850	Bruce Castle School

No:	CCD #	Chapter	Christ's/other #	Address	Date	Description
88	1396	5	Christ's 78a	Malvern	27 March 1851	Fox's father's death
89	1425	5	Christ's 79	Down	29 April 1851	Cruel loss of Annie
90	1476	5	Christ's 80	Down	7 Mar 1852	Bug-bears are Californian & Australian Gold
91	1489	5	Christ's 81	Down	24 Oct 1852	Oh the profession oh the gold & oh the french
92	1499	5	Christ's 82	Down	29 Jan 1853	Second volume on Cirripedes
93	1522	5	Christ's 84	Eastbourne	25 July 1853	Seahouses, Eastbourne; death of Mudge boys
94	1527	5	Christ's 85	Down	10 Aug 1853	Loss of Fox's daughter, Louisa Mary
95	1651	5	Christ's 87	Down	19 Mar 1855	For & versus the immutability of species
96	1656	6	Christ's 88	Down	27 Mar 1855	Pigeons, Call Duck
97	1675	6	Christ's 89	Down	26 April 1855	Asks for young of several poultry breeds
98	1678	6	Christ's 90	Down	7 May 1855	Ducks, Seeds
99	1683	6	Christ's 91	Down	17 May 1855	Lizard Eggs!!!
100	1686	6	Christ's 92	Down	23 May 1855	Rule of three
101	1698	6	Christ's 93	Down	11 June 1855	fair night after the murder
102	1704	6	Christ's 94	Down	27 June 1855	Young ducking
103	1728	6	Christ's 95	Down	22 July 1855	Cyanide! Murders
103A	1733	6	Christ's 65	Down	31 July 1855	Mr Dixon of Poultry notoriety
104	1745	6	Christ's 94A	Down	22 Aug 1855	"splendaceous Dorking"
105	1766	6	Christ's 96	Down	14 Oct 1855	Game chick
106	1815	6	Christ's 86	Down	3 Jan 1856	Great skin/ skeleton call duck
107	1646	6	Fox DAR 164:174	Delamere	8 Mar 1856	Scotch Deer Hound
108	1843	6	Christ's 97	Down	15 Mar 1856	Tegetmeier
109	1887	6	Am Phil Soc	Down	4 Jun 1856	I write one line
109A	–	6	Am Phil Soc	Delamere	<Fri 4 Jun 1856	Ellen Sophia Fox
110	1895	6	Brit Columbia	Down	8 June 1856	Lyell and his advice
111	1901	6	Christ's 98	Down	14 Jun 1856	Mention of Essay
112	1967	6	Christ's 100	Down	3 Oct 1856	Annies grave
113	1978	6	Christ's 99	Down	20 Oct 1856	Owls & Eagles
114	11799	6	Fox DAR 77:170	Delamere	19 Dec 1856	Peas
115	2049	6	Christ's 110	Down	8 Feb 1857	Momodes (Mormodes?) & Helix pomadice

No:	CCD #	Chapter	Christ's/other #	Address	Date	Description
116	2057	6	Christ's 101	Down	22 Feb 1857	Sweet Peas, Malvern
117	2085	6	Christ's 103	Down	30 April 1857	Pigs
118	2161	6	Christ's 104	Down	30 Oct 1857	Your pear-trees
119	2187	6	Christ's 105	Down	17 Dec 1857	Rabbits
120	2202	6	Christ's 108	Down	14 Jan 1858	Hewitson
121	2208	6	Christ's 109	Down	31 Jan 1858	Woods proposed election to Athenaeum
122	2219	6	Christa 111	Down	22 Feb 1858	Return of Hewitson's book
123	2229	6	Christ's 112	Down	28 Feb 1858	Turkeys & Kite
124	2256	6	Christ's 112A	Down	16 April 1858	Dun horses
125	2270	6	Christ's 113	Down	8 May 1858	Moor Park & Turkey chicks
125A	7815	6	Fox DAR164:194	Delamere	11 June 1858	Sir Francis Darwin's daughters
126	2293	6	Christ's 114	Down	24 June 1858	Dr Lane's case
127	2296	6	Christ's 115	Down	27 June 1858	Scarlet fever. . . .Charles Waring dangerously ill
128	2300	6	Christ's 116	Down	2 July 1858	Death of Charles Waring
129	2304	6	Christ's 117	Down	6 July 1858	Plans for IOW
130	2312	6	Christ's 118	Sandown	21 July 1858	Wallace's letter
131	2360	6	Christ's 119	Down	13 Nov 1858	William at Christ's
132	2412	6	Christ's 106	Moor Park	12 Feb 1859	Lyell staggered
133	2436	6	Christ's 120	Down	24 Mar 1859	Last illness of Fox's mother
134	2451	6	Christ's 121	Down	10 April 1859	Death of Fox's mother
135	2493	6	Christ's 122	Down	23 Sept 1859	Revision of proofs of *Origin*
136	2502	6	Christ's 123	Ilkley Wells	5 Oct 1859	Ilkley Wells House/ Homeopathy
137	2533	6	Christ's 124	Ilkley Wells	16 Nov 1859	Weariful book on spp.
137A	2604	6	Christ's 125	Down	25 Dec 1859	My book...very successful in the ordinary sense
138	2733	6	Christ's 127	Down	22 Mar 1860	Muriatic acid diet
139	2809	6	Christ's 128	Down	18 May 1860	Reviews of "*Origin*"; my bigger Book
140	2836	6	Christ's 129	Down	18 June 1860	Musk Duck reverting; H. Galton
140A	S06319	7	Fox DAR 164:188	Southampton	Aug 16 1860	White cats with blue eyes stone deaf
141	2953	7	Christ's 130	Eastbourne	18 Oct 1860	Gigantic correspondence on *Origin*
142	3075	7	Christ's 130A	Down	17 Dec 1860	Photographs of CD, Robert & Erasmus

No:	CCD #	Chapter	Christ's/other #	Address	Date	Description
143	3046	7	Christ's 126	Down	9 Jan 1861	Inkstands/white sows
144	3204	7	Christ's 131	Torquay	8 July 1861	Torquay/Henslow's death
145	3544	7	Christ's 132	Down	12 May 1862	Turkeys
146	3555	7	Christ's 133	Leith Hill Place	17 May 1862	Possible visit by Fox
147	3717	7	Christ's 134	Bournemouth	12 Sept 1862	Hairy Ears; Leonard ill
148	3732	7	Christ's 135	Bournemouth	20 Sept 1862	Orchids/Variation
149	3970	7	Fox DAR 164: 176	Delamere	6 Feb 1863	Sons & daughters
150	3975	7	Christ's 136	Queen Anne St	10 Feb 1863	Invitation to visit Queen Anne St or Down
151	397A	7	Fox DAR 164: 177	Hampstead	11 Feb 1863	Reply to invitation
152	4033	7	Christ's 138	Down	9 Mar 1863	Muscovy/Musk ducks
153	4037	7	Fox DAR164.178	Delamere	12 Mar 1863	Sheep & Ducks Gully
154	4044	7	Christ's 137	Down	16 Mar 1863	The Ram; Malvern
155	4057	7	Amer Phil Soc 292	Down	23 Mar 1863	Authentic particulars of lambs
156	4178	7	Fox DAR 164: 175	Delamere	16-22 May 1863	"Gullys Brain"
157	4181	7	Christ's 139	Down	23 May 1863	Illustrated Times & Squib
158	4193	7	Fox DAR 164: 179	Delamere	29 May 1863	Ben Rhydding, Ilkley
159	4292	7	Christ's 140	Malvern Wells	4 Sept 1863	Annie's grave?
160	4296	7	Fox DAR164.180	Delamere	7 Sept 1863	Whereabouts of Annie's Grave
161	4294	7	Emma Darwin	Malvern Wells	6-27 Sept 1863	Animal traps
162	4312	7	Emma Darwin	Malvern Wells	29 Sept 1863	Annie's gravestone found
163	4355	7	Emma Darwin	Down	8 Dec 1863	Royal Soc for Prevention of Cuelty to Animals
164	4484	7	Fox DAR 164: 181	Hillfield	5 may 1864	Proposal of visit to Down
165	4487	7	Emma Darwin 143	Down	6 May 1864	CD sick-not seeing visitors
166	4497	7	Emma Darwin 144	Down	16 May 1864	RSPCA competition & enclosure
167	4483	7	Fox DAR 164.182	Delamere	28 Nov 1864	Copley Medal
168	4685	7	Christ's 145	Down	30 Nov 1864	Thanks re Copley Medal
169	4741	7	Fox DAR164:183	Delamere	6 Jan 1865	3 Forms of *Lythrum salicaria*
170	4903	7	Fox DAR164:204	Delamere	<26 Oct 1865	Charles Woodd Fox at Oxford
171	4924	7	Christ's 146	Down	25-26 Oct 1865	Dr Bence Jones/Geo "Little go"
172	13808	7	Fox DAR 164: 205	Delamere	<1 Mar 1866	Dr Brushfield, superintendant, Woking

No:	CCD #	Chapter	Christ's/other #	Address	Date	Description
173	5195	7	Fox DAR 164: 184	Down	20 Aug 1866	Susan's sufferings: Lyell at Whitby
174	5197	7	Smithsonian	Down	24 Aug 1866	Illness of Susan Darwin/"*Variation*"
175	5388	7	Fox DAR 164: 185	Delamere	1 Feb 1867	No less than 3 Free Martins
176	5392	7	Christ's 147	Down	6 Feb 1867	Selling of The Mount
177	5837	8	Fox DAR164:186	Delamere	3 feb 1868	George Darwin 2nd Wrangler
178	5842	8	Christ's 148	Down	6 Feb 1868	There is so much detail in my book
179	5929	8	Christ's 148a	Down	25 Feb 1868	CD wants *info* on Magpies, Turkeys, etc
180	5762	8	Fox DAR 86:83/84	Delamere	>25 Feb 1868	Info on Magpies, Pheasants, etc
181	6172	8	Christ's 148b	Down	14 May 1868	Magpie Marriage/Black cocks/pheasants
182	6187	8	Fox DAR 86:187	Hillfield	19 May 1868	Can't visit
183	6426	8	Christ's 149	Down	21 Oct 1868	More on Magpie Marriage/Sedgwick
184	6436	8	Fox DAR 164: 189; 193: 112; 83: 187: 84.1: 128-30; 86: A87-9	Delamere	29 Oct 1868	Long, *long* letter!!!
185	6447	8	Christ's 150	Down	4 Nov 1868	Swan - Goose; stay with Eras
186	6455	8	Fox DAR 84.1: 126-7; 85: B36-7	Delamere	9 Dec 1868	More info for *Descent*: Talk on Sheep & Cows
187	6500	8	Amer Phil Soc	Down	12 Dec 1868	Short: has received Fox's letter
188	7107	8	Fox DAR104:190	Broadlands	15 Feb 1870	Temple of the Winds
189	7113	8	IOW Sandown	Down	18 Feb 1870	News on the Darwins, Geo William Bessy
190	7353	8	Fox DAR 164: 191	Delamere	28 Oct 1870	Address of Governess
191	7370	8	Amer Phil Soc	Down	15 Nov 1870	Correcting proofs of Descent of Man
192	7376	8	Fox DAR 164: 192	Sandown	18 Nov 1870	Descent of Man: Monkeys Ape & Apesses
193	7505	8	Fox DAR 164: 193	Sandown	21 Feb 1871	Thanks for receipt of "*Descent*"
194	8408	8	Fox DAR 164: 195	Sandown	13 July 1872	On our way to Cheshire
195	8413	8	Amer Phil Soc	Down	16 July 1872	*Expression of emotions in Man & animals*
196	8577	8	Fox DAR 164: 196	Sandown	25 Oct 1872	Nice old Lady. Visit in 1846
197	8585	8	Christ's 151	Down	29 Oct 1872	Staying at seven Oaks, *Expression*
198	9023	8	Fox 164 199	Delamere	22 Aug 1873	Receipt of *Expression*/PSS

No:	CCD #	Chapter	Christ's/other #	Address	Date	Description
199	9040	8	Christ's 152	Down	1 Sep 1873	Ugly head symptoms. Glorious 1st
200	9446	8	Fox DAR 164: 197	Sandown	8 May 1874	Package of Mr Merrion's/Kitchen Garden
201	9454	8	Christ's 153	Down	11 May 1874	Rec'd Package: newsy letter
202	9499	8	Christ's 154	Down	18 June 1874	Utricularia and Pinguicola
203	9507	8	Fox DAR164:198	Sandown	22 June 1874	Cannot collect U & P; vipers
204	10073	8	Fox DAR164 200	Sandown	16 July 1875	Fox leaving house for summer; insectivorous plants
204A	S13810	8	Fox DAR 210.14: 36	Sandown	[1876]	My dear Mother's death
205	10515	8	Brit Columbia	Dorking	26 May 1876	Cannot see Fox/Caroline's Fall/ William Fall
206	10923	8	Fox DAR 110.2: 62	Down	3 April 1877	Pulmonaria
207	11261	8	Fox DAR 164: 201	Sandown	29 Nov 1877	William's marriage/ Smoking
208	11266	8	Christ's 155	Down	2 Dec 1877	William marriage/Geo hereditory weakness
209	11355	8	Fox DAR164:202	Sandown	12 Feb 1878	CD's Birthday wishes
210	11358	8	Brit Columbia	Down	14 Feb 1878	Reply to Birthday wishes
211	11598	8	Brit Columbia	Down	10 July 1878	Death of Dora in Times
212	11625	8	Fox DAR 164: 203	Sandown	22 July 1878	Reply to CD over Dora's death; Hewitson
213	11913	8	Fox DAR 99: 172-74	Sandown	3 Mar 1879	Long gossipy letter; harvest mice
214	11995	8	Fox DAR 99: 175-6	Sandown	15 Apr [1879]	To Geo Darwin: information on Erasmus Darwin
215	12006	8	Fox DAR 99 177-8	Sandown	21 Apr 1879	To Geo Darwin re Erasmus Darwin
216	12554	8	Brit Columbia	Down	29 Mar 1880	Fox grievously ill
217	12569	8	C. Fox DAR 164	Sandown	8 Apr 1880	Charles Fox reports the death of W D Fox
218	12752	8	Brit Columbia	Down	10 Apr 1880	Condolences to family over the death of Fox

List of Figures

Notes on Abbreviations and Transcriptional Procedures

Abbreviations

Alum. Cantab. – Venn, J. A. (1922-1954). Alumni Cantabrigienses, 1752 – 1900, 6 Vols, Cambridge University Press, Cambridge, UK.
Alum. Oxon. – Foster, J., (1891-2). Alumni Oxonienses, 1715-1886, 2 Vols. Parker & Co., Oxford, UK.
CD or Darwin – Charles Robert Darwin
CCD – Correspondence of Charles Darwin – see Burckhart and Smith (1985-2008), References
Darwin Pedigrees – see Freeman, R. B. (1984), References.
Fox – William Darwin Fox.
LL (Vol i, ii & iii) – Life and letters of Charles Darwin – see Darwin, F. (1887), References.
Darwin's *Diary* – see *Darwin Online* or in volumes of CCD.
Emma Darwin's *Diary* – see *Darwin Online*.

Transcriptional procedures

Addresses

In writing the address, where this was done on separate lines, all the parts are put on one line separated by commas.

Italicised and bolded words

The usual convention has been followed of italicising words that are underlined. Doubly underlined words are bolded.

Marginalia

Darwin scored or double scored interesting passages with black, brown or blue pencil and sometimes made notes. These sections are noted after the letter and before the footnotes and in a few cases the reader is referred to the CCD version for extra details.

Numbering system

The letters are numbered in sequence. Where a letter is problematical as to date or is by a person other than Darwin, Fox or Emma Darwin an "A" has been added to the number of the previous letter. The CCD number and the LL page number is given in the heading to each letter. Provenance is given after each letter: in the case of the Christ's College, Cambridge, Darwin/Fox collection, the number given by Francis Darwin is listed there, too.

Transcription differences

Where there is a small difference between my transcription and that of the CCD version, the difference is listed in Appendix 6 (Corrigenda).

Unclear or obscured words

The following system has been adopted for words that are unclear or are obscured by age, tears (mine!), etc.: *correct*, the word is obscure or missing but can be reasonably asserted; t***t*d, these letters of a word are decipherable; ***** a word of approximately this length cannot be deciphered; **** **** ****, three words are estimated to be missing or indecipherable.

Use of "//".

"//" is used to denote the end of a page. In earlier letters where no paragraphs were used, "//" are used to make an arbitrary break in the text. In later letters where paragraphs are regularly used, "//" are generally inserted within paragraphs.

Use of the spacer "—".

This spacer was a common feature of letters of the time. In the transcriptions it has generally been kept within paragraphs; but in extra long passages it has been used, arbitrarily, to mark a new paragraph.

Split words "="

Where a word ran on to the next line Darwin and Fox used an "=" sign between the split parts and this has been retained in the text.

Figures

Fig. 1 Erasmus Darwin (1731-1802, physician and polymath; grandfather of Charles Darwin. Painting by Joseph Wright of Derby, in 1770, when Erasmus Darwin was 38; Copyright, Darwin College, Cambridge, with permission

Fig. 2 Robert Waring Darwin (1766-1848), physician and father of Charles Darwin. Engraving by T Lipton after a painting by J Pardon. Copyright English Heritage Photo Library, with permission

Fig. 3 The Mount, Shrewsbury, the home of Robert Darwin and family, from ca1800 to 1848 (and later by Charles Darwin's sisters, Susan and Catherine, until it was sold in 1867, see Letters #175 & #176). (Recent photograph by A W D Larkum)

Fig. 4 Osmaston Hall, near Derby, the home of Samuel Fox III and family, from 1813 to ca 1850; photographed in 1862, by Richard Keene

Fig. 5 The main entrance to Christ's College, Cambridge. From an etching of a water colour painting by W Westal in 1815

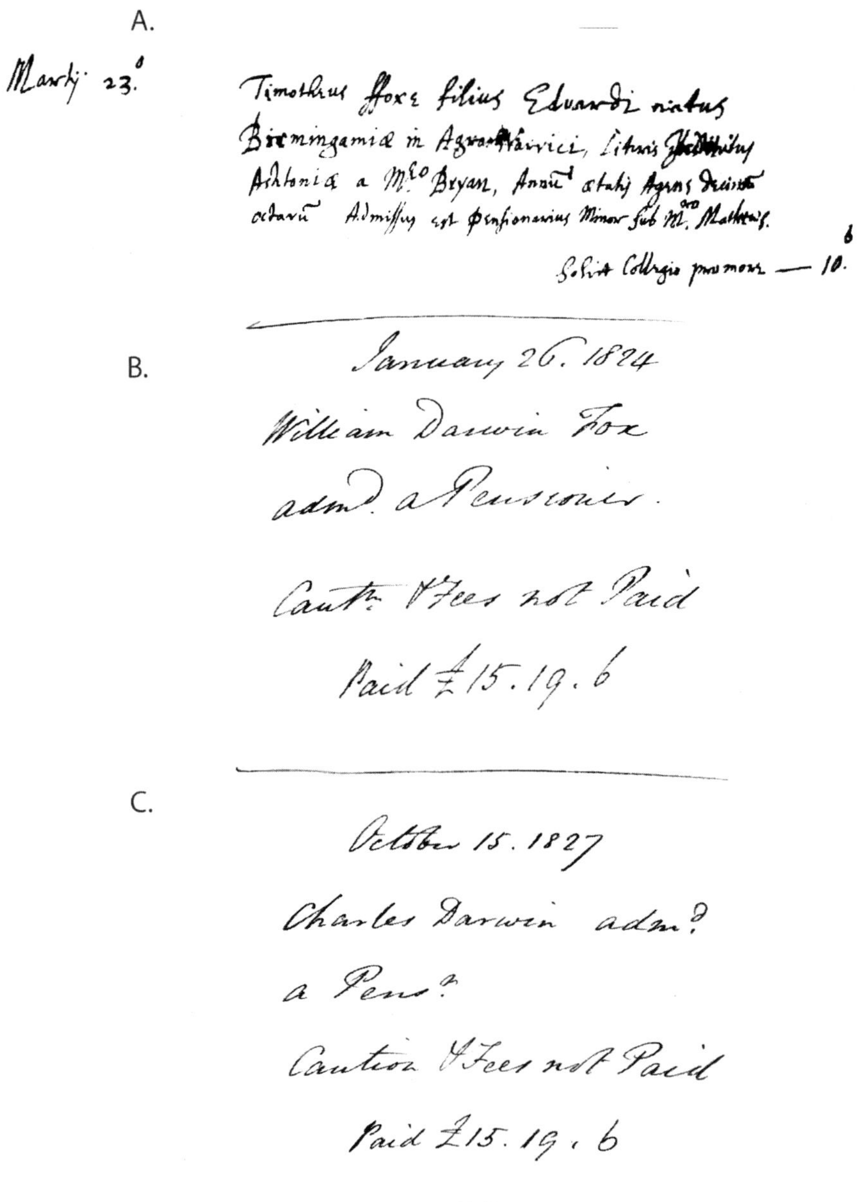

A.

Martij 23°

Timotheus ffoxe filius Edvardi natus Birmingamiæ in Agro Warwici, literis [illegible] Ashtoniæ a Mro Bryan, Annu ætatis Agens decimu octavu Admissus est Pensionarius Minor sub Mro Mathews.

Solvit Collegio pro more — 10. 6

B.

January 26. 1824

William Darwin Fox

admd. a Pensioner.

Cautn. & Fees not Paid

Paid £15.19.6

C.

October 15. 1827

Charles Darwin admd.

a Pensr.

Caution & Fees not Paid

Paid £15. 19. 6

Fig. 6 Entries in the Admissions Book of Christ's College, Cambridge, for Timothy Fox (23rd March 1646), William Darwin Fox (24th January1824) and Charles Robert Darwin (15th October 1827). With permission of the Master and Fellows of Christ's College

Fig. 7 A view of the south-east side of Second Court of Christ's College, Cambridge. W D Fox's rooms were on the left side of the staircase on the right hand side of the picture, on the first floor. (Recent photograph by A W D Larkum)

Fig. 8 A view of the south-east side of the First Court of Christ's College, Cambridge. Darwin's rooms were on the right side of the middle staircase on the first floor. (A recent photograph by A W D Larkum)

Fig. 9 Darwin's living room on the centre staircase, First Court, Christ's College, Cambridge; as reconstructed for the Centenary celebrations in 1909. Copyright: Christ's College, Cambridge, with permission

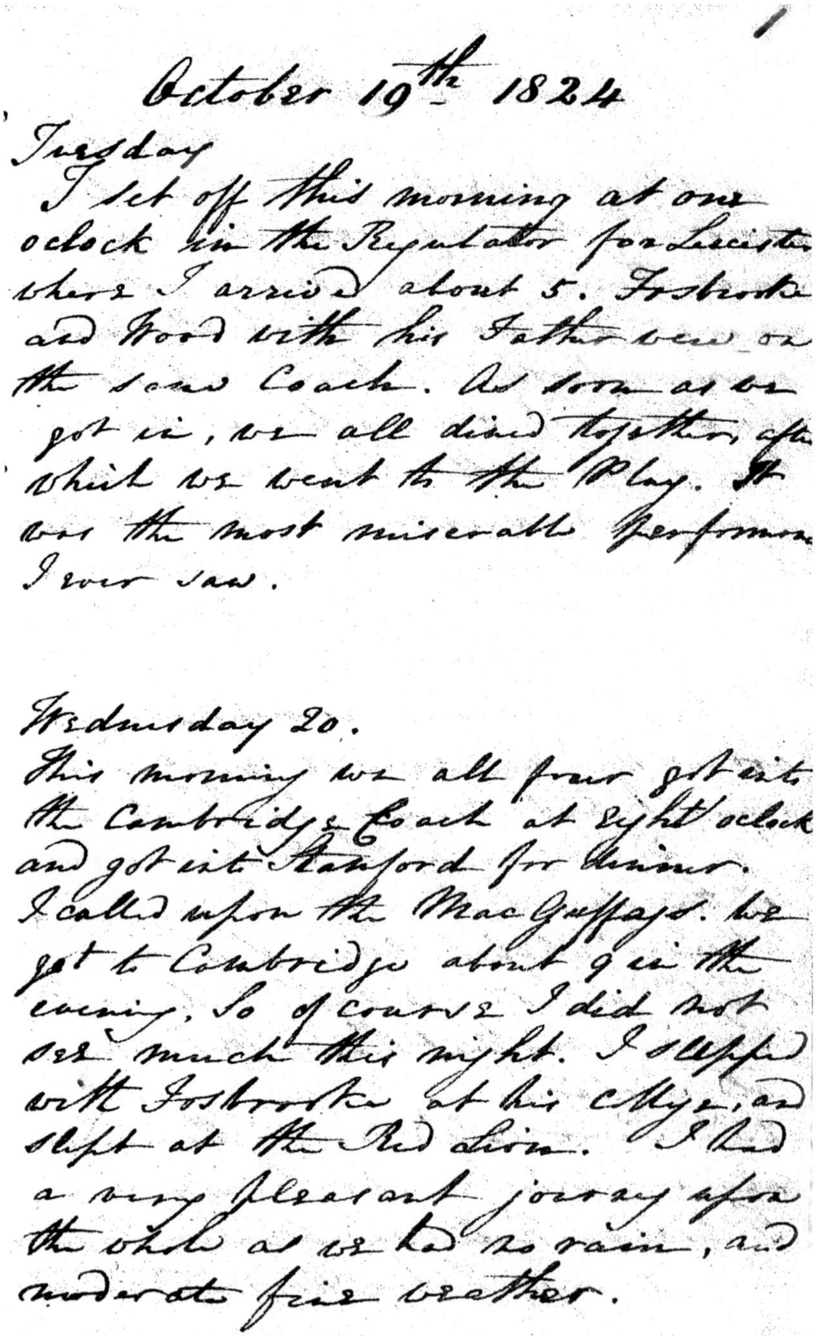

October 19th 1824

Tuesday

I set off this morning at one oclock in the Regulator for Leicester where I arrived about 5. Fosbrooke and Wood with his Father were on the same Coach. As soon as we got in, we all dined together, after which we went to the Play. It was the most miserable performance I ever saw.

Wednesday 20.

This morning we all four got into the Cambridge Coach at eight oclock and got into Stamford for dinner. I called upon the MacGuffogs. We got to Cambridge about 9 in the evening. So of course I did not see much this night. I supped with Fosbrooke at his College, and slept at the Red Lion. I had a very pleasant journey upon the whole as we had no rain, and moderate fine weather.

Fig. 10 The first page of W D Fox's journal (1824–1826)

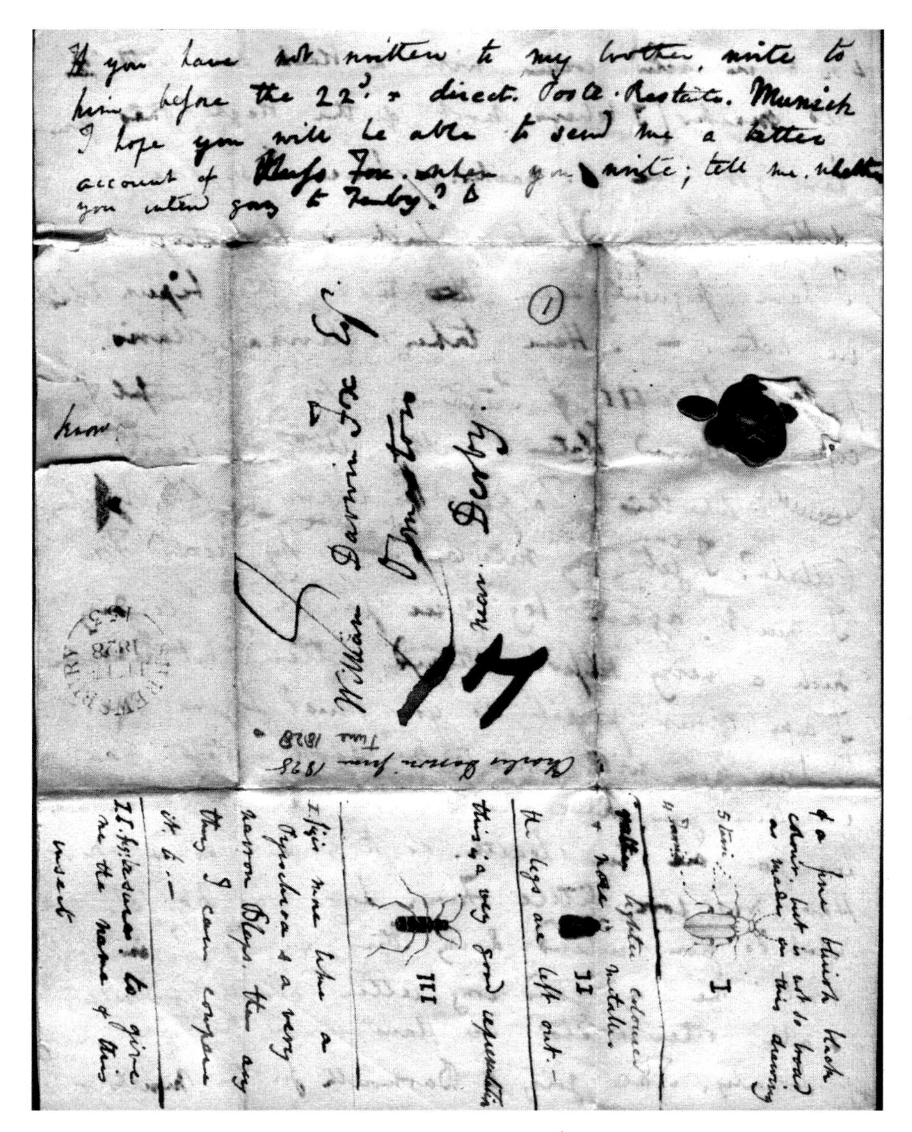

If you have not written to my brother, write to him before the 22d & direct. Poste Restante. Munich
I hope you will be able to send me a letter
account of Mrs Fox. when you write; tell me whether
you intend going to Tenby?

William Darwin Fox Esq
Osmaston
near Derby

Fig. 11 Letter #1, page 4: Charles Darwin to W D Fox, 12th June 1828. This sheet demonstrates the method used for correspondence at the time: the sheet was folded four times and the address was placed on the front; the back was sealed with wax, and often received the postmark stamp of the sender's post office. If two sheets were used, the first sheet would have writing on both sides and the second sheet would have writing on one side, and on the free flaps not used for the address and the seal. If the letter was unfinished at this point then writers often continued on the first sheet by writing perpendicularly to the initial lines of writing (see Fig 12)

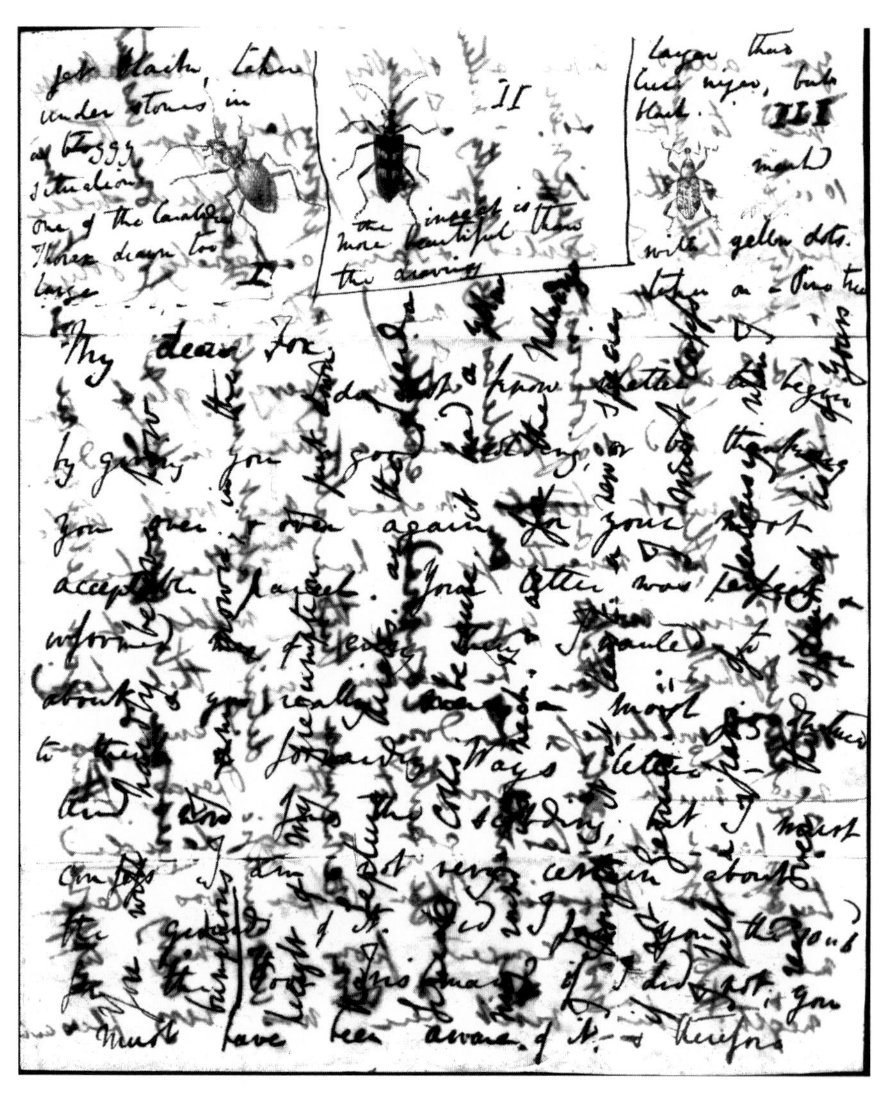

Fig. 12 Letter #2, page 1: Charles Darwin to W D Fox, 29th June 1828

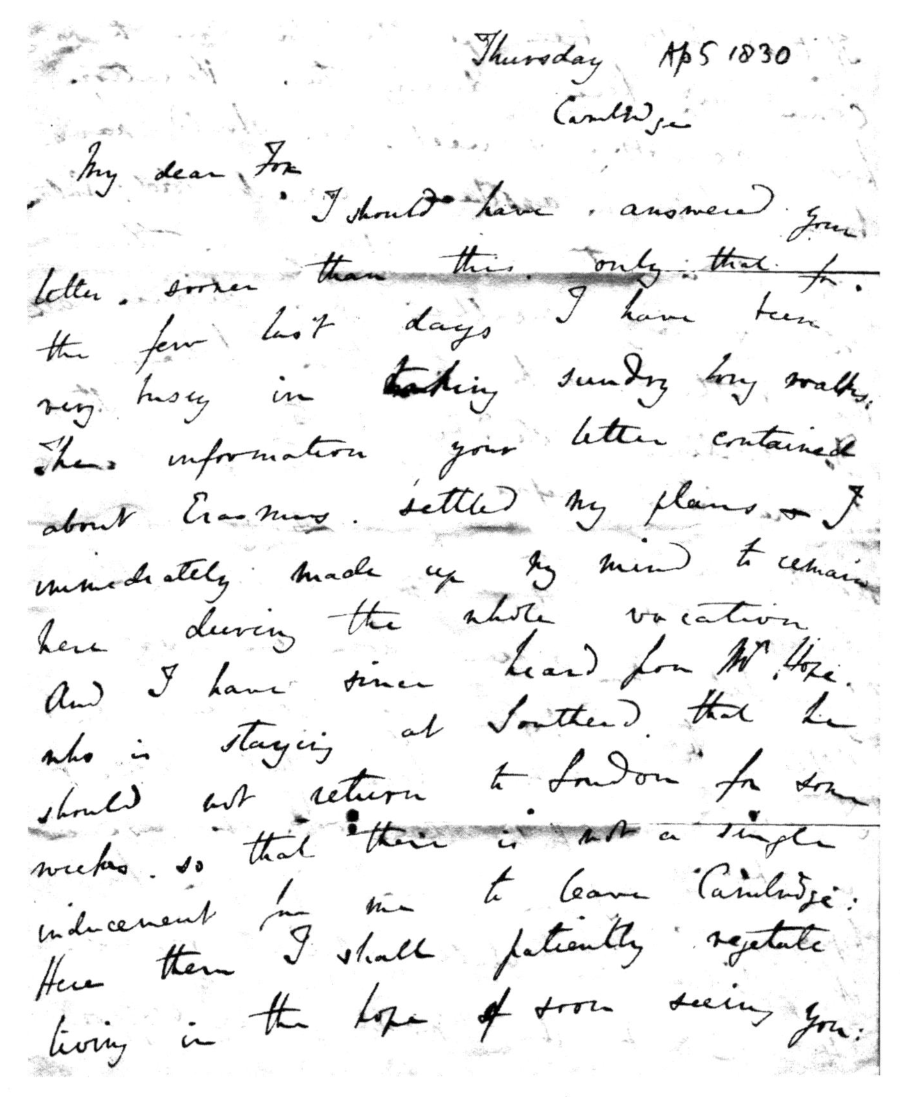

Thursday Ap 5 1830
Cambridge

My dear Fox
I should have answered your letter sooner than this, only that for the few last days I have been very busy in taking sundry long walks. The information your letter contained about Erasmus settled my plans & I immediately made up my mind to remain here during the whole vacation. And I have since heard from Mr. Hope who is staying at Southend that he should not return to London for some weeks, so that there is not a single inducement for me to leave Cambridge: Here then I shall patiently vegetate living in the hope of soon seeing you:

Fig. 13 Letter #28, page 1: Charles Darwin to W D Fox, 25th March 1830. Darwin announced to Fox that he had passed his "Little Go" - the Previous Examination, at Cambridge University, at this time; Darwin took his examination at the end of January 1830, after two years at the University. The insertion at the top of the letter "Cambridge March 1830" (in blue pencil) is not original and was probably inserted by Francis Darwin in the 1880s

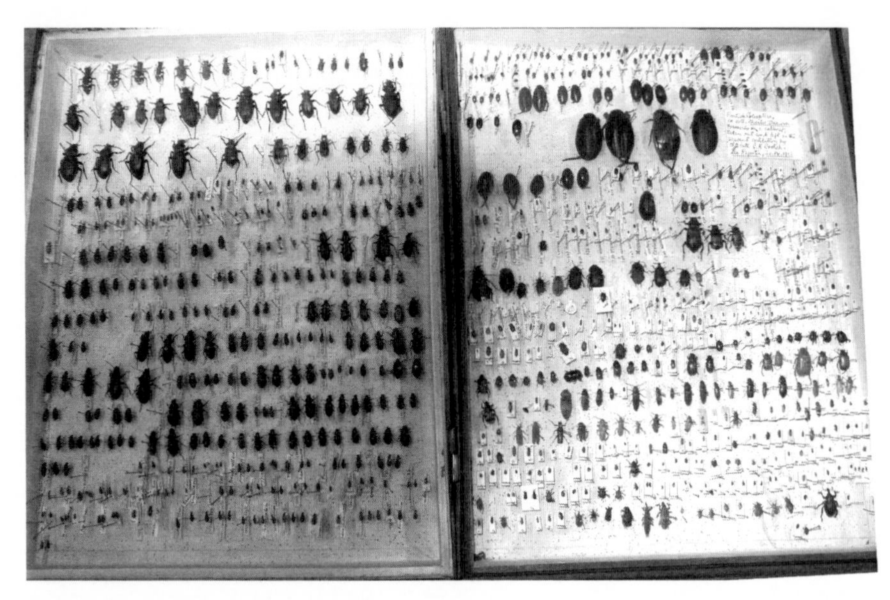

Fig. 14 Part of Charles Darwin's Insect Collection (not in the original cabinet) held in the Zoology Department, University of Cambridge. Copyright Zoology Department, Cambridge, with permission

Fig. 15 Cartoon drawings by Albert Way, illustrating Charles Darwin's enthusiasm for insect collecting in 1828, while they were students at the University of Cambridge. Copyright: University Library, University of Cambridge, with permission

Fig. 16 Coloured illustrations of beetles and butterflies, painted by one of Fox's sisters, possibly Julia Fox. Dated: 1828. 1. Gryllus Grillotalpa - Mole cricket (top left), 2. Papilio Io – Peacock butterfly (bottom left), 3. Chrysalis of ditto (centre), 4. Caterpillar of ditto (centre), 5. Ichneumon Persasorius (ichneumon wasp, bottom right), 6. Cerambyx Violaceus (longhorn beetle, top right). The identifications and transcriptions taken from labels on the mount of the original drawings by W D Fox; words in parenthesis not in the original. Copyright: Gerard Crombie, with permission

Fig. 17 Coloured illustrations of beetles and butterflies, painted by one of Fox's sisters, possibly Julia Fox. Dated: 1829. Chrysomela Fastuosa (top), Tipula Rivosa (Crane fly, middle right), 3 Phalaena Podella - Small Ermine Moth (middle left), 4. Papilio Machaon - Swallow Tail butterfly (bottom left), 5 Caterpillar and chrysalis of Small Ermine Moth (centre). The identifications and transcriptions taken from labels on the mount of the original drawings by W D Fox; words in parenthesis not in the original. Copyright: Gerard Crombie, with permission

Fig. 18 Drawing of William Darwin Fox, possibly drawn by Mr Cruikshank on September 14. 1825 at Osmaston Hall (see *Diary* entry for September 14th 1826, Appendix 4)

Fig. 19 Portrait of Emma Darwin (1840). Chalk and water-colour drawing by George Richmond (1809-96). Copyright English Heritage Photo Library, with permission

Fig. 20 Down House, Downe, Kent. Rear view of the house from the garden, taken in the twentieth century. Copyright English Heritage Photo Library, with permission

Fig. 21 Photograph of William Darwin Fox taken in the 1850s

Fig. 22 Photograph of Julia Fox, ca 1855

Fig. 23 A recent photograph of St Peter's Church of England Church and graveyard, Delamere. The Church was consecrated in 1817 on Crown Land of Delamere Forest. William Darwin Fox was the fourth Rector (1838-1874). The Rector possessed a glebe of some 50 acres, which W D Fox farmed, and which became known as Rectory Farm. Kindly supplied by Gerard Crombie

Fig. 24 A contemporary photograph of Delamere Rectory, taken ca 1865. Kindly supplied by Marion Williamson

Fig. 25 A contemporary photograph of Delamere Church School, taken about 1865. The school was built in 1846 with the active involvement of W D Fox in its foundation and its daily affairs, so that until quite recently it was known as Fox's School. Kindly supplied by Marion Williamson

Fig. 26 The headstone of the grave of Anne (Annie) Darwin in the churchyard of the Church of England Priory Church, Great Malvern, Worcestershire. The inscription reads: "IHS Anne Elizabeth Darwin Born March 2 1841 Died April 23 1851 A Dear And Good Child". See Letters #112, #160 & #162. (Photograph by A W D Larkum)

ILLUSTRATED TIMES 317

THE ORIGIN OF SPECIES, DEDICATED BY NATURAL SELECTION TO DR. CHARLES DARWIN.

NO. I.—LIKE A BULL AT A GATE.—(DRAWN BY C. H. BENNETT.)

Fig. 27 Cartoon from the *Illustrated Times (London)*, 2nd May 1863, p 317. The caption was "*The origin of species, dedicated by natural selection to Dr. Charles Darwin.*" No. 1. — *Like a Bull at a gate* (drawn by C H Bennett)"; referred to by Fox in Letter #156

Fig. 28 Cartoon of female Andaman monkey with pipe, which appeared in the *Illustrated London News* (Supplement) on 18th September 1869; referred to by Fox in Letter #192

Fig. 29 Portrait of Charles Darwin, painted in oil in 1875. A copy of the original by the artist, W W Ouless (signed and dated 1883). Held in Christ's College, Cambridge. The original is in Darwin College, Cambridge

Chapter 1
Grandfathers and Fathers

Ere Time began, from flaming Chaos hurl'd
Rose the bright spheres, which form the circling world;
Earths from each sun with quick explosions burst,
And second planets issued from the first.
Then, whilst the sea at their coeval birth,
Surge over surge, involv'd the shoreless earth,
Nursed by warm sun-beams in primeval caves,
Organic Life began beneath the waves.
The Temple of Nature, Canto I. Erasmus Darwin

On the 12th November 1756, a young physician, not long out of medical school at Edinburgh University, armed with letters of introduction, set up his shingle at the cathedral town of Lichfield, in the Midlands of England, in the time of George II, King of Great Britain and Ireland. This was Erasmus Darwin's second attempt to establish himself as a practising physician, as the first, at Nottingham, had only yielded one patient in 4 months (King-Hele, 1999). Lichfield was at that time a thriving town, not yet under the shadow of Birmingham, then a small provincial centre of 24, 000 inhabitants[1], which was soon to be the driving force of the industrial revolution, in no small measure as a result of the actions of this new and penniless doctor. Luck and good judgement was on his side and as a result of curing a seemingly incurable case, Erasmus soon had enough patients so that he could relax in his new home, find a wife and start a number of intellectual pursuits over the next 20 years, which would bring him international fame[2].

Erasmus Darwin was one of the greatest polymaths of the eighteenth century. The only Englishman of equivalent stature at this time was Samuel Johnson, who is also associated with Lichfield, having been born and educated there. Much has been written on these two men, both at the time and ever since (for Erasmus Darwin, see King Hele, 1999). Erasmus Darwin came from a line of distinguished lawyers and doctors and an ancestral home at Elston Hall near Newark, in Nottinghamshire[3]. The youngest of four brothers, he decided to study medicine at Edinburgh. Becoming established in Lichfield, he bought a house in the Cathedral grounds and married his first wife, Mary Howard[4], just one year after arriving, in 1757. He had 4 children by Mary, three sons and a daughter. The last of these sons was Robert Waring Darwin, the father of Charles Darwin, one of the two subjects of this book.

A. W. D. Larkum, *A Natural Calling*, DOI 10.1007/978-1-4020-9233-6_1,

England at this time was a very different land from the one that exists today. The Industrial Revolution, which was just beginning, had not left its stamp across the land, increasing the population many times, building entirely new cities and changing the grid of transport connections. Just as important as this growing commercial activity at home was the continued development of the East India Company (formed in 1600), especially after the defeat of the French in the seven years war (1756-1763), which opened up new markets in India and the East, alongside those already established in the Americas. A secondary effect of these overseas developments, was the effect at home of giving an outlet to an expanding population, and, for our story, also allowing the development of the influential Public School system of England, providing new opportunities for the sons of the ruling classes and the new colonialists, and providing the intellectual base for the Industrial Revolution and the development of the British Empire.

On a more prosaic note, this was a land in which the horse-drawn stage coach ruled supreme. Roads were not macadamised and, while some turnpikes were well maintained, many roads were rutted and dangerous. In his practice, Erasmus Darwin covered the astounding distance of up to 10,000 miles in a year, and a serious accident on the road a contributed life-long injury to his large frame, and in the end curtailed his travelling[5]. It is important to note that the grid of staging posts then does not coincide well with current arterial roads. The main centres then were largely the diocesan towns and cities, which possessed a cathedral, often going back to Norman times, or even to Roman Britain. Having a cathedral meant that many advantages were conferred on a town: it would have special markets and other privileges conferred by the Crown, such as choristers, which usually signified a local school of high quality. Today, Lichfield, with a population of just 28,000 inhabitants is hedged in on all sides by towns and cities of the Industrial Revolution, and it is difficult to see it as the provincial hub that it once was. And even as events were developing then in the middle of the eighteenth century, Birmingham, and other cities, were beginning to drain it of power and influence. Perhaps, the best description of the effect of these dramatic changes on provincial towns is in the novels of Anthony Trollope.

Both Charles Darwin and William Darwin Fox, the subjects of this book, began their careers by taking stage coaches across the country, particularly to Cambridge. Soon, however, railways were built rapidly between the towns and across the rural countryside. The earliest of these was the Liverpool and Manchester Railway in 1830, built 1 year before HMS "Beagle" sailed off on its 5 year circumnavigation of the globe; but by 1854 every major town in England was connected by a railway. Fox was an avid supporter of railways, as his diaries make clear. Yet Darwin too made great use of them in his family migrations to places all over England and he chose his home partly on the basis of a handy rail link. And as a final observation we should note that Wallace, the co-originator, with Darwin, of the Theory of Evolution, began his working life as an assistant surveyor for railways being developed in England and Wales.

During his time at Lichfield, and later in Derby, Erasmus Darwin became not only a successful doctor, whose services were requested by George III, but also, firstly, an inventor of renown (the discoverer, for example, of adiabatic heat changes), secondly, one of the foremost propagandists for the Linnean system of botany in England, thirdly, and a poet, who at one time was regarded as the leading poet in England[6]. Perhaps his most lasting achievement was to be one of the founders of the Lunar Society. This group of early industrialists and inventors, including James Herriot Watt, can justifiably be claimed to have sparked the Industrial Revolution, in the Midlands of England. It was the enthusiasm of Erasmus Darwin that held the group together. When he moved to Derby with his second wife (Elizabeth Chandos-Pole, 1747-1832) in 1780, the group began to disintegrate. In Derby, Erasmus formed the Derby Philosophical Society, and a founding member was an S. Fox, probably Samuel Fox III, the son of Samuel Fox II, and father of William Darwin Fox.

Thus our story begins in the early part of the eighteenth century, with two members of the newly-emerging professional classes: Erasmus Darwin (1731-1802), grandfather of Charles Robert Darwin, and Samuel Fox III (1715-1769), grandfather of William Darwin Fox, the second cousin of Charles Darwin.

Much less is recorded of the Foxes than of the Darwins. The North Midland town of Derby, on the edge of the Peak district, was an early centre for the industrial revolution. Coal mining was carried out in the hills to the northwest and weaving was carried out both as a cottage industry and also in the newly emerging mills (see below). It was perhaps most well known in the mid eighteenth century as the town where Bonney Prince Charlie housed his army and then retreated on his ill fated attempt on the Capital in 1845 (William Darwin Fox's step-grandfather Jedediah Strutt owned the house in which Charles Stuart stayed - see Diary entry of 27th Jan 1825[7]).

The historical record of the ancestors of William Darwin Fox in the eighteenth century is sparse. Derby in this century was plagued by Foxes, of unrelated families. However, we do know, from the "*Darwin Pedigrees*" and other sources[8] that the family goes back to a dispossessed puritan minister of the Restoration, the Revd Timothy Fox[9], who read for a degree at Christ's College, Cambridge (see Fig. 6). His fourth child, Samuel Fox I (1676-1755), was a draper, probably through his marriage to Elizabeth, the daughter and sole heiress, of Gilbert Wagstaff, draper of Derby[10]. It is likely that this was the S. Fox who was Mayor of Derby in 1741-42[11]. This would have made him 65 at the time. Samuel Fox II, the son, is mentioned as being a soap-boiler[12]. Soap-boilers were in fact the industrial chemists of the eighteenth and nineteenth century[13] and would have had much to do with the Industrial Revolution and the industrialisation of the Midlands. In Derby, new textile mills were rapidly erected towards the end of the eighteenth century[14], under the leadership of such eminent visionaries as Jedediah Strutt (1726-1797), the inventor, with William Woollatt, of the ribbed stocking machine, and Richard Arkwright (1732-1792)[15], the inventor of the water frame, a significant development over the of the spinning jenny of James Hargreaves. Both these inventions were used in Derby[16]. Jedediah Strutt enters our story in a much more personal role, however, because he was W D Fox's step-grandfather and, from the Fox diaries, Strutt's son clearly played an important place in the early life of Fox.

The talents of a soap-boiler would have been in great demand since many of the chemicals needed in the textile industry, such as caustic soda and dyes, would have been the stock-in-trade of soap boilers. It is also probable that Samuel Fox, II, kept up some activities in the drapery business. Both silk and cotton were woven in Derby at this time and his son, Samuel Fox III (1765-1851) first appears in the records, before 1813, as living in Thurlston Grange in the grounds of Elvaston Castle, owned by the Earl of Harrington[17]. An even earlier record exists in the cash book of Jedediah Strutt, for 1786, of 6 guineas to Samuel Eyre "for learning Mr Fox to work in frame &c", a clear reference to the hosiery business and possibly Samuel Fox III's initiation. And not long after this, in 1891, Fox married Strutt's second daughter, Martha.

Derby, a Midlands town of some 7,000 inhabitants in the Eighteenth Century, was the site of the first production of silk in England, as early as 1702[18]. And, it was John Lombe who set up a successful Silk Factory in 1717, on the banks of the Derwent, trading on knowledge, which up to that time had been kept secret in Italy, to where in turn it had been transferred from China. This factory underwent a number of vicissitudes throughout the eighteenth century, but by 1820 the mill was still active and had been joined by 12 other silk factories in the town and up to 6 elsewhere in the district (Glover, 1829). This activity led eventually to the weaving of silk, especially for silk stockings. Glover (1829) says "*The weaving of pieces of goods in silk was first introduced in Derby by Mr William Taylor, at his factory in Bog Lane, about seven or eight years ago*". Glover also stated that "*There are now about two hundred and twenty looms at work*". It is clear that silk hosiery was a large part of this activity and that the rib machine of Jedediah Strutt was adapted to this task, to produce ribbed silk stockings. In 1843, Glover reported "*Silk hosiery is not the least important branch of the silk trade; there are many excellent mechanics connected with this branch, and the manufacturers are Samuel Fox, jun, Esq., Mr Langden, Mr Lewis, Mr Morley and others*". Samuel Fox, jun, Esq. was probably Samuel Fox IV, Samuel Fox III, being close to retirement.

As already stated, Samuel Fox III married Martha, the second daughter of Jedediah Strutt, in 1791,[19] after a protracted affair. Samuel Fox III's silk hosiery business was successful enough for him to lease Osmaston Hall, just outside Derby, in 1814[20], a very large country house by any standards (Fig. 4). This suggests that business was very good at this time. The business premises for this activity were at 14 North Parade, Hardwick, Derby[21].

It is very likely that Erasmus Darwin and Samuel Fox II knew each other before Erasmus moved to Derby in 1781 and after that they certainly had close connections. Upon leaving Lichfield, Erasmus could no longer easily keep up his connections with the Lunar Society in Birmingham. After acquiring a house for his practice in Derby[22] (in addition to Radburn Hall, four miles from Derby, which belonged to his new wife, Elizabeth), he quickly reorientated his intellectual pursuits and formed the Derby Philosophical Society in 1783 (Sturges, 1979), and Samuel Fox III was a founding member. There were certainly many eminent men of the industrial revolution in Derby at that time and the Society was successful initially. However, this offshoot, like its sister, the Lunar Society, ran into the troubled times leading up to, during and beyond the French Revolution. In Birmingham the house and church of Joseph Priestley, the discoverer, amongst other things, of the element oxygen, and a member of the Lunar Society, was burnt down by a riotous mob in 1791 (Schofield, 2004). Priestley escaped shortly after, to London, and then in 1794 left for the United States of America, never to return. The actions of the mob were condemned and restitution for Priestley was sought but was never achieved. In Derby, at this juncture, the Philosophical Society passed a motion deploring these actions and expelled the Revd Hope over the issue, although we have few details[23]. Evidently there were divided opinions over the matter. It seems clear that Samuel Fox III allied himself with the reformist element in the Society, along with William Strutt, Peter Crompton and Erasmus Darwin[24]. At this time, too, Erasmus was writing polemical poetry in support of the French Revolution, and also on anti-slavery, on the education of women and on other liberal causes (King-Hele, 2006). While these actions and views were tolerated even by the reactionary elements of society up to the time of the French Revolution in 1789, thereafter they certainly were not. King-Hele (2006) relates how William Ward, the editor of the "Derby Mercury" was put on trial for (fairly mild) views expressed in his Address of 1792. The verdict was "not guilty" but if convicted, Ward, like Thomas Muir of Edinburgh, could have been transported to Australia for fourteen years. Furthermore, had that trial been "successful" it is possible that others, including Erasmus Darwin, would have been put on trial also[25].

The actual situation was somewhat more subtle in the case of Erasmus; George Canning, later the Prime Minister, whose politics swung drastically to the right at this time, penned a poem entitled *The Loves of Triangles,* a parody of Erasmus's poem *The Loves of Plants,* and backed up by long notes ridiculing his scientific ideas, particularly the "absurd notion" that human beings had evolved from lower forms of life[26]. Of course, many of the great poets have suffered from parodying of their work, including William Shakespeare, and it is a matter of personal taste how one regards the poetry of Erasmus Darwin today. At the time, and prior to these events, Erasmus was, highly regarded, even by the young poets, William Wordsworth, Samuel Coleridge, Mary Shelley and Lord Byron[27]. Yet as the country swung more and more to the right, Darwin's star began to fade. Part of the problem here was his strong support for what would now be called "evolution" although the term was not used until much later. In his book "Zoönomia" he devoted a large part to a theory of organic evolution and said,

"*All vegetables and animals now living were originally derived from the smallest microscopic ones.*"

He also had verses of a similar nature in his last poetical work "The Temple of Nature", which was published posthumously:

> *"Organic life beneath the shoreless waves*
> *Was born and nurs'd in Ocean's pearly caves*
> *First forms minute, unseen by spheric glass,*
> *Move on the mud, or pierce the watery mass;*
> *These, as successive generations bloom,*
> *New powers acquire, and larger limbs assume;*

Whence countless groups of vegetation spring,
And breathing realms of fin, and feet, and wing."

In the climate of the time this was seen as verging on heresy and sedition, a viewpoint upheld not only by the Church, but also by many of the scientific establishment; and the public attacks saw a sudden eclipse in his popular support. Sadly, Erasmus Darwin died, untimely, in 1802 at the age of 70, seemingly at the height of his powers.

One small example of the times is the tale of how Erasmus added to the family emblem of three scallop shells, the words "*E conchis omnia*" ("out of shells everything"), which he placed on his coach in 1770[28]. This clear reference to organic evolution was noticed by the Cannon of Lichfield, the Revd Thomas Seward, an influential friend, who observed in satirical verse that this was offensive and intimated that if he continued to display it he would be in danger of losing several wealthy patients[29]. The sign was removed (but Erasmus Darwin continued to use it on his letterhead, and so did his son, Robert Waring).

Robert Waring Darwin was setting up as a doctor at about this time (1787) in Shrewsbury after he gained his M D at Edinburgh University, and is said to have been horrified by the vilification of his father; so much so that he kept his religious scepticism and evolutionary beliefs to himself, even keeping them from his own son, Charles[30]. Shrewsbury at this time was a rich county centre, on the River Severn, near the Welsh border and surrounded by agricultural communities. Shrewsbury is remembered for one of the bloodiest battles in English history, the Battle of Shrewsbury, immortalised in the play, Henry IV Part 1, by William Shakespeare. It had been offered a cathedral by Henry VIII but turned it down[31] and owed its prosperity to the wool trade of the fourteenth and fifteenth centuries. This quiet rural centre suited Robert, who was not active politically or scientifically, although through his father's help he did become a Fellow of the Royal Society. Therefore it is difficult to say how much of his father's views carried over to him, or how much of these ideas he transmitted in turn to his sons. It is known that he also had letterhead with the three scallop shells and the inscription, although it is not known how much he used it. In his "*Autobiography*", Charles Darwin says (speaking of Dr Grant in Edinburgh). *"He one day, when we were walking together, burst forth in high admiration of Lamarck and his views on evolution. I listened in silent astonishment, and as far as I can judge without any effect on my mind. I had previously read the 'Zoonomia' of my grandfather, in which similar views are maintained, but without producing any effect on me. Nevertheless it is probable that the hearing rather early in life the views maintained and praised may have favoured my upholding them under a different form in my 'Origin of Species'. At this time I greatly admired the 'Zoonomia' "*. This seems to suggest that his father had passed on his admiration of Erasmus' ideas to his son, in some shape or form. Erasmus Darwin was clearly not a devout church-goer, but he would have observed decorum in this regard. Probably Robert shared these views. Robert's wife, Susannah, neé Wedgwood (1765-1817), was brought up in the Unitarian Church, which was the faith of Wedgwoods. In the "*Autobiography*" of Charles Darwin[32], his son, Francis, relates how his father was sent to the day-school of Revd Case, minister of the Unitarian Chapel[33], but that both he and his brother, Erasmus, were "*christened and intended to belong to the Church of England*". From later remarks in letters, we know that Charles Darwin supported the left-of-centre party, the Whigs and this was almost certainly true for Robert Darwin, too.

At a deeper level still, religion was to play an important and changing role in the zeitgeist of the age into which the young Darwin and his second cousin were thrust. Their grandfathers grew up in the Enlightenment, with its liberal views (for the upper classes) and a rather lax Church of England attitude to religion, which has been described as a comfortable social institution rather that an evangelical movement. Catholics, until 1829, and even beyond, were disenfranchised, only allowed to go to special schools and not allowed access to state positions or the institutions of higher learning. However, a sea-change was under the way with, primarily, the rise of the Methodists and other protestant faiths, which, together with the swing to the right, caused by the French Revolution, put pressure on the Church of England to act in a much more evangelical way and to overhaul its pastoral care, resulting, for example, in the Oxford Movement. For two people intent on entering the ministry of the Church of England, these effects must have come at a very personal level, estranging them from the attitudes of their fathers and grandfathers. And at a broader level it led to all those attitudes, which have subsequently been labelled "Victorian", and flavour the literature of the period, in such books as those of Dickens, Trollope and Hardy. And it was this swing to the right in terms of religion and politics that damned up the forces of rational belief, and made the reception to the Darwin-Wallace theory of organic evolution so much of a traumatic event when it occurred in the 1860s[34]. Indeed, echoes of this reception are still being heard today, in the early twenty-first century.

Meanwhile in Derby, Martha Strutt, the first wife Samuel Fox III, died, in 1798[35], leaving a son, Samuel Fox IV, who carried on the family silk business. Samuel Fox III remarried, this time to a Darwin. She was Ann, daughter of William Alvey Darwin of Elston Hall in Nottinghamshire, the older brother of Erasmus Darwin. How this connection came about we have no recorded facts, but it was almost certainly through the friendship of Erasmus and Samuel Fox III. From this second marriage of Fox came five daughters and a son[36]. The son was William Darwin Fox (1805-1880), the second cousin of Charles Darwin and the main subject, with Charles Darwin, of this story.

The connection to the Darwins was further cemented in about 1814, when the widow of William Alvey Darwin, Jane Darwin, moved to Thurlston Grange. This house had previously been the residence of Samuel Fox III, but in 1814 was freed up when he moved his family to Osmaston Hall. All the children of the marriage of William Alvey's son, William Brown, and Elizabeth de St Croix were born at Thurlston Grange (*Darwin Pedigrees*), with the births almost certainly overseen by the grandmother and by Aunt Ann Fox (see Appendix 4, note 123).

The political persuasion of the Fox family is less certain than that of the Darwins, but it is almost certainly on the side of reform and emancipation, given the family descent from the puritan clergyman, Timothy Fox, in the late 1600s. Samuel Fox

III, became a JP in 1821, was a Whig as he proposed John Ponsonby, a Whig, as a candidate for the House of Commons seat of Derby in 1835 (Derby Mercury, 7 Jan 1835), an election won by Edward Strutt[37]. We know that Samuel Fox did a great deal of good works in Derby from the 1820 to1850. He is also acknowledged by Glover (1829), who says "*This gent. used great exertions to relieve the distress of the frame-work knitters in the silk hose branch in 1820 and collected upwards of £400 for them*". We also know from the correspondence and diaries (see, e g, Letter 30 June 1832) that W D Fox was a keen supporter of the Whigs and the Reform Bills and voted Liberal, successors to the Whigs, in 1859 and 1865.

It is difficult from our perspective of the early twenty first century, to appreciate the great diversity of political and social opinion at the beginning of the nineteenth century. Like any society, it over-reached itself in some directions and was very backward in others. The eighteenth century is known as the Age of Enlightenment and certainly men like Erasmus Darwin fitted this ideal; but general practice fell far behind. The practice of press-gangs and flogging in the navy went on. Slavery was not outlawed in England until 1807, and even this act took many people by surprise, and was not outlawed overseas until 1832. We now look back on slavery and this period with a somewhat sanctimonious air. Yet, it is as well to remember that in 1859, when Darwin's *On the Origin of Species* was published, the United States of America was gearing up for a bloody Civil War, over slavery, which would last four years. We know that Charles Darwin was vehemently opposed to slavery, just as his grandfather had been. The new working class, personified here by the weavers of Derby, had no rights at all. Wages for weavers steadily dropped over the period 1770–1840. When the economy nearly failed during the French and Continental Wars of 1789–1815, wages were drastically reduced and many people were thrown out of work. This, together with the enclosures of the late eighteenth century, caused unrest right up to the time of the Tolpuddle Martyrs of 1834, and even beyond. There were various attempts to organise the workers but these were often repressed with fear and brutality. In and around Derby conditions were especially bad and are aptly illustrated by the "Luddite" Riots of 1811, where stocking-frame workers caused damage to machinery and organised protests. This led on in 1817 to the infamous trial of forty workers in the so-called "Pentrich Rising", the leader of which was Jeremiah Brandreth, a stocking-frame worker. Brandreth and two others were found guilty of high treason and were hung, drawn and quartered, although in a slightly a modified manner, outside the gaol of Derby, the last such execution outside London[38]. The crowd at the execution was very angry and had to be controlled by cavalry with drawn swords and militia. The poet Shelley attended the execution and wrote an account of it. The scene is well described in a letter by J D Strutt to his uncle, later Lord Belper, Edward Strutt[39], in which he condemned the proceedings roundly. The Grand Jury in this case had on it such people as Lord George Cavendish, Henry Cavendish, Sir Robert Wilmot, Richard Arkwright and John Crompton[40]. The latter two were both leaders of the textile industry in the Midlands and Sir Robert Wilmot was the owner of Osmaston Hall, which the Fox's rented from 1814 until 1865. So Samuel Fox and his son, who was twelve at the time of the execution, could hardly have been unaware of the brutal repression going on around them. It may be that the Foxes had little sympathy with this brutal stance but the social fabric was hard to break. The Derby "Mercury" from 1800 up to 1835 carried accounts of the local assises at which many workers were accused of sedition or worse and were either imprisoned, deported or sentenced to death; and Samuel Fox III was a JP from 1821[41] and is listed as either assistant to the judge or a member of the Grand Jury on many occasions.

The history of the weavers' strikes in Derby from 1820 – 1830 is dealt with in detail by Butterton (1996, 1997).

Political discontent throughout England in the 1820s, 30s and even 40s was high, and Derby, with its advanced industrial activity, was a centre of political activity from the end of the eighteenth century. The miners in the Peak district near to Derby were particularly active and on at least one occasion raided a local jail and set free prisoners; and after the defeat of the First Reform Bill of 1831, this happened again in Derby itself. The political system at this time, and up to the Third Reform Bill of 1832, and beyond, was corrupt with no fair balloting system and certainly no vote for the working classes. Even those who had a vote had difficulty voting in a government of their choice because of the system of "rotten boroughs" and tied constituencies and widespread bribery[42]. This must have made rising families such as the Strutts and Foxes very frustrated. We know from the correspondence between Charles Darwin and William Darwin Fox, that Fox was scrupulous in leaving time to go to his domicile (Derby) to cast his ballot, and was very pleased when the Reform Bill was voted in, finally, in 1832 (see Letter #48).

This is some of the background to the upbringing of young Charles Darwin at Shrewsbury and young William Darwin Fox at Derby. In both cases we know that they went to good Public Schools, nearby. Darwin went to Shrewsbury Grammar School (whose head at that time was the famous Dr Samuel Butler, 1798-1836) and often ran across the fields to home at lunchtime[43]; Fox did the same, although the distance was considerably longer, from Repton School, several miles south, to his home at Osmaston Hall, just outside Derby[44].

Another family that should be mentioned in this introduction is that of the Wedgwoods of Etruria, since Robert Darwin married Susannah Wedgwood, the daughter of Josiah Wedgwood I (1730-1795), and Charles Darwin married Emma Wedgwood, his cousin, the daughter of Josiah Wedgwood II (1769-1843)[45]. Josiah, father and son, were members of the Lunar Society. Josiah Wedgwood I, who came from a long line of potters, formed a very successful pottery, first at Burslem and then at Etruria, near Newcastle under Lyme, in the Potteries district of Staffordshire. He was a very inventive potter and introduced a number of lines that made his chinaware very popular and therefore very profitable. Josiah Wedgwood II continued in this mould, although he was more of a patrician. He produced the company's first bone china ware. He married Elisabeth Allen, of Cresselly, Tenby, Wales, in 1792 and had four sons and two daughters, one of whom, Emma, as we have seen married Charles Darwin, her cousin. In 1807 the family bought the residence Maer Hall, a 17th century stone-built manor house also near Newcastle under Lyme, and the family lived there until Josianh's death in 1843. This lively family, with a Unitarian

Church background and much free-thinking, was to play a major role in the lives of Charles and Emma Darwin, and, thus, also of William Darwin Fox. It is quite certain that the Foxes and Darwins would have known each other quite well[46], a fact that certainly would have impelled the young Charles Darwin to seek out and befriend William Darwin Fox at Christ's College, when he first went there in January 1828. It is also likely that the Foxes knew the Wedgwoods, since the latter's son, Hensleigh Wedgwood, also went to Christ's College, Cambridge, at the same time as W D Fox, and became a Fellow there (see Letter #10).

Thus our story should naturally begin at Christ's College in 1828, when Charles Darwin and W D Fox were there together, and Fox introduced Charles to beetles, natural history and the life, at that time, of an undergraduate at Cambridge. However, several written records go back before this and bear on our story. In particular we have the diaries of W D Fox from 1821 to 1828. And even more significantly, we have the *incomparable* diary of Fox when he was an undergraduate at Christ's College, just before he was joined there by Charles Darwin; this is given in Chapter 2 (and Appendix 4).

Notes

1. See Hutton, W. 1836. The History of Birmingham. J. Guest.
2. King-Hele, 1999. p. 25.
3. R B Freeman, 1978.
4. Mary Howard (1740-1770), daughter of Charles Howard, a lawyer. She was called Polly by Erasmus (see King Hele, 1999).
5. a) King-Hele, 1999, pp 64 and 82; b) DAR 227.5: 12. c) Hopkins, 1956, p 96.
6. See a) King-Hele, 1999, p 264; b) Horace Walpole on "The Loves of Plants" Part II The Botanic Garden: "the most delicious poem on Earth . . . all, all; all is the most lovely poetry. . .. The author is a great poet." Letters of Horace Walpole, ed W S Lewis (1737-1783) xi 10-11.
7. Fitton and Wadsworth, 1858.
8. R B Freeman, 1978.
9. Oxford Dictionary of National Biography (*DNB, ODNB*), 1917, 2004.
10. Derby Poll Book, 1710.
11. Derby Local Studies Library, Archives.
12. Derby Poll Book, 1765.
13. See details on soap boilers in the biography by Julia Blackburn, (1989). *Charles Waterton (1782-1865): Traveller and Conservationist*. Bodley Head, London.
14. a). Butterton H E, (1996). "*The Old Derby Silk Mill and its Rivals*". ISBN 0952910909. b). Page (1970).
15. Fitton and Wadsworth, 1958.
16. See, e g, Glover 1843. A History of Derbyshire.
17. Farey J, 1816. Agriculture and Minerals of Derbyshire. London, 1816.
18. a. Butterton 1996. b. History of Derbyshire, Glover, 1827, 1829 and 1843. c. Page, W. (1970).
19. See Fitton and Wadsworth, 1958.
20. Craven M and Stanley M, 2001. The Derbyshire Country House. Ashbourne. Pp 168-170.
21. Glover, 1846. But in 1857, White lists his address as 3, Regent Terrace, London Road.
22. King-Hele, 1999. p 188.
23. Sturges, 1979.
24. Sturges, 1979.
25. King-Hele, 1999. p. 277.

26. See King-Hele, 1999, pp 315-6.
27. King-Hele, 1999
28. King-Hele, 1999; Darwin C R 1878, Erasmus Darwin, Murray, London.
29. King-Hele, 1999. Erasmus Darwin: A Life of Unequalled Achievement.
30. King-Hele, 1999; King-Hele, 2003.
31. Victoria County History, Shropshire. A catholic cathedral was consecrated in 1856 on Town Walls and celebrated its 150th anniversary in 2006: see Shrewsbury Cathedral Web Site.
32. The original Autobiography was published in 1887 in the three-volume Life and Letters of Charles Darwin by Francis Darwin Nora Barlow, CD's granddaughter published an unexpurgated version with appendix and notes in 1958; more recently King-Hele (2005) published an unexpurgated version with notes.
33. In his diary (Darwin Online) CD relates (for 1817) "*Went to Mr Case's school in Spring. (8 years old)*". However the following Midsummer (1818) the diary relates "*Went to Dr. Butler's school.*" (That is, to Shrewsbury School.) So the young CD was only at Mr Case's school for just over 1 year. Before that presumably CD's sisters were in charge of both Erasmus' and Charles' education.
34. Interested readers are referred to Desmond, 1989; Glass, Temkin and Strauss, 1959; and numerous other books on the impact of the "Origin of Species".
35. Glover, S. 1833. A History and Gazeteer of Derbyshire. 2nd E, Vol 2, Pp 596-7, Derby. Supplemented by a pedigree compiled by W J C Waddington, given to Maxwell Craven in 1982.
36. See previous note.
37. Edward Strutt, son of William, and grandson of Jedediah Strutt, represented the borough of Derby as a Liberal from 1830-1847, then Arundel, 1851-2 and Nottingham, 1852-6. He was created Lord Belper of Belper in 1856.
38. a) Derbyshire, Vol 2 in the series "The Victoria History of the Counties of England." Publ. for the University of London, Institute of Historical Research. Ed, William Page, FSA. Reprinted from the original edition of 1905 by Dawsons of Pall Mall, London, 1970, pp 152-154. b) Derby "Mercury", 6th November 1817. c) Trials of Brandreth and others for High Treason. 1817. Vol 1 & 2. Taken in shorthand by William Brodie Gurney. Butterworth & Son, Fleet Street. Copy at Derbyshire Records Office, Matlock.
39. The Victoria History of the Counties of England, 1970, see note 38.
40. "Derby Mercury".
41. Justice of the Peace Oaths, 1st May 1821. Derbyshire Records Office, Matlock.
42. See, e g, Evans (1983).
43. "Autobiography"; refer to note 32.
44. Repton School records; and diaries of W D Fox
45. a) "Darwin Pedigrees" b) Dolan B, 2004, c) Wedgwood, B and Wedgwood, H, 1980, d) Hedley, 2002.
46. Letters of CD's sisters to CD, Vol 1 of *The Correspondence of Charles Darwin* (1985).

Chapter 2
Christ's College, Cambridge (1824-1826)

> *"We have often wondered how many months' incessant travelling in a post–chaise it would take to kill a man; and wondering by analogy, we should very much like to know how many months of constant travelling in a succession of early coaches, an unfortunate mortal could endure. Breaking a man alive upon the wheel, would be nothing to breaking his rest, his peace, his heart —"*
>
> Sketches by Boz. Charles Dickens

On October 19th 1824 a young man of nineteen made his farewells to his family on a country estate just outside Derby and set off in a horse-drawn mail coach to become an undergraduate at Christ's College, Cambridge. This was William Darwin Fox, second cousin to Charles Darwin, who would be joined by his cousin in the same College in 1828. The young Fox had attended Repton School, just outside Derby, and like several other school contemporaries chose to round off his education by studying for a Batchelor of Arts degree at the University of Cambridge, before entering the Church of England. For the son of a wealthy businessman with connections by marriage to many of the county's established families, including the Darwins, this was a fairly common sequence of events. Fortunately, on this very day W D Fox decided to begin a new diary, not using the small pocket diaries that were so common at the time (and used by Charles Darwin), and which he used previously and subsequently, but a half-quarto volume with ample room for detailed entries (see Appendix 1 and note 1). This remarkable document traces the daily life of the young Fox from October 1824 to June 1826. This allows us to follow the course of a fairly typical undergraduate of the time and, because he led such a similar early life to Charles Darwin, and went to the same College, it acts as a trustworthy mirror onto the college life of Charles Darwin himself. A large part of the Diary is given at the end of this Chapter and the rest is given in Appendix 4.

Christ's College, Cambridge, was formally established in 1505, under the initiative of Lady Beaufort, the mother of Henry VI. Later, in 1625, it was the college of John Milton and, just a few years later than this and despite being fairly broad in its fellowship, it was accused of harbouring "Latitude Men"[2] and "purged" of 11 fellows during the Civil War from 1642-1651. Nevertheless, the beautiful New

A. W. D. Larkum, *A Natural Calling*, DOI 10.1007/978-1-4020-9233-6_2,

Building (known today as the Fellows Building) was constructed in the troubled times of 1640-1642 and completed during the Civil War[3].

By the time that Fox, and then Darwin, arrived, Christ's College was a small, but for all that a fairly typical, Cambridge College, with 14 Fellows and approximately 60 undergraduates[4]. Like most Colleges it was essentially an ecclesiastical institution with strong links to the Crown, the Church of England and Parliament.

Cambridge University at this time was a loose association of Colleges, which were largely autonomous; however, academic degrees were organised and awarded under a centralised university body controlled by the University Senate. Academic standards were in general low, undergraduates proceeding to a general Batchelor of Arts degree, awarded on the basis of a single exam at the end of 3 years (or completion of 11 terms). Through this system were squeezed all undergraduates, whether they were to be high-flying mathematicians, historians and poets or professional men such as lawyers, doctors, and clergymen. Most Fellows of colleges were expected to enter the Church and thus most took up a benefaction in mid life and then juggled the responsibilities of a parish elsewhere with tutoring or teaching duties in the college. Teaching duties were not onerous, however, and some professors never delivered a single lecture!

Other professors at the time, such as Adam Sedgwick, Woodwardian Professor of Geology, took their duties very seriously (Desmond and Moore, 1991). For example, John Stevens Henslow, who became Darwin's most significant mentor at Cambridge (Walters and Stow, 2001), was a Fellow of St John's College, having achieved a high place in the mathematical honours list (and undergoing ordination into the Church of England in 1824). His bent, however, was towards the natural sciences and he first became Professor of Mineralogy in 1822, but not without a controversy, which caused him to change to a Crown-appointed Chair of Botany in 1825. Henslow was a key figure with Adam Sedgwick in founding the influential Cambridge Philosophical Society. He also gave intensive courses in mineralogy and botany. Later, however, the financial imperatives of family life meant that he became a vicar, far away from Cambridge, in 1832, and then a rector in 1847, and these responsibilities kept him away from Cambridge for increasing periods of time. Nevertheless, he was pivotal in the intellectual development of the young Darwin, firstly because he inspired the young man so profoundly and secondly because he was instrumental in obtaining for Darwin the position as companion to Captain Robert FitzRoy on HMS Beagle in 1831.

Change was in the air; by the 1820s, the monastic, inward-looking perspective of the University was changing rapidly. Younger members of the University were clamouring for change. Innovators such as George Peacock, from 1810-1820 and Whewell in the 1820s were urging radical changes to reform the curriculum and the examination system. The single examination in the eleventh term, which examined mathematics, classics and theological subjects, was now seen to be inadequate; but change came slowly. In 1822, two examinations were brought in, a compulsory one in fifth term, called the Previous Examination (or "Little Go") and a final examination, the Senate House Examination, in the tenth term, modified from the previous final examination[5]. The Previous Examination tested knowledge of the classical languages, latin and greek and christian theology. The latter was largely based a book loosely referred to as the "*Evidences*" written by the Rev William Paley, a Fellow of Christs's College in the late eighteenth century. The Senate House Examination was largely based on mathematics, testing a rudimentary knowledge of some Euclidean geometry, arithmetic and algebra[6]. This examination was held both in written and oral form. Scholars who passed the basic level were allowed to proceed on to an optional part, the Tripos, and the highest performers were awarded special honours, such as Senior and Junior Wrangler for that year. However, most undergraduates did not attempt this extended examination and were called "Poll" men. It was said that many Poll men were often able to memorise most propositions, rather than understand them! However, even here some difficulties were encountered. Brooke (1997) records how "*on the arithmetic paper, it was said, the examiners were 'obliged to be satisfied with less than one-third of the paper', otherwise they would have been reduced to the necessity of refusing degrees to a large proportion of the Candidates*". There were attempts to tighten up on the standard throughout 20s and in CD Letter #118, Darwin wrote to Fox "... *but I must read for my little Go. Graham smiled & bowed so very civilly, when he told me that he was one of six appointed to make the examination stricter, & that they were determined they would make it a very different thing from any previous examination that from all this, I am sure, it will be the very devil to pay amongst all idle men & entomologists.*" and in several other letters Darwin refers to anxiety amongst his fellow undergraduates studying for their exams. Many young men of a scholarly bent could do well in these exams. However, the situation was particularly dire for "Arts" students, especially poets. For T B Macaulay, it is recorded that "*in his Tripos in January 1822 he withdrew after two days knowing he would not achieve honours; he had not been able, wrote his father 'to sacrifice his miscellaneous reading to Mathematics'. Macaulay was gulphed (i.e. failed to achieve a sufficient standard), given a pass degree but not made to suffer the mortification of seeing his name in the list of Poll men*"[7].

It is not known for sure why Christ's College, Cambridge was chosen for the two Darwin sons in the early nineteenth century. As the Master of Christ's, A E Shipley, said at the time of the Darwin Centenary in 1909[8], "*Their grandfather Erasmus, of the "Loves of Plants", was at St John's, a college imminently connected to Shrewsbury, but at that period the life of an undergraduate at the sister Foundation seems to have been a somewhat troubled one, whereas, to judge from the expressions of contemporaries of Charles Darwin, Christ's was in their day 'a pleasant fairly quiet college, with some tendency to horsiness'* ". This may well be true but it should not be forgotten that the great, great grandfather of William Darwin Fox, Timothy Fox, went to Christ's in 1647. When Erasmus Darwin moved to Derby in 1781[9], he must soon have come into contact with Samuel Fox III (see Chapter 1); and when the latter married his niece, Ann Darwin in 1799[10] the connection must have been strengthened and the link with Christ's would have come to the surface (see Fig 6). This may explain why Erasmus Darwin sent one of his younger sons, John, to Christ's in 1809, having earlier sent his son, Erasmus, to St John's[11]. Charles Darwin's brother went to Christ's in 1822. Then William Darwin Fox, who was four years older than Charles Darwin, entered there in 1824 (see Fig 5) to study theology, which involved taking a general Batchelor of Arts degree (see note 8, Chapter 3). There is another connection too. Robert Darwin was closely connected with the affairs of the Wedgwoods and their pottery; he lent considerable sums of money to Josiah Wedgwood II and married Josiah's sister Susannah, Charles Darwin's mother. Josiah Wedgwood sent three of his sons to Cambridge, two of them, Frank and Hensleigh to Christ's College. Hensleigh the more scholarly of them all, was made a fellow of Christ's and was there when Charles attended in 1828. Thus Robert Darwin would have sought the view of his brother-in-law, Josiah Wedgwood, on the matter of where to send his sons and the name of Christ's College would very likely have come up.

As we will see in Chapter 3, there is very little written evidence on what Charles Darwin did at Cambridge from 1828 to 1831. The letters of CD to W D Fox are the main source on these years and most of these letters were exchanged when the two correspondents were away from Cambridge. Apart from letters to the Owen sisters and CD's brother or sisters, there are just 7 other letters of CD still extant for this period[12]. To be sure we have several pages devoted to this period in the "*Autobiography*"; however, this was written by CD in old age and therefore may be selective and inaccurate in detail. CD, at some stage after he had returned from the *Beagle* voyage, decided to keep a brief diary of the major events in his life. Retrospectively, he entered important events of his early life: the years covering the time that he was at Cambridge give few details about his time there. In 1831 he records: "*Christmas passed my examination for B.A. degree, & kept the two following terms.*" And later he records "*During these months lived much with Prof. Henslow, often dining with him, & walking with. became slightly acquainted with several of the learned men in Cambridge. which much quickened the little zeal, which dinner parties & hunting had not destroyed. In the Spring, paid Mr Dawes a visit, with Ramsay, Kirby & Henslow & talked over an excursion to Teneriffe.—In the Spring, Henslow persuaded me to think of Geology & introduced me to Sedgwick.— During midsummer geologized a little in Shropshire*". Prior to the time of publication of "*Life and Letters of Charles Darwin*", in 1887, his son, Francis Darwin, attempted to contact fellow students of CD, and several of their letters appear in "*Life and Letters*". Again, we have no guarantee that these are not selective versions of the truth. As a result, the 42 letters to W D Fox during this period, together with Fox's diary from 1824-26, constitute the most reliable source of information on how Darwin spent his time at Cambridge.

Fox seems to have kept a regular diary from 1818 to the year of his death in 1880. Unfortunately not all these diaries are still in existence[13], if indeed they did exist[14]. In particular the diary for the year 1828, when he was at Christ's with Darwin, is largely missing[15]. However, the diary for 1824-26 is extant and gives a wonderful description of the life of an undergraduate at Christ's College at the time. The first part of this diary is transcribed verbatim, with notes, in this chapter, and the remainder of this diary is given in Appendix 4. This account can also be augmented by other accounts of undergraduate life at Cambridge at the time (e g, Desmond and Moore, 1992; Brown, 1995; Brooks, 1997).

Rules at the twenty one colleges (at this time) varied greatly. At Christ's College gates were closed at 8 pm and undergraduates were not normally allowed out again before 6 am[16]. Each undergraduate was in the charge of a college tutor, who was responsible for his (there were no female undergraduates until 1870) moral welfare. Outside at night, the University employed Proctors[17] to police the streets. These were allowed to arrest both students and "certain women" and imprison them overnight in the University lock-up, if there were suspicious circumstances. Drinking and gambling were frowned upon. The students were also expected to be either in college or at lectures for a certain number of days per term, and this necessarily restricted them from taking time out to visit London or the races at the nearby Newmarket racecourse, etc.

As can be seen from his diary of 1824-1826, Fox and probably a good number of undergraduates broke these rules flagrantly. There was much card playing and drinking by students at night, and this inevitably involved late, or even early morning, entries into colleges by strategic external windows and other entry points. In Fox's case these "flings" were often followed by hangovers or even periods of considerable pain and lethargy. Often Fox would take some medicament or potion to help him over such patches. Yet, as these prophylactics included mercurials or arsenicals and other substances that are now known to be poisonous[18], it is possible that these alone would have caused long-term suffering. On at least one occasion, Fox with friends rode off illicitly to Newmarket to attend a race meeting[19]. Fox kept detailed records of his expenditure, and so we find him giving regularly 1 shilling to a beggar, which is clearly code for one further activity, to which he did not want to own up.

Nevertheless, the diary from 1824-1826 is a gem, not only in the context that it indicates what Darwin's life was like broadly at Cambridge just three years later, but also as a factual account of the life of a fairly typical undergraduate of the time. The diary for these years is set out in full after the Introduction (and in Appendix 4). It begins with the departure of Fox from Derby and his arrival in Cambridge. Fox stays the first night in Cambridge at an Inn and the reason for this becomes clear on the second day when he has to buy a bed for his room in Christ's College. Thereafter we see, in the daily entries, the life of an undergraduate, who begins to make friends and have parties, break college rules, attend lectures, employ a private tutor and attend chapel or church up to twice a day. Entries such as, "*I dined to day with Breynten of Maudlin, and after had a large supper party in my own rooms. Just before we sat down, Charles St Croix, (to my surprise) made his appearance. He would not stay supper, so I shewed him to an inn, and then returned. My supper went off very well except that Walton got very drunk and broke a good many Glasses* (*May 27. 1825)* make it clear that this was fairly typical for the time, despite the strict rules of the college. The entry, "*This morning I went with three men to see the Newmarket coursing meeting and did not get home till nearly seven, when I went and dined with F. Hall, and found a large party there. About 10 we began playing at Whist & as I won, I was obliged to stay on, till it got too late for me to return to College and I stayed there till Chapel time after which I lay down for an hour or two" (March 2nd 1825)* bears out the remark of The Master of Christ's College, Arthur Shipley in 1924, that Christ's College was known at the time for a good deal of "horsiness" (Shipley, 1924).

The diary is exasperating in the way in which Fox sets out menial details, with both too much detail, where it does not matter, e g, "*I attended Chapel twice and both lectures to day*" and not enough detail where it does matter, e g, "*There was a grand row tonight with the Snobs*[20]; *Fosbrooke and myself were once in the affray but luckily did not get hurt. There were several men very much hurt with brickbats, etc. I got to bed about ½ past 11, after taking some pills*". Who were the Snobs, who was Fosbrooke and what were the pills that Fox was taking? Sometimes detective work can answer these questions (and the reader is here referred to the notes appended to the diary entries), but on many occasions we are left tantalisingly in the dark. What subjects was Fox studying and what did he think about them ? Where did he struggle and where did he excel? What books was he reading? And what was the subject of general conversation at his parties? Perhaps it is unfair to ask of him to be another Samuel Pepys, yet what we do have is a wealth of original detail that we did not have before. And we should not forget that Darwin's personal journal for his three years at Cambridge takes up less than two pages and does not refer to life in detail at Cambridge at all![21] In fact, as mentioned above, Darwin did not write about this until late in life in his *Autobiography*, which was published posthumously[22].

One thing is clear from the Fox *Diary*, he was committed to natural history long before he met Darwin. His interests as shown in the Diary were varied and rich. They included mammals, birds insects and flowers. The pastime of collecting insects was all the rage at Cambridge at this time (Desmond and Moore, 1991) and Fox was caught up in this pursuit by at least 1825. So by the time he met Darwin in Cambridge in 1828 he was the local expert, who inducted Darwin into the new science of entomology. In later life Darwin spoke of this enthusiasm as mere "*stamp collecting*", but it was definitely more than this, for both of them, and can rightly be seen as the trigger for Darwin's scientific career. On a more problematic note the question has to be asked: how Darwin, ostensibly reading theology at Cambridge, could become the proponent of organic evolution just 30 years later? The thought of evolution was anathema to the conservative society of Cambridge, as evidenced by Adam Sedgwick's pained reaction to the *Origin of Species* in 1859. This is shown up starkly in a comment in Fox's *Diary* for 13th December 1826, "*The celebrated Monsieur Mazurier performed the part of a monkey afterwards in a most amazing manner. I never saw a man so active, or one that had such perfect command of his limbs. This performance was on the whole, tho' very extraordinary, one which led one to disagreable reflexions, on our similitude to the monkey tribe.*" Clearly, such thoughts were proscribed in the would-be clergy of Cambridge - and right into old age this was the one sticking point that Fox had with Darwin's ideas (see Chapters 7 & 8). The obverse of this is the more interesting conundrum – of how Darwin threw off these mental shackles and approached the "Holy of Holies": how one species transforms into another? The reader will find some clues to the answer in Chapters 4, 5 & 6. Much more has been written elsewhere.

The greatest mystery of the diaries is the affair that jumps out of the pages at Fox's entry for New Year 1825. Here he is saying "*My Follies of this Year have indeed been dreadful, now far more than a year ago I could have believed I ever should have been guilty of. They will serve to cast a shadow over, if not embitter, the remainder of my life.*" Is this the youthful hand-wringing of an idealist full of higher thoughts and bent on entering the Church? Or were there deeper undercurrents to an apparently normal life of a young man from the privileged middle class of that time? Who was "Fanny" (who had to be written in Greek - see note 63)? Was she just an early love or were his troubled thoughts the sign of some youthful sowing of wild oats? Perhaps we will never know. The diary preceding this one is for 1823, so the first nine months of 1824 are unrecorded. In 1823 Fox was at Repton School and he records an existence where he had little schooling and a good deal of shooting and high-spirited activity. There is reference here also to "Fanny", with references going as far back as to the diary of 1822. Many of the relevant entries are written in Greek, as if in code[23]. Also several pages have exclamatory statements and a symbol that recurs later: ※. So we have few clues as to what may have occurred in Derby, or at Repton, in the period leading up to Fox's departure for Cambridge on October 19th 1824. Then from the autumn of 1825 and through to May 1826, there is an unaccounted break in Fox's studies at Christ's College. At first during this period Fox was physically active, pursuing an almost daily round of game shooting; but later he became quite sick and confined to bed with swellings in his groin. Although he specifically denied, to his brother-in-law, Samuel Ellis Bristowe, that this was any kind of venereal disease (13-14 Feb 1826), the suspicion is still there that Fox was either hiding the truth from himself or from others; and it could, of course, have been some other kind of affliction such as glandular fever or incipient tuberculosis; nevertheless, a careful reading of the diaries suggests that Fox was sexually active at this time. Nevertheless, Fox does return to College for a few weeks in late May, 1826, before heading off for a recuperative spell by the sea side. This is where we leave him in person for the time being, as there are no more complete diaries until 1833 (some fragments suggest that Fox did keep a diary in the intervening years - see Appendix 5). Fortunately, the tale is taken up, in June 1828, by the letters of Charles Darwin to Fox, which begin in Chapter 3, and then go on right through to the death of Fox in 1880 (Chapter 8). In some of the intervening years diaries for Fox do exist and these allow for a greater understanding of the Darwin letters (Appendix 1). On some occasions, too, we actually have the reply of Fox to a letter from Darwin; unfortunately, these replies are all too few.

Fox went on to become a respected clergyman and family man, marrying 2 wives and having sixteen children, despite a life wracked with illness. The earlier sexual peccadilloes, if such they were, seem to be left behind, apart from some passionate letters to his sister-in-law after the death of his first wife (Appendix 5). While we may never know the real truth behind these records and our suppositions, they add, to say the least, a dramatic tension to the story of two men going up to Cambridge to study theology, with the intention of becoming clergymen.

With a note of irony it may be recalled that not long after this time in Fox's life, Charles Darwin himself was also involved in an early relationship with Fanny Owen, the daughter of Major William Mostyn Owen, of Woodhouse, near Shrewsbury. There is no clear evidence that this was more than "puppy love", as the correspondence of the time seems to bear out[24]. However, the continuing correspondence between CD and Fanny's sister, Sarah, long into adult life, indicates that this was no passing relationship.

Other parallels between Darwin and Fox can also be drawn from the *Diary*. Both young men loved to ride horses and go shooting, at this stage in their lives, pastimes that neither men pursued later on. This was clearly a passport to an exalted level of society for sons of professional or business men. Cobbett in his book "*Rural Rides*" of 1830[25], describes how the landed gentry talked about little else than shooting and shooting parties, which of course incorporated horse riding. As we have seen, Darwin spent much time at the estate of Major William Mostyn Owen at this time, but the attraction was not just Owen's daughters. Clearly to the chagrin of Owen's daughters, Darwin visited there even when they were away, and one can conclude that the shooting may have been an equal attraction[26]. In his Autobiography, Darwin stated "*How I did enjoy shooting! But I think that I must have been half-consciously ashamed of my zeal.....*"[27]. In the diary of 1824-1826, Fox records how much of his time at home was spent shooting and riding, either locally or on estates of family friends.

"*Last night there was a terrible fire at Linton eleven miles from Cambridge*", Darwin recorded in letter #9, "*seeing the reflection so plainly in sky, Hall, Woodyeare, Turner & myself thought we would ride & see it we set out at ½ after 9, and rode like incarnate devils there, & did not return till 2 in the morning: altogether it was a most awful sight.*" And again in letter #34 he says "*I left Shrewsbury on Tuesday & slept in Daventry, where I overtook my horse & rode him myself the two last days journey. The poor beast was so tired that he hardly knew whether he stood on his heels or his head. - & it will be sometime before I undertake to ride a young horse a long journey again.*"

A further similarity between the two men was that both came from fairly large families with a predominance of girls in their generation. Darwin had four sisters (3 older and 1 younger) and an elder brother. Fox had four sisters (2 older and 2 younger) and an elder step-brother. In both families the elder brother seems to have been amiable but unsuccessful in life and did not marry. In both families, the father was a dominant figure, but in the case of Darwin his mother died relatively early and had little influence on his development, whereas Fox's mother, Ann Darwin, was a strong partner to her husband, Samuel Fox III and certainly made her views felt in the later life of Fox (see Appendix 5).

Illness in mature life was another common bond. The afflictions and illnesses of the young Fox have already been recorded. Yet this poor health may have been partly hereditary. From the diaries we learn that Fox's father was very sick during the mid 1820s. Later in a letter from Caroline, Darwin's sister to CD on the *Beagle* in 1835[28], she says "*I heard from W^m. Fox last week. He talks of spending the summer at Barmouth or Beaumaris & paying us a visit on his road – his poor wife has been exceedingly ill all winter, she had a dead child & never recovered her strength – what a pity in that sickly family that M^rs. Fox should prove more delicate than any of them –*". In later life, Fox records medical advice that he was afflicted with poor lungs and poor blood supply from the heart (Letters #175, 177 & 184). These conditions seem to have impacted on his glandular problems in 1825-27. Subsequently he seems to have overcome his glandular condition, but was susceptible to lung infections exacerbated by overwork, and this seems to be the major reason why he gave up his curacy at Epperstone, in 1833 (Letter #48). Admittedly, Darwin showed few signs of illness prior to the *Beagle* voyage, except for palpitations and sore lips while waiting for embarkation. Yet for the rest of his life Darwin was beset by continuous ailments and ill health, of which much has been written[29]. Perhaps we should let Darwin have the last word on this, in his Autobiography: "*Even ill-health, though it has annihilated several years of my life, has saved me from the distractions of society and amusement.*"

As a coda to the diary of Fox and to this chapter, we may note the detailed financial records made by Fox (set out over 26 pages at the end of the diary – and not transcribed here). These detail the daily expenditure throughout this period, and not only present an invaluable record for historians, but also testify to significance that such records played in the life of Fox. We have no such evidence in the case of Darwin, either in the maintenance of daily records or financial records. However, we see in the letters to Fox during the Cambridge years, e g, letter #16, a considerable attention to financial accountability. It may reasonably be concluded then that Fox was the model that Darwin followed in this regard. It is also possible that the record taking of Fox influenced the young Darwin. When one looks at the journals that Darwin kept on the voyage of the Beagle, one might see the diaries of Fox as trivial and inconsequential. However, it is likely that Darwin became aware at Cambridge of the daily records that Fox kept and was impressed. In a letter dated 30th June 1832 (Letter #48), to Darwin on the *Beagle*, Fox wrote, in his typical verbose style, "*I have often regretted one trait of your character which will I fear prevent your making so great an advantage as might do from your present travels, and which I regret also very much on my own account, as I might perhaps get the perusal of it; - I allude to your great dislike to writing & keeping a daily methodical account of passing events, which I fear (tho' I have also hope the other way from the overwhelming influence of every surrounding object) will prevent you from keeping a Regular Journal. – If you do not do this, the vast crowd of Novelty which will surround you, will so jostle ideas, that to say nothing of the many that will be lost altogether, the vivid reality & life which a memorandum taken at the moment gives to every passing event & thing, is done away with. – With this one exception (which I dare say you have overcome) I know of no one so fitted altogether for the expedition you are engaged in.*" With hindsight one can smile at the irony of this remark and wish that Fox's letter had some of the vivacity of a Darwin's. However, at the time it was probably exceptionally percipient advice, which was well taken.

Part 1 of the Diary of William Darwin Fox from Oct 1824 to June 1826 (Transcribed from DAR 250: 5[1])

(Note that Venn (1944) *Alumni Cantabrigienses, Vol 2, 1752 – 1900* and Peile (1913) *Biographical Register of Christ's College, Vol 2, 1505 –1905* have been used together with the Dictionary of National Biography and other source material to establish the identity of persons named in the Fox *Diary* and the Darwin letters. If a person cannot be identified it probably means that, if they were an undergraduate at Cambridge, that they did not go on to take out a degree, and were therefore not always recorded, although some, such as Henry Charleton (see note 54),were.

Note, also, that the use of italics for the days, dates and page numbers has been done for ease of reading the text; they were not underlined in the original text.)

William Darwin Fox
Christ College
Cambridge

Osmaston Hall
Derbyshire

October 19th 1824[30]

Tuesday

I set off this morning at one oclock in the Regulator[31] for Leicester where I arrived about 5. Fosbrooke[32] and Wood[33] with his Father were on same Coach. As soon as we got in, we all dined together after which we went to the Play. It was the most miserable performance I ever saw.

Wednesday 20.

This morning we all four got into the Cambridge Coach at eight oclock and got into Stamford for dinner, I called upon the MacGuffays. We got into Cambridge about 9 in the evening. So of course I did not see much this night. I supped with Fosbrooke[32] at his College and slept at the Red Lion. I had a very pleasant journey upon the whole as we had no rain and moderate fine weather.

p 2. October 21 1824

I breakfasted at ½ past 8 with Fosbrooke after which I waited upon the Tutor[34] who shewed me my rooms in College[35]. I was very busy ordering Furniture &c and walked about the Town with Fosbrooke[32]. I found out Boston[36] who was just come up, I dined to day in Hall and afterward wined and drank tea with Fosbrooke. I managed to get a bed put into my room and slept to night.

22$^{\underline{nd}}$

I breakfasted this morning in my rooms tho' they were in quite confusion. I met Brewin[37] and F Hall[38] to day and was with Foster[39] most of the day, I dined in Hall and attended Chapel twice[40] today. I supped and stayed late with Brewin; and met there Wood[41] an old Reptonian[42].

23$^{\underline{rd}}$

Power[43] called on me this morning and took me about with him. I dined in Hall and attended Chapel twice to day. I had Wood[40] and Hall

p 3.

junior[44] to tea and wine with me; I wanted to get acquainted with the latter. I found Roberts[45] in the Town to day.

24th.

I attended Chapel to day, and St Mary's in the afternoon. I wined with Power and got another cut. I came home and got to bed about 11.

Monday 25th
I attended Chapel twice to day. We had lectures to day for the first time, at both of which I attended[46]. I wined to day with J Brewin[37]; My box arrived to day from home. So I began to get more settled.

Tuesday
~~I attended Chapel twice to day and lectures. I had Wood and Frank Hall~~[43] ~~to tea with me. I got to bed pretty early.~~

~~Wednesday 27th~~
I attended Chapel in the morning and lectures; and I afterwards went snipe shooting to the Fens with Fosbrooke[32]. He killed 1 snipe and I nothing. I dined with him at//

p 4. Tuesday Oct 26th. 1824
his rooms on our return. I got to bed about 11.

27th
This morning I had Power to breakfast with me early, after which we set off together to the Fens shooting. We killed 5 snipes and a partridge. I killed 2 snipes and the partridge. Power dined with me and I after supped with Hall at St Johns. I did not go to bed till ½ past 12. For missing lectures this morning I had a week of my term taken from me[47]. I attended Chapel at night. Mr J. Theobald called at my rooms while I was out.

Thursday 28th
I was dreadfully stiff this morning with being wet yesterday. I attended Chapel twice and lectures. I supped and spent the night with Power. We played a rubber at Whist. I won 14s. I got to bed about 12. I wrote a long letter to my Father to day, and received two newspapers from home.

p 5. Friday October 29th 1824
I attended lectures and Chapel once today. It was a rainy day and I did not stir out much. Power, Fosbrooke and Beecher[48] wined with me. I got to bed early.

Saturday October 30th 1824
I attended Chapel twice and both lectures to day. I began Euclid[49] to day with Joseph Power as my Private Tutor, and liked him well. I spent an hour to day with Frank Hall who was very *unwell.* I brought to day a celeret and paid £1″1 for it upon delivery. I also had my sopha in to day. I supped to night with Wood on Oysters and Snipe. Got to bed about ½ past 12.

Sunday. October 31st
I attended Chapel twice to day; I read all morning. I had Wood and a man of Christ named Evans[50] to wine with me and drank tea. I wrote to Jo Walker[51] and Jas Holden[52].

Saturday Nov 1.
I attended Lectures this morning, and went to my private Tutor.

p 6. November 1 1924
I also attended Chapel once. I was in my room most of the day, except an hour that I sat with Frank Hall[53] who was confined to his bed. I drank tea with Fosbrooke and got to bed very early.

Tuesday Nov 2nd
I attended lectures &c as usual, and went to Chapel once. Chawner[54] called on me to day. I made acquaintance to day with a man of the name Charleton[55] of this College. I wined with Wood, but drank nothing as I was not at all well. I received 6 dozen wine from London to day. At night I was entered into the Union club and attended its opening for the term. There was some very good speaking by Hilyard Pread &c[56]. – Fosbrooke returned with me and we had coffee in my rooms.

Wednesday. Novr 3, 1824
I to day attended all as usual and chapel twice. I felt very unwell all day, and was out but a short time with Power, and with F. Hall at his//

p 7. November 3rd. 1824
room; he was still confined almost to his bed. I ordered a new waistcoat today of Mr Peter, tailor. I sat up rather late reading Euclid.

Thursday November 4th.
I this morning received a long letter from home, and one from Bristowe[57]. All were quite well. It was a beautiful day. I went to my Tutor and lectures and Chapel once. I began to starve myself today, as I did not feel *well*. I supped with Wood and spent a very pleasant morning there, but did not drink anything.

Friday. 5th
This morning was a beautiful day, a sharp frost, all the men were turning out hunting and shooting but myself. I attended Chapel twice and my Tutor. I walked about with Power and Fosbrooke. I dined with former on Mutton broth. I went this afternoon and consulted Mr Ficklin[58] and had some medicine from him. There was a grand row with the snobs; Fosbrooke//

p 8. Friday November 5th 1824
and myself were once in the affray but luckily did not get hurt. There were several men very much hurt with brickbats &c. I got to bed about 1/2 past 11, after taking some pills.

Saturday. 6th
As I was at breakfast this morning Frank Wilmot[59] came in. He had come up the night before. I took an immense deal of physic today which kept me home most of the day. I did not attend lectures or Chapel to day.

Sunday 7.
I continued taking medicine to day which kept me at home. Fosbrooke drank tea with me, and Chawner, Brewin, and Fry[60] joined us at it. I got to bed rather early.

Monday 8th
I saw Ficklin again to day, and continued my Physic. I got an ægrotat[61] to day, so that I continued keeping my term, although I attended nothing at all. I had some broth for my dinner.

p 9. Monday November 8th
I received a letter from Bristowe this morning, saying he had sent some wines for me. I got to bed early.

Tuesday 9th

I this morning went to my Private Tutor, but was in all the rest of the day. I sent Bristowe 1 brace of teal a brace of wigion and 3 couple of Snipes to day. I took a great deal of Physic still, and got no meat. I wrote at night to Bristowe, Miss Woodcock[62] and φαννι[63] and sent the Former off to night. Wilmot[64] drank tea with me.

Wednesday. 10th

I was in my rooms all day. And took a good deal of medicine. I went in the evening to sit with Evans. We played at van John[65] and I lost 2£. I made a resolution never to play loo or van John again in Cambridge[66]. I did not get to bed until 2 oclock.

Thursday 11th

I had my rooms stained this morning. I went to my private Tutor.

p 10. November 11th. 1824

I was out a little in the morning and in the afternoon Wood came and drank with me. At night I read thro Medwin conversations[67] of Lord Byron, and was very much disappointed in them, as I had seen all the most interesting parts in the Papers before.

Friday Novr 12th

I went to my tutor this morning and he afterwards took me to the FitzWilliam Museum[68], with which I was much pleased. I received a letter from J. Holden to day. At night Power Wilmot and Fosbrooke came and played a rubber in my rooms, and Evans and Williams joined me at supper. I lost 13s. I still kept on with my Physic and got to bed about 1 oclock.

Saturday Novr 13th

I heard this morning from Miss Woodcock[62]. This morning was Matriculation day and accordingly we i.e. the freshmen of Chr went about 1 oclock to the Senate House and//

p 11. November 13th. 1824

after writing our name down took the oath. I had nobody with me to day at all. I kept up taking Medicine. I wrote at night a very long letter home.

Sunday Novr 14

I was unwell this morning. Mr Ficklin came and gave me some fresh medicine which soon made me feel better. Fosbrooke came and sat with me most of the day. –

Monday. 15.

This morning I felt pretty well again and went on taking physic as usual. I received to day a dozen Red wine from Bristowe. I went at night and sat with Charleton and played at van John. I neither lost nor won. I went to bed about 12. –

Tuesday. 16

Early this morning I sent home a dozen very fine Snipes. I was in my rooms most of the day. I had Charleton, Bond[69], Williams[70] & St John[71] to wine with me. I did not get to bed till late.

p 12. November 17. 1824

This morning I went to my Private Tutor and afterwards went to see some Wild Beast Show. I gave St John 8£ today for liberty to buy 5£ worth of Prints in a shop which I accordingly did this morning[72]. I reflected seriously this afternoon about my present life and hope to improve it[73]. I went to bed in good time. I began to take Cabebs[74] &c I heard from Miss Woodcock.

Thursday 18th.

I went to my private Tutor this morning, and to day gave up my Ægrotat, as I felt rather better. I drank tea with Fosbrooke, and I went with him to an Auction, where I bought a large latin herbal, for 5/6[75]. I went to bed early.

Friday 19th

This morning I went in to lectures as usual. I afterward took a row with Evans up the River for an hour. I attended Chapel twice to day. Got to bed early.

p 13. November 20th. 1824

I attended lectures and my Private Tutor to day as usual. The rest of the day I was chiefly in my rooms. Wilmot asked me to sup with him but I declined.

Sunday Novr 21st.

I attended Chapel and Church to day and except walking a little with Fosbrooke was in my rooms much of to day.

Monday. Novr 22

I attended lectures this morning and did not go out much, as I felt very unwell. I went to bed in good time. I was with my private tutor. I heard from φαννι[63]. & . 1λ[76] well.

Sunday. Nov 23.

This morning after lectures saw my Tutor. I went walking and bought a young Poodle Dog[77]. At night I went to wine with Wilmot, and afterwards went to a concert here, Miss Straus, Braham and Lindley were the Chief performers[77a]. I supped with Brewin after and did not get to bed till late. Was not at all well.

p 14. November 24th. 1924

I this morning received a long letter from home. All well, except my father, who had been rather unwell, but who was now better. This morning I attended Lectures, and was the rest of the day in my room. I saw Ficklin again this morning and as I was not so well again as I had been, he gave me some physic. I went to bed early.

Thursday 25th

I went to Lectures, Tutor Chapel &c as usual. I heard to day from Fosbrooke, that he was unwell. I went to see him, and found him very much so. I got to bed in good time. I dined with F Hall.

Friday 26th

I was much better to day. I attended lectures &c as usual. I had a party at my rooms to breakfast this morning, and went to Charletons to wine. Fosbrooke was much better today.

November 27th – 1824

I was not at all well this morning having had a very bad headache. I dined with Fosbrooke who was much better and afterwards went to Wilmots Rooms to play a Rubber. I lost at Shilling Points 3 sv. I left his rooms early and got to bed.

Sunday 28th
I breakfasted this morning with Charleton after Chapel. I was in my room most of the rest of the day, as I did not feel at all well.
Monday 29th
I this morning attended lectures, Chapel and my Tutor as usual. I wined to day with Williams, and afterwards went to play a rubber and sup with Chawner & won about a Pound. I was rather late in evening into College, but I heard nothing of it afterwards.
Tuesday 30th
I was in my rooms reading all day, except when at lectures.

p 16. Wednesday December 1st. 1824
I attended lectures and my Tutor as usual. At night I had Power, Wilmot, Brewin, Wood, Hall, Chawner, Evans and Charleton to wine, cards and supper. I kept my term today. I won 1£
Thursday, 2
This morning I went to lectures &c as usual. I dined with Power to meet James Jackson, who had come over from Oxford. There was a grand concert to night but I did not go.
Friday 3rd
I was not at all well this morning and was in my room a good deal. I attended lectures and my private Tutor as usual. Got to bed early.
Saturday 4th
I was very busy to day, as after attending lectures I got my Exeat[78] from Shaw, and paid him his bill. Charleton left Cambridge to day. I was buying things to take home all day, except the time I was with my private Tutor. I wined with Wood.

p 17. Sunday December 5th 1824
This morning after Chapel, I was packing up most of the day. I had Holden and some others to wine with me, and did not finish packing till late.
Monday 5th
This morning I got up at ½ past 5 and breakfasted with F. Wilmot, who came to see me off. I then set off home in the Leicester Coach. I got safe to Leicester at about 9 oclock.
I got to bed about 11.
Tuesday. 7.
This morning I left Leicester early, and got home about 10. I found all at home, and all well except my Father who had been very ill, but is now I am glad to say much better. All appeared glad to see me, and every thing seemed much as I left it except perhaps that my long absence now made things appear more delightful than before.

p 18. December 8th. 1824
This morning I rode over to Derby and called on Mr Strutts[79] and Mr. Hadens. The frost was breaking up fast.
December 9. Thursday
I went out shooting this morning in Osmaston and killed a hen and a rook. Holden returned from Oxford today.

Friday 11
I was out shooting this morning but killed nothing.
Sunday 12
I heard to day from J Holden. All was going well at present.
Wednesday 15th
This morning I went shooting at Trusley, but never got a shot all day. At night I received a letter from φαννι. She mentioned the death of Miss Chortus[80] in it. It shocks me dreadfully tho' as yet I think it//

p.19. Decr 16th. 1824
must be Mary C. if not it is indeed a dreadful visitation. I found out today it was Miss H. Chortus[81]. This morning I set off by Nottingham Mail, to go to the Cricket Ball. I drank tea with M A[82] and Bristowe and we afterwards went to the ball. There were nearly 400 people present. It was dreadfully hot before Supper but pleasant afterwards. I did not leave the ball Room till nearly six oclock when I got to bed.
Wednesday 22nd
I did not get up this morning till 12, when I went and breakfasted with Bristowe. Afterwards Bristowe returned to Beesthorpe, and M A Boteler[83] and myself returned to Osmaston in his carriage. Upon the whole the Ball certainly was a pleasant one.
Friday. 24.
I received a long letter from φαννι I dispair of getting any circumstantial account, without by language.

p 20. 25th Dec. 1824.
This morning being Christmas day we had Mr Bligh[84] to dinner.
** (28th Tuesday)**
This morning I set off in the coach to Mr Jedediah Strutts[85] to stay a day. I got there to breakfast and found M^{r} and Mrs Henry Walker staying there.
**(29th Wednesday*) vice versa*
This morning I was at home reading.
30 Thursday
I slept at Belper and about 1 oclock called at Bridge Hill, after which I returned to Osmaston with Mr Miss and Mrs Strutt in their carriage.
31. Friday.
I was at Derby this morning with Mr J Strutt. I wrote φαννι a long letter. Sent her £5. At night my Mother, Emma[86], F.J.[53], Julia[53] and Bristowe with Mr and Mrs Strutt and myself went to the Derby Ball. We got there much earlier than was pleas//

p. 21
-ant. It was a most excellent Ball and one of the pleasantest I ever was at in Derby. We did not leave till nearly four oclock........
Tho I ended this year so pleasantly, for the moment, yet, the remembrance of it, will I fear, be ever painfull to me; If ever any one was miserable I have been so in this (to me) eventful year. May I never pass such another. I hope the resolutions which I have now formed I may be enabled to keep. I will (if possible) pay off my Debts. And if I can get round, which by a course of economy I may yet, I hope for the

future to live regularly within what I have. My Follies of this Year have indeed been dreadful, now far more than a year ago I could have believed I ever should have been guilty of. They will serve to cast a shadow over, if not embitter the remainder of my life.

p. 22 blank
p. 23 blank

p. 25 Jany 1st 1825
This morning I went to Derby and saw Frank Hall and engaged to go to his house next week.
Monday 3rd
This morning I left Osmaston for Beesthorpe in Bristowes Carriage. We arrived there to a late dinner, and found all well and right.
Tuesday
Bristowe and myself were out shooting most of the morning, but had very poor sport. I only killed one hare.
Wednesday.
We were shooting again this morning and had much better sport, as we killed 2 partridge and 8 hares, of which I killed 1 bird and 3 hares.
Thursday. 6th
This morning I got up at 4 oclock and went by coach to Mansfield from where I walked to Park Hall, and found Frank Hall there. The family consisted of his Mother, a Sister and another brother, both younger than himself//

p. 26
January 6. 1825
Upon the whole I liked them almost as soon as I saw them. We went a shooting a short time but killed nothing. About 8 in the evening we set off to Col Needs, to a dance given there. It was a very pleasant one, and I never spent a more agreable evening. I met one or two I know but the rest were all strangers, but I was introduced to everybody. It was kept up till 4 oclock when Hall and I returned and got to bed.
Friday 7$^{\underline{th}}$.
I got up about 10 oclock, and I a short time after breakfast left Halls and went in his Gig to Mansfield, from where I returned to Beesthorpe per coach; I found Bristowe just at dinner with Becher. We afterwards went to Newark Ball, but I was too tired to enjoy it much. Tho' I danced a good deal.

p. 27 Jany. 8th. 1825
I found myself a good deal tired this morning. In the evening Bristowe and myself went to the Wood and killed 3 pheasants. I killed 2 of them. When we got back we found the Edward Galtons[87] and Edw[88] and Mr. J Strutt[89] arrived to stay at Beesthorpe some days.
Sunday.
We walked about and attended Church today. Miss Bakewell came at night to dine and stay.

Monday
This morning after breakfast I went to Newark and from thence to Osmaston per mail. I got there to tea and found all well. I had a sore throat when I left Beesthorpe in the morning, which increased towards night, and I felt so ill that I got to bed almost as soon as I got home, feeling very unwell and shivering.

p. 28 Jany 11th. 1825
This morning I found myself worse and was in bed most of the day.

Friday 14th
I this morning for the first time this week got up to breakfast, and was rather better but had headache all day.

Monday 17.
As my headache still continued, &c, I was very far from well. My mother sent this morning for Mr Haden, who came, and said I was weak and must be quiet, and take strengthening Medicine which I accordingly did.

Friday 21
Bristowe was here to day; he came yesterday and left M A well. This headache which I had had all the week began to night to leave me and I felt much better.

Monday 24
This morning I felt so much better that I rode over to Derby, where I saw Mr Haden. I still continued//

p. 29 Jan$^{\underline{y}}$ 24th. 1825
to take Medecines. I took a letter this morning for Miss φαννι. How I do wish I could see her, tho' for a moment, but this attack will make it impossible. I paid Chapman on acct 4£ to day.

25 and 26
I continued getting better both these days and kept at home.

27 Thursday.
I this morning went to Derby to Mr Robertson the dentist, and had what was necessary done to my teeth. I was almost well, to day but towards evening found I had caught a cold which on the following morning proved a bad one, and threatens to lay me up some days longer.

Saturday 29.
M$^{\underline{r}}$ Haden came this morning; I was much surprised at not hearing from φαννι to night. I sent a letter this morning to Frank Hall to tell when I hoped I should be able to go to Cambridge, if I continued fairly well, as I now hoped.

p. 30 January 31. 1825
I went to Derby in the carriage this morning, and walked back as I was got nearly well again. I received a letter to night from φαννι, and found all was right and well in that Quarter. I paid this morning at Derby Mills and Brigger's bills.

Feby. 1 Thursday
I was at home all day busy getting ready for Cambridge.

Wednesday 2.
I went to Derby this morning, and breakfasted with Mr Haden. I saw Willoughby Mundy there, who was going to Sandhurst per coach.
Thursday 3.
I rode over to Thurlston[90] this morning, and stopped dinner there. There was a good deal of snow fell in the course of this night and it froze very sharp.
Friday 4.
There was more snow this morning. I was over to Derby.

p. 31 February 4. 1825
At night my Mother and Eliza and Frances went to a Ball at Derby. It froze very hard indeed.
Saturday 5th.
I was at home all to day packing up my things for Cambridge.
Sunday 6.
The frost still continued. We had morning service.
Monday 7.
This morning early I rode to Derby and saw M$^{\underline{r}}$ Haden. I then returned home, and after making an early dinner set off for Leicester by the Regulator, where I arrived safely about ½ past 5 after a rather pleasant ride. I got some dinner at the Crown and slept there.
Tuesday 8$^{\underline{th}}$
This morning I went on to Cambs where I got rather late. I had some pleasant companions most of the day, and found my room quite ready for me on my arrival.

p. 32 February 9$^{\underline{th}}$ 1825
I got up late this morning, and called upon several men after breakfast. It was a beautiful day. I wined with Fosbrooke and got to bed in good time as I felt rather unwell.
Thursday 10.
I attended Lectures this morning and afterwards walked about Town with Evans and Williams. I bought a few coins this morning. I had 6 men to wine with me to night, and we spent a very pleasant evening.
Friday 11$^{\underline{th}}$
I attended Chapel and Lectures this morning and afterwards took a row in a boat for a short time with St. John and Hall.
Saturday 12th.
I found this morning that I had a bad cold and a slight sore throat, owing I supposed to my going on the water. I kept in all day and got to bed early.

p. 33 February 13th.1825.
I lay in bed till late this morn. I wined with Wilmot, and went afterwards to Trinity Chapel to hear the singing.
Monday 14.
I found my cold still very troublesome tho' not quite so bad. I walked over to Chesterton this morning with Power and I stayed there several hours while he played billiards. As we should have been late for Hall we dined there. I wined with S$^{\underline{t}}$ John. I wrote home to day.

Tuesday 15.

This morning, Evans, Williams, St John and I got some Pigeons to shoot at, and after lectures went to Chesterton where we shot them. We could none of us kill more than 2 out of six from the trap at 15 yds.

Wednesday 16.

I was in most of the morning and wined with F. Hall. I afterwards plaid a Rubber in Williams Room.

p. 34 February 17. 1825

I began to day an University set of Lectures by Professor Farish on Experimental Philosophy. I wined with Wood to day and spent the evening.

Friday 18th.

I attended Farishes Lectures to day as I hope to do every day.

Saturday 19th.

After breakfast this morning I was weighed and was 10 stone 9 lbs. I set off at ½ past 11 this day to ride with Williams, Evans & St John. We went and saw Dr S Cottons and Sir G Leeds Houses. They were not worth going so far for. We went so far we did not get home till near six, when we dined at the Greyhound and wined in my rooms. As I had not ridden lately and as we had run about 40 miles very hard, besides <leaping>, I was dreadfully tired and got to bed early.

p. 35 February 20. 1825.

I found myself so stiff this morning that I could scarcely walk so kept rather quiet. I attended Chapel and also went to Kings in the morning. I drank tea with Fosbrooke. I heard to day from home, they were all well.

Monday 21.

I had a large wine and card party to day, and sat up rather late. St John and Robertson slept on my sopha as they did not go home.

Tuesday 22.

I found myself so unwell and bilious that I sent for Ficklin, who came and gave me some physic. There was row in Christs and Yarborough was expelled for striking the Porter. I was much better towards night.

Wednesday 23

I was very unwell this morning and was unable to attend any Lectures. I was in room all day.

p. 36 February 24th 1825

I was still nearly confined to my room but was better than I had been.

Friday 25th

I attended lectures &c as usual today again.

Sunday 27.

After Chapel this morning I sat down and wrote a long letter to Julia. I sat with Fosbrooke tonight till about 8 and got to bed early.

Monday 28

This morning after lectures I rode out with St John, Williams and Bond to try to meet with the Hounds but could not. I thought my ride did me a great deal of good. Wined with Robertson.

Tuesday 1
This morning I went with three men to see the Newmarket coursing meeting and did not get home till nearly seven, when I went and dined with F. Hall, and found a large party there. About 10 we began playing at Whist &//

p. 37 March 2nd 1825
as I won, I was obliged to stay on, till it got too late for me to return to College and I stayed there till Chapel time after which I lay down for an hour or two.
Wednesday 2nd.
I was very stiff and tired this morning and got to bed very early at night, as I felt very sick and ill.
Thursday 3rd
I did not feel at all right this morning, so after lectures I took a ride with William Evans and St John to the Coursing Meeting again at New Market, and found my self at night tho' rather tired much better for it.
Friday
I took a constitutional ride this morning and was much better after it.

p. 38 Sunday 6th
I had several Men to breakfast with me this morning after Chapel. I called upon several Men to day and was walking most of the day.
Monday 7th
I after lectures this morning got a horse and went with Evans and Williams to meet the Harriers. We had a pretty run with a hare which we killed and afterwards run a fox for about half an hour. I was much better for my ride
Friday 11th
This morning I got a good hound & went with Evans and Williams to meet the Gamsden Hounds at Caxton. It was a miserable wet day, and we went all the course and country round till about 4 oclock without finding, when we returned home. It was my first days hunting, and a very unsatisfactory one. We all dined at the Greyhound.

p. 39 March 12th. 1825.
I was rather stiff this morning and kept in my rooms most of the day.
Monday 14.
I took a Hack this morning after lectures and found the Hounds at Madingly. I stayed with them some time but they did not find. After I left them they had a very good run.
Friday
I this morning got my exeat from Shaw and packed my things. I had about half dozen of Men to wine with me.
Saturday 19
This morning early I set off by Stamford Coach to Leicester where I arrived very safe and comfortable about 8 oclock. I got some dinner and coffee and then went to bed.

Sunday 20.
This morning I left Leicester about 9 oclock and got to Osmaston just as they were finishing breakfast.

p. 40 March 20. 1825
I found all well. My father very much better than when I came to Cambridge. Eliza and my Mother were with M. A.
Tuesday 22.
I rode over to Derby this morning and went to the Assizes, where I heard a few trials, and heard one man condemned. I called on Ed Strutt whom I found very well. The Miss Strutts were still very unwell indeed.
Wednesday 23.
I went for a short time this morning with Julia, and spent the rest of the day at home. I wrote to φαννι.
Friday 25
I went out a shooting this morning before breakfast to the old Derwent and killed a Snipe and a Waterhen. The rest of the day I was at home.
Saturday 26
I shot another Snipe this morning before breakfast. Bristowe returned to Beesthorpe this morning. My Mother came from B. this morning, and I think looking//

p. 41 March 26. 1825
uncommonly well. Eliza still remained with M. A. whom my mother left much better.
Sunday 27
We had Evening Service to day, and Sam and Mr Bligh dined here.
Monday 28.
I went to Derby this morning, and paid several bills. I dined with the Hadens, and after a dull evening returned home about 11. Ed Strutt returned to London to day.
Friday. April 1.
We had morning service this morning, and Sam dined here. The weather was beautiful to day, just as it had been for the last fortnight, quite like Summer.
Sunday 3
We had morning service to day. Sam dined here.
Monday 4.
I called to day at Darley[91], where I found F. Wilmot was staying, tho' he was not at home then.

p. 42 April 5. 1825
I dined to day at Darley[91]. F. Wilmot was there who told me that all the Men I know at Cambridge had got thro' their Little Go[5] well. I returned home early.
Friday 8th
I walked over to Derby this morning and found it very warm walking indeed. It was Easter Fair, and I went thro' it with Wilmot and the Holdens. I visited Trombwells Wild Beast Show which was at Derby.
Sunday 10.
Mr Bligh preached in the afternoon and dined here, with Sam.

Monday 11.
This morning I breakfasted at Darley[91] and from there went with Frank Wilmot in a hired Gig to Mr Woods. We found only Mrs Wood at home who invited us to stay dinner, which we did. Hugh came home to dinner. We returned about 10 oclock after having spent a very pleasant day on the whole.

p. 43 April 13. 1825.
I rode over to Derby this morning, and on my return found Mr Mrs and Miss Galton at Osmaston, who came to spend a few days with us and go to the Hunt Ball. I liked their first appearance much, and thought Miss Galton much improved since I last saw her.

Thursday 14.
I was at home all to day except that I took a walk to Derby with Mr Galton. Mr Wood came to Dinner. Mr Mrs Miss G., My Father and Mother dined with Mr J. Strutt[92]. About 9 we, i.e. Eliza (who came to Osmaston from M As on Tuesday), Frances J., Wood and self set off in our carriage to the Hunt Ball; we waited at Mr J. Strutts till they were ready, and then proceeded to the Ball Room. It was very well attended, as there were nearly 400 people present. I danced all night and a good deal of it with Miss G. who improves very much as you get to know her. I was on the whole very//

p. 44 April 14. 1825
much pleased to night, and I enjoyed myself excessively, as I think my Sisters did also. We did not return home till 4 oclock.

Friday 15.
This morning I got up about nine and breakfasted with H. Wood, who afterwards set off to Cambridge on his horse. About 11 after bidding all goodbye, I walked over to Derby and took Coach from thence with F Wilmot to Leicester, where I arrived about 6. We found H Wood at Leicester and all dined together there. I was very tired, but could not get much sleep, as I was very feverish.

Saturday 16.
We came on to Cambridge this morning by old Nevills Coach: at Stamford I called and saw Mr MacGuffay, who was very well. We got to Cambridge in pretty good time, and Wilmot had a stake with me. I found all Men I know at Christs up, and well.

p. 45 April 17. 1825
I was walking about all day with Evans, Charleston and Bond: I had several men to wine with me.

*Monday 18th*
This morning I attended lectures and afterwards I went with Charleston to Newmarket Races in a gig. It was a beautiful day and I enjoyed it very much. We dined at Newmarket.

Tuesday 19
It was a beautiful day and I walked about almost the whole of it, paying some bills &c. I wined with Hand and afterwards dined with F Wilmot, when I met Crews who was going from Cambridge the next day.

Wednesday 20th
I was walking most of this morning and had a party in my rooms in the evening to supper.
Saturday 23rd.
I took a horse this morning, and went to Newmarket. It was not a fine day on the whole for//

p. 46 April 23. 1825
the sport, but there was some very good running, and the rain held off till evening, when I got wet thro. I dined to day with Brewin, but did not drink any wine, as I did not feel well. I got to bed very early. I went to J. Power for the 1st time.
Sunday 24.
When I got up this morning I found myself far from well, so got some salts and kept in all day. I got some broth for dinner.
Thursday 28.
I still continued my broth diet and medicine and was on the whole very little better. But the Rifleton Club dined to day and as I was much pressed to go, I did, and came away early. Of course all were very drunk. I did not take any wine.

p. 47 April 30th. 1825
I fear to day was a fearful blank on my existence.
Sunday May1
I was in my rooms almost all to day. St John souped with me I got to bed early.
Monday 2.
I attended lectures Private Tutor &c to day and dined in Hall. I had a large wine party to day, and we played at cards in the evening. I was obliged to sit up with them till nearly 4 o clock, and was dreadfully tired.
Tuesday 3.
I went to J. Power, attended lectures as usual. Wined at Coombes and stayed the evening.
Wednesday 4.
I wined to day with St John and we sat up till ½ past 1 playing at cards. N. B. the last time I intended doing so at present.

p. 48 May 7 1825
I hired a horse this morning and rode over to 6 Mile bottom. There were many inferior races, but not many good ones. I had 9 men to dine with me in the Evening, and we sat up late.
Sunday May 8. 1825
I had a letter from MA. and another from Miss W. this morning. I was in my rooms most of the day. I wined with H. Wood where I met a large party.
Monday 9.
I was in all day writing letters and preparing algebra for J. Power, whom I attended as usual.

Wednesday 11.
I was busy all morning, and at night dined with Robertson who had a very large dinner party. I went afterwards walking with several men and amusing ourselves till near 3 when we came in and got to bed very tired indeed.

p. 49 May 12. 1825
This morning I did not get up in time for lectures, but attended J. Power and had lesson from Mr. Wood. I drank tea and played a rubber in St Johns rooms.
Saturday 14th
There was a row today about Men meeting to run horses at 6 Mile bottom, and many Men were found out riding there.
Monday 16.
I dined to day with Frank Hall and met several Men who were going to Epsom Races for the day, & as several men I knew were among them I agreed to go with them.
Tuesday 17.
I attended Lectures &c. I dined in a quiet way with F. Wilmot, and afterwards supped with Phillips, who was to take his degree in a few days. The Men I met there got most of them very drunk indeed. It was rather late before I got to bed.

p. 50 May18th. 1825
I got an Exeat to day from Mr Shaw to go out for two days. I had a cold dinner with F. Wilmot. At 11 oclock at night I set off with Brewin in one of the Coaches for London, where we arrived at 6 the next morning, and had great difficulty getting a room to drop in at any of the Coffee Houses. We at length got one at Joys in Covent Garden, where we lay down an hour or two, and then dressed ourselves, and breakfasted with Goodwin, Walton, Birch, Wilmot, Genner, Robertson and one or two others, and immediately afterward set off to Epsom Races in an open Barouch[93] and four with two boxes. It was a very fine day, and one of the most curious of scenes I ever beheld. All the way going the road was full of Carriages Gigs, and Horsemen, and we had to go Fast Pace a great part of it. The course was the busiest scene I ever saw. A complete Mass of Carriages, Horsemen, and Foot people. There but one good race which was the Derby, and which excited//

p. 51 May 19th. 1825
great interest. The favourites were all beaten and the race was won by the Webb Colt[94]. As soon as the running was over, we got off the course as well as we could, and saw a many accidents on our way home, among which, several people were much hurt. We did not get into Town till 7 oclock, when Brewin and myself went to Drury Lane[95], with which I was very much struck, as I had never seen it before. The performance was Der Frischutz[96], which I had long wished to see. The Entertainment[97] was a very good one too, and Miss Foote[98] performed her favourite character, Maria Darlington[99]. On the whole I was very much pleased with the scenery, performance, &c. I went also to see the conclusion of Convent Garden, which is certainly much the most handsome in the interior of the two. I was very tired at Joys, after supper, which, as I had had no dinner was very necessary. There was a beautiful girl in the box I was in with whom I was very much pleased.

p. 52 May 20th 1825
I breakfasted this morning at 10 with the Men who went to Epsom, and afterward Brewin, and myself set out, and walked all over the West End, and saw all we could. We took places in the Cambridge Coach, and after getting a chop at Offerys set out about six, in the evening to Cambridge where we arrived about 3 oclock, and found great difficulty in getting in to Brewins rooms, where I got to bed

Saturday 21.
We did not get up till 12, when we found ourselves still rather tired. I dined to day with Hand, who had a large party; - I got home as early as I could. I found a letter when I got to my rooms, from my Mother with a £55 bill in it which was very acceptable. I was very sorry, when reading it, that I had been to Town, as I had spent more there, than I had intended on setting out.

p. 53 May23rd. 1825
I attended J. Power as usual today, and lectures which finished this morning for this term. The Men began to drop off for the Long Vacation very fast.

Tuesday 24
I paid J. Power 28£ to day for tutoring me. I supped to night with St John. I wrote home to day.

Thursday 25
I supped to day with Charleton who left Cambridge the day after. Most of our Men are now gone. I heard from Sam, inviting me to join him in Town in a very kind manner, but as I am very poor, I fixed not to go.

Saturday ~~Friday~~ 27
I dined to day with Brenten of Maudlin, and after had a large supper party in my own rooms. Just before we sat down, Charles St Croix, (to my surprise) made his appearance. He would not stay supper, so I shewed him to an inn, and then returned. My supper went off very well except that//

p. 54 May 27. 1825
Walton got very drunk and broke a good many Glasses.

Sunday 29.
Charles St Croix breakfasted with me and I did northing all day, but walk him over the Town, shewing him as much as I could. We attended service at King's Chapel and St Marys. He dined with me and after ten went to sleep at his room to my no small relief, as he had quite walked me off my legs.

Monday 30
I was occupied all day in shewing Charles St Croix the different things he had not seen yesterday, and packing up. He left me about 9 in the evening and appeared much pleased with Cambridge.

Tuesday 31.
This morning at 7 oclock, I left Cambridge in the Stamford Coach, in the inside, with Chawner; the Coach was quite full of Cambridge Men.

p. 55 May 31. 1825
We got into Leicester in very good time where I got my dinner; I met the two Clearks, and Hand in Leicester, and we all supped together at Bishops. I got to bed soon after in a double bedded room with Hand.

June 1. Wednesday

This morning I got up before 5 oclock, and set off to Osmaston in the Telegraph, and found them all waiting breakfast for me. I was very much pleased to find all looking well, but especially my Father, who was looking quite as he used to do, before his illness. Osmaston looked beautiful. My Grandmother[100] I was sorry to find not so well as usual, and looking much older since I last saw her.

(*The diary of Fox from 1824 -1826 is continued in Appendix 4*)

Notes

1. Held in the Manuscript Room, University Library, University of Cambridge, under DAR 250: 5.
2. Nicolson, M. N. 1929. Christ's College and the Latitude Men. Modern Philology 27, 36-53.
3. Reynolds (2004).
4. The total number of undergraduates at Christ's College is not recorded at this time. Reynolds (2005) records "*Admissions were running at over twenty a year, part of a surge across the University which now welcomed more than four hundred new undergraduates annually*" (pp.108-109). The Fellowship at this time remained steady at 13-15 (p. 115).
5. The Previous examination (or "Little go") and the Senate House (or final B A) examination. The academic year consisted of 3 terms: Michaelmas Term, 10th Oct – 16 Dec; Lent Term, 13th Jan – 10th day before Easter; Easter Term, 11th day after Easter to the Fri after the first Tues in July. The number of terms required to matriculate decreased throughout the eighteenth and nineteenth centuries. At the time of Fox and Darwin it was 10 terms. There appeared to be some flexibility in the length required for each term (see Fox *Diary*, 1824-26) and what formally was required to satisfy termly attendance. For example, CD took his Senate House Examination in late January, 1831. He claimed that he then had to attend for two more terms, viz. Lent and Easter Terms 1831. However, he was not officially awarded his degree until January 1832.
6. In 1831 the examination consisted of six parts: Homer, Virgil, Euclid, Arithmetic and Algebra, Paley's *Evidences of Christianity* and *Principles of moral and political philosophy*, and Locke's *An essay concerning human understanding* (*Cambridge University calendar*, 1831). However, it was probably assumed that the Previous Examination accounted for most of the students acquired knowledge on Paley, Locke Homer and Virgil.
7. Pinney 1974.
8. Shipley, 1924.
9. Chapter 1.
10. *Darwin Pedigrees* (Fig 2.2).
11. King-Hele, 1999.
12. See *Correspondence of Charles Darwin*, Vol 1.
13. Diaries exist for the years 1820,1821,1822, 1824-1826, 1828 (fragment),1833. Excerpts exist for the years 1846 -1852, 1857, 1858,1861-1865, 1878,1879. In addition there are three travel diaries from Italy in 1833, a tour of Wales in 1835 and Antwerp to Italy in 1842 and one notebook of Latin verse in WDF's hand for 1819. These are held at the University Library, Cambridge under the call number DAR 250.
14. Several torn-out pages of what appear to be diary entries exist for 1831 in the Woodward Library, University of British Columbia, indicating that WDF probably did keep diaries for the lost years. For more information on the Woodward Library, see Appendix 1.
15. Diary notes for June 1828 exist (Cambridge University Library, DAR 250.6) and is transcribed in Appendix 5.
16. Brooke, 1997.
17. Proctors were younger Fellows of Colleges, especially appointed for this purpose.
18. See entries for 10, 12, 13, 14, 17 & 24th November 1824.
19. Date of Newmarket foray.

20. Snobs; Cambridge slang: "*anyone not a gownsman; a townsman*": The Shorter Oxford English Dicionary.
21. See individual volumes of CCD and *Darwin Online* for Darwin's personal '*Journal*' (1809-1881)] or University Library, Cambridg, DAR158.1-76.
22. See Note 31, Chapter 1.
23. Some of the Greek words are real Greek words. Others are English words spelled with Greek letters, such as λατε probably meaning "late".
24. Browne, Vol 1, 1995. See Desmond and Moore, 1991, for a view that the letters were coded, in more than just a familiar patois.
25. Cobbett, 1830.
26. Darwin Correspondence, Vol 1. Fanny Owen [8 September & 4 October 1830].
27. Autobiography, p 16.
28. CCD, Vol 1. Letter 253. March 30th 1835.
29. See Colp, 1977; Barloon T. J. and Noyes R., Jr, 1997, *Charles Darwin and panic disorder.* J Am Med Assoc, 277, pp 138–141 and many other attempts to diagnose CD's affliction.
30. Across the top of the first page of the Diary, W. D. Fox had later written:

Weight						
1826		st		lb	Feb 19 1825	
April	28.	9	"	9	after very ill	
May	7.	9	"	11	I was	
-	27	9	"	12	st	lbs
June	6 -	9	"	13	10	9
August	21 -	9	"	2		
Sep^{r}	11 -	9	"	7		

W D Fox's weight for 1826 (again) & 1827 are given at the end of the diary (Appendix 4).

31. A coach linking Derby with Leicester.
32. Leonard Fosbrooke was admitted to Pembroke College on Feb. 26 1823, one year ahead of WDF. He went to Rugby School and would have known WDF since he lived at Shardlow Hall, not far from Osmaston Hall, just outside Derby. The Fosbrookes of Shardlow Hall, Ravenstone, were an upper class family of longstanding in Derbyshire and were linked by marriage to the Foxes. Leonard Fosbrooke became a barrister. (*Alum Cantab*).
33. Frank Wood (see below) a member of St John's College, admitted in 1823. He went to Repton School and so knew WDF (see second part of Diary, Appendix 4, for further contacts). The Woods were an upper class family of longstanding in Derbyshire. *Alum Cantab.*
34. The Tutor at this time was Mr Joseph Shaw (1786–1859), also the Tutor of CD in 1828. Later he held most of the positions in the College, even becoming the Master for a short time; see, Shipley, 1924, Chapter VIII, Charles Darwin (1809-1882).
35. Christ's College. WDF was admitted pensioner on 15 Dec 1823 according to the Christ's College admissions book: see Fig 6. Erasmus Alvey Darwin, CD's brother was admitted nearly 2 years before on 9 Feb 1822. By the time of WDF's arrival in Cambridge it seems clear from the diary entries that Erasmus was no longer at the College and was pursuing his M. B. degree (with CD in tow) at Edinburgh.
36. Not identified.
37. J. Brewin, admitted pensioner at St John's College, Dec 25 1822, son of John Brewin of North Deighton, Yorks. School: Repton (and would have known Fox at school). He became a lawyer.
38. Probably Frank Hall of St Johns – see entry for 27th Oct 1824. Frederick Hall. Admitted pensioner St Johns, March 29 1822. Son of J Cressy Hall of Alfreton, Derbys. Repton School. Became a conveyancer and a JP.
39. Not identified.
40. College regulations at this time were for attendance at Chapel once per day: see Rackham, 1927.
41. Hugh Wood. Born 1804 at Swanswick Hall, Derby. Eldest son of Rev John Wood. School: Repton (and so would have known Fox at School). Admitted pensioner at St Catharines, Feb 20 1824, BA

1828. Rector, Blore Ray, Staffs. Wood visited Osmaston Hall, 31 Aug - 2 Sep 1825, and Fox visited Wood at Swanswick with Jas Holden, 5th Sept 1825 (Appendix 4). (*Alum Cantab*)

42. WDF attended Repton School from 1816 to 1823. The Head Master at this time was Dr William Boultbee Sleath, who raised the standard of the School and increased its student numbers significantly: see Page, 1970 (*A History of Derbyshire*. Vol II).
43. Joseph Power. Adm pensioner, Clare College, Cambridge, Mar 21 1817. Son of John Power M D of Market Bosworth, Leics. 10th Wrangler in 1821. Later Cambridge University Librarian.
44. Probably Frank Hall of St Johns – see note 38 and entry for 27th Oct 1824.
45. John Roberts. Admitted sizar at Trinity, May 21 1822. BA 1826. School: Repton. No further details. (*Alum Cantab*)
46. Few records exist to indicate the number of lectures given per day and by whom. This entry implies that there were two per day. Lectures were usually given by University Lecturers and Professors. From 1822 onwards the B A degree system involved two examinations, one in fifth term, the Previous Examination or "Little Go" and the Senate House Examination in the eleventh term. The Previous Examination, for which WDF was studying, tested classical tongues (Latin and Greek) and William Paley's "*Evidences of Christianity*", see *Introduction* to this chapter and Chapter 3, Letter #9 (n11).
47. That is, one week extra was added to his term, which at this period was supposed to be 10th Oct to 16th December (Brooks, 1997), although WDF seems to have arrived after this and left earlier – see below.
48. Possibly John Drake Becher. Admitted pensioner, St Johns, June 5 1824. Son of John Becher, vicar of Southwell, Notts. BA 1828. Became a vicar at S Muskham, Notts.
49. "*Candidates for the ordinary degree had to attempt just three papers: on arithmetic and simple algebra, the first three books of Euclid and six propositions with simple examples – at a generous estimate, no more than would have been known by a bright pupil of fourteen, when in the 1960s Euclidean geometry was last taught in British Schools*." Brooke, 1997, p 176.
50. Possibly David Evans, admitted Sizar to Christ's July 6 1820.
51. Not identified.
52. James Holden. Repton School. Adm pensioner Christ's, Oct 13 1825. Son of Rev Charles Holden (formerly Shuttleworth – but took the name of Holden from Robert Holden of Aston by his third wife). Entered the Church of England. (*Peile, 1913, Alum Cantab*)
53. See note 38, above.
54. Richard Croft Chawner. Admitted pensioner, Clare June 24 1823. Son of Rupert Chawner M D of Burton-on-Trent, Staffs. Became a lawyer.
55. Henry Wilmot Charleton. Adm pensioner, Christ's College, 12 Dec 1823; kept 2 terms in 1824 but did not graduate. Joined the Dragoon Guards and retired Lt-Colonel in 1850. The second name "Wilmot" suggests a connection with Henry Wilmot, who was also at Christ's College at this time and lived in the same village in Gloucestershire – see note 59, below. (*Peile, 1913*)
56. The Minute Book of the Union Society, I, Nov 4 1823 to Dec 14 1824 (Cambridge University Archives) records as follows: "*October Term, First Meeting on Tuesday Nov 2*," then goes on to record the election of officers including the President, Mr Haughton, followed by a ballot for 64 new members, amongst whom was "*Fox, Christs*" nominated by "*Fosbrooke, Pembroke*". These were all unanimously elected. There followed a debate on "*Whether the Constitution of England or that of America was to be considered most favourable to the Liberties of the Subject*". It was defended by Mr Hildyard (St Catherines), Hall and Reid (Trinity) and opposed by Mr Thornton (Trinity) and Kennedy (St John's). The debate was adjourned at 1 minute before 10 pm to the 9th Nov.
57. Samuel Ellis Bristowe of Beesthorpe Hall, Notts, who went to Repton School and then to Emmanuel College, Cambridge in1818 and married WDF's sister, Mary Ann, in 1821 – see Bibliographic Register. WDF would eventually marry, as his second wife, Ellen Sophia Woodd, who was connected to the Bristowes; note that Bristowe was supplying wine to Fox and Basil George Woodd, Ellen's father, who ran a wine merchants company (see letter from Woodd to Fox (5th Dec 1835), concerning the overpayment on a consignment of wine; in the Fox-Pearce (Darwin) Collection, Woodward Library, University of Vancouver).

58. College physician. There were several Ficklins in Cambridge at this time. This may have been George Ficklin, whose son, Thomas John, practised at The Grove, Cambridge, in 1855.
59. Frank Sacheveral Wilmot son of Rev Edward Wilmot. The Wilmots were an ancient family in the Derbyshire. The most successful line of the family, Sir Robert Wilmot, I, II and III, lived at Catton Hall and rented the ancestral home, Osmaston Hall, to Samuel Fox III from 1814 until his death in 1851. At this time there was another Wilmot at Christ's: Henry Wilmot, admitted pensioner Dec 12 1823, the son of Dr R J Wilmot of Oveston and Alveston, Gloucs. It is not clear, later on in the diary, which Wilmot is referred to since only the surname is given. See also Letter #77 n(3).
60. Charles Gulliver Fry. Admitted pensioner at St Johns College, Jan 21 1824. From Wimborne in Dorset. Became Vicar of Eltam, Kent.
61. Ægrotat; medical certificate to excuse a student from attending lectures, etc.
62. Miss Woodcock was probably a tutor or governess to the Fox sisters at Osmaston. She appears to have held the affections of Fox and left abruptly on 31st August 1825, to live with her sisters in London, although Fox continued to correspond with her after that, and visited her in June 1828 (see Appendix 5). While this "affair" was going on there was also an association with the unknown φαννι (see next note).
63. φαννι (literally translated as "Fanny") is an unidentified woman, in the life of Fox. Fox records, in his diary, sending several letters, and significant amounts of money, to her, and at one point (22 Nov 1824) may imply that there is a child (note 76). Fox's diary of 1823 has several references in Greek, which imply an affair. She may have lived at or near Repton.
64. Henry Wilmot? See note 59, above.
65. The card game of "Twenty one" or "Vingt et un", from which the name van John derived.
66. A resolution broken three days later.
67. Thomas Medwin, 1824. "*Conversations of Lord Byron noted during a residence with his Lordship at Pisa, in the years 1821 and 1822*". Henry Colburn, London.
68. This was the old Fitzwilliam Museum, which was housed in The Free School in Free School Lane before the current design of George Basevi, which took effect from 1834-1848. The same building later became, amongst other things, the Cavendish Laboratory for Physics.
69. Not identified.
70. Not identified.
71. Not identified.
72. On the face of it this sentence does not make perfect sense and one wonders whether there is some coded language involved. This supposition is re-enforced in the next few lines.
73. This seems innocent enough, but when seen in the light of the statements of December 31st they take on an added significance.
74. It is not easy to decipher what is written here and it may be a code.
75. Presumably 5s 6d.
76. It is not clear what is referred to here: possibly a child.
77. Dogs and cats were not allowed as pets by either Fellows or students of Christ's at this time (or since).

77a. There were two major works: a Symphony by Haydn and the "Grand Symphony" of Mozart, as well as many songs. The Cambridge Chronicle reported on the 26th Nov 1824: "*Two great concerts have this week been performed at the Town Hall, under direction of Professor Clarke Whitfeld. The principle attractions were Miss Stephens and Mr. Braham, Mr. Nicolson and Mr. Lindley and seldom have we witnessed such general satisfaction as was evinced by numerous and genteel audiences. To criticise the merits of these eminent performers would be superfluous, and to notice those pieces which more particularly elicited approbation, is difficult, the applause being always universal. We may, however, remark that on Tuesday evening, Mr. Braham gave "Deeper and deeper still" in a manner which no other singer in existence can equal: and Miss Stephens sang "Sweetbird" in the most enchanting style, which Nicolson in his accompaniment on flute, as well as in a previous concerto, displayed his inimitable powers in an astonishing degree. At the close of the concert there was a general call for "God Save the King" (which was announced as the finale of the following evening) and it was admirably sung by Miss Stephens and Mr. Braham, the audience joining in the chorus with enthusiastic loyalty."*

78. Permission to leave College: possibly required because WDF was leaving early: see note 47, above.
79. This would have been the household of Jedediah Strutt, Jr (see notes 85 & 88, below).
80. This is an unusual name, not common in England. It may be an Irish name, as one family of immigrants to New York from Ireland, in the potato famine of 1846-1851, had this name. In Latin the name means "grass".
81. Added in between lines.
82. M A Bristowe, neé Fox, Fox's sister, Mary Ann.
83. Boteler: Samuel Boteler Bristowe (1822-1897), son of Samuel Ellis and Mary Ann Bristowe (see notes 57 & 82, above) of Beesthorpe Hall.
84. Mr Bligh was the local clergyman at Osmaston.
85. Jedediah Strutt, Jr (1785-1854) was the son of George Benson Strutt and grandson of Jedediah Strutt, Sr (1728-1897). Jedediah Strutt, Sr, was Fox's step-grandfather, since Samuel Fox III married Strutt's daughter, Martha, in 1791. She died in 1799. Jedediah Strutt, Jr was therefore Fox's step-uncle.
86. Emma (1803–1885), Frances Jane (1806-1852) and Julia Fox (1809–1889) were three of Fox's five sisters.
87. Not identified. Not closely related to Samuel Tertius Galton, who married Violetta, the daughter of Erasmus Darwin, and gave rise, among others, to Bessy Wheler, née Galton, and Francis Galton, FRS.
88. Edward Strutt, was the son of William Strutt and grandson of Jedediah Strutt, Sr (1728-1897). Represented the Borough of Derby, 1830-1847, Arundel, 1851-52 and Nottingham, 1857-56. He became Baron Belper of Belper in 1856. He did not take an active part in the family business.
89. Brother of Edward Strutt, note 88, above.
90. Thurlston Grange had been the previous residence of the Foxes before they moved into Osmaston Hall in 1814. It was in the grounds of Elvaston Castle - see Chapter 1 (note 17) and Appendix 4, note 55. At this time it seems to have been occupied by relatives, including Jane Darwin (see note 100, below) and retainers of the Fox family (see Appendix 4).
91. Darley Abbey and Darley Hall, a fine Georgian mansion, lay on the River Derwent on the outskirts of Derby; they belonged, about this time, to the mill-owning Evans family, but the Strutt family may also have owned a part, as two of Jedediah Strutt's offspring (William and Elizabeth) married offspring of Thomas Evans (see Appendix 4, note78).
92. Probably Joseph Strutt, a son of Jedediah Strutt Sr, but could have been Jedediah Strutt Jr, the grandson of Jedediah Strutt; however, Fox usually distinguishes the later as Jed Strutt.
93. A barouche was normally a four-wheeled coach drawn by two horses, fashionable in the nineteenth century. It had double seats at the back, facing back and front, and a box in front for the driver. The "four" may imply that it was a large coach drawn by four horses.
94. Middleton was the winner of the Epsom Derby in 1825. He was sired by Phantom and the dam was Web.
95. The Theatre Royal, in Drury Lane, opened in 1812.
96. *Der Frieschütz*, a seminal work of German romantic opera, by Carl Maria von Weber was first performed in Berlin in 1821. The Dimond-Cook version in English ran for several weeks from 4th April 1825, and Washington Irving had a major role in transcribing the libretto (G. R. Price. 1948 Music and Letters Vol. 29, 348-355). There had been a previous production at Covent Garden in 1824.
97. This would have been an additional performance, of a lighter kind than traditional opera.
98. Maria Foote, who in 1831 married General Charles Stanhope, the 3rd Earl of Harrington, who lived at Elvaston Castle just outside Derby and close to Osmaston Hall (Fox records in his diary visiting Thurlston Grange, which was in the grounds of Elvaston Castle, and had been the Fox family residence before they moved into Osmaston Hall – see note 90, above). She was a successful actress at the time, in partnership with Lucia Elizabeth Vestris, who ran the Olympic Theatre in Drury Lane and who is mentioned elsewhere in the Fox diaries.

99. A comic character in a farce, from this time, called "*A Rowland for an Oliver*". A "*Rowland for an Oliver"* was a proverb from earlier times, whose meaning was "to meet one tall story with another".
100. Jane Darwin (1746–1835). Mother of Ann Darwin, Fox's mother. In 1814–15 she moved to Thurlston Grange (see note 90, above) to be near to her daughter at Osmaston Hall, Derby, where she died in 1835.

Chapter 3
Charles Darwin at Cambridge: The Letters to William Darwin Fox

Letters #1 to #43; 12th June 1828 to 1st Aug 1831

That punctual servant of all work, the sun, had just risen, and begun to strike a light on the morning of the thirteenth of May, one thousand eight hundred and twenty seven, when Mr Samuel Pickwick burst like another sun from his slumbers; threw open his chamber window, and looked out upon the world beneath. Goswell-street was at his feet, Goswell-street was on his right – as far as the eye could reach, Goswell-street extended on his left; and the opposite side of Goswell-street was over the way. "Such," thought Mr. Pickwick, "are the narrow views of those philosophers who, content with examining the things that lie before them, look not to the truths that are hidden beyond".

Pickwick Papers, Charles Dickens

Introduction

Just over three years after the arrival of William Darwin Fox in Cambridge, another young man of nineteen arrived, in early 1828, who was destined to cause one of the biggest upheavals of the Nineteenth Century. Charles Darwin did not come directly from school as Fox did; he had left Shrewsbury School at the age of sixteen and had joined his elder brother Erasmus in Edinburgh, with the intent, largely on his father's side, of becoming a physician, like his grandfather, his father and his brother. However, young Charley, as his sister's called him, was honest enough to own up that the blood and gore did not suit him for that kind of life, and as it was determined that that also excluded a military career, the remaining profession for a gentleman of the age was chosen, that of a clergyman.

By the time Charles Darwin arrived in Cambridge in January 1828[1], Christ's College was an obvious choice and by the time Charles Darwin his cousin had already been at Christ's for four years and Erasmus Darwin II, his brother, had recently left to take up his medical studies at Edinburgh University. Furthermore, Hensleigh Wedgwood, CD's cousin, was also an undergraduate at Christ's and would later become a Fellow of Christ's in 1829. Fox, it appears from the letters,

A. W. D. Larkum, *A Natural Calling*, DOI 10.1007/978-1-4020-9233-6_3,

quickly introduced Darwin to many of the good and, probably, "bad" aspects of Cambridge (see "*Autobiography*" and *Diary* of W D Fox, Chapter 2). Clearly Fox was already a committed amateur naturalist and had formed a circle of friends at Cambridge, including the brilliant Prof John Stevens Henslow[2], who was appointed Professor of Mineralogy at the University in 1822 and then to the Chair of Botany in 1825 (see Chapter 2). It appears from the correspondence of Darwin to Fox, but not from Fox's diary (see below), that Fox was not only a collector of birds and animals which he hoarded at Osmaston Hall[3] just outside Derby, but was also a keen beetle hunter.

We know this from Darwin's letters to Fox and also from an admission in his *Autobiography* that: "*... in my life at Cambridge, my time was sadly wasted, there and worse than wasted. From my passion for shooting and for hunting, and, when, this failed, for riding across country, I got into a sporting set, including some dissipated low-minded young men. We used often to dine together in the evening, though these dinners often included men of higher stamp, and we sometimes drank too much, with, jolly singing and playing at cards afterwards.*" To which Francis Darwin added the footnote, "*I gather from some of my father's contemporaries that he has exaggerated the Bacchanalian nature of these parties*". Clearly this was a sensitive family issue and may account for why the diary of Fox for 1828 is missing. After Fox died in 1880 his effects were passed on to his wife and four of his sons[4]. Darwin himself died two years later, and at that point, Francis Darwin, his son, began to assemble "*The Life and Letters of Charles Darwin*", which was published in 1887. The letters of Darwin to Fox form a strong component of this book and it is clear that the letters were made available to Francis Darwin, at an early stage by Charles W Fox (Letter, 1 Nov 1887). Therefore, it is likely that the diaries were also known to Francis Darwin, although he does not acknowledge them (see also Appendix 1).

In 1909, at the time of the Darwin Centenary celebrations, a set of letters was left to Christ's College in a deed signed by Fox's son, Gilbert Basil Fox. These letters are a prize-possession of the College. However, they do not constitute the entire collection of letters, which must have been in the collection of W D Fox at his death, together with other letters and memorabilia which have turned up elsewhere (see Appendix 1 and note 5). What is not extant is any letter of W D Fox to Darwin at this stage. In fact, this does not matter unduly for two reasons. Firstly, the letters of Darwin are self-explicit and there is very little that needs explanation in the form of a reply: one can read them rather like a diary of events. Secondly, we have the diary of Fox from 1824-26 that unmasks the persona and life style of Fox at Cambridge, perhaps as well, if not better, than the letters in reply. Nevertheless, we may well ask why no letters exist? The immediate reason seems to be that Darwin did not keep them! However a number of letters of this period are extant (see Correspondence of Charles Darwin); furthermore it is unlikely that Darwin threw the letters away as he would want them to refer to information on beetles, etc., that he was working on. More likely they were lost during or after the voyage of the *Beagle*. Some later letters of Fox do survive and give us a flavour of him as a letter writer (and more of that later in Chapter 4). Furthermore, after the *Beagle* period Darwin came more and more to cut out interesting parts of letters from scientific correspondence, including many of Fox's. Perhaps, CD was selective in his storage of letters and threw away many of Fox's letters; Fox on the other hand kept all of Darwin's letters!

As we can readily see the letters of Charles Darwin at this period are the expressions of a young man who is enjoying life to the full in a perfectly normal manner; and, unlike his grandfather, there is no indication of excessive sexual passion. The letters show Darwin on a sharp learning curve; for example, in terms of beetles, he came from a starting point of zero knowledge to becoming an expert and discovering new species within a single year. Thus, in this regard, he overtakes Fox very quickly but does so in a way that does not diminish the role of Fox as a close adviser and friend. In 1828 he writes to his brother Erasmus[6] "*I left Fox up in Cam. in great awe and tribulation about his degree*[7]*, which is to be made very much more strict, so that they give out that at least 50 will be plucked: I live almost entirely with Fox and Entomology goes on most surprisingly.*" There is no doubt that Darwin had a very close friendship with Fox at Cambridge over one year that sparked his interest in natural history, especially entomology. He even visited Osmaston Hall (see Appendix 2), where the Fox's sisters were an added attraction in addition to natural history, shooting and music[8]. Yet he turned down further offers to visit and the two rarely met in later life, despite the lively correspondence between them. The stance that Darwin took with Fox, is a recurring pattern throughout his life. Of course at the time Darwin was preoccupied with his studies and then in learning geology with Prof Sedgwick[9]. However the strong impression one gets is of an active mind scrupulously saving time for the important matters at hand, while trying to maintain good relations: a lesson that stood Darwin in great stead throughout his life, particularly during the stormy times after the publication of the "*Origin of Species*" in 1859.

There is clearly also the development of professional rivalry, not so much with Fox, who, it is apparent from the correspondence, is not going to be a serious rival but with the entomologist F W Hope, who later endowed the Hope Chair of Zoology at the University of Oxford, and with the naturalist-parson Leonard Jenyns[10]. The rivalry with Jenyns almost got out of hand, and indeed this animosity may have been the reason that the post of naturalist and companion to Capt Fitzroy on the *Beagle* was offered to Jenyns, before Darwin. And perhaps it was just as well that Darwin patched up this nascent rivalry, since J S Henslow, his early mentor, was married to Jenyns' sister, Harriet, and their daughter, Frances Harriet married Joseph Dalton Hooker, the great botanist, who played the unlikely role of youthful mentor in Darwin's middle years.

Overall the correspondence reveals Darwin as a young man with a zest for life and particularly for natural history. He is interested in art and music as well as hunting and shooting. At no point can one discern the mind of a man who will change the way the world thinks. He goes up to Cambridge a failure, in the eyes of his family, and with the modest ambition of becoming a clergyman. His friends, particularly J S Henslow, but also Adam Sedgwick, and the mysterious donor of a microscope (who was Herbert) and even Fox, all saw something in this young man. With this encouragement, Darwin began to see something in himself and even planned an expedition to the Canary Islands, an expression of this developing self-confidence. It is fair to say, however, that as yet we see no hint of the man who was to give his name to what has been called the "*Darwinian Revolution*".

At this time we do not have letters of Fox, only his diary. We see from the reflection of Darwin's letters and from Fox's diaries, a man with many of the same interests as Darwin, but someone with less ability to think clearly and express his thoughts in a lively fashion.

So, when we first meet the young correspondent, Darwin is a breath of fresh air: excited, exciting, vigorous, informed and blessedly vivid in his letters. In fact, the letters start after the two cousins have been at Cambridge for two terms, and have parted, CD to go home, and Fox to visit London (see Appendix 5). The first letter to Fox starts "*I am dying by inches*"; not an original expression as it was used by the Owen sisters in their letters to Darwin (see CCD, Vol 1), but so aptly reused by Darwin. And so it goes on in brilliant letter writing for some 43 letters. These fill in the burgeoning activities of Darwin as a student of theology at Cambridge and as a student of natural history, especially beetles, in his spare time.

These letters need no further introduction, and the letters follow on sequentially from Darwin's first year at Christ's College to his final year in 1831 (and the events leading up to his voyage on the *Beagle*). These letters form an incomparable record of Darwin's life at this time. We can only smile when we read Darwin's comment at the end of Letter #2, "*What a pity it is that you do not collect letters, like insects, birds & beasts*", and be thankful that Fox did indeed collect these letters.

Notes

1. CD was admitted to Christ's College in Michaelmas 1827 (see Fig 6), but studied Latin and Greek at Shrewsbury during the rest of 1827; see "Autobiography".
2. John Stevens Henslow (1796-1861). Clergyman, botanist mineralogist. Professor of Botany at the University of Cambridge (1827-1861), teacher and mentor to CD; See Walters and Stow, 2001.
3. See Letter #5 re "Osmaston Museum"; in Letter #6, CD writes "*he had not seen such a rich case collected: . . ... yours must be still richer*" in relation to a meeting with F. W. Hope.
4. WD Fox's will left all his effects to his second wife, Ellen Sophia. She died in 1887 and gave instructions to her executors to distribute "*plate, jewels pictures, books, furniture, etc according to lists of distribution*" and in default to the judgement of her executors. Clearly by 1909 the papers were in the possession of one of the last surviving sons, Rev Gilbert Basil Fox (see Chapter 2 and Appendix 1). The Rev Gilbert Basil Fox died in 1941 and apparently the materials were passed to Anthony Basil Darwin Fox, who, like several later members of the family, lived at Southwold, Sussex. However, in his will Gilbert Basil Fox also left some effects to Erasmus Pulleine Fox and then to Marjory Pearce, and it may be through this line of inheritance that Capt Christopher Pearce obtained some papers (see Note 15). The main line of inheritance then followed through to Agnes Letitia Darwin Fox (died 1988), Susan Frances Crombie (died 1991) and her husband Col Richard Crombie (died 1992) and currently to Gerard James Crombie (see Appendix 1).
5. Clearly other members of the Fox family gained access to the collection of letters, diaries and memorabilia and kept small portions of them. How this came about we can only guess, as there are no written records. The letter at the Sandown Museum, Isle of Wight can reasonably be seen as

a gift by one of WDF's offspring [most probably Robert Gerard Fox (1851-1909)] to his beloved local town[5]. Those at the American Philosophical Society, Philadelphia (six letters) and that at the Smithsonian Museum may be the result of a gift by Francis Darwin to a member of those institutions. The letters of the Fox-Pearce (Darwin) Collection in the Woodward Library, University of Vancouver (5 letters plus one from Emma Fox to WDF and one from ED to Mrs W D Fox, plus other memorabilia) are known, from available letters, to have been sold by Capt Christopher Pearce to the Woodward Library in 1970; how they got into his hands is not clear (but see Appendix 1 and note 14). Christopher Pearce was the grandson of Alexander Pearce, who was the son-in-law of W D Fox. Christopher Pearce served in the Royal Air Force until 1956. After visiting Canada in 1957 and selling the WDF letters and memorabilia to the Woodward Library, he entered the United States of America, and has not been traced since. For the main line of inheritance of the WD Fox materials see Appendix 1 & Note 14.

6. Darwin Correspondence, Vol 1, letter 53.
7. WDF took his final exam in Jan. 1829, see Letter 9, 25-29th Jan 1829 (Letter 56, CCD) re "*I think your place a very good one....*". In fact Fox ranked 88th on the list of 160 candidates (*Cambridge Chronicle and Journal ...*, Friday, 23 January 1829, 2d edition; repeated, Friday, 30 January 1829) which is not nearly as good as Darwin's 10th out of 178 candidates, two years later.
8. WDF had four unmarried sisters of eligible age in 1828: Eliza (27), Emma (25), Frances Jane (23) and Julia (19). The oldest sister, Mary Ann, married Samuel Ellis Bristowe, Fox's close friend (see Diary for 1824-26).
9. Adam Sedgwick (1785-1873). Woodwardian Professor of Geology at the University of Cambridge, 1818-1873; Canon of Norwich, 1834-1873. (*DNB*)
10. Leonard Jenyns (1800-1893). Naturalist, expert in insects and fish (described the fish specimens collected by CD on the *Beagle*). Vicar of Swaffham Bulbeck, Cambridgeshire. Later. inherited Bottisham Hall, just outside Cambridge. (Desmond 1977; *DNB*)

#1 (CCD 42). To: William Darwin Fox Esq, Osmaston near Derby.

[Shrewsbury]
Friday [12th June 1828]

My dear Fox

I am dying by inches, from not having any body to talk to about insects: — my only reason for writing, is to remove a heavy weight from my mind, so now you must understand, what you will perceive before you come to the end of this; that I am writing merely for my own pleasure & not yours. — I have been very idle since I left Cambridge in every possible way & amongst the rest in Entomology. I have however captured a few insects, about which I am much interested. My sister[1] has made rough drawings of three of them: I. fig: is I am nearly sure, the same insect as Hoar[2], of Queens took in a Willow tree, & which Garland[3] did not know. I took ∗it∗ under the bark of sail, was very active, striking looking insect, took 3 specimens

<I think this is an admirable ~~priz∗∗~~ prize.>//

II. fig: is an *extremely* common insect; of the family of scarabidæ. Do you know it's name? —

III. fig: A *most* beautiful Leptura[4] very like the Quadrifasciata[5], only the body is of the same size throughout. — I ~~ask~~ <tell> you all these ~~questions~~ particulars, as I am anxious to ~~hear~~ know something about these little g∗ ∗ ∗s[6]. — I was not *fully* aware of your extreme value before I left Cambridge. I am constantly saying "I do wish Fox was here". — And I again say, I hope you will come & pay me visit before the summer is over — My Father desired me to say, that he should be at anytime most happy to see you. — I have taken 3 species of Coccinellæ[7], one, the same as Hoar took in the Fens, which you said was rare, & another with 7 *white*! ~~spots~~ <marks> on each elytron. — I will mention, as I believe you are interested about it, that I have seen the Cocc: bipunctata[8] (or dispar)//

4 or 5 in actu coitus with a black one with *4* red marks (I believe most of the black ones you have got have *6*[9] marks, & hence I suppose a different species) also, which is very singular, I have frequently seen two of the bipunctata's in actu. — I Have taken Clivina Collaris, fig 3[10] Plate III of Stephens[11] also a beautiful *large*[12] copper-coloured Elater (with Antennæ

pectin<ated>  like this. Do you want any of the Byrrhus Pillula? I <can> get any number. — My dear Fox I must again beg your pardon for sending such a very selfish stupid letter: but remember I am your pupil, so you must forgive me. — I hope you will write to me soon, & tell me every thing you have been doing, & more particularly how you are in health, as to your eyes, & body[13]. — How was poor little Fan[14]: how was No 16!? what do you intend doing this summer? in short write me a good long letter about yourself & all other insects: My plans remain the same as formerly. I am going to Barmouth[15] for two months. — If you have not written to my brother, write to him before the 22^{d}. & direct, Poste Restante, Munich I hope you will be able to send me a better account of Miss Fox[16], when you write; tell me whether you intend going to Tenby[17]?
(written cross-wise on the first page)
I should not send this very shamefully stupid letter, only I am very anxious to get some *crumbs* of information about yourself & the insects. Believe me my dear Fox

Yours most sincerely
Charles Darwin

Shrewsbury

(on fly leaf)
of a fine bluish black colour, but is not so broad as made in this drawing

[*The following figures and labeling were done on one part of the second side of the last page, of which the middle was reserved for the address (see Fig 11).* The transcription of the labeling is given on the right hand side.]

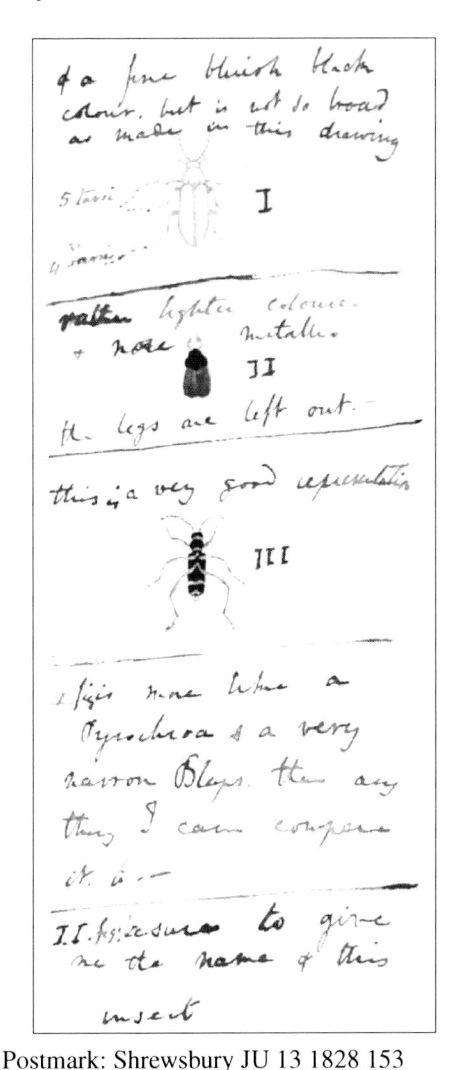

of a fine bluish black colour, but is not so broad as made in this drawing

I

rather lighter coloured & more metallic

II

the legs are left out. —

this is a very good representation

III

I. fig is more like a Pyrochroa or a very narrow Blaps, than any thing I can compare it to. —

I.I. fig: be sure to give me the name of this insect

Postmark: Shrewsbury JU 13 1828 153
Christ's College Library, Cambridge, #1.

Notes

1. Which sister is referred to is not specified. It was most likely to have been Susan or Catherine Darwin. Fox's sisters also made colours drawings of insects for him (see Figs 16 & 17).
2. Probably William Strong Hore (1807-1882), admitted to Queens College, Cambridge, May 1826. Became a clergyman and botanist. The name is obscured but in later letters CD refers to Hore (see Letters #8 and #8A). Mr Hoare, Queens College, Cambridge is mentioned as a subscriber to J F Stephens' "*A systematic Catalogue of British insects*" (see note 1, below) (*Alum. Cantab.;* Desmond, 1977). J F Stephens (1792-1852) was a naturalist and entomologist, employed by the Admiralty office.
3. Possibly Lewes (or Lewis) Garland (d. 1844). B A Trinity College, 1827. (*Alum. Cantab.*)
4. A long-horned beetle. In both Stephens and Samouelle (1819) but no annotations in CD's copies.
5. Note that at this time it was unusual to italicise species binomial names (as it is today) and that it was common to begin the specific epithet with a capital.
6. Illegible due to sticking of wax from postmark.
7. Coccinellidae – Ladybird beetles.
8. Two-spotted Ladybird.
9. Obscured.
10. Obscured.
11. James Francis Stephens. *A systematic Catalogue of British insects: being an attempt to arrange all the hitherto discovered indigenous insects in accordance with their natural affinities. Containing also the references to every English writer on entomology, and to the principal foreign authors. With all the published British genera to the present time.* London, XXXIV + 416 ss, 1829. This book is referred to, subsequently, as Stephens' *Catalogue Br insects.* J F Stephens (1792-1852) was a naturalist and entomologist, employed by the Admiralty office, and responsible for arranging insect collections at the British Museum. He also produced another book which CD used: *Illustrations of British entomology; or a synopsis of indigenous insects containing their generic and specific distinctions; with an account of their metamorphosis, times of appearance, location, food, and economy, as far as practicable embellished with coloured figures of the rarer and more interesting species.* 11 Vols (1827-46), London, [see Letter #24 (n11)].
12. Obscured by tear.
13. The health of Fox in 1828 is not known. A short journal entry for June (Appendix 5) suggests that at this stage his health was reasonable. Previously, in the later part of 1826, Fox was convalescing from a glandular fever (Chapter 2 and Appendix 4) and this may have affected his health in 1827, for which no diary exists. CD's enquiry seems to suggest that Fox still had health problems.
14. One of Fox's dogs [see Letter #17 (n4)]; another, a bitch, was Sappho.
15. On Cardigan Bay, Gwynedd, North Wales, overlooked by the mountains including Cader Idris. CD took a reading tour in the summer vacation of 1828 with some undergraduate friends and George Ash Butterton, of St John's College, a private tutor in mathematics (see *LL* 1: 166, with reminiscences from John Maurice Herbert to Francis Darwin, now in DAR 112). See also CCD, Vol 1, letter to J M Herbert, 13th September 1828.
16. Fox's married sister, Mary Anne, died on the 19th April 1829, so it is likely that the reference is to her (see next letter, #2).
17. Tenby is on Carmarthen Bay, in Pembrokeshire, S Wales.

#2 (CCD 43). To: Wm Darwin Fox Esq, Osmaston, Derby

[June 29th 1828]
[Shrewsbury]

(*across the top of page 1*)[1]

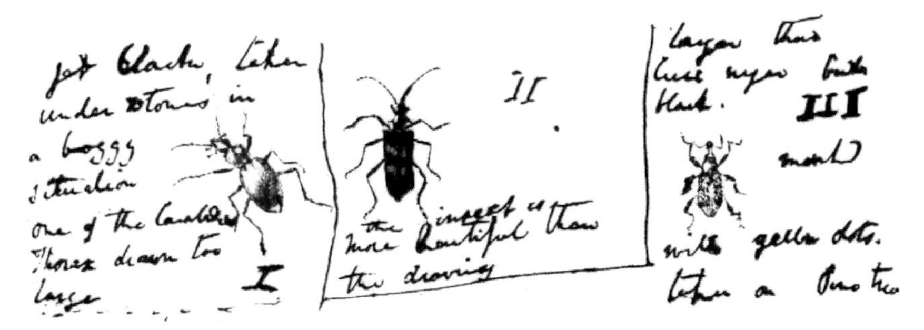

jet black taken under stones in a boggy situation One of the Carabidæ Thorax drawn too large I	II the insect is more beautiful than the drawing	larger than Curc niger but black, III marked with yellow dots. taken on a Pine tree

My dear Fox,

I do not know whether to begin by giving you a good scolding, or by thanking you over & over again for your most acceptable parcel. Your letter was perfect & informed me of every thing I wanted to hear about & you really were most goodnatured to think of forwarding Way's[2] letter. — And now for the scolding; but I must confess I am not very certain about the grounds of it. Did I pay you the 10"6^{d} for the Poor Irishman?[3] if I did not; you must have been aware of it, — & therefore//

you acted like a shabby fellow, in allowing me to be so. — I will pay you the 10"6^{d} & the 1£ for the pins, (which were just what I wanted & I am extremely obliged for you thinking of me) when we meet at Cambridge. — I am very glad you have had so pleasant a time in London[4]. it almost ~~tempts~~ makes me wish to be there: but nevertheless I had <much> rather be wandering with you in the wilds of Derbyshire; than be idling in that horrid smoky wilderness London. — I envy you very much in going to the Opera[5], for I have at last got a very decided taste for Music; When Roper[6] was here we had a concert from morning to night: his visit here was very pleasant//

as it is quite delightful to hear anyone sing with such spirit, & excellent good taste. — And now for Insectology, Way[24] need not doubt my ardour as it is I think redoubled; but my success does not equal what it did in Cambridge.
Fig: (I.) is this a Cychrus? I make it so by Lamarck[7]. (N.B. when you recognize an insect by my description, always say whether you possess it). fig II is a most splendid Cerambyx[8] —what is it? fig III, do you know this insect? I have also take another very large grey looking Curculio? Also a Cerambyx twice as big as that common one, with horn coloured elytra & the stomach marked with bands something like a Cockchafer. About 19 new Carabi — several Coccinellæ[9], & other insects which

I cannot describe. — Taken in plenty, the little beetle that destroy Scotch firs[10]. <What is its name. —>//

The Crysom: (fulgida(?)[11] <the same as your one specimen>, is the *greatest* profusion & am *quite* convinced it is a different insect from the little one you took in plenty in the Fens, as it never lives but on one plant, some sort of dead nettle. — of course I procure duplicate specimens for you, of all insects that I can.[12]

I have adopted Way's[2] plan of having a store box, & only having a ~~sm~~ few specimens of each insect in my regular case: —[12]

We heard from Erasmus[13] the other day just when he was on the point of leaving Vienna he is now on his road to Munich or Milan all letters must be sent to the latter place. I do not think he much admires Vienna, for it has been literally a perfect state of solitude for him, latterly he has got to know one single family Vivenot, a German Doctor, very pleasant people but who the devil would not be pleasant, when one has been silent for nearly a month? — *If* you have written to him either at Vienna or Munich, you need not be alarmed about their arrival, as he has left orders that all letters may be forwarded. — On Wednesday I set out on my Entomo-Mathematical expedition to Barmouth[14]; by the blessings of Providence I hope *the science* will not drive out of my poor//

noddle the Mathematics. — Talking of *the science* I must tell you, that since beginning this letter, I think, sir, upon my soul sir, I will take my oath sir, (as Way[24] would say), that I have discovered, that I posses a valuable insect, viz Melasis Flabellicornis, as the description & habits in Samouelle[15] & Lamarck *perfectly* agree. — look and see, what Samouelle says, "once taken[16]

(*written in middle of 4th page*)

(continued from the cover)

in Norfolk, by J. C. Curtis"[17] &c &c; also I have taken 2 new Elaters; one with reddish brown elytra, the other with yellow elytra, red thorax & black head. — Also, have taken Tritoma Bipustulatum, vide Samouell. Pl. II.[18]; also a small black Byrrhus[19], with brownish band across the back; also a small Silpha[20], with a red spot on each elytron; also have twice !*seen*! the Bombylius Minor[21], but curses on my clumsiness missed catching ~~him~~ them.[22] — //

(*Written cross-wise the 1st, 2nd and 3rd pages*)

You will hardly believe how *bumptious* I am grown, in the height of my presumption, I put down the Leptura Arietes, as the Saperda Lineato Collis[23], because it had a yellow ring round its neck! & as for the Melandrya[24] I thought it at least a new species if not Genus. — I must confess I felt a pang of jealousy, when I read over the splendid list of yours// & Way's[24] Captures. — What a pity it is that you do not collect letters, like insects, birds & beasts[25]. What a glorious specimen this would be of an interesting, neatly written & elegantly composed epistle: but my dear Fox

remember this, although my prosing unscientific details about insects may be very tiresome to you, do not for one instant suppose, that your letters are so to me,// for you cannot conceive, with what great pleasure I look out for an Entom: Letter now that I have nobody to talk to. I return Way's letter & must again thank you for it, when you write remember me most kindly & tell him, if he has a fit of idleness, he knows how glad I should be to hear from him. —
When you write, & I hope it will be soon, direct to Shrewsbury, as usual

Believe me my dear Fox
Yours ever sincr
C. D.

(*written under seal*)
Shall I or will you write to *** M.A.[26], if you say nothing about *** I shall, as soon as I hear from y*** alls well Wed.[27]

Postmark: Shrewsbury JU 30 1828 153
Christ's College Library, Cambridge, #2.

Notes

1. Note that the background on these images has been lightened and extraneous marks removed.
2. Albert Way (1805-1874), B.A., Trinity College, 1829. Antiquary and traveler. A founder of the Society of Antiquaries. He and Fox introduced CD to insect collecting. See also Fig 15. (*Alum Cantab, DNB*)
3. Not identified. Probably the same as 'Our old Irish friend' referred to in Letter #13. It may also have been a code name for some other activity. In Fox's accounts there are several records of payments to "poor man", "beggars" or "Irishman"; mostly of 1 shilling or so. In this case it seems that the recipient was genuine.
4. For Fox's trip to London from 24th June 1828, see Appendix 5.
5. In his diary entries from 24th June 1828 (Appendix 5), Fox records going to the Opera on 10th July. So CD is probably commenting on a coming event. It is possible that Fox had gone to the opera before the 24th June, at which date he was staying with family friends, the Bridgemans at Cheshunt, in Hertfordshire, north of London. However, it is more likely that he called on the Brigdemans *en route* to London from Cambridge.
6. Probably Henry W. Roper (1806-1883). St Johns, B A,1829. Major, 8th Regiment. (*Alum Cantab*)
7. Lamarck, 1815-22. *Cychrus* is described in Vol IV, pp 516–17. Volumes two and four, with CD's annotations, are preserved in Darwin Library (CUL), but there are no marginalia on these pages. However, in Darwin's copy of Stephens' *Catalogue of British insects* (1829) (see note 2) there is the marginal note "*Maer, 1828*" against *C. rostratus.*
8. A longhorn beetle (Cerambicidae). Several are mentioned in Samouelle (1809) and Stephens' *Catalogue of British insects* (1829).
9. Lady-bird beetles.
10. CD does not refer to where this beetle is found. Beetles in the genus *Tomicus* feed on pine needles; beetles in genus *Hylobius* (family Curculionidae) feed on pine roots. In Letter #4 (n6) CD refers to a beetle in the genus *Nosodendrum*, which is feeds on pine roots, but this is not a modern genus.
11. *Chrysomela fulgida* (Samouelle 1819, p. 416). CD's copy in CD's Library Collection, University Library, Cambridge; CD's copy of Stephen's *Catalogue of Br Insects* is asterisked at this species.
12. This sentence is written across the bottom of the Address page, which was then used as the 2nd page.
13. Erasmus Alvey Darwin, CD's brother.
14. See Letter #1 (n15).

15. George Samouelle (1790-1846) was curator in the British Museum (Natural History) and in 1819 brought out his book *The Entomologist's Useful Compendium, or an Introduction to the Knowledge of British Insects, etc.* Published by Thomas Boys, London, 496 pp. Samouelle 1819, p 415; describes Melasis flabellicornis under "July" as "woods, Norwich Windsor – page 160" (see note 21).
16. CD had left the mid-section of page 2, side 4, blank for the address, but, after adding a third sheet to his letter, he returned to the unused portion of page 4.
17. Samouelle, 1819, p 160: "*In England it has been once taken by Mr. J. Curtis, of Norwich, an excellent artist and an industrious entomologist; and several times near Windsor where it was first observed by Mr. Herschel.*"
18. Samouelle, 1819, Plate II, fig 9, and 'Explanation of plates', p 445.
19. A pill beetle.
20. Carrion beetle.
21. Heath bee fly; bee flies victimise burrowing beetles.
22. Written in address area of letter.
23. In Stephen's *Catalogue Br Insects, Saperda Lineatocollis* but in Samouelle *Cerambyx lineato-collis.*
24. False darkling beetle; mentioned in CD's copy of Stephens, but no annotation.
25. This is an ironic statement, indeed! We are indebted to W D Fox for preserving the letters of CD, which form an incomparable record of his life. "What a pity it is" that CD did not preserve the replies of Fox!
26. Mary Ann Bristowe; see note 16, Letter#1.
27. 'Shall . . . . Wed.' written on the outside fold of the cover.

#3 (CCD 45; LLi, fragment[1]). To: William Darwin Fox, Osmaston, Derby

Barmouth
[July 29th 1828]
Tuesday

Dear Fox

What excuse have you to offer for not having answered my letter long before this? I hope it is nothing worse than idleness; or what would be still better, I hope it arises from your being 10 fathoms deep in *the Mathematics*, & if you are God help you, for so am I, only with this difference I stick fast in the mud at the bottom & there I shall remain in statu quo[2]. — But you see *I* can afford time to write a single letter. — but I will not mention, how long I have been ~~wishing~~ hoping in vain to receive a letter from my old master; more especially at this present time, when, if you//

remember; we talked of meeting on the top of some Welch mountain between Barmouth & Tenby. I will not flatter you, by saying how much I wish to hear what you yourself are doing & how ~~insectology~~ <entomology> & your 101 other pursuits are going on? After this aweful Philippic I hope I shall soon receive a letter. — I have been at Barmouth ever since the first of July, & like it very much; the scenery & therefore the walks are quite delightful. I only wish you would make a trip here, & I would Cicerone you up & down the mountains, untill you had not a particle of wind left in your lungs. — I go on very badly with Mathematics; neither have I succeeded so well as I expected in entomology: but I must mention a few of the most conspicuous insects//

or rather those which I can at all describe. Amongst the Lepidoptera, the Pterophorus Pentadactylus (Lamarck)[3], a most extraordinary moth, — several large moths, one of which is figured in Samouelle; brown with a white band & spots; a beautiful <small> copper coloured butterfly, with a little tail like the Machaon[4]: allso a sphinx Zigena (Lam:)[5] Inter Coleoptera —. A carabus, larger than the Violaceus, jet black & smooth. — Of the Carabidæ[6], one very odd one; black polished; very long; thorax, orbicular, & sep*arate* from the abdomen. <of this length> caught near sea side. — A Donacia[7], with calves as big as mine; several large Cimices[8], one most beautiful, scarlet & black; — a red & black Ptinus[9], with a ~~thor~~ thorax like a night-cap; et multa alia. — How go all the Chrysalises *If* you answer this letter, tell me of your success in *the science*, (of *names*, I am afraid). — We have heard several times from Erasmus, he has changed his plans so often; that to follow him in his course//

would be to pursue a Machaon[10] on a windy day. I have nothing more to write about, but I hope soon to write again. — Heaven knows this effusion will not tempt you to write again: but I must take my chance. — My dear old Fox I long to see you again, but I suppose it will not be before Term begins

Believe me your most sincere & loving $\frac{\text{cousin}}{X^2}$

Charles Darwin

(on, being an unknown quantity[11], this formula will exactly express our relationship)

Endorsement: Charles Darwin Augst 4th 1828 WDF
Christ's College Library, Cambridge, #2.

Notes

1. *In Life and Letters* (1888), Vol. 1, p. 171, F. Darwin wrote "*The greater number of the following letters are addressed by my father to his cousin, William Darwin Fox. Mr. Fox's relationship to my father is shown in the tree given in Chapter I. The degree of kinship appears to have remained a problem to my father, as he signs himself in one letter "*$\frac{\textit{cousin}}{n^2}$*." Their friendship was, in fact, due to their being undergraduates together. My fathers's letters show clearly enough how genuine the friendship was. In after years, distance, large families, and ill health on both sides, checked the intercourse; but the warm feeling of friendship remained.*"
2. In his *Autobiography*, p. 58, CD comments on his lack of mathematics: "*I attempted mathematics, and even went during the summer of 1828 with a private tutor (a very dull man) to Barmouth, but I got on very slowly. The work was repugnant to me, chiefly from my not being able to see any meaning in the early steps in algebra. This impatience was very foolish, and in after years I have deeply regretted that I did not proceed far enough at least to understand something of the great leading principles of mathematics; for men thus endowed seem to have an extra sense.*' From Chapter 2 we see that Fox also struggled with mathematics and had a private tutor.
3. White plumed moth; Lamarck Vol IV (new edn.), p 184; in Darwin's copy of Stephens' *Catalogue Br insects*, p 229 but no annotation.
4. *Papilio machaon*, the Swallow Tail butterfly; Samouelle, 1819, p 235 and Plate 6, Fig 1; Stephens' *Catalogue Br insects* Vol 1 p 145 states, "*Plentiful throughout the fens between Ely and Cambridge. "Rev Leonard Jenyns*" ".
5. Probably the Burnet moth; Lamarck Vol IV pp 6 &7; the French spelling is "Zygèna"; the genus is in the Zygaenidae, or sphinx, family.
6. In CD's copy of Stephens' *Catalogue Br insects* two species of Carabidae have "Barmouth" against them: *Bradytus ferrugineus* and *Calathus mollis*, another two have "B" against them, *Calathus latus* and *Cincindela aprica*.

7. A water beetle.
8. Not a recognisable genus. Possibly referring to the family Cimicidae (Hemiptera), where there is the genus Cimex of which the plural would be "cimices".
9. Family Ptinidae; common name for one species of *Ptinus* is the "white-marked spider beetle". Several species are mentioned in Samouelle, 1819 and in Stephens' *Catalogue Br insects*.
10. See note 4, above.
11. Whether CD used an "x" here or an "n" is difficult to decide. Certainly in Letter #27 he uses "n" in the sense of the nth degree.

#4 (CCD 46). To: Wm Darwin Fox, Esq, Osmaston, Derby.

[Barmouth]
Tuesday Evening [19th Aug 1828]

My dear old Fox

How very unfortunate it was that our letters crossed on the road, more especially as mine contained such very severe remarks. I should indeed be most ungrateful, if after reading your letter, I did not repent of them. To give the devil his due, I must say you are, (excepting always punctuality) a perfect pattern for a correspondent. You answer so distinctly & satisfactorily all one's question's. You see I am already beginning to harp on Entomology. — But before that I must thank you for your most kind invitation to Osmaston[1] for the Music Meeting[2]. I shall be extremely happy to come, my only//

difficulty at present is how to get there as on the 7th I shall be in remote part of Staffordshire. Talking of that, upon my soul it is only about a fortnight to the first[3]. And then if there is bliss on earth that is it. — I suppose the Music Meeting will be very glorious as this will be my first, but I must say I expect much greater pleasure in seeing you, & ~~your~~ all your beasts, & last, but not least all the insects. — I suppose I had better bring my gun as I hope we shall have one or two shots together. — Looking over some insects the other day I found one of my *quondam* - Nebria[4], & eheu eheu[5] I believe it turns out to be a (Nosodendrum of Lam:[6]) as what I thought were //

the Palpi, I now believe to be very short thick Antennæ. Your description however agrees pretty well. — I have taken both varieties of the Pine destroyer. — Since I wrote, have taken a dirty-purple coloured Cicindela[7], with squarish white markings, is it Sylvatica? Also, a *most splendid* **Elater** not Buprestis, by Marsham[8] I make it out clearly to be the Cyaneus, but he gives no references[9], from which I infer it v∗ery∗[10] ∗r∗are. Elytra & thorax metallic purple blu∗e,∗ abdomen greenish blue, legs yellow, taken at a great altitude. — I hope, if you can, that you will answer this letter directly, as on the 27th. I leave for Shrewsbury, direct Post Office Barmouth: Look over my former letters, & answer my Entomolo: questions. It is quite absurd how interested I am getting about the science. — The great black Carabus, considerably bigger than Violaceus//

which I mentioned in former letter was taken at great altitude. — I am going to mention a few insects, which are very doubtful. — 2 most beautiful insects, Cryptocephalus sericeus[11], & Lebia Cyanocephalus[12], a Carabidæ. — allso, a

Bembidion[13], (Littorale). 3 sorts of dung beetles like the Vaccæ[14]. — a beautiful, square built Donacia[15], &c &c //

&c &c. I mean to take all my insects to Cambridge, & then you will see all these *wonders.* — Give me some instructions about keeping Crysalises, as I possess some of ditto. — I have got so much to say & so much to see <at Osmaston[1]> that it will be God's Mercy if//

(written cross-wise on 1st and 2nd pages)
I go away alive. nevertheless I much hope <much> for the experiment ~~will~~ to be tried A part of your letter has given me a Panic, you say you do not know when you shall return to Cambridge I *most sincerely* hope it only means at what time in *October*. I should be quite lost without my good old Entomological Tutor. — I have been abusing you for unpunctuality when I forget you might apply the argumentum ad hominem. The reason I delayed answering is that I have been //
on an expedition for a few days. For you must know that I am become a "Brother of the Angle"[16] under the superintendence of Mr Slaney[17] (M.P. for our town of Shrewsbury), who pronounces me a very flourishing Pupil. —
Do write soon, & believe me, dear Fox

Yours most sincerely
Chas Darwin

I need not tell you to mention to your Father how very much obliged I am for his kind invitation, & how happy I shall be to accept it. —
(written diagonally across the 3rd page)
Perhaps you had better direct to Shrewsbury as there is hardly time to receive a letter before the 27th.

Endorsement: C. Darwin Aug 23, 1828 WDF
Postmark: Barmouth 223
Christ's College Library, Cambridge, #4.

Notes

1. Osmaston Hall, the country mansion rented by Fox's father, Samuel Fox III, from 1814 to about 1850 (see Chapter 1, and Fig 4).
2. This was a Triennial Music Meeting in Derby, from the 9th – 12th September, 1828 (*Derby Mercury*); see also Letter #5 (n6). Fox also records, in his diary, attending the previous Music Meeting in Derby from 4 to 7th Oct 1825 (Appendix 4). Birmingham also had a Music Meeting to which CD went in 1829 (see Letter #24).
3. A reference to the start of the game (grouse) shooting season.
4. *Nebria* is a Ground Beetle in the genus of Caribidae. *Nebria brevicollis* is the commonest species found in shaded hedgerows and woods. Not mentioned in Samouelle, 1919, but is mentioned in Lamarck, Vol IV, p 519 and Stephens, *Catalogue Br insects*, Vol 1.
5. Latin for "alas".
6. Nosodendrum fasciculare is in Lamarck Vol IV, p 546-7 and in Stephens' *Catalogue Br insects* (but no notes by CD). It feeds on tree roots, hence the reference to "the pine destroyer". In Letter #2 (n10) CD asks Fox for the name of the beetle that destroys "Scotch pine". It would appear that Fox answered the question in an intervening letter, but it is not clear whether he indicated the genus *Nosodendrum*.

7. The tiger beetle, a very fast running beetle. CD's copy of Stephens' *Catalogue Br insects*, Vol 1 has the annotation "*Barmouth Aug 1828*" against *C. hybrida.*
8. Marsham, T, 1802. *Entomologia Britannica, sistens Insects Britanniae indigena, secundum methodum Linnaeanum disopsita, 1. Coleoptera.* London. xxxi + 548 pp. This volume is in Darwin's Library Collection, University Library, Cambridge. It is rarely referred to by CD as it was obviously being superceded by Stephens' volumes. *Bupestris* is in Stephens' *Catalogue Br insects*, Vol 1, p 119, but no annotation appears in CD's copy.
9. Marsham (1802) notes on p 388 "*32. Elater cyaneus. Habitat* —".
10. Some smudging occurs here.
11. The leaf beetle, in the family Chrysomelidae. In Samouelle (1819, p 213) and Stephens' *Catalogue of British insects*, Vol 1 but no annotations by CD in his volume.
12. A ground beetle. Found in the books of both Stephens and Samouelle, but no annotations in CD's copies.
13. Should be Bembidium, but the handwriting certainly suggests that CD wrote Bembidion. A ground or shore beetle found on sand or near the water's edge. In CD's copy of Stephens' *Catalogue Br insects* there is the annotation "*Barmouth*" against *B. pubescens.*
14. Dung beetles. Both Marsham (1802) Vol 1 and Stephens' *Catalogue Br insects*, Vol 1, p 105 mention Onthophagus Vacca. Scarabaeus Vacca is a synonym.
15. See Letter #3(n7).
16. "Brother of the Angle". The origin of this reference is unclear. The Gentleman's Magazine of 1827 refers to "A Brother of the Angle" in regard to the author of a book on angling. Therefore CD may have been on a fishing expedition.
17. Robert Aglionby Slaney, of Walford and Hatton, Shropshire; also the father of Elizabeth Frances, his co-heir, who married Thomas Campbell Eyton, a naturalist acquaintance of CD at Cambridge [see Letters #31 (n2) & #32] and later frequent correspondent.

#5 (CCD 48). To: William Darwin Fox Esq, Osmaston Hall, Derby

[Shrewsbury]
[1st or 7th October 1828]
Tuesday

My dear Fox

I think I mentioned that I had a few stuffed birds, & as they would be of much more use to you than to me, I have taken the liberty of sending them to the "*Osmaston Museum*"[1] & hope they will arrive safe. — I have also sent a few insects, a Carabus with 6 punctures — taken at Maer, & another Leistus[2] of a light brown colour. — Tell me what you think of these insects, also of a common black (but new to me) carabus. — N.B. The Terne was shot on Maer pool last September. — //

So much for Natu History, & excepting that I have been doing little else, & therefore my letter must be as stupid as I myself am. I staid two days at Maer, where I left orders about your birds, & on Monday returned to sweet home. Home is doubtless very sweet, but like all good things one is apt to cloy on it; accordingly I have resolved to go to Woodhouse[3] for a week. This is to me a paradise, about which, like any good Mussulman I am always thinking; the black-eyed Houris however, do not merely exist in Mahomets noddle, but are real substantial flesh & blood. Formerly//

I used to have two places, Maer[4] & Woodhouse[5], about which, like a wheel on a pivot I used to revolve. Now I am luckier in having a third, & I hope I need not say that third is Osmaston[6]: I must say, although for the 10th time, & although you doubtless would elegantly term it humbug, I do not know when I have spent 3 pleas*anter* weeks[6]. Would you be so kind as to present *to* your sister Emma[7], a few franks, which I have rummaged out, & I hope some few of them will be new. Mr. Joseph splendid example was before me, he indeed gave original sonnets, I alas must be contented with my minor contribution[8]//

I hope, when you write, you will give me a most minute account of every thing alive & dead about Osmaston. Remember me to your sister Julia & tell her how much amusement I received from her permission of peeping, & far from being able to forestall her in the article of news her letter afforded me as much pleasure as it did Catherine
(written cross-wise on the RHS of first page)
Remember me most kindly to Mr & Mrs. Fox & most tenderly (it is your own term) to the rest of your family & believe me my dear old Fox

Yours sincerely
Chas. Darwin

(written cross-wise on the first page)
I want to know the name of a butterfly, which you have got, its wings are *most wonderfully jagged*, & of a reddish colour, after an immense chase with all the servants in the house I at last captured it —
Look over your butterflies & you will soon perceive what I mean —

Endorsement: Charles Darwin Oct 1828 WDF
Christ's College Library, Cambridge, #5

Notes

1. Osmaston Hall was the Fox family home (see Chapter 1 and Fig 4).
2. A ground beetle. CD's copy of Stephens' *Catalogue of British insects* (1829) has "*Maer and Shrewsbury*" against *L. fulvibarbis.*
3. Home of Lieutenant William Mostyn Owen and his two daughters, Sarah and Fanny (see correspondence between these two sisters and CD in CCD, Vol 1).
4. CD was attracted to Charlotte Wedgwood as well as the two Owen sisters of Woodhouse (see note 3, above). However as this present letter makes clear he had also become attracted to Fox's sisters of whom 4 were unmarried at this time and who ranged from 19 to 27 years in age. CD singles out Emma (25) and Julia (19) for mention.
5. See Note 3, above
6. Osmaston Hall (see note 1). It is clear that CD had visited Osmaston Hall in his recent trip and attended the Music Meeting in Derby (see Letter #4). A triennial music meeting was held in Derby during these years. In 1828 it was held from 9th – 12th September (*Derby Mercury*). In his diary for 1825, Fox reports on attending the previous Music Meeting at Derby, in early October, 1825. In his letter to J M Herbert on 3rd October 1828 (CCD, Vol 1), CD reports. "*I have been enjoying myself very much in Derbyshire, the Music meeting went off very well. I shot pretty well: my Entomological pursuits succeeded, & the Miss Foxes are very pleasant girls, & so altogether I have spent a very pleasant month*".
7. Emma Fox (1803-1885), one of Fox's four sisters.
8. Clearly a reference to the Music Meeting in Derby, referred to in Letter #4 (n2) and note 6, above.

#6 (CCD 52). To: William Darwin Fox, Osmaston Hall, Derby

[Shrewsbury]
[29th October 1828]
Wednesday

My dear Fox

I should have answered your letter before this, but when you know the reason, you will say it is a good one. — But first, I must grieve over the loss both of your hawk & stuffed birds the first is the most serious one, but I cannot help grudging the fate of the others, as one gets a sort of affection for anything procured by one own self: I go to Cambridge on Thursday, & shall arrive there on Friday evening, & I hope it will not be long before you are there[1], & now for the glorious news. I have been introduced, & if I may presume to say so, struck up a friendship with Mr. Hope[2]: I met him at dinner, & I find he knows all my Scotch friends,//

& we had so much entomological talk, that he asked me bring over all my insects to Netley[3]. When we meet I will tell you the result, but I must ~~tell~~ mention, what he said, that for 4 or 5 years back, he had not seen such a rich case collected in one year <.·. yours must be still richer>: I hope you had all the duplicates of ~~carabus~~ <carabi> taken at Barmouth, as most are very rare, & some new he believes//

he is a perfect specimen of an Entomologist, so generous, strait-forward. I could not prevail upon him to have any new ones, except 2 flies, he believes both new to England. My head is quite full of Entomology. I **long** to empty some information out of it into yours. — He thinks he can give me 3 or 400 speci∗es∗ at Christmas. the other day he sent 700 to Dr Fleming[4]. In the spring he wants me to come on expedition, all over the Welch mountains & he insures me to find many *new* insects. I could write all day about him: but I long to see you! he has given me a great many water beetles.

Yours affectionate,
Ch Darwin

Catherine[5] <sends> something more than her love to Julia[6], & entreats she will write soon. I suppose you will soon be up at Xst. Coll: as the *delightful! mornings*! at the Priory[7] are come to end. —

Postmark: Shrewsbury OC 3<0> 1828 153
Christ's College Library, Cambridge, #6.

Notes

1. Fox probably went to Cambridge for the most of Michaelmas Term, 1828 – see Letter #7 (n1).
2. Frederick William Hope (1797-1862). Entomologist and clergyman (see Bibliographic Register), who endowed the Hope Chair of Entomology at the University of Oxford.
3. Netley Hall, the Hope family seat, about 5 miles south of Shrewsbury.
4. Probably John Fleming (1785-1857), zoologist and geologist; professor of natural philosophy at King's College, Aberdeen. (*DNB*)
5. Emily Catherine Darwin (1810-1866), CD's youngest sister.
6. Julia Fox (1809-1894), Fox's youngest sister.

7. The Priory was the home of Elisabeth Darwin and her daughter Frances Anne Violetta, see Letter #7. Frances Anne Violetta married Samuel Tertius Galton, who lived at the Larches, Sparkbrook, Birmingham; Fox was at one time attracted to their daughter, Bessy. It is likely that Frances and Bessy were visiting the Priory at this time and Fox was drawn there, not far from Fox's home, by the presence of Bessy. In Fox's dairy for 1825, he was clearly very attracted by Bessy (see 4th Oct 1825, Appendix 4).

#7 (CCD 54). To: William Darwin Fox, Christ's Coll: Cambridge[1]

[Shrewsbury]
[24th December 1828]
Wednesday

My dear Fox

I am sorry I did not write to you, as I promised, on Monday, but we have had people in the house so could not. — I & Mr. Dash[2] arrived quite safe here on Saturday morning He rises in my opinion hourly, & I would not sell him for a 5£ Pound note. — it would have excited your envy & spleen to have seen him on the scent of a covey of Birds, & the style in which he went down when I held up my hand. — I have begun to entomologize & have taken thirty specimens of Platynus Angusticollis[3]: the best silpha[4]: Dromius agilis[5], & a few other insects. — //

You cannot imagine how pleased my Father was with the Death-Head's[6]; to use his own words "if he himself had thought for a week he could <not> picked out a present so acceptable". He came in very tired from Nottinghamshire, & the sight of them acted like a charm — I hope it is not the Entomological Mr. Boothby[7], (who married Lord Vernons[8] daughter), who is so very ill: My Father in his way back again, called at the Priory: where Mrs Darwin[9] gave a magnifiquè account of you & your character to my Father. — Catherine had a letter the other day from Bessy Galton[10], which contained a flaming account of Erasmus[11] entry//

in the Navy, ushered in by right honourable Lords & gallant Captains. — This letter is a most unconnected tissue of facts; but whilst I recollect, I must put you in mind that you have my snuff box, & I have yours; doubtless yours is the most valuable: but mine was the gift of Mr. Owen, & he is the Father of Fanny, & Fanny[12], as all the world knows, is the prettiest, plumpest, Charming personage that Shropshire posseses, ay & Birmingham too, always excepting the blooming Bessy[13], & now that I know a most pleasant train of ideas are excited in your mind, I will not interrupt them by writing any more

But believe me my dear Fox
Yours most sincerely
Chas: Darwin.

//Do write soon & tell me how Entomology goes on. — Catherine is ruminating some message to you, which will express a proper proportion of kindliness & decorum mingled in due measure.
I cannot wait, so frame one according to your own liking. — I most cordially hope your degree goes on well[14]. write a most minute & particular account of all your doings.

Postmark: Shrewsbury <DE> 25 1828 153
Endorsement: Decr. 27, 1828. — C. Darwin
Christ's College Library, Cambridge, #7.

Notes

1. There is a large gap between Letters #6 and #7, i e, between 30th Oct and 25th Dec 1828. This meant, probably, that either one or more letters were lost or that WDF made an extended visit to Cambridge during this period, and was in daily contact with CD. The latter is the more probable since from the following letter (#8) we know that Fox was in Cambridge at CD's departure on 20th Dec and was still there in January, 1829 to take his B.A. examination (see Letters #8 and #9) and would have needed to have things in order. This accounts for the familiar reference to Dash, the dog and the lack of other details.
2. A small dog that CD and Fox acquired.
3. A ground beetle. In CD's copy of Stephens' *Catalogue Br insects*, Vol 1 (1829), there is an annotation against this species "Shrewsbury Jan 1829"
4. Silpha is a genus of the family Silphidae, or carrion beetles.
5. In CD's copy of Stephens' (1829) there is the marginal note "*Cam*" against this species.
6. Death's Head moth (*Acherontia stropos*).
7. Brooke Boothby (1784-1829). Prebendary of Southwell and Rector of Kirkby-in-Ashfield, Nottinghamshire. Married Louisa Henrietta Vernon in 1816. (*Alum. Oxon.*).
8. Henry Vernon, 3rd Baronet (1747-1829) of Sudbury Hall, Derbyshire (Burke's Peerage, 1980).
9. Elizabeth Darwin (1747-1832) (formerly Chandos-Pole, née Collier), the second wife of Erasmus Darwin.
10. Elizabeth (Bessy) Ann Galton (1805-1906), cousin of both CD and Fox. She was the daughter of Samuel Tertius Galton and Frances Anne Violetta, the daughter of Erasmus and Elizabeth Darwin. The Galtons, at this time, lived at the Larches, Sparkbrook, Birmingham. Bessy later married Edward Wheler and as Elizabeth Wheler wrote an interesting but unpublished account of the lives of the Darwins and Galtons, "*Memorials of my Life*" (see King-Hele, 1999).
11. Erasmus Galton, Bessy Galton's brother (see note 10, above).
12. Frances (Fanny) Owen, one of the daughters of Lieutenant William Mostyn Owen, Squire of Woodhouse, near Shrewsbury, with whom CD has a close relationship at this time (CCD, Vol 1; Desmond and Moore, 1991).
13. Bessy Galton, see Note 10, above.
14. Fox was preparing for his final B A exam, which he took in late January 1829. He came 88th in a list of 160 candidates – see Letter #9 (n2).

#8 (CCD 55). To: William Darwin Fox, Christ's Coll:, Cambridge

[Shrewsbury]
[7th Jan 1829]
Wednesday

My dear Fox

You must have thought me a most ungrateful wretch to have taken not the slightest notice of your offer of getting up the morning of my departure from Cambridge; the reason was, (& I told Impey[1] particularly to mention it to you) that I was called so late that in the hurry I forgot all about it, & in despair I rushed to your rooms, but found the staircase so dark, that I had not time to crawl up them[2]. — I am much obliged for your giving me so minute an account of your Entomological proceedings, although they //

have excited no small degree of envy & spleen: I literally can take nothing, excepting one small insect, something like the Trufle beettle, which Henslow[3] gave us. — The

Chrysalis goes on very well, it is much more lively, so that if touched, it will roll itself about. — & it appears to me the parts about the head are very much more distinct than they formerly were. — The process is, — a flower pot - full of dampish mold, & over that a stratum of not very dry sand, on which the Chrysalis is placed; an inverted glass vessel, which I suppose prevents too much evaporation, is placed over it, — the whole kept in a warm room. — I suppose you//

have heard poor Mr Boothby[4] is dead he expired whilst my Father was in the house. — Lord Vernon[5] also is there he is quite paralytic on one side. — What an excessively nice old gentleman he appears to be, it makes one pity him so much more. — From the bottom of my heart I pity your dismal state in Cambridge[6]: I suppose there is scarcely any body up there, & wit<h> this odious weather, & still worse degree, how it must make you long to be at dear old Osmaston[7]. — How very odd it is you do not hear from Chapman[8] — You can scarcely wish, half so much as I do, for it to be settled favourably, as I should think a curacy in Derbyshire would be pleasanter//

to you, but to me it will make all the difference whilst I remain at Cambridge[9] My Father has given me 20£ towards going to Edinburgh, but I still remain in doubt. it is so very long a journey to take in winter[10]. — Remember me kindly to Hore[11] & Jem. Holden[12] — I long to hear of your being AB[13]. as I have no doubt I soon shall. —

I remain dear Fox Yours sincerely
Chas. Darwin

Postmark: Postmark: Shrewsbury JA 8 1829 153
Christ's College, Cambridge, #8a

Notes

1. Impey was CD's 'gyp' (college servant). See Lichfield, 1915, *Emma Darwin* 2, p. 166 and Letter #131 (n7). Gyps looked after a number of students, usually on the same College staircase, making their beds, bringing meals, tidying the rooms and running errands.
2. The rooms occupied by WDF at Christ's College have been identified as Second Court, middle staircase, second floor (E7) (pers com, J van Wyhe, Christ's College, Cambridge).
3. John Stevens Henslow (1796-1861). See Bibliographic Register and Introduction to Chapter 2. Henslow became CD's mentor in his last two years at Cambridge and is mentioned in many subsequent letters (see for e g, #8 (n9), #26 (n6), #27 (n11), #29 (12), #30 (n11), #35 (n2 &7), #36 (n1), #39 (n4), #40 (n3), #41 (n8), #43 (n4), #44 (n2), #49 (n24). (*DNB*)
4. Mr Boothby, see Letter #7 (n7).
5. Lord Vernon, see Letter #7 (n8).
6. It is clear that Fox had stayed in Cambridge over Christmas to prepare for his final B A examination, traditionally at the end of January.
7. Osmaston Hall, the Fox family home just outside Derby (see Fig 4).
8. Possibly Benjamin Chapman (1798-1871). Christ's College, B A, 1829. Rector of Westley Waterless, Cambs, 1829-1836; vicar of Leatherhead, Surrey, 1837-1871. (*Alum Cantab*)
9. As well as taking his final B A exam (see result, next letter), WDF was looking for a curacy, which he obtained in the early part of 1830 at Epperstone, near Lowden, Nottinghamshire.
10. This journey never took place. See Letter #9.
11. William Strong Hore (1807-1882), B.A. Queens College, Cambridge, 1830. Clergyman and botanist. (*Alum Cantab, Desmond, 1977*)
12. James Richard Holden, a school friend of WDF from Repton School, see Chapter 2, note 52.
13. CD of course meant B A (Batchelor of Arts), the only undergraduate degree available at the University of Cambridge at this time, so this may have been a private joke.

#9 (CCD 56; LLi, part[1]). To: W. D. Fox, Esq., Christ's Coll, Cambridge

[25-29th Jan 1829]
[Shrewsbury]

My dear Fox

I waited till to day for the chance of a letter, but I will wait no longer: I must most sincerely & cordially congratulate you on having finished all your labours. I think your place a *very good* one[2], considering by how much you have beaten many men, who had the start of you in reading — I do so wish I were now in Cambridge (a very selfish wish however, as I was not with you in all your troubles & misery) to join in all the glory & happiness, which dangers gone by can give. — how we would talk walk & entomologize — Sappho[3] should be the best of bitches & Dash[4] of dogs: there should be "peace on earth — good will to men" //

(which by the way, I always think the most perfect description of happiness that words can give) I was very sorry to see Holdens name amongst what I suppose to be the plucked men[5], & amazed to see Pulleins name no where[6]. — It would be superfluous to thank you for the Newspaper, as I am sure you must know how very anxious I was hear the issue. I believe (?) we have some bets pending about Mr. Philpot[7] (who it seems has astonished the knowing ones) & about yourself, which must remain in doubt till I go to Cambridge & look at my betting book. — I received the swan about a week ago, & should have written to acknowledge & thank for it before; only that I thought it better to wait till the examination was over. We have not yet eat it & it is probable it will keep sometime longer. — //

My Father has had a bad fit of the gout together with a good deal of fever. he has been confined to his bed for a week. — he was very much pleased with the Hooper[8] & as he said himself, if we had thought for a month we could not have made two more acceptable presents than the Deaths head & swan. — I think it is probable we shall get for you both the common & Pine Marten, the latter alive, which I should think you would like, as th*ey* are very interesting animals to keep tame, so an Irish gentleman tells me. — Now for Entomology. — a beautiful Agonum with dark blood red elytra. — a small Elaphrus; an insect like the Pederus ripalis[9], only with a white mark on each elytron. — cum paucis aliis. — in your next letter tell me how you get on in the science. — My life is very quiet & uniform, & what makes it more so, my lips have lately taken to be bad[10], which will prevent my going to Edinburgh. — my Studies consist of Adam Smith & Locke, in the latter of which I suppose you are an adept, & I hope you properly admire it — //

About the little Go[11] I am in doubt & tribulation. I have had very little shooting. I went to Woodhouse for a week, & on the first day & first shot one of the young Owens[12] cut his eye so badly with a Copper Cap; that he has been in bed for a week. — I think I never in my life time was half so much frightened: I am sure I have written enough about myself & my own concerns: make a handsome return tell me every thing about yourself. I am exceedingly anxious to know what steps you have taken about the Curacy? where you are going? how long you intend staying at Osmaston & the Larches[13]? in short do give me an outline for the 2 or 3 <next> months. I hope you will write to me directly Erasmus[14] is in Rome & likes it very much, he wants me to join him in Geneva. he intends wintering next year in Paris. Remember me kindly to Whitley[15] if you see him, Pullein[6] & Hore[16], & believe me My dear Fox   Yours most sin

C. Darwin.

if you do not get the Curacy I do not know when we shall meet again
Qære.
How much of this is legible?

Postmark: Shrewsbury JA 2< > 1829 153
Christ's College, Cambridge, #8

Notes

1. Francis Darwin published the first part of this letter up to "perfect description of happiness that words can give" in *Life and Letters*, Vol 1, p 174 (F Darwin, 1888).
2. Fox was ranked 88th in a list of 160 candidates (*Cambridge Chronicle and Journal*, Friday, 23 January 1829, 2d edition; repeated, Friday, 30 January 1829).
3. Sappho was a bitch owned or minded by CD (see letter #13).
4. Dash was a dog that had previously been owned by Fox.
5. James Richard Holden (see Chapter 2, note 52). CCD, Vol 1, notes: "*James Richard Holden's name appears fifth in a group of fourteen names, without numbers, following the 160 that are ranked. In the list printed 30 January 1829 his name along with others appears with an asterisk, which, a footnote explains, designates men who 'have one or more terms to keep previous to being admitted to their degrees, although they passed their examination in the above order of arrangement' (ibid., 30 January 1829). No explanation is given of the separate unnumbered listing of 23 January.*" As Holden is not listed as having a B A in *Alum Cantab* it must be assumed that he never gained his degree and settled for the army.
6. Robert Pulleine (1806-1868). Emmanuel College; sat for his examination in March 1829 (see Letter #10]). Curate of Spennithorpe, Yorkshire, 1830-1845; rector of Kirkby-Wiske, 1845-1868. Later, he was well known as a judge at poultry shows [see *Cottage Gardener* 15 (1855–6): 208 and 227]. Fox kept up a friendship with him in later life (see Fox *Diaries*) and CD mentions him in Letter #106 (n6). (*Alum Cantab*)
7. Henry Philpott (1807-1892). St Catherine's College, B A, 1829. He was Senior Wrangler, that is, the student taking first place in the examinations for mathematical honours. Fellow of same college, 1829-1845; Master, 1845-1850. Bishop of Worcester, 1860-1890. (*Alum Cantab*)
8. Hooper swan.
9. No such species of *Paederus* exists. CD probably intended to write 'riparius'.
10. CD had a tendency to develop sores on his lips and hands [see also Letter 18 (n)]. In regard to his sore lips, CCD, Vol 1 notes "John Maurice Herbert, in a letter to Francis Darwin, stated that in his youth CD took small doses of arsenic for it. '*He told me that he had ment*[d]*. this treatment to his Father —& that his Father had warned him that the cure might be attended with worse consequences — I forget what he said the risk was, but I think it was of partial paralysis.*' (Letter from J M Herbert to Francis Darwin, 12 June 1882, DAR 112: 61v)". For an account of illnesses throughout CD's life, see Colp 1977. Interestingly, Fox also seems to have taken arsenicals, see Chapter 2.
11. "Little-go" was the popular name for the first examination at Cambridge University for the degree of B A, officially called "The Previous Examination" (*OED*). See Introduction, Chapter 2 (note 5) and Letters #10 (n13) & #28 (n2).
12. CD refers to Arthur Mostyn Owen, son of William Mostyn Owen, of Woodhouse, near Shrewsbury, Shropshire. As seen in letters #3 & #5 CD visited Woodhouse for shooting, especially in the game shooting season, and to see Fanny and Sarah Owen, sisters of Arthur.
13. Fox was looking for a curacy. The Larches, Sparkbrook, Birmingham, was the home of Samuel Tertius Galton, Deputy Lieutenant of Warwickshire, and Violetta Galton (née Darwin) until 1832.

(See Francis Galton *Memories of my life*; London: 1908, pp. 2-3 & 18). Their daughter, Bessy Galton (Elisabeth Ann), is mentioned in Letter #7 (n10).

14. Erasmus Alvey Darwin, CD's brother.
15. Charles Thomas Whitley (1808-1885). St John's College, B A, 1830). Attended Shrewsbury School and became a reader in natural philosophy and mathematics at Durham University. (*Alum Cantab*)
16. William Strong Hore (1807-1882). See Letter #8 (n11).

#10 (CCD 57; LLi, most[1]). To: W. D. Fox, Esq., Osmaston Hall, Derby

Cambridge
Thursday
[26th Feb 1829]

My dear Fox

When I arrived here on Tuesday, I found to my great grief & surprise, a letter on my table which I had written to you about a fortnight ago; the stupid porter never took the trouble of getting the letter forwarded[2]: I suppose you have been abusing me for a most ungrateful wretch: but I am sure you will pity me now, as nothing is so vexatious as having written a letter in vain. — Last Thursday I left Shrewsbury for London & staid there till Tuesday, on which day I came down here by the Times[3]: The two first day I spent entirely with Mr. Hope[4]. — & did little else but talk about & look at insects: his collection is most magnificent & he himself is the most generous of Entomologists//

he has given me about 160 new species, & actually often wanted to give me the rarest insects of which he had only two specimens: He made many civil speeches, & hoped you will call on him some time with me, whenever we should happen to be in London. He greatly compliments our exertions in Entomology & says we have taken a wonderfully great number of good insects On Sunday I spent the day with Holland[5], who lent me a horse to ride in the park with. — On Monday evening I drank tea with Stephens[6]: his cabinet is more magnificent than the most zealous Entomologist could dream of: He appears to be a very good humoured pleasant little man. — I am glad to find that neither he nor Mr Hope[4] have much opinion of Mr Jennynys[7]; they seem to think him very selfish & illiberal. — Whilst in town, I went to the Royal Institution//

Linnean society, & Zoological gardens, & many other places where Naturalists are gregarious. — If you had but been with me, I think London would be a very delightful place; as things were it was much pleasanter, than I could have supposed such a dreary wilderness of houses to be. — I shot whilst in Shrewsbury a Dundiver, (female Goosander as I suppose you know). Shaw[8] has stuffed it, & when I have an opportunity I will send it to Osmaston: There have been shot also 5 Waxen Chatterers, 3 of which Shaw has for sale, would you like to purchase a specimen? I have not yet thanked you for your last very long & agreable letter: It would have been still more agreable had it contained the joyful intelligence, that you were coming up here; my two solitary breakfastes have already made me aware how very very much I shall miss you — How goes on Entomology? You ought to take a vast number of Mycetophagos[9] insects in the Ash Wood//

Your new Dyticus is a most extraordinary one: I have ordered a 15£ Cabinet[10], & when we get the insects all arrayed together; we shall be able to make them much better out. I have taken this Vacation the Diaporis Anea[11], an insect which <only> Stephens & Hope ~~only~~ posses. —

By Grahams[12] decided advice, I do not go in for my little Go[13]. — Hoar[14] left Cambridge for Devonshire this morning. Pulleine[15] goes in for his degree today, in appearance he is in very good spirits, but his manners are quite different, & I think he has not at all//
(written cross-wise on head of first page)
recovered his most unfortunate accident. — Wedgwood[16] is elected a fellow. — & this is all the news I can scrape together; Cambridge is in statu quo I left it, rain, sleet & cold winds, alternating. Believe me My dear old Fox

Most sincerely Yours
C. Darwin

Write soon, for I long to hear what the Bishop[17] says.

Postmark: Postmark: Cambridge FE 26 182<9>
Christ's College Library, Cambridge, #9

Notes

1. In Life and Letters, Vol. 1, pp. 174-5 (F Darwin, 1888), the extract of this letter is up to "how very very much I shall miss you".
2. This information indicates that CD had written a letter to Fox, addressed to Christ's, which the porter had put on Darwin's desk, instead of forwarding it to Fox at Osmaston, presumably because the porter was uncertain where Fox was. This is probably Letter #9, as it is unlikely that CD wrote another letter in between #9 and the present letter (#10), and also unlikely that Fox did not keep it.
3. The 'Times' mail coach left London for Cambridge at 3 pm each day (*Cambridge University calendar*, 1829).
4. Frederick William Hope (1797-1862). Entomologist and clergyman. Founder of the Chair of Entomology at Oxford University. For further details see Letter #6 (2). (*DNB*).
5. Edward Holland (1806-1875). CD's second cousin. B A Trinity College, 1829. M P for East Worcestershire, 1835-1837, and Evesham, 1855-1868. President of the Royal Agicultural Society. (*Alum Cantab, DNB*)
6. James Francis Stephens, author of "*A systematic Catalogue of British insects*", 1829. See Letter #1 (n11).
7. Leonard Jenyns (1880-1893) [Note the misspelling, probably deliberate (cf. Letters #19 (n8) & 27 (n13)]. Naturalist and vicar of Swaffham Bulbeck, Cambridgeshire - see Bibliography and Introduction to this Chapter. In his *Autobiography*, pp 66-7, CD writes of Leonard Jenyns, '*At first I disliked him from his somewhat grim and sarcastic expression ... but I was completely mistaken and found him very kindhearted, pleasant and with a good stock of humour.*' Jenyns described fish specimens from the *Beagle* voyage. He is referred to frequently in later correspondence (see Index). (*Desmond, 1877, DNB*)
8. The brothers, Henry (Harry) Shaw (1812-1887) and John Shaw (1816-1888), originally from Tarporley, Cheshire, became famous animal collectors and taxidermists in Shrewsbury (Shrewsbury Museum Service; http://www.taxidermy4cash.com/Victorian.html). They started in their father's shop, so it is likely that CD was dealing through the father at this time
9. Meaning insects that feed on fungi. In fact, there was a family of beetles, at this time, the Mycetophagidæ, whose members lived in fungi and under the bark of trees. So CD may have had them in mind.

10. Cabinet: this is a notable sum of money and is certainly when CD acquired his first insect cabinet. However, reference later to a new cabinet [Letter #27 (n7)] suggests that CD had a least two cabinets built [for details of the second see Letter #43 (n7)]. The fate of these cabinets is still in doubt. Part of the insect collection itself is in the Zoology Museum, University of Cambridge; see Fig 14.
11. In Darwin's copy of Stephens' *Catalogue Br insects* (1829) there is no marginal note against this species.
12. John Graham (1794-1865). CD's tutor: Fellow, later Master, of Christ's College. Vice Chancellor of the University (1831, 1840). Bishop of Chester (1848-1865) (*Peile, DNB*)
13. "Little-go" was the popular name for the first examination at Cambridge University for the degree of B.A., See Chapter 2 (note 5). Also see Letters #9 (n11) & #28 (n2)
14. Probably William Strong Hore (1807-1882), see Letter#1 (n2).
15. Robert Pulleine (1806-1868), see Letter #9 (n6).
16. Hensleigh Wedgwood, son of Josiah Wedwood, II (see Bibliographic Register) had taken his B A (8th Wrangler) at Christ's College in 1824, and was elected a Fellow of the College in 1829. CCD, Vol 1, notes: "*Because he was last in the Classical Tripos, that place became known as the 'wooden wedge', corresponding to 'wooden spoon', the traditional designation for the last place in the Mathematical Tripos.*" (*Alum Cantab*)
17. Probably a reference to the curacy that WDF was seeking.

#11 (CCD 59) (LLi, part[1]). To: W. D. Fox, Esq., Clifton Post Office, Bristol; readdressed to Seaforth Parade, Bath.

[Mar 18th 1829]
[Cambridge]

My dear Fox

I am afraid if you judge from the time I take in answering your letters you will think I am not very impatient to receive them. Your last letter <however> was at no time more acceptable to me, as our correspondence, through the Porters stupidity had nearly come to an untimely end; — I am much grieved to find how likely it is my prophecy will be verified, & am still more so to hear the cause. I hope however by this time that Mrs. Bristoe[2] is quite recovered. — I am glad to hear that you are reading divinity[3]: I should like to know what books//

you are reading, & your opinions about them; you need not be afraid of preaching to me prematurely. — J. Price[4] is reading very hard on the same subject, & seems to find it no ordinary labour. — I suppose you have heard by this time both from Holden & Pulleine[5]: the latter, as you will perceive is half mad with joy: I saw him the next morning & he literally could not sit still in any posture for five minutes together. — I am anxious to hear how you like Clifton[6], & what you are doing there? I suppose your unsettled life will prevent you doing much of anything, & amongst the rest of Entomology: I had such a dose of "the Science" in London; that//

I have hardly recovered from it even to this time: this sort of scientific seediness after a nights debauch joined with Mss. Polo & Habours[7] idleness, has, *at present*, entirely put a stop to my Entomological proceedings: I am leading a quiet every-day sort of a life; a little of Gibbons history in the morning & a good deal of *Van. John*[8], in the evening this with an occasional ride with Simcox[9] & constitutional with Whitley[10], makes up the regular routine of my days. — I see good deal both of Herbert[11] & Whitley[9], & the more I see of them, increases every day the respect I have for their excellent understandings & dispositions. They have been giving some very gay parties — nearly sixty men there both evenings. — You ask me about the money? I know no more than the man in the moon what the sum is, indeed, I had entirely forgotten all//

about it till I received your letter. I thought I had *occasionally* heard you mention the regularity with which you keep your accounts[12]. I must confess, their *utility is most striking*. — I have paid Clayton[13] for the memorable Swan. After all it was hardly kept long enough: it was pretty good, but tasted like neither flesh nor fowl, but something half way, like Venison with Wild Duck — Have you heard anything more of the old sinner the incumbent[14], not that I exactly understand how his//

not paying for dilapidation influences you, as if I understood right, you were to live in a farm house <situated> some miles from any other human habitation. I hope you will write soon, & give me an account of all your "outgoings" & "incomings". — Give my most kind respects to M^rs^. Fox & the rest of your family & believe me my dear old Fox.

Yours most sincerely
C. Darwin

Recollect: that Deo Volente whether your Parsonage boast of a roof or not I shall pay you a visit this summer.
Sunday Evening

Postmark: Cambridge 1829; C 17 MR 17 1829
Christ's College Library, Cambridge, #10.

Notes

1. In LLi, p 176 (F Darwin, 1888) from "I am leading a quiet every-day sort of a life" to "nearly sixty men both evenings".
2. Mary Ann Bristowe, Fox's eldest sister (see Chapter 2, Bibliographic Register and Fox Family Tree, Appendix 7). In fact Mary Ann died on the 19th April 1829. See Letter #12. CD became more and most resentful for Fox's lack of response until a letter in April told him of the death, and then he is full of remorse (see Letter #15)
3. Fox returned to Cambridge, for short periods in 1829 and 1830, to read for divinity. This meant taking extra courses and lessons from dons and tutors in Cambridge. It did not lead to a formal degree, but probably helped in getting promotion in the Church of England. These visits did not coincide with the times that CD was in Cambridge (see Appendix 2).
4. John Price (1803-1887). St John's College, B A, 1826. He attended Shrewsbury School and was an assistant Master there from 1826-1827. He was tutoring at Cambridge in 1829 and studying for ordination (Desmond and Moore, 1991). He was a Welsh scholar, naturalist and schoolmaster. This is the "old Price" referred to in Letter # 203 (n4). (*Alum Cantab, Modern English Biography*).
5. James Holden and Robert Pulleine, see Letters #8 (n12) & #9 (n6).
6. The readdressed destination of this letter suggests that Fox was in Bath and not Clifton at the time, perhaps having visited Clifton previously. However, his sister Mary Ann Bristowe died in Clifton (see Letter #15).
7. Messrs Polo and Harbours sold specimens of insects to undergraduate entomologists (see Letters #12 and #14). Polo was apparently a nickname. CCD, Vol. 1, notes "*In an anecdote about his father, George Darwin says: 'Amongst his Cambridge expeditions I remember his speaking of going down*

to the fens ... with a sporting sort of guide who went by the name of Marco Polo, because he carried a leaping pole with a flat board fastened at the bottom for leaping the ditches' [DAR 112 (ser 2): 17]."

8. The card game also known as "21", a corruption of 'vingt-et-un'. As shown by the diary of Fox in Chapter 2 (see note 65) it was a popular card game amongst Cambridge undergraduates at the time.
9. As no Simcox is listed in *Alum Cantab* or in the *Cambridge University calendar*, this is possibly a familiar name for George Simpson (see Letters #16 ('Simpson') and #26, where 'old Simpcox' is referred to). George Simpson, Christ's College, B A, 1830; became a C of E clergyman. (*Alum Cantab*)
10. Charles Thomas Whitley (1808-1895). School friend of CD at Shrewsbury School. B.A. St John's College, 1830. Became a clergyman and Reader in natural philosophy at Durham University. (*Alum Cantab, Modern English Biography*)
11. John Maurice Herbert (1808-1882). St John's College, B A, 1830. Became a barrister and county court judge (*Alum Cantab, Modern English Biography*)
12. Clearly CD had observed that Fox kept accounts, and probably a diary, too; see comments in the Introduction to Chapter 2.
13. Robert Clayton was a Cambridge fishmonger and wild-fowl dealer.
14. CD seems, once more, to be referring to a curacy that Fox was seeking.

#12 (CCD 60; LLi: most[1]). To: W. D. Fox, Esq., Clifton Post Office, Bristol; readdressed to 9 North Parade, Bath.

Xst. College [Cambridge]
April 1st [1829

My dear Fox

In your letter to Holden[2] you are pleased to observe, "that of all the blackguards you ever met with I am the greatest": Upon this observation I shall make no remark, excepting that I must give you all due credit for acting on it most rigidly: And now I should like know, in what one particular are you less of Blackguard than I am? You idle old wretch why have you not answered my last letter which I am sure I forwarded to Clifton nearly 3 weeks ago? If I was <not> really very anxious to hear what you are doing: I should//

have allowed you to remain till you thought it worth while to treat me like a gentleman. — And now having vented my spleen in scolding you, & having told you, what you must know, how very much & how anxiously I want to hear how you & your family are getting on at Clifton, the purport of this letter is finished. If you did but know how often I think of you & how often I regret your absence, I am sure I should have heard from you long enough ago: I find Cambridge rather stupid, & as I know scarcely any one that walks, & this joined with my lips not being quite so well[3], has reduced me to//

sort of Hybernation, which almost equals "poor little Whitmores" melancholy case[4]. — Old Whitley[5] has begun to take your place, & we have just commenced a regular series of constitutionals. — Entomology goes on but poorly: a few Dromius[6] & Agonum's[3], together with the Poecilus[7] (with red thighs) make the g∗reat∗ part of what I have collected this ter∗m∗. I have caught M^{r}. Harbour[8] letting Babington[9] have the first pick of the beetles; accordingly we have made our final adieus, my part in the affecting scene consisted in telling him he was a d——d rascal, & signifying I should kick him down the stairs if ever he appeared in my rooms again: it seemed altogether mightily to surprise the young gentleman. — I have no news to//

tell you, indeed, when a correspondence has been broken off like ours has been, it is difficult to make the first start again. — Last night there was a terrible fire at Linton eleven miles from Cambridge; seeing the reflection so plainly in sky, Hall[10], Woodyeare[11] Turner[12] & myself thought we would ride & see it we set out at <1/2 after> 9, & rode like incarnate devils there, & did not return till 2 in the morning. altogether it was a most awful sight. — I cannot conclude, without telling you, "that of all the blackguards I ever met with, you are you are the greatest & the best

C. Darwin

On the fly leaf: 1 o'clock — going on very well

Postmark: Cambridge AP 2 1829
Christ's College Library, Cambridge, #11

Notes

1. LLi, pp 176-7 (F Darwin, 1888): the extract lacks the part from the words "which almost equals" to "I have caught Mr Harbours".
2. James Richard Holden, a school friend of Fox, see Letter #8 (n12) and Chapter 2 note 52.
3. This is the second statement we have that CD suffered from soreness of the lips [cf Letter 9 (n10)]. He also complained of this while waiting for the *Beagle* to embark (Desmond and Moore, 1991). A letter from J M Herbert to Francis Darwin, also recalls that CD had trouble with his lips at Cambridge (Letter #9 (n10).
4. Ainslie Henry Whitmore (1801-1843). Christ's College, B A, 1830; became a clergyman; he may have been 'rusticated'. (*Alum Cantab*, Peile, 1913)
5. Charles Thomas Whitley (1808-1895). See Letter #11 (n9).
6. In CD's copy of Stevens *Catalogue Br Insects*, Vol 1, there is an annotation of "*Cam*" against Dromius agilis, and several species of Agonum have an annotation "Cam Spring 1829".
7. In CD's copy of Stevens' *Catalogue Br Insects*, Vol 1, there is an annotation of "*Cam*" against Poecilus rufifermoratus and P. punctatostriatus.
8. Harbour: mentioned in previous letter (#11), a collector.
9. Charles Cardale Babington (1808-1895). B A, St John's, 1830. Professor of Botany at Cambridge, 1861-1895. J. Browne (*Charles Darwin, Voyaging*, Vol 1, 1995) says that he was called "Beetles Babington" and states that there was early rivalry between CD and Babington (p132). They were both keen students of J S Henslow. (*DNB*)
10. Jeffry Brock Hall (1807-1886). Christ's College, B A, 1830. Lived in Canada. (*Alum Cantab, Peile, 1913*)
11. John Fountain Woodyeare né Fountain (1809-1880). Christ's College, B A, 1831. He became a distinguished clergyman and chaplain to Dowager Countess of Cavan. (*Alum Cantab, Peile, 1913, Freeman, 1978*)
12. James Farley Turner (?-1841). Christ's College, B A, 1831. He became a C of E clergyman. (*Alum Cantab*, Peile, 1913).

#13 (CCD 61). To: W. D. Fox, Esq., Clifton Post Office, Bristol; readdressed to 9 North Parade, Bath.

[Cambridge]
Friday [11/12 Apr 1829]

My dear Fox

The very day I received your letter, I had given up all hopes of receiving one, & had written to you again at Clifton, in order to receive some account of all your proceedings: I suppose, as you are now at Bath, you will not have the pleasure of receiving <it>: that will not however be a very serious loss to you: as it contained nothing that might not as well have been left unsaid, excepting the old burthen of all my letters, that I wished to hear what you are doing. — I have been in such a perfect & absolute state of idleness, that it is enough to paralyze all ones faculties; riding & walking in the morning,//

gambling at Van John[1] to a most disgusting extent in the evenings, compose the elegant & instructive routine of my life. Lord help me, with this basis, how is it likely that I should be able to write a letter: for six weeks past have I been intending daily to write to my injured brother[2], & I cannot screw my courage to the sticking point[3], & so you ought to be much flattered at my writing at all to you. Therefore you must take the intention for the deed, & excuse my abominably stupid scrawls: Markham[4] has not sold Sappho accordingly I have given her to him: There are no letters for you: Pullein[5] wrote to you at Clifton & wants an answer direct//

Rev. Dr. Burrows, no 30 Fitzroy Sqr. London where he is going to reside for 6 months in order to read divinity: The only thing that is talked about in Cambridge, is little Go[6], which has been unusually strict. most men are in the first class, except Woodyeare[7], but a wonderfull number were called in a second time: Your friend Mr. Wale[8], the proctor, was most gloriously hissed, & pelted with mud: he was driven so furious that even his own gyp dared not go near him for an hour. — I shall go up to Town for a few days with old Whitley[9] next week, & I wish I could afford to put your very pleasant scheme into execution. But all my allowance is spent, & 2 Tutors bills may be expected. — in due time. — I think it is probable that you may see Wilmer[10] there sometime this Easter Vacation. — Polo[11] has not brought me//

any insects lately, but the Pieman[12] has brought me a great many, inter alia Chlænius holoriseus[13]. — I am sadly in want of somebody to entomologize with. I wish A. Way[14] was here. Holland[15] saw him in Switzerland from whence he was going into Italy. Who do you think walked into my room the other day? Our old Irish friend[16], dressed most respectably, he said Mr. Wynne's father had not helped him but that he had got into a very good place at Oxford, & was on his way to Ireland. he said he had written to you, he appeared very grateful & begged me most particularly to express his thanks to you: I do not think we have often spent a//

small sum of money more usefully. Chapman[17] tells me that he intends writing to you very soon: I cannot comprehend your plans. Whose Curacy is it you wish to get? I am very sorry to hear that Mrs Fox has been unwell, but I most sincerely hope she will soon be recovered

Yours most affectionately
C. Darwin

Postmark: Cambridge <A>P 11 <1829>: C 13 AP 13 1829
Christ's College Library, Cambridge, #12.

Notes

1. "Vingt et un", see Letter #11 (n7) and Chapter 2, note 65.
2. According to Desmond and Moore (1991), CD fell out with his brother Erasmus, in London, on his visit in mid February (Letter #10) about advice on how CD should behave and spend his time in Cambridge.

3. This misquotation of a line of Lady MacBeth's bears out the point made by Desmond and Moore (1991) that CD was suffering from ennui and the typical student lifestyle for this time, cf. Fox diary for 1824-25 (Chapter 2). It is reported by Francis Darwin in LLi (1888), p 170, "*Besides a love of music, he had certainly at this time a love of literature; and M^{r}. Cameron tells me that he used to read Shakespeare to my father in his rooms at Christ's, who took much pleasure in it.*"
4. Not identified; presumably a dealer in dogs. Sappho was a bitch, who belonged to Fox initially, see Letters #1 (n14) and #9 (n3).
5. Robert Pulleine (1806-1868). See Letter #9 (n6).
6. "Little-go" or "Little Go" was the popular name for the first examination for the degree of B A, officially called "The Previous Examination" at Cambridge (*OED*). See also Letter #9 (n11).
7. John Fountain Woodyeare né Fountain (1809-1880). See Letter #12 (n11).
8. Alexander Malcolm Wale (1797-1884). B.A., St John's, 1819. Fellow, 1821-1831; Senior Proctor, 1828-1829. "*In Apr. 1829 he was mobbed by undergraduates who pursued him from the Senate House to the gates of St John's, hissing and groaning, and though some of the offenders were summoned before the Vice-Chancellor's Court, the Proctors were not satisfied and sent in their resignations.*", Desmond and Moore (1991, p. 68), who give a more detailed account. (*Alum Cantab, Desmond and Moore, 1991*)
9. Charles Thomas Whitley (1808-1885), see Letter #9 (n15).
10. Bradford Wilmer (1808-1886). Christ's College, B A, 1831. Peile (1913) states that he commenced in 1822 and kept 8 terms, "*his name being removed in 1826 or 1827*". (*Alum Cantab, Peile, 1913*).
11. Polo was a collector used by CD. See Letter #11 (n6).
12. Not identified.
13. In CD's copy of Stevens' *Catalogue Br Insects*, Vol. 1, there is an annotation against Chlaenius holosericeus: "*Cam May 1829*".
14. Albert Way, see Letter #2 (n2).
15. Probably Edward Holland (1806 -1875), see Letter #10 (n5).
16. Probably the same "Poor Irishman" as mentioned in Letter 2 (n3).
17. Benjamin Chapman (1798-1871), see Letter #8 (n8).

#14 (CCD 62). To: W. D. Fox, Esq., Post Office, Clifton.

Sunday
[April 12th 1829]

My dear Fox

I received your letter yesterday morning, & it was with the most sincere grief I read its contents: I am well aware how illtimed all consolation must at present appear. But yet knowing how very unhappy you must be at your sisters[1] most alarming illness; I cannot help troubling you with these few lines, if it were merely to thank you for writing to me: I should indeed be sorry, if anybody, for whom I have such a strong friendship & respect as I must always feel, my dear Fox, for you, should be suffering under the most poignant grief, & I not aware of it. — I most fervently hope, by the time this letter reaches Clifton, that your sister will be better[2]. When you have a *spare* moment do write to me & tell me how she is going on. — I wrote to you at Bath only the day before I received your last letter, but little did I think, when I wrote it, that it would find you in such a melancholy situation

Believe me my dear Fox
Yours most affectionately
Charles Darwin

Cambridge
Postmark: Cambridge AP 14 1829
Christ's College Library, Cambridge, #13.

Notes

1. Fox's eldest sister, Mary Anne Bristowe (1800 -1829).
2. Mary Anne Bristowe died on 19th April 1829 at Clifton (*Darwin Pedigrees*).

#15 (CCD 63; LLi[1]). To: W. D. Fox, Esq., Post Office, Clifton.

Thursday
[23rd April 1829]

My dear Fox

I have delayed answering your last letter for these few days, as I thought that under such melancholy circumstances my writing to you would be probably only giving you trouble. — This morning I received a letter from Catherine[2] informing me of that event, which indeed from your letter I had hardly dared to hope would have happened otherwise[3]. — I feel most sincerely & deeply for you & all your family: But at the same time, as far as anyone//

can, by his own good principles & religion be supported under such a misfortune, you, I am assured, well know where to look for such support. And after so pure & holy a comfort as the Bible affords, I am equally assured how useless the sympathy of all friends must appear, although it be as heartfelt & sincere, as I hope you believe me capable of feeling. —
At such a time of deep distress, I will say nothing more, excepting that I trust your father & Mrs. Fox bear this blow as well as under such circumstances //

can be hoped for. — I am afraid it will be a long time, my dear Fox, before we meet. till then believe me, at all times

Yours most affectionately
Charles Darwin

Cambridge
Postmark: Cambridge AP 23 1829
Christ's College Library, Cambridge, #14.

Notes

1. In *Life and Letters*, Vol. 1 (F. Darwin, 1888), pp 177-8: verbatim (with minor punctuation changes).
2. Catherine Darwin, CD's younger sister, was clearly writing to CD at this time, as, no doubt, were his elder sisters. However, unfortunately, these letters, for the period that CD spent at Cambridge, have been lost, as have those of Fox. The first letters from Fox or family appear after the official start of the voyage of the *Beagle* in late 1831.
3. Fox's sister, Mary Ann Bristowe, who died on the19th April 1829 (*Darwin Pedigrees*).

#16 (CCD 64). To: W. D. Fox, Osmaston Hall, Derby.

[Cambridge]
Monday Morning [May 18 1829]

My dear Fox

I was exceedingly glad to receive your last letter, as I had been for some time past very anxious to hear how you & the rest of your family were after having undergone such a painful affliction. — I should have written <before> only I waited to get the

Bills you mentioned. — Claytons amounts to 2£"7"10, but Bakers' has not arrived yet[1]. —
I have procured for you from Polo[2] a beautiful specimen of the Ring Owzel[3], & gave 5 shillings, also a young male ring tail, with half the plumage blue & half like the <female> ring tail <price 5s>. — also 2 fine specimens of the black tern, price 3s. — also a Royston Crow[4]. — Now <being> upon Nat: history, I will go on: All the Crysalis you left in the deal Box, I never discovered till too late, & found all the moths utterly destroyed. The Cossus[5] has eat its way out & is lost. — //

Aiken[6] has got all the skins of your birds: he also had my Crysalis, which go on very well, & to which I have added a great many more. — I have taken, or rather had collected for me vast numbers of insects: a good many new ones to Cambridge, both large & small. — have since Christmas filled nearly three Standish Boxes. — But Hoare's[7] success is *most* brilliant: Carabus intricatus[8], an insect fit for Paradise; a beautiful Lamprias[9]: Badister cephalotes[10], & multa alia: My success also has been very good amongst the water beettles. I think I beat Jenyns[11] in Colymbetes. — I want much to hear what you have done in Entomology? I heard from old Pulleine[12] the other day. he wrote to enquire about you. I told him he had better direct to Osmaston so I daresay before long you will hear from him.//

Before I tell you my plans for the next 2 or 3 months, I must most cordially thank you for your very kind invitation to Osmaston. You used accuse me of *making speeches* about Osmaston. This accusation although proved to be most palpably absurd by the length of my former visit, shall be made doubly so, by the pleasure with which I accept your invitation. On the 5th of next month I go to Shrewsbury, either through London, in order to chaperon home my sisters, or else direct[13] through Birmingham. A *few* days of home, & then I go a three week trip with Mr. Hope[14] either in S or N Wales: after that home again (& Woodhouse & Fanny[15]), then to Maer[16], & then I hope to have the pleasure, my dear Fox of seeing you again, after our long absence: this till August will fill my time up pretty well. — As to what I have been & am now doing, the less that is said about it, the better: my time is solely//

occupied in riding & Entomologizing. Simpson[17] or Hall[18] are both industrious when compared with me. I cannot help thinking what a great deal of good advice I should receive were you but here. — I hope you will write soon & tell me all about every thing: I am sorry, although I think you are very wise in not going into orders at present Your mind at such a time must be very unfit for <so> serious an undertaking: Make any use of me & my rooms in future arrangements that may please you: Give my most kind remembrances to your father, & I hope that he & Mrs. Fox are going on pretty well:

Believe me my dear Fox
Yours most sincerely
Chas. Darwin. —

Postmark: Cambridge MY 18 1829
Christ's College Library, Cambridge, #15

Notes

1. Clayton and Baker were probably collectors employed by CD and Fox. They are referred to again in the list of accounts in the next letter (#17).
2. Polo was a collector, see Letter #11 (n6).
3. Ring Ouzel.
4. At this time the Royston Crow was a name for the Hooded Crow, *Corvus commix*, because they were so common around Royston, Hertfordshire.
5. A moth.
6. Not identified; presumably an animal collector.
7. Probably William Strong Hore of Queens College. See Letter #1 (n2).
8. In CD's copy of Stephens' *Catalogue Br insects* (1829), Carabus intricatus has no annotation against it.
9. In CD's copy of Stephens' *Catalogue Br insects* (1829), Badister has the annotation "p" against the species, B. chlorocephalus. See Letter #57 (n9). In Curtis' *British Entomology*, B. cephalotes is said to be a very rare insect.
10. In CD's copy of Stephens' *Catalogue Br insects* (1829) this has the annotation "*Waterhouse*" against it.
11. Leonard Jenyns (1880-1893); see Introduction to Chapter 3 (n10) and Letter #10 (n7).
12. Robert Pulleine of Emmanuel College. See Letter #9 (n6).
13. Obscured by tear.
14. Frederick William Hope (1797-1862). See Letter #6 (n2) and Biographical Register.
15. Fanny Owen of Woodhouse. See Letter #7 (n12).
16. Maer Hall: the home of Josiah Wedgwood II, and family, at this time.
17. George Simpson (?-1888). B.A Christ's College, 1830. He became a C of E clergyman. (*Alum. Cantab, Peile, 1913*, see also Letter #11 (n8).
18. Jeffry Brock Hall (1806 -1886). B A, Christ's College, 1830. He later lived in Canada. (*Alum Cantab, Peile, 1913*).

#17 (CCD 66). To: W. D. Fox, Osmaston Hall, Derby.

[Cambridge]
Sunday. June 7th. [1829]

My dear Fox

You did not give me time in your last letter, in order to see about your commissions, then write to you & wait for an answer: accordingly I have been obliged to act to the best in my power, & the imperfections must be supplied next October: I booked & sent of per waggon, a ~~trunk~~ portmanteau & deal box yesterday, containing I hope & trust every thing except your skins of birds & prints; Pulleine[1] had left such imperfect directions with Aiken[2] that we did not know yours from his: & we agreed that//

they had better be kept in my rooms till either Pulleine[1] or you could see them: they are all safely packed up together with the ring owzel: the other skins I have sent with the boxes (but one of the terns was too far gone). About the prints I thought they would go better some other time than with the boxes & tea chests. — You will find all the Keys in the deal box, which I had well nailed up & corded: & now for money matters; Accordingly I have paid for you altogether, 8£"12"6 & as you sent me 10£, I owe you, 1£"7"6, which I will pay you when I see you in the Summer.

//I have got all the receipts safe in Cambridge: — The man from whom you bought poor little Fan[3] told me he saw the other day in Bedfordshire, in very good keep & a very great favourite under such circumstances, I thought it better to leave her where she was. —
So much for your affairs[4]. I am afraid I have executed them imper∗fect∗ly but it is to the best of my power. — And now for myself. I leave Cambridge tomorrow morning, for London — & on Tuesday I go to Shrewsbury, from thence I go into Wales with Mr. Hope to Entomologize[5]. I will write to you again, whilst on our expedition: my success has been very splendid in the science, as you will say, when you next see my insects. I will write soon again. Yours affectionately

	£	s	d.	
Clayton	2 ”	7 ”	8	
Baker	2 ”	3		
<Brett		10 ”	6’>	
Dos for plate	”	16 ”		
”	5 ”	17 ”	6	
Paid for you for skins				s
& &c &c	”	15 ”	11	
Debt.	2 ”			
£	2 ”	15		
£	5 ”	17 ”	6	
”	8 ”	12 ”	6	

Chas. Darwin

Christ’s College Library, Cambridge, #16.

Notes

1. Robert Pulleine of Emmanuel College. See Letter #9 (n6).
2. Not identified; probably a collector, also mentioned in Letter 19 (n3).
3. Fan was the dog that Fox passed on to CD. See Letter #1 (n14).
4. In the list of accounts, Clayton and Baker were probably collectors [see Letter #16 (n1)]. Brett was also likely to have been a collector [see Letter #19 (n6)]. Clayton’s bill probably should have been *£2 7s. 10d* (see Letter #16).
5. See letter #16 (n14).

#18 (CCD 67; LLi, part[1]). To: W. D. Fox, Osmaston Hall, Derby.

Shrewsbury
Friday
[3rd July 1829]

My dear Fox

I should have written to you before, only that whilst our expedition lasted I was too much engaged, & the conclusion was so unfortunate, that I was too unhappy to write to you. this weeks quiet at home, & ~~Woo~~ the thoughts of Woodhouse next week has <at last> given me courage to relate my unfortunate case. — I started from this place about a fortnight ago to take an Entomological trip with Mr Hope[2] through all North Wales: & Barmouth was our first destination. the two first days I went on pretty well, taking several good insects, but for the rest of that week, my lips[3] became suddenly so bad, & I myself not very well, that I was unable to leave the room, & on the//

Monday I retreated with grief & sorrow back again to Shrewsbury. The two first days I took some good insects, Cillenum latterale, (Stephens, 16th. No[4]) Heterocerus 2.

species[5]. Calathus latus[6], (bad specimen of), & crocopus[7]: Opatrum tibiale[8]: Cerambyx nebulosus[9]: & many others: But the days that I unable to go out, Mr. Hope did wonders several <rare> species, of the genera Helobia[10], Carabus[11], Omaseus[12], Elater[13] Nothiophilus[14], & to day I have received another parcel of insects from him, such Colymbetes[15], such Carabi[16], & such magnificent Elaters[17], (2 species of <the> bright scarlet sort). I am sure you will properly sympathise with my unfortunate situation; I am determined I will go over the same ground that he does before Autumn comes, & if working hard will procure insects I will bring home a glorious stock. — //

It is so long a time since I have heard from you, that I am very anxious to receive a letter I want to know what you have been doing lately, & what your plans are for the rest of the summer. Catherine wants to know whether Mrs. Fox & Julia are returned ~~from~~ <to> Osmaston? I am most dreadfully stupid here, so you must excuse this equally stupid letter. I hope you will write soon, & then I will promise to write a longer one. — Catherine sends her kind *remem*brances[18] to you & the rest of your family, in which I beg leave to join & believe me My dear Fox, Yours most sincerely,

Chas. Darwin

I hope the boxes have arrived safe: keep an account of how much money I am indebted to you, as I have already forgotten//

I forgot to tell you, that my Father has got two Martens, I believe both species, ready mounted at Mr. Shaws[19], & I have got a Dun-diver[20], which can be sent anytime you like

Postmark: Shrewsbury JY 4 1829
Christ's College Library, #17.

Notes

1. In *Life and Letters*, Vol. 1, pp. 178-9 (F. Darwin, 1888) up to "It is so long a time....", but also omitting the line "But the days that I unable to go out, Mr. Hope did wonders".
2. Frederick W Hope, see Letter #6 (n2). The expedition to North Wales is mentioned in Letters #16 & #17.
3. Probably the eczema that CD suffered from, see Letter #9 (n10).
4. In CD's copy of Stephens' *Catalogue Br Insects* this has the annotation "*Barmouth Jun 1829*". Stephens' *Illustrations of British Entomology* was published monthly from 1st May 1828. No. 16 was published on 1st July 1829 and contained Cillenium laterale as the only species in this genus on p 4.
5. In CD's copy of Stephens' *Catalogue Br insects* (1829) there is the annotation "*N. W. 1829 June*" against H. marginatus (where N.W.= North Wales).
6. In CD's copy of Stephens' *Catalogue Br insects* there is the annotation "*Barmouth 1829*"
7. In CD's copy of Stephens' *Catalogue Br insects* there is the annotation "*Barmouth July 1829*" against C. crocopus.
8. Not in Stephens' *Catalogue Br insects*. However it is in Marsham (1802), *Entomologia Britannica, sistens Insecta Britanniae indigena secundum Linneum deposita. Coleoptera*, Vol 1, p 142 [see Letter #4 (n8)].
9. Appears in Marsham (1802), Vol 1, p 142. In Stephens' *Catalogue Br insects* it is called Pogonocerus nebulosus.

10. In CD's copy of Stephens' *Catalogue Br insects* there is the annotation "*Hope*" against H. Marshallana.
11. In CD's copy of Stephens' *Catalogue Br insects* there are annotations against 9 species as follows: C. cancellatus: "*Cam & Staffordshire*", C. purpurascens: "*N.W.*", C. arvensis, "*Dyffor & Cader Idris*", C. glabratus: "*Dyffor 1829 July*", C. nitens: "*Hope and Thomson*"; the rest have the annotation "*P*".
12. In CD's copy of Stephens' *Catalogue Br insects* there are three annotations as follows: O. aterrimus: "*Hope*", O. orinomum: "*Cader July 1829*", O. affinis: "*Cader*".
13. In CD's copy of Stephens' *Catalogue of British insects* (1829) there are no annotations against Elater.
14. In CD's copy of Stephens' *Catalogue Br insects* there is the annotation "*from Wales*" against Elater. However, this is a very large genus of click beetles.
15. Water beetles.
16. Ground beetles in the family Carabidae.
17. See Note 13, above.
18. Slight tear at this point.
19. See Letter #10 (n119).
20. The Dun-Diver or Sparling Fowl (*Mergus castor* L.) is closely related to the Goosander (Bewick et al. 1816. *A History of British Birds*).

#19 (CCD68; LLi: part[1]). To: W. D. Fox, Osmaston Hall, Derby.

[Shrewsbury]
Wednesday
[15th July 1829]

My dear Fox

Many thanks for your letter, it was very good of you to write so soon, the reason of my writing now is not to fulfill my promise of a long letter, for that is, I am afraid impossible, but to tell you, that, yours & Pulleins[2] birds skins are in my room. Mr. Aiken[3] seemed to be rather troubled with the size of the Boxes, for indeed they quite filled his little room, so I believe he has one box, & I have two, (in one of which is the ring-Owzel[4]), but I am sorry to say the Key is locked up in one of my drawers. But if you were to write to Impey[5], he could send the boxes (first getting them broken open & packed up by Mr Brett[6]) to Osmaston[7] — I am very sorry//

I have given so much trouble. But I did not think it fair to leave such Leviathan-like boxes in Mr. Aikens house. — I am also sorry to hear that some of your insects got loose. I thought I had packed them pretty tight: — It is terrible news to me to hear that you have been idle in the good cause, but I hope soon that your zeal will be renewed. You will see my name in Stephens' last number[8]. I am glad of it if it is merely to spite Mr. Jenyns[9]. — I shall like very much looking over all your insects again, & I think I shall be able to name a great number of the *genus* Colymbetes[10]: I have a great number of insects at Cambridge for you, & you shall have the duplicates of every thing I take this summer. whenever I take any thing good, you may rely upon it, that you always are uppermost in my mind: I shall not soon forget my first entomological walks with you, & poor little Way[11]://

I cannot help often wishing to hear something about him. — I have some thoughts of going for a week into Wales again this summer: would it be impossible for you to come here & then go on with me. We should do wonders amongst the Carabi. — I told Polo[12] to collect, but I am afraid he will never do much more as an Entomologist he is grown far too idle. You had better try your hand at him, for I despair. — You shall receive before very long, the two Martens, and my dundiver. — Recollect what the sum of money is that I am indebted to you, as I do not keep account, but you pretend to do, so put it to the proof, and do not forget. — I am going to Maer[13] next week in order to Entomologise & shall stay there a week: for the rest of this summer I intend to ~~live~~ lead a perfectly idle & wandering life, always taking care to have as little of home & as much of Woodhouse[14] as possible. You see I am much in the same state that you are, with this difference you make good resolutions & never keep them. I never make them, so cannot keep them: it is all very well writing <in> this//

manner, but I must read for my little Go[15]. Graham[16] smiled & bowed so very civilly, when he told me that he was one of the six appointed to make the examination stricter, & that they were determined they would make it a very different thing from any previous examination that from all this, I am sure, it will be the very devil to pay amongst all idle men & Entomologists. Erasmus[17] we expect home in a few weeks time; he intends passing next winter in Paris: — Be sure you order the 2 lists of insects, published by Stephens[18], one printed on both sides, & the other, only on one. You will find them very useful in many points of view.

Dear old Fox,
yours
C. Darwin

Postmark: Shrewsbury JY 16 1829
Christ's College Library, #18.

Notes

1. In LLi, p 170, from "I am going to Maer" to the end.
2. Robert Pulleine. See Letter #9 (n6).
3. Mr Aiken is presumably an animal collector, see Letters #18 and #17. We learn here that he lived in a very small house.
4. See Letters #17 and #18.
5. CD's college servant or gyp at Christ's College, Cambridge (1828-1831) - see letter #* (n1).
6. Mentioned in Letter #17.
7. Osmaston Hall, the Fox family home from 1813 - ca1850 (see Fig 4).
8. The issue of Stephens' *Illustrations of British Entomology* (1827-1846) for 1 June 1829 on *Haustellata*, Vol 2, has the entry (p 200) '*Graphiphora plecta "Cambridge". — C. Darwin, Esq.*' *G. plecta* is a moth. As pointed out in CCD, Vol 1 (p 98): " *The report of the moth, if it appeared early in June, would be the first time CD's name appeared in print, but in the Autobiography, p 63, where he writes of his delight at first seeing the magic words, "captured by C. Darwin, Esq.", his memory is of beetle collecting. Unfortunately the 'magic words' do not help in making a clear decision about the date or the publication; they do not occur in either list, nor have they been found elsewhere in Illustrations of British entomology. A total of 35 species of insects, 34 of them beetles, are listed as taken by CD. (See, also, Freeman, 1978, pp 19-20.)*".
9. Leonard Jenyns (1800-1893). Naturalist and Vicar of Swaffham Bulbeck, Cambridgeshire; see Bibliographic Register, Letter #10 (n7) and Introduction to this Chapter.
10. In CD's copy of Stephens' *Catalogue Br Insects* (1829) there are 40 species of Colybetes listed and CD has annotations against 26. Note that there is some confusion over Colymbetes and Colybetes, since these names were used interchangeably (see previous Letter #18).
11. Albert Way (1805-1874), see *Autobiography*, Letters #2 (n1) & 195 (n8) and Fig 15.

12. A natural history collector, see Letter #11 (n6).
13. Maer Hall: see Letter #16 (n12).
14. Woodhouse: the home of Fanny Owen, see Letter #7 (n10).
15. "Little Go" was the examination taken in the second year, known as the Previous Examination: see Introduction, Chapter 2 (Introduction, note 5, & Letter #9 (n11).
16. John Graham (1794-1865). CD's tutor at this time and later Master of Christ's College and Vice-Chancellor of the University. He was also on the committee revising the examinations at this time [see Letter #10 (n12)].
17. Erasmus Alvey Darwin (1804-1881), CD's brother.
18. See Note 5.

#20 (CCD 69). To: W. D. Fox, Osmaston Hall, Derby.

[Shrewsbury]
Wednesday 29th. — [July 1829]

My dear Fox

I take this opportunity of writing a few lines to you. It was by a mistake of Mr. Shaw[1] that the beasts & birds were not sent to you before this time. — I hope you will approve of their stuffing & that they will arrive safe & sound. The one with the smaller tail is from Ireland, & the other from Wales. Which is the Pine & which is the common Marten appears to me hard to ascertain. — I returned from Maer[2] on Saturday evening, & was most exceedingly grieved to hear from Catherine that there is no hope of seeing you & your family this summer at Shrewsbury//

I want very much to hear from you as I do not quite understand your plans. You are going into Yorkshire, sometime in the end of August, & I want to know how soon you will return? for I am very sorry to say that I do not see how it is possible for me to come to Osmaston[3], between this time & your intended trip to Yorkshire. The reason of this is that my sister Mrs. Parker[4] has the misfortune of having two unchristened children, & I am going after the first week in August to stand Godfather to them: I was thinking (& I hope you will excuse the coolness of my offering myself) that if you were at home: I could pay you a visit//

on my road to Cambridge. — When you write tell me what you think of this scheme. — I hope it will answer: I do hope that when you are in Yorkshire, you will redeem your character in Entomology, it is a long time since you have mentioned taking any good insects. — I think you would be surprized if you were to see all my insects. — Pray mention in your next letter how Uncle Edward[5] is, we were all very sorry to hear he had been so unwell: Catherine[6] begs that you will give her love Julia[7] & many thanks for her kind wish of seeing her at Osmaston[3], but is very sorry that it is impossible.

Believe me dear Fox
Yours most sincerely
Chas. Darwin

Endorsement: C. Darwin July 29, 1829.
Christ's College Library, Cambridge, #19.

Notes

1. Mr Shaw of Shrewsbury, probably refers to one of the brothers, Henry (Harry) Shaw (1812-1887) and John Shaw (1816-1888); see Letter #10 (n8).
2. From the previous letter it is clear that CD made a visit to his uncle Josiah Wedgwood II, and family, at Maer hall, for a week during the period 15th-25th July 1829.
3. Osmaston Hall was the home of Samuel Fox III and family (see Fig 4).
4. Marianne Darwin (1798-1858), CD's eldest sister, married Henry Parker in 1824.
5. Edward Darwin (1782-1829) of Mackworth, Derbyshire. Eldest son of Erasmus Darwin and Elisabeth Chando-Pole (née Collier). He was therefore the "Uncle" of CD by the second marriage of his grandfather Erasmus to Elisabeth Chandos-Pole, and also related to Fox since the latter's grandfather was William Alvey Darwin of Elston, a brother of Erasmus Darwin.
6. Emily Catherine Darwin (1810-1866) was CD's youngest sister.
7. Julia Fox (1809-94) was the youngest sister of Fox.

#21 (CCD 70). To: W. D. Fox, Osmaston Hall, Derby.

[Shrewsbury]
Wednesday 26th. — [August 1829]

My dear Fox

I returned from Black Game shooting at Maer on Monday, & was most exceedingly glad to find your letter, when I came home: I am very much obliged for your most kind & welcome invitation to Osmaston[1]. And, as by good luck your visit into Yorkshire has been delayed, if it will suit you, I will come there early in September. — I intend going to Maer[2] for the first, & shall stay there about a week, after which I will come through Newcastle[3] to Derby. — I will write & let you know when I can find out about coaches &c &c.//

But first, to know whether it is perfectly convenient I must have a letter directed to me at J. Wedgwood Esq. Maer Hall *Newcastle under Lime*[2]. You must write nearly as soon as you receive this. — I am afraid it is almost too late for Entomologizing, but I shall like a quiet week at Osmaston better than anything else: I had so very pleasant a visit the last time I was there; that every thing has a charm to me from Pig Park[4] to your Sanctum Sanctorum[5]. — The Wedgwoods[6] whom you saw at Kingscote[7] are staying at Maer. they have been for the last year living with Mrs. Sismondi[8] near Geneva & there they saw the Ways[9] who have a house there. — They did not see little//

Albert[9]. his direction is Les Dèlices[10], Près de Genève. — If you have any virtue & are not *too much* engaged, you will of course immediately write to him. — You will be surprized to hear that I have been again to Barmouth. my sisters took a sudden freak[11] to go to the sea-side so I chaperoned them there. The weather was dreadfully bad so I did not do much amongst the beetles. — We will work gloriously, when I am at Osma∗ston.∗ I shall extremely like going regularly through your collection, & I think I shall be able to name a good many. — I will bring you the duplicates of this summer work, such as they are. — My Father of course meant to send the Martens as they were as I also did the Dundiver, and we are glad they arrived safe. — My Father is glad to hear that Mrs. Darwin[12] is tolerably well: it must have been a terrible blow to her[13]. — //

The Owens of Woodhouse[14], the idols of my adoration, have been staying at Buxton[15], & are now returned for the 1st of September[16]. They give a most amusing account of the state decrepitude to which the Beau Garcons are there reduced. — But if I begin to talk about Woodhouse & la belle Fanny, I never shall conclude this letter, so believe me my dear Fox,

Yours sincerely
Chas. Darwin

My Father sends his best regard to you. — Be sure write soon, & direct to Maer.

Postmark: Shrewsbury AU 27 1829
Christ's College Library, Cambridge, # 20

Notes

1. Osmaston Hall was the home of Samuel Fox III, Fox's father (see Fig 4). CD was to visit Osmaston Hall in the next week or so (see approx dates in Appendix 2).
2. Maer Hall, in the village of Maer, near Newcastle under Lyme, Staffordshire, was the home, at this time of Josiah Wedgwood, II, and family.
3. Newcastle-under-Lyme (see note 2, above).
4. "Pig Park": this suggests that Fox was already interested in breeds of domesticated animals in 1829.
5. Presumably Fox's rooms at Osmaston, which housed his extensive animal collection, to which CD had been adding stuffed specimens (see letters #18-#20 and the "Martens" and "Dundiver" mentioned in this letter).
6. This would be the family of John Wedgwood from the reference to Kingscote (Note 7).
7. A small village in the Gloucestershire region of the Cotswolds. John Wedgwood and family rented "The Cottage" (in fact the local country house) there in 1828 and probably 1829 (see Litchfield, 1915, *Emma Darwin: A Century of Family Letters* 1: 180 & 197; and the diary of Alan Gardner Cornwall (http://www.owlpen.com/cornwall_diary.pdf).
8. Jessie Sismondi (née Allen), wife of the historian, Jean Charles Léonard Simonde de Sismondi and sister-in-law of Josiah Wedgwood II.
9. The family of Albert Way, see Letter #2 (n24).
10. CCD, Vol 1, notes: "Les Délices was one of Voltaire's houses, near Geneva. He bought and named the property in 1755. '*It is the palace of a philosopher with the gardens of Epicurus: it is a delicious retreat*' (Besterman 1969, p. 338)".
11. "A sudden causeless change or turn of the mind . . ." (*Oxford English Dictionary*).
12. Elizabeth Darwin, née Collier, the widow of Erasmus Darwin, who lived at The Priory, Breadsall, five miles from Derby.
13. CD is referring to the loss of Elizabeth Darwin's son, Edward Darwin, who died on 30th July 1829. He was in the 3rd Dragoons. (*Darwin Pedigrees*)
14. The home of Fanny Owen; see Letter #7 (12).
15. On the edge of the Peak District, in Derbyshire.
16. The official start of the grouse shooting season, "the glorious first" - see Letter #4 (n3).

#22 (CCD 71). To: W. D. Fox, Osmaston Hall, Derby.

[Maer]
[Friday] September 4th. [1829]

My dear Fox
I was very glad to receive your letter & to find that you can receive me next week at Osmaston[1]. — I write these few lines, to tell you that I shall be in Derby on Tuesday Evening[2], but I am afraid my visit must be very short. Erasmus[3] came home on

Tuesday, & intends only staying a fortnight or 3 weeks at Shrewsbury & then goes for the winter to Paris; so that I shall//

scarce be able to see anything of him. — I long to Entomologize with you, & have brought my duplicates with me. — There are literally no Partridges in this part of the country. I have been out two days & have only killed 3 birds a day[4]. —

Yours affectionately
Chas. Darwin

Postmark: Newcastle SE 5 1829
Christ's College Library, Cambridge, # 21.

Notes

1. Osmaston Hall, near Derby, the home of Samuel Fox III, Fox's father (see Fig 4).
2. Tuesday, 8th September 1829.
3. Erasmus Alvey Darwin, CD's brother.
4. CD was at Maer for the start of the game or grouse shooting season ["the first" of September, see Letter #4 (n2)].

#23 (CCD 72). To: W. D. Fox, Osmaston Hall, Derby.

[Shrewsbury]
Tuesday [22nd September 1829]

My dear Fox

I had hoped to have been able, before writing to you, to have seen Mr. Hope[1], and then I should have had something to write about, as it is I am in my usual predicament of having nothing to say. — When at Osmaston[2], Erasmus[3] plans were, you are aware, to go to Paris for the winter. accordingly I hurried home like a dutiful brother; I might have spared myself that trouble; as ~~his~~ it was not likely his plans would continue the same for a whole month: He now intends to give up Doctoring for the present & take rooms & live in London[4].//

He is come back very agreeable & industrious, & foreign parts have not all spoilt him. — he means to go the Birmingham Music meeting[5], from thence to London, making a detour to Kenilworth, Warwick & Oxford. — & before Xstmas he is coming to me in Cambridge[6]. — All this is excessively pleasant, & he means to have an air-cushion in his rooms for me to sleep on, which will make London a delightful place. — I myself am beginning to doubt about going to Birmingham[7] on account of the expence, but if you will make up your mind to go, so will I. — Mr. Hope[1] is gone up to London, so that I shall not be able to get anymore information about the insects we took together. — //

but I trust you have been to the Moor (take your net & sweep), & that you pretty often look over & shake the Fungi. — I had a letter from Turner[8] yesterday & he says he has actually collected a case full of insects. Do not let yourself be beaten by such a novice as he is. — I intend going to Cambridge about the 15th. of October, & I do hope you will contrive to be there before very long. — You must write *soon*

& tell me all your plans. — Give my love to the little Puss & remember me most kindly to all your family: & Believe me dear Fox

Yours most sincerely,
Charles Darwin

Write to me directly.

Postmark: Shrewsbury SE 23 1829
Christ's College Library, Cambridge, # 22.

Notes

1. Frederick William Hope (1797-1862). Entomologist and clergyman; see Letter #6 (n2) (*DNB*).
2. CD was at Osmaston Hall, near Derby, Fox's family home, from 8th September. It is not clear how long he stayed; possibly 5 – 7 days.
3. Erasmus, Alvey Darwin, CD's brother.
4. This the first indication that Erasmus Darwin would give up being a physician and live in London, which is, in fact, exactly what happened. This must have been done with the sanction of their father, Robert Darwin, and must have been an indication to CD that he himself might not have to become a clergyman.
5. The annual Music Meeting held in October at Birmingham, which attracted international musicians and singers (see Letter #24). It must have been similar to the Derby Music Meeting (see Appendix 4, 4th October 1825 and Letters #4 & 5 (n6)).
6. See Letter #25 (n5).
7. For the annual Birmingham Music Meeting (see note 5, above, and Letter #24). CD did attend this meeting but Fox did not (see Letter #23).
8. James Farley Turner (?-1841). B A Christ's College, 1831. A Shrewsbury school friend of CD. Became a C of E clergyman. (*Alum Cantab, Peile, 1913*).

#24 (CCD 73; LLi, part). To: W. D. Fox, Osmaston Hall, Derby.

Christ College [Cambridge]
Thursday [15th Oct 1829]

My dear Fox

I am afraid you will be very angry with me for not having written during the Music Meeting[1], but really I was worked so hard that I had no time: I arrived here on Monday, & found my rooms in dreadful confusion, as they have been taking up the floor, & you may suppose that I have had plenty to do for these two days. — The Music Meeting was the most glorious thing I ever experienced: & as for Malibrand[2], words cannot praise her enough, she is quite the most charming person I ever saw. — We had extracts out of several of the best Operas, acted in character, & you cannot imagine, how very superior it made the concerts to any~~thing~~ I ever heard before. — //

J. de Begnis[3] acted Il Fanatico[4] in Character, being dressed up, an extraordinary figure gives a much greater effect to his acting. He kept the whole theatre in roars of laughter. — I liked Madam Blasis[5] very much, but nothing will do after Malibrand; who sung some comic songs, & persons heart must have been made of stone not to have lost it to her. — I lodged very near the Wedgwoods[6] & lived entirely with them, which was very pleasant, & had you been there it would have been quite perfect. — It knocked me up most dreadfully, & I never will attempt again to go to two things the same day . — I passed through London on Sunday, but M^{r}. Hope was not there: Weaver[7] in Birmingham has a very good collection, but does not know much about//

them, & I bought 15^{s}. worth of insects of him: one insect, if it is British, quite surpasses my knowledge. — Would you believe it, that I have the Endomychus coccineus[8], a bad specimen however, given me by M^{r}. Hope? — I hope you have found something in the Fungi[9]: any how tell me when you, what you have done in Entomology. — Cambridge is most wonderfully empty. Wilmer[10] & the Rev — *Robinson*[11] are the only men up that you know: Poor Robinson[11] has got to go in again for his degree as he had not entirely kept one of his terms. — Your conjecture was perfectly right about the letter, when I came home I found it in my great big top-coat, to my sisters no small disgust & indignation. — I suppose you have sent Stephens Catalogues[12] to Shrewsbury, from//

whence I have ordered them to be sent here: it is a pity, as they have come so late, that they were sent for at all. — Do write soon & tell me your plans. I earnestly hope that you mean to come up early & stay for some time[13]: it will be so very snug being together once more. —

Yours most sincerely
Charles Darwin

Postmark: Cambridge <O>C 16 1829
Christ's College Library, Cambridge, # 23.

Notes

1. Birmingham Music Meeting; the programme (Birmingham Central Libraries) shows that it started on the 8th October 1829, when the opera *Il fanatico per la musica* by Antonio Sacchini was performed and the aria *I violini* was sung by Giuseppi de Begnis; see, also, Letter #23.
2. Maria Felicia Garcia Malibran (1808-1836). Mezzo-soprano opera singer, especially famous for her roles in Rossini operas.
3. Giuseppe de Begnis (1793-1849). Italian opera singer.
4. *Il fanatico per la musica*, an opera by Antonio Sacchini, see note 1, above.
5. Virginia de Blasis (1804-1838). Italian soprano.
6. Josiah Wedgwood II and family. It is not known who were present.
7. Not identified. The name is Weaver in LLi, Vol 1. However, it is not easy to decipher.
8. In CD's copy of Stephens' *Catalogue Br Insects* there is no annotation against Endomychus coccineus; synonym of *Chrysomela coccineus* Linnaeus.
9. CD refers to "Mycetophagos" insects, i e, insects living in or on fungi. Previously he referred to "Mycetophagos" insects in Ash Wood at Derby (see Letter #10 (n9).
10. Bradford Wilmer of Christ's College; see Letter #13 (n10). Ordained a deacon in 1829.
11. Torn out by seal but it is clear from the next passage that it is George Alington Robinson (?-1888); Christ's College, B A,1830. Robinson as we see was a clergyman taking orders. He matriculated, i e, began his studies, in 1825 and so must have had an interrupted history at Christ's, but was clearly respected by CD.
12. James Francis Stephens (1792-52) brought out a *Catalogue of British Insects*, in 2 volumes and 2 parts each, in 1829; see Letter #1 (n11). Stephens also published *Illustrations of British entomology* in 11 volumes and supplements between 1827 and 1846. CD could have been referring to either of these publications and both appear in his library. The publication with the majority of his annotations is the *Catalogue of British Insects,* Vol 1.
13. Fox, having passed his B A exam [Letter #9 (n2)], was now studying for his ordination which involved taking extra courses at Cambridge [see Letter #11 (n3)].

#25 (CCD 74). To: W. D. Fox, Osmaston, Derby.

[Cambridge]
Tuesday [3rd Nov 1829]

My dear Fox

I am ashamed at not having written to you before, & more especially as Aiken[1] gave me some very important messages to you, about Cocks & Hens[2], no less than ten days ago: Imprimis[3], there are two hatches: no mother: price of larger 3s each & of the lesser chickens 2"6. He is afraid you will think them very dear, but could not get them cheaper. — I am sure you will be very sorry to hear, that my Father has been very ill, but is now I hope steadily but slowly getting better: The cause of his illness is a deep seated//

Erysipelus[4] in his neck, attended with great deal of Fever. I am afraid he has undergone a great deal of suffering: & that it will be some time before he quite recovers the effect of it. I should have gone down to Shrewsbury, only that his Medical attendants particularly insisted that he should be kept as quiet as possible: My sisters appear to have been terribly alarmed, as I suppose on Wednesday he was in imminent danger. — Erasmus[5] has been staying here for a few days, & we passed the time very pleasantly chiefly in the Fitzwilliam[6] & he would have stayed some time longer, only that he went to London, in order to be in readiness to go down to Shrewsbury if//

my Father should be worse: None of the Cam: Entomologists have done anything to speak of, but I am going to Bottisham to see Mr. Jenyns[7] cabinet, & I believe he is coming to see mine: Professor Henslows[8] parties go on very well. the last was the pleasantest I was ever at:

I have been very idle since I have come up, but have had some good hunting: Rob∗in∗son[9] has been riding to the ad∗mir∗at∗ion∗ ∗of∗[10] every body. I think he is the best rid∗er∗ I ever saw. — I am very sorry that you are so undecided as to when you will come up to Cambridge. I do hope it will eventually take place. I shall go down very early in order to see my Father. — & return early in the next Term to read for my little Go[11].

Yours very sincerely
C. Darwin

Postmark: Cambridge NO 3 1829
Christ's College Library, Cambridge, #24.

Notes

1. Probably a natural history collector: see Letters #16 (n6), #17 (n2) & #19 (n3).
2. Perhaps the first indication we have of Fox's interest in animal breeding, which later letters show was an important motivation for the continuing correspondence between CD and Fox [but see Letter # 21 (n 4)].
3. Meaning "in first place" or "most importantly".
4. Erysipelas. A cutaneous inflammatory disease causing very red skin and leading to localised eruptions. Now known to be cause by streptococci bacteria. A common name is St Anthony's Fire.
5. Erasmus Alvey Darwin, CD's brother.

6. The collections of the Fitzwilliam Museum were temporarily housed in the Old Free School in Free School Lane at this time. The present Fitzwilliam Museum building was begun in 1837. See Chapter 2, note 68.
7. Leonard Jenyns (1800-1893), CDs rival at this time. See Bibliographic Register, Introduction to this chapter (n10) and Letter #10 (n7).
8. John Stevens Henslow had been appointed to the Chair of Botany at Cambridge University in 1825 following his earlier appointment to the Chair of Geology. CD had previously ignored him but Fox was an admirer and induced CD to attend lectures from 1829 (Browne, 1995). For more information see Bibliographic Register and Introduction to Chapter 2. Henslow is referred to frequently in later correspondence: see Index.
9. From the Fox *Diary* (1824-1826) (see Chapter 2 and Appendix 4), we know that illegal horse racing took place around Cambridge at this time and that several students were punished for their part in it. It is possible that Robinson was one of these. For details on Robinson, see Letter #24 (n11).
10. The page is torn at this place due to the seal. However, the meaning is clear. The Rev Robinson is referred to in the previous letter [#24 (n11)].
11. For "little Go" see Letters #9 (n11) & #10 (n13) and Introduction to Chapter 2 (note 5). It is possible that Fox did go to Cambridge for some time in December 1829 but he did not coincide with CD, who went up to Cambridge very early in January 1830, in preparation for his examination (see Letter #27).

#26 (CCD 75). To: W. D. Fox Esq, Osmaston Hall Derby

Christ Coll: [Cambridge]
Sunday Evening [3rd Jan 1830]

My dear Fox

I shall think myself very lucky if this letter is not returned unopened, for my shameful conduct well deserves it. I should have written long enough ago, only that when in Cambridge, I thought I should have had more to say in London, & when there I thought it better to put it off till I came here. So owing to these two weighty reason, it is nearly six weeks since I received your last, & it would serve me right, if it were six weeks till I received another, but I hope better things. — Cambridge is very empty[1], but for a change, I daresay it will be very pleasant: any how the bachelors dinners, I have no doubt will be very good. — I came here on Friday evening, having been staying with my brother[2] for the//

last three weeks in Town. I never was there for so long a time before & enjoyed it much more than I expected. There is literally nothing at all going on, & I cannot mention a single thing that I did. The time however glided away, much in the same way as it does at Osmaston[3]: I performed one Herculean task, having nearly finished Clarissa Harlowe, the most glorious novel ever written, & I advise you begin it as soon as you can[4]. — Mr. Hope[5] was in Town for the last week, & I worked away very hard at entomology. I went two evenings to Mr. Stephens[6], & I begin to like him very much. — a Mr. Waterhouse[7] gave me some most glorious insects, inter alia, 2 species of Trichius[8]: Your Tarus is basalis[9]. — I hope you have not thrown away the Fungi: but I believe we collected it rather too late. You can have no conception what magnificent insects some of the Fungiverous are. — Mr. Jenynys[10] has been looking at my insects, & he//

was very grateful for a good many insects I gave him. — I ~~have written~~ wrote to Baker[11] about the Shovellers & tracheæ: Mrs Field bill[12] is: 5£"7·6 & Bretts[13] for packing up a case, 12^{s}"6. & I paid 2 "6 for a grebe that he has got to skin: so you must keep an account of how we stand together, when I have paid Brett. — The men are all reading at a most wonderful pace. Old Simpcox[14] counts every minute in the 24 hours. — I am afraid Whitley[15] has a very poor chance of being Senior W.[16] but I trust he will be second. — Herbert[17] I believe will be in the first six. — I hope you will write soon & tell me what your future plans are. — for from the weekly variations, I can generally make a guess what they really will be. This indeed with my Brother[18] is my only way of finding out: — I shall if possible withstand temptation & not ride this term. I had one or two most glorious days hunting at the end of last. But the last day I had two such aweful rolls as nearly knocked my lungs out. This puts me in mind to give you a good scolding for imagining me such a blackguard//

as to forget my old friends & do nothing but poke about the stable & look knowing. Forgive my negligence as I do you for making so unjust a supposition & believe me dear Fox

Yours very sincerely
Chas Darwin

I saw old Pulleine[19], he looked very fat & rosy.
I forgot to mention, I dined with Sir J. Mackintosh[20] & had some talk with him about Phrenology[21], & he has entirely battered down the very little belief of it that I picked up at Osmaston[22]. He says, as long as Education is supposed to have any effect in decreasing the power of any organ of the brain, he cannot see how it ever can be proved true[23]

Postmark: Cambridge JA 3 1830
Christ's College Library, Cambridge, #25.

Notes

1. CD came to Cambridge to prepare for his "Little Go" [see Letters 10 (n11) and #28 (n1)]. The students taking their final examinations (B A) also took their exams at the same time; so Cambridge would have been very 'quiet" at this time. The "batchelors dinners" were, no doubt, given by students after their finals.
2. Erasmus Alvey Darwin; he had recently given up his career as a physician [Letter #23 (n4)] and decided to live in London; he stayed with CD stayed in Cambridge in early November [Letter #25 (n5)].
3. Osmaston Hall, near Derby, the home of Samuel Fox, III, Fox's father (see Fig 4).
4. Clarissa Harlowe was written by Samuel Richardson in 1747-8; it is a very long novel, in 3 parts, and, for the average reader, would take a reasonably long time to read. Just why CD would be reading it at this stage, just before examinations, is not known, and why Fox should read it "as soon as he can" is not clear. Clearly CD was very taken with the novel, which is about a virtuous young lady abducted by a libertine, who submits the heroine to a series of trials.
5. Frederick William Hope (1797-1862). Entomologist and clergyman. See Letters #6 (n1), #10 (n6), 18 (n1), #28 (n3).
6. James Francis Stephens (1792-1852). Entomologist and zoologist; published works on British insects [see Letters #1 (n11) and #24 (n11)]; employed by the Admiralty office and assisted at the British Museum. (*DNB*)
7. George Robert Waterhouse (1810-1888). Naturalist and founder of the Entomological Society; employed by the British Museum. (DNB).
8. Trichius has no annotations against in CD's copy of Stephens' *Catalogue Br insects*.

9. This is a new species introduced into Stephens' *Illustrations of British Entomology*. Mandibulata. Vol. 1 (1828), Addenda et corrigenda, p. 177.
10. Leonard Jenyns (1880-1893). [Note another misspelling of the name; cf. Letter #10 (n7).] Naturalist and vicar of Swaffham Bulbeck, Cambridgeshire - see Bibliographic Register, "Introduction" to this Chapter, and many references in subsequent letters (shown in Index).
11. Baker was probably a natural history collector in Cambridge; see letters #17 (n4), #30 (n7), #38 (n1), #39 (3), #40 (n2), #41 (n6).
12. Mrs Field; not identified, but mentioned in subsequent correspondence, where CD says "she says she never received any from Graham" (Letter #30) and "negotiations between Mrs Field and Graham" (Letter #35). John Graham [see Letter #10 (n12)] was CD's tutor at this time and, a little later, Master of Christ's College, so it may have been a bill for house-keeping services, etc. See also Letter #38 for what seems to have been the final settlement and an actual amount.
13. Possibly a college servant or house keeper, mentioned in Letters #17 (n4) & #19 (n6).
14. "Simcox" probably refers to George Simpson; see Letter #11 (n8).
15. Charles Thomas Whitley (1808-1895); see Letter #11 (n9). He was Senior Wrangler (see note 16, below) and achieved 2nd Class in Classics.
16. Senior Wrangler: the student coming highest in his Final B A Examination, including the optional Mathematical Tripos.
17. John Maurice Herbert (1808-1882). See Letters #1 (n13), #11 (n10). He was 7th Wrangler.
18. Erasmus Alvey Darwin. CD has already referred to his brother's changeability, likening him to "*a Machaon on a windy day*" [Letter #3 (n10)].
19. Robert Pulleine of Emmanuel College. See Letter #9, (n6).
20. Sir James Mackintosh (1765-1832), Philosopher, historian and politician. He was a part of the Wedgwood circle, through his wife, Catherine Allen, and their daughter, Fanny, married Hensleigh Wedgwood, whom CD would have known well, as he was elected to a Fellowship at Christ's College, in 1829 [see Letter #10 (n16)]. CD later called Mackintosh "the best converser I ever listened to" (*Autobiography*, p 55).
21. Phrenology was the doctrine, as elaborated by Franz Joseph Gall, in which the shape of the skull reflects the faculties of the brain and the qualities of the personality. Phrenology was very popular in England at this time. CD was no supporter of phrenology or of paranormal phenomena such as mesmerism and homoeopathy (see Letters #76 & #86). As indicated in this letter, Fox, like many Victorians, seemed to lean much more to esoteric approaches and paranormal phenomena (see, also, Letters #76 & #86).
22. Osmaston Hall, near Derby, the home of Samuel Fox III, Fox's father (see Fig 4).
23. The argument seems to have been that if a man's nature and habits can be modified by learning then these are not innate, and cannot be related to the gross structure of the brain or the shape of the skull.

#27 (CCD 76). To: W. D. Fox Esq, Osmaston Hall, Derby.

[Cambridge]
Wednesday [13 Jan 1830]

My dear old Fox

I observe that it has happened once or twice before, that at the same time that my guilty conscience has goaded me into writing, your stock of patience has been worn quite thread bare. — Owing to this exact proportion our letters unfortunately clashed on the road, & to show my real penitence I make this fresh start. — I opened your letter with trepidation & awe, & as all good books say, that a forgiving spirit is the truest revenge, so I found it; for your very kind letter made me more ashamed of myself than any well deserved & severe speeches could have done. — I wish it had so happened that I had gone to Shrew:[1] this Vacation, if it had merely been to have paid you a visit, for idle as I am, I would not have let you come all the way to Birmingham merely to see me, (always//

with the supposition that there was no "metal more attractive"[2] to be found in that metalliferous[3] district) I hope you had a pleasant visit at the Larches[4]. You do not say much about it. I hope you will mention in what sort of state you brought your heart back again to Osmaston[5]? in a sad raggy condition I am afraid[6]. — I forget whether I mentioned that I have ordered a Cabinet[7]. I long to begin about arranging & naming my insects; amongst the Carabidæ, I think I have a third in number. — I have lately taken Demetrias: imperialis!!![8] & Endomychus coccineus[9]. — What do you mean by saying Miller[10] has sent you Saperda Carcharias; why you have already got that, it is the insect you bought of Finch, large & mottled with brown. — There is a much better entom: than Miller in Bristol, of the name of Millard[11], but I hear very stingy. — N.B. M^{r}. Miller is suspected of having got a good many foreign insects in his//

collection from Germany, where he formerly collected. — My Father has been for some time quite well, but has since had a little gout. — I thought I had mentioned, that both the Deaths Heads[12] died in the Summer. — I rode over the other day to pay M^{r}. Jenyns[13] a visit. he has a nice snug little house, & lives very comfortably, & was altogether very civil, but not particularly liberal. I gave him an awesome lot of insects, & he gave me ∗2∗ good ones, & 2 or 3 more common ones, ∗he∗refused∗ me a specimen of the Necroph. sepultor[14], (the Fen one of which he can get plenty) although he has 7 or 8 specimens by him at that time. — There is a perfect specimen of liberality for you: a true disciple of the Curtis school[15]: My Dytici[16] & Colym:[17] astonished his weak mind: I do most sincerely hope you will at once fix a time for coming up here, & do let it be soon. I long to see you: & to go through my insects will be no small fun. — I shall call you irresolute & weak, if in your next letter//

you do not fix a definite time. — The men are all in a dreadful plight, from fear & anxiety[18], but nevertheless Cam: is remarkably pleasant, & if you were here, to have our old breakfasts together it would be delightful. — I never saw Whitmore[19] in such good spirits as he is at present: he goes walking about quite independently, & drinks brandy to the nth[20]. — It is quite curious, when thrown into contact with any set of men, how much they continue improving in ones good opinion, as one gets ackquainted with them. This was an argument used, in a religious point of view, by a very clever Clergyman in Shrews. to encourage sociability (he himself being very fond of society), for he said that the good always preponderates over the bad in every//

persons character, & he thought, the most social men were generally the most benevolent, & had the best opinion of human nature. I have heard my father mention this as a remarkably good observation, & I quite agree with him[21]. — Chapman[22] & many other men often ask after you. My dear Fox,

Yours sincerely
C. D

Postmark: Cambridge J< >30
Endorsement: Charles Darwin January 16, 1829. [Corrected to 1830 in purple crayon.]
Christ's College Library, Cambridge, #26.

Notes

1. Shrewsbury.
2. A quotation from Shakespeare (*Hamlet*, 3. 2, line 106).
3. Metals were not mined around Birmingham, but it was a centre for the metals industry since the seventeenth century. CD is referring light-heartedly to another attraction in Birmingham: Bessy Galton (see note 4, below).
4. The Larches, Sparkbrook, Birmingham, was home of the Galtons and especially Bessy Galton: see Letters #6 (n7), #7 (n10) and #9 (n13). However, CD deduces from Fox's letter that the latter's attraction to Bessy is fading (see note 6, below). The Larches was a large Georgian-style house, demolished in 1871. It was built on the site of the house of Joseph Priestley, the discover of oxygen, which was burnt to the ground in the Birmingham riots of 1791.
5. Osmaston Hall, near Derby, was the home of Samuel Fox, III, Fox's father (see Fig 4).
6. This is the last reference to Bessy Galton, so it appears that CD was right and the affair was over.
7. This seems to be a different insect cabinet from the one that CD reports he had ordered in early 1829: see Letter #10 (n10); the details of this later cabinet are given in Letter # 43 (n7).
8. In CD's copy of Stephens' *Catalogue Br insects* there is no species "D. imperialis". D. monostigma has the annotation "*Cam*" against it.
9. CD had already mentioned this species in Letter #24 (n8).
10. Not identified. Clearly an insect dealer in Bristol. This reference shows that both Fox and CD obtained specimens from collectors, as did many naturalists at this time. Most of the insects with annotations against them, in CD's copy of Stephens' *Catalogue Br insects,* Vol. 1 (1829), appear to be specimens collected by CD himself, but some rarer specimens may have been obtained from collectors.
11. Not identified.
12. Hawk Moth, *Acherontia atropos.*
13. Leonard Jenyns (1880-1893),: see Bibliographic Register, Introduction to this chapter (n10) and Letter #10 (n7). CD seems here to be mellowing in his antipathy to Jenyns, possibly under the influence of the growing relationship of CD with J S Henslow, who is married to Jenyns' sister.
14. In Stephens' *Catalogue Br insects* (1829) as Necrophorus sepultor. However, there is no annotation against this species in CD's copy.
15. John Curtis (1791-1862) was a consummate artist and author of one of the more important and beautiful entomological texts of the nineteenth century. His *British Entomology* was published in 193 parts and 16 volumes between 1823 and 1840. There was an unfortunate rivalry between Curtis and J F Stephens, whose book *Catalogue of British insects* CD used frequently. Both men had their supporters and CD is clearly referring to this here. In fact CD rarely refers to the works of Curtis, although these were clearly available and relevant to him.
16. CD had used the latinised plural for *Dyticus*, a genus of diving beetle. The more general name was *Dystiscus*. In his copy Stephens' *Catalogue Br insects* Vol 1, CD had annotations of "*Cam*" or "*P*" against several species.
17. Water beetles.
18. The end of January was the set time for examinations.
19. Ainsley Henry Whitmore (1801-1843): see Letter #12 (n6).
20. See Letter #3 (n11).
21. Certainly both CD and Fox seem to have taken this principle to heart in conducting their lives; both remained very sociable people despite ill health, large families and hard work.
22. Benjamin Chapman: see Letters #8 (n8), 13 (n17) & #28 (n6).

#28 (CCD78; LLi[1]). To: W. D. Fox Esq, Osmaston Hall, Derby.

[Cambridge]
Thursday
[25 Mar 1830]

My dear Fox

I am through my little Go[2],!!! I am too much exalted to humble myself by apologising for not having written before. — But, I assure you before I went in & when my nerves were in a shattered & weak condition, your injured person often rose before my eyes & taunted me with my idleness. But I am through through through. I could write the whole sheet full, with this delightful word. — I went in yesterday, & have just heard the joyful news. — I shall not know for a week, which class I am in. — The whole examination is carried on in a different system. It has one grand advantage, being over in one day. They are rather strict; & ask a wonderful number of questions://

And now I want to know something about your plans: *of course* you intend coming up here: what fun we will have together, what beetles we will catch, it will do my heart good to go once more together to some of our old haunts: I have two very promising pupils in Entomology, & we will make regular campaigns into the Fens; Heaven protect the beetles & Mr Jenyns[3], for we wont leave him a pair in the whole country. My new cabinet is come down & a gay little affair it is[4]. — And now for the time, I think I shall go for a few days to Town, to hear an Opera & see Mr. Hope[5]; not to mention my brother also <whom> I should have no objection to see. — If I go pretty soon, you can come afterwards//

but if you will settle your plans definitely, I will arrange mine. So send me a letter by return of post: — And I charge you let it be favourable, that is to say come directly. — Holden[6] has been ordained, & drove the Coach out on the Monday, I do not think he is looking very well. — Chapman[7] wants you & myself to pay him a visit, when you come up, & begs to be remembered to you. You must excuse this short letter, as I have no end more to send off by this days post. —

I long to see you again, & till then
My dear good old Fox
I am yours most sincerely
C. Darwin

Postmark: Postmark: Cambridge MA 25 1830
Christ's College Library, Cambridge #27

Notes

1. In *Life and Letters* (F. Darwin, 1888), Vol. 1, pp.180-1: verbatim (excluding a few punctuational changes).
2. For details on the "Little go" see Chapter 2 (note 5) and Letter #10 (n11). The results of the Previous Examination or "Little go" were published in two lists: those who passed with credit, and those to whom the Examiners had *only not refused* their certification of approval. CD's name appeared in the first list (Cambridge University Archives, Pass Lists, Previous Examinations 1824-83, Exam. L. 9).
3. Leonard Jenyns (1800-1893): see Bibliographic Register, Introduction to this chapter (n10) and Letters #10 (n119), #19 (n6), #26 (n5).
4. Se Letter #27 (n7).

5. Frederick William Hope (1797-1862). Entomologist and clergyman: see Bibliographic Register and Letters #6 (n2); #10 (n6); #18 (n2); #26 (n5).
6. James Richard Holden. See Chapter 2, note 52 and Letters #8 (n12), #9 (n5) & #12 .(n2)
7. Benjamin Chapman: see Letter #8 (n8), #13 (n17), #27 (n22).

#29 (CCD 79). To: W. Darwin Fox, Osmaston Hall, Derby.

Cambridge
Thursday
[1st April 1830]

My dear Fox

I should have answered your letter sooner than this, only that for the few last days I have been very busy in taking sundry long walks: The information your letter contained about Erasmus[1], settled my plans & I immediately made up my mind to remain here during the whole vacation. And I have since heard from Mr. Hope[2], who is staying at Southend, that he should not return to London for some weeks, so that there is not a single inducement for me to leave Cambridge: Here then I shall patiently vegetate living in the hope of soon seeing you: If it had but been possible for you to have come up here during this Vacation, how very snug & quiet we should have been, but in another respect, vid: Entomology, it is better as it is, as this cold & rainy weather has driven all the beetles back to their homes. — Do continue steadily to try to start for Cambridge, before the end of this month, for if you have many more delays, all the time will slip by. Let me hear from you some little time before you come; And I most fervently hope that time is near. — but there is no use in hoping, as I am sure, if you find it possible, we shall meet before the end of the month. — I suppose//

you do not know much more of your plans for the summer, than you do of those for the spring. Mr. Bristow[3] told Erasmus[1] that you intended going to the Sea side with your sisters: Is this quite fixed? & where is the place? — Mrs. Field[4] does not appear to be any great hurry for her money. she says she never received any from Graham[5]. — I think I told you the amount of the bill. — I want to know did you receive with your birds the ring Owzel? — Mr. Stephens[6] has been very ill with a fever, which is the reason that he has not been publishing lately. I have sent up all my Nitidulæ[7] &c for him to examine. — I find I get on very slowly with my cabinet, & shall be very glad of your assistance. I have only yet got to the Amaræ[8]. — Barker[9], whom we met at Sir Francis[10] is come up as Fellow Commoner[11] to Downing. is his Mother alive? for he talked one day as if Derbyshire was no longer a home to him. — I have been seeing a good deal lately of Prof: Henslow[12]; I took a long walk with him the other day: I like him most exceedingly, he is so very goodnatured & agreeable. — This letter is strange rag-tag & bob tail affair, & a poor return for your agreeable ones, but with it go my best wishes for you my dear Fox & believe me

Yrs. most sincerely
C. Darwin

Postmark: Cambridge AP 5 1830
Christ's College Library, Cambridge, #28.

Notes

1. Erasmus Alvey Darwin, CD's brother.
2. F W Hope, entomologist: see Letter #6 (n2), #10 (n6), #18 (n2), #26 n(5), #28 (n5), #29 (n2).
3. Samuel Ellis Bristowe, the husband of Mary Ann Bristowe, who died 19 April 1829: see Letter #15 (n3). Samuel Bristowe was involved with a wine merchant business in London.
4. Probably involved in housekeeping facilities at Christ's College [see Letter #26 (n12)].
5. Presumably John Graham, examiner, in 1829 for the "Little go", at this time CD's tutor and, a little later, Master of Christ's College [see Letter #10 (n12)].
6. J F Stephens, the entomologist and author of *Catalogue of Br insects*; see Letter #1 (n11).
7. Nitidula was a genus of beetles (no longer used), deriving from "nitidus" (Latin), meaning shining. In CD's copy of Stevens' *Catalogue Br insects*, Vol 1, there are no annotations against Nitidula.
8. In CD's copy, of Stevens *Catalogue Br insects*, Vol 1, there are annotations for "*Cam*", "*Shrews*", "*Maer*" and "*NW*" and the date "1829" against the species here.
9. Thomas Alfred Barker (1808-1891). M D Edinburgh, 1829. M B Downing College, Cambridge, 1835. CD may have known him at Edinburgh in 1826. (*Alum Cantab, Modern English Biography*).
10. Sir Francis Sacheverel Darwin (1786-1859). Physician and traveller. 4th child of Erasmus and Elizabeth Darwin. CCD, Vol 1, notes: "*Francis Galton mentions him as 'originally a physician, but for many years living in a then secluded part of Derbyshire, surrounded by animal oddities; half-wild pigs ran about the woods, tamed snakes frequented the house, and the like' (Galton 1874, p. 47).*" CD and Fox clearly had visited him at Sydnope Hall (see Appendix 2) and CD mentions him again in Letter #121.
11. Fellow Commoners (aristocrats and other 'men of family') were the first of three ranks in which students were matriculated, the others being Pensioners (ordinary paying students, like CD and WDF) and Sizars, who received scholarships or aid for which they had to perform services.
12. J S Henslow (1796-1861), Professor of Botany, University of Cambridge: see Bibliographic Register, Introduction to this chapter and Letter #8 (n3) & #25 (n8). This is the first indication of the special relationship, which developed between these two men.

#30 (CCD 80). To: W. Darwin Fox, Osmaston Hall, Derby.

[Cambridge]
Sunday [9th May 1830]

My dear Fox

I am very sorry to find that all our plans are likely to vanish into air. — It is most unfortunate your being obliged to go with your Father & Mother to Cheltenham, for the weather is so fine, the beetles so numerous, our zeal so ardent that the Science would have received a benefit never to be forgotten. — But it is a shame to talk about would have taken place: let us think of the future. — I am very much obliged for your invitation to Osmaston, but I do not see how it is possible for me to accept it.//

My brother has returned from Shrewsbury to London, & I have agreed to pay him a visit, which if I do not do, I shall not see him again for a good while. — And after that I really must proceed to sweet home. — for my unfortunate relations have been bereaved of my presence, for eight long Months. — But I have a plan which will remove a good many difficulties. — Why, in the name of providence, not pay me a visit in Salop[1]??? You know how glad we all shall be to see you. — You can start early from Derby, so through Lichfield to Birmingham, & then from there//

Wonder[2] to Shrewsbury, where at 11 oclock the same <night> you will arrive safe & sound, & make me glad by your sight. — What plan can be more easy or natural? It is too unreasonable even to hope, that you will come to Cambridge for so short a

time as four days, so that I shall not see till we meet at Shrewsbury. If you are half as eager as I am for that occurrence, you will not make any sort of excuse. — As soon as you return to Derby go to Mr Hey[3], & find the habitat Odontomyia[4] & Dromius[5] & promised specimens. —
My plans for the summer are totally undecided. — write soon be sure do not say no to Shrewsbury scheme else I will never forgive you. — till then, My dear old Fox

Yours very sincerely
C. D.

Wilmer[6] has left Cambridge, & begged to be most kindly remembered to you. — he has sent to Baker[7] for you a Falcon out of Norfolk. — I have got your prints. — Chapman[8] send Ditto. — he is going abroad. — I took the other day Elater sanguineus[9]. What do think of that? I have 19 species in genus Amara[10]. — I have seen a good deal of Henslow lately & the more I see of him the more I like him I have some thoughts of reading divinity with him the summer after next[11]. Adieu
(Across first page)
As far as to the Bembididæ, this in toto 339 species[12], out of which I have 208 of them — What do you think of that?

Postmark: C <1>0 MY 10 1830
Christ's College Library, Cambridge, #29.

Notes

1. The old name for the county of Shropshire and here meant to designate the county town of Shrewsbury.
2. The London, Birmingham, Shrewsbury mail coach.
3. The Rev. Samuel Hey of Ockbrook, Derbyshire, the father of William Hey [see Letters #35 (n9) and #36 (n5)]. CCD, Vol 1, notes that he is: "mentioned in Stephens 1827-46, *Mandibulata* 3: 373 as having rediscovered the rare *Chrysomela cerealis* and as having directed 'his friend and neighbour Mr. Fox to the locality' ".
4. CCD, Vol 1, has Odontomyia but this is not in Stephens' *Catalogue Br Insects*. Odontonyx does occur there and in CD's copy, Vol 1, there is the annotation "Maer and Cam" against O. rotundatus.
5. In CD's copy of Stephens' *Catalogue Br Insects*, Vol 1, D. agilis, D. quadrinolatus, D. melanocephalos have the annotation "Cam" against them and D. quadrimaculatus has "P" against it.
6. Bradford Wilmer of Christ's College; see Letter #13 (n10) and Letter #24 (n9). Ordained a deacon in 1829.
7. Baker: a natural history collector – see Letters #16 (n1) & #17 (n4).
8. Benjamin Chapman (1798-1871): see Letters #8 (n8), #13 (n17), #27 (n22) & #28 (n6).
9. There are not annotations in CDs copy of Stephens' *Catalogue Br Insects*, Vol 1, against this species.
10. In CD's copy of Stephens' *Catalogue Br Insects*, Vol 1, there are 15 species checked off. However, Darwin may have been counting species, in his collection, which he had not collected himself.
11. This is further evidence of the strengthening relationship between these two men (see Letter #29 (n12)) and perhaps a strengthening of CD's religious convictions. CCD, Vol 1, notes: "*If John Maurice Herbert's memory was accurate, CD's plan to take orders represented a change in his religious convictions. Earlier, at about the time of the reading tour at Barmouth (Letter #1), according to Herbert, CD had serious doubts about this career. Herbert remembered 'an earnest conversation' he had with CD 'about going into Holy Orders; & I remember his asking me with reference to the question put by the Bishop in the Ordination service: "Do you trust that you are inwardly moved by the Holy Spirit &c" whether I could answer in the affirmative; & on my saying "I could not", he said, "neither can I, & therefore I can not take orders."' (Letter of reminiscences to Francis Darwin, 2 June 1882, DAR 112 (ser 2): 63-4).*"

12. In CD's copy of Stephens' *Catalogue Br Insects*, Vol 1, there are 339 species. However, only 162 species have annotations against them. So, as with Note 10, CD may have been counting species, which he had acquired from collectors and friends, such as Leonard Jenyns [See Introduction to Chapter 3 (n10) and Letters #10 (n7), #19 (n8), #25 (n7) #27 (n11), #28 (n2)].

#31 (CCD 81). To: W. Darwin Fox, Osmaston Hall, Derby.

[Cambridge]
Monday [31st May 1830]

My dear Fox,

I write these few lines to give you an outline of my plans, to apologise for not having written long enough ago, & lastly to say how very glad I am to hear that my plan of your paying me a visit in Shrewsbury, meets with your approbation[1]. — I shall go up to Town on Friday, or Saturday, stay a few days there, & then on Shrewsbury: from whence I will again ~~wite~~ write to you. — //

Mr. Hope[2] has paid me a visit but went back disgusted at the weather. he wants me to make an expedition to the New Forest. — but I do not think I shall: after I once I get to Shrewsbury the sooner you come the better, the only thing at all likely to interfere is Mr Hope & Eyton[3] want me to come an Entomological trip into N. Wales. Cannot you perhaps join us? Think of it. — Of course you have heard of the new species of wild Swan[4], discovered in England, by Mr. Yarrell[5]. I have//

bad stuffed specimen of it, for 10s. — bad as it is, you may think yourself, lucky in getting. Yarrell himself, has pronounced it to be the new sort, so there can be no doubt. —

I have had a letter from old Simpson[6], & he begs to kindly remembered to you: I am packing up Pulleins[7] birds: I ∗have∗ <have> got for you a dusky Grebe.

I will write again when I have more time. I only send this scrawl that you may know how very happy I am at the thoughts of seeing you at Shrewsbury. —

Dear old Fox
Yours
C. Darwin

Mr Hope has taken P. Podalirius[8] in Shropshire, & seen it several times

Postmark: Cambridge MY 31 1830
Christ's College Library, Cambridge, #30.

Notes

1. Visit to Shrewsbury by Fox: we have little information on this, but from what CD said in Letter #34, it probably took place - and may have included a short visit to the Wedgwoods at maer Hall. The date of the visit is likely have been a few days in July or August (1830).
2. F. W. Hope, the entomologist. See Bibliographic Register and Letters #6 (n2), #10 (n6), #18 (n2), #26 (n5), #28 (n5), #29 (n2).
3. Thomas Campbell Eyton (1809-1880). A naturalist with independent means from Eyton, Shropshire. Studied at Cambridge, 1827-28, where he was an acquaintance of CD, but apparently did not gain a degree. Author of many books, including ducks and rare birds. Married Elizabeth Frances Slaney, daughter of the MP for Shrewsbury [Letter #4 (n17)]. On inheriting the family estate in 1855, established a large museum of birds at Eyton; opposed the Darwin-Wallace theory of organic evolution. (*DNB*)

4. Bewick's swan.
5. William Yarrell (1784-1856). Naturalist and bookseller. Wrote standard works on British birds and fishes (DNB). In 1830 he first described Bewick's Swan (*Cygnus bewickii*), which today is thought to be a sub-species of the Tundra Swan (*C. columbianus*). See Bibliographic Register. (*DNB*)
6. George Simpson (?-1888): see Letters #11 (n9), #16 (n17) & #49 (n24).
7. Robert Pulleine (1806-1868): see Letter #9 (n6).
8. *Papilio podalirius:* the "scarce the swallowtail butterfly". In CD's copy of Stephens' *Catalogue Br Insects*, Vol. 1, there is no annotation against this species. CD had collected many specimens of the common swallowtail butterfly (*Papilio machaon* - see Fig 17) in the fens and even used it as analogy to describe his brother's erratic nature: "*he has changed his plans so often; that to follow him in his course would be to pursue a Machaon on a windy day*" [Letter #3 (n10)].

#32 (CCD 82; LLi[1]). To: W. D. Fox Esq, Christ's College, Cambridge.

Barmouth.
Wednesday 25th. — [Aug 1830]

My dear Fox

I have been intending to write every hour for the last fortnight, but *really* have had no time: I left Shrewsbury this day fortnight ago, & have since that time been working from morning to night in catching fish or beetles. This is literally the first idle day I have had to myself: for on the rainy days I go fishing, on the good ones Entomologizing. You may recollect that for the fortnight previous to all this, you told me not to write, so that I hope I have made out some sort of defence for not having sooner answered your two long & very agreeable letters: I am delighted that we agree so well about all the Wedgwoods, & as you say it is so much pleasanter talking about any place or persons that both parties have seen. — //

I trust that Charlotte[2] has driven out of the field your musical enchanter at Ashby[3]: Cannot you imagine how very pleasant I must find Maer[4], not one drawback. I have not even the least dread of the *Parish*[5] a sensation, which it appears, you were not quite void of. I should enjoy most exceedingly to spend a week there with you: it is more probable now than that originally you should have been at Maer at all. — I suppose you are now living in Cambridge with Prof: Henslowe[6]. I shall enjoy hearing from you some gossip about him & his wife[7] & I hope you will return good for evil & write soon to me. — I am afraid you must find Cambridge rather stupid. it must be such a contrast from your former life there. — of course every thing my rooms contain is most entirely at your service: you do not know how it would please//

me if there is any chance of your being there in the first week in October, by which time I would be up, if it were merely to eat our breakfasts together as we did in good old times. — I shall be at Maer for the first week or 10 days in September, & write to me there, (direction is Maer, Newcastle, *Staffordshire*) & tell me your plans. — And now I give you some account of our Welch trip. Hope 2 Eytons (father & son[8]) & myself arrived at Capel Curig on a Thursday. On Monday we went up Snowdon, & were disgusted with it we took 3 Curculio[9] a few glabrati[10] & nothing else. Hope & Eyton went off next day to Barmouth, Old E & myself staid a few days longer & had some pretty good trout fishing. I joined Hope & young E. here on Friday, but they returned to Shrewsbury on Monday for a cricket match, & here I am enjoying my solitude extremely: I am quite disgusted with Hopes egotism & stupidity: how I wish you were here, the very thought of a days entomologizing with you, is quite refreshing. — Try & prolong your stay in Cam. to October. I long to spend a little//

time with you quietly. I do not call your visit to Shrewsbury anything, there were so many people making a row & confusion. — The Entomologists did not do a very great deal here, took some good Colymbetes[11], Cicin: hybrida[12], & Eyton[8] took a most beautiful Crysomela[13]. I have taken some good Leiodes[14] by the way ask Prof: Henslow, what is the time to get Leiodes cinnamomia[11]. I want some very much. — You ought also to take some good water beetles this time of the year. — When you write tell me what you read. Remember me most kindly to Prof: Henslowe — Excuse this patch-work letter & believe me dear Fox.

Yours sinc,
C. D.

Postmark: Barmouth; C 27 AU 27 1830
Christ's College Library, Cambridge, #31.

Notes

1. In *Life and Letters* (F. Darwin, 1888), Vol. 1, pp.181-2: verbatim (except for slight variations in puntuation).
2. Charlotte Wedwood (1797-1862), daughter of Josiah Wedgwood II.
3. Not identified. Fox had previously been interested in Bessy Galton who lived at The Larches, Sparkbrook, Birmingham. However this attachment seems to have been broken in early 1830; see Letter #27.
4. Maer Hall, in the village of Maer, near Newcastle under Lyme, Staffordshire, the home of Josiah Wedgwood, II, and family.
5. Maybe a reference to the somewhat parochial atmosphere at Maer Hall (see note 3, above); for life at Maer Hall, see *Emma Darwin* by Edna Healey, 2001.
6. J. S. Henslow (1796-1861). See Bibliographic Register, Introduction to Chapter 2 and Introduction to this chapter, and Letters #8(n3), #25 (n8), #29 (n12), #30 (n11). He was friends with both CD and Fox, and is mentioned frequently in subsequent correspondence;
7. Harriet Henslow, sister of Leonard Jenyns (see Introduction to this chapter, note10).
8. Presumably Thomas Eyton (1777-1855), barrister, and Thomas Campbell Eyton (1809-1880), naturalist (for the latter, see Letter #31). The situation is confusing because there were two other sons, Charles James and William Archibald, both of whom were undergraduates at Cambridge at this time. (*DNB, Alum Cantab*)
9. Curculio was a genus of beetle at this time, but these insects were also placed in a number of other genera.
10. glabrati: it is not clear what CD was referring to. This was usually a specific epithet for a known genus.
11. Colymbetes: water beetles. No annotations for Barmouth or Wales in CD's copy of Stephens' *Catalogue Br Insects*, Vol 1,
12. Cicindela hybrida; in CD's copy of Stephens' *Catalogue Br Insects*, Vol 1, there is an annotation "*Barmouth Aug 1828*" against this species.
13. Crysomela: no annotation for Barmouth or Wales against this in CD's copy of Stephens' *Catalogue Br Insects*, Vol 1.
14. Leiodes cinnamomea in Stephens' *Catalogue Br Insects*, Vol 1; in CD's copy there are no annotations against this species.

#33 (CCD 83). To: W. D. Fox Esq, Osmaston Hall, Derby.

Maer.
Wednesday [8th Sept 1830]

My dear Fox

I cannot help writing a few lines to say how very sorry I am to hear such bad news about Mrs. Fox[1], & to beg of you to write <to me> very soon to me again to say how she is going on. — And I most sincerely hope to receive a favourable account. — I return to Shrewsbury on Friday: soon after which I shall expect a letter. — //

I have had very bad shooting this season, the first day I killed 10 brace, since which I have not on any day bagged three. — But I find Maer[2] just as delightful as ever & Charlotte[3] just as agreeable as ever: id est, as agreeable as possible. — I have done a good deal in Entomology, & to day have worked so hard, that I can hardly see to write. — //

I am very much obliged for the tin Box. The Enicoceri[4] are invaluable. — But beetles, partridges & every thing else, are as nothing to me, now that I have got a horse[5]. — I am positively in love with him, stands 16 hands & one inch, and I think will make a very good hunter. — But now I have got on the subject of horses I had better wish you good night. — I shall be very anxious for a letter so I hope you will not disappoint me I sent your letter on to home, & they therefore will be equally anxious to hear.

Good night dear old Fox
Yours most sincerely
C. D.

I am quite knocked up[6]

(*On the fly leaf, in pencil, there is a what appears to be a medical prescription. It is not in CD's hand.*)

Postmark: Newcastle SE 9 1830Postmark
Christ's College, Cambridge, #32.

Notes

1. There are no records of an illness of Mrs Samuel Fox. She lived to a good of age of 81 (*Darwin Pedigrees*). In the next letter [#34 9n1)], she is reported to have recovered.
2. The home of Josiah Wedgwood II and family.
3. Charlotte Wedgwood (1797-1844) the second daughter of Josiah Wedgwood II and who married Revd. Charles Langton in 1832. CD was clearly attracted to Charlotte at this time, although she was 11 years older, see Letter #32 (n2).
4. Enocerus was a genus of beetles. CD has made the Latin plural. See Stephens' *Illustrations of British Entomology*, Mandibulata II (1829), p 196, plate XV f 6.
5. Probably the horse on which CD rode to Cambridge from Daventry [Letter #34 (n2)].
6. CD means that he is exhausted, but from what is not clear.

#34 (CCD 86). To: W. D. Fox Esq, Osmaston Hall, Derby.

Christ College [Cambridge]
Friday Evening [8th October 1830]

My dear Fox

When I received your last letter our house was full of company, & we had somebody there every day till I left home, and I always feel it quite hopeless to think of writing a letter if the house is not quite quiet. — We were all very glad to hear so good an account of Mrs. Fox[1]. — & I most sincerely hope she continues daily to gain strength. — I arrived here in my most snug & comfortable rooms yesterday evening, after having had//

a most comfortless journey up here. — I left Shrewsbury on Tuesday & slept at Daventry, where I overtook my horse[2] & rode him myself the two last days journey. The poor beast was so tired that he hardly knew whether he stood on his heels or his head. — & it will be some time before I undertake to ride a young horse a long journey again. — It will be very pleasant having horse up here. Moreover I think he will make a splendid hunter, from a specimen I had of him with Eytons[3] hounds. — There is not an individual up whom I know, & therefore I have had plenty of time to regret your absence. — How I wish you had been able to have//

stayed up here[4]. we should have suited so well, each of us reading all morning & being idle all evening. — But it is not only when I am solitary that I regret your absence. Many many times do I think of our cozy breakfasts & even wish for you to give me a good scolding for swearing, & being out of temper or any other of my hundred faults — But is no use regretting; what cannot be altered. — I have not seen Prof Henslow[5], but am going to a Party there to night; you have not told me half enough what you think about Mrs. Henslow[6] She is a devilish odd woman. I am always frightened whenever I speak to her, & yet I cannot help liking her. — I suppose you cannot at all//

tell me what your plans are, & whether there is any hope of your coming up to Cambridge. — but that of course entirely depends on your Mothers health[7]; I trust you will write soon & tell me how she is going on as it is sometime since I last heard. — This is a very stupid letter, but I will try not to write many such

& Believe me dear Fox
Yrs. sincerely
C. D.

Postmark: OC 8 1830
Christ's College Library, Cambridge, #34

Notes

1. Ann Fox née Darwin, Fox's mother, who we learnt had been sick in the last letter (#33).
2. CD's recently acquired horse, see Letter #33.
3. One of the Eyton brothers; see Letters #31 (n4) & #32 (n8).
4. CD and Fox have played cat and mouse once more. However they may have met in Shrewsbury for a few days in July or August of this year (1830) – see Letter #33 (n1). The visit may also have included a short visit to the Wedgwoods at Maer Hall (see Letter #32).

5. J S Henslow (1796-1861). Professor of Botany, Cambridge University, and mentor to both CD and Fox. See Letters #8 (n3), #25 (n8), #29 (n12) & #30 (n11).
6. Harriet Henslow, the sister of L Jenyns [see Introduction to this chapter, note 10, and Letter 10 (n7)].
7. Fox completed his B A in 1829 (Peile, 1900, Letter #9) and he was then studying for his ordination, it seems under the guidance of Prof Henslow. The health of Mrs Ann Fox is unknown.

#35 (CCD 87). To: W. D. Fox Esq, Osmaston Hall, Derby.

[Cambridge]
[5th Nov 1830]

My dear Fox

I have so little time at present, & am so disgusted by reading, that I have not the heart to write to any body. I have only written once home, since I came up. — This must excuse me for not having answered your three letters, for which I am really very much obliged. — In the last I received a 5£ note. I have not yet heard the end of the negotiations between Mrs. Field & Graham[1]. — I send your things per waggon as the box is very large, it would cost a fortune per coach. — In your insect box Henslow[2] has put in a Cassida[3] & Donacia[3]. — I have taken out of your bottle 2 or 3 specimens of a black Cryptocephalus[4] which you took in profusion. — //

I have not stuck an insect this term & scarcely opened a case[5]: if I had time I would have sent you the insects which I have so long promised, but really I have not spirits or time to do any thing. Reading makes me quite desperate, the plague of getting up all my subjects is next thing to intolerable. — Henslowe is my tutor[6], & a most *admirable* one he makes. the hour with him is the pleasantest in the whole day. I think he is quite the most perfect man I ever met with[7]. — I have been to some very pleasant parties[5] there this term. — his good nature is unbounded. — I saw your friend Mr. Hey[8] once or twice & liked him very much he appears to be a very sinsible man. — //

I have called on the young one[9], but in this term our communication will go no further. — I am sure you will be sorry to hear poor old Whitleys[10] father[11] is dead: in a worldly point of view it is of great consequence to him, as it will prevent him going to the Bar for some time[12]. — (Be sure answer this) What did you pay for the iron hoop you had made in Shrewsbury[13]? Because I do not mean to pay the whole of the Cambridge mans bill. — You need not trouble yourself about the Phallus[14], as I have brought up both species. — I have heard men say that Henslowe has some curious religious opinions[15]; I never perceived anything of it. Have you? I am very glad to hear, after all your delays, that you have at last heard of a Curacey[16], where you may read all the commandments, without endangering your throat. — I am also still more glad to hear that your mother continues steadily to improve. I do trust you will have no further cause//

for uneasiness about her. — with every wish for your happiness my dear old Fox
believe me yours most sincerely
Chas Darwin

Friday Nov. 5th. —

I heard from old Simpson[17] the other day, he is going on very prosperously, & sends numerous kind speeches to you. — Do you in one of your letters call M^{r}. G. Jenyns[18] a good or a grand fellow? I am curious to know. —

Christ's College Library, Cambridge, #33.

Notes

1. Mrs Field was perhaps involved in housekeeping at Christ's College. John Graham (1794-1865) was CD's tutor at this time and, a little later, Master of Christ's College. See Letters #10 (n12), #19 (n16) & #29 (n5).
2. J S Henslow: see Note 7, Bibliography, Introduction to this Chapter and Letter #8 (n3).
3. No annotations in CD's copy of Stephens' *Catalogue Br Insects*, Vol 1.
4. Fox records C. labiatus in his diary for 24th Aug 1830.
5. Is this evidence of CD's waning interest in insects because 1) Fox is losing interest and 2) J S Henslow is opening up CD's mind to other sciences?
6. CD is speaking in the general sense, no doubt, as Graham was his College Tutor; see Letter #10 (n12).
7. The friendship between these two men gathers pace [see Letters #29 (n12) and #30 (n11)]. The parties that CD now refers to are not the usual student gatherings but gatherings of naturalists and geologists at Henslow's house.
8. Samuel Hey (d 1852), see Letter #30 (n3). Vicar of Ockbrook, Derbyshire.
9. William Hey (1811-1882). St John's College, B A, 1834. Clergyman, teacher and entomologist. Son of Samuel Hey, note 8, above. (*Alum Cantab*)
10. Charles Thomas Whitley (1808-1895); see Letter #9 (n15) and note 12, below.
11. John Whitley (?-1830). Lived in Liverpool.
12. It would appear that Whitley did not follow this course of action as he became a reader in natural philosophy and mathematics at Durham University (*Alum Cantab*).
13. This probably refers to the visit of Fox to CD in Shrewsbury, which seems to have taken place in July or August (1830), see Letter #31 (n1).
14. *Phallus* is a genus of stinkhorn fungus. Possibly used by CD to attract insects.
15. This speculation seems to be wide of the mark, especially in light of CD's remark that he was thinking of reading divinity with J S Henslow; see letter #30 (n11). It does seem to bear out the point made by J M Herbert (in Letter #30 (n11) that CD was wrestling with religious doubts at this time. In the *Autobiography*, pp. 64-5, CD's memory of John Stevens Henslow is that he was "*deeply religious, and so orthodox, that he told me one day, he should be grieved if a single word of the Thirty-nine Articles were altered.*" In the background also is Hensleigh Wedgwood, who was a Fellow at Christ's College. Although Hensleigh is only mentioned once in all the previous letters to Fox [Letter #10 (n16)], he must have had conversations with CD, his cousin, at this time, concerning matters of religion, and, indeed, in 1831, Hensleigh resigned his Fellowship at Christ's College because he could not, in clear conscience, believe in the Thirty-Nine articles of the Church of England and because he had other religious doubts (Wedgwood and Wedgwood, 1980). Later still, in 1837, he resigned his government post as a Police Magistrate as he could not administer oaths with a clear conscience [Wedgwood and Wedgwood, 1980 and Letter #63 (n8)]. Everyone, including CD, thought that he was being unwise in making such an anguished decision, because he gave up a position, which brought in £800, when he had no other means to support a wife and three children. As we see later, Hensleigh Wedgwood was a strong influence in developing Darwin's mind after the return of the *Beagle*, when he lived in London just prior to, and immediately after, his marriage to Emma Wedgwood (see Inroduction to Chapter 5).
16. This seems to be a reference to the Curacy that W D Fox obtained at Epperstone, Leicestershire at about this time.
17. George Simpson (?-1888). See Letters #11 (n8) & #16 (n17). See also letter from Simpson to CD on 26th Jan 1831 (CCD, Vol 1, p 113).
18. The Revd George Leonard Jenyns of Bottisham Hall, Cambridgeshire, father of Harriet Henslow and Leonard Jenyns.

#36 (CCD 88). To: W. D. Fox Esq.

[Cambridge]
[27th Nov 1830]

My dear Fox

After a good deal of hunting they found at Prof. Henslows[1] house a box containing the various articles you mention, & if they will all go in, I will send them off to day. Your prints must be kept till some other time, also the beetle stick. —
Our accounts stand thus,

&c &c left unpaid at Henslows } including Lens }	14 . 6
Brett[2]. case & packing up	13 . 6
Mrs Field[3]	2 " 3 . 6
	£3 " 11 . 6
	5
Balance, on my side[4],	£1 " 8 " 6//

I am reading very hard[5], & have spirits for nothing. I actually have not stuck a beetle all this term. — Young Hey[6] does go to Henslow evening parties. — I am very glad to hear that there is some chance of your coming up to Cambridge. — I shall be delighted to see you, but I hope it will be after examination is over, as I am far too much plagued to enjoy any thing at present. — I trust after your numerous delays that you really will be in orders by Xst.mas, & that I shall really pay you a visit in the Spring. — I have nothing more to write about. You must put down my short letters to the right account. — stupidity

& believe me, My dear Fox
Yours most truly
C. Darwin//

You have no occasion to make speeches about giving me trouble I shall be most ready at all times to do anything I can, for you.

Endorsement: C. Darwin Novr 27 1830
Christ's College Library, Cambridge, #35

Notes

1. J S Henslow (1796-1861). See Bibliographic Register, Introduction to this Chapter and Letters #8 (n3), #25 (n8), #29 (n12) and #35.
2. Brett: see Letters #17(n4) & #19 (n6).
3. Mrs Field: see letter #29 (n4).
4. The balance is determined from the £5 that Fox sent to CD with the last letter (#35).
5. For his final examination in Jan 1831: the Senate House Examination (see Letter #37).
6. William Hey (1811-1882). See Letter #35 (n9).

#37 (CCD 89; LLi[1]). To: The Rev. W. D. Fox Esq., Epperstone, Nottingham[2]

[Cambridge]

Sunday [23rd Jan 1831]

My dear Fox

I do hope you will excuse my not writing before I took my degree[3]. — I felt a quite inexplicable aversion to write to any body. — But now I do most heartily congratulate you upon passing your examination[4]; & hope you find your curacey[5] comfortable; if it is my last shilling (I have not many) I will come & pay you a visit. — I do not know why the degree should make one so miserable, both before & afterwards; I recollect you were sufficiently//

wretched before, & I can assure I am now; & what makes it the more ridiculous, is I know not what about. — I believe it is a beautiful provision of nature to make one regret the less leaving so pleasant a place as Cambridge. — & amongst all its pleasures, I say it for once & for all, none so great as my friendship with you. — I sent you a Newspaper yesterday, in which you will see what a good place I have got in the Polls[6]. — As for//

Christ[7] did you ever see such a college for producing Captains & Apostles[8]. — There are no men either at Emmanuel or Christ plucked[9]. — Cameron[10] is gulfed[11], together with other 3 Trinity scholars! — My plans are not at all settled, I think I shall keep this term, & then go & economize at Shrewsbury, return & take my degree[12]. — A man may be excused, for writing so much about himself, when he has just passed the examination. So you must excuse. — And on the same principle do you write a letter brim-full of yourself & plans. — I want to know something about your//

examination[13] : tell me about the state of your nerves. — : what books you got up, & how perfect? I take an interest about that sort of thing as the time will come, when I must suffer. — Your tutor Thompson[14] begged to be remembered to you, & so does Whitley[15]. — If you will answer this; I will send as many stupid answers as you can desire. — Believe me dear Fox

Chas. Darwin.

Postmark: Cambridge JA 23 1831

Christ's College Library, Cambridge, #36.

Notes

1. In *Life and Letters* (F Darwin, 1888), Vol 1, pp 183-4: verbatim (with some punctuational changes).
2. This is the first letter addressed to Fox at Epperstone, a village near Nottingham, and suggests that he took up his curacy there early in 1831.
3. CD took his Senate House Examination for the B A without mathematical Honours. He was placed tenth on the list of 178 who passed (see note 5, below). Because he took residence only at the beginning of the second (Lent) term of 1827/28, he had to stay in residence for one more term and was officially listed among the B As of 1832. See *Autobiography*, p 68 and *LL* 1: 163.
4. W D Fox has obviously passed his clerical examinations.
5. CD refers to Fox's curacy at Epperstone, near Nottingham.
6. This was the Senate House Examination, sometimes called the Poll (OED), for those students who read for a "pass' " degree, i e., without attempting mathematical Honours (the latter known as the Tripos). CD was placed tenth on the list of 178 who passed (*Cambridge Chronicle*, 22 January 1831,

2d ed.). In his *Autobiography* (LLi, p 47) CD says: '*By answering well the examination questions in Paley, by doing Euclid well, and by not failing miserably in Classics, I gained a good place among the οι πολλι or crowd of men who do not go in for honours*'.
7. Christ's College.
8. The '*Captain*' headed the 'Poll' list; the Senior Wrangler headed the Mathematical Tripos list and the last twelve in the Mathematical Tripos list were called the '*Apostles*'.
9. That is, who failed.
10. Jonathan Henry Lovett Cameron (1807-1888). B A, Trinity College, 1831. Shrewsbury school friend of CD. Became a clergyman. See letter written by Cameron after CD's death (and referred to in CCD, Vol 1, p 112) in DAR 112: 14).
11. 'The position of those candidates for mathematical honours (at Cambridge) who fail to obtain a place in the list, but are allowed the ordinary degree.' (*OED*).
12. See Note 1.
13. At this time CD was considering reading for divinity with Prof J S Henslow (see Letter #36).
14. Thompson, Fox's tutor, but he has not been identified. There were several Thompson's in Cambridge at this time, but none at Christ's College.
15. Charles Thomas Whitley (1808-1895); see Letters #9 (n15), #11 (n9) & Letter #35 (n12).

#38 (CCD 92). To: The Rev. W. D. Fox Esq., Epperstone, Nottingham

[Cambridge]
Wednesday 9th Feb 1831

My dear Fox

I should have answered your last letter earlier, as indeed was necessary to redeem my character. — I waited till I could hear from Baker[1]; but the rogue has not yet been to my rooms; although I wrote to him. — I shall be very glad to be of any use in paying your bills & will make enquiries as you may direct me. NB. I owe you at present £1"4"6[2]. — I paid Markham Bennet[3] for Porterage, which makes a shilling or two difference from my last account to you. — I do not quite understand your last letter is Orridge[4] & Aikens[5] account a different affair?//

& now for giving you a scolding. I should like to know what you mean by such expressions as "begetting a horror of my handwriting" &c, is it a very refined species of irony? does it mean, if your very agreeable letters (I am forced to flatter in order to excuse myself) are horrible, that mine must be à fortiori, most intolerably horrible[6]? — To use your own expression, no more humbug. I am always very glad to receive your letter & you know it, & in Johnsons[7] language "there is an end of it." — I shall leave Cambridge a little time after division, my duty draws me away; my inclination would keep me all the next terms, & *if* <I> possibly//

can, most certainly I shall stay up the greater part of next term. — I have so many friends up here (Henslow amongst the foremost[8]) that it would make any place pleasant. — (NB you always spell Henslow with an e). — Is there any hope of your coming up next term, if there is we will most certainly meet. — I will write again, when I have settled your affairs here. — You ask after Eras: we never correspond excepting on business. — He is in London, & now you know as much as I do. — I could not write a long letter even <to> Charlotte Wedgwood[9] or Fanny Owen[9], so must excuse this short one. —

& Believe me dear old Fox
Your most sincerely
C. Darwin

Simpson's[10] direction is Feversham Kent. — do write to him, he will be so glad to hear from you.

Postmark: Cambridge FE 9 1831
Christ's College Library, Cambridge, #38.

Notes

1. Baker, a collector frequently referred to in previous and subsequent correspondence, see Letters #16 (n1), #17 (n4) & 30 (n7).
2. See Letter #36, where the sum is 1"8s"6d, but the "*shilling or two difference*" accounts for the difference.
3. Not identified.
4. A local carrier of goods?
5. Probably a natural history collector; see Letters #16 (n4), #17 (n2), #19 (n2) and #25 (n1).
6. Anyone who has to decipher the letters of CD and Fox will award Fox the winner! However even Fox's hand is difficult at times!
7. Samuel Johnson (1709-1784), writer and lexicographer. (*DNB*)
8. J S Henslow, Professor of Botany and mentor to both CD and Fox; see Bibliographic Register, Introduction to Chapter 2 and Letters #8 (n3), # 26, #27, #29 and #35.
9. Charlotte Wedgwood (of Maer Hall) and Fanny Owen (of Woodhouse); see Letters #32 (n2), #33 (n3), for Charlotte, and Letter #7 (n12) for Fanny Owen; the latter was fading in attraction to CD and Charlotte was gaining
10. George Simpson (?-1888). B A Christ's College, 1830. Became a C of E clergyman. (*Alum Cantab; Peile, 1900*); see Letters #11 (n8), #16 (n17) & #31 (n5).

#39 (CCD 94). To: The Rev. W. Fox, Epperstone, Nottingham

[Cambridge]
Tuesday [15th Feb 1831]

My dear Fox

I am going out this evening & have only time to write about business — I saw Baker[1] this morning & told him your message about writing, which he did not seem to like, so I do it for him. — He has for you the following Birds, pair of Hen Harriers — 3 ash coloured Falcons. — Woodpecker — pair Bearded Titmice. — Shrike. — Swan. (I sent to him to Price[2], but have not heard whether he is up) The Swan is in very bad condition. — I have sent also a dusky grebe, which I procured sometime ago. — His bill amounts//

to 5£"3^{s}"0. which includes a packing case 14^{s}. & a bushel Ribston pipins[3] 12^{s}. — I have not yet paid Aiken[4] but will see about it. — You will then be indebted to me to some small amount. — If you will give me directions I will tell Baker[5] to send your birds off. (You had perhaps better have the engraving with them) Henslows[6] former servant has left his service some time ago. —

Yours sincerely
Chas Darwin

I shall go away in the course of 10 days or there abouts. —
Adieu
If you have not read Herschel[7] in Lardners Cyclo[8] — read it directly.
1"16. —
3"17 —

Postmark: Cambridge FE 15 1831
Christ's College Library, Cambridge, #37.

Notes

1. Baker: a natural history collector: see Letters #16 (n1), #17 (n4), #30 (n7) and #38 (n1).
2. John Price (1803-1887). See Letters #11 (n3) & # 203 (n4).
3. Ribston pippins; a variety of apple.
4. Probably a natural history collector: see Letters #16 (n4), #17 (n2); #19 (n2), #25 (n1), #38 (n5).
5. A natural history collector with connections in Norfolk; See Letters #30 (n7) and #38 (n1).
6. J S Henslow; see Bibliographic Register, Introduction to this Chapter and Letters #8 (n3), #25 (n8), #29 (n12), #30 (n11), #35 (n2), #36 (n1), #39 (n4), #40 (n3) & #41 (n8). 43/5, 44/3, etc.
7. John Frederick William Herschel (1792-1871). Astronomer, mathematician and philosopher. Author of many learned works and books including the one cited by CD in Note 8, below. CD met him in Cape Town in 1836 (CCD Vol 1, Letter to J S Henslow, July 9th 1836).
8. Sir John Herschel's *Preliminary discourse on the study of natural philosophy* was published in 1831 in Dionysius Lardner's *Cabinet cyclopaedia*. It anticipated John Stuart Mill, and surpassed William Whewell, in the formulation of the methods of scientific investigation. Whewell was the dominant figure in the field of the philosophy of science in Cambridge at the time. This indicates that CD was probably dissatisfied with the latter's approach and was seeking out new authorities.

#40 (CCD 96). To: The Rev. W. Fox, Epperstone, Nottingham

Shrewsbury
Thursday [7th April 1831]

My dear Fox

Do you mean to cut the connection; why do you not write?[1] I sent the last letter; so by the laws of nations you ought to have written. — I was in such bustle, when I last wrote to you: that I really forget how our various money transactions go on. — I will state them. — I have in my possession your 5£: I have paid Aiken[2] 2"13"6. But have not paid//

Bakers[3] bill, (not having seen him) which amounts 5£"3s"6 including 12s. for apples for Henslows[4] — I shall start for Cambridge tomorrow week: but shall stay a few days in London to hear Operas &c &c. — Let me have a letter from you waiting at Cambridge, or before I go there: I will settle all your affairs for you. — I expect to spend a very pleasant Spring term: walking & botanizing with Henslow: I suppose it is out of//

the question, your snatching a Parsons week[5] & running up to Cambridge. I think you would enjoy; I am sure I should; — Think of it. — At present, I talk, think, & dream of a scheme I have almost hatched of going to the Canary Islands[6]. — I have long had a wish of seeing Tropical scenery & vegetation: & according to Humboldt[7] Teneriffe is a very pretty specimen. — Looking over your letter I find there is a bill Orridges[8], is it distinct from the 2£"13"6?//

If you are not busy, you had better write to me before tomorrow week, & give me circumstantial account of every thing that you can think of. — How all your family are? &c &c.

Believe me dear old Fox
Most sincerely

Chas Darwin

PS. tell me how, where &c &c, you are living?

Endorsement, April 10/31
Christ's College Library, Cambridge, #39.

Notes

1. This is a rather testy, fractious sentence. The reason for any lack of response by Fox at this time is probably due to ill health. We do not have diaries for these years (the next extant diary is for 1833). However we do know that Fox suffered a serious lung illness about this time, which forced him, eventually, to give up his curacy and recuperate on the Isle of Wight (see Letter # 49 and CCD Vol 1, Caroline Darwin to CD, 12-28th June 1832; and Susan Darwin to CD 15 [-18] August 1832).
2. Aiken: a natural history collector; See Letters #16 (n4), #17 (n2); #19 (n2); #25 (n1); #38 (n5) & #39 (n4).
3. Baker: a natural history collector with connections in Norfolk; See Letters #16 (n1), #17 (n4), #30 (n7); #38 (n1); #39/3 & #40 (n2).
4. J S Henslow; see Bibliographic Register, Introduction to this Chapter and Letters #8 (n3), #25 (n8); #29 (n1)2, #30 (n11), #35 (n2), #36 (n1) & 39 (n4).
5. Parson's week: 'the time taken as a holiday by a clergyman who is excused a Sunday, lasting (usually) from Monday to the Saturday week following' (*OED*).
6. This is the first indication of the scheme to visit the Canary Islands, which developed over the late spring and summer of 1831 and on which J S Henslow, initially, was to be one of his companions (LLi, p 190).
7. See Chapter two of Humboldt (1814): *Personal narrative of travels to the equinoctial regions of the New Continent during the years 1799 – 1804 (translation)*: see also CD's *Autobiography*, p 68. CD's six volumes (of various editions, 1814-1829) of the English translations of Humboldt's book are in the University Library, Cambridge.
8. A local carrier of goods; see Letter #38 (n4).

#41 (CCD 100). To: The Rev^{d}. W. Fox, Epperstone, Nottingham

Cambridge
Wednesday [11th May 1831]

My dear Fox

Cambridge has been in such a state of bustle & excitement for the last week[1], that I have done nothing but go gossiping about the town. — But thank goodness we are once again quiet: & I have had time to think about my plans. — I am very, very sorry to say I cannot pay you a visit at present. I really have no right to travel so many miles (& ∴ cost so many shillings) for my own amusement: the //

Governor[2] has given me a 200£ note to pay my debts, & I must be economical. — Independent of paying you a nice quiet, snug visit, I should have put a scheme into practice which I have long wished to do to see the Pictures at Stamford:[3] But both schemes must die the same death from inanition. — On the per contra side of the question Henslows lectures[4] come into play, & I should have been sorry to have missed even one//

of them. — And now for our eternal accounts. — I find I made a mistake in my last letter. — I have of your money, *6*"4"6 Orridge's bill[5] was 13^{s}. & what ∗I∗ have paid for you amounts in toto, 3£"6"6 ∴ remains now 2£18^{s}. — I have written to Baker[6] & given him proper directions, & will pay ∗h∗im his bill next time he comes to Cam.: — Some goodnatured Cambridge man has made me a most magnificent anonymous present of a Microscope[7]: did ever hear of such a delightful piece of luck? one would like to know who it was, just to feel obliged to him. — //

My time here is very pleasant. I am very busy at 3 or 4 λογοι[8], & see a great deal Henslow[9], whom I do not know, whether I love or respect most. — M^rs^. Henslow is hatching a young professor[10]. — She will be confined very soon. — As for my Canary scheme[11], it is rash of you to ask questions: My other friends most sincerely wish me there I plague them so with talking about tropical scenery &c &c. — Eyton[12] will go next summer, & I am learning Spanish. — How I wish we could meet. You would soon be tired of the subject

Good Bye
Chas Darwin

PS. Aiken[13] is not ill.
John Day[14] wants to know what to do with your wine as the Hampers are rotting & something must soon be done.

Endorsement: May 12, 1831
Christ's College Library, Cambridge, #41.

Notes

1. General Election polling took place in Cambridge on 3, 4, 5, and 6 May 1831. The Whig government of Lord Grey went to the country over the first Reform Bill. The University of Cambridge was allowed to elect two members of Parliament at this time. The constituency was the Fellows of Colleges - and the names of how these voted was published. So the Election would have been a very public and much-debated event in Cambridge; the Whig candidates were William Cavendish and Viscount Palmerston, the Foreign Secretary, and the Tory candidates were, firstly, Wellington's Chancellor of the Exchequer, the Rt Hon Henry Goulburn and, secondly, the brother of Sir Robert Peel, William Yates Peel (see *Desmond and Moore*, 1991, and *Parliamentary Records*). Both Tory candidates won, which must have caused quite a commotion.
2. CD's father, Robert Waring Darwin.
3. The collection of the Marquis of Exeter at Burghley House, Stamford. Stamford is quite as far from Cambridge as Epperstone would be.
4. J S Henslow gave a series of lectures every summer on botany and natural history, which was very popular. See Introduction to this Chapter. CD did not attend them in his first two years and it was probably Fox who inspired him to go initially, since Fox's name appears on lists in 1827 and 1828.
5. Orridge: local carrier of goods; see Letter 38 (n4) & #40 (n8).
6. Baker: a natural history collector with connections in Norfolk; See Letters #16 (n1), #17 (n4), #30 (n7), #38 (n1), #39 (n3) & #40 (n2).
7. Microscope: although anonymous it was later established that it was a gift from J M Herbert (1808-1882) of St John's College (CCD, Vol 1, p 123, note 1): see Letter #11 (n10).
8. Ancient Greek word meaning "prose writings".
9. J . Henslow: see Bibliographic Register, Introduction to Chapter 2 and Letters #8 (n3), #25 (n8), #29 (n12), #35 (n2), #36 (n1), #39 (n4) & #40 (n3), etc.
10. This was Leonard Ramsay Henslow (1831-1915), who became a C of E clergyman. (*Alum Cantab*)
11. Canary Islands scheme - See Letter #40 (n6).
12. Presumably Thomas Campbell Eyton; see Letter #31 (n2).
13. Aiken: a natural history collector who lived in a small house in Cambridge; see Letters #16 (n4), #17 (n2), #19 (n2), #25 (n1), #38 (n5), #39 (n4) & #40 (n)2.

14. John Day: not identified: a College functionary? This confirms what we knew already, from Chapter 2, that Fox had a liberal store of wine – probably supplied by his brother-in-law, Samuel Ellis Bristowe.

#42 (CCD 101). To: The Rev^d^. W. D. Fox, Epperstone, Nottinghamshire.

Shrewsbury
Saturday [9th July 1831]

My dear Fox

I arrived at this stupid place about three weeks ago, but have had no quiet time for writing till now. — I am staying here, on exactly the same principle, that a person chooses to remain in the Kings bench[1]. — Talking of poverty puts me in mind to give you a scolding: in your answer to my letter containing the reasons I could not come to Epperstone, you say you do not wonder at my not choosing to come to so stupid a place. now treating//

the thing logically, 1st you *must* have know what you call stupid, is just what I like, & 2nd you *might* have know, that if I could, I would most assuredly have come if it were merely for the pleasure of seeing you ∴ you have no excuse, & are (as we say in Spanish) un grandisimo bribon[2]. — I hope your prints arrived safe. I was in a perfect whirlwind of dust & confusion, when I sent them off, else I should have written with the box. — The Canary scheme goes on very prosperously[3]. I am working like a tiger for it, at present Spanish[4] & Geology[5], the former I find as//

intensely stupid, as the latter most interesting. I am trying to make a map of Shrops: but dont find it so easy as I expected. — How goes on Entomology with you? you are in a capital situation, that is if Sherwood forest[6] is at all like the New forest. — Hope & Eyton[7] did wonders the∗re∗ (I did not go propter pecuniam[8]). Your imagination cannot fancy the number of red elater, melasis[9], Cerambycidous[10] insects, without end. — I am just beginning Diptera[11]. — L. Jenyns[12] started[13] me, what an excellent naturalist he is. I have seen a good deal of him lately, & the more I see the//

more I like him. — I feel just the same way towards another man, whom I used formerly to dislike, that is Ramsay[14] of Jesus, who is the most likely person (I dont know whether I told you before) to be my companion to the Canaries. — How much do you know of the particulars of our plans? Shall you be at Epperstone in the Autumn, & if you are, would be convenient, if I could manage to pay you a visit — answer this sincerely. — Cannot you pay me a visit at Shrewsbury, is it impossible? This letter is all about myself. Do thou likewise,

good Bye dear old Fox.
C. Darwin

Postmark: Shrewsbury JY 10 18<3>1
Christ's College Library, Cambridge, #40.

Notes

1. The King's (Queen's) Bench was a division in the English superior courts system that heard civil and criminal cases. On the other hand, CD may have referred loosely to the King's Bench Prison, which was for debtors or people convicted of libel. The meaning is clear, that CD was only there at Shrewsbury because of the benefits, economically, of living at home.

2. Spanish for "a great rascal". CD is "showing off" his Spanish and was clearly learning Spanish for his projected trip to the Canary Islands (see note 3, below). This knowledge base would have stood him in good stead for the *Beagle* voyage, and would have been a further recommendation to Commander FitzRoy, when he was looking for a companion for the voyage.
3. For Canary scheme see Letters #40 (n6) and #41 (n10).
4. This reinforces note 2, above, that CD was learning Spanish. From his remarks, he does not seem to have found this easy. About this, time he also tried to learn German and did not make great progress there either, since he later used T H Huxley with his German translations.
5. CD wrote retrospectively, in 1838, in his *Journal* for 1831 '*In the Spring Henslow persuaded me to think of Geology & introduced me to Sedgwick.*' See also notes in CCD, Vol 1, p 125 and LLi, pp 56 & 57.
6. Sherwood Forest is north of Nottingham in the same region as Epperstone, where Fox was a curate.
7. Frederick William Hope (1797-1862). Letter #31 (n1) reported that Hope planned to make an expedition to the New Forest, to which CD was invited. See also Letter #6 (n2).
8. "propter pecuniam": Latin for "for reasons of lack of money".
9. Melasis: a synonym for Ceratophytum in Stephens' Catalogue Br Insects; it is in the Elateridae and so close to Elater; CD mentioned Melasis Flabellicornis (= M. bupestroides) in Letter #2.
10. Cerambicidous insects: the writing here seems clear; CD was using an anglicism to describe insects in the family Cerambicidae or longhorn beetles [see Letters #2 (n8) & #18 (n9)].
11. Diptera, an order of insects; common name "flies".
12. L Jenyns (1800-1893); See Bibliographic Register and Introduction to this Chapter (n10); see also Letters #10 (n7), #19 (n9), #25 (n7), #27 (n13), #28 (n2), #32 (n4) & #34 (n5), etc.
13. The writing is smudged here. It could be "started" or "startled". Francis Darwin and the CCD have "started". "Started" fits best if it is taken to mean "to be moved from a fixed position", that is, CD has reversed his previous poor opinion of Jenyns (see Letters #10 (n7) and #19 (n9).
14. Marmaduke Ramsay (?-1831). Fellow and tutor of Jesus College, Cambridge. (*Alum Cantab*)

#43 (CCD 103). To: The Rev^{d}. W. D. Fox, Epperstone, Nottinghamshire.

[Shrewsbury]
August 1st. [1831]

Dear Fox

I received your letter yesterday & sit down, according to your desire, to answer to the best of my power your questions. — Hope[1] will not be in town during the time you mention, his direction is 37 Upper Seymour St. & I am sure he would be very glad to show you his cabinet, if you should at any time be in Town when he is. — I am sorry to say I utterly forget both Stephens[2] Christian name & direction but I should <think> you could find out//

his direction from his bookseller, which latter you can find from the covers of the British Entomology[2]. — I have generally gone somewhere about 8 oclock: He is a very civil little man: there is one proper evening in the week, but I believe he has no objection to people out of the country calling at any time. — I forget whether I mentioned to you that I have 2 or 3 insects from, either M^{r}. Rud, or Wailes of Newcastle[3], Henslow gave them in my charge for you, & I cannot recollect//

for certain the Donor. — I will send these together with my long promised lot (which I am afraid will not be so numerous as you might expect) in the course of a few days. — The Canary scheme does not take place till next June. I am sorry to hear[4] that Henslows[5] chance of coming is *very* remote. I had hoped it was daily growing less so. — I shall in probability to go to Cam to pay my bills somewhere about the end of October, but I do not know for certain. — I cannot end this letter without adding, how *grieved* I am to find that you think me capable of telling base, hollow & deliberate falsehoods, in no other possible way can I interpret your letter[6].

Yours sincerely,
Chas. Darwin//

The man who made my cabinet is W. Edwards 29 Wilton St. Westminster. I advise to get one bigger than mine. — Mine cost 5£"10. & contained 6 drawers, depth, 1^{f}" 3. breadth 1^{f}" 7. — & whole cabinet stood in height 1^{f}"4[7]. —

Postmark: Shrewsbury AU 1 1831
Christ's College Library, Cambridge, #42.

Notes

1. F W Hope, the entomologist; see Letters #6 (n2), #10 (n6), #18 (n2), #26 (n5), #28 (n5) & #29 (n2), #31 (n1), 42 (n6).
2. J S Stephens, entomologist and author – see Bibliographic Register and Letter #1 (n11), etc.
3. Both L Rudd, Esq., of Marton Lodge, Yorkshire, and George Wailes, Esq., of Newcastle-upon-Tyne, are mentioned often in Stephens' *Illustrations of British Entomology* (1827-46)[see Letter #6 (n2)], as supplying specimens.
4. This suggests that Fox had recently seen Henslow, probably in Cambridge.
5. John Stevens Henslow (1796-1861). See Bibliographic Register, Introduction to this Chapter and Letters #8 (n3), #25 (n8), #29 (n12), #30 (n11), #35 (n2), #36 (n1), #39 (n4), #40 (n3) & #41 (n8). It is clear from this that the plan for Henslow to join the Canary expedition was failing and that the date had been postponed by a year.
6. It appears that these remarks were made seriously, and came on top of the strong remarks in Letter #40 "*Do you mean to cut the connection; why do you not write*?" In letter #44, CD says, "*You cannot imagine how much your former letter annoyed and hurt me. – But thank heaven, I firmly believe that it was my own entire fault in so interpreting your letter*." Francis Darwin (*LL* 1: 205) says "*He [CD] had misunderstood a letter of Fox's as implying a charge of falsehood.*" Fox's letter has not been found. It appears that CD was liable to the charge of becoming overbearing to his friend, for whatever reason.
7. These details are of interest in ascertaining any current identification of CD's cabinets and in distinguishing this cabinet from the one bought earlier [Letter #10 (n10)].

Conclusion

Although we have no letters from W D Fox to Darwin during this period, we have sufficient information from the letters to reconstruct the lives of the two young men quite accurately. Darwin starts off as the young student in unfamiliar surroundings and hanging on to every word of Fox. Fox is the mentor in general and teacher in matters of entomology. However, this phase quickly passes. By the end of 1828 (Letter #7), his first year at Cambridge, Darwin is confident enough to become an equal partner in their entomological enterprise. In his second year he begins to take over and even to hector Fox, when his enthusiasm seems to be waning. In his third year, Darwin is clearly gaining further in confidence, both in passing his "Little Go"

(Letter #28) and in his connections in the entomological world, where he is acknowledged by the leading entomologists such as Stephens, Hope and Jenyns. And in his second year, Darwin finally makes contact with John Stevens Henslow (Letter #25) and quickly forging a strong bond, which was to have a transformative effect on Darwin's life, and lasted until Henslow died in 1861 - not long after the publication of the *Origin of Species*. This association is key to Darwin's further development from insect hunter, natural history collector and geologist to the instigator of the "*Darwinian Revolution*". A crucial step here was the scheme, at the suggestion of Henslow, to accompany Adam Sedgwick on a geological expedition to North Wales, and it is during that tour that Darwin received the invitation, orchestrated by Henslow, to accompany Commander Robert FitzRoy, on the second surveying trip of HMS Beagle.

William Darwin Fox, on the other hand, took his finals in early 1829, and then had to seek a curacy. We do not have Fox's diaries for this period, although some few pages exist to show that he did continue to keep a diary during these years (Appendix 5). In fact, Fox's position for his B A was not very good, despite what Darwin says (Letter #9) and it took two years for his curacy to eventuate. He had to return to Cambridge to take his clerical examinations (Letter #32) and perhaps he was ill for some of this time. He began at Epperstone, just outside Nottingham, as a curate, in early 1831, just as Darwin was taking the final examinations for his B A (Letter #27). The two do not see each other thereafter for another six years (Letter #63). Furthermore, the curious fact about their friendship is that while they continue corresponding for the rest of their lives they rarely meet (Appendix 2). The total number of documented meetings is seven! Despite the fact that there may have been rare meetings at other times, it seems probable that the wonderful record we have, set down in their correspondence, represents the only major contact between these two friends after their glorious student days – which was a subject of great nostalgia for the rest of their lives.

At the end of this period, Darwin was clearly a young man of expanding interests. His enthusiasm for insects, in which he attained a professional level for this time, attracted the attention of some influential men of science in Cambridge. This in turn led to the offer of the passage on the second voyage of the *Beagle*. And this six-year voyage would launch Darwin on the unique career that would, 28 years later, culminate in the formulation, with Alfred Russell Wallace, of the first, viable theory of organic evolution, and the publication of the *Origin of Species*.

For this period, the correspondence with Fox forms a unique window into the life and times of this remarkable young mind. It also shows us the life of a typical undergraduate, Fox, of this time. Fox became a clergyman in the Church of England and lived a fairly normal life, bringing up a large family in the English countryside - a calling which up to this time, and beyond [see Letter #50, (n6)], beckoned Darwin as well. This, after all, was the natural profession to pursue, and there must have been several people, who, like Robert Darwin and Hensleigh Wedgwood (Wedgwood and Wedgwood, 1980, p 214), shook their heads as they heard of the venture, upon which Darwin was about to engage.

Chapter 4
Darwin's Voyage on the Beagle

Letters #44-63 (20); covering 1831 to 1837

"So much for the grand end of my voyage; in other respects things are equally flourishing, my life when at sea, is so quiet, that to a person who can employ himself, nothing can be pleasanter. — the beauty of the sky & brilliancy of the ocean together make a picture. — But when on shore, & wandering in the sublime forests, surrounded by views more gorgeous than even Claude ever imagined, I enjoy a delight which none but those who have experienced it can understand."

C Darwin to W D Fox, May 1832, Letter #47

The *"Voyage of the Beagle"* is now as iconic an event in the life of Charles Darwin, as it is in the history of science. However, this is only with the hindsight of history and one should not forget the fact that there was a previous voyage and many subsequent voyages of this surveying ship of the British Navy, right up till the time that HMS *Beagle* was sold off for scrap in a Chatham yard in the 1860s. Nevertheless, the offer to Charles Darwin to act as a companion to Captain Robert Fitzroy on a voyage of HMS Beagle, his final acceptance and the voyage itself is now so famous that the original account (as Volume 3 of the Narrative of the surveying voyages of his Majesty's ships Adventure and *Beagle*, and its frequent retelling fills whole library shelves. For our story the letter of 6th Sept 1831 (#44) begins this new phase of both Darwin's and William Darwin Fox's life very succinctly: Darwin has accepted the *Beagle* offer and will sail in October 1831. What will become a theme of the continuing correspondence starts here: Darwin is too busy to contemplate meeting Fox but desires the correspondence to continue. In fact the *Beagle* did not sail until late December and Darwin had unexpected periods of waiting when he could have arranged to meet with his second cousin, William Fox. However, as with a number of personal connections at this time Darwin is too caught up in events and too anxious about the voyage ahead to allow time for old friends and relatives. But from this time on we have a daily record of the voyage, by Darwin himself with his movements and impressions.

Unfortunately, the diaries of Fox are missing for this critical period and the first complete diary is for 1833; for the missing years we have only a few diary scraps (Appendix 1). On the other hand, we do have the first letters of Fox to Darwin. Here, at last, we are able to observe and compare their individual styles. The letters of Darwin that we have come to know are readable, even racy, full of 'bon mots',

A. W. D. Larkum, *A Natural Calling*, DOI 10.1007/978-1-4020-9233-6_4,

and above all are short enough to carry the interest. By contrast, Fox's letter writing style turns out to be very "Victorian": long, verbose, full of unnecessary general detail and lacking in the personal touch. However, now we can see the relationship between these two men more clearly.

Both men are embarking on their respective careers, although for Darwin the possibility of a career in natural history probably came gradually, and the early letters of his voyage reflect a nostalgic and romantic ideal of the country parson surrounded by his family around a warm hearth: *I hope my wanderings will not unfit me for a quiet life, & that in some future day, I may be fortunate enough to be qualified to become, like you a country Clergyman* (see Letter #50). And in truth his sense of isolation and separation must have been underscored by the fact that two of his romantic attractions had married: Fanny Owen to Robert Biddulph and Charlotte Wedgwood to Rev Charles Langton. And, as if to further emphasise the vicissitudes of life, the attractive Fanny Wedgwood died suddenly in 1832, probably of cholera. As his sister Caroline wrote, it was probable that all the Wedgwood sisters could by married by the end of the year.

Darwin was not a good sailor and despite the fact that Commander Robert FitzRoy took every opportunity to allow him shore leave, he found the sea voyages very tiring and the length of the trip almost intolerable. On the other hand he must have compared his circumstances against Fox as the months and years went by. Fox had obtained a curacy in 1831, at Epperstone, just outside Nottingham. However, sickness, in the form of a lung infection, forced him to withdraw in 1832, first to his father's home at Osmaston, and then to the Isle of Wight. Fox's diary exists for 1833, so his life is well documented for this period. It was in April 1833 that he met his wife-to-be, Harriet Fletcher. She was one of five daughters of Sir Richard Fletcher who had died heroically during the Peninsula war, where he was an important officer under the command of General Wellesley (later the Duke of Wellington). After his death, a small pension and a baronetcy enabled the family to live comfortably on the Isle of Wight. By 1834 Fox was married, but still recuperating from his lung infection. One way in which the newly-married couple did this was to take a pony and trap and visit various parts of England, first in the North (Letter #54) and then in the West and in Wales. As engaging as this "beau ideal" might seem, Darwin's sisters quickly brought him down to earth. Early on in the marriage his sisters reported that Harriet had miscarried near to term. And then later, Darwin's sister Catherine wrote to him, "*I dare say you will like his Lady better than we did; for one reason, she was in a bad state of health when we saw her, which I dare say accounted for much of her crossness*" (CCD, Vol 1 29th Jan 1836). So for Darwin suffering at sea, the rosy glow of the hearth at home must have been misted over to some degree. And for him there was the added excitement of new ideas.

Just what was in the air is difficult to discern from the distance of time. Darwin left England in 1831, a largely rural nation, totally dependant for transport upon the horse. When he returned, in late 1836, he would take most of his long journeys by rail. Politically, too, there was a sea change going on. While he was away the Reform Bill was passed, giving a wider franchise, but more significantly tacitly acknowledging that the old order must change. "*By the time of your return*" Fox wrote, in his long-winded fashion, *"we shall be better judges of the happy effects of our Reform Bill, at least if it is allowed to have its natural course in the correction of the abuses of Church and state. — Party Spirit runs now very high indeed — The Tories are merely (to use their own words) endeavouring to prevent the vessel from altogether sinking, & the Whigs & Radicals all alive. — For some days we certainly were on the very verge of Revolution. The excitement in the Country was quite extraordinary among the lowest orders. All still but evidently all prepared for anything that might turn up — We have however now I trust safely passed this grand Corner, upon which so much hung & shall proceed steadily & prosperously, though much remains to be done that is formidable to look into."* (Letter #48). The cities were growing and factories were being established as the industrial revolution got into full swing. And markets overseas, such as America and India, supplied these factories with burgeoning outlets. This was the soil that nurtured new ideas: new ideas such as evolution, as we now call it. It is no coincidence that the Vestiges of Creation published anonymously a little later, in 1844, would be a Victorian best-seller. Already at Edinburgh, Darwin had been exposed to ideas of transmutation by Robert Grant, who would move during this time to the new University College in London. Yet something was missing, a new approach was needed. Darwin on the periphery of the known world and seeing things afresh may have caught the mood and then embodied it. The second volume of Charles Lyell's *Principles of Geology* was delivered to Darwin when he arrived in Rio de Janeiro. The early adventures on the East coast of South America and the testing time of the sea voyage around Cape Horn may have restricted his access to this book. However, in the calmer waters of the Pacific and during the extended stay on the West coast of South America, Darwin had time to read in depth. And what he read was all about transmutation of species. Lyell wrote this second volume like the devil's advocate. He virtually took up the cause of transmutation and gave it flesh and blood. However, in Chapter 11, where he discussed the evidence for new species he seems to have lost his nerve and concluded that evidence for new species would be very difficult to obtain. Coincidentally this second volume also put forward Lyell's view on how coral reefs were formed, which Darwin must have mulled over until the *Beagle* reached the Keeling Islands (now the Cocos Keeling Islands) and Darwin obtained evidence to overturn Lyell's, and everybody else's, ideas. Thus, Lyell had thrown up two challenges that Darwin thought he might be able to tackle.

Darwin must now more than ever have looked anew at the world around him. Evidence for change was all too abundant on a geological scale. "*An old piece of ambition of mine has been gratified, viz finding the remains of large extinct animals I think some of them are new; I have teeth & fragments of about 7 kinds.*" Darwin wrote to Fox (Letter #50). And that was before the visit to the Galapagos where "*one might fancy that from an original paucity of birds in this archipelago, one species had been taken and modified for different ends*" as Darwin later wrote (Journal of researches, 1839). Then, from his experiences in Australia, he was led to observe, "*surely two distinct Creators must have been [at] work; their object however has been the same & certainly the end in each case is complete*" (Journal of researches, 1839).

The excitement, which this ferment of ideas would have produced, was enough to keep Darwin on the voyage and prevent him from giving in to the attractions of home. "*This same returning to dear old England is a glorious prospect; I wish it was rather nearer; but it is sufficient to make up for a thousand vexations*." Darwin wrote to Fox from Valparaiso, in 1835, "*Five years is a sadly too long period to leave ones relations & friends; all common ideas must be lost & one returns a stranger, where one least expects or wishes to be so*." Darwin then significantly went on to remind himself (and us) of the importance of Fox to his present travels, "*I think the recollections of the snug breakfasts & pleasant rambles at Cambridge, will make us remember each other. You are one of the indirect causes of my coming on this voyage: by taking me as your dog in the grand chace of Crux Major you made me an Entomologist & introduced me to Henslow. I am very glad I have come on this expedition, but like a Sailor I have learnt to growl at all the details*" (Letter #55).

From the point of view of the correspondence between Darwin and Fox, we may observe that the period during the voyage of the *Beagle* is less significant than other periods because we have many other letters from this time, both to and from Darwin. There are only 6 letters from Darwin to Fox, while there are many others from Darwin to other people. Moreover, the *Beagle* voyage is documented in great detail, both in the ship's log and in Darwin's *Beagle* diaries. On the other hand, we have for the first time extant letters from Fox to Darwin. These are interesting because they give us a flavour for the first time of Fox as a correspondent; and they show Fox's letters as much less readable than Darwin's. Fox's later letters prove it to be true in the future too. However, the significance for the correspondence of these two men is that it did survive this period, when it was threatened by their busy lives and competing interests - and survived to fulfil a much more important role in revealing the later lives of these two men.

#44 (CCD 120; LLi, part[1]). To: The Rev[d]. W. D. Fox, Epperstone, Nottinghamshire.

17 Spring Gardens London
Tuesday 6th. — [Sept 1831]

My dear Fox

When you read this you will understand why I have not answered your letter earlier. — I returned from a geological trip with Prof: Sedgwick[2] in N Wales on Monday 29th of August. — I found your letter there, & with a joint one from Henslow[3] & Peacock[4] of Trinity, offering me the place of Naturalist in a vessel fitted for going round the world. — This I at first, (owing to my Father not liking it I refused) but my Uncle, M[r]. Wedgwood[5], took every thing in such a different point of view, that we returned to Shrewsbury on 1st of September, & convinced my Father. — On 2[d] started for Cam: then again from a discouraging letter from my captain I again gave it up. But yesterday every thing was smoother. — & I think it most probable//

I shall go: but it is not certain, so do not mention it to any body. — I have had a most tremendous hard week of it. — Cap Fitzroy[6] seems every thing I could wish: the most serious objections are the time (3 years), & the smallness of the vessel. Fitzroy is determined to get over the latter for me. — Our route is Madeira, Canary, Rio de Janeiro, 18 month in S America chiefly S. extremity, we shall see every principle city in it. — then through South sea islands, Australia, India, Home. — Of course nothing is quite certain. we sail on 10th October. — of course I will write to you again. — Your letter gave me great pleasure. — You cannot imagine how much your former letter annoyed & hurt me[7]. — But thank heaven, I firmly believe//

that it was *my own entire* fault in so interpreting your letter. — I lost a friend the other day, & I doubt whether the moral death, (as I then wickedly supposed) of our friendship did not grieve me as much as the real & sudden death of poor Ramsay[8]. — We have known each other too long to need I trust any more explanations. — But I will mention just one thing; that on my death bed, I think I could say I never uttered one insincere (which at the time I did not fully feel) expression about my regard for you — One thing more. — The <sending *immediately* the> insects on my honor was an unfortunate coincidence, I forgot how you naturally would take them. — When you look at them now, I hope no unkindly feelings will rise in your mind. —//

& that you will believe that you have always had in me a sincere & I will add, <an> obliged friend. The very many pleasant minutes that we spent together in Cambridge rose like departed spirits in judgement against me: May we have many more such, will be one of my last wishes in leaving England. — God bless you dear old Fox

May you always be happy. —

Yours truly

Chas Darwin

I have left your letter behind so do not know whether I direct right

Postmark: C.H 6 SE 1831 X

Christ's College Library, Cambridge, #43.

Notes

1. In *Life and Letters* (F. Darwin, 1888), Vol. 1, pp. 205: from "Your letter gave me great pleasure.—" to end.
2. Adam Sedgwick (1785-1873). Geologist and clergyman. Woodwardian Professor of Geology at Cambridge University (1818-1873). He promoted the early geological interests of CD but was a strong opponent of "*On the Origin of Species*". (*DNB*)
3. John Stevens Henslow (1796-1861). See Bibliographic Register, Introductions to Chapters 2 & 3, Letters #8 (n3) and many subsequent letters.
4. George Peacock (1791-1858). Mathematician. Tutor, Trinity College (1823-1839); Lowndean Professor of Geometry and Astronomy, University of Cambridge (1837); Dean of Ely (1839-1858). (*DNB*)
5. Josiah Wedgwood II (1769-1743).
6. Robert FitzRoy (1805-1865). Commander of the HMS Beagle (1828-1836). Hydrographer and meteorologist. See Bibliography. (*DNB*)
7. See Letter #42 (2).
8. Marmaduke Ramsay (1831). Jesus College, Cambridge; B A, 1818; Fellow and Tutor, 1819 -31; died suddenly at Perth, Scotland, at this time (*Alum Cantab*). See Letter #42 (n11).

#45 (CCD 132; LLi, most[1]). To: The Rev^{d}. W. D. Fox, Epperstone, Nottinghamshire.

Monday 19th [Sep 1831]
17 Spring Gardens
(& here I shall remain till I start)

My dear Fox

I returned from my expedition to see the Beagle at Plymouth on Saturday & found your most welcome letter on my table. — It is quite ridiculous what a very long period these last 20 days have appeared to me, certainly much more than as many weeks on ordinary occasions. — this will account for my not recollecting how much I told you of my plans, therefore I will begin a novo. — The expedition, under the command of Cap FitzRoy is fitted out principally for completing a survey of the S. parts of S America: The western shores of these parts have been well done by Cap King[2], under whom Fitzroy went out second in command[3]. We accordingly shall principally work on the Eastern coast of Patagonia from Rio de Plata to St^{s} of Magellan. — The second object is to ascertain the longitudes of several places, more accurately than they are at present, & to carry a series of them round the world. — The expedition is entirely a government affair.//

My appointment is not a very regular affair, as the only thing the Admiralty have done is putting me on the books for Victuals, value 40£ per annum. — I have some thoughts of having it taken off again. I should certainly do so, if I thought it would give me a more absolute disposal of my collection, when I return to England[4]. — But on the whole it is a grand & fortunate opportunity; there will be so many things to interest me. — fine scenery & an endless occupation & amusement in the different branches of Nat: History: then again navigation & metereology will amuse me on the voyage, joined to the grand requisite of there being a pleasant set of officers, & as far as I can judge this is certain. — On the other hand there is very considerable risk to ones life & health, & the leaving for so very long a time so many people whom I dearly love, is oftentimes a feeling so painful, that it requires all my resolution to overcome it — But every thing is now settled &//

before the 20th of Octr I trust to be on the broad sea. — My objection to the vessel is its smallness, which cramps one so for room for packing my own body & all my cases &c &c. — As to its safety I hope the Admiralty are the best judges; to a landsmans eye she looks very small. — She is a 10 gun 3 masted brig. — but I believe an excellent vessel. — So much for my future plans, & now for my present.— I go to~~morrow~~<night> by the mail to Cambridge, & from thence <after settling my affairs> proceed to Shrewsbury <(most likely on Friday 23^{d} or perhaps before)>: there I shall stay a few days & be in London by the 1st of October, & start for Plymouth on the 9th. — And now for the principal part of my letter. — I do not know how to tell you how very kind I feel your offer of coming to see me before I leave England. — Indeed I should like it very much; but I must tell you decidedly that I shall have very little time to spare[5], & that little time will be almost spoilt by my having so much to think about: & 2nd I can hardly think it worth your while to leave your Parish for such a cause. — But I shall never forget such generous kindness. — Now I know you will act, just as you think right, but do not come <up> for my sake. Any time is the same for me. — I think from this letter you will know as much of my plans as I do myself, & will judge accordingly the where & when to write to me. —//

Every now & then I have moments of glorious enthusiasm, when I think of the date & cocoa trees, the palms & ferns so lofty & beautiful — every thing new everything sublime. And if I ~~life~~ live to see years in after life how grand must such recollections be. — Do you know Humboldt[6]? (if you dont, do so directly) with what intense pleasure he appears always to look back on the days spent in the tropical countries: I hope, when you next write to Osmaston[7], tell them my scheme, & give them my kindest regards & farewells.—

Good bye my dear Fox.
Yours ever sincerely
Chas Darwin

Postmark: C.H 19 SE 1831 X
Christ's College Library, Cambridge, #44.

Notes

1. In *Life and Letters* (F. Darwin, 1888), Vol. 1, pp. 211-3: with the exception of the part from "therefore I will begin a novo.—" to "disposal of my collection, when I return to England.—".
2. Philip Parker King (1793-1856). Born on Norfolk Island, Australia. Naval officer, hydrographer and landowner in Sydney. Commander of HMS *Adventure* and HMS *Beagle* on the first surveying trip to South America, 1826-1830. (*DNB*)
3. See CCD, letter from George Peacock [*c*. 26 August 1831] n 1, p 130. In fact, FitzRoy had been with Rear Admiral Sir Robert Otway on HMS *Ganges* and in 1831 sailed the *Beagle* as its new commander, from Rio de Janeiro to England, after the suicide of its captain, Pringle Stokes.
4. CD apparently decided not to remove himself from the ship's books. CCD, Vol 1, discusses the details concerning this (p 164). FitzRoy later stated (*Narrative* 2: 19) that "*an offer was made to Mr. Darwin to be my guest on board, which he accepted conditionally; permission was obtained for his embarkation, and an order given by the Admiralty that he should be borne on the ship's books for provisions. The conditions asked by Mr Darwin were, that he should be at liberty to leave the Beagle and retire from the Expedition when he thought proper, and that he should pay a fair share of the expenses of my table*." In fact, disputes did arise later as to the disposal of some of the *Beagle* specimens: Darwin was allowed to dispose of his personal specimens freely. When Darwin wanted to inspect the Galapagos finches, collected by other members of the crew, access was impeded. Also the wording of the Preface to Darwin's Volume of the *Narrative* (Volume 3, CD's contribution - which later became popularly known as "The Voyage of the Beagle") - led to some bitter words between CD and FitzRoy about CD's position on the Beagle and his obligations to the ship's crew (CCD; FitzRoy to CD: 15th Nov 1837 & 16th Nov 1837, p 57-58).
5. CD plays cat and mouse with Fox, as he does for the rest of their lives. And to some extent this tactic is reciprocated by Fox so the two men meet very infrequently, hereafter (Appendix 2).
6. Friedrich von Humboldt (1769-1859). Naturalist and traveller. See Introduction to this chapter and Letter #40 (n7). (*DSB, DNB*)
7. Osmaston Hall, the family home of Samuel Fox II, Fox's father (see Fig 4).

#46 (CCD 149). To: The Rev^{d}. W. D. Fox, Epperstone, Nottinghamshire.

4 Clarence Baths
Devonport 17th [Nov 1831]

My dear Fox

I daresay you will be surprised to see my hand-writing, you I suppose thought that I was far on the high seas. — We have had delay after delay & now we do not sail till the 5th of next month. — I always had intended writing to you before making my final exit. — But as that period is so far distant. — I made up my mind to do it earlier. — I do not think I can do better than give you an outline of the Instructions, which came down yesterday. — They leave a great deal to the Captains judgement, & I am sure they could not leave it to a better one. — We first go to Canary//

or Teneriffe, then to Cape Verde, Fernando Norunha[1], Rio de Janeiro, staying about a week in each of these places & rather longer at Rio. — I am very glad that, for I hear there is no view in the world at all equal to it. — Then Monte Video, which will be our headquarters for some time, or rather I should say, the point to which we often return for fresh provisions &c. — From Monte Video, we begin our regular work & go down the coast of Patagonia: parts of Terra del Fuego. — Falklands Islands: — After this done, we have some work about island of Chiloe, & then we are to proceed as far Northward as Captain likes (I daresay to California) so as to leave time to make a good traverse amongst Islands to New S Wales, after which take short cut//

amongst E Indian Islands, Cape of Good Hope, so home. — Everybody, who can judge, says it is one of the grandest voyages that has almost ever been sent out. — Everything is on a grand scale. — 24 Chronometers. The whole ship is fitted up with Mahogany, she is the admiration of the whole place. — In short everything is as prosperous as human means can make it — Time, which no one can alter, is the only serious inconvenience. — Why, I shall be an old man, by the time I return, far too old to look out for a little wife. What a number of changes will have happened; I suppose you will be married & have at least six small children. — I shall very much enjoy seeing you attempting to nurse all six at once. — & I shall sit by the fire & tell such wondrous tales, as no man will believe. — When I think//

of all that I am going to see & undergo, it really requires an effort of reasoning to persuade myself, that all is true. — That I shall see the same land that Captain Cook <did>. I almost doubt the truth of the old truism, that man may do, what <man> has done. — when I think that I, an unfortunate landsman, am going to undertake such a voyage, — I long for the time, when sea sickness will drown all such feelings & that time I do suppose will be the 5th of next month. — Will you write to me//

(written cross-wise at the top of the 1st page)

& tell me, how you are getting on. — I heard of you at the Music Meeting[2] & that you sat by the incomparable Charlotte[3]. Report progress. I have hooked her into a correspondence[4].

Good bye
Yours affectionately
Chas Darwin

Postmark: Dev*onpo*rt NO 1<7> 1831
Christ's College Library, Cambridge, #45.

Notes

1. Ilha Fernando de Noronha. An archipelago in the Atlantic Ocean, about 350 km off the coast of Brazil.
2. Presumably the Music Meeting at Birmingham, which CD attended previously in 1829 [see Letter #24 (n2)], although music meetings were also held in Derby [see Letter #5 (n6) and Appendix 4].
3. Charlotte Wedgwood (1797-1862). Married Charles Langton in 1832. See Letters #5 (n4) #32 (n)2, #33 (n3).
4. See CCD, Vol. 1, letter from Charlotte Wedgwood to CD, 12th Jan – 1 Feb. 1832.

#47 (CCD 168; LLi part). To: The Revd W D Fox, Epperstone, Nottingham; re-addressed to Wyndhams House, Ryde, Isle of Wight.

Botofogo Bay, near Rio de Janeiro
May 1832. —

My dear Fox

I have delayed writing to you & all my other friends till I arrived here & had some little spare time. — My mind has been since leaving England in a perfect *hurricane* of delight & astonishment. And to this hour scarcely a minute has passed in idleness. — I will give you a very short outline of our voyage. We sailed from England after much difficulty on the 27th of December & arriv'd after a short passage to St Jago. — I suffered exceedingly all the first part, the snowy peak of Teneriffe by convincing me I was well on the road to see the world first put fresh life into me. — At St Jago my Natural Hist: & most delightful labours commenced. — during the 3 weeks I collected a host of marine animals, & enjoyed many a good geological walk. — Touching at some islands we sailed to Bahia, & from thence to Rio, where I have already been some weeks[1]. — My collections go on admirably in almost every branch. as for insects I trust I shall send an host of undescribed species to England. — I believe they have no small ones in the collections, & here this morning I have taken minute Hydropori, Noterus Colymbetes, Hydrophilus, Hydrobius, Gyrinus, Heterocerus Parnus, Helophorus Hygrotius, Hyphidrus, Berosus &c &c, as a specimen of fresh-water beetles[2]. — I am entirely occupied with land animals, as the beach is only sand; Spiders & the adjoining tribes have perhaps given me//

from their novelty the most pleasure. — I think I have already taken several new genera[3]. — But Geology carries the day; it is like the pleasure of gambling, speculating on first arriving what the rocks may be; I often mentally cry out 3 to one Tertiary against primitive; but the latter have hitherto won all the bets. — So much for the grand end of my voyage; in other respects things are equally flourishing, my life when at sea, is so quiet, that to a person who can employ himself, nothing can be pleasanter. — the beauty of the sky & brilliancy of the ocean together make a picture. — But when on shore, & wandering in the sublime forests, surrounded by views more gorgeous than even Claude[4] ever imagined, I enjoy a delight which none but those who have experienced it can understand — If it is to be done, it must be by studying Humboldt. — At our antient snug breakfasts at Cambridge, I little thought that the wide Atlantic would ever separate us; but it a rare priviledge, that with the body, the feelings & memory are not divided. — On the contrary the pleasantest scenes in my life, many of which have been in Cambridge, rise from the contrast of the present the more vividly in my imagination. — Do you think any diamond beetle will ever give me//

so much pleasure as our old friend Crux Major[5]. — Can we ever forget our few days at Whittlesea Meer[6] with little Albert[7]? It is one of my most constant amusements to draw pictures of the past, & in them I often see you & poor little Fan[8] — Oh Lord, & then old Dash[8] poor thing! — do you recollect how you all tormented me about his beautiful tail. — I am now living here by myself, as the Beagle has returned to Bahia to settle a longitude question, & about the middle of next month we shall sail for Monte Video. — I rely upon your writing to me there (it will be *our* head quarters for a terrible long time) direct M*** HMS. B*eagle* Monte Video, S. America: do as I have done, & tell me all about yourself how contrive to live on such a stationary, slow sailing craft as a Parsonage: what you are, have, & intend doing. — Remember minutiae become more not less interesting, as the distance increases. —

I suppose I shall remain through the whole voyage, but it is a sorrowful long fraction of ones life; especially as the greatest part of the pleasure is in anticipation. — I must however except that resulting from Natur. History; think when you are picking insects off a hawthorn hedge on a fine May day (wretchedly cold I have no doubt) think of me collecting amongst pineapples &//

orange trees; whilst staining your fingers with dirty blackberries, think & be envious of ripe oranges. — This is a proper piece of Bravado, for I would walk through many <a> mile of sleet, snow or rain to shake you by the hand. My dear old Fox. God Bless you. Believe me

Yours very affectionately
Chas Darwin

Remember me most kindly to M^{r} & Mrs Fox & to all your family: Once more good night & good bye.

Postmark: J AU 11 1832
Christ's College Library, Cambridge, #46.

Notes

1. Botofogo Bay, now heavily polluted, lies near the centre of modern metropolitan Rio de Janeiro.
2. These early forays after beetles and insects became less conspicuous but still continued.
3. After the voyage, CD's insect specimens were described by Charles Cardale Babington, Frederick William Hope, Francis Walker, George Robert Waterhouse, and Adam White. See *Collected papers* 2: 295-300.
4. Claude Lorrain (1600-1682). 15th century French landscape painter popular in Italy. His works are held in many national galleries.
5. *Panageus (Panagaeus) crux-major*, a carabid beetle, which Wollaston reported in "immense numbers' at Wicken Fen, Cambridgeshire, in 1842 (Wollaston, M, ed, 1933. *Letters and Diaries of A F R Wollaston*. Cambridge, University Press). See also Letter #55. Now on the endangered species list. CD, Fox and Albert Way had clearly collected it at Whittlesea Meer (see note 6, below) and

it is a recurrent theme in letters right into the 1870s (see, e g Letter #195). CD mentioned it in his *Autobiography*: "*Panagæus crux-major was a treasure in those days*".

6. Whittlesea Meer or Whittlesey Meer lay some 14 km east of Peterborough in the Isle of Ely, and was, until drained by Dutch engineers in the 18th century, one of the largest bodies of water in the region [see also Letter 48 (n17)]. The excursion to Whittlesea Meer with CD, Fox and Albert Way is recalled many times in subsequent letters (see Letters #47 (n4 & 5), #48 (n16), #62 (n7), #90 (n17), #96 (n10), #71 (n8) & #115 (n6) and in the *Autobiography* (Darwin F, 1887 and Darwin, F, Life and Letters, 1888, Vol 1, p 51).
7. Albert Way (1805-1874). Antiquary and traveler [See Letters #2 (n2), #13 (n14), #19 (n11), #48 (n18), #62 (n15) & #195 (n8)]. See Fig 15.
8. Dogs of Fox & CD. See Letters #1 (n12) for "Fan", #1 (n12), #9 (n3) for "Sappho" and #9 (n3) for "Dash".

#48 (CCD 175). To: Charles Darwin Esqe, on board his Brittanic Majestys Ship Beagle, Rio Janeiro

Epperstone near Nottingham.
June 30. 1832

My dear Darwin

It is now I believe a month since your Sister[1] was so kind as to send me word that you were at length heard of, and where I could write to you. — I commenced a letter at the time but was prevented finishing it, and ever since I have purposed writing from day to day, and as constantly put it off, sometimes owing to illness, sometimes idleness, & frequently from feeling that I had nothing in the world to tell you of that you will not hear from Shrewsbury; and I am now once more commencing (with a determination to finish it) merely that I may put you in mind of my existence & prevent your totally forgetting me in the midst of the wonders of Creation you are now surrounded by, & will behold previous to your return to England. — I can scarcely realize the idea sometimes of your being at such a distance, and revelling in the midst of scenes I have always longed intensely to see, and hope to have a sort of idea of sometime secondhand from your description. — I had often wondered where you were and how going on, and was very anxious to hear of you, when your sisters letter gave me the welcome news. From what she says of you, you seem most happily situated in every respect; your health, ship & Companions all remaining as perfect as you hoped they would prove previous to your departure. From all I hear of South America the Climate is very little to be feared with proper precautions. I cannot help a little fearing that the ardour (which I remember your shewing in Chase of Machaons[2] in Bottisham Fen[3]) may, to compare great things with small, lead you into difficulties, & into disregard of dangers of various kinds when in pursuit of Nat: History where all is new & all glorious to the last degree. — I often long so to be with you & join in//

your happiness, and think over the difference of our lots & the ridiculousness of my pursuits in Nat Hist: compared with yours. In consequence of a severe Inflammation of the Lungs[4] I had early in April, I have been more or less Invalided ever since, and have amused myself in santering about the Fields on horseback studying the Small <summer> Birds of Passage, their nidification[5] &c, and when thus employed the thoughts of you and your occupation most forcibly & frequently struck me. I pottering in a Hedge Rows to watch the proceedings of a Whitethroat & you surrounded by the Noble Trees of a S. American Forest with every luxury of vegitation & life ~~surrounding~~ around you. — You must have much regretted your not seeing the Madeira & Canary Islands, tho' perhaps the time thus saved will be abundantly recompenced hereafter and as they are pretty well explored, at least the former certainly, the harvest will be richer gathering where you now are & will be. — The extreme novelty of every thing around you, must now be rather wearing off, and you are becoming more used to the Intoxication of feelings, the Country you are now in, must produce. — I have often regretted one trait of your Character which will I fear prevent your making so great an advantage as you might do from your present travels, and which I regret also very much on my own account, as I might perhaps get the perusal of it; — I allude to your great dislike to writing & keeping a daily methodical account of passing events, which I fear (tho' I have also hopes the other way from the overwhelming influence of every surrounding object) will prevent you from keeping a Regular Journal[6]. — If you do not do this, the ~~******~~ vast crowd of Novelty which will surround you, will so jostle about ideas, that to say nothing of the many that will be lost altogether, the vivid reality & life which a memorandum <taken> at the moment gives to every passing event & thing, is done away with. — With this one exception (which I dare say you have overcome[7]) I know of no one so fitted altogether for the expedition you are engaged in. We have had many extraordinary//

changes in England since you went, even in this short six months, what may occur before your return therefore in three years? — You have of course heard of the Incomparable Charlotte Wedgwood[8] changing her name. From what I hear from all that know her husband or rather have seen him, her choice seems a very happy one, indeed she is not one that would readily be taken in. You would I think much regret to hear of Sir J. Mackintoshs[9] death, as I have often heard you speak of him as one you much esteemed. Mrs Darwin[10] of the Priorys Death will not much affect you. By the time of your return we shall be better judges of the happy effects of our Reform Bill[11], at least if it is allowed to have its natural course in the correction of the abuses of Church and state. — Party Spirit runs now very high indeed — The Tories are merely (to use their own words) endeavouring to prevent the vessel from altogether sinking, & the Whigs & Radicals all alive. — For some days we certainly were on the very verge of Revolution. The excitement in the Country was quite extraordinary among the lowest orders. All still but evidently all prepared for anything that might turn up — We have however now I trust safely passed this grand Corner, upon which so much hung & shall proceed steadily & prosperously, though much remains to be done that is formidable to look into. — The Cholera[12] is now spread all over England, and tho' not the very dreadful scourge we were led to expect it to be, is very awful in many places. — It began I fear at Derby last week, & is now in ~~all~~ <many> of the large Towns slowly making progress in England Scotland & Ireland. — I never remember such a season as the present. Every kind of crop promises to be most abundant. It has been an extraordinary year for Insects, but I have not been able to go in search of any[13]. I have not seen any of your Family since you went, but hear very flattering accounts of all. — Your

Father is uncommonly well. — All at Osmaston[14] I am happy to say are much as usual. — My Father has been poorly but is much better. — Of Cambridge Friends I have not heard for some time. — I hope however very shortly to hear of Henslow[15]. I ought to have taken my Masters degree as next week, & should have rather gloried in having my vote at the Commencement of a Reformed House of Commons, but I have been obliged to forego it. — Pulleine[16] is to spend some days with me next week — Do you remember our excursion to Moncks Wood[17] <& Whittlesea[18]> at this time of year with Albertus Way[19] <At Whittlesea[20] the Cholera has killed 48 & there are 130 new cases last report>. — I have never heard of him, whether he is still at Leamington or not. Did you ever see Old Mr Galton of Dudston[21]. — He is just dead after a lingering illness. — I must now give you a few lines about my own dear self. — I have as I told you before, been unwell which has inca=pacitated me from taking any duty for the last 3 months and I am only returned to Epperstone for a short time[22], as I fear I can be of no use at present//

I am now very much ~~st****ed~~ stronger than I have been, in fact comparative well, but as is always the case with Chest Complaints[23], vary very much in health & spirits. I did at one time think I should never meet you again in this world, but trust now to do so & see you in full vigour after your wanderings are over. — I often look forward to the time of your return with great delight, and regret I did not see you before your departure. I had no idea that you had stayed so long in England — You scarcely left us in time to say it was 1831. — I hope you will not be disgusted at my very stupid letter. — You who abound in novelty must not censure we plain housekeepers for having nothing to communicate. I do not ask you to write to me as you must have plenty to occupy your time, and I shall hear of you from Shrewsbury when you write there, as a few lines from them will give the information I want as to your welfare. — And now my Dear Darwin with every wish for your welfare and success in all your undertakings & that I may again see you in health & happiness in Old England which after all is the prettiest & best Island in the world

Believe me your attached friend
William D. Fox. —

Postmark: PAI* 3 JY 1832
Provenance: University Library, Cambridge; Manuscript Room, DAR 204.6.2

Notes

1. Caroline, Susan and Catherine Darwin were all corresponding with CD at this stage so it is not possible to say which sister wrote to Fox.
2. Machaon: *Papilio machaon*, the Swallowtail butterfly; See Letter #3 (n4 & 10).
3. Bottisham is a village 10 km north-east of Cambridge, where George Jenyns, Leonard Jenyn's father, lived, at Bottisham Hall.
4. This is the first indication of the cause of Fox's recent ill health; Fox's Diary for 1824-1826 (Chapter 2 and Appendix 4) do not suggest that he had chest problems apart from coughs and colds, but there may have been a family weakness: severe illnesses of his father and mother are recorded but their nature is not.
5. Nidification was a common term at this time to describe nesting in birds.

6. See Introduction to Chapter 3 for a discussion of Fox's influence on Darwin. The letters during the Cambridge period show how CD ridiculed Fox's detailed account taking; however, he may have secretly been impressed and determine to do better.
7. The current scholarly industry in the interpretation of CD's notebooks shows that CD did indeed overcome his earlier deficiencies in this regard.
8. Charlotte Wedgwood married Charles Langton on 22nd March 1832. Since CD greatly admired Charlotte [see Letters #5 (n4), #32 (n2) and #33 (n3)] this must have been a further disappointment in addition to learning that Fanny Owen had been married (on 31st May 1832: see letter of Caroline Darwin to CD, 12 -28 June 1832, *CCD Vol 1*). Furthermore, Susan Darwin wrote to CD, "*I expect Fanny and Emma to follow this mad m*arrying* example before this year is over.*" (*CCD, Vol. 1*, 12 Feb. [- 3 Mar.] 1832).
9. James Macintosh (1765-1832). Philosopher, historian and member of the Whig government in the late 1820s. He married Catherine Allen, whose sister, Elizabeth, married Josiah Wedgwood, II. They were frequent visitors at the Wedgwoods, both at Maer Hall and in London. (*DNB*)
10. Elizabeth Darwin (née Collier)(1747-1832), second wife of Erasmus Darwin. Lived at Breadsall Priory, just outside Derby (see Letter #6 (n7); #7 (n9); #21 (n12)
11. The Reform Act of 1832 was passed finally by the House of Lords on the 7th June 1832. The manoeuverings from the time that the House of Commons passed the Act in late 1831 had brought the country to the brink of revolution (as testified by Fox's remarks). The passage was acclaimed with much rejoicing by a majority and especially by Whigs, which included CD and Fox.
12. Cholera first came to England in epidemic proportions in 1831-32. See Morris R. J. (1976) *Cholera 1832 -The Social Response to An Epidemic.* Holmes & Meier Publishers, New York,.
13. This is the ultimate admission, that without CD's influence, Fox has given up entomology. On the other hand he carries on with his own interests, which later are very helpful to CD.
14. Osmaston Hall: the country mansion rented by Fox's father, Samuel Fox III; See Chapter 1 and Fig 4
15. J S Henslow; see Autobiography and Letters #8 (n3), #25 (n8), #29 (n12), #30 (n11), #35 (n2), #36 (n1), #39 (n4), #40 (n3), #41 (n8), 43 (n5) & 44 (n3). It is clear from this and other indications that Fox kept up communications with Henslow independently of CD.
16. Robert Pulleine (1806-1868). See Letters #9 (n6), #10 (n15), #11 (n4), #13 (n5), #16 (n12), #17 (n1), #19 (n) and #31 (n6),.
17. Monks Wood, 3 km west of Cambridge.
18. Whittlesea Meer or Whittlesey Meer lay some 14 km east of Peterborough in the Isle of Ely, and was, until drained by Dutch engineers in the 18th century, one of the largest bodies of water in the region. Fox was responding to CD's mention of it previously: see Letter 47 (n5 & 6); and see also Letter #63.
19. Albert Way (1805-1874). Accompanied CD on trips to Barmouth. He is also mentioned in the Autobiography: "*Afterwards I became well acquainted, and went out collecting, with Albert Way of Trinity, who in after years became a well-known archaeologist*". See Letters #2 (2), #13 (n14), #19 (n11), # 47 (n6 & 7) and Fig 15.
20. Whittlesey, near Peterborough.
21. Samuel John Galton FRS (1753-1832) of Duddeston House. Father of Samuel Tertius Galston and grandfather of Francis Galton.
22. Probably Fox has been at Osmaston Hall recuperating; see letter by Caroline Darwin to CD, 12 June 1832, CCD Vol. 1, p. 243 and from Susan Darwin to CD, 155 Aug 1832, CCD Vol. 1, pp. 256 & 257. Later, as shown by Letter 49, he recuperated on the Isle of Wight.
23. This is the first indication we have that Fox has had a lung infection. Problems with his lungs are not reported earlier in his diaries, although he frequently reported having colds and coughs–see note 4.

#49 (CCD 184). To: Charles Darwin Esq, HMS Beagle, Monte Video, S America

Ryde. Isle of Wight.
August 29. 1832
[Posted 28 Sept. 1832]

My dear Darwin

The sight of your well known hand upon a letter back gave me no little pleasure as I began to doubt whether I should hear from you during your journeyings[1]; it seemed so long since I had done so, I had however heard excellent accounts of you tho' not from you & that was next to it, tho' far from being so pleasurable as hearing of your happiness in your own words. — I wrote a very stupid letter some three months since which you have probably received ere this, & if not you have no loss, as I felt very stupid & owlish when I wrote it & almost repented sending such rubbish so far. — Your letter I got a fortnight since when I was just on the point of leaving home for this Island in consequence of another attack upon my Lungs, which tho' very much slighter than my former one in March, was still bad enough to make a change to Warmer Air desirable. I gained ground surprisingly for a week after leaving home, indeed was nearly well to my feelings, when an imprudent walk on a hot day gave me another attack which has lasted me now a week & will not depart while the very cold weather & rain we now have, last. — I have intended every day commencing this, but not felt equal to it, but trust now I have once begun, to go on with it & send it across the Atlantic tomorrow. — You tell me to give you a particular account of my life &c &c. — You would have a very monotonous description indeed, were I to give you one. Since my last letter I have vibrated between Epperstone[2] & Osmaston[3], well enough to enjoy myself, but not sufficiently so to do any parochial Duty, or Entomologise, or in fact do any thing but amuse myself. — I shall not therefore inflict upon you the Diary of an Invalid, but sometime before Winter sets in I will write to you again & tell you how I then go on & how the World wags in these parts. I do not at present know where my Winter will be spent. I believe we (ie My Father, Mother & 2 younger sisters with my little niece[4]) stay here for about 6 weeks when it will depend upon my health whether I return to Osmaston[3] or go somewhere where the Air is milder than Derbyshire to pass the Winter. As far as I understand myself, I have no actual affection of the Lungs[5] but they are in such a delicate & excited state as to be extremely liable to all the Diseases liable to them, without great care &c. And unless I gain more strength & am very careful about them, Consumption would most probably ensue. — At present however I trust I am in a fair way of recovery & mean to leave no caution on my part wanting to ensure it.//

September 18. — When I wrote the first page I fully intended sending this off immediately, but the next day I was so unwell that I could not finish it, & so continued for some days when finding I was losing strength very fast I applied to a D^{r}. Barrett (here at present to recover from an attack of Cholera) who ordered me Blisters[6] &c &c which kept me *very tame* for some time longer, and until lately I have had no inclination to perform the mechanical part of writing, tho' very often thinking of you. Indeed there is seldom half a day passes that I do not wonder what you are about & wish I was with you. — I am now very much better, as well or very nearly so, as before my last attack and enjoy myself much. — All my plans for the Winter are however altered and it is now fixed that I & my two younger sisters stay here during the Winter, and we have already got into our Winter Quarters, a very comfortable House which seems to be very warm & snug; I much wish you were likely to be an inmate of it for a few weeks during our stay; tho' I am a great Beast to wish you here when you seem so very delightfully occupied where you are. I saw by the Paper a few days since that a Ship from Rio left your little Beagle in safety there and of course you were then well or we should have heard; indeed from your account of yourself & from what I know of the climate of S. America from books & travellers, I should hope there is every chance of your having your health there as well as in England with common precautions. — September coming in must have brought England very much to your mind, & the Partridge shooting you used so much to enjoy here, & I trust will do again. I have now not heard from Shrewsbury for a month. At that time your Father & Sisters were in good health. Poor Fanny Wedgwoods death[7] had just been a great shock to them, as it will be to you I am sure. Her death seems to have been very sudden, as she was not ill more than a week & then was not thought in any danger. How changed is that Family party since the last time (the only one) I saw them together. — Three sons & a daughter married & one gone[8].

September 28. — Again was I stopped in my progress, and been so engaged since with one thing or another that I have never resumed my letter tho' every day resolving I would do so and send it off. I have often wished you were with me in my little walks here when I have seen Insects that in your infantine days of Entomology you would have thought much of. I have seen one Colias Hyale[9], and had I been able to go after them believe I might have seen all the species of that beautiful Genus in the Island. Not one capture have I made since I came except a few Coleoptera that have crossed me, & one Sirex juvenens[10], tho I every day see something new to me. — I often think when I see our minute crustacea or mollusca or any insect, of the gloriously beautiful & curious creatures that are lying in profusion round you, & fancy you revelling among them, with sky, ocean, Trees, in fact every thing grand & sublime. You must sometimes//

almost feel bursting with your feelings, tho' of course by this time the extreme novelty of Tropical Climes is somewhat softened down. — I met with a case of South American Insects here some days since which I was extravagant enough to buy merely because they came from where you are collecting. — There are among them some magnificent Cerambycidæ[11], Geotrupidæ[12], Cicadæ[13] &c &c but I have at present no book by which I can make them out. There is also among them the Walking stick Mantis and a huge Nepa[14], or nearly allied to it. — From the list you give of the Water Beetles taken the day you wrote to me, your favorite Family seems to be very abundant. — I looked anxiously in Stephens for the Cryctocephalus[15] Darwinii, but it never appeared, nor do I think your name has ever been mentioned by him since your departure — You will however have your revenge in this I hope some time, as you will have plenty to tell the Lovers of Natural History when you return. I quite agree with you in what you say about minute insects not having been commonly collected abroad[16] —I have very seldom seen any but those remarkable for size or beauty in any Cabinets. — Stephens is now quite out of my books. He has kept prevaricating so egregiously about his No^{s} as to disgust many very much, and now he has crowned all but getting an injunction in Chancery against a book of Rennies[17] as infringing upon his work that has nothing to do with it, and has discontinued his N^{os} for the last 2 months, because forsooth others may copy out of it, if he does so. He is sadly in want of cash I hear, which may extenuate his conduct. A new Entomological Monthly M*agaz*ine is just come out[18], which I have not yet seen, but *hear* it is likely to be well supported by the first Entomologists and *n*o contests are to be allowed. — Are you aware that there is now no Duty upon Shells coming into this Country[19]. — I mention it as it may induce you to bring more than you otherwise might do over, as the Duty used to forbid a large collection. — You never told me whether you were to be allowed your own collection of Nat History or whether as you feared, our Government would swallow all. You surely will be allowed your duplicates, though after all the stores you will lay up in *the mind* is the great thing, in comparison with which, the Cabinet is of little moment. — Of course you see our Papers whenever a ship comes out and must devour them with no little interest. — Now the Reform Bill[20] has passed away the Cholera & coming Election are the great topics. — The Cholera is now everywhere almost, but is not by any means so dreadful a visitor to those of our rank in Society as it at first threatened to be[21]. — Among the poor & needy it is often very fearful in its ravages and sometimes also among the better classes, as in London where very many enjoying every comfort of life have been cut off. — In this Island there has been only one doubtful case though it has been at Portsmouth some weeks, and I believe many more have been attacked with it there, than is generally known. I say this from my own enquiries there as I often go over. To shew you how little is thought of it when it once has made its appearance, I yesterday went to the launch (with all our party of the Neptune[22] 130 Gun ship, and I do not think an inhabitant of the country within 20 miles that could come, was left behind — It was a glorious sight to see that fine Harbour a mass of living souls all looking happy. — I do not know//

whether you ever were here. It is a beautiful place commanding Spithead and all that fine anchorage opposite Portsmouth, and generally enlivened, as at the present time, with Ships of War. We have now three English & one French there & are hourly expecting the French Fleet in, when they are going to give the last decision of their countries to Holland on the Belgic Question & if not acceded to, blockade all her ports[23]. — It is somewhat curious to see a French Frigate with the Tricolor, abreast a 74 taken from them, in our roads. I do hope that the antipathy of the 2 nations so long cherished is now about to cease, would that your favorite Text, "peace on Earth, & good will to men" generally might spread abroad over the whole earth. It has just struck me that from my maritime situation this winter, I might be able to be of some use to you in forwarding any stores you may want. If there is anything that I can do of any sort, do not hesitate to make use of me as nothing would give me more pleasure (as I hope you know) than to be of use to you. Of course you have agents in London for common things but I thought I might of some utility in sending any Nat. Hist: stores you may now want. I can easily clear out any packages for you here, get store cases, any thing. Pray use me if you can for my pleasure. — I have gone on scribbling nonsense till my paper is gone. You will be glad to hear that I am now quite convalescent & hope a Winter here will quite restore me. — My Father Mother & sisters all in their best way and all wishing you every kind wish for success in all your undertakings, good health &c &c. I have just heard from Simpson[24], who tells me that Jeffry Hall[25] is actually settled in Canada. Simpson is gone into Church. — Of Henslow[26] I have not heard some months. I conclude Erasmus is alive as I have never heard to the contrary but have not heard of him in any way for some months past, I might almost say years— I must now my dear Darwin bid you farewell for a time. I shall certainly torment you again before very long. If you can find time to write me 6 lines any time, they will be thankfully & joyfully received; just tell me how & where you are & what doing — no more — Goodby. May you have a prosperous voyage & both of us live the one to hear & the other relate Your Travellers Wonders

God Bless you & Believe me
Ever yours affectionately & faithfully
W D. Fox

Postmark: Ryde
Provenance: University Library, Cambridge; Manuscript Room, DAR 204.6.2

Notes

1. Fox is probably referring to CD's letter of May 1832 (#47), which may have been a long time in passage.
2. Epperstone, near Nottingham: the parish where Fox was a curate.
3. Osmaston Hall was the country mansion rented by Fox's father, Samuel Fox III, near Derby (see Fig 4).
4. Anna Maria Bristowe (1826-1862). Daughter of Mary-Ann, Fox's sister who died in 1828 [see Letter #15 (n3)] and Samuel Ellis Bristowe.
5. Fox is referring to consumption, or tuberculosis, (as named directly in the next sentence) that was the scourge of the upper and middle classes in Europe in the nineteenth century. In a letter to CD, Susan Darwin also alludes to consumption "*After all our plans being settled to go to Osmaston we had a letter yesterday to tell us poor William was again so unwell they c*[d] *not receive us I suppose it is Consumption they fear as his chest is affected.*" (15 Aug 1832, CCD).
6. Probably blood letting. In a letter to CD, Caroline Darwin says "*We heard a report that William Fox has been very ill, in letter from Miss Bent to one of the Wedgwoods she says 'he is now so much better that he walks out every day, but his doctor bled him so violently that it will be long before he recovers his strength'*" (12-28 June 1832, CCD, Vol. 1). However that referred to an earlier bout of illness than to that mentioned here.
7. Fanny Wedgwood (1806-1832), fifth child of Josiah Wegwood II, died on 6th Aug 1832. She died of "*inflammation of the bowel*", which may well have been cholera (Wedgwood and Wedgwood, 1980).

8. With the "incomparable Charlotte" (Letter 46) married [Letter 48 (n7)] and Fanny dead, this left only two unmarried Wedgwood daughters, the eldest, Sarah Elizabeth, and the youngest, Emma. With Fanny Owen also married, CD must have felt that his chances for marriage to any of his acquaintances were lessening by the day.
9. Pale clouded yellow butterfly.
10. Probably *Sirex juventus*, a wood boring wasp in the family Sirenidae. Placed by Samouelle, 1819 in the genera *Cephus, Urocerus and Xiphaedria.*
11. A family of longhorn beetles. See Letter #2, Note 28
12. A family of earth-boring dung beetles.
13. Cicadas.
14. Water scorpion.
15. Fox misspelled "Cryptocephalus". Apparently CD had sent a new species for inclusion in Stephens' *Illustrations of British Entomology*, 1827—46. See Letter 19 (n5). "Ch. R. Darwin, FRS" is acknowledged in the supplement of 1846. Freeman (1978) mentions 11 beetles that have the specific epithet 'darwini', but none is in the genus Cryptocephalus.
16. See Letter #47.
17. James Rennie (1787-1867). Professor of natural history at King's College, London. Emigrated to Australia in 1840. Published "*Alphabet of Insects*", London, 1832.
18. *The Entomological Magazine*, edited by Edward Newman (London 1833-8).
19. The excise on shells and other natural history objects had been removed in July 1825. For further details, see CCD, Vol 1, p 268.
20. See Letter #48 (n11).
21. Ironically Fanny Wedgwood may well have died of cholera, see Note 6, above, and Healey, 2001.
22. HMS *Neptune*, famous for taking the HMS *Victory* in tow to the safe haven of Gibraltar, after the Battle of Trafalgar in 1805. Later used to transport convicts to Australia.
23. In October-December 1832 the British and French cooperated in a blockade to force the Dutch to surrender Antwerp to newly independent Belgium.
24. George Simpson (1888). See Letters #11 (n8), #16 (n17) & #31 (n5).
25. Jeffry Brock Hall (1807-1886). Christ's College, B A, 1830. Emigrated to Canada and resided at Guelph, Ontario. (*Peile.* 1913; *Alum Cantab*).
26. John Stevens Henslow (1796-1861). See Bibliographic Register, Introductions to Chapters 2 & 3 and Letter #8 (n3) and many subsequent letters.

#50 (CCD 189). To: The Rev^{d}. W. D. Fox, Osmaston Hall, Derby.

Rio Plata.
[12-13th] November 1832[1]

My dear Fox. —

I am going to take you at your word, & send you a very short letter sooner than none. — It may appear very odd to you, but I never found so much difficulty in writing letters, as at present. — There is so much to say, or might be said, that I am quite overwhelmed & generally finish by saying very little. — After leaving Rio de Janeiro, we sailed to M: Video. — from whence we had a surveying cruize to the South. — (at Rio I wrote to you[2]). — All this part of S America is wonderfully stupid; sand hillocks & undulating plains makes a poor change after mountains & vallies glowing with the rich vegetation of the Tropics. — At M: Video we had not heard from anybody for four month: & as you may suppose, like a Vulture I devoured yours & other letters. I was very sorry, my dear old Fox, to hear from yourself & home, that you had been so unwell; I trust that when you receive this you will be sitting, well & cheerful, by a blazing fireside. — Eheu Eheu[3] how long will it be before I enjoy that pleasure. — In about a weeks time, we sail down the coast for//

the Falkland islands & then for Cape Horn. I suspect there will be some difference between this & an English fire side: it would all be very tolerable, if there was some moderate limit steadily to look forward to. — But the Captains plans enlarge as the time advances[4], & I see no end to the voyage. — We return from Cape Horn & winter in the Plata; from thence we go to the other side & coast up to Panama, & then our voyage may be said again to commence. — During these last month, the only source of enjoyment, & it has been a large one, has been from Nat: History. — I have principally been lucky in Geology & amongst pelagic animals. — An old piece of ambition of mine has been gratified, viz finding the remains of large extinct animals I think some of them are new; I have teeth & fragments of about 7 kinds[5]. Even this does not reconcile me to leaving the golden//

regions of ye Tropics. — My peep at these climates has quite spoiled me for any other; I must however except the English autumnal day, the clearness of the atmosphere of which will stand comparison with anything. — Poor dear old England. I hope my wanderings will not unfit me for a quiet life, & that in some future day, I may be fortunate enough to be qualified to become, like you a country Clergyman[6]. And then we will work together at Nat. History, & I will tell such prodigious stories, as no Baron Monchausen[7] ever did before. — But the Captain says if I indulge in such visions, as green fields & nice little wives &c &c, I shall certainly make a bolt. — So that I must remain contented with sandy plains & great Megatheriums[8]: — At this present moment we are beating against a dead foul wind, on our return from B. Ayres to M Video. — Buenos Ayres is a fine large city & has an Europæan appearance, with the exception of a few wild Gauchos, with their bright coloured ponchos, riding through the streets. — Do you know Heads book[9]? it gives an excellent account of the manners of this country. —

I hope you will write to me again: & recollect that all details about yourself gain in interest instead of losing from the distance: S. America, by itself will be the best direction for the Beagle: I shall be very busy next week in packing specimens. Henslow[10] most kindly has undertaken to receive them. It is one more to the many many obligations I owe to him. — I always consider it one of the luckiest days in my life, when you introduced me to him[11]. — The friendship of such a man, is indeed worth gaining: — Good bye, dear Fox.
God bless you & keep you as happy, as you deserve to be, & Write to me again.

Yours very affectionately
Chas Darwin. —

Christ's College Library, Cambridge, #46a.

Notes

1. Dated from the time of the Beagle's voyage from Buenos Aires to Montevideo (*Beagle diary,* p 113).
2. Letter #47.
3. CD twice uses the latin word meaning "alas" or "oh! no!".

4. Robert FitzRoy decided to extend the expedition to allow more time to finish surveying the Patagonian coast and the Falkland Islands (see *Narrative of the surveying voyages of his Majesty's ships Adventure, and Beagle,* 1839, Vol 2, part 2).
5. See *Narrative* 3.4; among these were the Megatherium.
6. It is interesting that in this early part of the voyage CD still had ideas of becoming a clergyman. After rounding the Horn two and a half years later he did not mention this again [cf. Letter #55 (n4)].
7. *Baron Münchhausen's Narrative of his Marvellous Travels* (1785) was published anonymously, but it is known to have been written by Rudolf Erich Raspe, who was born in Hannover. The tales were "tall" stories but were widely popular and based on stories told by a real person, Karl Friedrich von Münchhausen, a retired army captain. The tales were much copied and extended at the time.
8. See Note 5.
9. Head, F. B. (1826) *Rough notes taken during some rapid journeys across the pampas and among the Andes*. London.
10. John Stevens Henslow (1796-1861). Professor of Botany at the University of Cambridge and mentor to both CD and Fox; see Bibliographic Register, Introductions to Chapters 2 & 3, Letter #8 (n3) and many subsequent letters.
11. This confirms the information from other sources, including the *Autobiography*, that Fox was the first to introduce CD to J S Henslow.

#51 (CCD 197). To: Charles Darwin Esq, HMS Beagle, Monte Video, S America

Ryde, Isle of Wight.
January 23. 1833.

My dear Darwin

As I remember promising that I would write to you again as Winter came on, I sit down to redeem it, fearing that you would almost rather I did not take the trouble, as I have nothing in the world to say that will interest you of any one excepting myself, and self is rather a dull subject to either write a letter upon, or receive one. — I had hoped to have heard a few lines from you in the last 5 months but I can easily imagine your time so taken up that your necessary letters home are quite sufficient for you, and perhaps in your case I should be as bad a correspondent. I have tried to instigate Julia[1] for some weeks to write to your Sisters in order that I might hear what they knew of you, but hitherto without success, & if I do not soon have better luck, I feel half inclined to do it myself; Erasmus[2] would not I conclude answer my letter if I wrote to him, nor do I know his direction. — I have seen so many vessels on the point of setting out to South America from Portsmouth & waiting for winds at the Mother Bank[3] that my erratic propensities have been often quite painfully excited and I have dreamt by the night that I was as busy as could be collecting with you, all around new, beautiful & strange. My destiny however is I fear quite fixed to the Continent at least, not to say (as perhaps may be much nearer the truth,) the country I was born in, and of tropical regions I must be content to hear from *Humboldt & Darwin*. — I hope your companion will be sufficiently great even for your enlarged ideas. But if I go on at this rate, I shall fill my paper without even telling you of myself, of whom I half flatter myself you will wish to hear, as I was in a very poor way when I last wrote to you. Since then I have become much stronger & better able to bear exertion, tho' I have had many attacks some of a more serious & others slight nature, and I still continue a good deal of an Invalid, and fear I shall do for some time to come[4]. I did dread the Winter very much indeed but by great precautions & with the very mild climate of Ryde, I hope now to get it over pretty well, and then I trust that next Spring & Summer may do a great deal to take away the remaining affection of my Lungs[5]. I often have great doubts whether I shall ever again be able to exert them for any continued length of time, as at present a few minutes quite oversets me without resting them. I must however hope for the best, at present I have very great cause for thankfulness that I am as I am. — I remain here with my two younger sisters & little Anna Maria[6], (who will have it that she has quite forgot you) for the Winter and most probably the Spring, when my health will determine what then becomes of me. You will be glad to hear that all at Osmaston[7] are quite well, My Father and Mother only left here two days ago. — I was much pleased a few weeks since by finding out in this town, a splendid Case of Insects from Rio de Janeiro, furnished by a Mr. Bescke Naturalist living on the Praca da Constituicao there — a Gentleman whom I daresay you visitted. What magnificent Lepidoptera there are there; There were several kinds of Mantis I never before saw & one of those Libellulæ which have such disproportioned long bodies. I fancied you in the height of Entomological Happiness & longed to be with you.

I have much enjoyed seeing our Navy constantly going & coming on account of this Dutch Blockade, and among all sizes have often fancied your little Beagle. You will have seen in the Papers, that we have had a French & English Fleet lying together at Spithead[8], & since cruizing together & now lying in Downs. I rejoiced much at it & hope the National antipathies may be done away, but I have been much amused by the annoyance it has given many of the Officers in our Ships — In several I visitted they could scarcely find names sufficiently bad for the French Officers, & the older the Officers the more bitter their hatred. I went over the finest French Ship & was much pleased with her & her Officers & Crew, all picked for the purpose of showing Englishmen what Nick Frog can do in his Navy. — Erasmus Galtons[9] Ship came here some months since & has just been paid off. He really is a very good specimen of a Midshipman & I am sure you would be much pleased with him. The whole Family came here & stayed some weeks while the ship was refitting for Holland. I wish much I could enter into your Geological Researches as well as your Ornithol: (which by the by you say nothing of, at which I marvel) Entomolo: and other Nat: Histy: Pursuits. I can easily imagine that the Spiders & adjoining tribes must be magnificent from the few I have seen. I have often thought that some parts of S. America would make a delightful Residence, & if one could get a few of ones most valued Friends to group together, I really think I could easily prevail upon myself to leave Good Old England, for a more sunny Clime. The Spartiate 74[10] is now lying in my sight just on point of coming to you with a New Admiral — I wish I could get to know some of the Officers that I might enquire when exactly they are going as it would be a great gratification to send you a *word of mouth* message. — There is something so freezing in sending stupid letters such a distance, & you scarcely having time to read them. — Our old Cambridge days often come over the mind like a dream. — They are hours gone by never to return I fear. I have been very busy here lately with the Pselaphidæ[11] and Scydænidæ[12] —I fancy your smile of contempt. I cannot help thinking that in other countries, larger genera allied to these will be found — perhaps it will be your lot, the S.A: is not the most likely place where you are at present. — I want much to know where you are likely to go after you leave S.A: — I fear I cannot again write to you, but that is of little consequence, you can sometimes send me a few lines, & I really hope you will. Mind I do not want more than 10 lines just to say how you are, where you are, & where going. You cannot think what pleasure a few lines of this kind would give me. — We all like Ryde[13] very much — there are many pleasant sensible people here, & the climate of the Island is delightful. I had hoped to have found many Insects that I knew were taken here, but have not been able to look after them much. The Gulls have been a great source of amusement to me. I thought when I came I could tell them pretty well, & was rather mortified to find out my ignorance of them. I set to *studying* them (for really I worked hard) and with what dead specimens I could meet with, & 5 live tame ones I have picked up at different times, I have now made out all that come here, with their several changes of plumage from nest to 5 years old. I could not have done this without your old Friend Fleming[14]. It is a great pity he was so run away with by his fondness for new names, as he is decidedly the best Naturalist generally of any who publish. — I can make out Fish Mollusca and Birds by him when I cannot by any other book I possess guess at them. Since you went there is a very respectable Entomological Magazine set up, published every 2 months, and Rennie has just commenced "A Field Naturalists Mag". monthly[15] at 1/. & others are talked of. I forget whether Hewitson[16] had commenced his eggs before you left but I think he had[17]. They are beautifully executed now, & sale of work encreases rapidly. I was amused with your seal of Cupid trimming the Sails of a vessel[18]. If any love trimmed the sails of your Vessel to encircle the world, it must have been the love of Beetles, spiders, or Rocks. — If you continue (as you purpose doing you say) in the Beagle during her whole voyage, your opportunities as a Geologist will be very great indeed. Is there not some danger of your becoming like Waterton[19], so much attached to Wandering, that the itch will again become irresistable when you have been home for a year or two. I wish I knew whether there are any Books or things of any kind I could send you; do remember that any thing I can do for you in any way, will give me very great pleasure indeed. I think it very likely I shall be here till middle of Summer & here I am on spot for doing any thing — At all events I now know plenty here that will superintend any thing for me when I am away. — I have now set up here a Pony, a terrier, 5 gulls of various kinds, Julia a Cockatoo I bought her — & A M a Cat, so that our house begins to look very sociable. When I last heard of your Father — he was in excellent health & all your Family well & in good spirits, but you have probably heard from them. I must now conclude this long, rambling, nonsensical letter — I fear you will think among my other Ailments, I am somewhat Insane. — & now my Dear Darwin Believe one with the most grateful & pleasing recollections of former days at Cambridge[20] & elsewhere, & ardent hopes for future long histories of men with tails & single eyes

Ever your affectionate Friend
William Darwin Fox.

University of Cambridge Library, Manuscript Room: DAR 204.7

Notes

1. Julia Fox (1809-1894), Fox's youngest sister. Artist (see Figs 16,17 & 22).
2. Erasmus Darwin (1804-1881), CD's brother. See letter from Erasmus to CD 18 Aug 1832 (CCD Vol. 1, p 258).
3. A sheltered anchorage between Spithead and the Isle of Wight (further descriptions in Letter #49).
4. Fox's *Diary* exists for 1833. He resided largely on the Isle of Wight with a visit to Osmaston Hall, the family home, between April and July. Fox recovered well, on the Isle of Wight; in April he met his wife-to-be, Harriet Fletcher, there.
5. This reinforces the information from the previous letter, that Fox has had a severe lung infection. This has not previously been referred to and seems to have been a new affliction for Fox, although it may have been prevalent in his family.
6. Anna Maria Bristowe, Fox's niece; see Letter #49 (n3).
7. Osmaston Hall, the home, near Derby, of Fox's father, Samuel Fox III; see Fig 4.
8. See Letter #49.
9. Erasmus Galton (1815-1909). 2nd son of Samuel Tertius Galton and Frances Anne Violetta Darwin; also the brother of Francis Dalton. See Letter #7 (n11).
10. HMS *Spartiate*, 74 guns, was originally a French ship captured during the Battle of the Nile. She fought in the Battle of Trafalgar and forced the surrender of the Spanish ship, Neptuno (Colledge and Warlow, 2006, Ships of the Royal Navy; Revsd. Ed. London: Chatham).
11. Now a subfamily of small beetles of very variable morphology.
12. Probably Scydmaenidae a family of small ant-like stone beetles.
13. Ryde was where Fox and family settled temporarily on the Isle of Wight, from August 1832 to April 1833. Later in life Fox retired near there, at Sandown (see Letters #198 and later).
14. Probably John Fleming (1785-1857). Zoologist and geologist. An annotated copy of Fleming,1828, *The History of British Animals*, 2 vols, Edinburgh, is in the Darwin Library Collection, University Library, Cambridge.
15. *The Field Naturalist* (London, 1833-5).
16. William Chapman Hewitson. The first part of his *British Oology* was published in 1831.
17. William Chapman Hewitson (1806-78). Fox knew, and had correspondence with, Hewitson: see Fox (Pearce) Darwin Collection, Woodward (Medical) Library, University of British Columbia. (see also Letters #120 (n2). & 212 (n5)
18. No doubt a seal for the wax seal on letters.
19. Charles Waterton (1782-1865). Naturalist and traveller. See Chapter 1, Note 13. In fact, after youthful excursions in the Amazonian forests, Waterton, like CD, settled down to a rural life in early middle age and did no more exploration (Blackburn, 1989). Later CD visited him and acknowledged his help over ducks in *Variation of Animals and Plants under Domestication* (1868); see Letter 125A (n5).
20. "Good old days at Cambridge" is a touchstone for the continuing correspondence between these two men.

#52 (CCD 207). To: W. D. Fox, Osmaston Hall, Derby

Maldonado. Rio Plata.
May 23d. 1833

My dear Fox

Upon our return from a cruize amongst the islands of Tierra del Fuego, — I received your two letters dated at the wide of interval of August & January[1]. — I am very much obliged to you for writing; your letters never fail to throw me into a pleasant reverie of past times & this is one of the highest pleasures I now enjoy. — I find the loss of society a great one[2]. — there is nothing on board the Beagle which can call to mind our evenings in Cambridge. — There is indeed a difference between one of your Coffee parties, with Whitmore[3] &c &c & an evening spent in the gun-room. — But it will be all the same in fifty years, as a boy says, who is going to be well flogged, & thus I have brought myself not to care much for anything which does not interfere with Natural Hist. — This summers cruize has not been a very profitable one; excepting some little in Geology. — I wish you would begin, like myself, to be a smatterer[4] in this latter branch. — she will soon be the favourite mistress & one easy to be wooed[5]. — I hope for better luck, when the happy day arrive of doubling the Horn & steering for warmer climes. — This whole East side is totally devoid of all picturesque//

beauty; & the coast not being rocky & there being no forests, it is bad for the greater part of Zoology. — We are passing this winter in this vicinity; when we shall say farewell to the R. Plata, I know not: I trust that next summer will complete the whole of this part of S. America. — The voyage is an immense one; how different from the first proposed two years. — it is, as you say, a serious evil, so much time spent in wandering. — I often conjecture, what will become of me; my wishes certainly would make me a country clergyman[6]. — You expect sadly more than I shall ever do in Nat: Hist:[7] — I am only a sort of Jackall, a lions provider; but I wish I was sure there were lions enough. — Now this morning I have collected a *host* of minute beetles; who, I should like to know, in England is both capable & willing to describe them[8]? — You ask me about Ornithology; my labours in it are very simple. — I have taught, my servant[9] to shoot & skin birds, & I give him money. — I have only taken one bird, which has much interested me: I daresay it is as common as a cock sparrow, but it appears to me as if all the Orders had said, "let us go snacks[10] in making a specimen". — I collect reptiles, small quadrupeds, & fishes industriously, especially the first: The invertebrate marine animals, are however my delight; amongst them I have examined some, almost disagreeably new; for I can find//

no analogy between them & any described families. — Amongst the Crustacea I have taken many new & curious genera: The pleasure of working with the Microscope[11] ranks second to geology. — I *strongly* advice you instanter to buy from Bancks in Bond St. a *simple* microscope, such as *the* Mr Browne[12] recommends. — & then make out insects scientifically by which I mean separate, examine & describe the trophi: it is very easy & exceedingly interesting; I speak from experience, not in insects[13], but in most minute Crustacea. I am very glad to hear in your last letter, that your health, after such struggles, is at last so much better, & that you are actually collecting the dear little beetles. — Your domestic arrangements at Ryde sounded exquisitely comfortable: it makes me envious to fancy them. I have told you nothing about our cruize to the South: because (you will say a very odd reason) I have too much to tell. — We had plenty of very severe gales of wind; one beating match of 3 weeks off the Horn; when it often blew so hard, you could scarcely look at it. — We shipped a sea — which spoiled all my paper for drying plants: oh the miseries of a real gale of wind! In Tierra del I first saw bonâ fide savages; & they are as savage as the most curious person would desire. — A wild man is indeed a miserable animal, but one well worth seeing. — Will you write again? I make a poor return: but indeed letter writing is not my fort: if you were but in hail, I would talk you//

deaf on the spot. — Once more I must thank you for your most kind letters. — I assure you I well know how to value & cordially be grateful for your friendship. Believe me. My dear old Fox

Yours affectionately
Chas Darwin

Remember me most kindly to M^{r} & Mrs Fox & to every one at Osmaston[14]. — Tell Miss A. Maria[15] I will remember her out of mere spite, because she wont me. — Direct to Valparaiso, per futuro. — I have collected in this place about 70 species of birds & 19 Mammalia: Your question: what I did in Ornithology? has done me good: I have watched the manners of the whole set:[16] —

Christ's College Library, Cambridge, #46b.

Notes

1. See letters #49 and #51.
2. CD was clearly nostalgic for his close circle of friends, chief amongst them Fox, and his family.
3. Ainslie Henry Whitmore (1801-1843); see Letter #12 (n2).
4. "*One who has only a smattering or superficial knowledge of a subject*" (*OED*).
5. Interesting sexual analogies? This may suggest CD's longing for suitable female company.
6. It is interesting that CD again states that he still intends to be a country clergyman: this is probably nostalgia and sympathetic support for Fox, who is one, albeit recuperating from sickness (cf. Letter #50 (n6).
7. In retrospect this is the understatement of the millennium.
8. Several entomologists, including CD's old rival C C Babington, identified insects from the Beagle voyage – see Letter #47 (n3).
9. CD contracted with Capt FitzRoy to take on Syms Covington as an assistant (see Narrative 3 and CCD, Vol 1, CD to Catherine Darwin, 22 may 1833).
10. To take equal shares. The bird is probably the same 'inosculating creature' of which CD wrote in his letter to J S Henslow, [*c.* 26 October] 24 November [1832], CCD.
11. This is a new interest for CD. He was given a microscope by Herbert, anonymously (see Letter #41 (n7).
12. Robert Brown (1773-1858). Botanist and librarian to Joseph Banks; discoverer of "Brownian motion"; advised Darwin on the use of simple microscopes (see LLi). (*DNB*).
13. It may be this experience, which steered CD away from his early work on insects and induced him to say in his *Autobiography*, "It was the mere passion for collecting, for I did not dissect them, and rarely compared their external characters with published descriptions, but got them named anyhow." (p. 21). See Letter #47 (n3).
14. Osmaston Hall was the Fox family home, see Introduction to Chapter 1 and Fig 4.
15. Niece of Fox; see Letter 49 (n4).
16. Patched seems clear. However, CCD has "watched" and says, "*The words transcribed as 'watched' and 'set' are not clearly legible. If correct, the meaning appears to be that CD observed the habits of the species collected. This is borne out by the Maldonado entries in his 'Zoological diary' (DAR 30.2: 77-86v.), which contain detailed observations of flight, nesting, feeding, and other habits. Most of these entries are reproduced in 'Ornithological notes', pp. 214-25".*

#53 (CCD 223). To: W. D. Fox, Esq, Osmaston Hall, Derby

Buenos Ayres
October 25th. 1833

My dear Fox

In less than a fortnight we shall be on our course to Tierra del Fuego. All this summer we shall be buried alive amongst the Barbarians[1]. — I send you this to wish you good night. — We shall not receive or write letters for the next six or eight months. I hope at Valparaiso (our future direction) to find one from you. — I have lately been a great wanderer. When the Beagle was at the R. Negro I left her & crossing the Colorado went to Bahia Blanca, from thence also by land, to this place[2]. — It is a long & most dreary ride, & till very lately never performed. — The government here sent out a large army against the Indians in their route, they left at wide intervals a troop of horses & five men, forming a line of Postas[3] to keep up some sort of communication with the Capital[4].//

I obtained an order for these horses & was of course very glad to profit by such good fortune: it was rough work many days living on nothing but Ostriches & Deer & sleeping in the open camp. — When the weather is fine nothing can be pleasanter than the Gaucho[5] fashion of travelling. — kill your game in the day, as soon as the sun sets, manacle your horses & with the cloths of the Recon[6] make your own bed. — Falkner the old Jesuit who resided so many years with the Indians has given a most accurate account of this country[7]. — One of the most interesting parts of this ride, was the ascent of the Sierra Ventana (or Casuahati of Falkner) a Mountain which rises in the campo like an island in the sea, to the height of between 3000 & 4000 feet. — In the greater part of the road the novelty & wildness were the chief charms, for one league did not differ an iota from//

another. — After arriving here, in a weeks time I started to St Fe. This country is comparatively civilized & true Pampas with all its characteristic features, Thistles &c &c. My object in all this galloping was to understand the Geology of those beds which so remarkably abound with the bones of large & extinct quadrupeds. I have partly succeeded in this; but the country is difficult to make out. — every thing in America is on such a grand scale, the same formations extend for 5 or 600 miles without the slightest change — for such geology one requires 6 league boots. — I thank Providence I have returned with an uncut throat; both Indians & these mis-called Christians, have most carotid-cutting faculties. We shall in a week leave for ever the muddy estuary of the Plata, & now the voyage may be said to have commenced. — I hope you will write to me, & as I do, give (but a longer) a history of//

yourself. Excepting my own family I have very few correspondents; & hear little about my friends. — Henslow[8] even has never written to me[9]. I have sent several cargoes of specimens & I know not whether one has arrived safely: it is indeed a mortification to me: if you should have happened to have heard, whether any have arrived at Cambridge, do mention it to me. — It is disheartening work to labour with zeal & not even know whether I am going the right road. — How is Henslow's family & what is he himself doing? I should so enjoy receiving ever so short a letter from him. — But patience is a fine virtue & there does not lack the opportunity of exercising it. — Remember me most kindly to all at Osmaston[10]; perhaps in five more years I may once again be there. — I hope to hear you are reestablished at your Curacy, & leading a White of Selbourne life[11]. — Good bye, my dear Fox. As the Spaniard says, God protect you for many years.

Yours affectionately
Chas. Darwin

Christ's College Library, Cambridge, #46c.

Notes

1. CD refers to the inhabitants of Tierra del Fuego. The *Beagle* was carrying three inhabitants to deliver back there, who were taken during the previous cruise: Jimmy Button, Fuegia Basket and York Minster (*Narrative 2*). So CD would have been alerted to what he was likely to meet, both by these persons and by tales of the sailors on the ship.
2. See *Narrative* 3.
3. Outposts.
4. Buenos Ayres.
5. A South American word deriving from Spanish and meaning a native of the Pampas of La Plata, of Spanish descent; the gauchos lived by cattle breeding and were noted for their skill in horsemanship and the use of the lasso and the bolas (*Lloyds Encyclopedic Dictionary*, 1895).
6. 'Recon' is not a traceable term; CCD suggested it may have been a gaucho term for saddle.
7. The physician and Catholic priest, Thomas Falkner, carried out expeditions in the Patagonian region of what is now Argentina, and initiated the base that became the town of Lobos (meaning 'wolves'). He spent 38 years as a missionary in the area learning Patagonian languages. His account was published in 1774 as '*A description of Patagonia and adjoining parts of South America. An account of the language of the Moluches, with a grammar and a short vocabulary*'. Printed in Hereford, England.
8. John Stevens Henslow (1796-1861). See Bibliographic Register and Introductions to Chapters 2 & 3; also Letter #8 (n3); and numerous subsequent letters.
9. Henslow had, in fact, written on 6th February 1832 and again on 15-21th January 1833 (see CCD, Vol 1). The letter of 15-21th January 1833 arrived at Valparaiso on 24th July 1834 (see letter to J S Henslow, 24 July -7 November 1834, CCD, Vol 1).
10. Osmaston Hall, Derby, the family home of Samuel Fox III, Fox's father (see Fig 4).
11. Gilbert White (1720-1793), The most famous of 18th century parsons who practised natural history. Wrote *The natural history and antiquities of Selborne in the county of Southampton* (1789), 2 volumes, London, which is one of the most published books in English. (*DNB*)

#54 (CCD 261). To: Charles Darwin Esq, Valparaiso, HMS Beagle

Osmaston
Nov 1. 1834

My dear Darwin

It is now so long since I have heard from you that I thought you most probably were really in the South Seas, your long promised Land, and tho' I much wished to write to shew you that I had not forgotten you, I hesitated to do so thinking that a letter directed to South America would not reach you. I have however written to your Sister Caroline for instructions, & today having heard from her, that if I direct Valparaiso, you will certainly get it, down I sit determined that this shall forthwith set sail after you. — My letters must all be sadly prosy I fear, but you say (& I believe you from my own feelings) that you like to hear from me, & therefore I write. I will commence by telling you, as it may be later information than you have when this reaches, that the Dr. Erasmus[1] & your Sisters are all quite well —the Dr. particularly so[2] & enjoying Gardening more than ever. You will be sorry to hear that Mr. Langton[3] is so unwell that he & his Wife are thinking of leaving England for the two next Winters — going to Madeira or the West Indies for the sake of the voyage. Your old friend Eyton[4] is going to be married to Miss Slaney — Has not a love of Nat: Histy been probably the means of this match, as Slaney[5] is himself I know an Ornithologist. — This is the first thing I have heard of Eyton since I remember going to his Fathers with you to entomologise & see his Cygnus Bewickii[6]. — I must now tell you all about our Party here. We are all very much as when you were here. My Grandmother, Father, Mother, & Sisters all in the same health and following their avocations as then. Indeed every thing at Osmaston has gone on just as it has for years. Our Landlord Sir Robert Wilmot[7] died after some weeks illness in July, & at the request of the Family was buried by myself—the first bit of duty I have done for two years. You will I dare say remember him well.//

I do not think that his death will make any difference as to our, or rather I should say, my Fathers living at Osmaston[7], as its present Possessor Wilmot Horton[8] is Governor of Ceylon & so deeply in debt that he will most likely remain at present where he is. I wrote to you about four months ago, and directed my letter to Valparaiso, so trust you have got it tho I dare say it was not worth much — It is always a comfort to me in writing foreign letters, that I pay the Postage. I think I wrote to you from the neighborhood of Doncaster where I was staying with my Wifes sister & her husband. After that I came here & spent a few weeks when we set out on a little tour into Yorkshire, where we were moving about the neighborhood of Harrogate York & Ripon till the middle of September. We were almost all the time at different friends & Relations Houses — amongst others, we were several days with Robert Pulleine[9], whom you must well remember at Cambridge. He is living in a most beautiful romantic part of Wensleydale; in a most comfortable Parsonage House, surrounded by dogs, Pigs & Poultry. I never was with a more truly hospitable kind hearted fellow. He fed us upon Venison & Moor Game & wanted us very much to stay as many weeks as we did days. Among many of our old friends and former times Charles Darwin was not forgotten, & we talked you over thoroughly. I wish you could have been with us — It was just at the commencement of Grouse shooting, which is excellent in that neighborhood, and thro' Harriets[10] friends I could have got as much shooting as we liked, & though I have never fired a Gun since I was in the Church, I should so much have enjoyed walking over this beautiful country with my much valued & old friend Darwin, and unless you are much changed, & have learnt to despise such small game since you took to Ostriches, you would I think have enjoyed it too. I hope we shall yet have this pleasure together, and as you bring down your Grouse or Partridge you can tell me prodigious tales of your sport in other climes.

We have this Summer travelled eleven hundred miles in a little Pony Carriage which with all our Luggage a stout Pony has conveyed. In this distance we have only had one break down, which was of no consequence as we soon got our Carriage mended, & no one untoward event of any kind. — I cannot tell you how much we have enjoyed it, tho' I am not sure that our enjoyment would not have been as great if we had been quietly located in some snug little Parsonage—my wife being quite as homely & domestic in her tastes & pursuits as your//

humble servant. We came here about six weeks since, intending to go to Ryde[11] late last month in our little Carriage, and to stay there the Winter, and I had hoped that a Winter there in addition to my two last, would enable me to take a curacy in the

South in the spring, as tho' very much better in my Lungs, I am still incapacitated from exerting them much: however things have fallen out otherwise — three weeks since my Wife was taken very suddenly extremely unwell with inflammation in the Peritonæum. We luckily had Medical Aid at hand immediately and after bleeding &c she was brought round, but requires extreme Care, as there is a great tendency to recurrence. This alone would have been sufficient to have made it extremely hazardous for her to have undertaken a long journey; but in addition to this, she expects to be confined at the end of December or beginning of the following month, and the two things together make it perfectly impossible for her to think of it. We therefore now purpose staying at Osmaston[7] thro' the Winter, and in Spring I hope we shall set out a trio, instead of a Duet. I begin now to think that your early prediction when you set off, of my having a family of children before you returned, may be realised, as at all events, unless something untoward occurs, there is a great probability of one child. I need not set forth to you the manifold advantages of Matrimony, as you were <always> a Philo*gyn*ist & purposed entering that state as soon as you could. — I will on*ly* say that from my present experience, I warmly recommend it to you & all I wish well to. I love my dear little wife more dearly now if possible, than when we were married, & have every reason to hope that we shall pass thro' life together, each adding much to the others comforts. But you will say I am writing a most pretious heap of nonsense — & so I am — but the truth is, that I am always apt to write to you as I should speak. — I wish you would do the same — It is now so long since I saw your handwriting that I cannot tell you the pleasure it would give me. I feel however that your time is valuable & mine worth nothing, which makes a vast difference. — My Uncle & Aunt Darwin[12] from Elston are now staying with us for a week. I was much amused a day or two since at my Uncles saying —"That he was quite sure no son of his would live to inherit Elston — That both his would die — that then the Estate would come to Erasmus, who he had understood never meant to marry — that you would be drownd in your present voyage — when it would pass to Sir Francis Darwin's children[13]."

Do you remember my being your first informant that Elston was entailed upon Erasmus & you after my Uncles sons? — His eldest Son is a dreadful invalid, having a complaint of the heart, but is a very nice dispositioned Boy. He has now had these attacks so long & always got over them, that I think he may be reared, but must always be an Invalid. His other son is a fine healthy Boy as ever was seen[14], so that I do not think your & Erasmus's chance is a very good one, but many more improbable events have happened. //

I heard a few days since a very poor account of Sir Francis Darwins[13] health. He fancies that he has some mortal disease and doctors himself for it. He either has used strong Medicines which have much brought him down, or is become very low spirited. He does not go out any where to his old friends & is very uncomfortable about himself. He is just expecting an increase to his Family from Lady Darwin, who had I imagined long since ceased from such expectations. I had fully arranged to have accepted a very kind invitation I had from Shrewsbury this Summer to go there & spend some time at your Fathers, but time slipped on so fast while we were moving about the country, that at last we were obliged to give up the idea, fearing we could not get to the Isle of Wight before the Winter set in[15]. I wish much I knew where you were at this time. I find from Caroline[16] that your last letter was in April from the Falkland Isles, and she mentions something of one Jemmy Button[17], thinking I know all about him, but she has quite mystified me as you never mentioned him to me. I hope you will experience as much delight in the South Sea as you anticipate — If you are not already on your way when this letter reaches Valparaiso, you must be on the point of setting off for that long promised land — Pray do not think my dear Darwin from this very dull letter that my feelings of friendship are correspondently blunted — You are never a whole Day absent from my thoughts & I think there are few of your friends who will more heartily rejoice to welcome you to England again than myself. But you will find me a strange dull fellow then I fear. All here unite in sending their kind love & best wishes for you in every way & believe me to remain

Ever your faithfully attached friend
William Darwin Fox.

Provenance: University Library, Cambridge, Manuscript Room: DAR 204.8

Notes

1. Erasmus Darwin, CD's brother.
2. Robert Darwin, CD's father.
3. Charles Langton, who married Charlotte Wedgwood.
4. Thomas Campbell Eyton (1809-1880). Married Elizabeth Slaney, co-heiress of Robert Aglionby Slaney, in 1835. See Letters (*DNB, Burke's Landed Gentry, 1879*).
5. Robert Aglionby Slaney (1791-1862). MP for Shrewsbury and later High Sherriff of Shropshire. (*DNB*)
6. Bewick's swan - see Letter #31 (n3). This is the only mention of a meeting of CD and Fox with Eyton, Sr., but see Letters #31 (n2), #32 (n8) & #34 (n3).
7. Robert Wilmot, 2nd Baronet (1834). Lived at Chaddesden Hall, just outside Derby and rented the family country mansion, Osmaston Hall, to Samuel Fox III. (*Burke's Peerage, 1927*).
8. Robert John Wilmot Horton (1784-1841). The son of Robert Wilmot (see Note 7). 3rd Baronet (of Osmaston). Active Tory MP for Derbyshire from 1818 and Governor of Ceylon (1831-1837). (*DNB*)
9. Robert Pulleine (1806-1868). Friend from Cambridge; curate of Spennithorpe, Yorkshire, 1830-1845. (*Alum Cantab*). See Letters #9 (n6), #10 (n15), #11 (n4) and many subsequent letters.
10. This is the first letter from Fox after his marriage to Harriet Fletcher, and Fox only refers to it obliquely. It seems that an intervening letter went astray and then has been lost: see letter #56 (n2). The marriage was at Ryde, Isle of Wight on the 11th Mar. 1834. Harriet Fletcher one of the five daughters of Sir Richard Fletcher, 1st Baronet, R E, who was greatly honoured for his death in the assault on San Sebastian, 1813, during the Spanish Peninsular War. CD first learns of the marriage from his sister Caroline (CCD, Vol 1, 9-28th Mar 1834) and in his next letter to Fox in March 1835 (#55), still has not received news of the marriage from Fox, himself.
11. Isle of Wight, where Fox first met his wife, Harriet.
12. The William Brown Darwins of Elston Hall, the Darwin family seat near Newark, Nottinghamshire, since the mid-seventeenth century (see *LLi*, pp 2-3). Fox's grandfather was William Alvey Darwin, the brother of Erasmus Darwin, who inherited the family seat of Elston Hall. At that time inheritance followed the male line and this was how Erasmus Darwin, CD's brother, was conjectured to inherit

Elston Hall. Ironically this all changed in the mid-nineteenth century and so Elston Hall was passed on through the female line to Francis Rhodes (Darwin) – see Note 13.

13. Francis Sacheverel Darwin (1786-1859). Son of Erasmus Darwin and Elisabeth (née Collier); Deputy Lieutenant of Derbyshire. This would have occurred if the entailment of Elston Hall clearly followed the male line. However, Elston Hall was inherited by William Brown Darwin's daughter Charlotte Maria Cooper, who married Francis Rhodes. Francis Rhodes later changed his name to Francis Darwin and took on the arms of Darwin under the will of Robert Alvey Darwin (see Note 14, below), in 1850. (*Alum Cantab*)
14. In fact the elder son, William Waring Darwin died the following year in 1835 and the younger son, Robert Alvey Darwin, died unmarried in 1847. So the predictions were not too wide of the mark. However, Robert Alvey left Elston Hall and the Darwin Arms to his sister Charlotte (Freeman, 1978), who married Francis Rhodes (see Note 13, above).
15. Fox's diary for 1834 is not extant. However, there is a document describing a horse and carriage ride with Fox's wife, Harriet, Fox's mother and his sister, Frances Jane, in the early Summer of 1835, during which they called on the Darwin house, the Mount, in Shrewsbury, for two weeks, between 28th May and 11th June. Fox gave no further details apart from, "*We arrived at Dr. Darwins at 7 oclock and received as kind and hospitable reception as we could wish for.*" The next entry is "*June 11th – Thursday. Shrewsbury After a fortnight spent very delightfully at Shrewsbury, we today left it with regret.*" They proceeded, eventually, to Harlech, Barmouth and North Wales.
16. Caroline Darwin (1800-88), CD's sister.
17. See Letter #53 (n1).

#55 (CCD 270). To: W. D. Fox, Esq, Osmaston Hall, Derby

Valparaiso.
March 1835

My dear Fox

Our correspondence seems to have died a natural death or rather I will say an unnatural death. I believe I wrote last to you, but it was before I heard the news of your marriage[1]. You have my most sincere congratulations, mixed however with some little envy: I hope you are now stronger in your health; & then I am sure you will be as happy as you well deserve to be. How changed every body & every thing will be by the time I return. You a married clergyman, ave maria, how strange it sounds to my ears. I wonder when I shall see you: If you continue to reside in the Isle of Wight perhaps it will be in Portsmouth. If a dirty little vessel, with her old rigging worn to shreds, comes into harbor September 1836 you may know it is the Beagle. You will find us a respectable set of old Gentlemen, with hardly a coat to our backs. This same returning to dear old England is a glorious prospect; I wish it was rather nearer; but it is sufficient to make up for a thousand vexations. Five years is a sadly too long period to leave ones relations & friends; all common ideas must be lost & one returns a stranger, where one least expects or wishes to be so. — I hope at least it will not happen//

with you & me.— I think the recollections of the snug breakfasts & pleasant rambles at Cambridge, will make us remember each other. You are one of the indirect causes of my coming on this voyage: by taking me as your dog in the grand chace of Crux Major[2] you made me an Entomologist & introduced me to Henslow[3]. I am very glad I have come on this expedition, but like a Sailor I have learnt to growl at all the details — What I shall ultimately do with myself — Quien Sabe[4]? But it is very un-Sailor-like to think of the Future & so I have done. — We leave for ever the coasts of America in the beginning of September, our route lies by the Galapagos, Marquesas, Society, Friendly I^{s}, New Zealand (1784), to Sydney. I hope, shortly after receiving this you will write to me at the latter place. I have heard nothing about you for a long time; excepting the one grand thing marriage[1]; this certainly is a host in itself, but I should like to hear some more particulars, what doing, where living, & infuturity? — Can you drink to Hopes[5] toast of "Entomologia floreat' ". in one of your letters you told me you had been collecting Pselaphidæ[6]. In the damp forests of Chiloe & Chonos Archepelago, I had the satisfaction of taking many small English genera: amongst them Pselaphus, Corticari's, minute Staphylini, Phalacrus, Atomaria & Anaspis. (Remember the Fungi at Osmaston[7])//

&c &c & Elmis beneath a stone in a brook. — Latterly however I have been paying more attention to Geology even to the neglect of marine Zoology. We are now making a passage from Concepcio>n: you will probably have seen in the Newspapers an account of the dreadful earthquake[8]. We were at Valdivia at the time; the shock was not quite so strong there, but enough to be very interesting. — The ruins of Concepcion is a most awful spectacle of desolation[9]. There absolutely is not one house standing. — I have thus had the satisfaction in this cruize both of seeing several Volcanoes & feeling their most terrible effects. It is certainly one of the very grandest phenomena to which this globe is subject. — As soon as the Beagle reaches Valparaiso, I intend going on shore & shall reside there till 1st of June, when the Beagle will *pick* me up on her road to Guyaquil[10]. — I am at present full of hope to be able to cross the Cordilleras & see the Pampas of Mendoza.— I am very anxious to connect the geology of the low country of Chili with the main range of the Andes.— I will keep this letter open for the chance of receiving one from you.— I should have written before sailing on the last cruize to the South; but I was very ill for 6 weeks & found all labor, even of writing too irksome[11]. Farewell dear Fox. God bless you. — I hope you are both well & happy.

Your affectionate friend.
C. Darwin. —

Postmark: JY 15 1835 M
Christ's College Library, Cambridge, #47.

Notes

1. Fox married Harriet Fletcher at Ryde, Isle of Wight on the 11th Mar 1834 [see letter #54 (n10)]. Fox had already written about his marriage in his letter of 1st Nov. 1834; however, a whole batch of letters had gone astray (see Letter #56 and Desmond and Moore, 1991). The marriage was reported first to CD by his sister, Caroline (CCD, Vol 1, 9 -28th Mar 1834).
2. *Panageus (Panagaeus) crux-major*, a carabid beetle; See Letter 47 (n5).
3. John Stevens Henslow (1796-1861). See Bibliographic Register, Introductions to Chapters 2 & 3 and Letters #8 (n3), #25 (n8), #29 (n12), #30 (n11) and many subsequent letters.
4. "Who knows" (Spanish). Significantly, CD no longer says that he still intends to be a country clergyman (cf. Letters #50 (n6) and #52 (n6).
5. F W Hope (1797-1862). Entomologist and clergyman. See Bibliographic Register and Letters #6 (n2), #10 (n6), #18 (n2), #26 (n5), #28 (n5) & #29 (n2), #31 (n1) & 42 (n6).
6. See Letter #51 (9).

7. Osmaston Hall, near Derby, the home of Fox's father, Samuel Fox, III (see Fig 4).
8. CD described the earthquake at Concepción in his paper, to the Geological Society in March 1838, '*On the connexion of certain volcanic phenomena in South America*' (*Collected papers* 1: 53-86) and in Narrative 3.
9. Captured as a drawing by J C Wickham, engraved by S Bull in *Narrative* 2, facing p 405. Reproduced in R D Keynes *The Beagle Record*, 1979.
10. Guayaquil, Ecuador.
11. This was the famous illness of CD, which he suffered from and necessitated his convalescence at Richard Corfield's house in Valpairiso (*Narrative*, 3). There are many theories as to what CD may have contracted, ranging from overwork to Chagas' disease from the Benchuca bug (Browne, 1995), and the influence of this on CD's ill health in middle age (Colp, 1977).

#56 (CCD 282; in LLi. part[1]). To: The Rev[d] W. D. Fox, Osmaston Hall, Derby

Lima
[9-12th Aug] July[2], 1835

My dear Fox,

I have *lately* received two of your letters, one dated June & the <other> November 1834[3]. (—They reached me however in an inverted order; —) I was very glad to receive a history of this the most important year in your life. Previously I had only heard the plain fact, that you were married. — You are a true Christian & return good for evil. — to send two such letters to so bad a Correspondent, as I have been. God bless you for writing so kindly & affectionately; if it is a pleasure to have friends in England, it is doubly so, to think & know that one is not forgotten, because absent. — This voyage is terribly long. — I do so earnestly desire to return[4], yet I dare hardly look forward to the future, for I do not know what will become of me. — Your situation is above envy; I do not venture even to frame such happy visions. — To a person fit to take the office, the life of a Clergyman is a type of all that is respectable & happy: & if he is a Naturalist & has the Diamond Beetle[5], ave Maria; I do not know what to say. — You tempt me by talking of your fireside, whereas it is a sort of scene I never ought to think about— I saw the other day a vessel sail for England, it was quite dangerous to know, how easily I might turn deserter. As for an English lady, I have almost forgotten what she is. — something very angelic & good[6]. As for the women in these countries they wear Caps & petticoats & a very few have pretty faces & then all is said. — But if we are not wrecked on some unlucky reef, I will sit by that same fireside in Vale Cottage[7] & tell some of the wonderful stories, which you seem to anticipate & I presume are not very ready to believe. Gracias a dios, the prospect of such times is rather shorter than formerly. — From this most wretched "city of the Kings' " we sail in a fortnight, from thence to Guyaquil — Galapagos — Marquesas — Society Is[d]., &c &c[8]. — I look forward to the Galapagos, with more interest than any other part of the voyage. — They abound with active Volcanoes[9] & I should hope contain Tertiary strata. — I am glad to hear you have some thoughts of beginning geology. — I hope you will, there is so much larger a field for thought, than in the other branches of Nat: History. — I am become a zealous disciple of M[r] Lyells views, as known in his admirable book[10]. — Geologizing in S. America, I am tempted to carry parts to a greater extent, even than he does. Geology is a capital science to begin, as it requires nothing but a little reading, thinking & hammering. — I have a considerable body of notes together; but it is a constant subject of perplexity to me, whether they are of sufficient value, for all the time I have spent about them, or whether animals would not have been of more certain value. — I have lately had a long ride from Valparaiso to Copiapò; in the Northern half the country is frightfully desert, & the sole source of interest was in the Geology. The scarcity of fossil shells is very inconvenient, as it will render any comparison of the formations with those of Europe nearly impossible[9]. The Andes, at the period when Ammonites lived, (which corresponds to the secondary rocks) must have been *a chain* of Volcanic Islands, from which copious stream*s of* Lava were poured forth & subsequently covered with Conglomerates. Such beds form the Cordilleras of Chili. — For the last months I have been shamefully negligent of all branches of Zoology; I hope to make up a little in the Pacifick; but all our future visits will indeed be flying ones. — The Captain talks about arriving in England September year. I doubt the possibility; but Heaven grant it may not be much after[11]. — Will you write to me once again, soon after receiving this & direct to the C. of Good Hope, & in answer to it you will see me in Person; Till that joyful day arrives, I must wish you a long Farewell. I shall indeed be glad once again to see you & tell you how grateful I feel for your steady friendship. — God bless you. My very dear Fox. Believe me,

Yours affectionately
Chas. Darwin—

Christ's College Library, Cambridge, #47a.

Notes

1. In part in LLi, p 262-3. The part omitted is from "I have lately had a long ride from Valparaiso to Copiapò" to "I must wish you a long Farewell."
2. Probably a mistake for August (see CCD, Letter 281, to Caroline Darwin, [19] July - [12 August] 1835, Note 1).
3. The letter of June has not been found; the other is Letter #54.
4. There are strong indications of homesickness and nostalgia in this letter.
5. 'Diamond Beetle', *Curculio imperialis,* of Brazil, so-called because of its sparkling elytra. Fox may have mentioned that he had acquired a specimen of this species in his former letter (see Letter 51).
6. CD is clearly nostalgic (see Note 4).
7. The reference here is unknown; maybe it is to something in the missing letter from June 1834, see note 3, above.
8. Only two of the four places named, the Galápagos and the Society Islands, were visited. CD mentioned Guyacil, now Guayaqil, Ecuador, in Letter # 55 (see Note 8).
9. CD must have been disappointed on this score since the islands of the Galápagos visited by the *Beagle* were only weakly volcanic. CD reported on a small jet of steam issuing from a crater on Albemarle (Isabela) Island (*'Beagle' diary*, p. 338).
10. Charles Lyell (1797-1875). His book *Principles of Geology* (1828-1833) had a profound effect on CD. (*DNB*). CD received Vol 3 of Charles Lyell's "Principles of Geology" in South America in early 1834. In his letter to J S Henslow (24th July 1834) CD says "*After leaving the Falklands, we proceeded to R. S. Cruz; followed up the river till within 20 miles of the Cordilleras: ....I think these facts will be interesting to Mr Lyell. – I had deferred reading his third volume till my return, you may guess how much pleasure it gave me*". Therefore, it is likely that CD received the third

volume on the Falkland Islands between 10th March and 7th April 1834. Volume 3 is concerned with tertiary strata and their identification through the presence of small marine fossil shells. Hence, CD's remarks to Fox. This is the first volume of Lyell to mention the "Miocene".

11. CD was being urged by his family to leave the *Beagle* and return to England (CCD, Vol 1, Letter from Catherine Darwin, 28 January 1835) and he was very tempted to do this in Valparaiso in Feb 1835 (Browne, 1995).

#57 (CCD 299). To: The Revd W. D. Fox, Osmaston Hall, Derby

Hobart Town. — Van Diemen's land —
February 15th. 1836

My dear Fox

On our arrival at Sydney[1], we all on board the Beagle were bitterly disappointed in not finding a single letter. — For the first occasion, the Beagle was before her appointed time; & hence the cause of our grief. I daresay otherwise I should have received a letter from you. — It is now a long time since I heard any news.— the last was, from home, of Mrs Fox's ill health[2]. — You have had much to endure in your own bodily suffering & if to this is superadded unhappiness from another & deeper source you will indeed have a heavy burthen to support. — But I sincerely hope, my dear Fox, I am croaking about calamities, which have passed away & that you are as happy as you ought to be from the bright picture you drew in your last letter. — I presume you heard from me at Lima[3]; since that period time has hung rather heavily on hand. — Not that the present is absolutely disagreeable, but I cannot refrain from thinking of the future. — I am sure, if a long voyage may have some injurious tendencies to a person's character, it has the one good one of teaching him to appreciate & dearly love his friends & relations — Now that the object of our voyage is reduced simply to Chronometrical Measurements, a large portion of our time is spent in making passages. — This is to me, so much existence obliterated from the page of life. — I hate every wave of the ocean, with a fervor, which you, who have only seen the green waters of the shore, can never understand. It appears to me, I am not singular in this hatred. — I believe there are very few contented Sailors. — They are caught young & broken in before they have reached years of discretion. Those who are employed, sigh after the delights of the shore, & those on shore, complain they are forgotten & overlooked: All think themselves hardly used, that they are not sooner promoted, I thank my good stars I was not born a Sailor. — I will take good care no one shall shall ever persuade me again to volunteer as Philosopher[4] (my accustomed title) even to a line of Battle Ship. — Not but what I am very glad I have come on the expedition; but only that I am still gladder it is drawing to a close. — I have had little opportunity, for some time past of doing anything in Natural History. — I draw up very imperfect sketches of the Geology of all the places, to which we pay flying visits; but they cannot be of much use[5]. Leaving America, all connected & therefore interesting, series of observations have come to an end. — I look forward with a comical mixture of dread & satisfaction to the amount of work, which remains for me in England. I suppose my chief *place* of residence will at first be Cambridge & then *Lon*don[6]. — The latter, I fear, will in every respect turn out most convenient. I grieve to think of it; for a good walk in the true country is the greatest delight, which I can imagine. — I shall find the different societies of the greatest use; judging from occasional glimpses of their periodical reports &c, there appears to be a rapidly growing zeal for Nat: Hist. — F. Hope[7] informs me, he has put my name down as a member of the Entomological Soc: — I do not know, whether you are one. — Formerly, when collecting at Cambridge, how very useful such a central Society would have been to us Beetle Capturers. The banks of the Cam, the Willow trees, Panagæus Crux Major[8] & Badister[9], which was not cephalotes, all form parts of one picture in my mind. To this day, Panagæus is to me a sacred genus. — I look at the Orange Cross[10], as the emblem of Entomological Knighthood. At Sydney I took a fine species, & long did I look at it, as compared to any other insect. — Poor little Albert Way[11], I wonder, what has become of him. I wish I could think he was well. —

I do not understand where you are now residing, in the last letters from home, (which was several months ago) nothing was mentioned[12]. Probably I shall not receive another letter, before reaching England, if it turns out so, there will be then a space of 18 months[13], of the events of which//

I shall be entirely ignorant. — God grant they may not be unfortunate. — I think it will be on a September night when we shall first make the Lizard lights. On such an occasion I feel it will be quite necessary to commit some act of uncommon folly & extravagance. School boys are quite right in breaking the binding of their books at the end of the half year & likewise Man of Wars men, when they throw guineas into the sea or light their tobacco pipes with Pound notes, to testify their joy. — The time is now so short, before, I trust, we shall meet, that I feel it is almost useless to describe imperfectly, what we shall have opportunities of talking over. Visiting Australia, which one day will rise the Empress of the South, was interesting. It has given me a grand idea of the//

(written crosswise on 1st and 2nd pages)
power & efficiency of the English nation. To see Colonies which in age, bear the proportion of tens of years to hundreds, so far outstepping in Civilization those of S. America, is really most astonishing. — Although full of wonder & & admiration at this Spectacle, I should be very loth to emigrate. The moral state of the lower orders is of course detestable; the society of the higher is rancorously divided by party feelings & the country itself is not to me pleasing. But with respect to money-making it is a very paradise to the Worshippers of Mammon. — It is an undisputed fact that there are Emancipists now living worth 15,000! pounds per annum.[14] — After touching at King Georges Sound we proceed to the Isle of France. — It will clearly be necessary to procure a small stock of sentiment on the occasion;// Imagine what a fine opportunity for writing love letters. — Oh that I Philo[15] had a sweet Virginia to send an inspired Epistle to[16]. — A person not in love will have no right to wander amongst the glowing bewitching scenes. — I am writing most glorious nonsense, so that I had better wish you good night, although at this present moment you probably are just awaking on a cold frosty morning. We are on opposite sides of the World & everything is topsy turvy: but I thank Heaven, my memory is in its right place & I can bring close to me, the faces of many of my friends. Farewell,

my dear Fox, till that day arrives, when we shall really once again shake hands. God bless you. —

Your affectionate friend
Chas Darwin.

Postmark: 7 JY * 1836
Christ's College Library, Cambridge, #48.

Notes

1. HMS *Beagle* arrived at Sydney on 12th January 1836, from the Bay of Islands, New Zealand.
2. See Letter #54.
3. Letter #56.
4. CD's nick name on the Beagle.
5. CCD, Vol 1. Notes "*For an account of CD's geological observations while in Tasmania, see Banks 1971*".
6. CD forecasts very accurately the events that at this stage were at least six months away. The letter is torn here and makes the word obscure but the sense is clear enough.
7. F W Hope (1797-1862). Entomologist and clergyman. See Bibliographic Register and Letters #6 (n2), #10 (n6), #18 (n2), #26 (n5), #28 (n5) & #29 (n2), #31 (n1) & 42 (n6).
8. *Panageus (Panagaeus) crux-major*, a carabid beetle; see Letter 47 (n5), #55 (n1); also Francis Darwin, Life and Letters, Vol 1.
9. *Badister*. A genus of ground beetles. See Letter #16 (n9).
10. *P. crux-major* has an orange cross on its back.
11. Albert Way (1805-1874). See Letters #2, Note 24; 13, Note 159; #19, Note 8; #47, Note 7. Albert way is invariably referred to as "little Albert Way". However, CD refers to him respectfully in his *Autobiography*; he became an eminent antiquarian and founder of the Society of Antiquaries (see also Fig 15). (*DNB*)
12. CD had heard last about Fox in a letter from Catherine Darwin written on 28th Jan 1835 and received in March 1835, in Valparaiso (CCD, Vol 1); CD wrote Letter #56 to Fox from Valparaiso (in March) and Letter #57 from Lima (in July).
13. CD last heard from Fox in a letter dated 1st Nov 1834. The Beagle was expecting to get back to England in the summer of 1836; so this would amount to about 18 months. In fact, CD was to receive letters at the Cape of Good Hope (June 1836), after thirteen months without a single letter, and at Bahia, Brazil (early August 1836; see CCD, letters to Catherine Darwin, 3 June 1836 and to Susan Darwin, 4 August 1836).
14. These words are reminiscent of what CD later wrote in *Narrative* 3 (*Journal of Researches*), "*Farewell, Australia! you are a rising child, and someday will reign a great princess in the South: but you are too great and ambitious for affection, yet not great enough for respect. I leave your shores without sorrow or regret.*"
15. "Philo": familiar shortening of CD's nick-name "Philosopher" (see note 4, above).
16. CD is making reference to *Paul et Virginie* (1787) by Jacques Henri Bernardin de Saint-Pierre.

#58 (CCD 319). To: The Revd W. D. Fox, Ryde, Isle of Wight[1]

43 Great Marlborough St[t2]
Novr 6th [1836]

My dear Fox,

I have taken a shamefully long time in answering your letter[3]. — But the busiest time of the whole voyage has been tranquillity itself to this last month. After paying Henslow[4] a short but very pleasant visit, I came up to town to wait for the Beagle's arrival. At last I have removed all my property from on board, & sent the specimens of Natural History to Cambridge, so that I am now a free man. — My London visit has been quite idle, as far as Nat: History goes but has been passed in most exciting dissipation amongst the Dons in science. — All my affairs indeed are most prosperous; I find there are plenty, who will undertake the description of whole tribes of animals, of which I know nothing. So//

that about this day month, I hope to set to work: tooth and nail at the Geology, which I shall publish by itself[5]. — I do not know how I shall be able to manage a visit to the Isle of Wight[6]. — My plans are on Thursday to go to Shrewsbury by the way of Maer[7] & Overton[8] staying at each place about a couple of days. — For Shrewsbury I hope to spare ten days, but I foresee, it will be a most uncomfortable visit, — a mere struggle how many people can be visited, to whom I am bound, & indeed, (if time, just at present was not so precious) am most anxious to see. — I shall return to London for two or three days & then go and reside in Cambridge for some months[9], & try if I can settle my jolted brains into some kind of order. — Have you during the winter any sort of business, or anything which will tempt you to pay Cambridge a visit[10]. — It is too long a journey in the//

depth of winter, and I fear for your health, to take for any one motive; though you have offered to come up all the way to London, merely for a flying visit. When I leave Shrewsbury, I will write again, as by that time, I shall know more certainly about the future. At present I am exceedingly doubtful between the merits of London and Cambridge. — It is quite ridiculous, what an immensely long period it appears to me, since landing at Falmouth. The fact is I have talked and laughed enough for years instead of weeks, so my memory is quite confounded with the noise. — I am delighted to hear you are turned Geologist. when I pay the Isle of Wight, a visit, which I am determined shall somehow come to pass[11] , you will be capital cicerone to the famous line of dislocation[12] . — I really suppose there are few parts of the world more interesting to a Geologist than your island. — Amongst the great scientific men, no one has been nearly so friendly & kind, as Lyell[13]. — I have seen him several times, &//

feel inclined to like him much. You cannot imagine how good-naturedly he entered into all my plans. I speak now only of the London men, for Henslow was just like his former self, & therefore a most cordial and affectionate friend. — When you pay London a visit I shall be very proud to take you to the Geological Society, for be it known, I was proposed to be a F.G.S. last Tuesday. It is, however a great pity that these & the other letters, especially F.R.S.[14] are so very expensive. — I do not scruple to ask you to write to me in a week's time in Shrewsbury, for you are a good letter writer, & if people//
(written cross wise at the top of the 1st page)
will have such good characters they must pay the penalty. I wish I had the pleasure of being acquainted with Mrs. Fox, that I might beg to be remembered to her, but I hope the time is not far distant, when I shall be so. — Pray tell me what relation I am your daughter, for I have been puzzling my brains & can come to no conclusion[15].

When formerly at Cambridge you explained to me my various relations, you little thought the next time you would be genealogist for your own daughter.

Good Bye Dear Fox
Yours C. D.

Postmark: D N0 7 1<83>6
Christ's College Library, Cambridge, #49.

Notes

1. Note that Fox was at Ryde on the Isle of Wight. There is no extant diary for this year. However, it is likely that Fox was still recuperating there, from his lung illness, and where his first wife, Harriet, had family connections. As indicated in Note 11, below, Fox was also at Ryde on the Isle of Wight the following year, on 21st Nov 1837, so he probably spent extended stays there during this period.
2. The address is the residence, at this time, of Erasmus Alvey Darwin, CD's brother.
3. This letter has not been found.
4. On returning from the voyage in the Beagle on 2nd Oct 1836, CD went home immediately to Shrewsbury and was there by 4th Oct. By the end of October he was at Greenwich helping to sort his specimens. In the interval it is clear, from this letter, that CD visited Cambridge to consult with J S Henslow (see Bibliographic Register, Introduction to Chapter 2 and Letters #8 (n3) and many subsequent letters).
5. *Narrative of the surveying voyages of his Majesty's ships Adventure, and Beagle,* 3 Vols & appendix to Vol 2, Colburn, London, was published in1839: and contained some geological observations of the voyage. Darwin's contribution, Volume 3 of the *Narrative*, was ready in 1838, but its publication was delayed by the preparation of Vols 1 & 2. Volume 3 was published simultaneously as *Journal of researches into the geology and natural history of the various countries visited by H. M. S. Beagle*, Colburn, London, 1839. It was republished later, with extensive revisions, as *Journal of researches into natural history and the geology of the various countries visited by H. M. S. Beagle round the world* (Murray, London, 1845). This volume was popularly known as the *Voyage of the Beagle* (and has been published as such since 1905). CD is referring here to his intention to publish a more detailed set of geological observations, which appeared later [*Coral Reefs*, 1842; *Volcanic Islands*, 1844; and *South America*, 1845 – see Letter #62 (n4)].
6. See note 11, below.
7. Home of Josiah Wedgwood II and family.
8. Home of CD's sister Marianne Parker at Overton-on-Dee, near Wrexham, Flintshire.
9. CD resided in Cambridge from 13th Dec 1836 to Mar 6th 1837, mainly in lodgings at 22 Fitzwilliam St, see Letter #59 (2).
10. CD did not seem to have sufficient motivation to visit Fox on the Isle of Wight and it must be fairly clear to CD (see following sentence) that Fox who is unwell himself and with a sick wife would not be in a position to travel. This is the pattern for the rest of their lives: it is a relationship largely sustained by their correspondence, their family ties and Cambridge memories, with a few exceptions (e g, Note 11, below).
11. It is known that CD visited Fox on the Isle of Wight: in his diary CD recorded for 1837 "*Novemb. 20th: Two days Isle of Wight to see Fox.*- " The next record of CD on the Isle of Wight is 12th Sept 1846 and then at Sandown and Shanklin, on a family holiday and writing the "*Origin of Species*", in July (see Letter #130) and August 1858.
12. "famous line of dislocation"; CD may be referring to the cliffs that separate the Solent and the rest of England from the Isle of Wight. This is where the major fossil deposits, including dinosaurs, were to be found most easily.
13. Charles Lyell (1797-1875). His book *Principles of Geology* in 3 volumes (1828-1833) had a profound effect on CD. See Letter #56 (10). (*DNB*)
14. CD was elected to the Royal Society, London on 24th Jan 1839.
15. CD and Fox were second cousins, i e, they shared a great grandfather [William Alvey Darwin (1726-1783) of Elston Hall]. Thus, Fox's daughter would be only slightly related to CD.

#59 (CCD 327). To: The Revd W. D. Fox, Ryde, Isle of Wight

Cambridge —
December 15 [1836]

My dear Fox

It is now sometime since I wrote to you at Ryde, since that time, I have been living a life of busy enjoyment, but now, at last, I see (at least) a short prospect of quiet tranquillity. I am at present staying in Henslow's house[1], (which by the bye, must be new since you were in Cambridge) it is not very large, but very comfortable. Henslow and Mrs Henslow are so kind and affectionate, that I quite feel to them as to my nearest relations. Their comfortable ways of life however do not suit hard work[2], and, in consequence, tomorrow I migrate into solitary lodgings in Fitz William St[3].//

(not far from where Hoare[4] lived). In the mornings I have to arrange my specimens into groups, and to examine all of the geological fragments one by one, which will be a most tedious task, and in the evenings, I shall write. As it is now Christmas vacation I hope to extract a good many solid hours out of each day. — I do not think excepting Henslow, that we have one single friend in common, now at the university. — It appears to me, most strange to stand in the court of Christ[5], and not to know one undergraduate: It was however some kind of satisfaction to find all the old "gyps". — John Dowly enquired much after you, as did some others[6]. — After packing up all my goods on board the Beagle, I went to Maer, where I staid two most uncommonly pleasant days. I found Mr Wedgwood[7] much aged, perhaps more changed, than any other individual. as for poor Aunt Bessy[8], she is only the wreck of her former self.— With all these//

changes, and the absence of Mrs Langton[9], yet the family party was most agreeable. — I returned home with Caroline, and staid at Shrewsbury about a fortnight. — My time was most grievously destroyed by visits to stupid people, who neither cared for me, nor I for them[10]. My father told me, next time I wrote, to remember him most kindly to Mrs Fox and yourself—and to say how very anxious he always is to hear how you are going on. — I do hope, when you write, you will be able to send good accounts. My direction for the next month will be Christ Coll: after that period or a little later, I fear I again shall be compelled to change my place of residence to London. — This will be necessary if Capt Fitz Roy & myself publish any joint stock concern[11], and indeed it will be necessary for all zoological information. — To complete my annals; from Shrewsbury I went to London, where I staid a week with my good dear old brother Erasmus, and the day before yesterday arrived here. — Whilst in London I disposed of the most important part of my collections, by giving all the fossil bones to the College of//

Surgeons. — Casts of them will be distributed, and descriptions published. — They are very curious and valuable: one head belonged to some gnawing animal, but of the size of a Hippopotamus! another to an Ant-Eater of the size of a horse! — When shall we ever meet. I cannot really spare time at present to go to the Isle of Wight. Already several people have attacked me, for not having sooner commenced. — And it is out of the question your taking the trouble of crossing so many miles of country, and leaving M^{rs}. Fox[12]. I do not see any earlier prospect than the summer,

without some good accidental circumstance should draw you up to town. But pray write to me; though we may not meet, let//
(*written crosswise at the top of the 1st page*)
tell to each other our respective annals. I truly hope your next annals will be good and comfortable ones. — Good Night. I am alone Henslow and Madam having gone to a Ball which I shirked.

Yours ever most truly
C. D.

Postmark: Cambridge DE 16 1836
Christ's College Library, Cambridge, #50.

Notes

1. John Stevens Henslow (see Bibliographic Register, Introduction to Chapters 2 & 3, and Letter #8 (n3) and many subsequent letters – also Walters and Stow, 2001). CD had arrived in Cambridge on the 13th. CCD (p 326) records a note by J Romilly, who met CD at the Henslows.
2. This is an interesting note on the difference in pace of scientific work between these two naturalists at Cambridge at this time. CD had developed the habit of long working hours on the Beagle and was intent on publishing his accounts of the voyage as quickly as possible.
3. The lodgings were at 22 Fitzwilliam Street, Cambridge.
4. William Strong Hore (1807-1882), admitted to Queen's College, Cambridge, May 1826. Became a clergyman and botanist. CD invariably misspelled his name. See Letters #1, Note9; #8/ Note 97; #9, Note 114. (*Alum Cantab; Desmond, 1977*)
5. Christ's College, Cambridge. CD had rooms on first court, Fox's rooms were on second court at E7.
6. John Dowly has not been traced.
7. Josiah Wedgwood II (1769 -1843). He had palsy (Parkinson's disease) at this time (Wedgwood and Wedgwood, 1980).
8. Elisabeth (Bessy) Wedgwood (née Allen) (1764-1846) wife of Josiah Wedgwood II. She probably had Alzheimer's disease and was unable to attend the marriage of CD and Emma Wedgwood, her daughter, in early 1839. Nevertheless, she lived on for nearly a decade in this state and considerably outlived her husband (note 7).
9. Charlotte Langton (née Wedgwood)(1797-1862). She is the "*incomparable Charlotte*" of earlier letters: see Letters #46 (n3) and #48 (n8). For the view of Erasmus Darwin of the effect on Charlotte of her marriage, see CCD Letter 182, 18th Aug 1832.
10. CD is at his most saturnine. Was this a coded message about Fanny Biddulph (née Owen) and the passage of time, exemplified by Josiah Wedgwood II, his wife and CD's father, all in obvious stages of decline?
11. CD refers to the projected *Narrative* in 3 volumes: see Letter #58 (n5).
12. See Letter #58 (n10). CD was here rather more considerate of Fox's situation.

#60 (CCD 348; LLi, part[1]). To: The Revd W. D. Fox, Ryde, Isle of Wight

[43 Great Marlborough Street]
Sunday Evening [12th Mar 1837]

My dear Fox

It is a long time since I wrote to you, from Cambridge, but I was determined to wait till I was fairly settled, which however I can hardly say I am yet, but on Tuesday I go into lodgings, at Nor 36 Grt Marlborought St. which I have taken for the year[2]. — I am at present in my brothers house no 43. — It is very pleasant our being so near neighbours[3]. — My residence at Cambridge was rather longer, than I expected, owing to a job, which I determined to finish there, namely looking over all ~~the~~ my geological specimens was finished. — Cambridge yet continues a very pleasant, but not so half so merry a place as before. — To walk through the courts of Christ Coll: and not know an inhabitant of a single room gave one a feeling half melancholy. — The only evil I found in Cambridge, was its being too pleasant[4]; there was some agreeable party or another every evening, and one cannot say one is engaged with so much impunity there as in this great city. —//

It is a sorrowful, but I fear too certain truth, that no place is at all equal, for aiding one in Natural History pursuits, to this odious dirty smokey town[5], where one can never get a glimpse, at all, that is best worth seeing in nature. — In your last letter you urge me, to get ready *the* book. I am now hard at work and give up every thing else for it. Our plan is as follows. — Capt. FitzRoy writes two volumes, out of the materials collected during both the last voyage under Capt. King to T. del Fuego and during our circumnavigation. — I am to have the third volume, in which I intend giving a kind of journal of a naturalist, not following however always the order of time, but rather the order of position. — The habits of animals will occupy a large portion, sketches of the geology, the appearance of the country, and personal details will make the hodge-podge complete[6]. — Afterwards I shall write an account//

of the geology[7] in detail, and draw up some Zoological papers[8]. — So that I have plenty of work, for the next year or two, and till that is finished I will have no holidays. — Do you recollect telling me, the new ostrich should be called "*darwinii*". By an odd chance Mr Gould has actually so named it! [9] — We are going to read a paper to the Zoological on Tuesday about it[10]. — I hope you will pay London a visit this spring; surely you will be able to find some good excuse, some necessary business, which will oblige you to come up. — I trust Mrs Fox is pretty well[11]. — My Father often makes enquiries through letters, and I have been ashamed lately, that I have been ~~able~~ not been able to send any tidings. — I will not make any excuse about this letter being almost every word about myself, for I want you do the same, indeed, we have been so long separated, that there is not much else to do. — My good dear old brother lives the same life of//

tranquility as usual. Going to Shrewsbury he considers a dreadful journey, only to be undertaken once a-year, and as far anything further, as altogether impossible. — He seems very well contented and happy, but for my own part, I would not care for a hundred years of life without a little more excitement. As for the good sisterhood at home, they remain, in statu quo, and long may they so remain. — You cannot imagine how gloriously delightful my first visit was to home. It was worth the banishment. How lucky I have been not to lose one near friend.

Good Bye - Dear Fox
Yours affectionately,
C. Darwin

Postmark: MR 13 1837
Christ's College Library, Cambridge, #51.

Notes

1. In LLi, pp 278-279 there is an abstract from "My residence at Cambridge was rather longer" to "engaged with so much impunity there as in this great city" (on p 278) and then, separately on p 279, from "in your last letter you urge me to get ready *the* book." to "and till that is finished I shall have no holidays." The penultimate two sentences of this letter are also abstracted on p 273: "You cannot imagine how gloriously delightful my first visit was to home. It was worth the banishment."
2. In the his '*Journal*' CD states that he entered his lodgings in Great Marlborough Street, London on the 13th, but that entry was made retrospectively in August 1838, when he started the '*Journal*'. Hence he probably made a mistake at the time of starting the journal.
3. In addition to his brother Erasmus, Hensleigh Wedgwood now came on to the scene. Hensleigh and family, including Mary Rich née Macintosh, a very pious woman who was at odds with Hensleigh (but much respected by Emma Wedgwood and Fanny Hensleigh), first rented a house at Notting Hill in London, at this time, and later bought a house next door to Erasmus Darwin in Great Marlborough St. This led to a great deal of socialising, one outcome of which was a crucial meeting between CD and his future wife, Emma Wedgwood (Wedgwood and Wedgwood, 1980). The interaction between CD, Erasmus and Hensleigh was clearly a fortunate happenstance in CD's intellectual development at this formative time (Wedgwood and Wedgwood, 1980).
4. CD again identifies a weakness of Cambridge life, at that time: it was a very comfortable and sociable life, which did not fit comfortably with the industrious lifestyle of a person focussed on a single goal. CD would have come quickly to the conclusion that Cambridge offered only a position in a college involving teaching and college duties and little opportunity for his researches, that is, the publication of the *Beagle* materials – which would take up the greater part of his waking hours for the next 3 to 4 years. In any case, it is unlikely that Christ's College, and other Cambridge Colleges of the time, would have seen CD's general B A degree as sufficient qualification for an academic position. One may contrast the life that CD now led with that of Fox, as an undergraduate (Chapter 2) and to some extent that of CD as an undergraduate (Chapter 3), with its round of many student parties and socialising. On his return from the *Beagle*, CD was clearly a very changed person [see also Letter #59 (n2)]. This probably accounts for the harsh remarks in the *Autobiography* "*During the three years which I spent at Cambridge my time was wasted, as far as the academic studies were concerned, as completely as at Edinburgh and at school.*"
5. CD refers to London. For the advantages of the various learned societies in London, see Letter #58.
6. For the rather byzantine nature of the publication(s), in retrospect, see Letter #58 (n5).
7. In fact, CD did publish some geology in Vol 3 of the Narrative [but see Letter#58 (n5)]. Moreover, the title of that first edition had geology before natural history in the title. It was not until the republication by Murray in 1845 [see Letter #58 (n5)], in a substantially changed account, that the place of geology and natural history were switched. CD must, therefore have changed his mind at this time, possibly because he foresaw that much of the natural history would be published as *The zoology of the voyage of H. M. S. Beagle ....1832-1836* in 5 parts (1838-1842). CD's geological observations were eventually published in greater detail, see Letter #58 (n5).
8. Most of the zoology was published in *The zoology of the voyage of H. M. S. Beagle* (see Note 4, above). However, other parts were published in scientific papers.
9. *Rhea darwini* (see The zoology of the voyage of H. M. S. Beagle, Birds, p. 123, Plate 47).
10. 'Notes upon the Rhea americana', *Collected papers* 1: 38–40.
11. CD is naturally hesitant here because he knows that Harriet Fox had a miscarriage in December 1834 (see CCD: letter of Caroline Darwin to CD, 28th Jan 1835 and Letter #54). Then after only 13 months she had a girl, Eliza Ann, on 10th Jan 1836 and now was expecting again (Harriet Emma was born on 24th April 1837).

#61 (CCD 364, LLi, part[1]). To: W. Darwin Fox, Ryde, Isle of Wight

36 Grt. Marlborough Stt—
Friday July 7th [1837]

My dear Fox

It is a very long time since I have heard any news of you; why have you not written to tell me how you are going on? — Are you turned idle, or do you think I am too full of South American bird beasts and fishes to care about old friends. — I returned last night from a flying visit of a eight days to Shrewsbury[2]; & I have now got a piece of news to tell you, which I am sure you will be interested about. Caroline is going to be married to Jos Wedgwood[3]. I do not know whether you recollect him, he is the eldest son. — He is a very quiet grave man, with very much to respect & like in him, but I wish he would put himself forward more. He has a most wonderful deal of information, & is a very superior person; but he has not//

made the most of himself. — I am very glad of the marriage for Caroline's sake, as I think she will be a very happy person, especially if she has children, for I never saw a human being so fond of little crying wretches, as she is. But I am an ungrateful dog to speak this way, for she was a mother to me, during all the early part of my life[4]. — And I forget I must not talk to *you* of; — crying little wretches — You will not guess that I mean such little angels as all children doubtless are. — This puts me in mind of your neice, who must have grown into a young lady[5]; I have never heard where she is, or anything about her, — she was a great ally of mine, when I was at Osmaston[6]. I gave myself a holiday, and a visit to Shrewsbury, as I finished my journal[7], I shall now be very busy in filling up gaps & getting it quite ready for the press, by the first August. — I shall always feel respect for every one who has written a book, let it be what it may, for I had no idea of the trouble, which trying to write//

common English could cost one. — And alas there yet remains the worst part of all correcting the press. — as soon as ever that is done I must put my shoulder to the wheel & commence at the geology. — I have read some short papers to the geological Soc[8], & they were favourably received by the great guns, & this gives me much confidence, & I hope not a very great deal of vanity; though I confess I feel too often like a peacock admiring his tail. — I never expected that my geology would ever have been worth the consideration of such men, as Lyell[9], who has been to me, since my return a most active friend. — My life is a very busy one at present, & I hope may ever remain so; though Heaven knows there are many serious drawbacks to such a life, & chief amongst them is, the little time it allows one for seeing one's natural friends. For the last three years, I have been longing and longing to be living at Shrewsbury, and after all, now in the course of several months, I see my good dear people at Shrewsbury for a week. — Susan & Catherine[10] have however been staying with my brother[11] here for some weeks, but they had returned home, before my visit. —

Pray write to me, and tell me how you are, — my Father always enquires after Mrs Fox[12] & yourself, — & I could tell him nothing. — I dont know when I shall be able to pay the Isle of Wight a visit; I am very curious to see its geolog. besides the very great pleasure//

my dear Fox, of seeing you once again[13]. — I often think of our old Entomological walks, —at this moment I can see a part of a wood (famous for big fungi & little jumping beetles (anaspis? <orchesia>[14])) near Osmaston[6], as plain as if we had been sweeping there a month ago instead of some seven long years. — And many a good day in Cambridge has left, & will I think ever leave, as bright a picture in my mind, as any I am capable of enjoying. —

How our friends are dispersed. — where is Whitmore[15]? Orlebar[16], I hear has become an extremely evangelical preacher. — Albert Way is grown quite steady & turned Antiquarian[17]; he is in town but I see nothing of him. I oftener recollect Whitlesea meer[18] than I actually see him. — When & where we shall ever meet is doubtful, but I treasure up the past with joy.

God Bless you dear Fox. — Ever yours affectionaly
C. Darwin.

Postmark: 8 JY 8 1837
Christ's College Library, Cambridge, #52.

Notes

1. Abstract in LLi, p. 280, from "I gave myself a holiday" to "but they had returned home before my visit."
2. CD recorded in his *Journal*, "*June 26th. Short visit to Shrewsbury.*" It is interesting that this is immediately followed by, "*In July opened first note Book on "transmutation of Species*". The only reference to this in the correspondence with Fox at this time is the short sentence in Letter #63 (15th June 1838), "*I am delighted to hear, you are such a good man, as not to have forgotten my questions about the crossing of animals. It is my prime hobby & I really think some day, I shall be able to do something on that most intricate subject species & varieties.*"
3. Caroline Darwin and Josiah Wedgwood III, first cousins, were married at Maer on 1st August 1837 (*Darwin Pedigrees*). Josiah was forty and Caroline thirty seven. Josiah had no inclination to take over his father's business; the couple moved to Leith Hill Place, near Dorking, Surrey, and produced three daughters, one of whom was the mother of Ralph Vaughn Williams, the composer.
4. Caroline Darwin had taken the major responsibility for looking after CD, after his mother, Sukey, died in1817, and when he was eight years old.
5. Anna Maria Bristowe was born in 1826 (*Darwin Pedigrees*). Fox's eldest sister, Mary Ann, married Samuel Ellis Bristowe, giving rise to a son, Henry Fox Bristowe and a daughter Anna Maria. Fox's sister died in 1829 (see Letter #15). However, this is not the end of the story since Fox's niece (note CD's mispelling), Anna Maria later became the second wife of Baron Dickinson Webster of Penns, Warwickshire and Captain in the Staffordshire Yeomanry. Fox clearly kept up a close connection with Baron Dickinson Webster and Anna Maria and the latters' son Baron Dickinson Webster of Newland Court, Great Malvern, married Fox's 7th daughter, Ellen Elizabeth, in 1875 (*Darwin Pedigrees*).
6. Osmaston Hall, near Derby, the family home of Samuel Fox III, Fox's father (see Fig 4).
7. This would become the third volume of *Narrative of the surveying voyages of his Majesty's ships Adventure and Beagle* (1839), which was later published separately and known familiarly as the *Voyage of the Beagle* [Letter 58 (n5)]. The publication was held up because the other authors, Robert FitzRoy and P Parker King, were not ready.
8. CD read papers on "Coral formations" and "deposits in Pampas" to the Geological Society in May (CD's *Journal*).
9. Charles Lyell (1797-1875), the author of *Principles of Geology* (1828-1833), which so impressed CD [Letter #56 (n10)], and who befriended CD on his return from the Beagle (Letter #58 (n13).
10. CD's sisters.
11. Erasmus Alvey Darwin, who resided at 43 Great Marlbrough Street, London, at this time (see Letter #58).
12. Harriet Fox, née Fletcher [see Letter #54 (n10)].
13. CD visited Fox on the Isle of Wight in November1837 , see Letter #63 (n1).
14. *Anaspis* is a genus in the Meladryrdae beetles, in Stephens' *Catalogue Br insects*, Vol 1.
15. Ainslie Henry Whitmore [see Letter #12 (n2)].
16. Cuthbert Orlebar (1808-1861). Christ's College, Cambridge, B A, 1831. Vicar of Podington, Bedfordshire, 1832-1836. (*Alum Cantab)*
17. Albert Way (1805-1874). Antiquary and traveler [see letter #2 (n2) and several subsequent letters, including #47 (n7) & #57 (n11)]. See Fig 15.

18. Whittlesea or Whittlesey Meer, near Peterborough [see Letter #47 (n5 & 6)]. CD, Fox and Albert Way apparently had a memorable excursion there in their undergraduate days and found the bombadier beetle and *Panagaeus crux-major*. This excursion is mentioned many times in succeeding correspondence and obliquely in CD's *Autobiography*: "The pretty *Panagæus crux-major* was a treasure in those days".

#62 (CCD 374). To: W. D. Fox (*No address*)[1]

36 Grt. Marlborough — St
August 28th [1837]

My dear Fox

I take shame to myself for not having sooner answered your letter, — and *such* a letter deserved more gratitude — I partly waited, that I might be able to tell you more definitely my plans. The proof sheets are beginning to tumble in; so that I shall be tied by the leg, hard at work as any galley slave during the next five weeks. — I then mean to run down to Shrewsbury for a few days, as actually since the middle of the winter I have only been able to pay them one visit of nine days. — It is a good joke, that during the whole five years I was longing most sincerely to be at Shrewsbury & now time glides by, & I am stewing in this great den of a place. — How I should like a good walk or a good days shooting, with you on a fine clear//

autumnal day, when every thing does look so very beautiful. In the whole world there cannot be anything more delightful than a wooded country in England during the Autumn. I recollect the day, I shot the ears of an unlucky pointer, (which always sticks in my conscience) at some farm to which we drove, was one of those glorious days. — But to go on with my plans. — After taking a peep at Shrewsbury & Maer[2], I come back to Marlborough St. & begin in earnest with pure Natural History. — I quite forget whether ~~I told~~ when I last wrote the Government had given me a 1000£ to illustrate Zoology of Beagle's voyage[3]. I shall not have much to do with it beside superintendence, but it will consume time; whilst this is going on I shall be trying to make progress with the Geology[4]. — But writing is most tedious & difficult work. Till lately I had not the slightest idea what a difficulty it was to

<last page missing >

(*across top of page 1*)
I suppose you heard long ago by the papers, of Henslow's good living[5]. — Is it not glorious. I think we shall live to see him, my lord Bishop, if so the Whigs will spoil a good naturalist, but make excellent bishop

Incomplete
Woodward Medical Library
University of British Columbia

Notes

1. This is the first of 5 letters from CD to Fox in the possession of the Woodward Library in Vancouver. The history of these letters and other material [The Fox-Pearce (Darwin), Collection 1821-1884] in the Woodward Library is given in Appendix 1. Fox's whereabouts are not known at this period but it is clear that he spent extended stays on the Isle of Wight, at Ryde, and CD visited him there on 20/21

Nov 1837 (see *Journal* of CD), and the next letter of 11th Dec 1837 was addressed there (Letter #62; as were Letters #58 & #59).

2. CD visited the Mount at Shrewsbury, the home of his father and unmarried sisters, and Maer Hall, Maer, near Newcastle under Lyme, Staffordshire, the home of Josiah Wedgwood II and family.
3. See CD's *Journal*, 1837: "*16th August. The Chancellor of the Exchequer granted me £1000 towards publication of the Zoology.*" See letter to J S Henslow, 20th June 1837 and J S Henslow to Thomas Spring Rice, 21st June 1837 (CCD, Vol 2). The official letter for the grant came on 31st August 1837.
4. *The zoology of the voyage of H.M.S. Beagle ...during the years 1832 to 1836,* edited by Charles Darwin, appeared in 5 parts from 1838 – 1942. As noted in Letter #58 (n5), the geology referred to by CD was published in the 3rd volume of the "*Narrative*" in 1839 (also published as *Journal of researches into the geology and natural history of the various countries visited during the voyage of H. M. S. Beagle* in 1839). Since CD had largely finished his part of the Narrative by this time he was probably looking forward to the four more books on the geology of the *Beagle*, which appeared as *The structure and distribution of coral reefs. Being the first part of the geology of the voyage of the Beagle, under the command of Capt. Fitzroy, R.N. during the years 1832 to 1836* (Smith Elder and Co, London, 1842), *Geological observations on the volcanic islands visited during the voyage of H.M.S. Beagle, together with some brief notes on the geology of Australia and the Cape of Good Hope* (Smith Elder & Co, London, 1844), *Geological observations of South America* (Smith Elder & Co, London, 1846) and, a combination of the last two, *Geological observations on coral reefs, volcanic islands, and on South America* (Smith Elder & Co, London, 1851).
5. This was the Crown Living at Hitcham, Biddeston, Suffolk, worth £1000 per annum, which was given to J S Henslow (see Biographical Register) in early 1837. See letter of CD to J S Henslow, [12 or 13 July 1837]

#63 (CCD 393). To: W. D. Fox, Binstead, Ryde, I.O.W.

36 Grt Marlbro' St
Monday
[11th Dec 1837]

My dear Fox

I have been shamefully idle in not having written to you some weeks since. But *you* can make allowance for indolence in this particular line: — why writing a letter should require a greater stock of energy, than almost any other act during the whole week, I know not, but so it is with me. — I had a very prosperous journey after leaving you at the inn at Portsmouth[1], although I did not see so much of the country as in coming down. — Is there any chance of your coming up: if you do, you ought to bring that lower jaw[2]. — I made Lyell[3] open his eyes//

at hearing about it. What a shame and pity it was, that the quarry was closed. — Since being in town Henslow[4] has been up about some London University business, — he is one of the Senators. — He enquired much after you. — I saw also little Albert Way[5], who expressed great delight at seeing a South American Brachinus[6]; which brought back to his mind the famous old expedition to Whittlesea Meer[7]. Way's chief occupation is hunting out old engraved brass-plates in churches, & poring over antiquities in the British Museum.—

I am sure you will be sorry <to hear> a piece of bad news, with respect to the Wedgwoods. Hensleigh W[8]. who married Fanny Mackintosh[9] (& who is the pleasantest of the whole family) has long had great scruples, about the profanation of taking oaths on//

trifling occasions. A week since it ******[10] for him to swear *42* oaths (!) under the new reign to qualify as magistrate. — this he could not bring himself to do, & consequently resigned his place <of 800£ per annum> & is now utterly thrown out of all employment. — He has three children & may probably have many more, & has scarcely anything to live on. — It is a most distressing case: many thousand people might be searched & not so excellent, clever & admirable <a> pair could be found as H. & his wife. — and now they are actually thinking of going to *America*! — thoug*h* I cannot but hope that something may turn up. — I hope before very long, you will summon up some of the energy, which I say is so difficult to do, & write to me & tell me how you are all going on. — Though most certainly this letter deserves no return. — I am very glad I paid you that flying visit, hurried as it was, for I like being able to form a picture in my mind of the place, whence a letter comes. —

Good bye, my dear Fox//
Yours[10] most sincerely,
Chas. Darwin. —

Pray remember me most kindly to all your family. —
T. Eyton[11] is in town, preparing a great book, a monograph on Ducks of the World[12]. - with drawing of skeleton of each genus. — Eyton will make a first rate naturalist, some day, I think. — and it is a *shame* that you are not so *now*. recollect that Sir —

Postmark: 12 DE 12 1837
Christ's College Library, Cambridge, #53

Notes

1. In his "*Journal*" CD notes for 1837 "Novemb. 20: Two days Isle of Wight to see Fox.-"
2. The lower jaw of a *Chaeropotamus,* an extinct member of the pig family, and other fossil remains found by Fox at Binstead, Isle of Wight, were described by Richard Owen at the Geological Society on 7th November 1838 (see CCD, Vol. 2). When Fox retired to the Isle of Wight in 1873, he was probably attracted to dinosaur fossils (as they were now called). However by that time another Rev W Fox, of Brixton, now Brighstone, had been working on them for ten years, and today this Fox has the most dinosaur species named after him of any Englishman.
3. Charles Lyell (1797-1875), palaeontologist and geologist. A close friend of CD after the return of the Beagle. (*DNB*)
4. J S Henslowe (1897-1861). See Bibliographic Register, Introductions to Chapter 2 & 3 and Letter #8 (n3); and many subsequent letters.
5. Albert Way (1805-1874). Antiquary and traveler. A founder of the Society of Antiquaries. He and Fox introduced CD to insect collecting and to Whittlesea Meer, see note 7, below. See letters #2 (2), #13 (n14), #19 (n11), # 47 (n6 & 7), #48 (n18), #62 (n15) & #195 (n8) – also Fig 15. (*DNB*)
6. This is the bombadier beetle and is referred to by CD in his *Autobiography*.
7. Whittlesea Meer is described in connection with Albert Way in Letters 47 (n6 & 7) & #48 (n17).
8. Hensleigh Wedgwood (1803-1891). Fellow, Christ's College, 1829-1830. Philologist and barrister. Resigned his position as Metropolitan Police Magistrate at Lambeth, London, in 1837. (*DNB*)
9. Fanny Wedgwood (née Macintosh) (1800-1889). Daughter of Sir James Macintosh. Married Hensleigh Wedgwood in 1832. (Burke's Peerage 1980)
10. Torn at this point.
11. Thomas C. Eyton (1809-1880). Cambridge contemporary of CD. Mentioned in several previous letters, see eg #31 (n2) (*DNB*)
12. T C Eyton. A monograph of the Anatidae, or duck tribe. London, 1838.

Conclusion

Darwin arrived home from the Voyage of the *Beagle* in October 1836. However, in some ways this was not the end of the voyage and in many ways it was just the beginning of Darwin's scientific life. Darwin had been changed for ever by his experience. In particular the long hours pouring over geological sites, biological specimens or books had given him a confidence in his abilities and an impetus to get things done. And those things were clearly to document and systematise the specimens collected on the five-year voyage – and beyond that to explore the exciting ideas that were swirling around in his brain and which he began to write down in his special notebooks. Darwin had thought about his return for a long time and had already written to Fox in 1835 that his plan would be to sort the specimens out in Cambridge, whence many of his specimens had been sent (the general collections from the Beagle went to the British Museum), "*I suppose my chief place of residence will at first be Cambridge & then London. —*", Darwin had written from Hobart, "*The latter, I fear, will in every respect turn out most convenient. I grieve to think of it; for a good walk in the true country is the greatest delight, which I can imagine. — I shall find the different societies of the greatest use; judging from occasional glimpses of their periodical reports &c, there appears to be a rapidly growing zeal for Nat: Hist.*" (Letter #57). So after making his social calls to family and friends in Shrewsbury, Maer and London he moved to Cambridge.

Cambridge now seemed a foreign place to him with the intervening of the years. Of course most of his friends, like Fox, were no longer there: "*To walk through the courts of Christ Coll: and not know an inhabitant of a single room gave one a feeling half melancholy*", Darwin wrote to Fox (Letter #60). But even more importantly he was now a professional. When John Stevens Henslow and his wife went off, in the evening socialising, Darwin stayed behind to catch up with his work (Letter #59). "*The only evil I found in Cambridge, was its being too pleasant; there was some agreeable party or another every evening*", Darwin later wrote to Fox (Letter #60). To be sure the Colleges offered a collegial atmosphere, but there would have been few with whom Darwin could converse on the subjects that interested him. Most such people were in London, where the many new societies beckoned. So it was not long before Darwin had his specimens crated up once more and sent off to his brother's house in London. "*It is a sorrowful, but I fear too certain truth, that no place is at all equal, for aiding one in Natural History pursuits, to this odious dirty smokey town*", Darwin wrote to Fox (Letter #60). Soon Darwin rented a house close to his brother's by and continued his sorting.

And here a further fortunate circumstance occurred; Darwin found a new mentor. John Stevens Henslow, in Cambridge, had been the major motivator for Darwin's professional development, up to this point. On returning to Cambridge, Darwin wrote to Fox, "*Henslow and Mrs Henslow are so kind and affectionate, that I quite feel to them as to my nearest relations.*" (Letter #59). Now, in London, he was befriended by Charles Lyell, the eminent geologist, "*Amongst the great scientific men, no one has been nearly so friendly & kind, as Lyell. — I have seen him several times, & feel inclined to like him much. You cannot imagine how good-naturedly* *he entered into all my plans*" (Letter #58). Through Lyell, Darwin entered into a number of important connections, such as the Horner family. "*I never expected that my geology would ever have been worth the consideration of such men, as Lyell, who has been to me, since my return a most active friend*", Darwin wrote (Letter #61). Lyell also proposed him for the Geological Society of London and soon Darwin was its general secretary. Darwin was now busy writing up his part of the *Narrative* on Journal of Researches of the biological and geological aspects of the voyage, "*I am to have the third volume, in which I intend giving a kind of journal of a naturalist, not following however always the order of time, but rather the order of position*" (Letter #60). Darwin also received a grant towards the cost of bringing out a series of books on the Zoology. "*I quite forget*", he wrote "*whether when I last wrote the Government had given me a 1000£ to illustrate Zoology of Beagle's voyage*" (Letter 62). The last rung in his professionalism was being forged, but this came at a cost, and he wrote to Fox, "*I shall always feel respect for every one who has written a book, let it be what it may, for I had no idea of the trouble, which trying to write common English could cost one*" (Letter #61).

As far as Fox is concerned the present correspondence sheds much light, which other sources do not provide. Only one diary of Fox survives for this period, that for 1833, the year leading up to Fox's marriage to Harriet Fletcher on the 11th March 1834. At this time Fox was still recuperating from his lung illness, and he was frequently on the Isle of Wight. His wife gave birth to a healthy girl in January 1836. Nevertheless, both Fox and his wife were sickly and there seems to have been a hiatus in Fox's clerical career: he must at some stage have relinquished his curacy at Epperstone. Now, however, he could be of use to Darwin, again, not through insects, which had flown out the door, but through the giant fossils that were being discovered in the chalk cliffs of the Isle of Wight. A meeting was arranged and Darwin journeyed down to the Isle of Wight to meet Fox, the only time that they met outside London for the rest of their lives. "*Is there any chance of your coming up: if you do, you ought to bring that lower jaw. — I made Lyell open his eyes at hearing about it.*" Darwin wrote to Fox after the visit (Letter #63).

Soon after this Fox obtained the living of a small parish in the Forest of Delamere, at Delamere, near Northwich in Cheshire (see Figs 23 and 24). The parish was very new having been started in 1812 under the Crown. Such parishes were rare. They were not under the control of a Bishop or Cambridge College and so could be more easily moved by influence. The details of how Fox obtained the Rectorship are not known. Nevertheless he was installed there in July 1837 and he stayed on until 1873, through two wives and sixteen children.

The correspondence between Darwin and Fox in this *Beagle* period is limited, especially in relation to the large number of other correspondents, but it is still instructive. It is also important in bridging the gap, from the earlier very important letters of Darwin to Fox, until the next period of their lives where the letters of Darwin to Fox again form an important window into their lives.

Chapter 5
Professions, Marriage, Families and Illness

Letters #64-95 (34), 15 June 1838 – 19 Mar 1855, 16 years

Genius does what it has to, talent does what it can.
Bulmer, Lord Lytton.

Who trusted God was love indeed
And Love Creation's final law —
Though Nature, red in tooth and claw
With ravine, shrieked against his creed.
Alfred Lord Tennyson, In Memoriam A.H.H. (1850), canto 56

Introduction

For Charles Darwin, 1837-38 was a crucial time. Something quite extraordinary but unrecorded had happened. It must have been at this stage in his life that he embarked on a mental program that was to direct him for the rest of his life. The Darwin that emerged in these years is a quantum jump away from the Darwin we saw at Cambridge up to 1832. Quite how this transition occurred is not known. One can guess that the young Darwin had a mind-expanding experience on the Beagle, the most important aspect of which was a grasp of the possibility that species were not stable. As discussed in Chapter 4, the young Darwin may have measured himself against Charles Lyell, who was not much older, and found that he could come off better in some aspects. One of these was coral reefs, which Darwin set about to write on at this stage (and which was, in fact, his first attempt at an evolutionary way of thinking – Ghiselin, 1969)), and the second was evidence that species were not stable. The latter, much more profound insight, was to transform the Victorian world. However, at this stage it must have been a hazy vision rather than a well-thought out thesis – and one that had been shared with a number of forerunners, notably Lamarck. It seems that Darwin did not discuss his vision with any one else. If he had done so, particularly with his father, he would have been reminded that his grandfather, Erasmus, had proposed ideas of a similar vein. As far a can be judged Darwin did not consciously appreciate this connection until much later in his life: in the third edition of "*On the origin of Species*" he put in a section on forerunners, including his grandfather, but it was perhaps as late as the 1870s, when he began his biography of Erasmus Darwin, that he fully realised his grandfather's

A. W. D. Larkum, *A Natural Calling*, DOI 10.1007/978-1-4020-9233-6_5,

contributions. Puzzling as it may seem, there is no indication that, apart from a preliminary skim, he ever read the poetical works of his grandfather, although he did take Milton's "Paradise Lost" with him on the Beagle; but it should also be remembered that Erasmus Darwin was almost forgotten during this period and was certainly dismissed by the poets of the time, as having written doggerel verse and having had heretical ideas.

What is clear, however, is that having sorted his specimens, Darwin decided on his career as a naturalist, and with, it is assumed, the blessing of his father, settled in London, where he took two pivotal steps (Desmond and Moore, 1992, Browne, 2002): the first was to open his notebooks on transmutation of species, which he did on 7th July 1837 [see Letter 61 (n2)]; the second was to get married.

While the voyage of the *Beagle* was a unique event, in triggering Darwin's career, we should not ignore the stimulus provided, at this time, by his choice of London to start his career in natural history, and the lucky coincidence that he had a close circle of family friends to turn to. The inner circle here were his brother, Erasmus, and his cousin Hensleigh Wedgwood, now married to Fanny Macintosh, who all lived just doors away from Darwin's rented house at 36 Great Marlborough Street. Darwin was in and out of their houses regularly and attended the dinner parties that each gave for their circle of friends, which included Thomas Carlyle, the historian, his wife Jane, and Harriet Martineau, radical essayist and free-thinker of this time. Hensleigh had resigned his Fellowship at Christ's College, Cambridge because he could not affirm the 39 Articles of the Church of England and more recently had resigned his position as a Police Magistrate in London, because he felt could not in all honesty administer oaths. So the conversations at these meetings would have been free-ranging and very different from anything that Darwin had experienced on the Beagle, in Cambridge or with Charles Lyell and other functionaries of the London scientific societies (Manier, 1978).

It was probably the confidence that the young Darwin gained from these stimulating social contacts that encouraged him to set down on paper his scattered thoughts on the question of the transmutation of species. The first of these was begun in July 1837 [see Letter #61 (n2)], although he had begun the Red Notebooks earlier than this in 1836 (Herbert, 1980). These contained his earliest notes on the possibility of the transmutation of species. However once he began these transmutation notebooks, his thoughts tumbled out in a rather free flowing pattern of thought, filling a succession of notebooks on various aspects of what we now call evolution (see Beer et al, 1960–1967; Browne, 2002). This stimulated a broadening in Darwin's interests, from pure natural history and geology to a range of other interests (Manier, 1978). It seems Darwin resolved at this point to read widely in philosophy and subjects of broad scientific interest, a resolution which he kept up for the next fifteen to twenty years (see reading lists, CCD, Vol 4). One of the results of this was to purchase a copy of "*An Essay on the Principle of Population*" by Thomas Malthus (1798) on 13th Sept 1838 (CD's *Journal*, September 1838). With a sudden shock of realisation, he saw that Malthus had spelled out the thesis for humans with breathtaking brevity; very simply, it was that human populations grew, naturally, at a logarithmic rate, while resources could not grow faster than an arithmetic rate. This would mean that in every generation of humans some must die of starvation or malnutrition unless the population was controlled. Darwin saw immediately that when applied to animals and plants this, too, would set in motion a competition for resources in which the best fitted would survive, i e, what Herbert Spencer, and later, Darwin, were to term "**survival of the fittest**".

At the time, this inspiration must have been dramatic and a major driving force in Darwin's life, yet we have no indication that he shared his thoughts with any one else. His excitement must have been hard to contain and many of his close friends must have realised that he was "on to something"; for example, he wrote to Fox "*It is my prime hobby & I really think some day, I shall be able to do something on that most intricate subject species & varieties*" (Letter #64).

As to marriage, Darwin famously wrote out a sheet with the pros and cons of marriage (CCD, Vol 2, Appendix IV) in which he came to the conclusion "*Marry, Marry, Marry Q.E.D.*". However, this was a theoretical decision, which had to be put into practice. As described by Browne (2002), Darwin had a wide choice before him. Here a crucial meeting seems to have decided matters: it was the return of the Wedgwood family from a visit to Paris in 1838, when they stayed in London and visited Hensleigh Wedgwood and family. Here Darwin was able to see Emma Wedgwood, the youngest Wedgwood daughter in a new light, fresh from an overseas adventure and full of vitality; while Darwin coyly did not relate the presence of Emma, he certainly reported the overall event to Fox in Letter #64 (n11). Darwin had known Emma all his life. She was one year older than him. Darwin had admired all the Wedgwood girls, in turn, starting with the eldest, Charlotte, who was ten years older than him [see Letter #32 (n2)]. Emma had always suffered in contrast with her sister Fanny, who was two years older and who most commentators said was more vivacious and self-possessed; but there was probably not a lot of difference and these two younger daughters were known as the "dovelies" (Wedgwood and Wedgwood, 1980; Healey, 2001). The sudden death of Fanny, in 1832, probably from a bout of cholera [see Letter #49 (n7)] was a family tragedy, but it meant that by 1838, Emma was the only eligible Wedgwood daughter remaining. Darwin decided that Emma was to be his choice and summoned up the courage to ask her on a visit to Maer Hall, in November, 1838. Josiah Wedgwood II was delighted to have another Darwin cousin in the family; Caroline Darwin had married Josiah Wedgwood III in 1833 [Letter #61 (n3)].

They were married on 29th January 1839; Emma was thirty and Darwin was just a few days short of his thirtieth birthday; their first child, William Erasmus Darwin, was born in late December of the same year. The immediate result of the marriage was a more than doubling in Darwin's income. The allowance that his father had given him was £400 per annum and with the marriage came a dowry of £500 per annum, and there were also assurances about help over buying a house. This together with some savings that Darwin had made, meant that he had an income in excess of £1000 per annum (Freeman, 1978). This income was quite enough to live on comfortably. For example it was the income that Hensleigh Wedgwood gave up when he resigned his magistracy and was much in excess of the income Hensleigh later received of £400 p a, as a Registrar of Cabs. For the time being the young couple stayed in London. Nevertheless their intention was to find a place in the country near London (Letters # 69-71). These plans came to fruition in September1842 when they bought Down House, at Downe, Orpington, Kent, which was served at that time by the Beckenham Railway Station, on a line from London Bridge. At that stage they had two children, William and Anne, and Emma was expecting their third child in December. This child died after 3 months. Thereafter Emma gave birth to a child just about every two years until the 9th child, Horace, in 1851. There was then a gap of five years until the last child, Charles Waring was born in 1856.

Emma Darwin was a very religious person, and much influenced by the Unitarian background of the Wedgwoods. Darwin at this stage had probably relaxed his beliefs in a literal interpretation of the Old Testament, and maybe under the influence of Hensleigh Wedgwood and his brother Erasmus, had fresh questions about the interpretation of the New Testament. Against this background, Darwin kept his heretical thoughts about the transmutation of species to himself. Nevertheless by 1842 he had hammered out a sketch, over a number of pages, of his ideas, which was a good summary of the argument for evolution. This sketch was polished in a new version of 1844. Notwithstanding his ultimate goal of doing something significant in the field of species and varieties, Darwin's course at this stage in his career, having published most of his *Beagle* work, was to settle down to a full taxonomic analysis of the barnacles, a group of animals that had hitherto been neglected. This work expanded as it developed and was to absorb the activities of Darwin for the next ten years, although as the letters to Fox reveal, he never wavered in his determination to follow through with his ideas on the transmutation of species.

In contrast to Darwin, William Darwin Fox held to his clerical calling, but nevertheless pursued his natural history interests. Like many another nineteenth century parson, it was a passionate but, generally, subordinate activity. Fox's *Diaries* of this period are largely, missing (see Appendix 1), except for one in 1833, when he was recuperating from a chest illness on the Isle of Wight. It was here that he met Harriet Fletcher, his wife-to-be. As we have seen in Chapter 4, they were married in 1834. Their life together, judged on the few surviving documents, was a very happy one, although both suffered from ill health. The survival of the one diary for this period does not seem to be accidental. Like the earlier diaries from 1820 to 1826, the retention of this diary appears to be a deliberate act of recording the key stages in Fox's life. There is a travel diary for a visit to Germany in 1842 the year that Harriet died. The next diary, only a fragment, is for 1844, two years after the death of Harriet, in 1842, and leads into W D Fox's second marriage, in 1846 (covered by a complete diary). Piecing together these fragments one can put together the following sequence of events.

By 1837 William Darwin Fox and Harriet had two girls. In the same year Fox was successful in obtaining the position of Rector in the Parish of Delamere. Delamere Forest had for long been a Royal Forest, and the Crown established a church in the village of Delamere in 1827. So nearly 30 years later, Fox was the fourth incumbent. Here he brought his wife and family. In 1842, Harriet died from "disease of the lungs" (possibly tuberculosis, a very common disease of this time). Despite this enormous blow, Fox carried on bringing up his five children, four girls and a boy, with help from his sisters, his sister-in-law, Miss Fletcher (see Appendix 4), and a governess.

In 1844 we find Fox at home with his sister-in-law, Miss Fletcher, and the governess. And in early January, the family went to stay with Dr Robert Darwin, Charles Darwin's father, in Shrewsbury for three days, and later Susan Darwin visited for five days, so the connection to the Darwin family was clearly strong. These facts and a full diary of commitments and events testify to an active life at this stage This was despite periods of melancholy: for example on Tuesday 19th November 1844, Fox wrote "*At home all day superintending workmen & feeling all the time I was seemingly taking great interest in what gave me none in any way.*"

In 1845, Fox was recovered enough to entertain the Woodds, a rich merchant family from Hillfield, Hampstead, London, at Delamere, and to visit them a good deal in their London home. On the 7th March 1845, Fox recorded in his diary, "*Called in Bond St.*", where he met many of the Woodd family including one of the daughters, Ellen, and recorded "*Was much pleased with the latter*". On 21st November, Fox further recorded "*Was in London with Woodds til 24th during which time affairs were settled, & I was made most happy*". He was engaged to Ellen Sophia Woodd. In his diary for 1846 Fox recorded "*Remarkable Events in 1846. May 20. I was married to my dear Wife Ellen Sophia Woodd at Hampstead Church.*" There followed a detailed description of the Wedding. After taking a honeymoon at the Needles Hotel on the Isle of Wight, the couple spent four days with Charles Darwin and his family at Down (from Friday 29th May to Monday 1st June). Fox wrote:

"*Left our delightful retreat at the Needles Hotel at 7 oclock. We have been so uninterruptedly, perfectly happy here. Went in Fly to Yarmouth & there to Gosport by Steamer. Saw Mr Goode at Ryde Pier – that Pier brought such remembrances & emotions.*

"*By rail to Town –Robert Wood met us at Station. Went in coaches to London Bridge & thence by Rail to Sydenham. Here we took Fly to Charles Darwins at Down. Found them very kind and friendly.*

"*Sat 30th May. Lovely Day. Were quiet at home with C. Darwin his wife and children all day. The day was intensely hot but very beautiful.*

"*Sunday May 31st (Whit Sunday) ****(torn) day*

Went to Church with Ellen and Mrs C Darwin. E & I received Sacrament. Mr Junes is Clergyman & sadly Pepseyistic. I have seldom felt the sun more powerful than today

"*Monday 1 June (Whitsunday) Very hot. Left Down soon after breakfast for Hillfield where we arrived to lunch. I had so much lumbago that I took Red Pill on the road.*"

This happy marriage led to a further twelve children, six boys and six girls, the last being born in 1864. And this was not all, because the family took in two young boys, relations of his first wife, Harriet. These two boys, Zachary and Frederick Mudge, died tragically of scarlet fever along with his daughter. Lousia Mary, in 1853 (Letters #93 & 94). Miss Fletcher his sister-in-law seems to have been a semi-permanent resident too, but complications did arise (see Appendix 5). In the end, Fox had sixteen surviving children, well up to and beyond his retirement in 1873.

Fox had a history of illness from his early College days, although the lung infection that caused him to retire from his curacy at Epperstone in late 1832 or early 1833 (Letter #49) appeared to be a new condition. Unlike Fox, Darwin had no early signs of illness apart from sores on the face during stressful periods and an illness of several weeks during the *Beagle* voyage. He began to show signs of chronic illness in 1838, which was not allayed by marriage the following year. Stress again may have been the cause. The symptoms were over-excitement, stomach cramps, vomiting, passing wind and irregular pulse (Colp, 1977). Thus he gradually withdrew from social life and from an active role in the scientific societies to which he belonged. He was General Secretary of the Geological Society from Feb 1838 to Feb 1841, but thereafter took on few official roles. Even visitors induced such symptoms, so visitors were kept to a minimum. On at least one occasion when Fox offered to visit Darwin at Down, he was told not to come as Darwin was too unwell (Letter #165) and this probably occurred on other occasions as well; on still other occasions Darwin was away recuperating from bad spells of illness (Letter #93). It must be admitted however, that this was not an illness in the sense that Darwin was laid up in bed for considerable periods. Even at its worst, Darwin was able to carry on the work of an able-bodied man, and the evidence from his *Journal* and estimations of the time that he must have spent preparing his books, carrying out background reading and keeping up a voluminous correspondence, lead one to wonder how even a healthy man could have achieved all that he did. The concept of that peculiar malady, the Victorian "creative" illness has been canvassed by Pickering (1974) - see also Colp, 1977.

Interestingly, the correspondence gives little information on the times. Queen Victoria was crowned in 1837; the Chartist "Revolution" was suppressed in 1848, the great Irish famine lasted from 1845 to 1849; the Crimean War was fought from 1853–1866; the Indian Rebellion (Indian Mutiny) was in1857 and the American Civil War was from 1861-65. These, and the fast pace of change in England, could have been points for discussion. Yet few were. Darwin voiced his concerns over his three "bug-bears", "*By the way my three Bug-bears are Californian & Australian Gold, beggaring me by making my money on mortgage worth nothing — The French coming by the Westerham & Sevenoaks roads, & therefore enclosing Down — and thirdly Professions for my Boys.*" (Letter #88), but these, by comparison, were fairly trivial qualms. Both men did have concerns about these major events, which they set down in their diaries or letters to friends, but their friendship was based on natural history and their families, and they did not waste space on chit-chat.

During these times, there was no such thing as a "scientist" or a professional scientist as they exist today. A very few persons (all men) secured posts in disciplines, which would now be known as "science"; just a few of such post were available for the very first time at this time, in the new universities, such as the newly established University College, London (to become the University of London) and at the University College, Durham, where Whitley (Letter #26) gained a lectureship in mathematics. These handful of men were the very first "scientists". One such was Darwin's old teacher at Edinburgh University, Dr. Grant who was appointed Professor of Natural History at University College, London, in 1828.

Interestingly few of the persons that we consider the great "scientists" of that age came from this group or from the established Universities. Charles Darwin was a man of independent means, as was Charles Lyell, the renowned geologist. Richard Owen, the anatomist and palaeontologist studied medicine before joining the Hunterian Museum, later becoming the Hunterian Professor of the Royal College of Surgeons and later still, in 1856, being appointed Superintendent of the Natural History Department of the British Museum in South Kensington, London. Joseph Dalton Hooker, the renowned botanist and Darwin's young supporter, was initially of an assistant surgeon of the Royal Navy, later taking on many expeditions overseas before first becoming assistant director of the Royal Botanic Gardens, Kew and then finally its Director. Thomas Huxley, too, who left school at 10, was initially an assistant surgeon on a Royal Navy ship, H.M.S. Rattlesnake and then had to eke out employment in various teaching and other positions until his appointment to the professorship of Natural History at the Royal School of Mines (later Imperial College) in 1864. Alfred Russel Wallace, the cofounder of the theory of natural selection, with Charles Darwin, left school at age of 13 to help his brother as an engineer and surveyor and later, from 1848 earned his living as a natural history collector in South America and the East Indies, until finally gaining a pension from the British Government in1881.

#64 (CCD 419). To: Rev[d] W. D. Fox, Delamere Rectory, Cheshire

Friday
[15 June 1838]
36 Grt. Marlbro'

My dear Fox

I have been in great trepidation for the last fortnight at the thoughts of receiving a letter from you: for the fossils[1] only came to my house three days since. The casts were to have come yesterday, but have not yet. I do not doubt, however, I shall have them directly. — My chief object in writing now is to tell you <than> in a week, <or 10 days>, or at most a fortnight I leave London for about five weeks, so that if you want the fossils & casts sent anywhere you must let me know at once, or leave them till the beginning of August, when I shall be back here. — Owen[2] was annoyed//

at having been the cause of the delay: but in fact he is worked out of life & soul. He has sent me a letter to forward to you, which I have not done, thinking I would wait, till I heard, whether you chose to have the fossils sent to you; or what plan was decided on. — Although I was in great fears at seeing your handwriting until I had the fossils safe & sound, yet I have at the same time been wishing *very much* to hear how Delamere gets on[3]. — You must give up an hour of one of your solitary evenings, (for I suppose M[rs] Fox has not yet joined you) & give me long account, of place & people. I have not been very well of late[4], which has suddenly determined me to leave London earlier than I had anticipated. I go by the steam-packet[5]. to Edinburgh. — take//

a solitary walk on Salisbury crags & call up old thoughts of former times[6] then go on to Glasgow & the great valley of Inverness, — near which I intend stopping a week to geologise the parallel roads of Glen Roy[7], — thence to Shrewsbury[8], Maer[9] for one day, & London for smoke, ill health & hard work. — Catherine[10] came back yesterday from Paris, where she has been staying with party of Wedgwoods[11]. for a month — She stays a week in London & then returns <to> Shrewsbury. — I am going today to dine at the great Herschel dinner, which you probably will have seen announced in the Papers[12]. — it will, I am afraid be stupid but I trust Sir J. will give us some account, of his discoveries. — My journal will not be out until the Autumn, I am crawling on with the geology, — but the Zoological parts[13] murder much of my time. — I am delighted to//

hear, you are such a good man, as not to have forgotten my questions about the crossing of animals. It is my prime hobby & I really think some day, I shall be able to do something on that most intricate subject species & varieties[14]. Good Bye my dear old Fox — make a noble return to this excessively stupid letter, & tell me all the news of Delamere.

Yours most Sincerely
Chas. Darwin

Postmark: JU 15 1838
Christ's College Library, Cambridge #54.

Notes

1. See previous letter #63 (n2).
2. Richard Owen (1804-1892). Anatomist and palaeontologist. Coined the name "dinosaur". Later strongly anti-*Origin*. (*DNB*)
3. As the address shows Fox was now at Delamere in Cheshire. He was appointed Rector of Delamere, a Crown appointment, in the Forest of Delamere, Cheshire, in April 1838, although he had probably taken up his position unofficially in 1837
4. This is the first report of the illness that CD suffered for the rest of his life. The cause has been much debated: see Introduction to this chapter. It is noteworthy that the ensuing letters to Fox over the next year are the first intimation of CD's illness. Because of their close friendship and shared illness they were able to discuss these and other personal matters openly. This is why the Darwin-Fox correspondence has been used so extensively in biographies of CD.
5. Steam-driven boats had become popular in the early 1800s, especially on rivers in America. In his diary of 1824–1826 (Chapter 2 and Appendix 4)), Fox reported taking a steam boat from London to Margate.
6. See *Autobiography*, p. 53; CD has unflattering recollections of Professor Robert Jameson's field lecture on the Salisbury Crags. The notes of CD's visit of 1838 are in DAR 5 (ser. 2): 33–8.
7. The parallel 'roads' (terraces) on the slopes on either side of Glen Roy were a well-known geological conundrum. CD set out on 23 June 1838 to investigate them. He proposed, from his South American experience, that they were formed by up-lift of (marine) beaches. However, it turned out that they were deposits formed around freshwater lakes, created by glaciers in former times. Darwin later recognized this as his greatest blunder (*Autobiography*, p. 33). Darwin published his results in the Philosophical Transactions of the Royal Society, 1839.
8. To his father's house, The Mount, at Shrewsbury.
9. The home of Josiah Wedgwood II and family.

10. Emily Catherine, CD's youngest sister.
11. CD reports on his sister Catherine's return from Paris, but does not report the more significant fact, that his sister was accompanied by Emma Wedgwood: "*Emma Wedgwood and Catherine Darwin came back from Paris where they had met the Sismondis. All the travellers, Charles, Ras and Thomas and Jane Carlyle, came to dinner. After the party Charles began to think seriously of marrying his cousin Emma.*" (Wedgwood and Wedgwood, 1980).
12. John Frederick William Herschel (1792-1871). Astronomer. Herschel had just returned from South Africa, where CD had met him, and, back home, was much honoured for his astronomical discoveries. In the same year, at Queen Victoria's coronation (28th June 1838), he was made a baronet. See Letter #39 (n3).
13. See Letter 61 (n3).
14. This is the first time that CD mentions his ideas on species, and it is a prescient remark. At this time CD had already completed the Red Notebook, and Notebooks A and B and much of C, all of which mentioned the "species question" (Herbert, 1980). So CD had been working on his ideas on transmutation of species for over a year.

64A (CCD 418). William D. Fox to Charles Darwin [Great Marlborough Street, London]

[Delamere Rectory, Northwich]
[c November 1838][1]

My dear Darwin

Before I enter into any chitchat, I shall give you the result of my enquiries about Dogs, Geese, &c, which you requested me to see[2] about. — First about H: Galtons[3] Dogs — I wrote to him & have the following answers. "I have found from breeding in & in that there is considerable difficulty in keeping up the breed. Many of the females have never exhibited any ~~slight~~ <sexual> appetite & those which do so at all, very rarely.

The Knot in the tail appeared by accident in one of the finest Dog puppies I had, so fine that I kept it notwithstanding this imperfection & all his descendants had it until at last I got a cross with one of Lord Aylesfords[4] Bloodhounds, since which time it has disappeared[5]. The knot was always in the same part of the tail. Another consequence of breeding in & in is that the animals become prema=turely old." I give you this in his own words — They agree pretty much with what I told you, do they not?[6]

Now as to the Geese. I went//

to Mr Strutts[7] & examined them myself & had the old man who always takes care of them to examine. They had at one time a *pair* of Canada Geese — The Gander however was so shy that he never would perform his Marital Duties, & a Common Gander undertook them for him. — The first year, there were five young produced from this cross — all *very much* more partaking of the make & plumage of the Canada Goose than the Common Gander. This year from the same pair (viz the Canada & Common Gander) there were 6 eggs produced, but 5 were addled there is now one young one & it is so like the Canada breed, that at first I doubted the mans word, till on getting nearer to it, I saw that the white spot on the neck was not so pure a white — that the plumage was generally duller & that there was a difference in the shape. However to a casual observer 50 yards off, it would pass for a Canada. I should tell you that the Canada Gander has been long dead, so that there can be no mistake as there are only common or Chinese Ganders, she could cross with. The Chinese & common breed he tells me better together, than if not crossed. There is also a ½ bred Goose (between these two) which has now 8 young //

by a Chinese Gander. This visit proves what I told you that the Chinese Canada[8] & Common Geese breed freely to=gether & also that the ½ breed are fertile also.[9] So much for my enquiries.

<rest of side 3 excised - about 14 cm >

I hope to hear you are going on as comfortably & industriously as when I last heard from you. When is your Book to come out[10] — I want much to see it? — I am scribbling this in a sad hurry but I am determined to send it off today & I have a Funeral immediately. — Tell me whether[11]

<rest of page excised, as with side 3 >

(Annotations, see Notes 2, 5 &10)
2. "My dear. . . . . .to see" *crossed pencil*
5 "First . . . . . . disappeared" '(Breeding In & In)' *added brown crayon*
10. I hope. . . . . .whether" *crossed pencil*

Across top of first page, all in ink:
'about presentation casts'[12]
'Paper Read to Geolog Soc'[13]
'exact locality of Jaw'[14]
'Delamere Rectory Chester? where to sen Journal'[15]

Provenance: University Library, Cambridge; Manuscript Room: DAR 164: 173

Notes

1. Dated on the basis of the relationship between this letter and the letter to W. D. Fox, [15 June 1838], and CD's annotation (see n. 8, below).
2. Annotation 1.
3. Howard Galton, brother of Samuel Tertius Galton (1783-1844), who married Frances Anne Violetta Darwin, daughter of Erasmus Darwin, and produced Sir Francis Galton. The Galtons were a well-established Midlands family. Samuel John Galton (1753–1832), the father of Howard and Samuel Tertius, was a Quaker armaments manufacturer, a member of the Lunar Society and a Fellow of the Royal Society (*Darwin Pedigrees*). William Darwin Fox was distantly related to the Galtons through his mother, Anne Darwin, the daughter of William Alvey Darwin (1726-1783) of Elston Hall. In addition he was connected through his step-brother, via the Strutts to the Daltons. From the Fox diaries it is apparent that the two families kept up close social ties. At one time CD in his letters suggested that Fox was keen on Bessy Galton; see Letters #6 (n7), #7 (n10), #9 (13) & #27 (n4): a view confirmed by the Fox Dairy for 1824–26, see, e g, 14th April 1825 (Chapter 2).
4. Heneage Finch (1786–1859), 5th Earl Aylesford and High Steward of Sutton Coldfield (1835-1859).
5. Annotation 2.
6. Fox probably discussed this with CD in November 1837 when CD paid him a visit on the Isle of Wight ('*Journal*'; CCD, Vol 2, Appendix II). CD made notes about Howard Galton's bloodhounds in *Notebook B*: 175–6, 182, and 183 (between December 1837 and January 1838), and later used the information on bloodhounds in *The variation of animals and plants under domestication*, 1868, Vol 2, p 121.
7. The Fox family were on very close terms with Jedediah Strutt II and family in Derby (see Chapter 2 and Appendix 4) since Samuel Fox III's first wife Martha, was the daughter of Jedediah Strutt I. In addition Isabella Strutt had married Howard Galton (see Note 3).

8. The word is unclear but "Canada" seems to fit.
9. CD made notes from Fox's information on the crossing of geese in *Notebook D*: 9, 13. Later CD used this information on hybrid geese in his 'big book' (see *Natural Selection* (Stauffer, 1975), p 453).
10. Fox is presumably asking after CD's 3rd Volume of *Narrative of the surveying voyages of his Majesty's ships Adventure and Beagle*, 1839, which came out as *Journal and remarks 1832-1836* [also as*Journal of researches, 1839* – see Letter #58 (n5)]. CD had said that it would be published by late 1838 (see Letter #63). In fact, due to the tardy preparation of Volumes 1 & 2, *Journal and remarks* was not published until early 1839.
11. Annotation 3
12. From letter #63 it appears that CD had asked Fox to make casts of fossils that he had discovered and send them to CD.
13. Probably a reference to the publication: Owen, R, 1841-2. Description of some fossil remains of Chæropotamus, Palæotherium, Anoplotherium, and Dichobunes, from the Eocene formation, Isle of Wight. Transactions of the Geological Society of London. 2nd Ser, Vol 6, pp 41–45 (read 7 November 1838).
14. A reference to the lower jaw collected by Fox on the Isle of Wight; see Letter #62.
15. Letters to Fox from CD were variously addressed to i) Delamere Rectory, Chester, ii) Delamere Rectory, Cheshire, iii) Delamere Rectory, Northwich and iv) Delamere Rectory, nr Northwich, Cheshire. The first address was used most often.

#65 (CCD 541; LLi, part[1]). To: Revd W. D. Fox, Delamere Rectory, Cheshire

12 Upper Gower St
Octob 24th — [1839]

My dear Fox

I have been intending for some time past to write to you, although I have little to say. My chief object is to get a letter from you, to tell me how you & Mrs Fox are, & whether there is any likelihood of your coming to London, as you told me at Bermingham[2] this was possible. I had fully intended, when at Maer[3], to have paid you the morning visit I talked off, but during my whole visit in the country I was so languid & uncomfortable that I had but one wish & that was to remain perfectly quiet & see no one. — I scarcely even enjoyed my visit to Shrewsbury. — I have been much brisker since my return to London, & I am now getting on steadily, though very slowly with my work[4], &//

hope in a couple of months to have a very thin volume 8vo on Coral Formations published[5]. Emma is only moderately well & I fear what you said is true "she wont be better till she is worse"[6]. — [1]We are living a life of extreme quietness: Delamere itself, which you describe as so secluded a spot, is, I will answer for it, quite dissipated compared with Gower St. — We have given up all parties[7], for they agree with neither of us; & if one is quiet in London, there is nothing like its quietness — there is a grandeur about its smoky fogs, & the dull distant sounds of cabs and coaches: in fact you may perceive I am becoming a thorough-paced cockney & I glory in thoughts, that I shall be here for the next six months. — I am afraid you will say if I had not anything better to//

write about than the advantages of London over country life, I had better have been quiet: but I have been thinking of writing for some time past, so write I would—do you follow my example & let me have a line from you — I heard from Shrewsbury a few days since, that Mr & Mrs Fox[8] were going to pay a visit there: my Father will enjoy very much having some family talk with Mrs Fox: I think our fathers have never seen each other[9]. Talking of family affairs, can you tell me from *memory* what the motto to our crest is <for I mean to have a seal solemnly engraved>. — I have put your letter carefully by & for the life of me I cannot find it — it was *aude* ? et — ?[10] If your memory does not serve you, do not trouble yourself with looking in your papers — I gave you too much trouble before. — When at Shrewsbury I had a regular hunt through some old papers & pedigrees relating to our most ancient family, which as you say is older than the heralds office[11]. — It has given me a great wish to see Elston[12], which some future year I will put into execution. — The pedigrees want filling up terribly; so ancient a family ought not to be neglected: My father//

gave me a curious old ivory box, which belonged to W. Darwin of Cleatham, who died in 1682[13]. — By the way Hensleigh Wedgwood made a curious discovery regarding our august family, which I must tell you, that a W. Darwin my great grandfather is described in the Phil. Transacts for 1719[14], as a person of curiosity, who discovered the remains of a giant, evidently an Icthyosaurus. — so that *we* have a right of hereditary descent to be naturalists & especially geologists. — I had a letter yesterday from Caroline, the first I have received for a long time: she appears in much better spirits, & even writes about baby-linen & such points, which shows she can now somewhat master her grief. — I do not believe the deaths of but few babies have caused more bitter grief than hers, & I fear it will be a great draw back to her happiness through life[15].
Goodbye my dear Fox, excuse this letter — I am very old & stupid

Ever yours,
C. Darwin

(*across head of letter*)
P.S. I shall direct to Delamere Rectory, though I dont know whether it is a Rectory or Vicarage, but one always ought to give the highest title[16].

Postmark: OC 24 1839
Christ's College Library, Cambridge, #58

Notes

1. In LLi, p 299, there is a short abstract drawn from this letter from "We are living a life of extreme quietness;" to "and I glory in thoughts that I shall be here for the next six months."
2. CD and Fox must have met at the meeting of the British Association in Birmingham on 26th Aug 1839. In his Journal CD records "*Aug 23d to Maer & thence on the 26th to Bermingham for the Meeting of the Brit. Assoc.*"
3. Maer Hall, near the village of Maer, Newcastle under Lyme, Staffordshire, the family home of Josiah Wedgwood II.
4. CD's illness continued soon after his marriage on 29th Jan 1839. In his *Journal*, CD records several periods of being unwell and unable to work in late 1838, in1839 and in1840. See Letter #63 (n7).
5. *Coral reefs* was not published until May 1842, mainly because of CD's illness (see '*Journal*', above Note 2; also see CCD, Vol 2, Appendix II).
6. Emma was expecting her first child. William Erasmus Darwin (1839-1914) was born on 27th Dec of this year.

7. This is interesting since the family letters report many interactions between CD and Emma and Erasmus Darwin, Hensleigh and Fanny Macintosh and the family of Sir James Macintosh, all of whom lived close to one another in, what is now, central London. CD probably used his illness to restrict his social life and get on with his books and his studies.
8. Samuel Fox III and his wife Ann, W D Fox's parents.
9. The lack of contact is surprising with the connection of Ann Fox née Darwin and Erasmus Darwin and the Derby Philosophical Society (see Chapter 1).
10. The motto was '*Cave et aude*' (Burke 1884; *Darwin Pedigrees* by R B Freeman, 1984). CD's sons William and George also used the motto. This is different to the motto used by Erasmus and Robert Waring Darwin ("*E conchis omnia*", Freeman 1978, p 70; see Chapter 1)
11. Gerard Crombie, a descendant of the Fox, has a family tree, which traces the Fox family ancestry to John of Gaunt in the 12th century (Copy in Christ's College Library).
12. See next note (13).
13. CD refers to William Darwin (1655-1682) who inherited the family home of Elston Hall, near Newark on Trent. See *LLi* : 2–3. See also letter from Susan Darwin (CCD), [*c.* 24 October 1839], n 4.
14. Refer also to Letter #65 (n14). CCD, Vol 2, notes "*Stukeley 1719, p. 963, refers to an account from 'my Friend, Robert Darwin, Esq; of Lincoln's-Inn, a Person of Curiosity, of an human Sceleton (as it was then thought) impress'd in Stone found lately at the Rev. Mr. John South's, Rector of Elston' etc. At the end of the notebook which CD used as his 'Journal' (DAR 158) he noted: 'Robert Darwin my great-grandfather is described in the Philosop. Transactions for 1719 as a person of curiosity who discovered near Elston a skeleton of some large animal.—'* "
15. Caroline Darwin (1800-1888), CD's sister, married Josiah Wedgewood, III, in 1837. Their first daughter, Sophy Marianne Wedgwood died on 31st January 1839. Their other 3 daughters lived long lives and Margaret (1843-1937) was the mother of Ralph Vaughan Williams (1872-1958), the composer.
16. Delamere Rectory was correct. The distinction is in terms of the historical founding of the position and its annual income. Delamere was at the time one of the newest parishes in England, and had been set up by the Crown in 1817. Fox was the fourth Rector.

#66 (CCD572). To: Rev[d] W. D. Fox, Delamere Rectory, Cheshire

[Maer]
[7th June 1840][1]

<about seven lines missing>[2]
old friend much to endure in the ill-health of your wife[3]. — There is a degree of anxiety about the health of
<about ten lines missing>
boy — (William Erasmus, — proper *family* names) — he is a prodigy of beauty & intellect. He is so charming, that I cannot pretend to any modesty. — I defy anybody to flatter us on our baby, — for I defy anyone to say anything, in its praise, of which we are not fully conscious. — He is a charming little fellow, & I had not the smallest conception there was so much in a five month baby: — You will perceive, by this, that I have a fine degree of paternal//

fervour. And now for myself, about whom I can now give a very good account. — I have during the last six weeks been gradually, though very slowly, gaining strength & health, but previous to that time I ~~have been~~ was for nearly six months in very indifferent health, so that I felt the smallest exertion most irksome. This is the reason I have been so long without writing//

I had no spirits to do anything. I have scarcely put pen to paper for the last half year[4], & everything in the publishing line is going backwards — I have been much mortified at this, but there is no help, but patience. — My Father, who has certainly in a quite unexpected degree put me on the course in getting well, (having put a stop to periodical vomiting to which//

I was subject) feels pretty sure that I shall before long get quite well. — At present I only want vigour — in wanting which, however, one wants almost all, which makes life endurable. — We came here the day before yesterday and are enjoying the delightful smell of the damp earth & plants. — I think we shall never//

be able to stick all our lives in London, & our present castle in the air is to live near a station in Surrey about 20 miles from Town[5]. — We shall remain here about a month[6]. — should my health continue getting stronger, I will, should all things be fitting with you, pay you a flying visit. — I want much to see your house. — it is always very pleasant to have a picture of a person's//

residence rise before one, when one thinks of the person. — At present, however, I am determined to obey implicitly my Father & remain absolutely quiet. — You speak in your letter of going with M[rs] Fox to the sea. — pray send me a short note before long, telling me your plans. — I hope M[rs] Fox came home from Osmaston[7] as well as could be expected after//

her confinement[8]. — What an awful affair a confinement is: it knocked me up, almost as much as it did Emma herself. — You ask about scientific news — I have none to tell, for I have of late literally seen no one except Lyell[9] occasionally. — He is going to start immediately to examine Fren*ch* Tertiary deposits[10] — I have not seen Henslow for a long time, though I have occasionally heard from him. He likes his living exceedingly, & is growing fat. — He is very active, giving scientific lectures to parish & whole neighbourhood: displaying fire works, & preparing for publication a continuation of Lindley & Hutton's fossil coal Flora[11]. — Pray remember me most kindly to M[rs] Fox, & believe me dear Fox. Your's affectionately

Chas. Darwin//

(*On the fly leaf*)
I really have not ever heeded of your recommendations that I ought to take account of the envelope & fold other smaller.

Incomplete
Postmark: Newcastle under Lyme JU 7 1840
Christ's College Library, Cambridge, #60

Notes

1. This date is deduced from the postmark and the age of CD's first child, William Erasmus, given as 5 months. The Correspondence of Charles Darwin, Vol 2, notes: "*CD's 'Journal' (CCD, Vol 2, Appendix II) for 1840 reads 'June 10th Went to Maer & paid a visit to Shrewsbury', but the entry appears to be a mistake made in retrospect.*"
2. Only the bottom few lines remain and the rest has been excised from the first and second pages. As this is the time of the first "penny black" of the new postal system of Rowland Hill (1st April 1840), it is possible that the first part of the letter was cut off to obtain the stamp. Later letters, around this time, were in envelopes and the stamps are missing. Penny blacks have always been plentiful so their price has not been great; but, nevertheless, a good stamp could fetch a reasonable price.

3. Fox's wife, Harriet, seems to have been sick again; we know from Fox's letter to his Sister-in-law, Miss Fletcher (Appendix 3, DAR 250: 54) that she was identified with an incurable disease, probably consumption, at about this time. Nevertheless she had recently given birth to their fourth surviving daughter, Julia Mary Anne (1840-1883) on 30th April of this year and would give birth to a fifth child, Samuel William, on 3rd July 1841. She died of "disease of the lungs" on 28th March 1842.
4. As noted in the last letter, #65 (n5), this delayed publication of "*Coral Reefs*".
5. This is exactly what they did, when they bought their home at Downe, a few months later.
6. See Note 1.
7. Osmaston Hall, near Derby, the family home of Samuel Fox III, Fox's father (see Fig 4).
8. It is clear from this, and from the *Darwin Pedigrees*, that Julia Maria Anne Fox was born at Osmaston Hall, the home of Samuel Fox III and family.
9. Charles Lyell (1797-1875). See Biographical Register and Letters #56, (n10), #58, (n13);
10. Lyell left for France in July [Wilson, L G (1972), *Charles Lyell: the years to 1841*, Yale University Press; pp 490–2].
11. Lindley J and Hutton W 1831–7. The fossil flora of Great Britain. 3 Vols. London.

#67 (CCD 586). To: Revd W. D. Fox, Delamere Rectory, Northwich, Cheshire

12 Upper Gower St
Monday [25th Jan 1841]

My dear Fox

It is a long time since we have had any communication, — I daresay you will be glad to hear how I am going on, & I wish to hear of Mrs Fox and yourself. — My strength is gradually, with a good many oscillations, increasing; so that I have been able to work an hour or two several days in the week — I have at

last to my great joy sent the last page of M.S. of the Bird Part to the printer[1]. — I am forced to live, however, very quietly and am able to see scarcely anybody & cannot even talk long with my nearest relations. I was at one time in despair & expected to pass my whole life as a//

miserable useless valetudinarian but I have now better hopes of myself — You see I treat you, like a very old good friend, as you have always been to me, & write at great length about my own poor carcase. As for news of any kind, I am not in the way to give any to any body — We have//

the pleasure, at present, of a visit from Susan[2], who is in a very flourishing state of health, which for a Darwin is something wonderful. — but I will say nothing more about health, & as a consequence I must say nothing more about any of my Family — I will just add that Emma expects to be confined//

in March[3] — a period I most devoutly wish over — Our little boy is a noble fat little fellow & my Father has christened him Sir Tunberry Clumsy[4]. — Pray let me hear soon how you are all going on. — I hope this very severe winter has not much affected Mrs Fox and that your lungs//

have stood it pretty well[5]. — My dear old friend you have much to support.

Yours affectionately
Ch. Darwin

PS. If you attend at all to Nat. Hist — I send you this P.S. as a memento, that I continue to collect all kinds of facts, about "Varieties & Species" for my some-day work to be so entitled[6] —the smallest contributions, thankfully accepted — descriptions of offspring of all crosses between all domestic birds & animals dogs, cats &c &c very valuable — Dont forget, if your half-bred African Cat should die, that//

I should be very much obliged, for its carcase sent up in little hamper for skeleton. — it or any cross-bred pidgeon, fowl, duck, &c &c will be more acceptable than the finest haunch of Venison or the finest turtle. — Perhaps all this will only bothers you — So I will add no more, except, should you ever have opportunity when in Derbyshire, do enquire for me, from some person you told me of whether offspring of male muscovy & female common duck, resembles offspring of female muscovy & male common[7] — How many hybrid eggs are produced. — //

P.S. I enclose a Pamphlet on Oaths by Hensleigh Wedgwood[8]. — Should any **fitting occasion** happen, would you object to ask Sir. P. Egerton[9] to look over it. — Hensleigh is very anxious for ~~every~~ as many M.P: as possible, to consider the subject. —

Postmark: JA 25 1841
Christ's College Library, Cambridge, #59

Notes

1. The final volume of *The Zoology of the voyage of H. M. S. Beagle .... during the years 1832 to 1836 (Birds)*, edited by Charles Darwin, was published in March 1841.
2. Susan Darwin (1803-1866), CD's sister.
3. Anne Elizabeth Darwin (1841-1851) was born on 2nd March of 1841.
4. Sir Tunbelly Clumsy is a character in two plays: John Vanbrugh's *The relapse* (1696) and Richard Brinsley Sheridan's *A trip to Scarborough* (1777). (*Oxford companion to English literature*)
5. This is one of the few references that tell us what affliction Fox suffered from: see also Letters #48 (n4); #49 (n5).
6. The book that CD planned (in three volumes) but never published is often referred to as "*Natural Selection*" (see Stauffer, 1975). However, CD used the original idea for that book in the book that he rushed into print following the Alfred Wallace bomb shell in 1858 (see Chapter 6). CD wanted this book to be called *An abstract of an essay on the origin of species and varieties through natural selection or the preservation of favoured races in the struggle for life* but eventually under pressure from the publisher it was shortened to *On the origin of species by means of natural selection, or the preservation of favoured races in the struggle for life* (Murray, 1859).
7. This is the first of many references in their future correspondence to many kinds of geese and ducks.
8. See Letter #62 (n8), also CCD, Vol 1, letters from Catherine Darwin, 27 November 1833; from Caroline Darwin, 30 December [1833] – 3 January 1834, and *Emma Darwin* 1: 257–8, 285–6).
9. Sir Philip de Malpas Grey-Egerton (1806-1881). Palaeontologist specializing in fossil fish. He lived at Oulton Park, Tarporley, about 10 miles away from Fox's parish of Delamere and represented South Cheshire as a Tory MP from 1835-1868. The Fox *Diaries* show that Egerton and Fox were well acquainted. (*DNB*)

#68 (CCD 606). To: The Rev W. Darwin Fox, Delamere Rectory, Chester

[12 Upper Gower St]
Monday [23rd Aug 1841]

My dear Fox

I was very glad to hear from you — I had thought of writing but purposely deferred it, till I learned how your family were. — You must have gone through fearful suspense & suffering — I hope Mrs Fox has not had more than her usual suffering to endure with her wonderful fortitude. — I am glad to hear the poor Baby's life is at last pretty secure[1] — I well remember//

your letter to me in the Beagle & your request, which I shall be most happy to comply with. — you did not tell me his proposed name — But before I say anymore I conceive myself bound to tell you, that we have not had Godfathers or Godmothers to our children, — not from any objection to their having such — but as we should in that case have been obliged to have stood proxies & we//

both disliked the statement of believing anything for another. I earnestly trust this will make no difference in your making me your childs Godpapa, but with your deep feelings on religion, I thought possibly you might much dislike having a Godfather who could[2] stand in propriâ personâ as such. — I feel sure you will prefer my having been open on this head. — //

We are all well here and our two babies are, I think, strong healthy ones, & it is an unspeakable comfort this — For myself I have steadily been gaining ground & really believe now I shall some day be quite strong — I write daily for a couple of hours on my Coral volume & take a little walk or ride every day — I grow very tired in the evenings & am not able//

to go out at that time or hardly receive my nearest relations —but my life ceases to be burthensome, now that I can do something. We are taking steps to leave London & live about 20 miles from it on some Railway — we are going in a few days to see house & land 6 miles north of Windsor: we shall not actually be transplanted till next spring[3]//

I had a letter two days since from Henslow[4] , who appears very happy & flourishing — giving lectures, displays of fireworks, initiating agricultural prizes, & I do not know what besides, for his Parishioners — As I tell him, I wish Botanists were his Parishioners, he would then do a little//

more scientific work — I presume your wife & children are at Osmaston[5] & that you are by yourself at Delamere, but I direct this there, as I hope it will catch you. — Pray let me hear from you before long about my being Godfather[6] . — You & I will be then, what all Spaniards who hold that relation to each other, delight to call//

each other "compadre" so my dear old expectant compadre farewell.

Your affect friend
Charles Darwin

Postmark: AU 24 1841
Christ's College, Cambridge, #61

Notes

1. Samuel William Darwin Fox (1841-1918), born 3rd July of this year.
2. This is the place for the critical "not" referred to in the next letter (note 2).
3. The search for a house was successful the following summer. CD and family moved to Down House, Downe, near Farnborough, Kent, in September 1842.

4. John Stevens Henslow (1797-1861). See Bibliographic Register, Introductions to Chapters 2 & 3, Letter #8 (n3) and many subsequent letters.
5. Fox's wife had gone to his father's (Samuel Fox, III) home again to deliver her baby, see Letter #65 (n7).
6. There is no record that CD was godfather to any of the Fox children. However, the next Letter (#68) seems to confirm this point, even though CD was not at the christening.

#69 (CCD 609). To: The Rev W. Darwin Fox, Delamere Rectory, Cheshire

12 Up. Gower St
Tuesday [28th Sept 1841]

My dear Fox

I was very glad to receive so much an improved account of your wife — I heard from Shrewsbury, a few days before I received your note, that she had returned home & was surprisingly better — I earnestly hope for all your sakes she may continue to suffer less, though, as you say at the expence of being an invalid. — It is a long//

time now since I saw <you both> at the Isle of Wight[1]. — Pray remember me very kindly to her, & as you seem to like the title of compadre, I hope she will not <dislike> being my commadre, which I have heard Spaniards, I know not whether in joke or earnest, call the wives of their compadres. — You must give Mss. W. D. Fox a kiss for me, that is if you are//

as fond of kissing babies, as I am — some fathers are more cleanly in their tastes. — I am very glad you did not reject me as god-father — I hoped & expected you would not. — Has the ceremony taken place? Our children are baptized. — You were right in thinking I had left out the critical not[2]. — I forget how long it is since I received your note, but I suspect I have left//

it rather long unanswered, which you would forgive, if you knew what a turmoil I have been living in house-hunting. — We have seen one near Chobham & Bagshot in Surrey, wh. we think will suit, but the vendor asks an exorbitant price[3]. — I long to be settled in pure air, out of all the dirt, noise vice & misery of this great Wen, as old Cobbett[4] called it — I am going to Westcroft (the name of one place) on Friday with a//

valuer & then mean to make an offer — We shall not move till next summer indeed I fear from requisite alterations not until Autumn. I envy your discovery of the Cheirotherium footsteps[5] — it must have been very interesting encountering these relics, which seem more real & recall the past far more vividly than old bones — the very emblem of the past. — //

I suppose I shall hear more about them in course of time. — Is it not singular how long obvious phenomena remain unobserved! I never cease marvelling at this. — I am very glad you mean occasionally to pay London a visit — nothing on earth will be easier than coming on to us, when//

we are settled near Railroad in Surrey. — If you come up to town this autumn or winter do let me know beforehand & I have no doubt we shall be able to give you a bed & then we could see more of each other far over with less fatigue to me. — But I am grown a dull old spiritless dog to what I//

used to be. — One gets stupider as one grows older I think[6]. —

Ever yours
My dear Fox
C. Darwin

Postmark: SP 28 1841
Christ's College, Cambridge, #62

Notes

1. 20–1 November 1837, see Letter #63 (n1).
2. See letter #67 (n5).
3. In the last letter #67 (n2) it was Windsor. Now it is Chobham and Bagshot in Surrey. Finally they chose Kent, where CD and family moved to Down House, near Farnborough (or Bromley), Kent, in September 1842.
4. William Cobbett (1763-1835). Writer politician and agriculturalist. Author of *Rural Rides*, 1830, in which he uses this term for London. (*DNB*)
5. A creature known from footprints found in beds of the Red Sandstone in Saxony and named by Johann Jakob Kaup. See Buckland W (1836) *Geology and mineralogy considered in reference to natural theology*, London; Vol 1: 263–6 n.
6. CD is now is now 32 years old!

#70 (CCD 624). To: The Rev W. Darwin Fox (Envelope missing)

12 Upper Gower St
March. 23^{d}[1842]

My dear Fox.

Your affecting account of the loss of your poor wife[1] was forwarded to me yesterday by Susan[2]. — The cessation of her sufferings, which she seems to have borne with such noble, & devoted patience, must be a great relief to all. — I truly sympathise with you, though never in my life, having lost one near relation[3], I daresay I cannot imagine, how severe grief, such as yours, must be, & how little the longest expectation can resign one to the blow, when it falls. Few, I am sure will bear it better than you, although from your most affectionate disposition I cannot fancy anyone feeling it more deeply. — How fortunate Miss Fletcher's arrival was[4].
Your children must be now an infinite comfort to you; I trust they are well & again cheerful, for it is their peculiar blessing to love sweetly & yet not to mourn long. —
— Do write to me, my dear old friend, as soon as you have leisure & some degree of inclination & tell me how you are in health & how your children are — believe me. My dear Fox

Your sincere friend
Charles Darwin

Christ's College, Cambridge, #57

Notes

1. Harriet Fox died, of "disease of the lungs", at the Rectory, Delamere, Cheshire, on 19th March 1842 (*Cheshire Archives, Darwin Pedigrees*).
2. Susan Darwin, CD's sister, who stayed with Charles and Emma Darwin in London, often, at this time.
3. Although CD's mother, Susannah Darwin, died on 15 July 1817 (*Darwin Pedigrees*), when CD was eight years old, CD, in later years, said that he could 'remember hardly anything about her' (*Autobiography*, p. 22). See also *CCD* Vol. 2, Appendix III.
4. Harriet Fox's older sister; see Fox's letters to her concerning Harriet's loss in Appendix 3.

#71 (CCD 625). To: The Rev W. Darwin Fox (Envelope missing)

12 Up. Gower St.
Thursday [31st March 1842]

My dear Fox

I have purposely delayed writing to you for some little time, until the last and most sacred ceremony had past[1]. Your note was most affecting to read — such tender love and so soon separated. — What a comfort it must be to you, — that is I think I should find it the greatest — the having children[2] — it must make the separation appear less// entire—the unspeakable tenderness of young children must soothe the heart & recall the tenderest however mournful remembrances —

My dear old friend, I do not know why I write, for I can offer only sympathy, & I can well imagine what small comfort that must be — I trust your bodily health has withstood your trial pretty well. You must have a good deal of occupation at present, which// I should think would be your best relief — My health has been as usual oscillating up & down, but I certainly gain strength though at a very slow pace. — my two dear little children[3] are very well and very fat. Emma however is uncomfortable enough all day long & seldom leaves the house, — this being her usual state before her babies come into the world[4] — I trust time & your own innate powers of cheerful endurance will eventually make your future prospects appear// brighter than can now seem possible. — Farewell. — Let me hear from you again.

My dear Fox farewell

Charles Darwin

Christ's College Library, Cambridge, #56

Notes

1. Harriet Fox was buried on 28 March 1842.
2. There were five surviving children (*Darwin Pedigrees*).
3. William Erasmus and Anne Elizabeth.
4. This was Mary Eleanor Darwin; born 23rd Sept 1842, died 16th October 1842.

#72 (CCD 654; LLi, part). To: The Revd W Darwin Fox, Delamere Rectory, Chester

Down near Bromley Kent
Friday [9th Dec 1842]

My dear Fox

It is a long time since we have had any communication. They forwarded to me from Shrewsbury your note of enquiry about all of us[1]. — I by this heard of your return[2] & I have since quite lately heard of your travelling with my cousin Eliza Wedgwood[3] in the Railway, accompanied by your two very pretty little girls[4]. I was very glad to hear, that you were looking pretty well — you will hardly believe yourself how very bad you looked when I last//

saw you in Gower St[5]. What a great tour you have taken; I hope you have enjoyed it to a certain extent; under other circumstances, I should think travelling wd just suit your activity of mind & body — I should like to hear a little more of your proceedings & what struck you most. — I always feel more inclination for Switzerland, than any other part. — It is very hard that I should never have ~~been~~, as Erasmus says I never have been, abroad[6]. We like our new purchase <of this place> very well: it is not perfection, but there will//

always be drawbacks about every place[7]. It is very retired, indeed I do not think I ever walked in so retired a country, hardly excepting those great grass fields on the road to Whittlesea meer, from the Inn, in which we slept one famous night[8]. — our house lies high on the chalk — is good, largish & was cheap.— there is one fine view, but the country is rather dreary & what is worse very slippery from the thin bed of clay, which every where overlies the chalk. — Our removal has answered very well; our two little souls are//

better & happier[9] — which likewise applies to me & to my good old wife.— I wonder when will you come & see us — if ever you are in town & can possibly spare a few days I charge you to let us know & you can be here in about 2 hours from London Bridge[10]. — I attend the councils at the Geolog. Soc. (where by the way we have lately had a fracas, about Charlsworth the late Editor of the Journal of Nat Hist.[11] — who was rejected as candidate for poor Lonsdale's place (who has retired from ill-health) on account of ill- ~~bad~~ temper//

and consequently Mr Charlsworth got a special meeting up to call us of the council over the coals — but he burnt his own fingers — for the meeting of a hundred & odd members, with not one dissentient voice — voted the council "their grateful thanks". I hope by going up to town for a night every fortnight or 3 weeks to keep up my communication with scientific men, & my own Zeal, & so not to turn into complete Kentish Hog — I am now preparing a very thin//

volume or pamphlet on the volcanic islands visited by the Beagle — it forms a second Part[12] to the First on coral Reefs, now published. —

You will have heard in what a lamentable state of health, Emma's father is sunk into[13]. It is melancholy what a changed house Maer is[13]; within a few years there was always a large party there, full of intelligence & activity, & now all are scattered, except Elizabeth Wedgwood[14], & both Mr & Mrs//

Wedgwood are bed-ridden & enfeebled in mind & body. My Father keeps pretty well; but his powers of movement fail him. — I trust Mr & Mrs Fox are well[15] & all your family. — if any of your sisters are staying with you pray remember me to them, & believe me my dear Fox

Most truly yours
C. Darwin

Postmark: De 9 * * to D 10 1842
Christ's College Library, Cambridge, #64

Notes

1. There were active lines of correspondence between CD at Down, CD's sisters and father in Shrewsbury, Fox and family at Delamere and the family of Samuel Fox III at Osmaston Hall, Derby. No letters or diary entries have survived for this time to indicate who was involved in this particular communication.

2. There is an account by W D Fox of a visit to Munich, southern Germany and Switzerland, in October 1842 in DAR 250, Cambridge University Library (see also letter from Fox to his sister-in-law from Heidelberg, etc, 1842; Appendix 3).
3. Sarah Elizabeth Wedgwood (1793-1880), who was at this time looking after her bed-ridden parents, Josiah (II) and Elizabeth Wedgwood at Maer Hall (see CCD, Vol 2, Elizabeth Wedgwood to CD, 25th Sept 1842).
4. At this time Fox had four girls and a baby boy, all by his first wife. The eldest girl, Eliza Ann would have been 6 ½ years old and the next, Harriet Emma, 4 ½ years old.
5. There is no record of a meeting at 12 Upper Gower St. However, Fox had relatives in London and could easily have visited in the period that CD and Emma Darwin resided at Gower Street (i e, early 1838 to Sept 1842).
6. CD visited continental Europe on only one occasion; this was in the spring of 1827 when he visited Paris with his uncle Josiah Wedgwood II and his sister Caroline (the last two on their way to Geneva to accompany Josiah's daughters back to England (Litchfield, 1915, Vol 1, p 100). Erasmus Darwin, CD's brother, may have being making a joke - that seeing Paris was not seeing Europe.
7. This is the first letter from Down House. CD purchased it in late 1841 (see letters in CCD Vol. 2) and they moved in, between 14 -17th Sept. 1842 (Freeman, 1978).
8. See *Autobiography* and Letters #47 (n5 & 6), #48 (n17), #61 (n18), #90 (n17), etc.
9. They had William Erasmus and Anne Elizabeth. Their third child, Mary Eleanor Darwin, born on 23rd Sept. 1842, had died on 16 October 1842, after the move to Down; the strain of the move on both mother and child may have contributed to the death of this three week-old child. CD fails to tell Fox about this recent loss.
10. By rail to Sydenham or later to Beckenham (see Letter #115 (n5). Two hours represents the total time: the train journey would be less than 1 hour and then the carriage would take approximately 1 hour.
11. Edward Charlesworth (1813-1893). Naturalist and palaeontologist. For details of this affair see: Jackson C.E and Davis, P. 2001, Sir William Jardine, W *A life in natural history.* p 152. Continuum International.
12. *Geological observations on volcanic islands visited during the voyage of H.M.S. Beagle, together with some brief notices on the geology of Australia and the Cape of Good Hope.* Smith, Elder and Co., London, 1844.
13. See Note 3.
14. See Note 3
15. Samuel Fox, III (1765-1851), (1765-1851), and his wife Ann (née Darwin) (1777-1859).

#73 (CCD 665; LLi, part[1]). To: The Revd W Darwin Fox, Delamere Rectory, Chester.

Down. Bromley Kent
Saturday [25th March 1843]

My dear Fox

I was very glad to get your letter some six weeks since, but grieved to the heart at its contents. Your's does indeed seem at present a hopeless case — I should have thought that time *inevitably* would have done you more good, than it seems to have done — I had hoped (for experience I have none) that the mind wd have refused to dwell so long & so intently on any object, although the most cherished one[2] — Strong affections, have always appeared to me, the most noble part of a man's character & the absence of them an irreparable failure; you ought to//

console yourself with thinking that your grief is the necessary price for having been born with (for I am convinced they are not to be acquired) such feelings — But I am writing away without really being able to put myself in your position — you have my sincerest sympathy & respect in your sorrow — I can only hope that the intensity of your grief may shorten (however little you may think it possible) their duration. — I will tell you all the trifling particulars about myself, that I can think of. — we are now exceedingly busy, with the first brick laid down yesterday//

to an addition to our house[3]; with this, with almost making a new kitchen garden & sundry other projected schemes, my days are very full. I find all this very bad for geology — but I am very slowly progressing with a vol: or rather pamphlet on the Volcanic islands, wh we visited[4]; I manage only a couple of hours per day, & that not very regularly. — It is up-hill work, writing books, which cost money in publishing & which are not read even by geologists[5]. — I forget whether I ever described this place: it is a good, very ugly house with 18 acres; situated on a chalk-flat, 560 ft above sea — There are peeps of far-distant country & the scenery is moderately pretty; its chief merit is its extreme rurality; I think//

I was never in a more perfectly quiet country: Three miles South of us the great chalk escarpment quite cuts us off from the low country of Kent, & between us & the escarpment, there is not a village or gent^{s}: house, but only great woods & arable fields (the latter in sadly preponderant numbers), so that we are absolutely at extreme verge of world. — The whole country is intersected by foot-paths; but the surface over the chalk is clayey & sticky, which is worst feature in our purchase. — The dingles & banks often remind me of Cambridgeshire & walking with you to Cherry Hinton, & other places, though the general//

aspect of the country is very different. I was looking over my arranged cabinet (the only remnant, I have preserved of all my English insects) & was admiring Panagæus crux major[6]: it is curious the vivid manner in which this insect always calls up my mind, your appearance, with little Fan[7] trotting after, when I was first introduced to you— Those entomological days were very pleasant ones — I occasionally hear from Henslow[8]: he is absorbed at present with making the Suffolk farmers experimenters in agricultural chemistry[9], & has commenced by doing wonders in exciting their zealous cooperation. He seems very//

busy with parish concerns — We are all well here — that is essentially so for Emma is as bad as she usually is in her present state[10].— I am much stronger corporeally, b*ut* am **very*[11] little better in being able to stand mental fatigue or rather excitement, so that I cannot dine out or receive visitors, except relations with whom I can pass some time after dinner in silence[12]. Farewell my dear Fox with my best wishes. —

Ever yours
C. Darwin

Postmark: 27 MR 1843
Christ's College Library, Cambridge, #66

Notes

1. In LLi p 321, from "I will tell you all the trifling particulars" to "These entomological days were very pleasant one —"; and the last sentence "I am much stronger corporeally, but am *very* little better in being able to stand mental fatigue or rather excitement, so that I cannot dine out or receive visitors, except relations with whom I can pass some time after dinner in silence."

2. No diaries are extant for Fox for these years. The first is for 1844 and in that diary Fox is clearly still grieving for his wife, Harriet, who died in 1842 (see Letters #69 & #70). In the transcription of this letter done by Francis Darwin there is a note: "*Mr Fox had recently lost his wife*" (DAR 144: 135-296).
3. Extra bedrooms and a school room were added and bay windows placed on the drawing room and main bedroom (Healey 2002, p 235).
4. *Geological observations on the volcanic islands visited by H. M. S. Beagle, together with some brief notices of the geology of Australia and the Cape of Good Hope.* Smith Elder and Co, 1844.
5. CD had discovered the joys of being an author.
6. A rare carabid beetle, which CD took with Fox and Albert Way at Whittlesea Meer [see Letters) #47 (n5 & 6), #48 (n17), #61 (n18), etc].
7. Fan was Fox's dog [Letter #1 (12)], which he apparently took to Whittlesea Meer (see Note 6).
8. J S Henslow (1797-1861). See Bibliographic Register, Introductions to Chapters 2 & 3, Letters #8 (n3) and many subsequent letters.
9. See Darwin's Mentor by Walters and Stow, 2001.
10. Emma was in the early stages of pregnancy. She delivered Henrietta Emma Darwin on 25th September 1843.
11. The page is torn at this spot.
12. CD gives further details of his illness, which was associated with sensibility to excitement (see Colp, 1977).

#74 (CCD 715). To: The Revd W Darwin Fox, Delamere Rectory, Chester (envelope extant but stamp removed)

Down. Bromley Kent
Monday [20th Nov 1843]

My dear Fox

I have been a neglectful dog, for I see in your last note, an age since, that you ask for one line to hear about Emma & her Babby — She has never had so good a recovery & there never was such a good little soul — as Miss Henrietta Emma Darwin[1] — she is beginning to smile & be very charming, — though how//

she has any idea, except whether the milk comes fast or slow is hard to conjecture— I have now nearly got this place in order, though there is much yet to do — & I think when you next see it, you will think it greatly improved — I have made a nice little orchard, all dwarfs, as you, for one, recommended. — Susan is staying here[2] & likes//

the quiet of the place much. She has leave of absence, as Caroline & Jos[3] have not yet removed from Shrewsbury, where Caroline produced a little girl, a few days after our affair[4]. — We continue in the same profoundly tranquil state, as when you were here[5], & have only got to know one person well, namely Lady Lubbock[6], who is a very nice person, absorbed with the education of nearly a dozen very nice children. — I wish, with//

all my heart that we were, as you say, rather nearer, but remember we are near London & that in effect is being near everyone. — They heard of you the other day at Shrewsbury from Col. Egerton[7], who spoke much of you. — my Father likes Col. Egerton particularly & he must be a very pleasant neighbour for you. — But I fear this last sentence & all that speaks of pleasure grates on your feelings[8] — I trust you are well bodily & believe me

Ever yours —
C. Darwin

Postmark: NO 21 1843
Christ's College Library, Cambridge, #68

Notes

1. Henrietta Emma Darwin, born 25th Sept 1843.
2. See Letter #69 (n2).
3. Caroline Wedgwood (née Darwin) and Josiah Wedgwood, who were married in 1837.
4. Margaret Susan Wedgwood (1843-1874).
5. There is no record, apart from this of Fox having visited Down, prior to his visit in 1846, with his newly-wed second wife (see Introduction to this Chapter). However, there is no record either of the meeting at Gower Street before the 9th Dec 1842. There are no extant diaries for Fox, or letters from Fox to Darwin, during these years. Presumably Fox visited the Darwins between the time that they moved into Down House in September, 1842 and Nov. 1843.
6. Harriet Lubbock (1810-1873). Married John William Lubbock, 3rd Baronet, in 1833. He was an astronomer and mathematician and, later on, a good friend of CD; Their son was John Lubbock, First Baron Avebury. (*Burke's Peerage, 1980*)
7. Charles Bulkeley Egerton (1774-1857). Colonel in the army and later General. Uncle of Philip de Malpas Grey Egerton (see Letter #66 (n9). Both are mentioned often in Fox's diaries. (*Burke's Peerage, 1980; DNB*)
8. Fox was still mourning the death of his wife Harriet, who died on 19th Mar 1842, see Letter #70.

#75 (CCD 801). To: The Revd W Darwin Fox, Delamere Rectory, Chester

Down near Bromley — Kent
Dec. 20th[1844]

My dear Fox

I was on the point of going to London when your note arrived, whence I returned yesterday, from visiting Erasmus with whom Susan[1] staying. She begs me to say, (which I cd have said) that my Father, she is sure will be very glad to see you at any time, whenever convenient to//

yourself, when you will write to propose yourself, & if my Father is not then very well, he will have no scruple in deferring your visit. My Father's health, I grieve to say, is now *very* uncertain: he has just lately been very well, but about 3 weeks since he had rather a severe attack upon his chest. He has been prevailed upon to sleep down stairs, but it was most//

painful to him, <thus> to give up one step in life. He was upon the whole very cheerful when I was there. — Illness with his figure & constitution is very dreadful[2]. With respect to mesmerism, the whole country resounds with wonderful facts or tales[3] : the subject is most curious, whether real or false; for in the latter case, what is human evidence worth? I am astonished at your zeal, with respect to Miss Martineau[4], for//

to my mind, the girls case[5] bears more plainly, than I sh^{d} have expected, the mark of deception, possibly involuntary. You are no doubt aware that many doctors (for example Dr Holland[6]) have <long ago> remarked how marvellous a diseased tendency to deception there is in disordered females. Sh^{d} your zeal still continue, I w^{d} write to Miss Martineau & propose your visiting her (my Brother wishes to avoid all communication with her on this subject)— When in London, I saw a letter from her (not to my Brother), in which she says **crowds** of people are//

coming to her from all parts of England; *she does not seem to dislike <this>*, but she says she is going very soon to leave Tynemouth for rest from visitors; so you w^{d} have to go at once if you do go. I have just heard of a child <3 or 4 years old> (whose parents & self I well know) mesmerized by his father, which is the first fact, which has staggered me. — I shall not believe, fully, till I see or hear from good evidence of animals (as has been stated is possible) not drugged, being put to stupor; of//

course the impossibility w^{d} not prove mesmerism false: but it is the only clear *experimentum crucis*, & I am astonished it has not been systematically tryed. If mesmerism was investigated, like a science, this c^{d} not have been left till present day to be *done satisfactorily*, as it has been, I believe, left. — **Keep** some cats **yourself** & do get some mesmerizer to attempt it. One man told me he//

had succeeded, but his experiments were most vague, as was likely from a man, who said cats were more easily done, than other animals, because they were so electrical.!! (Miss Martineaus case was tumor, I presume, of the womb or Ovaria. the tumor is reduced greatly & her deafness decreasing[7]: she is in a most excited state. My Father has often known Mania relieve incurable complaints.)—
As long as you like to receive the Athenæums[8], I shall have *pleasure* in sending them.//

Shd you <happen to> call this winter at the House, in which the "Rippons & Burtons (R. 389) Vesta Stove"[9] is, w^{d} you be so kind as ask whether they continue to like it. 2^{d} whether it often wants feeding. 3^{d} whether it gives out much heat, enough to warm a **cold** House. I must get a new one: & I am confounded with so many recommendations. —
Pray excuse this very untidy note, written in a Hurry & believe me

ever yours
C. Darwin

Postmark: DE 20 1844
Christ's College, Cambridge, #70

Notes

1. Erasmus Alvey Darwin (1805-1881), CD's brother, and Susan Darwin (1803-1866) CD's third sister.
2. On his 80th birthday (30 May 1846), Robert Waring Darwin weighed 22 stone 3½ pounds ('*Weighing Account*' book, Down House MS). He died on 13th Nov 1848.
3. Mesmerism, or animal magnetism, was popularised in Vienna and Germany in the late eighteenth century by Franz Mesmer. A developed form of mesmerism became popular in England, and especially in London, in the 1840s, when séances were very fashionable, to call up spirits, for example, of the

deceased. These continued up to the turn of the century and beyond. It is interesting to note CD's very scientific approach to the subject. Alfred Russel Wallace, the cofounder of the theory of evolution with CD, was a strong believer in mesmerism (Slotten, 2004).
4. Harriet Martineau (1802-76). Author and reformer. A friend of Erasmus Darwin and the Wedgwood family. Her health had broken down in 1839. In 1844 she undertook a course of mesmerism, after which her health was dramatically restored. She published an account of her cure in the *Athenæum* and subsequently published it in a collection *Letters on mesmerism* (1845)(see Logan and Sanders, 2007; and also Arbuckle, ed,1983, p 78).
5. A partially blind 19-year-old girl, Jane Arrowsmith, according to Harriet Martineau, while in a mesmeric trance at a séance on 15 October 1844, predicted the safe return of the crew of a wrecked ship, see Note 4, above.
6. Henry Holland ((1788-1873). Physician. Distant relation of both the Darwins and the Wedgwoods. He treated CD, Emma Darwin and Queen Victoria. He was made a baronet in 1853. (*DNB*)
7. Harriet Martineau (note 4, above) suffered from 'prolapse of the uterus and polypous tumours' and had been deaf from the age of 20 (R K Webb, 1960).
8. This refers to the publication, "*Athenæum*" of the Athenæum Club, to which CD belonged. CD passed this publication on to Fox.
9. A form of slow-combustion stove; see F. Edwards, 1870 (Longmans Green and Co. 2nd Ed.); however, in the next letter [#76 (n7)], CD says that he now prefers the Arnotts stove.

#76 (CCD 827). To: The Rev^d W Darwin Fox (no envelope exists for this letter).

Down Bromley Kent
Thursday.[13th Feb 1845][1]

My dear Fox

I am very glad to be able to tell you that my Father is at present very well; he has, however, till lately been suffering a good deal from a cough. — When do you intend to pay your Shrewsbury visit? Thanks for your letter & all your news of your Noah's Ark[2]; it w^d really be a curiosity to see the place, besides other pleasures. Emma would have as much pleasure I think, as myself, in accepting//

your very cordial invitation, but I fear there is a reason against it, which even you will admit to be paramount, viz her confinement in the middle of July[3] & as soon as the Baby can travel viz Sept 1st we must visit Maer[4] & Shrewsbury. — It really is one of my heaviest grievances from my stomach, the incapability & dread I have of going anywhere: I literally have not slept, I believe, out of inns, my own, Erasmus' Shrewsbury & Maer houses, since I married. — My stomach continues//

daily badly, but I think I am decidedly better than one or two years ago, as I am able with rare exceptions to do my three hour's morning work, which at present is on the Geology of S. America[5], & very dull it will be. — You ask about Down, the house is now very comfortable; & the garden will be tolerably so, when the evergreens are grown up; I continue to like it very well; & its thorough rurality is invaluable. — By the way, do not ask about the Vesta stove[6] (& thanks for your remembering it) for after hearing it from other//

quarters, also, highly praised, I *saw lastly* an Arnotts[7] stove in action, & that struck the balance, & I have got one, & it answers admirably, requiring feeding only twice in 24 hours, & once will do. — I am very glad to hear you are coming to London, & do, if you possibly can, come on here; we shall be heartily & cordially glad to see you, & you now know, how quietly & in the eyes of most, dully we go on[8]. — I have not heard anything//

lately about Mesmerism; except Sidney Smiths[9] dream, (who is said to be dying[10]) that he dreamt he was in a madhouse & that he was confined in the same cell with Miss Martineau & the Bishop of Exeter[11]. It is said that the remarks on Miss. M. in the Athenæum, were by Brodie[12]; I thought them much superior to the general writing in that paper. — I hear Miss. M. is in so excited a state about Mesmerism, that she can hardly keep on peace with her old friends, who are unbelievers; I wonder how she & Erasmus will get on[13]. — I simply feel,//

that I *cannot* believe, in the same spirit, as it is said, that ladies do believe on all & every subject. I read aloud Arnolds Life[14] to Emma & liked it very much; I wish he had had rather a more lightsome & humorous spirit: as Carlyle would say, he was no "sham"[15].

Farewell my dear Fox.
I hope we shall soon meet here.

Ever yours
C. Darwin

P.S. I sh$^{\underline{d}}$ like very much <sometime> a few of my potatoes[16], & chiefly to get true seed from them, & see whether they will sport or not readily. —

Postmark: FE ** 1845
Christ's College, Cambridge, #69a

Notes

1. CCD based the date on an endorsement, not by Fox, of 'Feb 14 45' and on the death of Sydney Smith, see n 5, below. From the reference to the Vesta Stove it also clearly follows Letter #73.
2. Fox kept a large number of different varieties of domesticated animals for observation and breeding purposes.
3. George Howard Darwin (1845-1912), CD's second son, was born 9th July of this year.
4. Josiah Wedgwood, II, died on 12th July 1843, but Emma's mother, Elizabeth (Bessy) Wedgwood, and sister, Sarah Elizabeth (Elizabeth) Wedgwood, still resided at Maer, and were joined by Charles Langton and his wife Charlotte, Bessy's daughter in late 1842.
5. *Geological observations of South America*. Smith Elder and Co. 1846.
6. See Letter #75.
7. A form of slow-combustion stove. Two versions are described in *Our domestic fire-places* by F. Edwards 1870 (Longmans Green and Co. 2nd Ed. P. 162, figs 132-137).
8. Fox did visit CD and family at Down House, but not until after Fox's marriage, in May 1846.
9. Sydney Smith (1771-1845). Clergyman, essayist and founder of the Edinburgh Review. He was a friend of Erasmus Alvey Darwin and his circle. (*DNB*)
10. Sydney Smith died on 22 February of this year.
11. This refers to Harriet Martineau (see "Introduction" to this chapter and Letter #75, Note 4) and Henry Phillpotts. Martineau was an outspoken radical; the bishop was well-known for his Tory views.
12. Benjamin Collins Brodie (1783-1862). Surgeon and Professor of comparative anatomy at the Royal College of Surgeons. (Baronet, 1834). The anonymous article (*Athenæum*, no. 896, 28 December 1844, pp 1198–9) was a reply to a series of articles by Harriet Martineau in the *Athenæum* testifying to her own recovery from illness, apparently as a result of mesmerism. See R. K. Webb 1960, pp 230–3 and Letter #75 (n4).
13. Erasmus Alvey Darwin, CD's brother, was a close friend of Harriet Martineau. For the reaction of Erasmus to her involvement with mesmerism, see Letter #75.

14. Thomas Arnold (1795-1842). Clergyman and famous headmaster of Rugby School. (DNB) The '*Life*' is that of Stanley, 1844, as recorded in CD's list of '*Books Read*' on 30th November (see CCD, Vol 3).
15. Thomas Carlyle (1795-1881). Essayist and historian. He and his wife Jane, were close friends of Erasmus Darwin and met CD and Emma when they lived at Upper Marlborough Street. Carlyle was critical of the standards of literary criticism of his time (see CCD, Vol 3).
16. The potato tubers were supplied by John Stevens Henslow, from plants that he raised from CD's seeds from the *Beagle* (see letter# 78 (n2)).

#77 (CCD 859). To: The Revd W Darwin Fox (No envelope exists for this letter)

Down Bromley Kent
Thursday [24th April 1845][1]

My dear Fox

It is some time since we have had any communication. I write now chiefly to say, that I heard some little time ago from Shrewsbury, in which they said they had wished to have asked you to have come to the Shrewsbury Agricult. Meeting[2], but some invited & more self-offerers have filled the house, more in my opinion than ought to have been allowed. I shall keep out of the way. Mr & Mrs Wilmot//

of Nott:[3] will be there & Mr & Mrs (Miss Gifford[4] that was) G. Holland & E. Holland.[5] By the way, was it not an odd & friendly thing, Mrs Darwin & young Mr D. <of Elston[6] > called on my Brother a few weeks ago; & it seems the young man, whose appearance my Brother liked, has called several times formerly at Grt. Marlborough St. They would not let my Brother return the call. — I have forgotten to add that they desire me to say that they shall be//

particularly glad to see you at Shrewsbury, if you are inclined to go there any other time either before or after July. My Father has been pretty well lately; but yesterday we heard that his leg has suddenly inflamed & was very painful I hope it will not last; I intend going there for a week very shortly[7]. We have had Ellen Tollet staying with us, & heard indirectly much of you: the Tolletts[8] & you seem//

to have many acquaintances in common. Our children are very well, notwithstanding this most cold-catching weather: poor Emma is as bad as she always is, when she is, as she is[9]. (this last sentence is quite Shakespearian) As for myself, my most important news is that I have agreed with Murray for a second Edition of my Journal in the Colonial//

Library[10] in three numbers; & thanks to the Geological fates, I have written my S. American volume the first time over[11]. The only other piece of news about myself is, that I am turned into a Lincolnshire squire! my Father having invested for me in a Farm of 324 acres of good land near Alford[12]. Have you read that strange unphilosophical, but capitally-written book, the Vestiges[13], it has made more talk than any work of late, & has been by some attributed to me. — at which I ought to be much flattered & unflattered.

Ever yours My dear Fox. —
C. Darwin

Postmark: "Down", obscured
Christ's College Library, Cambridge #69

Notes

1. The date is based on the completion of CD's first draft of *South America* (see note 10, below).
2. In 1845 the annual meeting of the Royal Agricultural Society took place in Shrewsbury and the main show day was on the 17th July.
3. Possibly Edward Woollet Wilmot of Worksop Manor, Nottinghamshire, a governor of the Society. One branch of the Wilmots were the owners of Osmaston Hall [see Letter #54, (n7)] and E W Wilmot was in the Nottinghamshire branch of the family (Burke's *Genealogical and heraldic history of the landed gentry*, 1847).
4. George Henry Holland (the younger brother of Edward Holland - see note 5, below) married a Gifford, but further details are not known.
5. Edward Holland (1806-1875) and George Henry Holland (see Letter #73 (n5), CD's second cousins. EH was at Cambridge in 1829; MP for East Worcestershire, and later was president of the Royal Agricultural Society. (*DNB*)
6. Elizabeth de St Croix Darwin (1790-1868) and her son Robert Alvey Darwin (1826-1847). These names had come up in earlier correspondence over the inheritance of Elston Hall; see Letter #54 (n13 & 14).
7. According to his '*Journal*', CD left for Shrewsbury on 29 April, returning home on 10 May.
8. George Tollett (1767–1855). Of Betley Hall Staffordshire. Agicultural reformer. He was a great friend of Josiah Wedgwood, II. His daughters Ellen and Caroline were great friends of Emma Darwin (née Wedgwood) and her sisters. CD has correspondence with George Tollet (see CCD, Vol 2).
9. Emma Darwin was pregnant with her fifth child, George Howard Darwin, who was born on the 9th July 1845.
10. CD was paid £150 by Murray for the copyright to republish the *Journal of Researches into the geology and natural history of the various countries visited by H.M.S. Beagle*, 1839 (The other title for the 3rd Volume of the *Narrative of the surveying voyages of his Majesty's ships Adventure and Beagle*, see Letter #61 (n7), which was popularly known from this time on as the *Voyage of the Beagle*. Significantly, the title was change to put "natural history" before "geology". Apart from the £1000, which CD received to defray the costs of publication of the Beagle collections, this was the first income that CD received (Freeman, 1977). However, as shown by Freeman (1977), his income from books rose steadily after this, as also from investments (*Account Books*, Down House).
11. According to his *Journal*, CD completed the first draft of *Geological observations of South America* on 24th April. For the agreement with John Murray, see letter to John Murray, CCD Vol. 3, 17th April 1845. This was supposed to be a reprint of previous volumes.
12. The Beesby farm at Alford, Lincolnshire, which CD had purchased for £13,592 with a loan from his father (Freeman, 1978). CD's Investment Book (Down House MS) shows that he paid £213 13*s*. 8*d*. interest to his father on 10th August 1846. For further details see CCD, Vol. 3. CD visited the farm in Sept. 1845 (CD's *Journal*, 1845).
13. The book was *Vestiges of the natural history of creation* by Robert Chambers (1844: London). This book caused a storm of controversy, not least because it was published anonymously. It sold many more copies than did the *Origin of Species* in its initial years. CD's copy (Darwin Library Collection, University Library, Cambridge) is carefully annotated and it is clear that CD used it to avoid many pitfalls when he came to write the *Origin of Species*. It is interesting that CD says that some people have claimed that he is the author, since his evolutionary views were at this time a secret, known only to a few close friends. For example, the correspondence with Fox indicates that CD had only suggested broadly that he was interested in the relation of varieties to species. Whether they had any personal conversations on this topic is unknown. Long after the *Origin of Species* was published, Fox was not fully supportive of the theory of Natural Selection (evolution) (see Letter #192).

#77A (CCD558). To: Charles Darwin from W. D. Fox (fragment)

[after 1st June 1846][1]

with creatures of all kinds. The Geese decidedly take more after the *Swan Goose* Gander than the common mother[2].

Pray remember me most kindly to Mrs Darwin — Remind the children that they have a cousin & Believe me — Ever yours affec^{ly} W D Fox

Verso of note: 'Subsequently Fox tells me that all his Hybrids are exactly like each other.—'

Provenance: University Library, Cambridge, Manuscript Room; DAR 205.7 (2)

Notes

1. The date is suggested by a slip in DAR 205.7 (2): 209, headed '*W. D. Fox, 1 June 1846*', in which CD noted the details of the crosses mentioned in the letter. CD had several years earlier asked Fox for information on geese hybrids, see Letters #63 and 64A. Fox and his young bride Ellen Sophia, visited Down from 29th May to 1st June 1846 ("Introduction" to this chapter; Fox *Diary* for 1846). It must have been at that meeting that Fox provided CD with the information. Much later CD corresponded with Fox about these hybrids, when preparing for the *Descent of Man* in 1868 (Letters #185 (5) & #186)
2. In his planned "species book", in a discussion of the hybrids of different breeds of geese, CD stated that "*the Rev W. D. Fox informs me that in some other hybrids which he had seen there was considerable diversity in the degree of resemblance to either parent goose*." (Stauffer, 1975: *Natural Selection*, p 453).

#78 (CCD 13809). To: The Rev^{d} W Darwin Fox (no envelope exists for this letter)

Down Farnborough Kent
Friday
[before 3rd Oct 1846][1]

My dear Fox

The potato seeds[2] were collected from ripe tubers in the Cordillera of central Chile, in a most unfrequented district, many miles from any inhabited spot, & where the plant was certainly in a state of nature— they were collected in the spring <(ie autumn of the S. Hemisphere)> of 1835// & shipped for England in May, & so probably were planted in the spring of 1836 by Prof. Henslow[3], from whom came the tuber which you had.— I see *since* I wrote to you someone has urged the necessity of sending to S. America for new seed![4] // I have sometime thought of calling the attention of the readers of the Gardeners Chronicle to the remarkable difference of climate of the Chonos islands & central Chile, in both of which places the Potato grows wild— if you think it worth while to allude to this, refer to the **1**st Edit. of my Journal, if you have it[5] .—

Many thanks for your answer//about the Potash[6]. I shall certainly set up a bottle & quill—

Very many thanks for your most kind invitation to us all, & I assure you it would give us much real pleasure to accept it; but the journey is fearfully long, & my wretched stomach hates visiting out, as much as the rest of the inward man enjoys seeing his old friends— I enclose list of songs for Mrs Fox, & I wish with all my heart, I c^{d} hear her singing// them[7]— pray give our kind remembrances to her & believe me My dear Fox

Ever Yours
C. Darwin

Christ's College Library, Cambridge, #107

Notes

1. This letter was dated by reference to Fox's article referred to in the letter and detailed Note 2, below.
2. The information in this letter was used by Fox in a letter to the *Gardeners' Chronicle and Agricultural Gazette*, no 40, 3rd October 1846, p 661 (see also Letter #76 (n16).
3. J S Henslow (1797-1861). See Bibliographic Register, Introductions to Chapters 2 & 3, Letter #8 (n3) and many subsequent letters.
4. At this time (1845-1852) in Ireland the Great Hunger or Irish Potato Famine (Woodham-Smith, 1990), exacerbated by British Government neglect, caused the deaths of over one million people, and the emigration of many more. There was much discussion about the best way to maintain current stocks of potato and how to prevent further devastation by the fungal rot. In his letter to the *Gardeners' Chronicle and Agricultural Gazette* (see note 2, above), Fox asserted that there was no point in growing from seed, since CD's potatoes were also diseased. This was an understandable but false assumption and the idea of obtaining resistant strains from South America underpinned potato breeding for the next one hundred years (Erwin and Ribeiro. *Phytophthora Diseases Worldwide*, American Phytopathological Society,1996). John Lindley, editor of the *Gardeners' Chronicle*, was an outspoken proponent of the view that potato rot was caused by climatic factors.
5. CD has already told Fox of the second edition of his book, *Journal of researches*, see Letter #77 (n10). The first edition, *Journal of researches*, pp 347–8, gives more details than the second, *Journal of researches* 2d ed, pp 285–6, although both refer to the difference in climate.
6. Neither CD's letter with the query, nor Fox's reply, has been found.
7. Fox's first wife, Harriet, had died in1842 (see Letters #69 & #70). Fox remarried, on 20th May 1846, Ellen Sophia Woodd (1820-1887), daughter of Basil George Woodd, from Hillfield, Hampstead, near London. They visited the Darwin's at Down House after their honeymoon, from 29th May to 1st June 1846 ("Introduction" to this chapter; Fox *Diary*, 1846). It seems that she sang songs during her visit to the Darwins, which impressed CD. B G Woodd was a London merchant with interests in the wine trade (see Introduction to this chapter). The connection to Fox possibly came through his brother-in-law, Samuel Ellis Bristowe, who also ran a wine merchant business in London [see letter #29 (n3)]. In the Fox-Pearce Collection, Woodward Library, University of British Columbia, there is a letter dated 5th December 1835 from B G Woodd to Fox, recording the overpayment of money on a shipment of wine.

#79 (CCD 1222). To: The Revd W Darwin Fox (No envelope exists for this letter. The first page of the note paper has a black surround)

Down Farnborough Kent
Feb 6th [1849][1]

My dear Fox

I was very glad to get your note[2]. I have often been thinking of writing to you, but all the autumn & winter I have been much dispirited & inclined to do nothing but what I was forced to[3]. I saw two very nice notes of yours on the ocasion of my poor dear Father's death[4]. The memory of such a Father is a treasure to one; & when last I saw him he was very comfortable & his expression which I have now in my mind's eye serene & cheerful.—//

Thank you so much for your information about the water cure[5]: I cannot make up my mind; I dislike the thoughts of it much — I know I shall be very uncomfortable there, & such a job moving with **6** children. Can you tell me (& I shd be much obliged sometime for an answer) whether either your cases was dyspepsia, though Dr Holland[6] does not consider my case quite that, but nearer to suppressed gout[7] . He says he never saw such a case, & will not take on him to recommend the// Water cure. — I must get Gully's Book[8]. —

We shall indeed be very glad to see you here at any time; though Miss Annie[9] is not *quite* ready to be married yet: the new handwriting is our Governesses[10]. — Do you remember recommending me to tie up our fruit trees like besoms — well I did so, & then wrote Gardeners Chronic to ask his opinion[11] — & he answered "cut loose at once or you will spoil all your trees"; but I kept 2 or 3// pears & plums — standard & wall, tyed for experiment, & they answered wonderfully this summer being the only trees of several of same age which bore fruit — & made hardly any wood — one little Pear on wall had 90. — Have you gone on with this plan? do tell us, ie Comfort[12] & myself, as we are both very curious to hear. — We are now trying Apricots & Peaches. —

Pray give our kind remembrances to Mrs Fox[13], & believe my dear old Friend
Yours most sincerely
C. Darwin

Christ's College Library, Cambridge, #71

Notes

1. The year of the letter is given by reference to CD having six children.
2. This the first letter since before October 1846 (see Letter #78). The first sentence and the rest of the letter suggest no such break in correspondence. It must be concluded, therefore, that some letters have not survived. This could be because Fox was exceptionally busy during this period with his new wife, Ellen Sophia, and their growing family. More likely some letters were lost at a later date.
3. CD's '*Journal*' (CCD, Vol 4, Appendix I) records that his health was 'very bad January 1st to March 10th'. In January 1849 CD began to keep a health diary in which he made detailed daily entries of his symptoms and treatment until 1855; the diary is held at Down House.
4. Robert Waring Darwin died on 13th Nov 1848 at the age of 82.
5. From the information further on in the letter (Note 7) this must be Dr Gully's establishment at Malvern. It seems that Fox had suggested it to CD. CD became a regular visitor there for some years, and his daughter Anne Elisabeth died there in 1851 (see Letter #89).
6. Henry Holland ((1788-1873). Physician. Distant relation of both the Darwins and the Wedgwoods. He treated CD, Emma Darwin and Queen Victoria. He was made a baronet in 1853. See also Letter #75 (n6). (*DNB*)
7. Holland may have been thinking of the gout of Robert Waring Darwin, when he diagnosed CD as suffering from '*suppressed gout*'. CCD, Vol 4, stated that Henry Holland believed gout to be a hereditary disorder brought on by an accumulation of unknown toxic materials in the blood (H. Holland 1839, p 117).
8. Both James Manby Gully and James Wilson set up hydropathic establishments in Malvern, Worcestershire, in 1842. These flourished in the succeeding two decades. Dr Gully's water cure (*The water cure in chronic disease. An exposition of the causes, progress & termination of various chronic diseases of the digestive organs, lungs, nerves, limbs & skin; an of their treatment by water and other hygienic means.* 1846, London) held that all chronic diseases were caused by a faulty supply of blood to the stomach and that the application by various means of cold water to the skin would return the circulation to normal (see, also, Metcalfe R (1906) *The rise and progress of hydropathy in England and Scotland*; Simpkin, Marshall, Hamilton, Kent & Co, London).
9. Anne Elizabeth Darwin (1841-1851).
10. Miss Thorley. In CD's *Account Book* (Down House MS) payments are noted until 1st January 1857.
11. John Lindley was the editor of the *Gardeners' Chronicle*. No letter has been found.
12. Comfort was the gardener at Down House.
13. W D Fox and his wife Ellen Sophia visited CD and family at Down house, soon after their wedding, on the 29th May to 1st June 1846 ("Introduction" to this chapter; *Diaries* of W D Fox).

#80 (CCD 1233). To: The Rev[d] W Darwin Fox (No envelope exists for this letter)

Down Bromley Kent

Tuesday [6th March 1849][1]

Dear Mr Fox

As Charles is recovering from a lingering attack he has desired me to answer your very kind note. We really are going to set out "bag & baggage" to Malvern as soon as he is well enough. It is a great trouble taking all the household, but we think he could not give Dr Gully's treatment a fair trial under 6 weeks or 2 months & that would be too long to leave the children even with their Aunts[2]. We both feel very hopeful that it will be of some use to Charles tho we do not venture to hope a cure. This has been//

a disheartening winter with respect to his health. Dr Gully[3] writes like a sensible man & does not speak very confidently[4]. Susan & Catherine are now with us & both pretty well & I am glad to see that Susan's Spirits are good enough to make her feel an interest in all that goes on. I mention her particularly as she was more constantly//

with her father & therefore would feel the blank more[5]. Catherine is going abroad for some weeks with a sister & an Aunt of mine[6] in the course of the spring. One of the things we shall most enjoy if it pleases Heaven to give Charles the blessing of better health would be to see more of you & Mrs Fox & that all our children should become intimate[7]. I have no doubt//

I have not thanked you half enough for your thorough interest in this step of ours. Mrs Fox & I should feel very jealous & spiteful at each others babies just now for I think they are about the same age[8]. Ours *is* a remarkably nice one, I must say.

With our kindest remembrances to Mrs F.

E. D.

Christ's College Library, Cambridge #72

Notes

1. The date, as determined by CCD, is the Tuesday before CD and his family left for Malvern, where they arrived on 10 March 1849 (*'Journal'*; CCD, vol. 4, Appendix I). According to Emma Darwin's *Diary* (University Library, Cambridge), she and CD left Down on 8th March and travelled via London. She noted that the children also arrived in Malvern on 10 March.
2. Catherine and Susan Darwin, see later in letter.
3. James Manby Gully (1808-83). Physician and proprietor of an hydropathic establishment in Malvern. See Letter #79 (n8). (*DNB*)
4. No letters from James Manby Gully have been found. In his book (Gully 1846, see Letter #78), Gully critically assessed his water treatment. He specifically said that for nervous indigestion, CD's complaint, a cure would take at least six months (p 167).
5. Robert Waring Darwin died the previous November cared for especially by Susan Darwin.
6. It must be remembered that Maer Hall, where several of Emma's brothers and sisters had lived, had been sold recently. By the end of 1847 all the family except Frank Wedgwood had left Staffordshire. Sarah Wedwood moved close to CD and family at Petleys, near Downe. The Langtons and Elizabeth Wedgwood moved to Hartfield Grove, Sussex. In Shrewsbury, although Robert Darwin had died in 1848 (see Letter #78), both Caroline and Susan Darwin maintained the house. It is likely that the sister

was Elizabeth and the aunt was Fanny Allen, and that they visited Jessie Sismondi in Switzerland (see *A Century of Letters* by Henrietta Litchfield, 1915).

7. Fox and his new wife had visited CD and family at Down House in 1846 [see Letter s #77A (n1) & #79 (n13)], so the sincerity is likely to be real despite the formal opening to the letter, "*Dear Mr Fox*", which Emma, no doubt, considered the proper way to address him.
8. Frances Maria Fox was born on 28th February 1848 and Francis Darwin on 16th August 1848 (*Darwin Pedigrees*).

#81 (CCD 1235). To: The Rev. W. D. Darwin Fox, Delamere Rectory, Chester (The first page of the notepaper has a black surround)

The Lodge Malvern
24th [March 1849]

My dear Fox

As you took such very kind interest about me, & as I owe to so considerable extent my expedition to you, I must report progress. We came here this day fortnight & have got a very comfortable house, with a little field & wood opening on to the mountain, capital for the children to play in[1] . — We shall stay here at least 2 months, ie to <latter end or> middle or <possibly only to> early part//

of May. — I much like & think highly of Dr Gully[2]. He has been very cautious in his treatment & has even, had the charity <to> stint me only <to> six pinches of snuff daily. — Cold scrubbing in morning, 2 cold feet bath & compress on stomach is as yet the only treatment, besides change of diet &c. — I am, however to commence tomorrow a sweating process.— I am already *certainly* stronger & perhaps//

my stomach somewhat better. I was in a much shattered condition before coming here; my hands were becoming tremulous & head often swimming. — I expect fully that the system will greatly benefit me, & certainly the regular Doctors cd. do nothing. We have a spare bed-room & shd be delighted to see you here, but I shd enjoy your visit more towards latter half of April, when I hope to be stronger & less absorbed with my eternal short, walks//

bathings &c. all of which make me excessively tired in evening so that I am forced to go to bed at 8½. The only disagreeable part as yet to me, has been the excessive irritation of skin which comes on every evening over whole body. — so that I cannot sit quiet one minute after six or seven oclock. — This no doubt will before long go off. Have you ever been here? It is a curious & nice place & in summer must be//

very pretty; we have had half our time foggy or hazy, but most fortunate in not having had any rain: I do hope you will come, & that you will find me a ruddy strong man. We have all our servants & governess[3] here. —

My dear Fox Yours most sincerely
C. Darwin

Endorsement "March 24/49"
Christ's College Library, Cambridge #73

Notes

1. Patients for the water cure, and their families, generally lodged in private houses in Malvern, although limited accommodation for patients was available at the hydropathic establishments.
2. James Manby Gully (1808-83). See Letter #79 (n8).
3. At this time the governess was Miss Thorley [see Letter #88 (n8)].

#82 (CCD 1240). To: The Rev[d] W Darwin Fox (No envelope exists for this letter)

The Lodge. Malvern
April 18th [1849]

My dear Fox

Many thanks for your most friendly note a fortnight ago. I did not write sooner, till I knew our plans more definitely. Dr G. [1] now says we may return home at *end* of May (it will be on Friday 1st of June that we shall arrive, I hope at dear old Down) but that I shall have to go on with the Aqueous treatment for *many* months at home//

under his direction. — Our remaining here all May will, I hope give us a better *chance* of seeing you here, but your account of yourself & gigantic family[2] shows that it is but a chance; nevertheless we shall hope for it — I am very sorry to hear that you have not been very well this winter. With respect to myself I believe I am going on very well; but I am rather weary of my present inactive life &//

the Water Cure has the most extraordinary effect in producing indolence & stagnation of mind; till experiencing it, I c[d] not have believed it possible. — I now increase in weight, have escaped sickness for 30 days, which is thrice as long an interval, as I have had for last year; & yesterday in 4 walks I managed seven miles! I am turned into a mere walking & eating machine. — Dr G.[1] however finds he is obliged to treat me cautiously, & during last//

week all my treatment has been much relaxed. There are many patients here even already: last summer I hear he had 120! — He must be making an immense fortune. — Lady Wilmot[3] lives here with her son Col. Wilmot[3]; I have not called, for I was frightened at this great Dandy of a son: if it had been summer I wd have called to have seen the flower garden. — You need not send Athenæum[4] or Glacier Paper[5] till our return to Down. —

Yours very affectionately
C. Darwin

For auld langsyne
I have looked for Beetles on the hills here, but cannot find one[6]. —

Christ's College Library, Cambridge, #73A.

Notes

1. James Manby Gully (1808-83). See Letter 79 (n8).
2. Fox had seven children at this stage, but the one page of his diary for 1849 Fox records that he had, in addition, Miss Fletcher (his sister-in-law, by his first wife), Zachary and Frederick Mudge (both children) and a Miss Harker (governess?).
3. Marianne Wilmot and Eardley Nicholas Wilmot of Malvern, the widow and son of Sir Robert John Wilmot who died in 1841 and had been the landlord of Osmaston Hall, the home of Samuel Fox III and family [see Letter #54 (n7)]. Lady Wilmot resided, at this time, at Rosebank, Malvern.

4. CCD, Vol 4 suggested that this was possibly the *Athenæum*, no 1116, 17 March 1849, in which Robert Chambers 1848 was favourably reviewed (pp 275–6); or *Athenæum*, no. 1117, 24 March 1849, in which there was an account of Edward Forbes's lecture at the Royal Institution on 2nd March, entitled, '*Have new species of organized beings appeared since the creation of man?*' (p 304).
5. CCD, Vol 4 suggests that this was possibly the paper of James David Forbes, 1849, in which he drew an analogy between the flow of mud-slides and the movement of glaciers. The article was dated 2 December 1848 and was published in the January 1849 issue of the *Edinburgh New Philosophical Journal*.
6. The earlier enthusiasm of Fox and CD, in their college days, for beetles seems now to have have cooled on both sides (compare, e g, Letter #1).

83 (CCD 1249). To: The Revd W Darwin Fox (No envelope exists for this letter)

Down Farnborough Kent
Saturday 7th [July 1849][1]

My dear Fox

You will see by this that we have returned home, having staid *16* instead of *6* weeks. We got home last Saturday. — I go on with the Treatment here, just the same as at Malvern, though in a somewhat relaxed degree, so as to avoid bringing on a crisis[2]. — I am building a douche, for//

Dr Gully tells me I shall have to follow treatment for a year. I consider the sickness as absolutely cured. And about 3 weeks since I had 12 hours without any flatulence, which showed me that it was possible that even that can be cured, as Dr G. has always said he could. — The Water Cure is assuredly a grand discovery & how sorry I am I//

did not hear of it, or rather that I was not somehow compelled to try it some five or six years ago. — Much I owe you for your large share in making me go this Spring. Remember your promise of coming here this summer, & if possible I hope you will bring Mrs Fox[3]. —
If able I shall be at Birmingham for Brit. Assoc, for I am honoured//

with being appointed one of the Vice Presidents[4], & Dr Gully said he shd like to know in autumn, how I shd stand leaving off treatment with some mental excitement. — Till then I mean to remain slave to treatment. How I have prosed about myself!

Yours affectionately
C. Darwin

Endorsement "July 1849"
Christ's College Library, Cambridge #74.

Notes

1. Dated by Fox's endorsement and by CD's reference to having just returned to Down from Malvern. He arrived home on Saturday, 30th June 1849 (CD's *Journal*)
2. CD is referring to treatment at Dr James Manby Gully's hydropathic establishment at Malvern (see Letters #81 & #82). In James Manby Gully's usage, a 'crisis' was brought about by the transfer of morbid action from the original seat of the disorder to some less important part where it was eliminated by secretions (Gully 1846, p 2).

3. CD's Journal for 1849 records for 2nd November "William Darwin Fox visited"; Emma's *Diary* records that Fox visited on 2nd November 1849. Fox's *Diary* for 1849 is missing except for one page (see Appendix 1).
4. *Report of the 19th meeting of the British Association for the Advancement of Science held at Birmingham in 1849*. CCD, Vol 4 recorded another assignment given to CD at this meeting.

84 (CCD 1292). To: The Revd W Darwin Fox (No envelope exists for this letter)

Down Farnborough Kent
Thursday [17th Jan. 1850]

My dear Fox

Will you be so kind as to tell Mr Wood[1], I shall be happy to answer honestly & to the best of my power any questions he may put — though any very general questions could hardly be answered in the compass of a letter. I have only been at Geology Meeting once this year, — Proh pudor[2]. — The day before yesterday Emma was confined of a little Boy[3]. Her pains came on so rapidly & severe, that I cd not withstand her entreating for chloroform & administered it myself which was nervous work not knowing from eye-sight anything about it or midwifery[4]. The Doctor got here only 10 minutes before the Birth. — I thought at the//

time I was only soothing the pains — but, it seems, she remembers nothing from the first pain till she heard that the child was born[5]. — Is it not grand? — You ask after water cure. — I go honestly on & had the douche 36^0 to 37^0 for 5 minutes & to shallow bath with water at 39^0 for 4 minutes this very morning: it is sharp work, but not half so sharp as you would think & admirably invigorating[6]. — My health is better than when you were here. — Your visit was not thrown away upon us, we both much enjoyed it[7]. — Pray remember me most kindly to Mrs Fox.

Ever yours
C. Darwin

I am now at work on *fossil* Cirrepedia[8]

Endorsement: "Jany 19 /50
Christ's College Library, Cambridge, #75

Notes

1. Possibly Charles Henry Lardner Woodd (?-1893), of Roslyn, Hampstead, London. Woodd was elected a fellow of the Geological Society in 1846. The speciific matter was to do perhaps with Woodd's election to office in the Geological Society or to do with being proposed as a member of the Athenæum Club [see Letter #119 (n4)].
2. Latin phrase meaning "for shame".
3. Leonard Darwin (1850-1943), born 15th Jan 1850.
4. This is the first occasion that Emma took chloroform in childbirth. She did so for her next two children (Hedley, 2001). At the time the use of chloroform was much debated, it being argued by some that women should undergo the pain of childbirth without relief. However, Queen Victoria had her 10th child under chloroform in1850 which put a royal stamp of approval on the new approach. See Letter #85 (n4).
5. CD may well have over-dosed Emma with chloroform.

6. CD's *Health Diary* (Down House) indicates that he continued the treatment until 1855 [see Letter # 85 (n1)].
7. Both CD and Emma record in their *Diaries* that Fox visited on the 2nd November 1849. The *Diary* of Fox in 1849 is missing except for one page.
8. *A monograph of the fossil Lepadidae, or pedunculated cirrepedes, of Great Britain. A monograph of the fossil Balanidae and Verrucidae of Great Britain.* Palaeontographical Society. Vols 5, 8 &12 1851, 1854 &1858. CD worked on these publications alongside the two-volume book on "*Living Cirripedes*" (or barnacles) published in 1851 & 1854 by the Ray Society.

85 (CCD 1323). To: The Revd W Darwin Fox (No envelope exists for this letter)

Down Farnborough Kent
Thursday [May 1850]

My dear Fox

I am extremely much obliged to you for your very kind interest regarding me & my curative processes[1]. I am <rather> glad to hear that people can be injured by the Water cure; for I have rather feared that I was getting inured to it. — I had, however, heard quite enough to have even prevented me, treating myself blindly without//

guidance: I report myself regularly to Dr Gully[1] & receive instructions. My treatment varies every 4 or 6 weeks — About 8 weeks since I left off the Lamp[2] for a month: on an average I have had it throughout the last 9 months ~~either~~ from 2 to 4 times a week & so with Douche. — For present fortnight I//

am having Douche daily, which is first time since leaving Malvern. Having gained weight ever since I commenced[3] , is a clear sign that I have not overdosed myself. — Dr Gully has had plenty of experience & with me has always been cautious.— He now rather wants to see me; but I hate//

the trouble of travelling. If you are led to converse with your Dr, I shd like to hear his opinion; but I conceive it is simply impossible for anyone to say how much an unknown person can stand; & one must be under the orders of one Doctor. I have lately been *very* well; but previously//

I had a somewhat retrograde period of 6 weeks — though even then *infinitely* better than before I commenced the W. Cure. — We are *all* well & flourishing — I am sorry to hear about Chloroform; from what Dr Simpson says I believe in such//

cases as you mention; the frightened & generally prejudiced doctors do not give enough[4]. — I saw a few weeks since Sir P. Egerton[5] & he gave me such an account of your Farm[6], as one of the seven Wonders of the world — & delighted I was to <hear> all he said regarding you & your Farm —

Ever Yours
C. Darwin

Endorsement "May 1850"
Postmark: "Bromley", the postmark obscured.
Christ's College, Cambridge, #76

Notes

1. CD refers to his Water Cure being conducted under the instructions of James Manby Gully - see Letters #81 & #82. See also CCD, Vol. 4, letter to Susan Darwin, 19th March 1849, for a further account of James Manby Gully's hydropathic treatment. On returning to Down from Malvern, CD made arrangements to keep up the treatment by building a small outhouse containing a douche and a shallow bath. See also Letter #84 (n5). CD's *Health Diary* (Down House) contains daily records for five and a half years, until January 1855.
2. CD alludes to Gully's introduction of the lamp bath into hydropathic practice, which he mentioned in a letter to J D Hooker, 28th March 1849 note 2 (CCD, Vol. 4): "*I am heated by a Spirit lamp until I stream with perspiration*".
3. CD weighed 10 st 7 lb 12 oz at the beginning of his Malvern visit and on 12th May 1850 he weighed 11 st 13 lb 10 oz (*Health Diary*).
4. Fox seems to report on some adverse effects of chloroform. However, on 1st – 8th Oct 1852, Fox recorded in his *Diary*, "*I suffered much from a carbuncle on stomach – which was cut out on 8th under chloroform by J. Harrison*". James Young Simpson (1811-1872), an army surgeon and Director General, Army Medical Department, was a pioneer in the use of chloroform as an anaesthetic (*DNB*).
5. Philip de Malpas Grey-Egerton (1806-1881). Palaeontologist, specializing in fossil fish. He was MP for South Cheshire. His estate, Oulton Park, was not far from Fox's parish of Delamere and he knew Fox well; see Letter #67 (n9). (*DNB*)
6. Fox's *Diaries* show that he ran an extensive farm and had a great variety of animals for breeding purposes. See CD's references to Fox's '*Noah's Ark*' in Letters #76 (n2) & #87 (n4).

86 (CCD 1352). To: The Revd W Darwin Fox Delamere Rectory, Chester.

Down Farnborough Kent
Sept. 4th — [1850]

My dear Fox

I was much pleased to get your very agreeable letter with all its curious facts on the female sex & their hereditariness[1]. Undoubtedly the periodical shedding of the nails almost by itself wd have convinced any naturalist that the individual was specifically distinct. I wonder whether the queries addressed to//

about the specific distinctions of the races of man are a reflexion from Agassiz's Lectures in the U.S. in which he has been maintaining the doctrine of several species[2], — much, I daresay, to the comfort of the slave-holding Southerns. — Your aphorism that "any remedy will cure any malady" contains, I do believe, profound truth, —whether applicable//

or not to the wondrous Water Cure I am not very sure. — The Water-Cure, however, keeps in high favour, & I go regularly on with douching &c &c[3] : I am much in the same state as I have been for the last nine months, & not quite so brilliantly well as I was in the dead of last winter[4] . To be as I am, though I never have my stomach//

right for 24 hours, is, compared to my state two years ago, of inestimable value. My wife & all my children are well; & they, the children, are now seven in number; to what I am to bring up my four Boys, even already sorely perplexes me. My eldest boy[5] is showing the hereditary principle, by a passion for collecting Lepidoptera. We are at//

present very full of the subject of schools; I cannot endure to think of sending my Boys to waste 7 or 8 years in making miserable Latin verses, & we have heard some good of Bruce Castle School, near Tottenham[6] which is partly on the Fellenberg System[7], & is kept by a Brother of Rowland Hill of the Post-office, so that on Friday we are going to inspect it & the Boys[8]. I feel that it is an awful experiment//

to depart from the usual course, however bad that course may be. — Have you, who have something of an omniscient tendency in you, ever heard anything of this school? — You speak about Homœopathy; which is a subject which makes me more wrath, even than does Clairvoyance: clairvoyance so transcends belief, that one's ordinary//

faculties are put out of question, but in Homœopathy common sense & common observation come into play, & both these must go to the Dogs[9], if the infinetesimal doses have any effect whatever. How true is a remark I saw the other day by Quetelet, in respect to evidence of curative processes, viz that no one knows in disease what is the simple result//

of nothing being done, as a standard with which to compare Homœopathy & all other such things[10]. It is a sad flaw, I cannot but think in my beloved Dr Gully, that he believes in everything — when his daughter was very ill, he had a clairvoyant girl to report on internal changes, a mesmerist to put her to sleep — an homœopathist, viz Dr Chapman[11]; & himself as Hydropathist[12]! & the girl recovered. — My dear Fox, I do hope we shall sometime see you here again.

Your affectionate friend
C. Darwin

(*across top of page 1*)
By pure accident a bundle of Athenæums have been *much* delayed.

Endorsement "Sep 6/50"
Christ's College Library, Cambridge #77.

Notes

1. Fox's interest in these subjects is well attested by later letters and his diaries. In *Variation of animals and plant under domestication* (1868), CD refers to a case, cited by Fox, of a woman who shed her nails [see Letter #198 (n10)].
2. CCD, Vol 4, noted: "At a meeting of the American Association for the Advancement of Science at Charleston, South Carolina, 15 March 1850, Louis Agassiz stated that, viewed zoologically, '*the several races of man were well marked and distinct*' and that '*these races did not originate from a common centre, nor from a single pair*' (Lurie 1960, p 260). See also Lurie 1954."
3. Apart from CD's *Health Diary*, which showed that he kept up the hydropathic treatment for over 5 years [see Letter #83 (n2)], a reminiscence by George Howard Darwin, his son, is given in note 12, below.
4. According to Emma Darwin's *Diary*, '*Mr Fox*' visited Down on Friday, 2nd November 1849. Fox's *Diary* for 1849 is not extant, apart from a summary page - see Appendix 1.
5. William Erasmus Darwin (1839-1914), at this time nearly eleven years old.
6. Bruce Castle School in Tottenham was a continuation of the famous Hazelwood School, founded by Rowland Hill (originator of the Penny Post; see Letter #66 (n2) and his brothers, Arthur and Matthew. It had a progressive program, which included student self-government and a curriculum emphasising modern languages and science, but allowed for individual development. See: Stewart and McCann (1967–8), *The educational innovators, 1750-1967*; 2 Vols; Macmillan, London (Vol 1: 98–123). In the event, CD did not use this school, instead sending William to Rugby School (see Letter #87) and his other boys to Clapham Grammar School [see Letter #112 (n8)].

7. Philipp Emanuel von Fellenberg (1771-1844. Swiss educationalist. His theory focused upon the development of the child but placed emphasis upon rigid discipline. He also considered that class distinctions to be inevitable and in consequence favoured different types of education for each class (Stewart and McCann 1967–8, 1: 141–6 – see citation in note 6). (*DNB*)
8. Emma Darwin recorded in her diary "*Bruce Castle*" on 6 September 1850, having gone up to London the day before.
9. It is interesting that CD should display such a logical (materialist) view in condemning clairvoyance (mesmerism) and homeopathy. We do not have Fox's letters on these subjects, but it looks as if he has sympathy for them, as did many "Victorians" of the time (if we can call them by that name).
10. Lambert Adolphe Jacques Quetelet (1796-1874). A Belgian astronomer, statistician and foreign member of the Royal Society, who was important in introducing statistical methods to the social sciences. The views expressed by CD derive from the book by Quetelet, 1849, pp. 228–36. As pointed put by CCD, Vol 4, CD may have been basing his account on F W Herschel's review of the book in the *Edinburgh Review* (1850) 92: 1–57, in which these views are set out on pp 54–5.
11. John Chapman (1822-1894). Physician and publisher. Editor of Westminster Review, which was so influential in establishing early views in support of evolution. Thomas Henr Huxley gained his first job on the Review and improved its standards. Ironically, CD was to turn to Chapman in 1865 for medical advice, when Chapman visited Down House and recommended ice treatment to the spine [see *Adrian and Desmond*, 1991 and Emma Darwin *Diary*: Saturday 20 May "*No sickness great flat nausea put on ice at 5 p.m. Dr Chap – good night no flat*"; Sunday 28th May: "*Dr Chap – very brisk eve*"]. (*DNB*)
12. For CD's opinion of the homoeopathic treatments of James Manby Gully's, see CCD, letter to Susan Darwin, 19th March 1849. George Howard Darwin later recalled (DAR 112: 49): "*Dr Gully was a spiritualist & believer in clairvoyance. He bothered my father for some time to have a consultation with a clairvoyante, who was staying at Malvern, and was reputed to be able to see the insides of people & discover the real nature of their ailments. At last he consented, to pacify Dr Gully, but on condition that he should be allowed to test the clairvoyante's powers for himself. Accordingly, in going to the interview he put a banknote in a sealed envelope. After being introduced to the lady he said 'I have heard a great deal of your powers of reading concealed writings & I should like to have evidence myself: now in this envelope there is a banknote—if you will read the number I shall be happy to present it to you.' The clairvoyante answered scornfully 'I have a maid-servant at home who can do that.' But she had her revenge for on proceeding to the diagnosis of my father's illness, she gave a most appalling picture of the horrors which she saw in his inside*."

87 (CCD 1362). To: The Rev^{d} W Darwin Fox Delamere Rectory, Chester.

Down Farnborough Kent
Oct. 10th [1850]

My dear Fox

I am very much obliged for your juicy, as my poor dear Father used to call an interesting, letter. — We were very glad to get the sentence about Bruce Castle school, for we are still in an awesome state of indecision between Rugby & it[1]. I knew you were just the man to apply to to get information upon any out of the way subject. — We have//

taken much pains in making enquiries, & upon the whole the balance is decidedly favourable; yet there is so much novelty in the system that we cannot help being much afraid at trying an experiment on so important a subject. At Bruce castle, they do not begin Latin, till a Boy can read, write, spell, & count *well*: they have no punishments except stopping premiums on good behaviour. I do not see how we are ever to come to a decision; but we//

must soon. — Willy is 11 this coming Christmas, & backward for his age; though sensible & observant. I rather think we shall send him to Bruce C. School[2]. — Your own system of Education sounds **capital**[3], & why you sh$^{\underline{d}}$ think I sh$^{\underline{d}}$ laugh at it, I cannot conceive: I believe a good deal of diversity an immense advantage. It is one good point at Bruce Castle: that no one subject exceeds an hour & if a Boy can do it quicker, he may go out before the hour is over. —//

You say you are teaching riding: we have been teaching Willy & we began without stirrups, & in consequence Willy got two *severe* falls, one almost serious; so we are thinking of giving him stirrups; more especially as I am assured, that a Boy who rides well without stirrups has almost to begin again when he takes to stirrups: Can you give me any wisdom on this head; pray do if you can? —//

I never heard anything half so wonderful as your stock of cows, pigs horses & children: well might Sir Philip tell me it was marvellous what your farm did[4] . — How earnestly I wish we were nearer to each other: I sh$^{\underline{d}}$ beyond measure like to see your Noahs ark[4]: next summer I will really try to pay you a visit of a few days[5]. — I fear Emma will not be able, for about midsummer we//

expect our 8th arrival, & then we shall have, as it may be said, almost four babies of the same age. — I often speculate how wise it w^{d} be to start off to Australia, or what I fancy most, the middle States of N. America[6]. —
You ask after my Pear trees; those against the wall continue to bear *very well* for young trees; but//

my standards have not borne ~~well~~ <at all>; & I doubt whether the besom system of tying up, answers for them[7]. — I have planted my Rivers Quince Dwarfs, which I got on your recommendation[8]. — Whenever you can do pray pay us a visit even for one day[9]. — Susan & Catherine's tour answered capitally[10]

Yours affectionately
C. Darwin

Do you intend to educate// your Boys altogether at home? — The first-rate tutor[11] at whom Willy now is, teaches nothing on earth but the Latin Grammar, & his charge is 150£ per annum! Bruce Castle is cheap *with extras* about 80£ .

Adios

Have you any curiosity to read a pamphlet descriptive of Bruce Castle[12]? —

Endorsement: "Oct 10/50"
Postmark: obscured
Christ's College Library, Cambridge, #78

Notes

1. See letter #86 (n6).
2. The final decision was to send William Erasmus to Rugby School. He entered the school in early February 1852 (*Rugby School register;* CCD, Vol. *5*, letter of CD to William E. Darwin, 24th Feb 1852). In letter #90 (n5), CD confesses that he had not had the courage to break away from 'the old stereotyped stupid classical education'. However, CD sent his other sons to Clapham Grammar School, and not one of several prestigious "Public Schools" in South London.

3. Fox taught his own children, including his daughters, in their early years (Fox *Diaries*).
4. CD refers to Sir Philip Egerton: see Letters #67 (n9) & #85 (n5 & 6).
5. There is no record of CD ever having visited Fox at Delamere, although Susan Darwin did (Fox *Diary*, 14th May 1847).
6. See Letters #88 & 89. As shown by this letter, about this time, CD had have anxieties about life in England and had pipe dreams about emigrating to either Australia or America (see CCD, Vol 4, letter to Syms Covington, 23rd November 1850, n. 4 and Desmond and Moore, 1991, Chapter 26).
7. See Letter #79 (n12).
8. See Letter #72.
9. The *Diary* of Fox for 1851 shows that Fox and family visited the Great Exhibition (Grand Exhibition of Industry of all Nations) in early May. CD and family visited in late July, the visit delayed because CD wanterd to finish his barnacle volume [see Letter #88 (n13)].
10. Susan and Catherine Darwin's tour. There are no records or family letters that record what this tour might have been. Both CD's and Emma's brothers and sisters were very mobile at this time and visited each other's houses and also the home of the Allens at Cresselly, Tenby, Wales. Since it was the summer period Susan and Catherine may have gone somewhere special. Emma's *Diary* for 1850 records that "*Cath came*" on 24th Aug. In 1851 there is a record "*Cath went*" on 22nd Jan. Whether Catherine stayed the whole 6 months is not clear.
11. Henry James Wharton, vicar of Mitcham, Surrey. See Letter #88 (n6). CD's *Account Book* (Down House MS) has an entry for 19th August 1850: '*Whartton Revd— Willy School 74 3 6.*' In *Emma Darwin* (Litchfield, 1815, Vol 2: 145), there is the following footnote, against the name Mr Wharton in the letter that CD wrote to William Darwin at Rugby School on 24th Feb 1852 (see, also, CCD, Vol. 5): "*The schoolmaster at his preparatory school*"
12. This is probably "A. Hill, 1833. *Sketch of the system of education, moral and intellectual, in practice at the schools of Bruce Castle, Tottenham, and Hazlewood, near Birmingham.* London".

88 (CCD 1396). To: The Revd W Darwin Fox Delamere Rectory, Chester.

Malvern[1]

Thursday [27th March 1851]

My dear Fox

In passing through London two days ago[2] I heard from Erasmus[3] with sorrow of your Father's death[4]. A few weeks since I had been much interested in hearing from Susan[5] an account of your Fathers equable & apparently happy state, & of the surprising manner in which he retained his faculties & interests. In a note from Susan she expresses how very glad she now is at her last visit. — I grieve to hear that your health prevents you attending the Funeral: this was my case[6], & though it is only a ceremony I felt deeply grieved at this deprivation & you no doubt will feel this more. You have my sincere sympathy & very sorry indeed I am to hear that your health is not so good, even as formerly.

Hereafter, when *at leisure*, do let me have a line from you, my dear old Friend. I often think of our happy days at Cambridge; almost if not quite, the happiest part of my life & much associated in my memory with you[7]. — Long continued ill-health has much changed me, & I very often think with pain how cold & different I must appear to my few old friends to what I was formerly; but I internally know that the inner part of my mind remains the same with my old affections.

Believe me, my dear Fox, I am & shall ever be your affectionate friend

Charles Darwin

I have brought my eldest girl here & intend to leave her for a month under Dr Gully[8]; she inherits I fear with grief, my wretched digestion. I return in a day or two home. —
When next you write to any of your Family pray express my sincere sympathy for Mrs Fox[9] & all your sisters.

Endorsement: "March 29151"
Christ's College Library, Cambridge, 78a

Notes

1. CD's *Journal* (CCD, Vol 5, Appendix I) records that he left Down for Malvern on 24 March 1851.
2. See letter in CCD, Vol 5, to John Wickham Flower, 23 March [1851], n 3.
3. Erasmus Alvey Darwin (1804-1881), CD's brother. CD kept up his friendship with Erasmus and visited him regularly, when visiting or passing through London.
4. Samuel Fox died on 20th March 1851 at Kendalls, Hertfordshire (*Darwin Pedigrees*; Fox *Diary*, 1851).
5. Susan Elizabeth Darwin, CD's sister, still residing at The Mount, Shrewsbury, was a close friend of Fox's sister and of Fox. Fox records in his diary for 1847: "*May 17 Susan Darwin came till 22nd*."
6. In fact, Fox did attend his father's funeral at Little Stanmore or Whitchurch on 27th March, as shown by his *Diary* for 1851. It is only strictly true that Darwin did not attend his father's funeral; he got to the Shrewsbury after the service had started and retired to spend the rest of the time with his sister, Marianne Parker, at their father's home, The Mount (Wedgwood and Wedgwood, 1980; see also letter of Catherine Darwin to CD, in CCD, Vol 4, 13th November 1848).
7. CD once again plays on the sentimental attachment to Cambridge days.
8. This was a visit to James Manby Gully's hydropathic establishment in Malvern (see letters #81-84) for both CD and his daughter, Anne Elizabeth Darwin, who died there on 23rd April. Annie's health had been failing since the summer of 1850, but probably longer (*Emma Darwin* 2: 132; Keynes,R., 2001), was to be left in Malvern under the care of her nurse, Brodie, and (later) the governess, Miss Thorley, and with Henrietta Darwin as a playmate. According to Emma Darwin's *diary*, Miss Thorley left for Malvern on 28 March. CD. The events covered in the next letters (# 86 - #88) are covered in great detail by Keynes, R. (2001). CD accompanied his daughters to Malvern by stage coach on 24th March and returned to London on 31st March. Annie was well for a few days but then quickly became seriously ill. CD returned to Malvern on 16th April and Annie died 7 days later.
9. Ann Fox (née Darwin)(1777-1859), widow of Samuel Fox, III.

89 (CCD 1425; LLi, part). To: The Revd W Darwin Fox, Delamere Rectory, Chester. (Black surround to first page)

Down Farnborough Kent
Apr 29th [1851]

My dear Fox

I do not suppose you will have heard of our bitter & cruel loss. Poor dear little Annie[1], when going on very well at Malvern, was taken with a vomiting attack, which was at first thought//

of the smallest importance, but it rapidly assumed the form of a low & dreadful fever, which carried her off in 10 days. — Thank God she suffered hardly at all, & expired as tranquilly as a little angel. — Our only consolation is, that she//

passed a short, though joyous life. — She was my favourite child; her cordiality, openness, buoyant joyousness & strong affection made her most loveable[2]. Poor dear little soul. Well it is all over. My dear Emma supports herself admirably & is calm & courageous. — It was//

a severe aggravation that Emma could not possibly join me in nursing our darling: she almost daily expects her confinement[3],. I have not yet thanked you for your most kind & interesting letter of the 10th I most truly hope that your health is better.

Yours affectionately
C. Darwin

Christ's College Library, Cambridge, #79.

Notes

1. Anne Elizabeth Darwin (1841-1851). See Letter #88 for details of the trip to Malvern.
2. Anne (Annie) died on 23rd April 1851. A week later, CD wrote a memorial (see CCD, Vol 5, Appendix II) in order that "in after-years, if we live, the impressions now put down will recall more vividly her chief characteristics." Randal Keynes (2001) has written a sensitive biographical account of Annie's life, her death, its probable causes (consumption) and the effect on Emma and Charles Darwin. The effects on CD and his health are also dealt with by Colp, 1987, Moore, 1989 and by his daughter Henrietta (Litchfield 1915, 2: 137).
3. Horace Darwin (1851-1928) was born on 13th May 1851

90 (CCD 1476; LLi[1]). To: The Revd W Darwin Fox, Delamere Rectory, Chester.

Down Farnborough Kent
March. 7th [1852]

My dear Fox.

It is indeed an age since we have had any communication, & very glad I was to receive your note. Our long silence occurred to me a few weeks since, & I had then thought of writing but was idle. I congratulate & condole with you on your *tenth* child[2]; but please to observe when I have a 10th, send only condolences to me. We have now seven children, all well Thank God, as well as their mother; of these 7, five are Boys; & my Father used to say that it was certain, that a Boy gave as much trouble as three girls, so that bonâ fide we have 17 children. It makes me sick whenever I think of professions; all seem hopelessly bad, & as yet//

I cannot see a ray of light. — I should very much like to talk over this By the way my three Bug-bears are Californian & Australian Gold, beggaring me by making my money on mortgage worth nothing[3] — The French coming by the Westerham & Sevenoaks roads, & therefore enclosing Down[4] — and thirdly Professions for my Boys.] & I shd like to talk about Education, on which you ask me what we are doing. No one can more truly despise the old stereotyped stupid classical education than I do, but yet I have not had courage to break through the trammels. After many doubts we have just sent our eldest Boy to Rugby[5], where for his age he has been very well placed. By the way, I may mention for chance of hereafter your wishing//

for such a thing for any friends, that Mr Wharton Vicar of Mitcham, appear to us a really excellent preparatory tutor or *small* school keeper[6]. — I honour, admire & envy you for educating your Boys at home[7]. What on earth shall you do with your Boys?[8] Towards the end of this month, we go to see Willy at Rugby, & thence for

5 or 6 days to Susan at Shrewsbury[9]; I then return home to look after the Babies; & Emma goes to the F. Wedgwoods of Etruria for a week[10]. Very many thanks for your most kind & large invitation to Delamere[11]; but I fear we can hardly compass it. I dread going anywhere, on account of my stomach so easily failing under any excitement. I rarely even now go to London; not that I am at all worse, perhaps rather better & lead a very comfortable life with my 3 hours of daily work, but it is the life of a hermit. My nights are *always* bad, & that stops my becoming vigorous. — You ask about water cure: I take at intervals of 2 or 3 month, 5 or 6 weeks of *moderately* severe treatment, & always with good effect[12]. Do you come here, I pray & beg whenever you can find time: you cannot tell how much pleasure it would give me & Emma. I have finished 1st vol. for Ray Soc. of Pedunculated cirripedes[13], which, as I think you are a member, you will soon get. Read what I describe on sexes of Ibla & Scalpellum. — I am now at work on the Sessile cirripedes[14], & am wonderfully tired of my job: a man to be a systematic//

naturalist ought to work at least 8 hours per day. — You saw through me, when you said that I must have wished to have seen effects of Holmfirth Debacle[15], for I was saying a week ago to Emma, that ~~I~~ had <I> been, as I was in old days, I would have been certainly off that hour —

You ask after Erasmus[16]; he is much as usual, & constantly more or less *un*well. Susan[17] is much better, & very flourishing & happy. Catherine[15] is at Rome & has enjoyed it in a degree that is quite astonishing to my old dry bones. — And now I think I have told you enough & more than enough about the house of Darwin; so my dear old Friend Farewell. What pleasant times we had in drinking Coffee in your rooms at Christ Coll. And think of the glories of Crux Major[18]. Ah in those days there were no professions for sons, no ill-health to fear for them, no Californian gold — no French invasions.//

How paramount the future is to the present, when one is surrounded by children. My dread is hereditary ill-health[19]. Even death is better for them. My dear Fox your sincere friend

C. Darwin.

Remember do if you ever can, come here. You can at any time send Athenæum Newspaper addressed to me at the Athenæum Club, Pall Mall which is my House of call for Parcels of all kinds —//

P.S. Susan[17] has lately been working in a way, which I think truly heroic about the scandalous violation of the act against children climbing chimneys[20]. We have set up a little Society in Shrewsbury to prosecute those who break the Law[21] It is all Susan's doing. She has had very nice letters from Ld Shaftesbury & the D. of Sutherland, but the brutal Shropshire Squires are as hard as stone to move. The act out of London seems most commonly violated. It makes one shudder to fancy one of one's own children at 7 years old being forced up a chimney — to say nothing of the consequent loathsome disease, & ulcerated limbs, & utter moral degradation[22]. If you think strongly on this subject, do make some enquiries — add to your many good works — this other one, & try to stir up the magistrates. There are several people making a stir in different parts of England on this subject. — It is not very likely that you would wish for such but I could send you some essays & information if you so liked, either for yourself or to give away. —

Emma desires me to give her very kind remembrances to Mrs Fox, in which I beg to join[23]. —

Postmark: MR 9 1852
Christ's College Library, Cambridge #80

Notes

1. Francis Darwin in *Life and Letters*, Vol 1, quotes this letter almost completely leaving out only the first postscript "Remember do if you ever can, come here. You can at any time send Athenæum Newspaper addressed to me at the Athenæum Club, Pall Mall which is my House of call for Parcels of all kinds—"
2. Ellen Elizabeth Fox (1852-), born 26th February of this year; this was the fifth child by Fox's second wife, Ellen Sophia. He now had three boys and seven girls.
3. The reference is to the 1849 Californian gold-rush and the 1851 gold-rush in New South Wales and Victoria, Australia. CD is obviously fearful that a sudden influx of gold would lower the gold standard. In reality CD had nothing to fear. He and Emma were among the very rich of the time due to astute investments, many in railway stocks (Desmond and Moore, Chapter 26, 1991).
4. At this time a resurgent Louis Napoleon, president of the French republic, made a coup d'état on 2nd December 1851 and engendered fears of another round of Napoleonic imperial aggression. CCD, Vol 55, notes that at this time, that "*Bartholomew James Sulivan held forth at a dinner party at Down House on the subject of 'how easily a small invading force might overrun our south-eastern counties ... Those present urged him to write to the papers on the subject.' This he did in letters to the Naval and Military Gazette (10 and 31 January 1852), proposing the establishment of a volunteer corps (Sulivan ed. 1896, p. 426).*"
5. Rugby School, a famous Public School, at Rugby, Warwickshire.
6. Henry James Wharton (1798-1859). M A, Emmanuel College, Cambridge (1823). Vicar of Mitcham, Surrey. Tutor of William Erasmus Darwin from autumn 1850 until he entered Rugby School in February 1852 and then for several other sons (see *Diary* of Emma Darwin, 3rd July 1852). Clearly, CD was trying to obtain further employment for him. (*Alum Cantab*)
7. Fox's sons were 10½, 5, and 2½ years old at this time. As noted elsewhere (Introduction, Chapter 8) he helped set up a local school which was still known as "Fox's School" in the 1980s (see Fig 25). He helped teach the children there and also taught his own children there. Whether he also taught his girls is unknown but it is likely.
8. Fox's seven boys grew up to become men in respected professions. Many went to Oxford University and became clergymen, lawyers and financiers. Robert Gerard Fox (1849 – 1909) went first to Kings College, London and only later to Oxford University, and became tutor to the future Kaiser Wilhelm II of Germany [see Letter #213 (n14) and Bibliographic Register].
9. According to Emma Darwin's *Diary*, they visited Rugby on 23rd March 1852. They then presumably traveled to Shrewsbury to see CD's sisters, Susan and Catherine Darwin, who continued to live at the Mount, the old family residence. After this Emma visited Frank Wedgwood (see note 10, below).
10. According to Emma Darwin's *Diary*, she went with Susan Darwin on the 2nd April to Barlaston, where Frank Wedgwood, Emma's brother, who had taken over the Wedgwood pottery, was living at Etruria Hall. Emma returned with her son Frank, now four years old, on the 10th April. CD returned earlier on the 1st April (*Journal*, CCD, Vol 5).
11. Fox's Rectory of Delamere, Cheshire.
12. See Letters #79 (n8), #80, #81 and #82. CD returned twice to Malvern for therapy after his visit in 1849. He also built a special bath house at Down and continued the treatment until 1855, see *Health Diary* (Down House MS). By1855, the water cure treatment had lost its effect and CD consulted Dr Chapman, benefiting from treatment of ice to the spine [see Letter #85 (n10)]. After the death of his

daughter, Annie, at Malvern, in 1851 (see Letter #89), CD did not return there for many years and when he did it was not under the care of Dr Gully (see Letters #161 & #162, in 1863).
13. CD delivered the first volume to the publishers on going up to visit the Great Exhibition on 31st July 1851 (CCD, Vol. 5, Letter to Edwin Lankester, 30th July 1851; Emma Darwin *Diary*). CD here refers to the (late) delivery of this volume by the Ray Society, which had an 1851 date on it.
14. The second volume of the book *A monograph of the sub-class Cirrepedia*, 1854.
15. The village of Holmfirth, in the West Riding of Yorkshire, had been destroyed when the reservoir dam burst on 5 February 1852 (*Annual Register* (1852): 478–81).
16. Erasmus Darwin (1804-1881), CD's brother.
17. Susan (1803-1866) and Catherine (1810-1866) Darwin, CD's sisters, living at the Mount, Shrewsbury.
18. *Panagaeus crux-major*, a rare beetle, found at Whittlesea Meer (see *Autobiography* and Letters #47 (n5 & 6), #48 (n17, #61 (n18) and subsequent letters).
19. It has often been discussed that CD realised from his studies that inbreeding was not beneficial to many species. In marrying his first cousin, he may have thought that he had increased the chance of ill-health in his children. Annie had died in Malvern the previous year (see Letter #87) in circumstances that must have reawakened the ever-present Victorian fear of consumption (R. Keynes, 2001). George Darwin was later to be become interested in the topic of cousin marriages (Letter #207 (n8).
20. This became a 'cause celèbre' in the 1850s and reached the popular literature in *The Water Babies* by Charles Kingsley (1863). Under the Parliamentary Acts of 1834 and 1840 it was illegal to employ boys under the age of sixteen as apprentices to chimney-sweeps. However, the laws were not enforced effectively until 1864 [see, e g, Strange, K (1982) *Climbing boys: a study of sweep's apprentices, 1773-1875*, Alison and Busby, London].
21. CD's *Account Books* (Down House MS) shows a contribution of £5 to the '*Chimney Sweep Society per Catherine*' on 26th June 1852.
22. CCD, Vol 5, notes that "*CD had read Henry Mayhew's London labour and the London poor soon after publication in April 1851*" (CCD, Vol 4, Appendix IV, 119: 23b), in which a detailed account of the chimney-sweeps' climbing boys is given.
23. There is no record of CD and his wife Emma having met Fox's wife, Ellen, since 1846. Fox visited Down on 2nd Nov. 1849 [see Letter #84 (n7)] but not, apparently, with his wife. It seems very probable that they all met in London during 1851, at the time of the Great Exhibition, or in early 1852. However, there are no records in any of their diaries/journals to this effect.

91 (CCD 1489; LLi). To: The Revd W Darwin Fox, Kendalls[1], Ellstree, Hatford, Herts

Down Farnborough Kent
24th [Oct 1852]

My dear Fox

I received your long & most welcome letter this morning, & will answer it this evening as I shall be very busy with an Artist drawing Cirripedia[2] & much overworked for the next fortnight. But first you deserve to be well abused, & pray consider yourself well abused, for thinking or writing that I could for one minute be bored by any amount of detail about yourself & belongings. It is just what I like hearing: believe me that I often think of old days spent with you, & sometimes can hardly believe what a jolly careless individual one was in those old days. A bright autumn Evening often brings to mind some shooting excursions from Osmaston[3]. I do indeed regret that we live so far off each other, & that I am so little locomotive: I have been unusually well of late, (no Water Cure) [4] but I do not find that I can stand any change hardly better than formerly.//

All excitement & fatigue brings on such dreadful flatulence; that in fact I can go nowhere. The other day I went to London & back, & the fatigue, though so trifling brought on my bad form of vomiting. I grieve to hear that your chest has been ailing[5]; & most sincerely do I hope that it is only the muscles: how frequently the voice fails with the Clergy. I can well understand your reluctance to break up your large & happy party & go a broad; but your life is very valuable, so you ought to be very cautious in good time. You ask about all of us, now 5 Boys (oh the professions, oh the gold, & oh the French, — these three oh's all rank as dreadful bugbears[6]) & two girls: Emma has been very neglectful of late & we have not had a child for more than one whole year[7]. She is, thank//

God, right well, & so are all the chicks; but another & the worst of my bugbears, is heredetary weakness[8]. All my sisters are well, except Mrs Parker[9], who is much out of health; & so is Erasmus at his poor average: he has lately moved into 2 Gower St[10]. I had heard of the intended marriage of your sister Frances: I believe I have seen her since, but my memory takes me back some 25 years, when she was lying down; I remember well the delightful expression of her countenance: I most sincerely wish her all happiness[11]. Tenby by all accounts is a delightful place[12]. — I see I have not answered half your Queries: we like very well all that we have seen & heard of Rugby, & have never repented of sending him[13] there: I feel sure schools have greatly improved since our days; but I hate schools & the whole system of breaking through the affections of the family by separating the boys so early in life, but I see no//

help & dare not run the risk of a youth being exposed to the temptations of the world, without having undergone the milder ordeal of a great school. — I see you even ask after our Pears: we have had lots of Beurre d'Alenbery, Winter Nelis, Marie Louise, Passe Colmar & Ne Plus Meuris but all off the wall[14]: the standard dwarfs have borne a few, but I have no room for more trees, so their names w^d^ be useless to me. You really must make a holiday & pay us a visit sometime: nowhere could you be more heartily welcome. — I am at work on the second vol. of the Cirripedia[15], of which creatures I am wonderfully tired: I hate a Barnacle as no man ever did before, not even a Sailor in a slow-sailing ship. My first vol. is out: the only part worth looking at is on the sexes of Ibla & Scalpellum[16]; I hope by next summer to have done with my tedious work[17]. Farewell, — do come whenever you//

can possibly manage it. I cannot but hope that the Carbuncle may possibly do you good: I have heard of all sorts of weaknesses disappearing after a carbuncle: I suppose the pain is dreadful. I agree most entirely, what a blessed discovery is Chloroform[18]: when one thinks of one's children, it makes quite a little difference in ones happiness. The other day I had 5 grinders (two by the Elevator) out at a sitting under this//

wonderful substance, & felt hardly anything[19].

My dear old Friend
Yours very affectionately
Charles Darwin

Christ's College Library, Cambridge #81

Notes

1. "Kendalls", near Hatfield, Hertfordshire, North of London, was the final home of Samuel Fox III, his wife and family. They had previously moved from Osmaston Hall, by at least 1844, to Meopham Bank, near Tunbridge, in Kent, according to the Fox *Diaries*. The first mention of Kendalls is on 2nd March 1845: "*My first visit to Kendalls. Mother & sisters met me at Watford. Catherine Darwin there.*"
2. Emma Darwin noted in her *Diary*, for 25 October 1852,: '*Mr Sowerby came*'. CCD noted that in November that CD recorded in his *Account book* (Down House MS) that "*George Brettingham Sowerby Jr was paid £10: 'for drawing & journey [to Down] £7. 14 giving 2 guineas extra, & on Engraving account 2. 6. 0'*" (CCD, Vol 5). These drawings were presumably for *A monograph of the sub-class Cirripedia, with figures of all the species. Vol 2* (1854).
3. Osmaston Hall (Fig 4), the home of Samuel Fox III and family, near Derby, until about 1844, when his son Samuel Fox IV took over the lease (see Chapter 1, and notes 1 & 5, above and below).
4. CD's *Health Diary* (Down House MS) indicates that CD discontinued his douche treatment on 22 August, noting: "*Six weeks treatment; not much good effect, extremely tired in Evening. I do not think last Treatment did me much good. —*". The *Health Diary* indicates that CD continued in relatively good health until the end of the year.
5. In his *Diary* for 1852, Fox wrote, "*Oct 1- 8th. I suffered much from carbuncle on stomach – which was on 8th cut under chloroform by J. Harrison*", and again, "*Nov 4. I consulted Dr Watson who said heart was weak*". Nevertheless during October and November, Fox carried on a busy life, attending to his parish duties, visiting his mother and sister in Hertfordshire, his father-in-law in Hampstead, London, his brother at Osmaston Hall (twice), attending the marriage of his sister at Tenby, Wales (see note 11, below), and watching the funeral of the Duke of Wellington. Perhaps it is not surprising that his last entry for November 1852 was "*Poorly rest of month*", on 24th Nov. By December, Fox was busy again and on 14th Dec he wrote, "*Set off with Mr Dear at 3 in morning to visit Birmingham Poultry and Cattle Show*".
6. See also Letter #88 (n3 & 4).
7. Horace Darwin was born on 13th May 1851. Emma's last child, Charles Waring Darwin was born on 6th December 1856. CD may have joked about Emma's child-bearing but it was no joke to her. Her last three children were born using chloroform as an anaesthetic. See Letter #84 (n4).
8. CD harks back to the same subject as in a previous letter (#90).
9. Marianne Parker (1798-1858) was CD's eldest sister.
10. The letter has "2 Gower St" but LLi and CCD, Vol 5 has Queen Anne Street, Erasmus Darwin, CD's brother lived in Great Marlborough St, and then Park Street, prior to his move to 6, Queen Anne Street (see Litchfield, 1901, Vol. 2, p 146-148).
11. Frances Jane Fox (1806-1884) was married on 28 October 1852 to John Hughes, vicar of Penelly, near Tenby, South Wales (*Darwin Pedigrees*). CD visited Osmaston Hall before the *Beagle* voyage but did not mention Frances. Susan Darwin visited there in 1833: see CCD, vol. 1, Susan to CD, 22–31 July 1833, for a description of Frances : "*Frances Jane is grown much stronger so that she comes down stairs & is able to sit up much more than formerly. I think she is very superior to the rest of the girls both in sense and agreeableness, besides being very handsome & interesting.*"
12. See Note 11. Frances Jane was to move to Tenby, South Wales. At this time several of Emma Darwin's Allen aunts and an uncle still lived at Cresselly in Tenby.
13. William Erasmus Darwin, CD's oldest son, was at Rugby School in early 1852 (see Letter #88 and CCD, Vol. 5, letter of CD to W.E. Darwin, 24th February 1852).
14. CD and Fox were discussing the best way to grow fruit trees in 1849 in letter #79 (n11) .
15. The first volumes of both *A monograph of the sub-class Cirrepedia, with figures of all the species* and *A monograph of the fossil Lepadidae, or pedunculated cirrepedes, of Great Britain* were published by the Ray Society in 1851. The second volume of each publication was published in 1854 (n b, Freeman, 1977, suggests that for the fossil volume the date should have been 1855).
16. CD emphasises Ibla and Scalpellum in another letter to Fox [see Letter #88 (n14)].
17. CD was over-optimistic and because of ill health did not see the proofs of the later barnacle volumes until 1854 (see '*Journal*' and CCD Vol 5, Appendix I).

18. Fox seems to have come around from his previous position [see Letters #84 (n4); #85 (n4)]. Fox noted in his diary for 1st to 8th Oct 1852 "*I suffered much from a carbuncle on stomach – which was on 8th cut under chloroform by J. Harrison*".
19. The Darwin Correspondence team have noted that CD recorded in his *Account Books* (Down House MS) for 25 June 1852, '*Teeth extracted & Journey to London [£]4 17[s.]*'. Also CD's *Health Diary* (Down House MS) indicated that this dentistry took place on 24th June. Another entry in the *Account book* records for 5th September 'Mr Waite. Dentist' was paid £1 1*s*. (Down House MS). The Oxford English Dictionary records 'Elevator' as 'an instrument used in dentistry to remove stumps of teeth'.

92 (CCD 1499; LLi, part). To: The Rev^{d} W Darwin Fox, Delamere Rectory, Chester.

Down Farnborough Kent
Jan. 29th — [1853]

My dear Fox

Your last account[1], some months ago, was so little satisfactory, that I have often been thinking of you, & should be really obliged if you would fly me a few lines, & tell me how your voice & chest are. I most sincerely hope that your report will be good; this wonderfully mild winter must be in your favour.//

As for myself I really have no news: just lately my stomach has been a little extra ailing[2]. All other members of the family are flourishing. My eldest Boy is now home from Rugby[3]: he is a thoroughily steady, industrious & good boy; I fancy, (though perhaps it is fancy) that I see the contracting effects on his mind of his very steady attention to classics: formerly//

I think he had more extended interests, & cared more for the causes & reasons of things[4]. Our second lad Georgie, has a strong mechanical turn: & we think of making him an engineer: I shall try & find out for him some less classical school, — perhaps Bruce Castle[5]. I certainly sh^{d} like to see more diversity in Education, than there is any ordinary school: no exercise of the observing or reasoning faculties, — no general//

knowledge acquired, — I must think it a wretched system: on the other hand a Boy who ~~will~~ <has learnt to> stick at Latin & conquer its difficulties, ought to be able to stick at any labour. — I sh^{d} always be glad to hear anything about schools <or education> from you.

I am at my old, never-ending subject, but trust I shall really go to press in a few months with my//

second volume on Cirripedes[6]: I have been much pleased by finding some odd facts in my 1st vol. believed by Owen, & a few others, whose good opinion I regard as final[7]. — I have this morning been dissecting a most abnormal cirripede, which after a good meal has to vomit forth the residuum, for there is no other exit[8]!

I heard yesterday from//

Dr Hooker, who married Henslow's eldest daughter, of the birth of a son[9] under Chloroform, at Hitcham. I wonder when we shall see you here again: it w^{d} give Emma & myself no common pleasure. Do write pretty soon & tell me all you can about yourself—& family//

& I trust your Report of yourself may be much better than your last. Catherine & Susan are at present staying with Erasmus in London, & perhaps I shall go up & see them next week[10]. I have been very little in London of late, & have not seen Lyell since his return from//

America: how lucky he was to exhume with his own hand parts of 3 skeletons of Reptiles out of the *Carboniferous* strata, & out of the inside of a fossil tree, which had been hollow within[11]

Farewell
My dear Fox
Your's Affectionately
Charles Darwin

Christ's College library, Cambridge #82

Notes

1. Fox notes in his diary for 24th Nov. 1853, "*poorly rest of month*". So it seems that the letter was written at this time. From his *Diary* entry on 14th Dec 1852, Fox was well enough to set out for the Birmingham Poultry and Cattle Show. See Letter #89 in which CD mentions Fox's chest ailment; see Letter #48 (n4), for the onset of the illness affecting his lungs; see, also, letter from Susan Darwin to CD, 22nd July 1833 (CCD, Vol 1).
2. See CCD, Vol 5, letter to G R Waterhouse, 18th January [1853], n 3. At the end of January, 1853, CD summed up that he had had 11 days on which he felt very well, compared with 24 days in December, 1852.
3. See Letter #91 (n13).
4. CD seems already to be regretting sending his eldest boy to a prestigious "Public School", with all that that entailed in terms of a classical education in Latin and Greek. His other boys at this time were going to Mr Henry Wharton [see Letter #88 (n5)] at Mitcham and later were sent to Clapham Grammar School in South London [see Letters #86 (n6) and #112 (n8)]. Judging from all their successful careers, one can conclude that the education at Clapham did them no harm.
5. See Letter #86 (n6).
6. The final proofs of *A monograph of the sub-class Cirripedia* Vol. 2 (1854) were not sent to the printer until July 1854.
7. The sexual relations of *Ibla* and *Scalpellum* (*Living Cirripedia* (1851): 281–93). See Letter #91 (n16) and CCD Vol 5, CD letter to Richard Owen, 17th July 1852.
8. *Alcippe lampas* has no rectum or anus (*A monograph of the sub-class Cirripedia* (1854): 546–7).
9. William Henslow Hooker (1853-1942), born 24th January of this year (Desmond, 1999).
10. CD recorded the expenses of a trip to London on 3rd February 1853 in his *Account Books* (Down House MS). His *Health Diary* (Down House MS) indicates that the visit was from 1 to 3 February.
11. See Charles Lyell, 1853. Edinburgh New Philosophical J, 55: 215-225; American J Sci 2nd ser 16: 33-41.

93 (CCD1522; LLi, part[1]). To: The Revd W Darwin Fox, Delamere Rectory, Chester.

13. Sea Houses Eastbourne
[15 July[2] 1853]

My dear Fox

Here we are in a state of profound idleness, which to me is a luxury; & we shd all, I believe, have been in a state of high enjoyment, had it not been for the detestable cold gales & much rain, which always gives much ennui to children, away from their homes. — I received your letter of the 13th of June when working like a slave with Mr Sowerby at drawing for my second volume[3], & so put off answering it till when I knew I shd be at leisure. I was extremely glad to get your letter: I had intended a couple of months ago sending you a savage or supplicating jobation to know how you were, when I met Sir P. Egerton[4] , who told me you were well, &, as usual, expressed his admiration of your doings, especially your farming & the number of animals, including children, which you kept on your land. — Eleven children ave maria! it is <a> serious//

look out for you[5]. Indeed I look at my five boys as something awful & hate the very thought of professions &c: if one could insure moderate health for them it wd not signify so much, for I cannot but hope with the enormous emigration professions[6] will somewhat improve. But my bug-bear is heredetary weakness[7] . I particularly like to hear all that you can say about education: & you deserve to be scolded for saying "you did not mean to *torment* me with a long yarn". — You ask about Rugby[8]: I like it very well, on the same principle as my neighbour Sir J. Lubbock[9] likes Eton, viz that it is not worse than any other school: the expence *with all &c &c* <including some clothes, travelling expences &c> is from £110 to £120 per annum: I do not think schools are so wicked as they were, & far more industrious. The Boys, I think, live too secluded in their separate studies; & I doubt whether they//

will get so much knowledge of character, as Boys used to do, & this in my opinion is the *one* good of public schools over small schools. I shd think the only superiority of a small school over home was forced regularity in their work, which your Boys perhaps get at your home, but which I do not believe my Boys wd get at my home. Otherwise it is quite lamentable sending Boys so early in life from their home. I think of bringing up my eldest Boy as an Attorney; & my second, who has a mechanical turn & is very active minded, as an Engineer[10]. — To return to schools, my main objection to them, as places of education, is the enormous proportion of time spent over classics. I fancy, (though perhaps it is only fancy) that I can perceive the ill & contracting effect on my eldest Boy's mind, in checking interest in anything in which reasoning & observation comes into play. mere memory seems to be worked. — I shall certainly look out for some school, with//

more diversified studies for my younger Boys[11]. I was talking lately to the Dean of Hereford[12], who takes most strongly this view: & he tells me that there is a school at Hereford commencing on this plan: & that Dr Kennedy at Shrewsbury is going to begin vigorously to modify that school; but I rather mistrust Dr K's judgment[13]. I have some fears whether any school will do for my second Boy, as his health has lately failed rather[14]; & a very irregular pulse, (though not resulting from any heart complaint) I fear shows that the weakness is deep-seated[15]. — I am *extremely* glad to hear that you approved of my cirripedial volume[16]: I have spent an almost ridiculous amount of labour on the subject & certainly wd never have undertaken it, had I foreseen what a job it was: I hope to have finished by the end of the year[17]. — Do write again before a very long time: it is a real pleasure to me to hear from you.

Farewell with my wifes kindest remembrances to yourself & Mrs Fox. My dear old friend. Yours affectionately

C. Darwin

I am reading F. Galton's book & like it very *much*[18]. —//

P.S. I had not sealed up this letter an hour, before I saw with the utmost concern & astonishment the deaths in your house: [19] I most deeply hope that your own children have escaped this most fearful illness. I did doubt about sending off this letter till knowing how your own children were; but it need not be read[20]. Do pray sometime tell me how far you have escaped. And I fear your wife must just have been confined[21]. —

Postmark: JY 18 1853
Christ's College Library, Cambridge, #84

Notes

1. LLi, p 386–388. With the exception of the following: i) the three sentences: "I think of bringing up my eldest Boy as an Attorney; & my second, who has a mechanical turn & is very active minded, as an Engineer.", "but I rather mistrust Dr K's judgment. I have some fears whether any school will do for my second Boy, as his health has lately failed rather; & "a very irregular pulse, (though not resulting from any heart complaint) I fear shows that the weakness is deep-seated.", ii) the postscript.
2. CD wrote this letter in Eastbourne where the family stayed from 14th July to 4th Aug 1953. It did not have a date and was dated from the postmark.
3. This was the second volume of *A monograph of the sub-class Cirrepedia* (1854).
4. Sir Philip de Malpas Grey-Egerton (1806-1881), palaeontologist specializing in fossil fish, who lived at Oulton Park near Fox's parish of Delamere in Cheshire. See Letters #66 (n9), #73 (n6) & #87 (n5).
5. The grand total, finally, in 1864, was to be seventeen children!
6. Echoes of Letters #90 & 91.
7. Also echoes of Letter #89 (n3); and see Letter #91 (n7).
8. CD's eldest son, William Erasmus, had now been at Rugby School for 1½ years.
9. John William Lubbock (1803-1865), 3rd Baronet (in 1833). Banker, astronomer and mathematician. First Vice-Chancellor of London University. A good friend and neighbour of CD. (*DNB*)
10. William Erasmus became a banker and George Howard Darwin was called to the bar but did not practice, later becoming Plumian Professor of Astronomy at Cambridge University.
11. With the exception of William, all of CD's sons attended Clapham Grammar School [see Letter #112 (n8)].
12. Richard Dawes was a fellow of Downing College at the time that CD was an undergraduate in Cambridge. In 1842 Dawes founded King's Somborne School in Hampshire, which was considered a model school at the time. He was interested in the establishing a Bluecoat school in Hereford after his appointment as dean in 1850 (*DNB*).
13. CD was a pupil at Shrewsbury School from 1818–25 under Dr Butler. CD and his family in Shrewsbury would have known Benjamin Hall Kennedy, Headmaster of Shrewsbury School, from 1836–66. Like Dr Butler, Kennedy strongly promoted a classical education. (*DNB*)
14. George Darwin (1845-1912), who became Plumian Professor of Mathematics and Astronomy at Cambridge University
15. Emma Darwin noted in her *Diary* on 17th June 1853: "*up in London tour - Dr H. for G. and Sir B. B. for me*". Dr. H was probably Dr Henry Holland. George Darwin suffered from poor health throughout his life.
16. Fox, as a member of the Ray Society, would have received a copy of *A monograph of the sub-class Cirripedia* Vol. 1, in 1852 [see Letter #90 (n13)].

17. CCD, Vol 5 contains, as appendix II, "Darwins's study of Cirripedia", which discusses in detail Darwin's barnacle work. CCD, Vol 5, in note 10 for this letter, observes: "*CD had originally planned to write only a paper on Arthrobalanus, the aberrant cirripede collected during the voyage of the Beagle (see CCD Vol 3). His work soon encompassed several other genera, and late in 1847, at the suggestion of John Edward Gray he undertook a systematic study of the entire sub-class (see CCD, Vol 5, Appendix II)*".
18. Francis Galton (1822-1911), a cousin of CD and distant cousin of Fox: scientific writer, statistician and founder of the eugenics movement. CD refers to *The narrative of an explorer in tropical South Africa*, London, 1853, which CD recorded having read on 20 July 1853 (CCD, Vol 4, Appendix IV, 128: 5). (*DNB*)
19. *The Times*, 16 July 1853, Supplement, p 1, announced the deaths of Frederick William and Zachary Granger Mudge, aged 8 and 10, on 8th and 13th July respectively, at Delamere Rectory, Cheshire. They had both died of scarlet fever. The boys were the grandchildren of Zachary Mudge, a prominent naval officer, who died in 1852 (*DNB*). Their father, Zachary Mudge Jr, was at this time a barrister, living in the family home of Sydney, Plympton, Devon (*Gentleman's Magazine*, n s 5 (1868) i: 120). Fox and family had cared for the two boys since 1848, when their mother, Mary, died. Zachary Mudge, Sr was Fox's uncle-in-law, by his first wife, Harriet. No *Diaries* of Fox are extant for the years 1853-1856 (see Appendix 1).
20. The postscript was written on a separate sheet of paper. CD presumed that Fox would see this first.
21. Ellen Sophia Fox gave birth to a daughter, Theodora (1853-1878), on 16th June 1853 (*Darwin Pedigrees*).

94 (CCD 1527; LLi[1]). To: The Revd W Darwin Fox, Delamere Rectory, Chester (redirected to: Huntingdon Lodge, Rhyl, N. Wales).

Down Bromley Kent
Augt 10th [1853]

My dear Fox

I thank you sincerely for writing to me so soon, after your most heavy misfortunes[2]. Your letter affected me much. We both most truly sympathise with you & Mrs Fox. We too lost, as you may remember, not so very long ago, a most dear child, of whom, I can hardly yet bear to think tranquilly[3]; yet, as you must know from your own most painful experience[4], time softens & deadens, in a manner truly wonderful, one's feelings & regrets. At first it is indeed bitter. I can only hope that your health & that of poor Mrs Fox may be preserved; & that time may do its work softly, & bring you all together, once again as the happy family, which, as I can well believe, you so lately formed[5].

My dear Fox
Your affectionate friend
Charles Darwin

Christ's College Library, Cambridge #85

Notes

1. LLi, p. 388, verbatim.
2. Fox's three-year old daughter, Louisa Mary Fox (1850-1853), died of scarlet fever on 29th July. This was in addition to the deaths, also from scarlet fever, of Zachary and Frederick Mudge, detailed in the last letter [#93 (n19)]. In addition to CD's loss of his daughter, Annie (Letter #89), it should be remembered that CD was also to lose a one-and-half-year old child, Charles Waring, from scarlet fever in 1858, at the time of the joint presentation by Darwin and Wallace, of their theories of organic evolution to the Linnean Society of London on the 1st July 1858 [Letters #127 (n12) & #128 (n1)].

3. Anne Elizabeth Darwin died on 23 April 1851; see Letter #89.
4. CD refers to the loss of Fox's first wife on 19th March 1842; see Letters #70 & #71.
5. The Fox *Diary* for this year, 1853, has not been found and nor those for 1854-56 (see Appendix 1). It seems likely that Fox started a *Diary* in 1853; whether he discontinued and later destroyed it, from the pain of events, is a matter for conjecture. The absence of *Diaries* for 1854-56 is more difficult to explain. These *Diaries* may have gone to a different line of the Fox family (see Chapter 2, note 13; Appendix 1).

95 (CCD 1651; LLi). To: The Rev^{d} W Darwin Fox, Delamere Rectory, ~~Chester~~, Northwich.

Down Farnborough Kent
March 19$^{\underline{th}}$ [1855]

My dear Fox

How long it is since we have had any communication[1] & I really want to hear how the world goes with you; but my immediate object is to ask you to observe a point for me, & as I know how you are a very busy man with too much to do, I shall have a good chance of your doing what I want, as it w^{d} be hopeless to ask a quite idle man. — As you have a noah's ark[2], I do not doubt that you have pigeons; (how I wish by any chance they were fantails!) Now what I want// to know is, at what age nestling pigeons have their tail feathers sufficiently developed to be counted. I do not think I ever even saw a young pigeon. I am hard at work on my notes, collating & comparing them, in order in some 2 or 3 years to write a book with all the facts & arguments, which I can collect, *for & versus* the immutability of species[3]. — I want to get the young of our domestic breeds to see how young, & to what degree, the differences appear. I must//

either breed myself (which is no amusement, but a horrid bore to me) the pigeons or buy them young, & before I go to a seller whom I have heard of from Yarrell[4], I am really anxious to know something about their development not to expose my excessive ignorance, & therefore be excessively liable to be cheated & gulled — With respect to the *one* point of the tail feathers, it is, of course, in relation to the wonderful development of tail feathers in the adult fan-tail.// If you had any breed of Poultry pure, I w$^{\underline{d}}$. beg a chicken <with exact age stated> about a week <or fortnight> old! to be sent in Box by Post, if c^{d} have the heart to kill one, & secondly w$^{\underline{d}}$ let me pay postage. Indeed I sh$^{d.}$ be very glad to have a ~~young~~ <nestling> common pigeon sent, for I mean to make skeletons, & have already just begun comparing wild & tame ducks[5], & I think the results rather curious, for on weighing the several bones very carefully, when perfectly cleaned, the *proportional* weights of the two have greatly varied; the foot of the tame having largely increased. How I wish I could get a little wild duck of a week old, but that I know is almost impossible. —

With respect to ourselves, I have not much to say: we have just now a terribly noisy house with the Hooping cough[6]: but otherwise are all well. Far the greatest fact, about myself is that I have at last quite done with the everlasting// Barnacles. At the end of the year we had two of our little Boys very ill with fever & Bronchitis & all sorts of ailments[7]. Partly for amusement & partly for change of air we went to London & took a House for a month, but it turned out a great failure, for that dreadful frost just set in when we went, & all our children got unwell & Emma & I had coughs, & colds, & rheumatism nearly all the time[8]. We had put down first on our list <(of things to do)>// to go & see Mrs Fox[9], but literally, after waiting some time to see whether the weather would not improve, we had not a day when we both could go out. I do hope before very long you will be able to manage to pay us a visit: Time is slipping away, & we are getting oldish. Do tell us about yourself & all your large family. I know you will help me, *if you can*, with information about the young pigeons: & anyhow do write before very long.

My dear Fox
Your sincere old friend
C. Darwin

Amongst all sorts of odds & ends, with which I am amusing myself, I am comparing the seeds of the vars. of plants. I had formerly some wild cabbage seeds, which I gave to some one. Was it to you? It is a *thousand* to one that it is thrown away, if not I sh$^{\underline{d}}$. be very glad of a pinch of it[10].—

(On envelope)
"Young Duck
March ~~20~~ 19 *1865*"

Postmark: obscure
Christ's College Library, Cambridge, #87

Notes

1. The last known letter is for 15th August 1853. This was at the time of the three deaths in the Fox family; see Letter #94. This long lapse in correspondence may be explained by the grief following these losses. However, the reference of a possible visit to Mrs Fox (Note 9) suggests, despite the opening line that there had been some kind of communication between the families.
2. A previous reference to the large variety of animals that Fox kept on his farm: see Letters #74 (n2), #85 (n6) & #87 (n4).
3. This is the first reference to CD's abiding idea after CD had hinted to Fox about his views on the transmutation of species in 1838 (see Letter #64): "...... *the crossing of animals. It is my prime hobby & I really think some day, I shall be able to do something on that most intricate subject species & varieties.*" It is natural to wonder how much was spoken between them in the intervening years on this subject. Fox was always very guarded in later years [see Letter #192 (n6 & 7)] and so CD may not have communicated his ideas in detail, knowing that he might be rebuffed.
4. This is the first information to Fox on the subject of pigeons, although CD had for some time been attending pigeon fancier meetings in London. William Yarrell (1784-1856): Naturalist and bookseller; wrote standard works on British birds and fishes (*DNB*). He was well known to both Fox and CD from their student days; see Letter #31 (n4). From 1855 onwards, CD's *Account Books* (Down House MS) are full of entries relating to payments to Baily for pigeons: this is probably the son of John Baily, poulterer and dealer in live birds, 113 Mount Street, Berkeley Square [see Letter #106 (n4 & 5)].
5. See CCD, Vol. 5, CD letter to J D Hooker, 11th December 1854: "*I have just been testing practically what disuse does in reducing parts; I have made skeletons of wild & tame Duck (oh the smell of well-boiled, high Duck!!) & I find the tame-duck wing, ought according to scale of wild prototype to have its two wings 360 grams in weight, but it has it only 317 or 43 grams too little or 17th of own two wings too little in weight: this seems rather interesting to me.*"

6. Illnesses of the Darwin children were recorded in CD's family Bible (Down House MS). This indicated that all except William had '*Hooping cough*' in March, 1855. Emma Darwin's *Diary* records, "*10th March Hooping cough began*".
7. Francis and Leonard Darwin (see '*Journal*'; CCD, Vol. 5, Appendix I).
8. Charles Darwin and family stayed at 27 York Place, Baker Street, from 18 January to 15 February 1855. CCD, Vol 5 noted that the *Annual Register* (1855), Chronicle, pp. 23–5, reported that a severe cold spell affected the country in January and February with the Thames iced over for periods.
9. Fox's *Diary* for 1855 has not been found. Since his second wife's father, Basil George Woodd, and family lived at Hillfield, Hampstead, and his wife made frequent visits to them in London, it is highly likely that a visit had been planned.
10. CD was interested at this time in observing the longevity of viable seeds when exposed to seawater: See CCD, Vol 5, CD letter to J D Hooker, 7th April 1855.

Conclusion

The correspondence of this Chapter takes us to the middle of 1855, a significant year for Darwin, for it is the year that he finished his work on barnacles and felt free to devote his energies, at last, to that vexed question of the transmutation of species. In Letter #64, in 1837, Darwin first admitted to Fox that "*It is my prime hobby & I really think some day, I shall be able to do something on that most intricate subject species & varieties*." Now, eighteen years later he is "*hard at work on my notes, collating & comparing them, in order in some 2 or 3 years to write a book with all the facts & arguments, which I can collect, for & versus the immutability of species*." (Letter #95). The year is also significant because there is a growing impatience of many natural historians with the current theological attitude to the idea that God created all the plants and animals of the world just as they were. Significantly meetings were held on the subject, even at Down, the home of Darwin (Desmond and Moore, 1992). And, even more alarmingly for Darwin, someone else had begun to write sensibly on the subject. In 1855, Alfred Russell Wallace published a paper, which was like a banner on the new battle lines. It was entitled "*On the law which has regulated the introduction of new species*" and was published in the Annals and Magazine of Natural History (London). Darwin was alerted to this paper, initially, by Edward Blyth. However, it is clear that even his friend Charles Lyell was worried by the situation, since Darwin wrote in 1858, after the receipt of the Wallace 'bombshell letter' (see Chapter 6), "*Some year or so ago you recommended me a paper by Wallace in Annals, which interested you*". Lyell urged Darwin to publish quickly an abstract of his ideas on transmutation, but Darwin's answer was the book referred to above, which would be written in 2 to 3 years, that is by 1858 at the latest.

Apart from a break from about the 3rd Oct 1846 (Letter #78) to 6th Feb 1849 (Letter #79) the correspondence is fairly complete. This missing period covers the early years of Fox's second marriage (the wedding was on 20th May 1846), but there is no indication that this would have restricted the correspondence. Fox and his wife visited the Darwins at Down House on their honeymoon (see "Introduction" to this chapter). The next visit is by Fox (alone) in November 1849. Since there is no evidence for a break in communication it is likely that some letters for these years were lost. Fox clearly kept diaries at this time, but only kept summaries of the main events; the next complete diary is for 1851 (CUL DAR 250).

These years, from 1844, or so, to 1854, should not be thought of as wholly devoted to barnacles (published in 1851 and 1854/1856 (see Letter #95, 19th Mar 1855, "*Far the greatest fact, about myself is that I have at last quite done with the everlasting Barnacles.*"). In addition to completing publication of the Beagle material (the last volume of which, *Geological observations on South America*, was published in 1846), Darwin devoted a considerable amount of his considerable energy to charting variation in species of plants and animals during this time. There is little of this in the Fox correspondence of these years, which is devoted to family and health matters. The course that Darwin took here is not well charted, because Darwin did not publish these ideas and they remain as scraps of information in notes, marginalia on books and in correspondence. Manier (1978) and Ospovat (1981) analyse Darwin's intellectual progress during these years. It is essentially one of rational scientific development, with the discarding of irrational ideas such as the Quinary system of Macleay and progressivism in evolution (Ospovat, 1981). Just as Darwin was ahead of his time in rejecting paranormal beliefs such as phrenology (letter #26), mesmerism [Letters #75 (n3) & #76, (n8)], clairvoyance (Letter #86) homoeopathy [Letter #86 (n9)] and probably spiritualism [Letter #86 (n11), although it is recorded that CD and ED attended a spiritualist séance with George Eliot at R D Litchfield's house in 1874 (Freeman, 1978).

Darwin quickly strove to find a mechanistic theory for what he called "natural selection". He first wrote out an abstract, or sketch, of his theory in 1842. This was edited and written out in a fair hand (not Darwin's!) in 1844 and he left instructions that if he were to die, that Emma, his wife, should arrange to have it published. This was the basis of his ideas, but there was much to fill in. Darwin had little knowledge of plants at this stage, especially of their breeding systems. His new friend and confidant, Joseph Dalton Hooker, was invaluable in this regard, as he was a botanist and his father, William Hooker, was the Director of The Royal Botanic Gardens at Kew. More than anyone else, through these years, Darwin found a kindred spirit with whom he could share the most intimate thoughts, much more so than with Lyell and, later, with Huxley.

In the same year, an anonymous book called the *Vestiges of the Natural History of Creation*, appeared and became a Victorian sensation (Secord, 2000). This book gave a racy account of how the solar system may have formed and how life on earth may have "evolved" (although this use of the word had not yet been invented - see Freeman, 1978), from simple forms to the most complex. It was a "good read" but was seriously flawed by the lack of biological knowledge of the author, the Scottish publisher, Robert Chambers, who did not admit to his authorship until 1884. Like other accounts of the time it lacked a credible mechanism for how organisms might change from one species to another. For these reasons, and because of the strong religious revival that permeated early Victorian life, the book was roundly condemned by all sides, including natural historians. It may be for this reason that Darwin postponed entering the fray. His copy of "*Vestiges*", now in the University Library, Cambridge, is carefully annotated and Darwin would have learnt much from the example - of what was palatable and what was not, as for instance any suggestion that man had evolved from the apes. Darwin wrote to Fox "*Have you read that strange unphilosophical, but capitally-written book, the Vestiges, it has made more talk than any work of late, & has been by some attributed to me. — at which I ought to be much flattered & unflattered*." (Letter #77). Clearly some word had got out that Darwin had unorthodox views.

As mentioned above, the correspondence between Darwin and Fox during these years is largely devoid of biological comment. Both men were clearly much concerned with bringing up families and the woes that this entailed. It is not until the last letter in the series (#95) that Darwin begins to draw upon Fox for biological information. This coincides with the completion of his barnacle work (see above) and the statement "*I am hard at work on my notes, collating & comparing them, in order in some 2 or 3 years to write a book with all the facts & arguments, which I can collect, for & versus the immutability of species*." Since Fox was a Church of England clergyman, this appears to be a provocative statement. We have no idea what Fox's stance was on this, at this time, but it appears that Darwin had made him complicit in his work from an earlier time. "*There are points in your unrivaled Book "The Origin of Species*" Fox writes, much later, "*which I do not come up to*" (Letter #192, November 1870). From 1855 on, however, there develops a thriving correspondence on varieties of domesticated animals and in this sequence the correspondence is much more frequent (see Chapter 6).

It was not that Fox had been neglectful of his natural history interests. He kept up a correspondence with Hewitson and wrote letters to the *Gardeners Chronicle* on rearing resistant strains of potato (Letter #78). There was even the odd letter between Darwin and Fox on variation (Letters #66 and 77A). Fox also developed a thriving farm on his property at Delamere, which Sir P Egerton and Darwin referred to as Fox's "*Noah's Ark*" (Letters #76, #87 & #95). From here on, it is more the preoccupations of Darwin that poise and guide the correspondence; his coaxing and cajoling of Fox reaches a new high, and the result was a burgeoning correspondence (Chapter 6).

Nevertheless these years were filled with family matters. For Darwin, his plant studies began to emphasise the importance of out-crossing in nature and he must have wondered what negative effect the inbreeding in the Darwin-Wedgwood family might have incurred (Letters #90, 91 & 93). In a series of letters in 1844-46 he harked on the responsibility of bringing up a family: "*Ah in those days there were no professions for sons, no ill-health to fear for them, no Californian gold—no French invasions.*" (Letter # 90) and notes especially "*my bug-bear is heredetary weakness*" [Letter #93 (n6)].

Both men were sick during this period, although in neither case did it prevent a fairly full and active professional life. Darwin clearly used his ill health as an excuse to cut back on social contacts and he even applied this to Fox. There are only four recorded visits during this seventeen year period, all initiated and executed by Fox (see Appendix 2). It was Fox who persuaded Darwin to take up the "water cure" which came into fashion at this time - and after his first experience in 1849 he wrote to Fox, "*The Water Cure is assuredly a grand discovery & how sorry I am I did not hear of it, or rather that I was not somehow compelled to try it some five or six years ago*" (Letter #83). For a while, that is up until about the end of this series of letters (mid 1855), the water cure provided positive effects on Darwin's health, but eventually it seems that he got "inured" to the treatment (Letter #85).

There was also sickness and death in the families, too. Anne Darwin's death in 1851, while on a water cure at Malvern (Letter #89) had a profound effect on both Charles and Emma Darwin (Randal Keynes, 2001). For Fox, scarlet fever brought about the death both of his three-year old daughter (Letter #94) and of two boys of the Mudge family, who were staying with the Fox family (Letter #93). Both families were relatively large: 8 at this stage for Darwin, 11 for Fox. So deaths of this kind were not unexpected at the time, although hard to bear.

As for the times, there is little to hint at the great transition that Britain was going through at this time. This reflects the attitudes of both men, who, although strongly liberal (Whig) in their attitudes were not active participants in the politics of their day. Darwin retained his hatred of slavery (Freeman, 1978) and no doubt Fox did too. The letter concerning the resistance of potatoes to blight (Letter #78) was no doubt brought about by concern for the Great Hunger in Ireland from 1845–1852, but like most people in England at the time the events in Ireland were probably little understood and therefore of little concern. The reference by Darwin to "no French invasions" (Letter #90) gives some hint of the concern at the time of Louis Napoleon and the English paranoia of a French invasion at the time of European revolutions in 1848-51. The diaries of Fox also give some more background information, but not a great deal. On the 26th Feb 1848 he noted "FRENCH REVOLUTION" and then on 10th April "CHARTIST REVOLUTION SUPPRESSED", but gives no more details. He notes the death of the Duke of Wellington on 16th September 1851. In 1851 Fox also visited the Grand Exhibition of Industry of All Nations in Hyde Park, and the Great Britain Ship at Liverpool, the great failure of Isembard Brunel. He noted, too, "General National Humiliation before God (India)" on 7th October 1857, which may have been in response to realistic reports filtering through of the British repression of the Indian "Mutiny" of 1857-58. However, more likely these were merely prayers for the British reoccupation of Northern India, since the national fervour over this event led to very jingoistic statements from the British establishment (as demonstrated by the many jingoistic sermons preached on the last National Fast Day, declared by Queen Victoria, for 7th Ocotber 1857 – see, e g, the sermon of Rev C H Spurgeon at Crystal Palace Sydenham).

On a more sombre note, both men lost their fathers during this period, within a short space of each other: Robert Darwin died, at the Mount, Shrewsbury, on 13th Nov. 1848, and Samuel Fox, III, died at Kendalls on 20th Feb 1851. In the case of Darwin, he inherited a fortune of more than £40,000 (Freeman, 1978), but for Fox there was nothing but further responsibilities, since his father left a widow and four unmarried sisters, who were, at the time, living in a large house, Kendalls, near Hatfield, north of London. From this point on Darwin's careful management of his affairs led from strength to strength. Presumably he learnt much from his father's careful financial management and business acumen. For Fox, however, his financial affairs never improved and in his will he left a bewilderingly small estate of less than £5000 (see Chapter 8). This compares with the estate of £282,000 (or more) calculated for Darwin (Freeman, 1978). Curiously, therefore, despite Fox's early use of accounts, a practice which he passed on to Darwin, and for all his father's and brother's business enterprises, Fox neither inherited much money nor did he show a great deal of acumen in using the little money that he had to increase the family fortune (see letter from his brother, Samuel, to Fox, 19th November 1857, Appendix 3).

As we shall see in Chapter 6, Darwin's ability to save up facts, little by little, was also nearing the time when it would pay dividends in a big way.

Chapter 6
The "Origin of Species"

Letters #96-140 (total, 49), 27 Mar 1855 – 18 June 1860, over 6 years

The Origin of Species.

"The other book you may have heard of and perhaps read, but it is not one perusal which will enable any man to appreciate it. I have read it through five or six times, each time with increasing admiration. It will live as long the "Principia" of Newton. It shows that nature is, as I before remarked to you, a study that yields to none in grandeur and immensity. The cycles of astronomy or even the periods of geology will alone enable us to appreciate the vast depths of time we have to contemplate in the endeavour to understand the slow growth of life on earth. The most intricate effects of the law of gravitation, the mutual disturbances of all bodies of the solar system, are simplicity itself compared with the intricate relations and complicated struggle which have determined what forms of life shall exist and in what proportions. Mr. Darwin has given the world a new science*, and his name should, in my opinion, stand above that of every philosopher of ancient and modern times. The force of admiration can no further go!!!"*

Alfred Russel Wallace in a Letter to George Silk, 1 Sept. 1860; in *My Life* by A. R. Wallace (1905) Vol. 1, pp. 372-3.

Introduction

The period of the letters in this chapter, early 1856 to mid 1860, covers not only a most significant period in the lives of Charles Darwin and his cousin William Fox but a watershed in the history of science and the effect of Darwin's ideas on society as a whole.

The period sees a burgeoning of the correspondence between Darwin and Fox, which mirrors the increase in Darwin's correspondence in general. This can be gauged easily by reference to the publications of the Correspondence of Charles Darwin (CCD). While there are only 5 volumes covering the period up to 1855, the series expands rapidly until after 1860, there is approximately one volume a year.

A. W. D. Larkum, *A Natural Calling*, DOI 10.1007/978-1-4020-9233-6_6,

Why and how this change came about is mapped fairly well in the correspondence at this time between Darwin and Fox. And for the first time we begin to get many of the replies by Fox, which have been missing from earlier chapters.

The first letter in the series (Letter #96) signals the change that has come about: it is long and almost completely devoted to discussions of natural history and requests by Darwin for information and specimens: "*Your offer is most kind & generous. It will really save me a great expence, as I sh$^{\underline{d}}$ have had to build places for the varieties of Fowls, not to mention the cost of the different breeds themselves; & the care would have been extremely irksome to me.*" After much discussion about ducks and fowl, Darwin elaborates what he had said in the last letter as the reason for his renewed activity: "*I forget whether I ever told you what the objects of my present work is, — it is to view all facts that I can master (eheu, eheu, how ignorant I find I am) in Nat. History, (as on geograph. distribution, palæontology, classification, Hybridism, domestic animals & plants &c &c &c) to see how far they favour or are opposed to the notion that wild species are mutable or immutable: I mean with my utmost power to give all arguments & facts on both sides.*"

It goes on, remorselessly, letter after letter, in quick succession; now on ducks, now on dogs, now on seeds, now on lizards, and so on. "*You will hate the very sight of my handwriting; but after this time I promise I will ask for nothing more, at least for a long time.*" states Darwin (Letter 99), and one can believe it to be true! But only a few weeks later he is writing again, this time about lizard eggs and chickens (Letter #100)! And there are another five letters in that year, making 11 altogether, whereas in the previous year there were none, and only three in the year previous to that. And so it goes on with relentless zeal, into 1856 and 1857. Then in 1858 it all changes as fate plays an unexpected hand and sets in motion the rapid publication "*On the origin of species*", amidst family tragedy.

This remorseless activity is not confined to Fox alone but is reflected in the correspondence in general, as noted in the second paragraph. What Darwin is doing here is what he has been practicing with Fox and others on a smaller scale up to this period, and that is locking in a correspondence with all the major, and some minor, naturalists of his time. A correspondence, which on the surface seems innocent enough, but which, when presented together, will force them, and the world, to make a decision on a much broader enquiry – whether the facts support the hitherto heretical statement that species are mutable. In the correspondence with Fox one can see the knot tightening, slowly and letter by letter. Darwin does not disguise his intent but he does it in such a way that one feels that Fox is like the prey of the snake, caught in an ineluctable dance to the end. Fox's attitude is not, unfortunately, revealed in this series of letters. If one can put it in to words for him it is probably very similar to that of Lyell in his various revisions of "*Principles of Geology*": "Yes! Of course one can see fascinating changes around and trace the effect of parentage on the offspring and the effect of this and the environment on their survival", but it is a step too far to remove the action of God from the scene and say that this all happens "mechanistically". Nevertheless, Fox and his other correspondents are forced into a discussion of very trivial points, but nevertheless the very points that Darwin will use to paint a broad canvas depicting just such a world view of "mechanistic" change and the mutability of species.

It is not until the late sixties that we have any kind of recorded response by Fox to the "*Origin*" and then he says "*There are points in your unrivaled Book "The Origin of Species" — which I do not come up to — but with these few expressions omitted, I go with it completely. I do not think even you will persuade me that my ancestors ever were Apes*" (Letter #192). Frustratingly, before this we have to pick through the chaff to get the odd grain of Fox's reflected opinion. Thus Darwin wrote, in June 1856, "*What you say about my Essay, I daresay is very true; & it gave me another fit of the wibber-gibbers; I hope that I shall succeed in making it modest. One great motive is to get information on the many points on which I want it. But I tremble about it, which I sh^{d} not do, if I allowed some 3 or 4 more years to elapse before publishing anything* (Letter #111). What did Fox say? Was he encouraging? Was he urging haste? It does not seem likely from the Darwin's next letter: "*I remember you protested against Lyells advice of writing a sketch of my species doctrines; well when I begun, I found it such unsatisfactory work that I have desisted & am now drawing up my work as perfect as my materials of 19 years collecting suffice, but do not intend to stop to perfect any line of investigation, beyond current work. Thus far & no farther I shall follow Lyell's urgent advice. — Your remarks weighed with me considerably. I find to my sorrow it will run to quite a big Book.*" (Letter #112). Probably, Fox was urging Darwin to be cautious and thorough. Unfortunately we do not know. He may also have gently suggested that Darwin might be driven to publish before he was ready, partly, for egotistic reasons: "*I am got most deeply interested in my subject; though I wish I could set less value on the bauble fame, either present or posthumous, than I do, but not, I think, to any extreme degree; yet, if I know myself, I would work just as hard, though with less gusto, if I knew that my Book w^{d} be published for ever anonymously.*" (Letter #116).

Both men were enmeshed in a social fabric, by their families, by Victorian propriety and by their respective illnesses. "*A man ought to be a bachelor, & care for no human being to be happy! or not to be wretched.*" Darwin wrote (Letter #118). Charles Darwin suffered from his stomach disorder all through this period. Even after the death of Annie on a water cure excursion in 1851 (Letter 89) he continued his water cure habits at home. However, by 1855 his health was deteriorating again and he was willing to try other means. "*I do not think I shall have courage for Water Cure again: I am now trying mineral Acids, with, I think, good effect. I am not so well as I was a year or two ago*" he wrote (Letter #115) but a little later added "*I must now take a sitz Bath, my treatment being, — daily Shallow, Douche, & Sitz*" (Letter #117).

By 1857 Darwin had discovered Moor Park and its proprietor Dr Lane. More precisely it seems that Fox, was responsible for alerting Darwin to this establishment, for Darwin wrote in October 1856: "*I sh^{d} not be surprised if I went for a fortnight to Moor Park Hydropathic establishment for a fortnight, some time; for I have great faith in treatment, & no faith whatever in ordinary Doctoring. It was very kind in Dr Gully to speak so of me: if you go there again, pray remember me most kindly to him, & say that never (or almost never) the vomiting returns, but that I am a good way from being a strong man.*" (Letter 112). Later, in early 1857, Darwin wrote "*Dr L. is too young, — that is his only fault — but he is a gentleman & very well read man. And in one respect I like him better than Dr Gully, viz that he does not believe* *in all the rubbish which D^r. G. does; nor does he pretend to explain much, which neither he or any doctor can explain*" (Letter #117). And Darwin visited there right up to the publication of the "*Origin*" with the exception of a visit to Ikley Wells House just prior to and during publication (Letters #136 & #137), also probably on the recommendation of Fox. And this is the irony of Darwin and his illness: he tried so hard to be rational about everything, but the treatment of illness, at least at that time, was any thing but rational. The microbial basis for many diseases was only just being discovered and psychotherapy was unheard of. Into this gap stepped the many wonder cures of the time. Having rejected paranormal phenomena like mesmerism, homoeopathy and spiritualism, Darwin was still forced to submit to a variety of unsubstantiated "cures", such as the water cure, acidic ingestants and ice-cold compresses.

Likewise Fox struggled with illness during this period. Only his diaries for 1857 and 1858 are extant and the rest we glean from Darwin's letters or the two letters of Fox (#107 & #114). From the diaries we know that Fox had lumbago for several months in late1858. Also from Darwin's letters we know that Fox also suffered from his chronic lung affliction: "*I am very sincerely sorry to hear that you have been ill, & that your chest has been the peccant part*" (Letter #105). In addition there was a severe impairment to the leg, which may have been related to the lumbago that Fox later reported: "*I am very sorry to hear how much you have been ailing. I had not heard of this affliction of the leg. I do most sincerely hope that the water-cure will complete the good work which it has begun: the loss of locomotion to a man so active & energetic as yourself would be grievous*" Darwin wrote in October 1856 (Letter #112).

Despite their various afflictions, both men continued a grueling pace of activity: Darwin on his "Species" book and Fox running his parish, his farm and his extended family. Maybe, as a noted above Darwin started a prodigious correspondence from 1855 onwards. As a result his health gradually deteriorated and he had to take several imposed rest cures at Moor Park hydropathic establishment from early 1857. For the years for which we have Fox diaries, we know that during this period (1857 and 1858) he too suffered lumbago and lung problems, yet managed to keep up a constant round of visits between his parish in Delamere, Cheshire, his birthplace, Derby, London and many other places, such as Ipswich, Malvern and Harrogate. He did this by his favourite form of transport, the new railways. On one of these excursions he managed to visit Darwin at Moor Park (Fox Diary for 1858; Letters #125 & #134).

Not surprisingly as the correspondence intensified there was less and less room for family gossip and no reference at all to national affairs.

96 (CCD 1656; LLii, part). To: The Rev^d W Darwin Fox, Delamere Rectory, Northwich.

Down Farnborough Kent
March 27th [1855][1]

My dear Fox

Your offer is most kind & generous. It will really save me a great expence, as I sh^d have had to build places for the varieties of Fowls, not to mention the cost of the different breeds themselves; & the care would have been *extremely* irksome to me. As I want to compare the young only, when there are several varieties, I shall not want, I am extremely glad to say, your precious varieties of the White China, the Khoutom / or Call Duck[2]; the others w^{d} really be of very great value to me, viz Spanish, Game, Cochin China, Dorking & Sebright Bantam: also wild & Aylesbury Duck. How on earth can you get the young wild Duck? With respect to age: all sh^{d} be <of> the same age; but the exact age, in which the <character of the young, as> the Downy feathers *are best developed*, I sh^{d} be much obliged if you w^{d} aid me in settling: I said a week at hazard: ~~so also about the 2 Ducklings~~ <N.B. The ducklings must be seven days, as I have had a note *this morning* from Mitchell[3] of Zoolog. Soc to say that he will send me a seven day Penguin Duck[4].> //

The way I shall kill other young <things will> be to put them under tumbler glass with tea-spoon of ether or Chloroform, <the glass being> pressed down on some yielding surface & leave <them> for an hour or two <(young have such power of revivication)> (I have thus killed moths & butterflies). The best way w^{d} be to send them, as you procure them, in <paste-board> chip-box by Post <on which you c^{d} write, & just tie up with string>; & you will **really** make me happier by allowing me to keep an account of Postage &c. Upon my word I can hardly believe that *anyone* would be so goodnatured as to take such trouble <& do such a very disagreeable thing as kill Babies>; & I am very sure I do not know *one soul* who except yourself would do so. I am going to ask one thing more. sh^{d} *old Hens* of any above Poultry (*not Duck*) die or become so old as to be *useless*, I wish you w^{d} send her to me per Rail addressed to "C. Darwin, care of Mr Acton, Post Office Bromley Kent". Will you keep this address? as shortest way for *parcels*. But I do not care so much for this, as I could buy the old Birds dead at Bayleys[5] to make//

skeletons. — I sh$^{d.}$ have written at once even if I had not heard from you, to beg you not to take trouble about Pigeons, for Yarrell[6] has persuaded me to attempt it & I am now fitting up a place, & have written to Bayley about prices &c &c. — *Sometime* (when you are better) I sh^{d} like very much to hear a little about your "Little Call Duck"[7]. Why so called? & whence you got it? & what it is like? Also about the Khoutom / grey lag goose. I was so ignorant I did not even know there were 3 vars. of Dorking Fowl, nor how do they differ. It is an evil to me that M$^{r.}$ Dixon is such an excommunicated wretch[8].

I forget whether I ever told you what the objects of my present work is, — it is to view all facts that I can master (eheu, eheu, how ignorant I find I am) in Nat. History, (as on geograph. distribution, palæontology, <classification> Hybridism, domestic animals & plants &c &c &c) to see how far they favour or are opposed to the notion that wild species are mutable or immutable: I mean with my utmost power to give all arguments & facts on both sides. I have a *number* of people helping me in every way, & giving me most valuable assistance; but I often doubt whether the subject will// not quite overpower me[9]. —

So much for the quasi business part of my letter. — I am very very sorry to hear so indifferent account of your health[10]: with your large family[11] your life is very precious, & I am sure with all your activity & goodness it ought to be a happy one, or as happy as can reasonably be expected with all the cares of futurity on one. One cannot expect the present to be like the old Crux Major days[12] at the foot of those noble willow stumps, the memory of which I revere. I now find my little entomology, which I wholly owe to you, comes in very useful. — I am very glad to hear that you have given yourself a rest from Sunday duties[13]. How much illness you have had in your life! Farewell my dear Fox. I assure you I thank you heartily for your proffered assistance

Your affectionate friend
C. Darwin

I will not forget about your wish for two Boys[14]: it is not likely that I sh^{d} ever hear of such a thing; but sometime odd chances turn up. —

(*on the envelope*)
"March 26 1855
Spanish Game C.C. Selbright Dorking Wild Aylesbury Duck"

Postmark: obscure
Christ's College Library, Cambridge #88

Notes

1. The date was assigned by F Darwin and determined by relationship to the other letters of this time.
2. Khoutom duck. See Letter #106 where it also called a Malay Duck.
3. David William Mitchell (1813-1859) was the first full-time secretary of the Zoological Society of London (1847–59). (*Modern English Biography*)
4. Penguin Duck: so-called because it walked in an upright position in a poor imitation of a penguin. See: Dixon E S, 1848. *Ornamental and domestic poultry: their history and management,* London. CCD, Vol 5 notes that "*CD recorded in his Questions & experiments notebook, p. 20 (Notebooks), under 'Zoological Soc': 'Young Chinese or Penguin Duck in very young state for skeleton'*".
5. Probably John Baily. See Letter #95 (n4).
6. William Yarrell (1784-1856); see Letter #31 (n4) and Biographical Register.
7. The Call Duck is described in the book by E S Dixon (note 4) as a race of white ducks, from Holland, which because of their resonant calls were often used as decoys for sporting purposes.
8. The use of the term '*excommunicated wretch*' is not known, nor why in a subsequent letter CD refers to him as '*Mr Dixon of Poultry notoriety*' (see Letter #103A). CCD, Vol. 5, for this letter, notes "*It is possible that whatever prompted CD's reaction also explains why Dixon published under a pseudonym from 1854 onwards (see letter to Edward Blyth, 4 August 1855, n.7).*"
9. Once again CD hints to Fox of his abiding passion - the species question - but, as in Letter #95 (n3), does not go into detail
10. Diaries of Fox for the years 1854-1856 have not been found, so his health in these years is unrecorded.
11. By the time of this letter in1855, Fox's family had increased to eleven children (*Darwin Pedigrees*).
12. CD refers once again to this hallowed expedition to Whittlesea (or Whittlesey) Meer and the collection of the rare carabid beetle *Panagaeus crux-major*; see Letters #47 (n5 & 6), #48 (n17), #61 (n18) & #90 (n17), etc.
13. It is possible that Fox was able to get a curate at this time, although diaries do not exist for this year (see Note 10).
14. It is not known what CD was referring to here. Earlier Fox had the two Mudge boys living with the household (see Letter #93), but it seems unlikely that CD's remark was in reference to them.

97 (CCD1675; LLii, part). To: The Revd W. Darwin Fox, Delamere Rectory, Northwich.

Down Farnborough Kent
Ap. 26th [1855]

My dear Fox

Very many thanks for telling me exactly what I wanted to know. —

With respect to infant Poultry, I meant to have explained, that I wished you to fix how many days old, so as best to show the first downy plumage, the proportions, size & general character of the **young** birds of the several Breeds. — What is most important is that the age shd be exactly the same for all; I first proposed a week at hap-hazard. The younger the bird the better in *the more important respect* of showing the characters of the young[1]; on the other hand, whilst **very** young, I fear there wd be great difficulty in measuring the cartilaginous bones. Will you fix for me, for really it would be simple guess-work in me, who am so totally ignorant of Poultry?
Secondly I had not thought of the greater difficulty of getting old <dead> Cocks than Hens[2]. — I shd like to hear sometime from you, whether my impression is right that old cocks are more characteristic, *in proportions of body*, than are old Hens? — Thanks for suggestion about measuring Feathers. — I shd state that it is not my present intention to think of giving description of each breed; But I look only to certain points or questions, as the relative variation at different ages. — the effect of disuse on different parts: the result of my one comparison in this latter respect of the Ducks skeletons surprised me much, — more especially the great increase in weight in bones of feet, but I probably told you all about this[3].
Should a Call Duck[4] ever die, I shd be very glad of one for skeleton: Can you tell me whether they breed *freely* with common tame Duck? How are the young of the Call Ducks: Perhaps you will compare the young with the young of the common for me. —
Have you wild Turkeys now: ~~if they were~~: I shd be glad of one for skeleton, shd one die naturally, more especially if they are *tamed* birds of the *wild* Breed. —
Lastly, there is no point I am <so eminently> curious about, shd you ever *hear* anything which wd help me, is whether crosses between very different breeds, as <bull->dogs ~~at~~ ~~but~~ & greyhounds &c are ever (or in successive generations), **in the least degree** less fertile than the pure parents[5]. — I know it is said commonly that they are *more* fertile, but one statement makes me in the least degree doubt this[6].
Can you forgive so much trouble?
To save you rereading this, & to give me, whenever convenient, a better chance of having answer, I have put on other side, an abstract of my queries. —

Most truly yours
C. Darwin

I return ~~the~~ <Mr> Galton<s>[7]. letter with many thanks.
(1) To fix yourself age of infant Poultry.
(2) Are old Cocks more characteristic <in form> than old Hens?

(3) Do Call Ducks breed <freely> with common? character of the Ducklings? a chance dead Bird for skeleton.
(4) <Chance> Dead wild Turkey
(5) Any information on fertility of mongrels *of very diverse races*. —

Endorsement: C. Darwin April 26 1855
Christ's College Library, Cambridge, #89

Notes

1. CD was interested in when differences in plumage begin to develop in domestic birds; see Letter #96.
2. Few cock-birds were kept for breeding purposes and the rest were killed early for the table, whereas, hens were generally kept for longer for their egg-laying ability.
3. See Letter #96.
4. See Letter #96 (n1).
5. CD was interested in whether crosses between distantly related varieties were more or less fertile than between animals of the same variety. In *The variation of animals and plants under domestication,* 1868, CD refers to a famous case of the offspring of bulldogs and greyhounds bred by Robert Walpole, Earl of Offord and described by Youatt, W 1845. *The Dog*. London.
6. For CD's scanty conclusions refer to the section on fertility and interbreeding in dogs in *The variations of animals and plants under domestication,* 1968, Vol 1, 29–33.
7. Probably John Howard Galton, who had corresponded with William Darwin Fox in 1838 in response to a query by CD about Galton's strain of bloodhounds (Letter #63A). Galton was a cousin of Fox's father, Samuel Fox III.

98 (CCD 1678; LLii, part). To: The Revd W. Darwin Fox, Delamere Rectory, Northwich.

Down Farnborough Kent
May 7th — [1855]

My dear Fox

My correspondence has cost you a deal of trouble; though this note will not. I found yours on my return home on Saturday after a week's work in London[1]. Whilst there I saw Yarrell, who told me he had carefully examined all points in Call Duck & did not feel any doubt about its being specifically identical[2], & that it had crossed freely with common varieties in St. James' Park[3]. I sh^d therefore be very glad for a 7 days duckling, & for one of the old birds sh^d one ever die a natural death. Yarrell told me that Sabine[4] had collected 40 vars. of the common Duck!! I am rather low today about all my experiments, — everything has been going wrong — the fan-tails have picked the feathers out of the Pouters in their Journey home[5] — the fish at the Zoological Gardens after eating seeds would spit them all out again[6] — Seeds will sink in salt-water — all nature is perverse & will not do as I wish it, & just at present I wish I had the old Barnacles to work at & nothing new. — Well to return to business, nobody, I am sure could fix better for me, than you, the characteristic age of little chickens: with respect to skeletons I have feared it w^d be impossible to make them; but I suppose I shall be able to measure limbs &c by feeling the joints. What you <say> about old Cocks just confirms what I thought; & I will make my skeletons of old cocks. — Sh^d an old wild Turkey ever die please remember me: I do not care for Baby turkey. Nor <for> a mastiff. Very many thanks for your offer. — I have puppies of Bull-dogs & Greyhound in salt[7]. — & I have had Carthorse & Race Horse young colts carefully measured. — Whether I shall do any good I doubt: I am getting out of my depth. —

Most truly yours.
C. Darwin

Endorsement: May 8th 1855
Christ's College Library, Cambridge, #90

Notes

1. According to Emma Darwin's *Diary*, this visit to London lasted from Monday 16th April to Saturday 5th May, but CD may have joined later. This trip fits with a description given by CD to his son William (in part of a letter in Litchfield (1915), Vol 2, dated "Down, 29th" [1855 or 1856]). "*I am going up to London this evening and I shall start quite late, for I want to attend a meeting of the Columbarian Society, which meets at 7 o'clock near London Bridge. I think I shall belong to this Society, where I fancy, I shall meet a strange set of odd men. Mr Brent was a very queer little fish; but I suppose Mamma told you about him; after dinner he handed me a clay pipe, saying "Here is your pipe," as if it were a matter of course that I should smoke. Another odd little man (N.B. all pigeon-fanciers are little men I begin to think) showed me a wretched little Polish hen, which he said he would not sell for £50 and hoped to make £200 by her, as she had a black top-knot. I am going to bring a lot more pigeons back with me next Saturday, for it is a noble and majestic pursuit, and beats moths and butterflies, whatever you may say to the contrary.*"
2. CD cited William Yarrell on this point in *The variation of animals and plants under domestication*, 1868, Vol 1, 279 n.
3. Repeated in *The variation of animals and plants under domestication*, 1868, Vol 2, 262.
4. Joseph Sabine (1770-1837), Barrister, horticulturalist and authority on British birds. Hon secretary of the Horticultural Soc, 1810-1830 and treasurer of the Zoological Soc, 1830, FRS. His younger brother Edward Sabine, was President of the Royal Society from 1861-1871. (*DNB*)
5. This tallies with events described in Note 1.
6. CD described his experiment in a note dated 5th May 1855 (DAR 205.2: 115): for details see CCD, Vol 5, this Letter, note 5.
7. Pickled in brine; See letter #97.

99 (CCD 1683; LLii). To: The Revd W. Darwin Fox, Delamere Rectory, Northwich.

Down Farnborough Kent
May 17th [1855]

My dear Fox.

You will hate the very sight of my handwriting; but after this time I promise I will ask for nothing more, at least for a long time. — As you live on Sandy Soil, have you Lizards at all common? If you have, shd you think it too ridiculous to offer a reward for me for Lizards eggs to the Boys in your school[1]; — a shilling for every half-dozen, or more if rare, till you get 2 or 3 dozen & send them to me. — If snake's eggs were brought in mistake it would be very well, for I want such also: & we have neither lizards or snakes about here. — My object is to see whether such eggs will float on sea-water, & whether they will keep alive thus floating for a month or two in my cellar. I am trying experiments on transportation of all organic beings, that I can; & Lizards are found on every isl[d] & therefore I am very anxious to see whether their eggs stand seawater. Of course this note need not be answered, without by a strange & favourable chance you can someday answer it with the eggs[2].

Your most troublesome friend
C. Darwin

Postmark: MY 18 1855
Endorsement: May ~~19~~ 17/55
Christ's College Library, Cambridge, #91

Notes

1. Fox helped found and run the Church School at Delamere, which until recently was known familiarly as "Fox's School"; see Letter #90 (n7) and Fig 25.
2. After sending this letter, CD realised that most British lizards are viviparous and do not lay eggs! So he included this information in the next letter (Letter #100). At this time CD was grappling with the biogeographical distribution of plants and animals. Without a knowledge of continental drift, this was an insuperable task at a global level.

100 (CCD 1686). To: The Rev[d] W. Darwin Fox, Delamere Rectory, Northwich.

Down Farnborough Kent
May 23[d] [1855]

My dear Fox

I thank you heartily for the way you take all my requests.
I think I will insert request in Gardeners Chron. about Lizard Eggs[1]; it seems cool, however, making such a request to the world at large. — Very many thanks to your Lady friend from Jersey. — I write now to say that I have been looking at some of our mongrel chickens & I should say *one week old* would do very well. The chief point which I am & have been for years very curious about is to ascertain, whether the *young* of our domestic breeds differ as much from each other as do their parents, & I have no faith in anything short of actual measurement & the Rule of Three[2]. — I hope & believe I am not giving so much trouble without a motive of sufficient worth. — I have got my Fan-tails & Pouters (choice Birds, I hope, as I paid 20[s] for each pair from Baily[3]) in a grand cage & Pigeon House, & they are a decided amusement to me, & a delight to Etty[4]. Both kinds have laid eggs. — My wife is away from home at Rugby, nursing Willy, who has had your dreadful enemy the Scarlet Fever, rather badly, but yesterday & today's letters make me perfectly easy[5]; but I think I shall go in a few days, & relieve guard. — We are in terrible perplexity about contagion, for here *all* the children have the Hooping cough, which Willy has not had[6]. — I am glad to see from your note that you keep up your vivid interest about Birds, & all living things. With so many children & parishioners I feared that you could hardly have had time to care for anything else. Farewell my dear old friend

Ever yours
C. Darwin

PS
I have sent a communication to the Gardeners' Chronicle on the means of distribution of plants by sea-currents[7], but I expect that it is too long for insertion at least of the whole. — I had quite forgotten when I wrote to you, that the very common

British Lizard is ovo-viviparous[8]! & the chance of getting ova of the L. agilis, I fear is small. Jersey is evidently the best chance. — I am going to try land-snail shells & their eggs also. in sea-water. —

Christ's College Library, Cambridge, #92

Notes

1. This letter to the *Gardeners' Chronicle* requests eggs of *Lacerta agilis*, a lizard which inhabits southern England, but is very rare. CD offered a reward of a few shillings; see actual letter in CCD, Vol 5, to *Gardeners' Chronicle*, [before 26 May 1855].
2. CD described his observations on this point in *The variation of animals and plants under domestication,* 1868, Vol 1, 249–51. The "rule of three" is the algebraic formula for finding, x, using the formula: $a{:}b$ as $c{:}x$.
3. Probably John Baily; see Letter #95, Note 4.
4. Henrietta Emma Darwin (1843-1827), who married Henry Litchfield. In her book (Litchfield, 1915) she mentions her father's activities with pigeon fanciers in Vol 2, p 157, but she does not mention any particular interest on her own part. It should be remembered that CD's mother, Susannah, kept doves (Healey, 2002).
5. According to Emma Darwin's *Diary*, she visited William Erasmus Darwin from 21st May to 28th May.
6. All the children living at home had whooping cough in March 1855; see Letter # 95.
7. In the letter to the *Gardeners' Chronicle*, 21 May [1855] CD reviews his experiments that out of 23 kinds of seeds, randomly chosen, only members of the pea family were at all susceptible to being killed in less than 42 days, In that time he calculated, ocean currents could have carried them 1300 – 1440 miles. The letter to the *Gardeners' Chronicle* is reprinted in CCD, Vol 5 for 21st May 1855.
8. Ovo-viviparous animals develop within eggs that remain within the mother's body up until they hatch or are about to hatch. The distinction between this and true vivipary is that the embryo is nourished by the egg yolk rather than by the mother's tissues.

101 (CCD 1698). To: The Revd W. Darwin Fox, Delamere Rectory, Northwich.

Down Farnborough Kent
June 11th — [1855]

My dear Fox

Your "very fair night after the murder"[1] made us laugh heartily. — I thank you cordially for this first instalment[2]: I can only repeat my firm belief that there is not another man, at least I am very sure I do not know one, who would take so much trouble for another. The results may not, very likely prove of value to justify my having asked you to do such disagreeable work, but I can say that it is no idle passing wish, but one of several years standing. I have put the Duckling into the strongest sea-salt Brine, & shall wait till cooler weather, & till I have all the specimens for full comparison. — I sh$^{\underline{d}}$ like sometime know about how many generations your tame wild-duck has been tamed. Perhaps next Spring you might get me a specimen <of the really wild:> though I sh$^{\underline{d}}$ be very glad of one of yours.

The Box was hardly strong enough in which you sent the Duckling, for it had been compressed, & the intestines had been forced out, but not much injured, which I am very glad of, for I intend to measure length of intestines in tame & wild old Ducks, & if there be, (as I sh$^{\underline{d}}$ not be surprised) any considerable ~~length~~ difference, I sh$^{\underline{d}}$ wish *particularly* to look to the intestines of the young. I shall be so curious

to compare the feet of the young wild & tame; for in the old skeleton I find that all the bones together of the tame weigh 1/3 more than all of the wild, but the bones of the feet *alone* weigh actually double those of the wild, which I think must be attributed to the more terrestrial habits of the tame; I hope, therefore, your *tamed* wild Ducks have kept their aquatic habits. The wing-bones of the tame are proportionally <*considerably*> lighter, than those of the wild. The lower mandible, also, has not increased in weight with rest of skeleton, which I presume must be attributed to their being fed. — I mention these particulars to show the sort of points, which I am attempting to observe.
Would it not be well to get on my account half-a dozen little Boxes made by some Carpenter?

Farewell & believe me my very dear Murderer
Your affectl^{y}
C. Darwin

You will think me an awful beggar, but if ever one of your old Aylesbury Ducks die, I sh^{d} like the skeleton: I do not ask to save myself buying one, but not knowing appearance I sh^{d} very likely not get a pure breed. The one I have got, was ordered merely as a white large Duck, — the man told me it was Aylesbury but I have not the least idea whether it was. — My demands will cost you a fortune in little Baskets.
Remember address for *parcels*
"C Darwin"
"care of Mr Acton"
"P. Office"
"Bromley
"**Kent**"

Endorsement; June 11 55
Christ's College Library, Cambridge #93

Notes

1. Fox was, no doubt, referring to his sacrificing of animals for CD and, perhaps, as suggested by CCD, Vol 5, alluding to the murder of Duncan by Macbeth and Lady Macbeth (Shakespeare's *Macbeth* II, iii).
2. In Letter #98, CD requested Fox send him the body of a seven-day old duckling.

102 (CCD 1704). To: The Revd W. Darwin Fox, Delamere Rectory, Northwich.

Down Farnborough Kent
27th — [June 1855]

My dear Fox

I have received **safely** the young Duckling, & he is now pickled. — How very different from the young Aylesbury! I am rather disgusted for I expected otherwise. — What trouble you will have to take, & have taken about the young poultry, having to go to so many quarters. I am going on with my salting experiments. Several seeds have now come up after 65 & 70 days immersion[1]. — What you

say about testing for the percentage is *very true*; I had intended it, but I found it so very troublesome actually to count all the seeds. — But I have sown all kinds <without salting> so as to test, as far as eyesight serves, how far they were good seeds[2]. —
I have just ordered Almond Tumblers & Rusets, so I shall have soon a grand collection of Pigeons[3]. —
Should you ever be able to give me history of any *mongrel* crosses of//

any animals whatever, I sh^d be very glad of them[4]. I am awfully deficient in exact information on mongrels, though pretty rich in regard to Hybrids[5]. I mean to cross Pigeons systematically, & see how the offspring go, how much they vary & which parent take after &c &c. &c &c

Ever most truly your's
C. Darwin

(*envelope missing*)
Christ's College Library, Cambridge # 94

Notes

1. See Letter #100 (n7).
2. Fox clearly points out that CD needs controls, i e, tests in fresh water vs salt water, etc.
3. There are literally thousands of varieties of pigeons today.
4. The meaning here is that CD wants information on crosses between hybrid animals not crosses between pure breeds. CD cites Fox in *The variation of animals and plants under domestication*, 1868, Vol 2, 30–1, & 40, for information about reversion in the offspring of crosses between different breeds of domestic animals.
5. CD distinguishes between mongrel and hybrid crosses. The difference is one that generally is no longer made – presumably a hybrid cross is one where the variety of each parent is well established, whereas this is not the case in crosses with mongrel animals.

103 (CCD 1728, LLii, part). To: The Revd W. Darwin Fox, Delamere Rectory, Northwich.

Down Farnborough Kent
22^d [July 1955]

My dear Fox

Many thanks for the 7 days old White Dorking & for the other promised ones[1]. — I am getting quite "a chamber of horrors"[2]. I appreciate your kindness even more than before; for I have done the black deed & murdered an angelic little Fan-tail & Pouter at 10 days old. —
I tried Chloroform & Ether for the first & though evidently a perfectly easy death it was prolonged; & for the second I tried putting lumps of Cyanide of Potassium in a very large damp Bottle, half-an-hour before putting in the Pigeon, & the prussic acid gas thus generated was very quickly fatal. — I find a lump of this substance kept with a little cotton-wool in small wide-mouthed Bottle, excellent for my Boys,

who are ardent Lepidopterists: the gas kills quickly, sometimes almost instantly, & they most rarely revive after 1/2 an hour immersion[3]. —
Thank you for your letter about mongrels; by the way you deserve a good scolding, by ending your letter with "I will spare you anymore" — I can truly say that I never in my life received a letter from you that did not interest me, — & this particular one on the very subjects, which interest me most!! — I am never weary at marvelling with you at heredetary mental habits, tastes &c. — I wonder whether it would be possible to get precise information on the cross of greyhound with Bull-Dog, whether as you *think*, there were traces of the Bull-dog, after 8 generations: I sh$^{\underline{d}}$ be *extremely* glad to get such facts. Can you think of any channel? Do you know ~~who~~ in the least who made this cross[4]? I suppose & fear that it is quite impossible that you sh$^{\underline{d}}$ have any precise *evidence* that the Setter is a mongrel[5], as likewise the brown Retriever &c. I observe many writers on these subjects will not believe that any mongrel is ever true to its kind[6]. — I remember when you were here you told me many curious facts about Retrievers &c, &c, which I have written down. —

—

One more question, to the above point, do you know *positively* that L$^{\underline{d}}$ Hill's Dorkings[7] were crossed with the Game, & then fetched high prices as Dorkings[8]?

If *I* were to apologise, & talk of sparing you, by Jove it would be to the point. —

My dear Fox
Most truly your's
C. Darwin

This morning Atriplex or Orache seeds germinated after 100 days immersion! —

Postmark: JY 24 1855
Christ's College Library, Cambridge #95

Notes

1. See Letters #96; #100; #101; #102.
2. CD is comparing his set-up for killing animals (see note 3) with the 'Chamber of Horrors' in Marie Tussaud's waxworks, in Baker Street (established in1833).
3. CD must have had a space set aside somewhere for this; hence his reference (note 2) to a "*chamber of horrors*". It may be noted that cyanide salts have been banned from all laboratories in most countries for some time, under Health & Safety laws; they must now be stored and used under strictly supervised and regulated conditions.
4. See Letter #97 (n5). In *The variation of animals and plants under domestication,* 1868, Vol 2, 95, CD described Robert Walpole's crosses between greyhounds and bulldogs in the formation of new races [see Letter #97 (n5)].
5. This is discussed in *The variation of animals and plants under domestication,* 1868, Vol 1, 41, where CD quotes Youatt, 1845 [see Letter #97 (n5)] for evidence that a setter is a modified spaniel with the ability to mark game.
6. In *The variation of animals and plants under domestication,* 1868, Vol 2, 95, CD made the suggestion that breeds of dogs produced by crosses did "*breed true*" over longer periods. CD cited Fox on the example of pointers crossed with a foxhounds '*to give them dash and speed*'.
7. The source of this information is not clear. CD refers to Sir Rowland Hill (1795-1879)(of Hawkestone and Harwick in the County of Salop) the inventor of the modern postal system and the "penny post" [see Letters #66 (n2) & #86].

8. In *The variation of animals and plants under domestication,* 1868, Vol 2, 95, CD stated: "*Certain strains of Dorking fowls have had a slight infusion of Game blood*".

103A (CCD 1733). To: The Rev[d] W. Darwin Fox, [Delamere Rectory, Northwich.]

Down Farnborough Kent
July 31st [1855][1]

My dear Fox
I have received the *young Cochin*, *Call Duck*, & *Sebright Bantam* all safe, for which hearty thanks, & I congratulate you on your horrid work being more than half done[2]. Your deeds are beyond thanks. —

When you have occasion to write again: will you tell me, whether the Sebright// Bantam is a single sub-variety; I fancy there are 2 or 3 differently coloured Bantams so called: *if so* please inform me which kind was parent of the chloroformed infant. — There is going to be a great show of Poultry & Pigeons at end of August at Annerly, which I shall go to[3]. —

Thank you for extract from Wood:[4] I shall get the Book. — I quite agree// in probability of what you say about the Breeds of Dogs[5]; but there are many who would not, & therefore I am anxious to get as many precise facts as I can about crossing, both for this object, & generally for the comparison of mongrels & Hybrids. The line of argument you put in your note falls without the least impression on some people, as on Mr.// Dixon of Poultry notoriety[6], who argued stoutly for every variety being an aboriginal creation, & seemed to entirely disregard all the difficulties on the other side[7]. And difficulties enough there are, as it seems to me, on all possible sides!

I did not know Bewick had written on Dogs[8]; I will see to it. —
With hearty thanks

Ever yours
C. Darwin

Onion: Rhubarb: Beet & *Orache* or Atriplex; have all come up after 100 days' immersion. (I send. 3 stamps of 6[d]. each.)

Christ's College Library, Cambridge, #125.

Notes

1. The letter is dated by relationship to earlier letters of 1855 (Letters #96 & #103). Francis Darwin had a date of 1842/3 for this letter.
2. CD is referring to the work of killing by chloroform 7-day old chicks – "a chamber of horrors" (Letter #103), as per instructions in Letter #96.
3. There was a poultry show at Anerley on 28th to 30th August 1855 (*Poultry Chronicle* 3 (1855): 505). Anerley is located near Sydenham, Kent, not far from Downe, where CD lived.
4. A reference to Wood, J G (1855), *Sketches and anecdotes of animal life*; 2nd ed, London and New York, in which the habits and breeds of domestic dogs were discussed on pp 131–178, and which Fox had clearly recommended. A passage on a cross between a grey hound and a bulldog is quoted in *The variation of animals and plants under domestication,* 1868, Vol 2, 95.
5. It seems that Fox had suggested that the domestic breeds of dogs were descended from a single stock.

6. Edmund Saul Dixon (1809-93). Rector of Inwood with Keswick, Norfolk; author of books on poultry. The book referred to here is *Ornamental and domestic poultry* (1848). See Letter #96 (n8). (*Alum Cantab; Modern English Biography)*
7. Dixon argued that varieties of domestic animals were independently created and that breeding experiments would prove this (E S Dixon 1848, pp ix–xiv – see note 6, above).
8. Bewick T (1753-1828). Wood engraver and naturalist, after whom Bewick's swan was named [see Letter #31 (n4)]. The book referred to here is *A general history of quadrupeds* (1790), Newcastle upon Tyne. (*DNB*)

104 (CCD 1745). To: The Revd W. Darwin Fox (no envelope for this letter)

Down Farnborough Kent
Aug 22^{d} [1855]

My dear Fox
Many thanks for the little Dorking received this morning. You sent before a White Dorking. I sh^{d} like to know in what respect the "*splendaceous*" Dorking received this morning differs from the white, except in colour: — You once said the breeds of Dorking were considerably different. I have had from you.

Aylesbury Call Duck. Wild	Duck
White Dorking "splendaceous" Dorking Cochin Bantam	Chicken

When I get the Game & Spanish I shall be complete. A neighbour, I find, has very true Spanish, & I have just sent to know whether I can get chicken from him, & will append P.S. By your letter you seem to think you have sent Spanish, but I have not received such. — But I have received a magnificent, stupendous old Spanish Cock, (& his bones are now whitening) from Mr *Wilmot*[1] (?); I did not write to thank for I did not know name & address, & I forgot to write to you, which I ought to have done. —

~~From what you say, I think~~ With respect to other old Birds, I must get old Dorking, Game, Cochin, & Sebright Bantam, & Call Duck. Some few of them you can perhaps help me to; the rest I can get somehow or Buy from Baily[2]. — I have had a pair of rumpless Fowls given me. And I have now got 4 splendid races of Pigeons, which amuse me *extremely*. —

Thank you very much for all your manifold assistance & believe me

Ever yours
C. Darwin

P.S.

I can get a Spanish Chicken & am truly glad to save you one murder, you best & kindest of Murderers. —

Christ's College Library, Cambridge #94a

Notes

1. This is one of the Wilmots, who were plentiful around Derby and in the Midlands. Sir Robert Wilmot, had been the landlord of Osmaston Hall, where Samuel Fox III lived with his family [see Letters #49 (n2), #54 (n7); #83 (n3)]. Fox often recorded in his *Diary* a visit to a Wilmot in the 1850s, although he did not specify who this was. Eardley Nicholas Wilmot, one of Sir Robert's sons, lived with his mother at Malvern at this time, and they were visited there by Fox in September 1857. Another son Frank was at Cambridge as an undergraduate with Fox. E E Wilmot of Congleton was referred to in *The variation of animals and plants under domestication,* 1868, Vol 1, 293.
2. Probably John Baily, see Letter #95 (n4).

105 (CCD 1766). To: The Revd W. Darwin Fox (no envelope for this letter)

Down Bromley Kent
Oct 14th [1855][1]

My dear Fox.

I received this morning the Game Chick, for which very hearty thanks & for all the very great & very disagreeable work in this line, which you have done for me. — I have now quite a grand collection of chickens, & shall be able to ascertain how far the young really do differ proportionally with the old[2]. — I did// write, immediately after I got your last note, to Mrs Wilmot[3] to thank her. The bones of the old gentleman will soon be cleaned. —
With respect to the old Birds, please observe that I can get excellent game, so that all I want are, *first rate*, old cocks.

1. Dorking
2. Cochin
3. Call Duck//

especially the last. —

I go slowly on accumulating facts; — what I shall do with, remains to be seen. —
I am very sincerely sorry to hear that you have been ill, & that your chest has been the peccant part. How most truly & sincerely do I wish that we did live nearer, or that either or both of us were more locomotive; it would be a very great pleasure// indeed to see you here again, & learn wisdom from you of all kinds. —

I really have no news: the only thing we have done for a long time, was to go to Glasgow[4]; but the fatigue was to me more than it was worth & Emma caught a bad cold. On our return we staid a single day at Shrewsbury & enjoyed seeing the old place[5]. — I saw a little// of Sir Philip[6] (whom I liked much) & he asked me "why on earth I instigated you to rob his Poultry yard?" The meeting was a good one & the Duke of Argyll spoke *excellently*[7]. —
I had a letter some 2 months ago from Hore[8], who is settled, an old Batchelor, in Devonshire, & has given up Natural History, as he tells me[9]. —//

Farewell, my dear old friend. I do hope when next I hear, that you may be stronger.

Your's affect^y
C. Darwin

I have now

Fan-tails
Pouters
Runts
!!! Jacobins !!!
Barbs
Dragons
Swallows
!!! Almond Tumblers !!!
!!!

Endorsement: Oct ~~16~~ 14 1855
Christ's College Library, Cambridge, #96

Notes

1. Dated from letter to Letter #104, in which CD mentioned receiving a Spanish cock from Mr Wilmot.
2. The variations between the young chickens of various breeds of fowl are discussed in *The variation of animals and plants under domestication,* 1868, Vol 1, 249 - 250.
3. Lady Marianne Wilmot, the widow of Sir Robert Wilmot who had been landlord of Osmaston Hall, the home of Fox's parents. Lady Wilmot lived at Malvern with one of her sons, Eardley Nicholas Wilmot; see Letters 82, (n2) & #104 (n1), and also #54 (n7).
4. CD and his wife Emma, went to Glasgow from 12th to 18th Sept 1855 for the Annual Meeting of the British Association (Emma Darwin *Diary*; Freeman, 1978).
5. At this time his sisters Susan and Caroline were living at the Mount, the house that Robert Waring Darwin built in Shrewsbury.
6. Philip de Malpas Grey-Egerton. His estate at Oulton Park, Cheshire, was close to Fox's home at Delamere. Both Egerton and CD were ordinary members of the Council of the British Association for the Advancement of Science, and had met at the recent meeting in Glasgow (see note 4, above).
7. George Douglas Campbell, Duke of Argyll (1823-1900). Lawyer and Liberal M.P. President of the British Association, 1854–5. (See *Report of the 25th meeting of the British Association for the Advancement of Science held at Glasgow in September 1855*, p. lxxxii).
8. William Strong Hore (1807-1882), B A., Queen's College, Cambridge, 1830. Clergyman and botanist. A fellow undergraduate with CD and mentioned in several letters for his entomological enthusiasm. The letter referred to by CD has not been found. See Letters #1 (n2), 8 (n11), #9 (n16), #10 (n14) & #16 (n7). (*Alum Cantab; Desmond, 1977*)
9. William Hore (Note 8) was perhaps being humble; in fact he had taken up botany in his parish of Shebbear, Devon; he discovered the rare flowering plant *Trifolium molinierii* and sent marine algae to the great phycologist W H Harvey, who named a genus of algae "*Horea*" after him; see Desmond, R (1994). *Dictionary of English and Irish Botanists and Horticulturalists*. CRC Press, Boca Raton; ISBN 0850668433. Unfortunately, today, this algal genus has been discarded.

106 (CCD 1815). To: The Revd W. Darwin Fox (no envelope for this letter)

Down Bromley Kent
Jan. 3^{d} [1856]

My dear Fox

Thanks for your letter: I had your name on my list to write to soon to tell you how I got on in the Cock & Hen line of business. I have got nothing but promises as yet, & there are few or none like you, who do what they promise, though I fully believe they intend at the moment to do so. I sh$^{d.}$ be very glad of old & good Cochin, Dorking & Malay <Call Drake>[1]; though I think it possible that I may get Cochin elsewhere;// if I succeed I will instantly write to you. — Indeed almost any cock w^{d} be very valuable to me, if of good breed; but I do not want to be unreasonable, after the immense trouble you took for me in regard to Chickens[2]. I ~~had the~~ have as yet not tried energetically for Poultry, but I have for <dead> Pigeons, & my success has been not great, so that I do not want throw away any chance of good Bird from any quarter.// I offer payment for <dead> Pigeons, & Ducks & Rabbits to Mr Baker[3]. —

<Young> Baily, from whom I have bought live Pigeons, I found not at all obliging about dead birds[4]. If I fail by other channels, I will apply to the Father Baily & use Pulleine's name[5], or get him to write a note to him. — I have since I wrote last// greatly extended my scheme; & I have now written above 20 letters to every great quarter of the world to professional skinners, & others to get me collection of Poultry & Pigeons skins[6]; if I succeed I think it will be a very curious collection. —

With respect to the Athenæum I have f$^{d.}$ it so dull that I have for some time left off taking it in; — I quite forgot to mention this to you[7]. —//

With respect to seeds, my few remarks were made, on account of Hookers, extreme obstinacy, (as it appears to me) in not believing that seeds can live even a few years in the ground[8]. Your Isle of Wight case w$^{d.}$ have been interesting if written at the time with certainty about the species; but without every particular given Hooker & Bentham, sneer at every account[9]. —

I am particularly obliged by the particulars on the eggs & colour of the Muscovy Ducks// & sheep: all such facts are *valuable* to me[10]. I have heard something analogous about crossing Cochin Chinas. — Did you ever see the Poultry Chronicle; one of the best contributor B. P. Brent, lives not far off, & is very kind in giving me information: unfortunately he has given up keeping many Birds[11]. —

Farewell, we are not in a very flourishing condition; as many of us are rather poorly, & my wife for a couple of months has been suffering much from headachs. — I congratulate & condole with you on your 12th child[12]: in my own case, I sh$^{d.}$ have wished only for condolences!

Yours very affectionately
C. Darwin

P.S. Would you blow for me an *average* egg of the White & Slate-coloured or other var. of Musk Duck. — I sh$^{d.}$ very much like to have these.

Postmark: JA 3 1856
Endorsement: Jan 4 3/56 185**6** VDP
Christ's College Library, Cambridge #86

Notes

1. These breeds of duck were all mentioned in the previous letter (#105). The 'Call Duck', which CD added above 'Malay', is a breed of domestic duck with a resonant call, mentioned earlier - see Letters

#96 (n1); #97 (n4); #104; #105. In Letter #96, CD also referred to the Call Duck as a "Khoutom or Call Duck".

2. See Letters #101 - #105.
3. CCD, Vol 6, suggests that this was either Samuel C or Charles N Baker, both dealers in poultry, Chelsea.
4. As noted before [see Letters #95 (n4): #96 (n4); #100 (n3); #194 (n2)] this is likely to be John Baily, poultry dealer, a regular contributor on poultry-keeping to the *Cottage Gardener*. From 1855 onwards, there are frequent entries for '*Baily Pigeons*' in CD's *Account Books* (Down House MS). As noted in *The variation of animals and plants under domestication,* 1868, Vol 1, 132 n 2, it appears that Baily eventually provided CD with skins, etc.
5. Robert Pulleine (1806-1868), Emmanuel College, Rector of Kirkby-Wiske, Yorkshire; had been a friend of both CD and Fox at Cambridge [see Letters #9 (n6), #10 (n15), #11 (n4), #13 (n5), #16 (n12), #17 (n1), #19 (n)2 and subsequent letters]. Fox visited him at Wensleydale on a tour of Yorkshire with his young wife in 1834 [see Letter #54 (n9)]. Later he was well known as a judge at poultry shows [see *Cottage Gardener* 15 (1855–6): 208 and 227]. The elder John Baily, of Mount Street, Grosvenor Square, London, was also an established judge of poultry.
6. In December 1855, CD made a memorandum about skins and skeletons of poultry pigeons, rabbits, cats and, even, dogs. He listed 20 correspondents around the world. See CCD, Vol 5, CD memorandum, [December 1855], p 310, and subsequent letters in December, 1855.
7. CD customarily passed on old issues of the *Athenæum* to Fox [see Letter #74 (n7)].
8. CD was referring to his recent letters to the *Gardeners' Chronicle* (see CCD, Vol 5, letter to *Gardeners' Chronicle*, 13 November [1855], and [before 29 December 1855]), concerning the longevity of old, buried seeds. Joseph Dalton Hooker had written to the same journal (*Gardeners' Chronicle*, 8 December 1855, pp 805–6) querying CD's report and stating: '*my objection to placing complete confidence in this case, is the want of evidence that Charlock seed will withstand the destroying effects of moisture for any number of years*' (p 806).
9. George Bentham (1800-1884), noted botanist, who published *Genera plantarum* with Joseph D Hooker; he had expressed a view similar to Hooker's (see Note 8, above) in an article on the longevity of charlock seed in the *Gardeners' Chronicle*, 10 November 1855, pp 741–2. Fox convalesced on the Isle of Wight in the second half of 1832 and in1833 (see Letters #49 & #51) and met his first wife there (Fox *Diary*, 1833). CD also visited Fox there in 1837 [see Letter #63 (n1)]. At some point it seems Fox carried out experiments on the longevity of seeds (placed in moist soil?).
10. See *The variation of animals and plants under domestication*, 1868 (Vol 1, 248–9).
11. Bernard P Brent then lived at Bessel's Green, Riverhead, Kent. In *The variation of animals and plants under domestication* Vol 1: 132 n, CD acknowledged the help of Brent. Brent was a member of the Columbarian Society (see letter #98). Three volumes of The *Poultry Chronicle* were published between 1854 and 1855 before it was taken over by the *Cottage Gardener, and Country Gentleman's Companion.*
12. Frederick William Fox (1855-1931) was born on 8th December of this year (*Darwin Pedigrees*).

107 (CCD 1646). To: Charles Darwin, [Down House, Bromley, Kent]

Delamere[1]
Northwich
March 8 [1856][2]

My dear Darwin

I have anxiously inspected My Dorking & Cochin friends Yards for an old Cock of Each[3], & written to Captain Hornby[4] — but I fear you have not received any yet. There must be some die before long I think.

Have you a Sebright Bantam yet?[5] If not I have an old Gentleman I will send you shortly[6]. You should have an old *White* Dorking also, as they are quite distinct from the other in form.

I forget whether I ever told you that I had long considered the Scotch Deer Hound a mongrel, par Excellence. //

Dont tell any Scotch so, or I shall be murdered. It has long been a pet idea of mine, & I have often said I could breed them without any Deer Hound blood in them. I have also always thought the Irish Deer or Wolf Dog, was merely a cross with the Scotch & a Mastiff.

Some months ago in a conversation on this head with a Mr Lister[7] near here, he told me to my great delight that he had a Bitch ½ Deer Hound & ½ Mastiff. On looking at her it is wonderful how little the ½ Mastiff is recognisable in her. On minutely examining however, you find her mastiff Blood in neck & shoulder. I much wished this Bitch crossed back with Deer Hound. This has been done, & the result, as shown in a splendid Bitch puppy, is to *completely* restore the Scotch Deer Hound. I dined there last week, & met a stranger who was// enthusiastic about Scotch Dogs — of which by the way he gave a pretty story as having happened to himself. Walking one day in Regent St he felt something cold in his hand, & on looking, found a Scotch Dogs nose there, who had been with him Deer stalking &c 2 years before in the Highlands, & was then walking in London with his Master.

Lister, rather spitefully introduced me to this Captain Warren[8] — as being one who believed in the Scotch Dogs being mongrels. Of course I maintained my ground, when, to my intense amusement he (after warning me not to go to Scotland & especially Badenoch[9], with such views) quoted *the* puppy as an Example of pure blood, as might be seen <by> any one, & which he said was well worth 40£. He was so enthusiastic that I was obliged to break the fact by degrees, "that her Grandfather was a Mastiff." I am trying now// to get the ½ breed Scotch & Mastiff Bitch put to a pure Mastiff — & I expect either the produce of that — or the next cross at all events, to be the Irish Wolf Dog[10].

I would defy any Scot to detect the false blood in this puppy Bitch — She is quite a perfect Scotch Deer Hound[11].

I see Tegetmeyer — or some such name, who doctors all the Fowls in England — says he is engaged with you in examining the anatomy of Fowls[12]. He seems to know a great deal about them from his letters in Cottage Gardener — but I often think his prescriptions rather foolish. You are not meddling with Geese I think yet, are you[13].

Tell me how Mrs Darwin & your little ones all are — also Susan Catherine & Mrs Wedgwood[14] & Believe me always Yours affecly

W D. Fox.[15]

Provenance: University Library, Cambridge; Manuscript Room: DAR 164: 174

CD Annotations

1. *brown pencil* "16" (Note 1)
2. *crossed brown crayon* from "My dear Darwin" to "I will send you shortly." (Note 6)

3. *scored brown crayon* from "I would defy" to "Scotch deer hound." (Note 11)
4. *crossed pencil* from "I see Tegetmeyer" to "W. D. Fox." (Note 15)

Notes

1. Annotation across top of first page in brown pencil: "16": the number of CD's notes on hybridism.
2. The letter is dated by its relationship to the Letter #108.
3. See Letter #106.
4. Windham W Hornby of Knowsley, Lancashire, was a well-known breeder of Dorking fowl. *Diaries* of Fox have not been found for 1853-1856 (see Appendix 1). In 1851, on the 10th October, Fox wrote "*At Knowsley Sale. Went again on 14th to see last of this Splendid collection. Captain Hornby most civil*"; again, on December 14th 1852, Fox wrote " *Set off with Mr Dear at 3 in the morning to visit Birmingham Poultry and Cattle Show, Captain Hornby sold 1 pen of 3 spanish for 50£ - 3 ditto for 30£.*"
5. CD had asked John Lubbock, his neighbour at Downe, for specimens of Sebright bantams (CCD, Vol 6, letter to John Lubbock, [14 January 1856]).
6. Annotation 1.
7. CCD, Vol 5, suggests '*E. Lister, esq., Marston, Northwich*' according to the *Post Office directory of Cheshire* (1857).
8. Captain Warren has not been identified.
9. A district near Inverness, Scotland. Perhaps Capt Warren came from there.
10. See *The variation of animals and plants under domestication,* 1868. CD discusses this in Letter #108 (n2).
11. Annotation 2.
12. The *Cottage Gardener* Vol 15 (1855) had two articles relevant to this statement. On p 383 it stated, in an article entitled "*Golden Pheasants versus Spangled Hamburghs*": "*Mr Tegetmeier is now engaged with Mr. Darwin, in such research respecting the different classes of fowls, and, he courteously informs me, that the widest difference exists, between so-called Pencilled Hamburghs and Golden Pheasants; - and that, to classify these birds together, is quite incongruous and inadmissible.*" Then, on p 301, it is stated in a report on the 'Annual grand show of the Philo-Peristeron Society' that "*Mr. Yarrell, whose name is a "household word" with all zoologists, and Mr. Darwin, whose "Naturalist's Voyage round the World" is known all over the world, were present, and with our old correspondent, Mr. Tegetmeier, were examining bird after bird, with a view to ascertain some of those differences on which the distinction between species or varieties depend.*" William Bernhard Tegetmeier (1816-1912), was a journalist and naturalist, specializing in chickens and pigeons, and was undoubtedly the author of both articles, as he was a columnist for this magazine. In the next letter (#108) in this series, CD says "*Mr Tegetmeier is a very kind & clever little man; but he was not authorised to use my name <in any way>, & we cannot be said to be working at all together; for our objects are very different, & he began on skulls before I had thought on subject*".
13. Of course CD is interested in geese! As later letters show; see, e g, Letter #186.
14. Fox is referring to CD's sisters, Susan Elizabeth Darwin and Emily Catherine Darwin, and Caroline Sarah Wedgwood. All CD's sisters were alive in 1856: Fox refers to CD's (at this time) two unmarried sisters (Catherine married Charles Langton in 1863). Mrs Wedgwood would be Darwin's sister Caroline, who was married to Josiah Wedgwood III.
15. Annotation 3.

108 (CCD 1843). To: The Revd W. Darwin Fox (no envelope for this letter)

Down Bromley Kent
March 15th [1856]

My dear Fox.

I was very glad to get your note & congratulate you on your triumph about the Scotch Deer Hound, which however I do not know by sight[1]. — How I wish I further knew (can you find out for me) whether <the real Scotch Deer Hounds> breeds true; but I suppose this must be the case, whether or no your mongrel would do so[2]. This seems to **very valuable** case for me, for it would be a most bold hypothesis to imagine that the real Scotch Deer Hound was a pure & distinct aboriginal race, but that your mongrel, though identical in appearance, was *essentially* different: I do not even know what a common Deer Hound is. — Many thanks for your continued remembrance of me & my poultry skeletons: I am making some progress & have been working a little at their ancient History & was//

yesterday in the British Museum getting old Chinese Encyclopedias translated[3]. This morning I have been carefully examining a splendid Cochin Cock sent me (but I shd be glad of another specimen) & I find several important differences in number of feathers in alula[4], primaries & tail, making me suspect quite a distinct species[5]. — I am getting on best with Pigeons, & have now almost every breed known in England//

alive: I shall find, I think great differences in skeleton for I find extra rib & *dorsal* vertebra in Pouter[6]. — I have just ordered the Cottage Gardener[7]: Mr Tegetmeier[8] is a very kind & clever little man; but he was not authorised to use my name <in any way>, & we cannot be said to be working at all together; for our objects are very different, & he began on skulls before I had thought on subject: I have not yet looked at our pickled chickens & hardly know when I shall, for I have my hands very full of work; but they will come in some day most useful, as will a large series of young Pigeons, which I have myself killed & pickled[9]. —
I shd be very glad of old Sebright Bantam. —
I have been in London nearly all this week,//

working at Books[10] & we had at Erasmus's[11] a very pleasant dinner & sat between Mr & Mrs Bristowe & was charmed with both[12]. Bristowe often so reminds me of you some 25 years ago in certain expression of face & manner. They told me you had been far from well: why did you not mention yourself? Do not I always prose at good length//

about myself & pursuits. So I will say that my stomach has been better for some months than average, & I am able decidedly to work harder. — My sisters are pretty well: you heard of Dr Parkers release about 2 months ago[13]. How I do wish I had you nearer to talk over & benefit//

by your opinions on the many odds & ends on which I am at work. Sometimes I fear I shall break down for my subject gets bigger & bigger with each months work. —

My dear old friend
Most truly yours
Ch. Darwin

Endorsement "C Darwin March 18/56" and the note by CD: "Scotch Deer Hounds". Christ's College Library, Cambridge # 97

Notes

1. See Letter #107, in which Fox stated that in his view the Scottish deerhound was a 'mongrel' breed.
2. The Scottish deerhound had been established as a distinct breed by the sixteenth century.
3. From the letter to Samuel Birch, CCD, Vol 5, [12 March 1856], CD sought his help at the British Museum about ancient breeds of pigeons and poultry.
4. The alula is a vestige in birds of the front limb, consisting, in chickens, of three or four large quill-like feathers.
5. In wild birds, the number of feathers is generally constant and is used as a character in classification. The Cochin China cock had been sent to CD by Bernard P Brent (see CCD, Vol 6, letter to W B Tegetmeier, 20 March [1856]).
6. The differences in the number of ribs is discussed in *The variation of animals and plants under domestication,* 1868, Vol 1, 165–6. CD recorded that he was not sure that he had designated the vertebrae correctly; he stated the dorsal vertebrae were always eight in number (*Variation* Vol 1,165).
7. *Cottage Gardener*, and *Country Gentleman's Companion*. See Letter #107 (n12).
8. William Bernhard Tegetmeier (1816-1912). Author, journalist and naturalist, specialising in domestic animals and birds (pigeons and fowls). Editor of the *Field* (1864-1907) and Secretary of the Apiarian Society of London. See Letter #107 (n12).
9. CD's evidence of changes in young pigeons was used in *On the Origin of Species* in Chapter 14 (*Mutual affinities of organic beings: morphology: embryology: rudimentary organs*), pp 445–6. It was also used to compare breeds of fowls in *The variation of animals and plants under domestication,* 1868, Vol 1, 248–50.
10. CD was referring to books that he was reading on the subject and which he recorded in his reading notebooks: see CCD, Vol 4, Appendix IV, 128: 14 &16, p 493
11. Erasmus Darwin, CD's brother, with whom CD was staying at the time.
12. Probably Samuel Boteler Bristowe [eldest son of Samuel Ellis Bristowe (see Chapter 2, Note 57)] and his wife, Albertine. He was Fox's nephew, a good friend from childhood, and a barrister in the Inner Temple. Alternatively, CCD, Vol 6 suggests that it could have been the second brother, Henry Fox Bristowe, also a London barrister, and his wife, Selina.
13. Henry Parker, physician at the Shropshire Infirmary, married to Marianne Darwin (1798-1858). He died in January 1856 (*Eddowes Salopian Journal*, 16th January, 1856, p 5).

109 (CCD 1887). To: The Rev[d] W. Darwin Fox (no envelope for this letter)

Down Bromley Kent
June 4th [1856][1]

My dear Fox

I write one line only (for I have been writing till I am quite wearied) to say that a Dorking Cock, more like an ostrich than a simple fowl, arrived this morning; & I presume from you. — It is a splendid acquisition, & I thank you much//

for it. — It will make a grand skeleton. —
I hope all is well with you. —

My dear Fox
Yours most truly
Ch. Darwin[2]

Provenance: The American Philosophical Society, Philadelphia, 139
S-130 1887

Notes

1. Dated by the reference to the Dorking Cock mentioned again in Letter #110.
2. The last half of the last page of this letter was excised. It was probably used by Ellen Fox to write to her husband: see Letter #109A.

109A Ellen Sophia Fox[1] to William Darwin Fox

Friday [>6th June 1856]

Darling Husband,

I am going to try to save <1^{d}.> by filling up the scraps of paper, I was sorry to send you on 2^{d} letter, I put in the Libry ***** message just at this <back> and forgot to put on another stamp. Thank you *very*, *very* much dearest for <your last> letter, it is the one sunshine spot in the day, & I can do and think of nothing till the post comes in; I have kissed the ***** letters many times, but it is very *cold* consolation, and makes me long still more for your dear face; and at night it is no easy matter to grow sleepy, for want of your arms around me. It seems as if you had been gone a month at least, & only 3 days have passed, Oh! how it seems a long dreary time in prospect, but like other things, I suppose I shall get over it, and only feel more happy & joyous to see you home again safe and well; it is indeed cause for real gratitude & I do feel *truly* thankful that you are better, I hope the pain in the ball of the toe is going off, and that you will be able to continue y$^{r.}$ walks. How exactly I can fancy you & yr medical attendant lounging about yr dogs, & I can fancy you drinking the odious waters too, I hope you go to the Old Well, the water is much better there; how do you go to the Baths[2]? It is <handy> for you to walk. What stories you Dr M[3] will tell each other! & what discussions about the [Plan?], it is so like Mrs Mudge, it does so amuse me. I c^{d} almost say "*thank you*" for liking the old house, the reason I like it, is because of the light staircase & cheerful large landing and I remember many many happy days there, in Uncle Mittons[4] time, he lived there & everything was so very nice, it was always holiday time at Harrogate, & we were sadly spoilt by being made a great deal of there, so of course it is very pleasant to me to have you say it is a nice old place, I hope, still, we shall some day be there together[5]

(*on a separate scrap of paper*)[6]

**** **** at the school all day yesterday, they amused themselves for some time with ringing the bell, to w^{h} a rope is now attached, I can't think what they c^{d} find to do all day – I went to Peter this morning with your messages, he & Clarke are weeding the small side beds in the farm garden, & Johnson & Gregory doing one of the borders; Peter says after today & tomorrow they must be in the fields, I told them I hoped they w^{d} get on with the Garden as much as they c^{d}.

Peter says

Have you done anything about some *gorse* *seed* you said you w^{d} write for?

Have you written to [Errington?] for some Cucumber plants?

Peter has 4 pretty good ones. There is a duck wanting to sit, & we propose to put the 5 call ducks eggs under her, what shall we fill up the number with?

Beckett (Mr Douglas partner) said that they were to have the tiles from the farm *floor*, did you make such arrangement. . . . .

(*last half of page torn off*)

Provenance: The American Philosophical Society, Philadelphia, 139
S-130 1887

Notes

1. Fox's second wife, Ellen Sophia, who would have been 36 years old at the time of this letter. The author of this letter is identified by the references contained in the letter and its association with Letter #109. It is likely that Ellen Fox was saving on paper and stamps by excising the last part of CD's letter (Letter #109) and wrote on that. The rest of the letter has not been traced (but see scrap of letter that follows the first sheet; see, also, note 6).
2. From a reference later in the letter, the place was Harrogate, a well-known resort for water cures in the eighteenth and nineteenth centuries. Two chalybeate wells, i e, with water high in iron, were known for their curative properties. The earlier well, known as Tewit Well, may be the "old well" referred to here. The other was St John's Well. CD and family went to Ilkley Wells near Harrogate in 1859 (see Letters #136 & #137).
3. It is likely that this was Dr Mudge. The "Mrs Mudge" referred to in the same sentence is probably his mother. Fox's first wife was the daughter of Elizabeth Mudge, who was the daughter of John Mudge, FRS, and whose brother was Zachary Mudge, a naval officer, best known for his exploits on the Vancouver Expedition. Zachary and Frederick Mudge, who died of scarlet fever in the care of Fox in 1853, were the grandsons of Zachary Mudge, Sr [see Letter #93 (n17)].
4. Ellen Sophia's mother was Mary Mitton. It is therefore likely that she was referring to Mary's brother. Both were the offspring of the Rev Robert Mitton.
5. Letter #111 from CD to Fox is addressed to the "Old Parsonage, High Harrogate", so it is likely that Fox and his wife did return together there. However, the diary of Fox for 1856 is not extant (see Appendix1).
6. This seems to be another note from Ellen Sophia Fox to her husband W D Fox. The reference to call ducks correlates well with letters of this period. However, it is not necessarily part of the previous letter.

110 (CCD 1895). To: The Rev W. Darwin Fox, Old Parsonage, Harrogate

Down Bromley Kent
8th [June 1856]

My dear Fox

I wrote just to say the splendid Cock had arrived, & now I have got your letter[1]. I am most sincerely sorry, my dear old friend, at all the great suffering you have undergone. I have always understood that Lumbago & sciatica cause extreme pain. — How much illness you have had in your life, & at different times how much misery! It is lucky, when young, that one does not foresee the future. I do hope that the waters are so nasty that they will do you //

good[2]. — We are all well here, except my poor dear wife who is *wretched* (perhaps in Family way) & has been wretched for weeks; & I fear that it will stop Tenby[3]. I have been working of late very hard & have now an enormous correspondence. — I think I shall make an interesting collection of domestic vars. for promises are coming in from all quarters[4]. This morning I heard from Rajah Brooke with promises of energetic assistance[5]. and Honbl. Ch. Murray says Pigeons <& Fowls> are on their road for//

me from Persia[6], — as are others from E. Africa[7]. — At this present moment I am most interested about domestic Rabbits; having just sent an Angora to be skeletonised, & having compared a P. Santo, common, & Hare Rabbit skeletons & found, I believe, most remarkable differences[8]. — Ducks too are my delight: pray do not forget a Call *Drake* <or Duck>, if you shd have the *misfortune* (oh what hypocrisy!)

to lose one. Can you tell me anything about differences in habits in *Rabbits*? Do you know any breed of Ducks//

besides tufted, hook-billed, (both of which I have had from Holland); Penguin; Call; & Black Indian or B. Ayres Duck[9]. — Even sub-vars wd. be of value to me. On account of doubt on origin I have come to care more about these & Pigeons than about Poultry; not that I shall give up poultry. — Sir C. Lyell was staying here lately, & I told him somewhat of my views on species, & he was sufficiently//

struck to suggest, (& has since written so strongly to urge me) to me to publish a sort of Preliminary Essay[10]. This I have begun to do, but my work will be horridly imperfect & with many mistakes, so that I groan & tremble when I think of it. By the way I want just to put, as an illustration of singular corelations in organisation. That "I have been assured by the Rev. W. D. Fox that he has//

never seen or heard of a blueish-grey Cat which was not deaf." May I quote you? or <must I> put in "friend" Do you still believe in the generality of fact? Might I say that you have seen or heard of as many as 6.[11] But please mind, do **not** answer this note, if it pains or troubles you. — I do most truly regret that you were not able to pay us a visit; I shd. have so much enjoyed it. —

My dear old friend
Yours most truly
Ch. Darwin

I have sent you a very long note all about myself. — And now to write to Borneo & the Cape of Good Hope[12]!

Postmark: JU 10 1856
Provenance: University of British Columbia, Woodward Library, "A"

Notes

1. The cock is referred to in the previous letter (#110). As indicated in Letter #109a, Fox had travelled to Harrogate to take a restorative cure for lumbago and sciatica, from where he had written the letter to CD, which is referred to here. As shown by Letter #109a and the readdressed destination, CD's letter, addressed to Delamere, had been sent on here by his wife. From extant diaries of Fox, from1857 onwards (1853-1856 are missing), it is clear that Fox suffered from lumbago.
2. Note that in Letter 109A, Ellen Sophia Fox mentions *"I can fancy you drinking the odious waters too, I hope you go to the Old Well, the water is much better there"*. In addition to the two chalybeate wells, with reasonably drinkable water, there was in addition a well with water high in sulphur, to which Ellen is presumably referring in writing the *"odious waters"*.
3. Emma Darwin was, indeed, expecting their tenth child, Charles Waring (born 6th December 1856). CD and family did not go to Tenby in July as planned (see CCD, Vol. 6, Letter to Tegetmeier, 11th May 1856 and Note 4).
4. This reiterates the point made in Letter #106 (n6), see also CD Memorandum referred to there in CCD, Vol 5.
5. James Brooke, Raja of Saráwak, Borneo, who sent CD specimens of pigeons and ducks (*The variation of animals and plants under domestication,* 1868, Vol 1, 132, n1 & 280.
6. Charles Augustus Murray (1806-1895) envoy to the court of Persia. See CCD, Vol 6, letter to W. B. Tegetmeier, 29th November 1856.
7. In *The variation of animals and plants under domestication,* 1868, Vol 1, 25 & 246 n, CD refers to S Erhardt, as having provided information on domestic animals in East Africa.

8. The bone characteristics of rabbits were recorded in *The variation of animals and plants under domestication,* 1868, Vol 1, 115–130.
9. Discussed in *The variation of animals and plants under domestication, 1868*, Vol 1: 276–7.
10. Charles Lyell and his wife had visited Down from 13 to 16 April 1856 (See Emma Darwin *Diary* and letters of CD in CCD, Vol 6, to and from Charles Lyell, May 1856). CD was working at this stage on his "*Big Book*", which was never published; parts have subsequently been published posthumously (Stauffer, 1975, *Natural Selection*), but the first two chapters, composed in 1856 and dealing with '*Variation under domestication*', have not been found and it is probable that CD adapted this material later in compiling *The variation of animals and plants under domestication* (1868). Lyell at this time was urging CD to publish a quick synopsis, advice that was to come home to haunt CD later (see Letter #130).
11. In *The variation of animals and plants under domestication,* 1868 Vol 2: 329, CD made the statement that '*white cats, if they have blue eyes, are almost always deaf*', and then wrote: '*The Rev. W. Darwin Fox informs me that he has seen more than a dozen instances of this correlation in English, Persian, and Danish cats; but he adds "that, if one eye, as I have several times observed, be not blue, the cat hears. On the other hand, I have never seen a white cat with eyes of the common colour that was deaf."*' This information also appears in *On the Origin of Species*, p 12, but without a reference.
12. CCD, Vol 6 suggests that this is a reference to James Brooke and Edgar Leopold Layard. No letter to Brooke is extant, but the letter to Edgar Leopold Layard, 8 June [1856] is given in CCD, Vol 6.

111 (CCD 1901, LLii, part[1]). To: The Rev W. Darwin Fox, Old Parsonage, High Harrogate

Down Bromley Kent
June 14th [1856]

My dear Fox

Very many thanks for the capital information on Cats[2]; I see I had blundered greatly, but I know I have somewhere your original notes; but my notes are so numerous during 19 years collection that it wd. take me at least a year to go over & classify them[3]. — I do not intend to attend systematically to cats, there are such great doubts on// origin & they have been crossed in so many countries with native Cats. — I have bespoken the *so-called* Himmalaya Rabbit in Zoolog. Gardens[4]. I am particularly obliged about Call Drake[5]. Have you ever crossed them with common Ducks: it would be a very valuable experiment for me to know whether the half-bred are// fertile inter se, or with some third breed; I am trying this extensively with Pigeons; but I am overwhelmed with subjects & work, & do so wish I was stronger. —

What you say about my Essay[6], I daresay is very true; & it gave me another fit of the wibber-gibbers; I hope that I shall succeed in making it// modest. One great motive is to get information on the many points on which I want it. But I tremble about it, which I shd not do, if I allowed some 3 or 4 more years to elapse before publishing anything.

My dear old friend
Yours affecty
C. Darwin

I am off in 10 minutes to a great Pigeon Fancier at Black Heath[7]. —

Postmark: First postmark obscure; second: Harrogate JU 15 1856 A
Christ's College Library, Cambridge, #98

Notes

1. In Life and Letters, Vol 2, p 71, from "What you say about my Essay" to "some 3 or 4 more years to elapse before publishing anything", and after (p 71/72) from "Very many thanks for the capital information on Cats" to "at least a year to go over & classify them".
2. See Letter #110. Fox is cited on cats in *The variation of animals and plants under domestication,* 1868,Vol 2, 329.
3. CD boasted of his cataloguing system. However, in the case of Fox, CD sadly neglected to keep Fox's letters in a systematic way. In a number of cases he cut out the interesting part of a Fox letter and kept it in a folder on a relevant topic. Several of these excised parts are now in DAR 84, 85 and 99 in the University of Cambridge Library. In his *Autobiography*, CD said '*I keep from thirty to forty large portfolios, in cabinets with labelled shelves, into which I can at once put a detached reference or memorandum*' (p 137).
4. The Gardens of the Zoological Society of London. CD had been a member of the society since 1839. See letter #110 (n8) for the outcome on rabbits in *The variation of animals and plants under domestication,* 1868, in which he cited information given by Fox on Himalayan rabbits.
5. See Letter #110.
6. This remark suggests that CD had given Fox an outline of his ideas. However, the later remarks in Letter #130 seem to make clear that Fox had no knowledge of the abstract of 1842 or 1844. It is more likely that CD had described his book in terms of an "Essay". For Fox's reaction to "*On the Origin of Species*" and CD's other books, see Letter #192 (n4 -6).
7. In CD's *Address Book* (Down House MS) the address of Matthew Wicking is given as 'Clifton Villa, Blackheath Park'. Wickings is cited in *The variation of animals and plants under domestication* (1868), concerning modifications in domestic animals living in mountainous regions, and on a numerical relation between the hairs and excretory pores of sheep.

112 (CCD 1967, LLii, part[1]). To: The Rev W. Darwin Fox, Delamere Rectory, Northwich

Down Bromley Kent
Oct. 3$^{\underline{d}}$ [1856][2]

My dear old Friend

I am very sorry to hear how much you have been ailing. I had not heard of this affliction of the leg. I do most sincerely hope that the water-cure will complete the good work which it has begun: the loss of locomotion to a man so active & energetic as yourself would be grievous. No one can wish more truly for your recovery//

than I do. — Thank you for telling me about our poor dear child's grave[3]. The thought of that time is yet most painful to me. Poor dear happy little thing. I will show your letter tonight to Emma. — About a month ago I felt overdone with my work, & had almost made up my mind to go for a fortnight <to Malvern>; but I got to feel that old thoughts would revive so vividly that it would//

not have answered; but I have often wished to see the grave, & I thank you for telling me about it[4]. I sh$^{\underline{d}}$ not be surprised if I went for a fortnight to Moor Park Hydropathic establishment for a fortnight, some time[5]; for I have great faith in treatment, & no faith whatever in ordinary Doctoring. It was *very* kind in D$^{\underline{r}}$ Gully to speak so of me: if you go there again, pray remember me *most* kindly to him, & say//

that never (or almost never) the vomiting returns, but that I am a good way from being a strong man. Do not mention Moor Park; but sh$^{\underline{d}}$ I ever go there, I sh$^{\underline{d}}$ *certainly* inform him. — My poor wife has been having a bad summer; for poor soul, we shall have another Baby this autumn[6]! Our second Boy, George (an enthusiastic *Herald*! & Entomologist) [7] has been for the last six weeks, at his first School; the//

Revd. C. Pritchard at Clapham, — a school which has been much patronised by scientific men, Herschel, Airy, Grove & Gassiott[8]: I like it very well, & we can have him home monthly for a day. — You know that our Aunt M[s] S. Wedgwood lives here[9]: a month ago she slipped on the road & fractured the head of thigh-bone & is utterly crippled, — a grievous accident for her with all her peculiarities, — far worse than death. — //

I remember you protested against Lyells advice of writing a *sketch* of my species doctrines; well when I begun, I found it such unsatisfactory work that I have desisted & am now drawing up my work as perfect as my materials of 19 years collecting suffice, but do not intend to stop to perfect any line of investigation, beyond current work. Thus far & no farther I shall follow Lyell's urgent advice[10]. — Your remarks weighed//

with me considerably. I find to my sorrow it will run to quite a big Book[11]. — I have found my careful work at Pigeons really invaluable, as enlightening me on many points on variation under domestication[12]. The copious old literature, by which I can trace the gradual changes in the Breeds of Pigeons has been extraordinarily useful to me. — I have just had Pigeons & Fowls *alive* from the Gambia! Rabbits & Ducks I am attending pretty carefully, but less so than Pigeons. I find most remarkable differences in skeletons//

of Rabbits[13] — Have you ever kept any odd breeds of Rabbits & can you give me any details? Your Call Drake is quite hearty: I have not watched it much, but have not noticed its loquacity; the beak seems short & breast very protuberant. If at any time you could spare time, I sh[d] very much like to hear any particulars about habits of Call Ducks. — Do they show any migratory restleness in Autumn? — One other question, you used to keep//

Hawks, do you at all know, after eating a Bird, how soon after they throw up the pellet? No subject gives me so much trouble & doubt & difficulty, as means of dispersal of the same species of terrestrial productions on to oceanic islands[14]. — Land Mollusca drive me mad, & I cannot anyhow get their eggs to experimentise[15] on their power of floating & resistance to injurious action of salt-//

water. — I will not apologise for writing so much about my own doings, as I believe you will like to hear. — Do sometime, I beg you, let me hear how you get on in health; *& if so inclined* let me have some words on Call Ducks.

My dear Fox
Yours affectionately
Ch. Darwin

Christ's College Library, Cambridge #100
(*The envelope is missing*)

Notes

1. In Life and Letters, Vol 2, p 84, from the sentence "I remember you protested against Lyells advice of writing a sketch...." to the end with the exception of two sentences in the middle of this section: "If at any time you could spare time, I sh$^{\underline{d}}$ very much like to hear any particulars about habits of Call Ducks. — Do they show any migratory restlessness in Autumn? — "
2. Dated by the reference to CD's aunt, Sarah Elizabeth Wedgwood (see Note 9).
3. Anne Elizabeth Darwin (1841-1851) had died at James Manby Gully's hydropathic establishment in Malvern, Worcestershire, on 23rd April 1851: see Letters #85 & #86. CD left before the funeral and so did not know any details of the grave site. Fox was clearly in Malvern in the summer of 1856 (see next Letter #113) and had visited the grave and was therefore able to give directions to CD and family when they visited Malvern and could not find the grave. The grave had become overgrown in the five years since (see Letter #114). No *Diaries* for Fox exist for 1855 or 1856 (Appendix 1).
4. This information must have occurred in a missing letter. This reference to Malvern and Annie's grave indicates that Fox had recently visited Malvern, as confirmed by the next letter (#113), although we know that he had also recently visited Harrogate for relief from lumbago (Letter #110 (n1).
5. Moor Park, was a hydropathic establishment, at Farnham, Surrey, run by Dr Edward Wickstead Lane. CD paid the first of several visits there in April 1857 [see CD's '*Journal*' and Letters #114 (n4), #118 (n3&4), #125 (n1), 125A (n 2), #134 (n4), #135 (n10)].
6. CD repeats information given in Letter #110 (n3). Charles Waring was born 6th December 1856.
7. George Howard Darwin (1845-1912), aged 11 at this time, followed his father's interests in collecting moths and butterflies, and also had an interest in drawing soldiers and coats-of-arms. Several of his detailed coloured drawings are preserved in the University Library, Cambridge (DAR 210.7).
8. The named persons are John Frederick William Herschel (1792-1871), astronomer; George Biddell Airy (1801-1892), mathematician and astronomer; William Robert Grove (1811-1896), lawyer and physical scientist, and John Peter Gassiot (1797-1877), businessman and pioneer with Grove in work on electrical phenomena; also Charles Pritchard (1808-1893), clergyman, educational reformer and astronomer. Pritchard had founded the Clapham Grammar School in 1834 and instituted a curriculum that included the sciences. It was situated at 95-97 High Street, Clapham, on the east side, and had a chapel erected in 1846 (*Gentleman's Magazine*, 1846). See J R Moore, 1977, for CD's opinions on the education of his sons.
9. Sarah Elizabeth Wedgwood (1776-1856) the only surviving child of Josiah Wedgwood I. In 1847 she had moved, with three servants, to Petleys, a house near Downe, that she rented from John Lubbock; the move was motivated by the desire to be close to Emma Darwin. On 3rd September 1856, Emma Darwin recorded in her *Diary*: '*Aunt S. fall*'. As a result of the fall she died a few months later in the same year.
10. CD reversed this decision on receipt of the "bombshell" letter and article (intended for publication) by Alfred Russell Wallace, which arrived sometime before 18th June 1858 [see Letter #130 (n8)].
11. It is not clear what Fox's advice was. It would appear to be "to publish an extensive account rather that an abstract"; if so then he should have approved of CD's plan for a "*Big Book*". Lyell seems to have advised CD to publish a short abstract as well as a larger book However, CCD, Vol 6 (1856-57) presents little evidence for such advice, although there is a letter from CD to Lyell (July 5th 1856) in which CD says "*I am delighted that I may say (with absolute truth) that my essay is published at your suggestion*". The use of the term "essay" is confusing, but it does refer to the "*Big Book*". CCD appends a note to this letter referring to an interesting entry in one of Lyell's notebooks under the title "*Letter Darwin Dedication*". See, also Note 10, above.
12. See Letter #108 (n8).
13. Chapter 4 of *The variation of animals and plants under domestication,* 1868, Vol 1, is on "*Domestic rabbits*" and contains a long section on "*Osteological characters*".
14. CD was interested in the ways in which remote islands might be populated with species. Birds of prey with long home ranges was clearly a possible means of introduction. See reply referred to in next letter (#113).

15. CD uses the same word "experimentise" in the same context in a significan letter to A R Wallace in 1857 (Dec. 22), CCD, Vol 6, "*I have experimentised a little on some land mollusca & have found sea-water not quite so deadly as I anticipated. You ask whether I shall discuss "man"; — I think I shall avoid whole subject, as so surrounded with prejudices, though I fully admit that it is the highest & most interesting problem for the naturalist. — My work, on which I have now been at work more or less for 20 years, will not fix or settle anything; but I hope it will aid by giving a large collection of facts with one definite end: I get on very slowly, partly from ill-health, partly from being a very slow worker. —*
I have got about half written; but I do not suppose I shall publish under a couple of years. I have now been three whole months on one chapter on Hybridism!"

113 (CCD 1967, LLii, part). To: The Rev W. Darwin Fox, Delamere Rectory, Northwich

Down Bromley Kent
Oct. 20 [1856][1]

My dear Fox

I am so sorry to hear of all your suffering: it is most grievous that you should so soon fall back after Malvern, & how patiently you seem to bear it. I suppose the pain is extremely severe: very oddly I had yesterday morning just a touch of it; my back feeling locked & rigid! I never felt such a thing before[2]. — I should very much like to have ever such brief a note before very long to hear how// you are getting on. — I had no idea that lumbago ever became so long-continued & severe as in your case.

I fear I sh<u>d</u> not possibly be able to join you, at Malvern; for I could hardly leave my wife now for the next two or three months[3]. You ask Georgy's age; he is just turned eleven: perhaps a little, & but very little earlier would have been better for his// going; as he is thrown back by his entire ignorance of Greek: he is very forward in everything except that & Arithmetick, of which much account is made[4]. Georgy was at first put under care of Penfold, & as he says that he (or his brother I forget which) is your God Child, he must be son of P. of Christ Coll[5]: — Many thanks for such answers as you could give to my diverse queries: I have since taken birds with seeds in crops to Eagles & Owls at Zoolog. Soc., & the pellets// with all seeds apparently perfect were thrown up in 18 & 16 hours; but the Keepers thought this *much* sooner than often happens. We have thus an effective means of distribution of any seed eaten by any Birds, for Hawks & owls are often blown far out to sea: I am trying whether the seeds will germinate.[6] —

If *when you are better* you do not dislike writing// to your nephew in Jamaica, it would be really an *important gain* to me, to know whether the eggs of Lizards, (or snakes) & Shells will float in **sea-water** say for a week (or 10 days); & if they will float for that time, if he would wash them in pure water & keep them to see whether they would hatch, it would be still greater// Service. If many could be obtained it would be well to try whether some would hatch after 5 days floating & some after 10 days. But if they sink, I sh<u>d</u> care less[7]. —

I have given up in despair trying to get lizards eggs in England[8]. — But I fear your nephew w<u>d</u> think the experiment too foolish. My dear old friend, no one can more sincerely wish for your recovery than I do. — With our very kind remembrances to Mrs. Fox

Your's most truly
Ch. Darwin

Christ's College Library, Cambridge #99

Notes

1. Dated by the references i) to Emma Darwin's forthcoming confinement (see note 3, below) and ii) to George Howard Darwin being 11 years old.
2. As indicated by Letter #112 (n4), it seems that Fox has visited Malvern recently. Thus it appears that the waters of Harrogate conferred no benefit on Fox's lumbago and that he turned to the hydropathic cure of Malvern, probably from the references, Dr Gully's establishment [see Letter #88 (n8)].
3. As indicated previously [Letter #110 (n3)], Emma Darwin was pregnant with her last child (seven months at this stage): Charles Waring was born on 6th December.
4. As indicated in the previous letter (#112), George had just entered the Clapham Grammar School. Since he became Plumian Professor of Astronomy and Experimantal Philosophy at the University of Cambridge, his deficiency in arithmetic seems to have done him no harm.
5. James Penfold. Admitted pensioner at Christ's College, Cambridge in Oct 1825, B A, 1830. Vicar of Thorley, near Yarmouth, Isle of Wight, 1841-1856. (*Alum Cantab, Peile, 1913*).
6. CD describes this experiment with owls' pellets in his Experimental book, p 15 (DAR 157a): '*Oct 19th Planted pellet from Snowy owl, charged with seeds. planted pellet whole — 16¼ in stomach. Such seeds wd be well manured — Birds wd float if died in sea. —*'. In another letter to J D Hooker (CCD, Vol 6, 19th Oct. 1856), CD described an experiment with hawks: "*The Hawks have behaved like gentlemen & have cast up pellets with lots of seeds of them.*" This experiment is also reported in CD's *Experiment Book* (p 15).
7. See Letter #123 (n1).
8. See Letters #99 and #100.

114 (CCD 11799). To: Charles Darwin [Down, Bromley, Kent]

Delamere Ry
Northwich
Dec 19 [1856][1]

My dear Darwin

I am got back from Malvern very much better for Dr Gully[2]. I sadly miss the Douche here & wish I had yours[3]. My substitutes are 3 pails of water thrown over me on rising before a shallow — & 6 minutes water poured down spine at Noon — but they do not fill the place of the Douche. It certainly is a wonderful system, & the class of *Incurables* getting well there is quite a study. All the Drs are making fortunes. I think there are eight there now, who have plenty to do with Gullys leavings[4].

I have today received the enclosed letter[5] in answer to my[6]

(*section excised? Or whole page missing*)

"Peas are grown extensively by the Messrs Sharp (Sleaford)[7] and as they are considered liable to mésalliances[8] considerable precautions are employed to secure separation" — Gard Ch Dec 15, 828[9]

I noticed this, the week after you had made enquiries about Leguminous Plants hybridising[10]. As "in the multitude of Counsellors is wisdom" I will add my experience upon the same plant, Pease. I have grown "The Queen Pea" — a great favorite with us some ten years past, with many other sorts by Garden by them[11], & never find the seed deteriorate in the least, & I do not therefore believe Messrs. Sharp.

Annotations
1) Delamere . . . to my *crossed pencil*
2) Peas are grown extensively] "Chapter 3." *added pencil*

Incomplete
DAR 77: 170

Notes

1. Dated from CD's inclusion of this information in his *Big Book*, which in 1857 morphed into *Natural selection* (never published in CD's life time, but see Stauffer, 1975), and from the article in the *Gardeners' Chronicle* (see note 9, below).
2. It had been clear from Letters #112 and #113 that Fox had visited Malvern, Worcestershire, for hydropathic treatment, probably at James Manby Gully's hydropathic establishment [see Letter #79 (n8)]. This was probably for treatment for lumbago. Fox subsequently visited Malvern on several occasions. It is interesting that CD visited Dr Gully on several previous occasions [see Letters #79 (n8), #80 (n3), #82 (n1), #83 (n2), #85 (n1), #88 (n8)] but not after the death of his daughter, Annie in, 1851 (Letter #89). CD had given up the douche by 1853, but clearly carried on with some kind of water treatment (see letter to Edward Cresy, 19th April 1853, CCD, Vol 5). In 1856 CD started to go to Dr Lane's Hydropathic establishment at Farnham, Surrey [#112 (n4), #114 (n4), #118 (n3&4), #125 (n1), etc].
3. CD had built a special out-house and douche, see Letters #83 and 85.
4. It is not clear what is meant by "leavings". There was much rivalry in Malvern at the time and Dr Gully took long sojourns overseas in later years (Mann, 1983). Dr Manby Gully retired in 1872 (Mann, 1983).
5. This letter has not been found.
6. End of annotation 1.
7. The nursery and seed-supplying firm of Sharpe and Co. owned by Charles Sharpe of Sleaford, Lincolnshire.
8. End of annotation 2. CD means that unwanted crosses are possible unless careful attention is given.
9. *Gardeners' Chronicle and Agricultural Gazette*, 13 December 1856, p 823. Fox got the page number wrong. CD cited this quotation in chapter 3 of his species book, '*On the possibility of all organic beings occasionally crossing*' (*Natural Selection*, p. 70). This chapter was begun on 13 October 1856 and completed on 16 December ('*Journal*'; Appendix II), although CD made subsequent changes. The nursery and seed-supplying firm of Sharpe and Co was owned by Charles Sharpe of Sleaford, Lincolnshire.
10. This implies that Fox wrote a letter to CD on the subject but it has not been found. CD may have been referring to adding notes to his Big Book.
11. The irony here is that, about this time, Gregor Mendel, the monk and great founder of modern genetics, worked on crossing peas to a very large extent. His work was published in 1866. See e g, Bateson (1902). *Mendel's principles of genetics*. Cambridge University Press, Cambridge, UK.

115 (CCD2049). To: The Rev W. Darwin Fox (no envelope[1])

Down Bromley Kent
Feb. 8th [1857][2]

My dear Fox

I was very glad to get your note; but it was really too bad of you not to say one single word about your own health. Do you think I do not care to hear? —

It was a complete oversight that I did not write to tell you that Emma produced under blessed Chloroform our sixth Boy almost// two months ago[3]. I daresay you will think only half-a-dozen Boys a mere joke[4]; but there is a rotundity in the half-dozen which is tremendously serious to me. — Good Heavens to think of all the sendings to School & the Professions afterwards: it is dreadful. —

I am very sorry to hear of your 3/4 child![5] // We shall be most heartily rejoiced to see you here at any time: we have now R^y to Beckenham which cuts of 2 miles & gladly will we send you both ways at any time[6].

But the other morning I was telling my Boys about some of our ancient entomological expeditions to Whittlesea meer &c[7]; & how we two used to drink our tea &// Coffee together daily. We had not then 20 3/4 children between us; & I had no stomach.

I do not think I shall have courage for Water Cure again: I am now trying mineral Acids, with, I think, good effect[8]. I am not so well as I was a year or two ago. I am working very// hard at my Book, perhaps too hard. It will be very big & I am become most deeply interested in the way facts fall into groups. I am like Crœsus overwhelmed with my riches in facts. & I mean to make my Book as perfect as ever I can. I shall not go to press at soonest for a couple of years.

Thanks about W. Indies. I have just had a Helix pomatia[9] withstand 14 days well in Salt-water; to my very great surprise[10]. I work all my friends: Are there any Mormodes[11] at Oulton Hothouses[12] or any of those Orchideæ which eject their pollen-masses when irritated: if so will you examine & see what// would be effect of Humble-Bee visiting flower: w$^{\underline{d}}$ pollen-mass ever adhere to Bee, or w$^{\underline{d}}$ it <always> hit direct the stigmatic surface? —

You ask about Pigeons: I keep at work & skins are now flocking in from all parts of world. —

You ask about Erasmus[13] & my sisters: the latter have been tolerable; but Eras. not so well with more frequent fever fits & a good deal// debilitated: Charlotte Langton[14] has been very ill with Asthma & Bronchitis; but I hope is recovering.

Farewell, my dear old Friend
Yours affect^y
C. Darwin

Are castrated Deer larger than ordinary Bucks? Do you know?

Christ's College Library, Cambridge, #110

Notes

1. The original envelope is not extant. The letter was placed in another envelope on which someone (probably Francis Darwin) wrote "1857", crossed out and changed to "1859" and then to "1858".
2. Dated by the birth of CD's eighth surviving child (see Note 3).
3. Charles Waring Darwin was born on 6 December 1856. CD had anaethetised Emma with chloroform during the two previous labours (see Letter #84 and letter to J. D. Hooker, 3 February 1850, CCD, Vol. 4).

4. Fox at this time had twelve children, five by his first marriage and seven by his second. Of these, four were boys (*Darwin Pedigrees*).
5. CD refers to the advanced stage of Ellen Sophia Fox's pregnancy. Edith Darwin Fox was born on 13 February 1857 (*Darwin Pedigrees*). Fox wrote in his diary of 1857 "*Feb 13 My dear Wife confined of a ~~Boy~~ Girl Edith Darwin – Mr MacGregor*"
6. CCD, Vol 6 reports "*In April 1855, a plan was drawn up for an extension to the South Eastern Railway to run from Lewisham to Beckenham. John William Lubbock was the chairman of the committee; CD was a shareholder (CCD, Vol. 5, letter to W. E. Darwin, [25 April 1855], n. 4). The Mid-Kent Railway was opened on 1 January 1857 (The Times, 9 February 1857, p. 7).*" The other route was from London Bridge to Sydenham [see Letter #72 (n10)].
7. Whittlesea, a village near Peterborough, was near the site of an extensive meer before it was drained in the eighteenth century [see Letter #47 (n6)] and the destination of a memorable entomological excursion for CD, Fox and Albert Way during their undergraduate days. This is where they found *Panagaeus crux-major*. See Autobiography and Letters #47 (n4 & 5), #48 (n16), #62 (n7), #90 (n17), #96 (n10) and #71 (n8).
8. For the possible treatment, see Colp, 1977. Colp suggested that 'Mineral acids' "*probably meant a mixture of muriatic (hydrochloric) acid and nitric acid*". The belief was that some cases of dyspepsia were caused by a lack of acids secreted by the stomach and that these could therefore be replaced. Colp describes the precise prescription on p 157.
9. This is the edible snail or Roman snail of continental Europe.
10. In his *Experimental Book*, p 16 (DAR 157a), CD recorded on 5 February 1857 that a snail '*moved distinctly after 14 days in salt-water —*'. See, also, letter to J. D. Hooker, 10 December [1856], and letter from T. V. Wollaston, [11 or 18 December 1856], CCD, Vol. 6.
11. A genus of orchids.
12. Oulton Park, Cheshire, the home of CD's and Fox's friend, Philip de Malpas Grey-Egerton [See Letters #66 (n9); #73 (n6); #87 (n5), #93 (n2) & #105 (n6)]. The *Cottage Gardener* (J W Johnston) Sept 1851 has an article by Mr R Errington, Gardener to Sir P. Egerton, Bart.
13. Erasmus Darwin (1804-1881), CD's brother.
14. Charlotte Langton née Wedgwood(1798-1862), the "Incomparable Charlotte" of earlier letters [see Letter #48 (n8)]. She and Charles Langton moved to Hartfield Grove, Hartfield, near Ashstead, Kent, some twenty miles from Down, in 1847. Charlotte was never very well after moving there.

116 (CCD 2057; LLii, part[1]). To: The Rev W. Darwin Fox, Delamere Rectory, Northwich

Down Bromley Kent
Feb. 22$^{\underline{d}}$ [1857][2]

My dear Fox

I am much obliged for your various enclosures, viz <(1st)> about yourself, & most heartily glad I am that D$^{\underline{r}}$ Gully has done you some good[3]. — Emma desires to be most kindly remembered to M$^{\underline{rs}}$ Fox & we are very glad that she & the little girl are both well[4]. —

I hope that your nephew may succeed in finding some lizard eggs; for it seems// that he will try his best to ascertain the point in question[5]. — By the way I have just had Helix Pomatia quite healthy after 20 days submersion in salt-water[6]. — Thanks about Pea case: it is a very great puzzle to me; for if I could trust to my observations on Bees, I cannot see how they can avoid being crossed; but the evidence certainly// preponderates on your side, & most heavily in case of Sweet Peas[7]. — I suppose the Queen Pea flowered at same time with adjoining Peas: are you sure of this?

With respect to Clapham School: I think favourably of it: the Boys are not so exclusively kept to Classics: arithmetic is made much of: all are taught drawing, & some modern languages[8]. — I was rather frightened by having heard that it was rather// a rough school; but young Herschel did not agree to this[9]; & Georgy[10] is rather a soft Boy & I cannot find out that he has anything to complain of, though of a very home-sick, disposition. I will at any time answer any queries in detail: I do not know, but could find out, whether Clergymen's sons are charged less. —//

My wife agrees very heartily with your preachment against overwork, & wishes to go to Malvern; but I doubt: yet I suppose I shall take a little holiday sometime; perhaps to Tenby[11]: though how I can leave all my experiments, I know not — I am got most deeply interested in my subject; though I wish I could// set less value on the bauble fame, either present or posthumous, than I do, but not, I think, to any extreme degree; yet, if I know myself, I would work just as hard, though with less gusto, if I knew that my Book w^{d} be published for ever anonymously[12]

Farewell, my dear Fox
Ever yours
C. Darwin

Christ's College Library, Cambridge #101-2

Notes

1. The part in LLii is from the sentence "I am got most deeply interested in my subject;" to the end.
2. Dated by the reference to the birth of Fox's daughter, Edith Darwin (see note 4, below).
3. James Manby Gully, see Letter #114 (n2) and previous correspondence, especially #79 (n8).
4. Ellen Sophia Fox had given birth to a daughter, Edith Darwin Fox, on 13 February 1857, see Letter #115 (n4).
5. There is no identification for this nephew. It is possibly one of the Bristowes, since Fox was on very good terms with his brother-in-law, Samuel Ellis Bristowe (1800-1855). His son, Samuel Boteler Bristowe (1822-1897), Fox's nephew, was a favourite of Fox, since he was the son of Fox's sister Mary Ann, who died in 1829 [see Letter #15 (n2)]. However, it could be one of Bristowe's other son's or another branch of this family (see *Darwin Pedigrees*). CD had met one of the Bristowes recently [see Letter #108 (n11)].
6. CD repeats information given in his last letter [#115 (n9)].
7. CD's hypothesis in *Natural Selection*, was that all sexual organisms had to cross-fertilise occasionally ("*On the possibility of all organic beings occasionally crossing*") and the sweet pea (*Lathyrus odoratus*, Leguminosae) was the greatest challenge to his hypothesis. CD discussed the evidence for and against cross-fertilisation in Leguminosae and particularly sweet peas (pp 68-71). Apparently Fox had sent him evidence against cross fertilization in sweet peas.
8. CD repeats information given in Letters #112 & #113.
9. Alexander Stewart Herschel, son of John Frederick William Herschel, also attended Clapham Grammar School.
10. George Howard Darwin (1845-1912), see Letter #113 (n4).
11. CD and family never did go to Tenby although they did visit Wales. On this occasion CD went to the hydropathic establishment of Dr Lane at Moor Park, Farnham, Surrey, from 22nd April to 6th May (see next letter #117). CD and Emma's daughter, Anne Elizabeth Darwin, died attending Dr. Gully in Malvern in 1851 (see Letter #89) and there may have been a natural aversion to returning there, although CD and family did go to Malvern Wells on one later occasion (see Letters #161 & #162).
12. CD enunciates clearly the attitude of most authors.

117 (CCD 2085). To: The Rev W. Darwin Fox, Delamere Rectory, Northwich

Moor Park
Farnham
Surrey
Thursday [30th April 1857]

My dear Fox.

I have now been here for exactly one week, & intend to stay one week more. — I had got very much below par at home, & it is really quite astonishing & utterly unaccountable the good this one week has done me. — I like Dr Lane & his wife & her mother[1], who are the proprietors of this// establishment very much. — Dr L. is too young, — that is his only fault — but he is a gentleman & very well read man[2]. And in one respect I like him better than Dr Gully, viz that he does not believe in all the rubbish which Dr G. does; nor does he pretend to explain much, which neither he or any doctor can explain[3]. — I enclose a paper for the strange chance of//

your ever knowing anyone in the S. in want of Hydropathy. — I really think I shall make a point of coming here for a fortnight occasionally, as the country is very pleasant for walking[4]. — But I, also, think it highly probable that we *all* shall move to Malvern this summer, not for my sake, but for Etty's, who has now been out of health for some six or 8 months. I hardly know yet when//

we shall go, if we do go; but I very much wish that we might meet you there. Etty is now & has been for some time at Hastings[5]. I am well convinced that the only thing for Chronic cases is the water-cure[6]. — Write to me either here or after Wednesday next to Down, & tell me how the world goes on with// you, & how, especially, the Sciatica has been, if it was sciatica, which caused you so much suffering[7]. —

I believe that I worked too hard at home on my species-Book, which progresses, but very slowly. Whenever you write, as you are, I know, very learned in Pigs, pray tell me, whether any breed, known to have originated or to// have been greatly modified, by a cross with the Chinese or Neapolitan Pig, whether any such crossed-breed, breeds true or nearly true[8]. — I am pretty sure I have read of some breed known to have been formed by a cross with one of the above Breeds, but I cannot remember// particulars. — I must now take a sitz Bath, my treatment being, — daily Shallow, Douche, & Sitz[9].

Farewell, my dear Fox
Yours most truly
C. Darwin

Postmark: 1 MY 1857
Christ's College Library, Cambridge #103

Notes

1. Dr Edward Wickstead Lane. Lane took over the lease of Moor Park, a large country house with associated parkland, from Thomas Smethurst, who had turned the house into a hydropathic establishment in or around 1850. As part of his cure, the patients were accommodated in the same building as Lane and his family [Lane (1857), *Hydropathy; or, the natural system of medical treatment. An explanatory essay*. John Churchill, London]. Prior to this Moor Park was at one time owned by the

Whig statesman, Sir William Temple in the late seventeenth century. Temple employed Jonathan Swift and Ester Johnston ("Stella") and the lasting love affair between Swift and Stella resulted in *'Journal to Stella'* (1710-11).
2. Lane was 34 years old.
3. As we have seen before, James Manby Gully advocated homœopathy, and clairvoyance in some cases, in his water-cure establishment in Malvern [see Letters #79 (n8), #80 (n3), #82 (n1), etc]. CD was very skeptical of both of these treatments in medical practice, see Letter #86 (n9, 10, 12 & 13); see also letter to Susan Darwin, 19th March 1849, CCD, Vol 4. For details on Gully's hydropathic establishment and practice, see Letter #79 (n8).
4. Dr. Lane believed that a change of scenery was necessary condition for a successful cure. Moor Park was very suitable, as CCD, Vol 6, points out "*the site of Moor Park having been chosen for its location in a 'picturesque district abounding in pleasant and varied walks, with a dry soil under-foot and the fresh breezes of health playing about . . . over-head from morning till night' (Lane 1857, note 1, above, p 43)*".
5. This plan did not eventuate: Emma Darwin took her daughter Henrietta Emma to Hastings on 9th April 1857 to see whether her health would improve at the seaside. Henrietta returned home on 12th May. Emma took Henrietta to Moor Park on 29th May where she remained until 7th August (Emma Darwin's *Diary*). CD returned to Moor Park for two weeks in June (CCD Vol. 6, '*Journal*'; Appendix II). Emma visited Malvern in 1863 [see Letter #161 (n2)].
6. CD's beliefs about the water cure were undergoing change at this time. His previous attitude was totally supportive (see Letter #83) but by1853 he had given up the douche [see Letter #91 (n4)]. By the 1860s, CD had largely given up the standard water treatment as ineffective and took to cold spinal treatment [see Letter #86 (n12)].
7. As stated in Letter #114, Fox had recently visited Gully's hydropathic establishment in Malvern for treatment. The treatment was probably for lumbago, see Letter #114 (n2).
8. At this time CD had completed a chapter "*Variation under Nature*" for "*Natural Selection*" and was embarked on "*The struggle for existence as bearing on natural selection*". CD discussed the subject of pigs and their origins in *The variation of animals and plants under domestication* (1868) Vol 1, 78–9, but Fox's name is not mentioned.
9. See Letter previously referred to (letter to Susan Darwin,19th March 1849, CCD, Vol 5), for a description of CD's previous treatment at Gully's establishment in Malvern. See also Lane, 1857, pp 51–78; for citation, see note 1, above.

118 (CCD 2161). To: The Rev W. Darwin Fox, Delamere Rectory, Northwich

Down Bromley Kent
Oct. 30$^{\underline{th}}$ — [1857]

My dear Fox

I was very glad to get your note with a very fairly good account of yourself. — I cannot say much for myself; I have had a poor summer, & am at last rather come to your theory that my Brains were not made for thinking, for twice I staid for a fortnight at Moor Park[1], & was so extraordinarily better that I can attribute the difference, (&//

I fell back into my old state immediately I returned) to nothing but to mental work; & I cannot attribute the difference but in a very secondary degree to Hydropathy. Moor Park, I like *much* better *as a place* than Malvern[2]; & I like Dr. Lane very much: by the way have you seen his Brochure on Hydropathy[3]; it seems to me very good & worth reading.//

Unfortunately for me, I believe Dr L. means to look out for some new place[4]. — We have had Etty[5] there all summer; but she comes home for good next Saturday[6]. — She has received much benefit, I think, from Hydropathy; but can walk very little

& is still very feeble. For the last month or two we have had trouble about Lenny[7], who has been the picture of strength & vigour, & now his pulse has become feeble & often very irregular//

like three of <our> other children: it is strange & heart-breaking. A man ought to be a bachelor, & care for no human being to be happy! or not to be wretched.
I make slow progress in my work, which is altogether too much for me; I have done only 2 chapters in rough, first copy during the last six months[8]!
I see you ask about Mr Pritchards school[9]: I have//

nothing to say against it; but were it not for the great advantage of having George home on monthly Sundays, & *short* Michaelmas & Easter holidays; I think I shd prefer Rugby; but if you ask me why, I declare I cd give no answer. You ask about all my Sisters & Eras: all are much as usual: Catherine is thinking of taking a house in London & living there//

at least during greater part of year, but I do not know how it will answer[10]. How you have been spinning all about the English world[11]: we have been all fixture, except Moor Park. — We have, however, been recreating ourselves with building a new Dining Room & large bedroom over it; for we found our party, when we had cousins had quite outgrown our old room. — *Your* Pear-trees have born very well this year for the first time; & we had lots on the wall; & pretty well off for Plums; but no other fruit succeeds with us. Louise Bonne & Marie Louise have been splendid[12]. Not only have we been//

nowhere; but we have hardly had any visitors, except Henslow for 2 or 3 days[13]; & he was, all what he always is, — than which I cannot give higher praise. — I am very glad your children are flourishing.

My dear Fox
Yours affectionately
C. Darwin

Emma desires her very kind remembrances.

Postmark: OC 31 1857
Christ's College Library, Cambridge # 104

Notes

1. CD spent a fortnight at Moor Park [see Letter #117 (n1)] from 22nd April to 6th May 1857 and again from 16th June to 30th June 1857 ('*Journal*'; see CCD, Vol 6, Appendix II).
2. It was at Fox's suggestion that CD had undergone hydropathic treatment in 1849 and 1850 under the care of James Manby Gully in Malvern, Worcestershire, see Letters 79 (n4), #82 (n1); #83 (n2), #85 (n1).
3. Edward Wickstead Lane (1823-1889) the proprietor of the hydropathic establishment at Moor Park. His book was "*Hydropathy: Or the natural systems of medical treatment. An explanatory essay*", John Churchill, London, 1857.
4. This move did not take place until 1860, when Dr Lane moved his establishment to Sudbrook Park, near Richmond, Surrey (see Colp 1977, p 68; Freeman, 1978). CD visited there from 28th June to 7th July 1860 (CD's "*Journal*").
5. Henrietta Darwin, CD's daughter.
6. Henrietta Darwin was at Moor Park for six weeks and returned to Down House on 31st October 1857 (Emma Darwin's *Diary*).

7. Leonard Darwin, CD's fourth son.
8. CD recorded finishing two chapters (*Laws of variation*) and Chapter 8 (*Difficulties on the theory*) of *Natural Selection* on 29th September 1857 (see CCD Vol 6, Appendices II and III). He started them at the beginning of April 1857. This would mean that they had taken approximately five months.
9. CD is referring to Clapham Grammar School, started by Rev. Pritchard, see Letter #112 (n8). CD's first son William Erasmus was sent to Rugby School, see Letter #88.
10. See letter to W E Darwin, Thursday evening, November 1857, CCD, Vol 6. CD reports that Catherine Darwin was intending to rent a small house in YorkTerrace, Regent's Park, London.
11. Fox's *Diary* from May to October,1857, indicates that he went to London (visiting his parents-in-law at Hampstead and the Crystal Palace at Sydenham Hill), Manchester (twice), Rhyl (Wales), Malvern and Harrogate; so CD's remark is apt.
12. For CD's earlier discussion with Fox of the value on securing fruit-trees to the wall, see Letters #79 & #87.
13. John Stevens Henslow (1796-1861). Clergyman, botanist & mineralogist. Professor of Botany at the University of Cambridge (1827-1861), teacher and mentor to CD; See Walters and Stow, 2001. At this time he was still Professor of Botany at Cambridge and on many national committees, but spent much time in his parish at Hitcham, Suffolk. Henslow visited Down House in August 1857 (see letter to J. S. Henslow, 10th August [1857], CCD, Vol 6; although Emma Darwin did not record this in her diary).

119 (CCD 2187). To: The Rev W. Darwin Fox, Delamere Rectory, Northwich

Down Bromley Kent
Dec. 17th [1857]

My dear Fox

I am very much obliged for your Rabbit Letter: I have forgotten your having before mentioned to me this breed: I presume it is that called at Shows the "Himalaya breed"[1]. — You call it black nosed & brown-tailed; I suppose the ears are black. If *not*, & you could get me an old accidentally dead specimen I sh$^{\underline{d}}$ be very//

glad, but I can hardly doubt it is the so-called Himalaya, one of which I have now skeletonising from Zoolog. Soc^{y} — I am *particularly glad to hear about this variety being so true*: its history has been published, & I have it somewhere, but for the life of me I cannot tell where[2]. It is was produced, **I think**, from a cross of two *sub*-breeds of the//

French Silver Grey Rabbit. — <I know it is *closely* connected with Silver Greys.> If the ears are **not** black, will you tell me: all that I have seen had *red* eyes. — There is one odd point in this breed, which will serve to identify it, the young are quite white & the black ears &c appear subsequently[3]. — I had forgotten that <the> **under** side of tail was brown. — By the way Silver-Grey Rabbits are quite black when born, a curious contrast with//

so called Himalaya. — I have heard from Mr Woodd & will second him & if possible attend the voting[4]. —
In regard to Douche you will find the *same* water will last sweet, especially with charcoal floating on it, for a *long time*[5].

Yours affectionately
C. Darwin

Postmark: 18 DEC 1857
Christ's College Library, Cambridge #105

Notes

1. CD mentioned the Himalayan breed of domestic rabbits in Letter #111 (n3). Note that in this letter CD refers to them as Himalaya rabbits. However, in "*Variation*" they are referred to as Himalayan rabbits.
2. In *The variation of animals and plants under domestication,* 1868, Vol 1, 108 n15, CD cited an article in the *Cottage Gardener* (1857), 18, 141–2, on the origin of the Himalayan rabbits. This author believed the breed resulted from a cross between a tame silver gray doe and a silver gray buck that had '*one-eighth of wild blood in him.*' (*Cottage Gardener* 18,141).
3. In *The variation of animals and plants under domestication,* 1868, Vol 1, p 109, CD cited Fox as: *having found that in Himalayan rabbits sometimes 'the young are born of a very pale grey colour'.*
4. Charles Henry Lardner Woodd, Fox's brother-in-law and a geologist with whom CD had earlier corresponded (see CCD, Vol 4, letter to C H L Woodd, 4th March 1850), was seeking membership of the Athenæum Club [see Letter #121 (n13) and CCD, Vol 7, letter to George Bentham, 27th January 1858]. Woodd is not listed in Francis Gledstanes Waugh, *Members of the Athenæum Club, 1824 to 1887* (London, [1888]). There is also mention of a Wood in earlier correspondence [see Letter #84 (n1)], so there may have been more than one attempt.
5. For CD's changing views on the water cure see Letter #117 (n6).

120 (CCD 2208). To: The Rev W. Darwin Fox (no envelope)

Down Bromley Kent
Jan. 14th [1858][1]

My dear Fox

Have you Hewitsons work on Eggs, in which you gave some observations <on variations> of Birds' nests?[2] If you have could you dare send it per post well done up & allow me to pay postage; for I do not know how else I cd get to see it. —
Have you ever heard that the *first* year that Birds build, they are not quite//

so adroit — LeRoy states this to be so, but I can hardly believe him[3]. Can you give me any thoroughly well authenticated facts on ever so little variations in nests; I do not mean such cases as the Water owzel habitually having a domed[4] or open nest — or difference of Sparrow's nest in tree & in hole; but rather any *slight* difference in degree of perfection of nest of same species in different districts or of any individuals of same species.[5] —
The smallest charity in this line would be gratefully accepted. —

Ever yours
In Haste
C. Darwin

Endorsement: "Jan 14 58"
Christ's College Library, Cambridge, #108

Notes

1. The letter has the year '1858' added to the date, but it is not clear whether this is in CD's hand. The endorsement by Fox confirms that the letter was written in 1858.
2. William Chapman Hewitson (1806-78). Naturalist and specialist in diurnal lepidoptera as well as birds and birds eggs. The volume referred to here is Hewitson (1831–44) *British Oology*; Newcastle upon Tyne. Fox corresponded with Hewitson personally (see 11 letters between W D Fox and Hewitson, 1830–35; The Fox/Pearce (Darwin) Collection, Woodward Library, University of British Columbia. See, also, Letter #121. Fox was thanked in the preface to Hewitson (1831-44) for supplying information and specimens of nests and eggs and was cited frequently in the text.

3. Leroy, C. P. 1802, *Lettres philosophiques sur l'intelligence et la perfectibilté desanimaux, avec quelques letters sur l'homme,* Paris. (pp 104–5). Cited in chapter 10 of CD's species book (Stauffer, 1975: *Natural Selection*), on the '*Mental powers and instincts of animals*', p 474, n 4).
4. CD's handwriting here certainly looks like "doomed", which is given in Francis Darwin's transcription and in CCD, Volume 6. However, the sense is "domed" and in chapter 10 of CD's species book (Stauffer, 1975: *Natural Selection*), CD wrote "*or that of the Water-owzel making or not making a dome to its nest, ought to be called a double instinct, rather than variation.*"
5. See Letter #121; Fox had long been interested in bird nesting [nidification: see letter from W D Fox, #48 (n5)].

121 (CCD 2208). To: The Rev W. Darwin Fox (no envelope)

Down Bromley Kent
Jan. 31st [1858][1]

My dear Fox

I am extremely sorry to hear of your Lumbago[2], & that I shd have troubled you at such a time. — I am, also, rather sorry you have had the trouble & risk of sending a precious Book to London, in as much as it does not contain, you say, *your own* observations. And now I am sorry to say//

I do not know Mrs Fox's address[3], so will you any time during next fortnight ask to have the Book[4] sent or left at my Brothers 57 Queen Anne St or at Athenæum Club if more convenient: I would have gone & called at Mrs Fox, when I go to London about middle of February, but the distance is so great, & I have so//

much to do, & so precious little strength to do it. I could look at Book in evenings, & would return it to you direct, or to Mrs Fox's whichever you might direct[5]. — <(A memorandum might be enclosed with Book)> But all this, I am ashamed to say is giving you trouble. Very many thanks for fact of Black-birds nests[6]; it is just what I wanted//

to know: I quite agree, also, with what you say about relation of intellect & reason[7]. I have applied to one great Canary Fancier, B. P. Brent[8], & he does not believe ~~that~~ in young Canaries building worse than old ones[9]. I find my chapter on Instinct very perplexing from not knowing what to choose from the load of curious facts on//

record[10]. — There are two *little* points, on which I want information, & if you can, *which is a mere chance*, I think you will not grudge, *sometime* (for I am in no hurry) giving me information; but if you cannot, then I beg you *not* to write merely to say that you cannot. Do <you know positively whether> chickens or young Turkeys scatter & squat when they **themselves** see large Bird in air, or perceive//

danger, or do they do this, only when their mother utters the cry of danger? <Secondly> I have been struck with surprise in comparing my memoranda how often crossed animals are said to be <very> wild, even wilder than *either* parent: I have thought I would just put a foot-note to this effect, giving my cases[11]: my memory, which I dare not trust, tells me//

that the cross from wild Boar & common pig at Sydnope was wilder than the wild Boar: do you remember anything of this: I refer to our, memorable visit to

Sydnope[12]. — Tomorrow night Mr Woodd's election for Athenæum comes on[13]. — I do not second him, as he found Sir F. Palgrave's proposal was <not> invalidated[14]; & as it is the last day of my//

two Boys holidays[15] & for other reasons (expected visitor amongst others) it would be excessively inconvenient for me to sleep in London. I have written to him to say how sorry I am I cannot attend. But I have written seven notes to various members, begging their assistance, & I hope this will do very nearly as well as my attendance.

My dear Fox
Yours affectionately
C. Darwin

Endorsement: "young turkeys Jan 31 1848"
Christ's College Library, Cambridge #109

Notes

1. Dated by reference to Hewitson's book "British Oology" (see note 4 below); see also Letter #120, in which CD asked to borrow Hewitson, 1831–44.
2. In his *Diary* for 1857 Fox recorded for 11th January, "*Lumbago till end of month*".
3. Mrs Ann Fox (née Darwin) appears to have moved to Kensington Park Gardens, with her daughters, from Kendalls, their previous home near Hatfield. This would have been a reasonable place to deliver the book (see Note 4, below) and have CD pick it up.
4. A reference to Hewitson's book "British Oology": see Note 1.
5. CD recorded reading Hewitson (1831–44) at the end of 1858 (CCD, Vol 4, Appendix IV,128: 22). However, CD incorporated information from Hewitson in the chapter on instincts of animals that CD he recorded finishing on 9th March 1858 (Stauffer, 1977, *Natural Selection*, p 463).
6. CD discussed nesting in birds (nidification) and variability in the nesting instinct in *Natural Selection* (Stauffer, 1977; pp 498–506). On p 505, CD reports '*The Rev. W. D. Fox informs me that one "eccentric pair of Blackbirds" (Turdus merulus) for three consecutive years built in ivy against a wall, & always lined their nest with black horse-hair; though there was nothing to tempt them to use this material. The same excellent observer (in Hewitson British Oology. Pl cx) has described the nests of two Redstarts, of which one alone was lined with a profusion of white feathers.*'
7. CD said "*Although I do not doubt that intelligence & experience often come into play in the nidification of birds, yet both often fail*" (*Natural Selection*, p 504) and he was certain that nest-building was predominantly instinctive (pp 503–5).
8. See Letters #106 (n11) & #108 (n5).
9. Neither CD's letter to Bernard Philip Brent, nor Brent's reply, has been found (but see CCD, Vol 6, 23rd Oct 1857 for another letter from Brent).
10. CD discussed this topic in *Natural Selection*, Chapter 10, pp 466–527: "*Mental powers and instincts of animals*". Small parts of this chapter were used in *The Descent of Man* (1871). The greater proportion was never used by CD but much of it was passed on to G Romanes, who published much of the material in two books (Stauffer, 1975).
11. This is discussed in *Natural Selection*, "*Mental powers and instincts of animals*", pp 486–7. However, the examples given are: a cross between musk duck and common ducks, pheasants and fowl, wild and tame goats, and Indian bulls and common cows.
12. Sydnope Hall, Derbyshire, was the residence of Francis Sacheverel Darwin, the son of CD's grandfather Erasmus Darwin by his second marriage to Elizabeth Chandos-Pole (née Collier). F S Darwin kept a large menagerie there (Galton 1874, p 47). CD and Fox visited Sydnope Hall when they were undergraduates at Cambridge [see Letter #29 (n10) and Appendix 2]. In his discussion of reversion in *Variation in Animals and Plants under Domestication* (1868), Vol 2, 45, CD stated that F S

Darwin had crossed a domesticated Chinese sow with a wild Alpine boar, the offspring of which were '*extremely wild in confinement*'.

13. Charles Henry Lardner Woodd (1821-1893). See preceding letter; Waugh, 1888, does not list Woodd as a member of the Athenæum Club.
14. Francis Palgrave was Joseph Dalton Hooker's uncle (CCD, Vol 6).
15. Emma Darwin recorded in her *Diary* that '*G. went to school*' on 1st February and that '*W^{m} went to tutors*' on 2nd February 1858.

122 (CCD 2219; LLii part[1]). To: The Rev W. Darwin Fox, Delamere Rectory, Northwich, Cheshire

Down Bromley Kent
Feb. 22^{d} [1858][2]

My dear Fox

I was in London last week & got Hewitson all safe at Athenæum Club[3]. It is indeed magnificently bound & I am not surprised that you did not like to send it per post, & am only surprised that you would send it anyhow. — I was headachy & half-knocked up during all my stay & I found it impossible to look it over//

whilst in London & so have brought it with me, & shall consult it deliberately for I have found more variability in Birds nests than I expected ~~when I~~<after I had> put my notes together[4]. I presume you are in no hurry, & so will leave it at Mr Woods[5], when next, in about 6 weeks, when I go to London. — Very many thanks for the loan. — In your last note, what//

a funny fact that it of the Turkeys & the frog: I think it will do to work in: I wish I knew about how old they were, but probably you will not recollect. How fine it would have been if they had squatted at the sight of the monster.[6] —

You say that if you can anyhow manage it, you will this Spring pay us a visit; it would really give Emma & myself great pleasure; so do//

not forget & come if you can. I am not likely to be from home, this summer, without it be sometimes for a fortnight to Moor Park[7]. — I think Etty[8] gets gradually rather stronger. Farewell my dear old friend. If the spirit moves you at anytime to send me a line about yourself, glad shall I be.

Ever yours
C. Darwin

Christ's College Library, Cambridge #111

Notes

1. The last paragraph of this letter is in LL ii, p 110; the date of the letter is given as Feb. 8th [1858].
2. Dated by reference to Hewitson, from Letter #121.
3. Hewitson's Book "*British Oology*", 1831–44. See Letters #120 & #121.
4. See Letter #121 (n6).
5. Either Basil George Woodd, Fox's Father-in-law, or Charles Henry Lardner Woodd, Fox's brother-in-law, both of whom lived in Hampstead. CD does not refer to the proposed election of Charles Henry Lardner Woodd to the Athenæum, referred to in Letter #121.
6. Clearly Fox had replied to CD's query about fear in young turkeys and chickens (see Letter #121). CD placed this information in *Natural Selection* (Stauffer, 1977), pp 487–8 n5. "*I may add that the Rev W. D. Fox saw a brood of his Turkeys with their mother in agonies of horror at a frog peeping out*

of a hole; as Mr. Fox remarked their instinct probably misled them to mistake the bright eyes of the frog for those of a deadly N. American snake."

7. Fox visited CD at Moor Park on 27th April 1858; see Letter #125 and Appendix 2.
8. CD's daughter, Henrietta (1843-1927), who was a delicate child, had been ill in the early part of 1858 and went to Moor Park to recuperate.

123 (CCD 2229). To: The Rev W. Darwin Fox, Delamere Rectory, Northwich

Down Bromley Kent
Feb. 28th [1858]

My dear Fox

I return you with thanks your nephew's letter: he has forgotten to mention one most important element, viz whether the eggs floated[1]; if you have any communication with him I particularly wish you wd ask this question, & tell him to open eggs, as you suggest, if he tries the experiment again. If the eggs do not float or are killed by salt-water it is marvellous how Lizards get on every oceanic island. —//

Westwoods Butterflies of Grt Britain1855, is a beautifully illustrated thin large 8vo. & I am almost certain costs 15s; but I cannot find out positively[2]. Stainton has published vol I. 12mo 1857. of his Manual of British Butterflies & moths[3]; this 1st vol. includes Butterflies & "stout-bodied Moths" — it is illustrated with many uncoloured woodcuts, & I believe is very good. — it is *very* cheap, I think only 4s, certainly not// more than 7s/. —

Very many thanks about Turkeys[4]; I shall be delighted if you can succeed in trying the experiment this summer with the young Turkeys; but how on earth will you get a Kite — you speak as if everybody had a live Kite. Oh, perhaps you mean a *paper* Kite! Thanks, also, for fact about Terriers — Jesse[5] has a very parallel fact about his own *Family* of Terriers, which grinned & protruded feet when caressed. — I shall try & quote your fact, but, as I before said, I am overfacted. [6] —

I have lately some facts given to me by you years ago, about birds carrying away egg-shells very useful, as illustrating *small* instincts.[7] — GoodBye my dear Fox —

Yours most truly
C. Darwin

Endorsement" "Feb 28 1856"
Christ's College Library, Cambridge #112

Notes

1. In Letter #113 CD had asked Fox to write to a nephew in Jamaica asking him to collect eggs of lizards and snakes and to find out whether they floated and survived in sea-water.
2. John O Westwood (1805-1893). Entomologist. Hope Professor of Zoology at Oxford University. Published *The Butterflies of Great Britain*, London, 1855. (*DNB*)
3. Henry T Stainton (1822–1892). Entomologist. FRS. Published *A Manual of British Butterflies and Moths*, London; in two volumes (1857 & 1859).
4. See Letter #122.
5. E Jesse (1780-1868). Deputy surveyor of royal parks and palaces and a writer on natural history. CD quoted him from *Nat History* (3 series), in *Natural Selection* (Stauffer, 1977), p 480/81, n2: "*Jesses*

in his Gleanings in Nat Hist. 3 series p 149, says he has had a breed of Terriers, which all shew their teeth & put out their paws when caressed".

6. In the same note referred to above (Note 5), CD cited Fox on the inheritance of the behavioural traits of his Skye terriers: "*The Rev W. Darwin Fox tells me that he had a Skye Terrier bitch which when begging rapidly moved her paws in a way very different from that of any other dog which he had ever seen: her puppy, which never could have seen her mother beg, now when full grown performs the same peculiar movements exactly in the same way.*"
7. In *Natural Selection* (Stauffer, 1977), p 522, n2, CD stated "*The Rev. W. Darwin Fox informs me that he has attended to the case of the parents removing the broken egg-shells, & that there can be no doubt of its truth.*"

124 (CCD 2256; LLii, most[1]). To: The Rev W. Darwin Fox, Delamere Rectory, Northwich.

Down Bromley Kent
Ap. 16th [1858]

My dear Fox

I want you to observe one point for me, on which I am *extremely* much interested & which will give you no trouble beyond keeping your eyes open, & that is a habit I know full well that you have. I find Horses of various colours often have a spinal band or stripe of different & darker tint than rest of body — rarely transverse bars on legs, generally on under side of front legs — still more rarely a very faint transverse shoulder stripe, like an ass. —
Is there any breed of Delamere//

Forest Ponies[2]. — I have found out little about Ponies in these respects. Sir P. Egerton[3] has, I believe, some *quite thorough* bred Chestnut horses: have any of these the spinal stripe. Mouse-coloured ponies or rather small horses, often have spinal & leg bars. So have Dun Horses (by Dun I mean real colour of cream mixed with brown bay or chestnut)<so have sometimes Chestnuts, but> I have not yet got case of spinal stripe in Chestnut *Race* Horse, or in quite heavy//

Cart-Horse[4]. — Any facts of this nature of such stripes in Horses would be *most* useful to me. — There is parallel case in legs of Donkey & I have collected some most curious cases of stripes appearing in <various> *crossed* equine animals[5]. — I have, also, large mass of parallel facts in the *breeds* of Pigeons about the wing-bars[6]. — I *suspect* it will throw light on colour of primeval Horse[7]. So do help me if occasion//

turns up. — I have not yet returned your Oology[8], though I have finished with it; for I have not been in London since, & I did not like to intrust it to Carrier; though perhaps I had now better do so. — My health has been lately very bad from overwork & on Tuesday I go for fortnights Hydropathy[9]. My work is everlasting.

Farewell — My dear Fox,
I trust you are well
Farewell
C. Darwin

Endorsement: "April ~~18~~ 16 1858 (*On envelope:* 'Horses "obliterated by mistake" ')
Christ's College Library, Cambridge #112A

Notes

1. This letter is reproduced in *Life and Letters of Charles Darwin* by F Darwin (Vol 2, pp 111-112) with the exception of the following sentence, "I have not yet returned your Oology, though I have finished with it; for I have not been in London since, & I did not like to intrust it to Carrier; though perhaps I had now better do so." It is possible that Fox collected the volume during his visit to see CD at Moor Park (see Note 7, below).
2. Fox was Rector of Delamere Parish, Cheshire, which lay in the crown lands of Delamere Forest, established during the Napoleonic wars (Frank A. Latham (Ed) Delamere - the history of a Cheshire parish 1991, published by Herald Printers [Whitchurch] Ltd, ISBN 0 9518292 0 3). The forest today is an extensive area of 5000 ha of mixed deciduous and pine forest managed by the Forestry Commission. CD had finished Chapter VII (*Laws of Variation*) for *Natural Selection* by July 1857 (Stauffer, 1975). CD discussed dun colouration and banding in horses, donkeys, quaggas and zebra on pp 328-332. It seems that in 1858 he was still updating these ideas. Ponies from the New Forest of Hampshire were well known to be a well-defined stock dating back to at least 1016 AD. CD seemed to be wondering if another breed of pony existed in the Delamere forest. We do not have Fox's reply but lack of any reference in any subsequent literature seems to suggest a negative reply.
3. Philip de Malpas Grey-Egerton, of Oulton Park, Cheshire, was a neighbour and friend of Fox's, sometime President of the British Association, Tory MP for South Cheshire, and well-known to CD [see Letters #66 (n9); #73 (n6); #87 (n5) & #93 (n2)].
4. See letters to W E Darwin, CCD, Vol. 7, 11th [February 1858] and 27th [February 1858].
5. See Note 2, above. CD also relied heavily on these aboriginal markings in discussing the evolution of the horse in "*On the Origin of Species*". On p 130 (Chapter V, *Analogous Variations*)(Sixth edition) he wrote, " *He who believes that each species was independently created, will, I presume, assert that each species has been created with a tendency to vary, both under nature and under domestication, in this particular manner, so as often to become striped like other species of the genus; and that each has been created with a strong tendency, when crossed with species inhabiting distant quarters of the world, to produce hybrids resembling in their stripes, not their own parents, but other species of the genus. To admit this view is, as it seems to me, to reject a real for an unreal, or at least an unknown, cause. It makes the works of God a mere mockery and deception; I would as soon believe with the old and ignorant cosmologist, that fossil shells had never lived, but had been created in stone so as to mock the shells living on the sea-shore.*"
6. CD is referring to the appearance of two black bars on the wings of pigeons of hybrid origin, which he believed represented a reversion to the ancestral characters of the aboriginal rock pigeon. In *Natural Selection*, pp. 321–3, CD said "*... in my discussion on Pigeons , in the fact that all breeds occasionally throw blue birds, & that these always have two black bars on the wing, generally a white rump & a white external web to the exterior caudal feathers, - all characteristic of the aboriginal Rock Pigeons.*"
7. CD discussed this point in Chapter V, *Analogous Variations*, of *On the origin of species* (pp 127-129)(sixth edition). On p 129, he said " *I am aware that Colonel Hamilton Smith, who has written on this subject, believes that several breeds of the horse are descended from several aboriginal species – one of which, the dun, was striped; and that the above-described appearances are all due to ancient crosses with dun stock. But this view may be safely rejected, for it is highly improbable that the heavy Belgian carthorse, Welsh ponies, Norwegian cobs, &c, inhabiting the most distant parts of the world, should all have been crossed with one supposed aboriginal stock.*"
8. See Letter #122 (n3). It is possible that Fox collected the volume during his visit to see CD at Moor Park (see Note 9).
9. CD wrote in his *Journal* for April 20th 1858 "*Moor Park. Returned May 4th*" (CCD, Vol 7, Appendix II, p 504,). For 27th April CD recorded "*William Darwin Fox came to Moor Park*". Fox recorded in his *Diary* for April 27th "*Went to Moor Park to see Charles Darwin, who is at Dr Lanes Hydropathic Establishment – Spent day there.*" See also Letter #125, where the visit of Fox is mentioned.

125 (CCD 2270). To: The Rev W. Darwin Fox, Delamere Rectory, Northwich.

Down Bromley Kent
May 8th [1858]

My dear Fox

I send by this Post Dr Lane's Book[1], which you can keep as long as you like. I enjoyed your little visit extremely[2]; & short as it was, you won golden opinions from the members of the Family there[3]. — After you went, I thought of half-a-dozen subjects, which I sh^{d}. have much liked to have talked over with you. — I was an ungrateful dog not to have thanked you for all your several letters on Turkey chicks &c &c: and this, moreover, was the most serious kind of ingratitude; for it was ingratitude for favours to come; for I shall be very curious to hear about the Kite & Turkey-chicks *by// themselves*[4]. If you make the experiment, <when the Hen is present,> pray describe how far off the chicks run <from her> before they squat.

I got splendidly well the few last days at Moor Park, & walked one day two miles out & back[5]; & I am now hard at work again as usual. —

Whenever you write, do tell me, whether you got// your Hoes. Farewell my good & dear Friend

Yours most truly
C. Darwin

Mem. if ever you see a Dun, even with only dorsal stripe, enquire colour of dam & sire: I have not been able even in *one* case to ascertain how Duns appear[6]. —

Postmark: 8 May 1858 London 8
Endorsement "May ~~9~~ 8 1858"
Christ's College Library, Cambridge, #113

Notes

1. Dr Edward Wickstead Lane (1823-1889). Owner of the hydropathic establishment at Moor Park [for the current and previous ownership of Moor Park, see Letter #117 (n1)].
2. See Letter #124 (n9).
3. Fox visited CD at Moor Park on 27th April 1858 [see letter #124 (n9)]. It is not recorded who these family members were, but from Emma's *Diary* and from Letter #134 it seems clear that CD was referring to members of the Dr Lane's family, including Dr Lane's mother-in-law.
4. See Letters #121, #122 & #123.
5. In his diary of 1858, Fox recorded for 27th April: "*Went to Moor Park to see Charles Darwin, who is at Dr Lanes Hydropathic Establishment — Spent day there*." Although CD does not refer to Fox's visit here, he does in Letter #131, Note 11. In DAR 205.11 (2): 108, (University Library, Cambridge), there is a note, giving Fox as the source, dated "*May 5/58*" describing the nest-building instincts of the golden wren (gold-crest).
6. See Letter #124, Note 7.

125A (CCD7815). To: Charles Darwin, Down House, Bromley, Kent

Delamere – Northwich
June 11 [1858][1]

My Dear Darwin

I have never thanked you for the loan of Dr Lanes little Book which I liked much, & will return soon[2]. I am so glad you have a plan of retreat to flee to when you get unbearably unwell, but I think you a very bad Husband & Father to let yourself get into such a state.

I have two little lots of Turkeys ready to try your experiment[3]. They have been kept in a room & never seen any thing to alarm them. I hope to tell you the result of the trial at the end of this[4]. By the way of a little example upon some subject. I hatched some call Ducks under a hen & they were not wilder than other young Ducks are.

Soon after a *wild* Duck hatched some eggs (*of the same Parents*) and I never saw any little wretches so wild from the moment they were out of shell//
They were just as real wild ducks young are — as wild as possible — & this before they ever were out of the nest, which was in the common Hen house[5]. —
Now this must have been caused by the Wild Ducks notes of alarm stimulating their organ of caution. The eggs were the same in both instances.

I have this week been paying a visit to Worsley at Platt[6]. He nursed one of Sir F Darwins[7] daughters, & I there met Mrs Wilmot, Violetta, & Millicent[6]. Did you ever have the history of our old friend Tiger (at Sydnope when we were there[8]) getting loose & biting off a great part of a Sows cheek? — she was a pig, & all her little ones were deficient were deficient in one cheek. Mrs Wilmot said she remembered the fact perfectly — & when the pigs were grown up & salted, she could remember Sir Francis showing the deficient cheek at Table[9]

Across top of p. 1: I shall probably be in town next week. If I can get away for a night I will write you word & offer myself possibly about 5 or 6th.[10]

Ever yours W D Fox

Incomplete, 2nd sheet missing
Provenance: University Library, Cambridge: Manuscript Room: DAR 164: 194

Notes

1. The year of this letter is determined by relationship to Dr Lane's book, see note 2, and by the record in Fox's *Diary* of a visit to the Worsleys on 7th June 1858 (see Note 6). Fox had visited CD at Dr Lane's hydrotherapy establishment at Moore Park (see note 2) on the 27th April 1858 (see Appendix 2).
2. See Letter #125 (n1). CD had posted Fox Dr Lane's book on hydrotherapy. Dr Lane was the proprietor of a hydrotherapy establishment at Moore Park, which CD attended regularly in the late 1850s [see Letters #112 (n5), #114 (n4), #118 (n3&4), #125 (n1), #134 (n4) & #135 (n10)].
3. In *Natural Selection*, the big book that CD was writing (the greater part of which was published by Stauffer, 1977) can be found the footnote, on p 488 (of the Stauffer book): "*I may add that the Rev. W. D. Fox saw a brood of his Turkeys with their mother in agonies of horror at a frog peeping out of a hole: as Mr. Fox remarked their instinct probably mislead them to mistake the bright eyes of the frog for those of a deadly N. American snake.*"

4. CD was in the habit of excising portions of letters with material for a book, and putting the scrap into a special folder. This is probably the reason that the last part of this letter is missing – and it probably contained the account of the young turkeys and the frog.
5. CD referred to Fox on p. 13 of *Variation of Animals and Plants under Domestication* (1868), Vol 2, but in relation to crossing of slate-coloured with wild musk ducks. In relation to the wildness of common ducks crossed with wild ducks (call ducks?) CD refers on p 20 of Vol 2 (*ibid.)* to information from Mr Waterton; see Letters #62 (n5), #96 (n1), #97 (n4), #104, #105, #111 (n4) & #127 (n4) for further correspondence on call ducks.
6. In his diary for 1858 Fox wrote on the 7th June: "*Ellen and I to Platt to the Worsleys*." Platt House near Manchester had been the family home of the Worsley family for many generations and Major General Charles Worsley was buried in Westminster Abbey in 1656. The fourth child of Erasmus Darwin by his second wife, Elizabeth Pole née Collier, married Charles Carill Worsley of Platt Hall (?-1864). Thus Fox was visiting three daughters of Erasmus Darwin, by his second wife, and a granddaughter; Violetta Harriot Darwin and Millicent Susan Oldershaw née Darwin were the other daughters of Erasmus Darwin who were present; Mrs Wilmot, was the granddaughter, being Emma Elizabeth Darwin ((1812-?), the daughter of Sir Francis Sacheverel Darwin: she married Edward Woollett Wilmot (1808-1864), the sixth son of Sir Robert Wilmot (*Darwin Pedigrees*). Sir Robert John Wilmot, who died in 1841, had been the landlord of Osmaston Hall, the home of Samuel Fox, III and family [see Letters #49 (n2). #54 (n7)]. So Fox had many family connections to the people he was visiting.
7. Francis Sacheverel Darwin (1786-1859) was CD's half-uncle, being the son of Erasmus Darwin by his second wife Elizabeth Pole née Collier. He was also Fox's great uncle, since they were both descended from William Alvey Darwin of Elston (1726-1783), Erasmus Darwin's father.
8. Sydnope Hall and CD's and Fox's visit there is detailed in Letter # 121 (n12) [see also Letters #29 (n10), #137 (n5 & 6)]; while Sir Francis Sacheverel Darwin had lived there in the 1820s and 1830s, he resided at Breadsall Priory, after the death of his mother Elizabeth Darwin née Collier in 1832 (*Darwin Pedigrees*).
9. This seems a far-fetched tale. If true it would indicate a Lamarckian factor in inheritance, to which CD, at this time, was steadfastly opposed.
10. No record of any meeting has been found (see Appendix 2).

126 (CCD 2293). To: The Rev W. Darwin Fox, Delamere Rectory, Northwich.

Down Bromley Kent
June 24th [1858]

My dear Fox

Will you read the enclosed & return it me soon — All those, whom I have asked think that Dr. L. is probably innocent[1]. — Mr Thom's (a very sensible nice young man) evidence[2]; the admitted coldness of Dr. Lane's letters[3], — the absence of all corroborative evidence — & more than all the unparalleled fact of a woman detailing her//

own adultery, which seems to me more improbable than inventing a story prompted by extreme sensuality or hallucination, — altogether make me think Dr. Lane innocent & that it is a most cruel case. — I fear it will ruin him. I never heard a sensual expression from him[4]. — I am writing this under//

much hurry, (but I will not miss a post) as poor dear Etty has been most seriously ill with an attack very like Diptheria; but thank God after much suffering is recovering; but last night our Baby commenced with Fever of some kind[5].

Yours affectionately
C. Darwin

Endorsement: June ~~25~~ 24 1858
Postmark: JU 24 58 London 8
Christ's College Library, Cambridge #114

Notes

1. CD had probably sent Fox the long reports in *The Times* (15 June 1858, pp 11–12; 16 June 1858, p 11; 17 June 1858, p 11; 22 June 1858, p 11) on the sensational case where Edward Wickstead Lane, the proprietor of Moor Park hydropathic establishment, had been cited as co-respondent in a divorce case brought by Mr Henry Robinson of Reading against his wife. Mrs Robinson kept a diary, which detailed her adultery with Lane during a visit to Moor Park in 1845, which was presented as evidence. The judge's summing up was reported, along with some sensational extracts from the diary, in *The Times* on 22nd June 1858. The case turned on the possibility that Mrs Robinson had imagined the whole affair. The suit was eventually withdrawn (*The Times*, 5 July 1858, p 11).
2. Mrs Robinson claimed in her diary to have also had an affair with a Mr Thom at Moor Park. Mr Thom had been called as a witness in the trial and denied any affair with Mrs Robinson (*The Times*, 17 June 1858, p 11).
3. Dr. Lane had corresponded with Mrs Robinson, but strictly on health matters, after the alleged adultery.
4. CD and family visited Moor Park again in 1859 [see Letter #135 (n4)], but in the lead up to and after the publication of *On the Origin of Species* CD's health deteriorated and the Moor Park treatment did not seem to have effect: he then tried spas at Ilkley Wells, Yorks (see Letters #136 & 7). At the same time, in 1860, Lane moved his establishment to Sudbrook Park, Petersham, near Ham, Surrey (Letter #133 (n6); Freeman, 1978).
5. Emma Darwin's *Diary* records for 18th June 1858 "*Etty taken ill*" and for 23rd June 1858 "*Baby taken ill*". Etty was Henrietta Emma Darwin. The baby was Charles Waring Darwin, who died of scarlet fever on 28th June 1858 (see Letter #128): CD's memorial is given in CCD, Vol 7, Appendix V, p 521.

127 (CCD 2296). To: The Rev W. Darwin Fox, Delamere Rectory, Northwich.

Down Bromley Kent
27th [June 1858][1]

My dear Fox

I am extremely glad to hear the view you take of Dr. Lane's case[2]. What extraordinary facts you tell me. The soul of some great physician has transmigrated into you. I am profoundly sorry for Dr. L. & all his family, to whom I am much attached[3]. — We shall, indeed, be delighted// to see you here if you can anyhow come. —

Your fact about Call-Ducks is *first-rate* for me, & I will quote it; as I particularly wanted such cases of influence of parent, independently of instinct[4].
The Sow-case would have been valuable, had it been more recent, so that I cd have ascertained, that the same cheek was affected in young, & had known how// many young pigs had same deficiency. I have generally been inclined to account for the several similar reported cases by coincidence & inaccuracy, or from *disease* of bone having been set up. As the sow was actually pregnant such case does utterly stagger me[5]. —

I had heard something of the Leicester sheep & am very glad to have more details: my doubt is, whether// in all kinds of sheep black are not sometimes dropped[6]. I thank you much for all the very kind trouble, which you have taken to get me information on all the above points; and about Horses. I have lately seen some splendid cases of barred legs; but I never can find out about colour of parents. — I hardly know what roan is. — I shall be very glad to hear// about young Turkeys, if you succeed; but in 3 out of 4 of my experiments, something, which one had not calculated on, interferes with the result[7]. —

I have lately been observing & experimentising with much care on the construction of Bees' cells & have been testing the accuracy of Huber's observation & on some points I do not think the blind man's observations stand the test very well[8]. — I think I have got theory, which greatly// simplifies the marvellous power of construction of all the wondrous angles & perfect cell[9]. —

You will be sorry to hear that we have had Etty most seriously ill with a modified form of that horrid new complaint, Diptheria; but all danger is over & she is slowly recovering[10]. We have the Baby, also, very ill with fever[11], but the Doctor declares not dangerously; We have been// much terrified as Scarlet Fever has been very bad. Our nurse, too, has sickened[11], so we have had much trouble, but I hope things are now clearing. —

Yours affectionately
C. Darwin

(*on a separate page*)
Since this written our Baby has become suddenly most ill. — it is Scarlet Fever, & the Doctor can only say there is yet some hope[12].
Endorsement: June ~~25~~ 27 1858
Postmark JU 28 1858 London 8
Christ's College Library, Cambridge #115

Notes

1. Dated by endorsement and reference to Dr Lane's case and illness of CD's baby.
2. See Letter #126.
3. CD had always liked Edward Wickstead Lane (see Letters #117, #118 and #126). He maintained a good opinion of him throughout his life. CD continued his visits to Moor Park in 1859 and early1860 but visited Sudbrook Park (see below) only once [see Letter #156 (n5) – see also Letter #133 (n6)]. From1860–79, Dr Lane moved his establishment to Sudbrook Park, near Ham, Surrey. Lane also practised in Harley Street, London, 1879–89. Lane's name was on the list of 'personal friends' for invitations to CD's funeral in 1882.
4. See Letters #120 and #121. There is no identifiable reference to this "fact" in any of CD's writings. CD finished writing Chapter 10 (*Mental Powers and Instincts of Animals*) of *Natural Selection* on 9th March 1858 (Stauffer, 1975 and CD's *Journal*). He may have made notes on this subject, which have been lost, or he may have been distracted at this time by a) the death of his son Charles Waring Darwin (one day later, on 28th June 1858) and b) the preparation and reading of his and Wallace's papers on organic evolution for the Linnean Society on 1st July 1858. (Note that CD does not relate the receipt of Wallace's letter and enclosed paper (*On the Tendency of Varieties to depart indefinitely from the Original Type*) to Fox until Sept 1858 [Letter #130 (n6)]. The official date for receipt of this letter is 18th June 1856, in a letter to Charles Lyell, CCD, Vol. 7; for a discussion of when Wallace's (missing) letter may have arrived, see Brooks, 1984.
5. CD apparently did not use this fact, which appears to suggest that an injury to the cheek of a pregnant sow that was passed on to the progeny.
6. CD used this fact in *The Variation of Animals and Plants under Domestication, 1868, Vol.* 2, p 3: "*Since the time of the famous Bakewell, during the last century, the Leicester sheep have been bred with the most scrupulous care; yet occasionally grey-faced, or black-spotted, or wholly black lambs appear (Footnote: "I have been informed of the fact through the Rev W. D. Fox, on the excellent authority of Mr Wilmot*").
7. See Letters #122 and #123.

8. François Huber (1750-1831). Swiss entomologist, who had lost his sight at fifteen. Author of *Nouvelles observations sur les abeiles*; Paris, 2nd edition, 1814.
9. In Chapter 10 (*Mental Powers and Instincts of Animals*) of *Natural Selection* (Stauffer, 1975) there is a section on "*Bees Comb*" (pp 513-517) with many observations on different types of combs and quotations from Huber (pp 513-516). There is no clear indication of what new hypothesis CD is referring to in this letter. Stauffer (1975) added the footnote: "*Darwin later added in pencil: 'Theory of the construction of the cells of Hive Bees. – Waterton has given theory very like that above hinted at*'. In the Darwin Manuscript Collection (DAR), University Library, Cambridge, there is much material about bees. In the section of vol 46.2 entitled "*Habits of Bees*", are many note scraps made by Darwin from July 1840 at Maer to Sept. 1862 at Bournemouth. At the end of vol. 48 is a collection of notes, diagrams and drafts regarding construction of bees' cells. In the relevant section in *Natural Selection*, CD refers to Pierre Huber. He was the son of Francois Huber, and the latter published independently on bees and ants (*Recherches sur les mœurs des formis indigenes*; Paris, 1810) and edited an appendix to the 2nd edition of his father's book.
10. Emma Darwin's *Diary* records "*Etty taken ill*" on 18 June 1858. This was Henrietta Emma Darwin.
11. Emma Darwin's *Diary* records "*Baby taken ill*" for 23rd June 1858 and "*Jane sore throat*" for 26th June 1858.
12. Emma Darwin's records the death of Charles Waring Darwin on the 28th June 1858. CD's memorial of Charles Waring Darwin is set out in CCD, Vol 7, Appendix V, p 521.

128 (CCD 2300). To: The Rev W. Darwin Fox, Delamere Rectory, Northwich.

[Down]
July 2^{d} [1858]

My dear Fox

Our poor Baby died on 28th at night[1]. — One of his nurses has caught Fever. We had resolved not to move the others, after consultation <& taking all sorts of precautions>; but your letter has been turning point & this very day they all go to Sussex to their Aunt, who is here & takes them[2]. We stay till Etty//

can move & I of course stay till nurse is out of all danger whatever[3]. You may believe we are terribly anxious, but fear has almost driven away grief. — What a miserable fortnight we have had. — Many thanks my dear Fox for your most kind & affectionate letter

Yours
C. Darwin

We have *all* had slight sore-throats, but I often have, & my wife thought it might be in some degree fancy in all. —

Postmark: illegible
Endorsed: July ~~3~~ 2 1858
Christ's College Library, #116

Notes

1. Charles Waring Darwin died of scarlet fever on 28 June 1858 (Emma Darwin's *Diary*; '*Journal*'; CCD, Vol 7, Appendix II). CD's memorial of Charles Waring Darwin is set out in CCD, Vol. 7, Appendix V, p 521.
2. Fox, had experienced the trauma of death, from scarlet fever, of his daughter, Louisa Mary Fox and two young relatives in 1853 (see Letters #93 and #94). It appears that Fox had advised CD to move his children away as soon as possible [see Letter #130 (n5)]. Sarah Elizabeth Wedgwood, Emma Darwin's

sister, arrived at Down House on 25th June 1858 and Emma records on 2nd July 1858 "*children and Eliz went to Hartfield*", the home of Elizabeth at Hartfield, Sussex (Emma Darwin's *Diary*).

3. CD and Emma remained at Down House until 9th July and then joined their children and Elizabeth at Hartfield (Emma Darwin's *Diary*).

129 (CCD 2304). To: The Rev W. Darwin Fox, Delamere Rectory, Northwich.

Down Bromley Kent
July 6th [1858][1]

My dear Fox

I write one line to thank you for your second most kind letter. All is going on well. The nurse has had it pretty severely, but the crisis is over. No one else// has taken it & I hope now will not, but we had a fear about our Governess, but she is out of House[2], & we are getting less frightened & in every way, more composed, I have been much knocked up & so has my poor dear wife// but we are now much better.

Etty is too weak to move yet: she has not even put on her clothes, but our Dr. is strong for her moving as soon as ever she can. — We go first to Elizabeth Wedgwoods & thence to the sea; but where is our puzzle[3]. We should// much like S. side of the Isle of Wight; but it is indispensable <on account of Etty> that the House shd be very near the sea, & I fear such does not occur on account of cliffs. — Shd you know anything on this head, will you let me have a line; but do not otherwise trouble yourself to write[3].

My dear Fox, with cordial thanks for all your affection.

Ever yours
C. Darwin

Provenance: Woodward Library, University of British Columbia
This letter has the label "117" in the top right hand corner of the first sheet. Letter #117 is missing from the Christ's College Library collection and it seems clear that the label "117" was placed on this letter by Francis Darwin[4].
On the side of the first page, vertically: "Part of a letter of C. Darwins Came with Parcel from Isle of Wight Gerard Fox -"

(*First page with a black surround*)

Notes

1. The events described here clearly relate to those described in Letter #128 and the death of Charles Waring Darwin on 28th June 1858.
2. The governess Miss Pugh, employed since April 1857, apparently took fright. Emma Darwin's *Diary* records that on 4th July 1858 '*Miss Pugh went*'. However, she did come back for short periods in late 1858 and early 1859 (Emma Darwin's *Diary*). Later she went insane and was confined to an asylum (Freeman, 1978).
3. As noted in Latter #128 (n3), Charles, Emma and Etty Darwin went first to Hartfield, and on the 16th July to Ryde, Isle of Wight (Emma Darwin's *Diary*; see also Letter #130), and later, on the 27th July, to Shanklin. Since Fox had recuperated and met his first wife, Harriet, at Ryde (see, e g, Letters #49, #51, #58, #59, #61 and #62) it is likely that he had recommended this resort. Fox later retired to Sandown, IOW.

4. If this conclusion is correct it suggests that all the letters were returned to Charles W Fox, Broadlands, Sandown, Isle of Wight, by Francis Darwin, and were divided up after that (see Appendix 1).

130 (CCD 2304). To: The Rev W. Darwin Fox, Delamere Rectory, Northwich.

King's Head Hotel — Sandown — I. of W.
21st [July 1858][1]

My dear Fox

As you said you would like to hear how we were going on, I write to say that we are all very fairly well, & after some house-hunting are settled here till next Tuesday when we go to Norfolk House Shanklin, where there is a single row of Houses on the Beach[2]. This place has evidently sprung up like a mushroom; & there are three hotels & many villas. Years ago you took me across the isld & I have a great notion that the solitary sandy bay which I then saw must this be place[3]. — It suits us very fairly & the little Boys are very happy.

We have just heard of my sisters Marianne Parker's death, — a blessed relief after long continued & latterly very severe suffering[4]. She was an admirable woman, & I thank God is at rest. We are all here, but my eldest son starts in a week's time abroad, for a little tour before going to old Christ College[5]. —

After all, I am now beginning to prepare an abstract of my Species Theory. By an odd coincidence, Mr Wallace in the Malay Archipelago sent to me an Essay containing my exact theory; & asking me to show it to Lyell[6]. The latter & Hooker have taken on themselves to publish it in Linnean Journal, together some notes of mine written very many years ago[7]; & both of them have urged me so strongly to publish a fuller abstract, that I have resolved to do it, & shall do nothing till completed: it will be published, probably, in Journal of Linn. Socy[8] & I shall have separate copies & will send you one. — It is impossible in abstract to do justice to subject. —

There has been another child die in village of Down; which makes the fifth; so we rejoice we acted on your advice & left home[9]. — We shall stay here to middle of August. — With very sincere thanks for all your most kind sympathy.

My dear Fox
Yours most truly
C. Darwin

Dr. Lane has his house full, I am glad to say[10].

Endorsement: July ~~24~~ 21 1858
Postmark: JY 22 58 Southampton H
Christ's College Library, Cambridge #118

Notes

1. Dated by relationship to Letter #129 and the place of writing (cf. letter to J. D. Hooker CCD, Vol. 7, 21st July 1858).
2. In Letter #129, CD specifies that the family need a place by the sea for Etty's health.
3. See CCD Vol 7, this Letter (2304), Note 1, for a description of Sandown in earlier times, when there was no development. Fox had stayed at Sandown in 1833-37 (see Fox *Diaries* and Letters

#58-#62) so it must have had some guest houses at that stage. CD visited Fox there on 21st Nov 1837 [see Letter #58 (n11)] and clearly, although a long time ago, it had changed a great deal since then.
4. CD's eldest sister, Marianne Parker (1798-1858), died on 18th July. She had been married, since 1824, to Henry Parker, MD, of the Shropshire Infirmary in Shrewsbury; he died in 1856. They left fours sons and a daughter, who, as grown-up children, were given special attention by Susan Darwin, CD's sister, until her death in 1866 (*Litchfield*, 1915, Vol 2, p 184, *Darwin Pedigrees*).
5. William Erasmus Darwin (1839–1914) was to begin his studies, in October, 1858, at Christ's College, where both CD and Fox had been undergraduates (see Chapters 2 and 3). In the meanwhile he went abroad from 29th July to 4th Sept. (Emma Darwin's *Diary*).
6. The letter from Alfred Russell Wallace (1823-1913) has not been found. The receipt of this letter is usually assumed to be about the 18th June 1858 (see letter by CD to Lyell, CCD, Vol 7, 18th [June 1858]; LLii, p 427, but see Letter #127 (n4)].
7. Note that this statement seems to suggest that CD had not shared his ideas or his previous abstract with Fox, despite reference to an "Essay" in Letter #111 (n5).
8. The papers of A R Wallace (*On the Tendency of Varieties to depart indefinitely from the Original Type*) and C. Darwin (*Abstract of a Letter from C. Darwin, Esq., to Asa Gray, Boston U.S., dated Down, September 5th 1857*) were read at the meeting of the Linnean Society on 1st July 1858 and published back-to-back, with CD's articles first, in 1859, in the *Journal of the Proceedings of the Linnean Society (Zoology)* 3, 45-62.
9. See Letter #128 (n2).
10. CD must have feared that the scandal arising from the divorce proceedings of a patient, in which Lane was named as a co-respondent, would damage the operation of Moor Park. In fact, Lane had been planning to move for some time [see Letter #118 (n4)] and did so in 1860, moving to Sudbrook Park, Petersham, near Ham, Surrey (Freeman, 1978).

131 (CCD 2360; LLii, part[1]). To: The Rev W. Darwin Fox, Delamere Rectory, Northwich.

Down Bromley Kent
Nov. 13th [1858][2]

My dear Fox

Hooker & Henslow both applied to me to suggest as many names of persons as possible to whom to send the circular about Mr Ralfs[3]; & I gave yours amongst many others[4]. From all that I have heard from Hooker I believe the case to be one deserving of aid; but with the many claims which you must have on// you, I, if in your place, would not give without perfectly convenient. — I have not much news to tell you about myself; I am working slowly & steadily at my Abstract & making progress & hope to print in the Spring. My stomach has been bad enough, & I have lately spent a very pleasant week at Moor Park, & Hydropathy & idleness did me// wonderful good[5] & I walked one day 4 1/2 miles, — a quite Herculean feat for me!

William, my son, is now at Christ Coll. in the rooms above yours[6]. My old Gyp. Impey[7] was astounded to hear that he was my son & very simply asked "why has he been long married?"[8] What pleasant hours, those were when I used to come & drink coffee with you daily! I am remin//ded of old days by my third Boy[9] having just begun collecting Beetles, & he caught the other day Brachinus crepitans[10] of immortal Whittlesea-mere memory[11]. — My blood boiled with old ardour, when he caught a Licinus, — a prize unknown to me[12].

You say nothing about yourself, so I hope that you are pretty well.//

I had heard of your Mother's severe illness[13]. My children are tolerably well for them, & this as much as I can ever say.

Farewell my dear Fox. I often think what a good man you were to come down & see me at Moor Park[14]. Farewell//

Yours most sincerely
C. Darwin

Henslow is coming here for a couple of nights on the 25th[15] —
If you have ever opportunity do not forget about parentage of Dun Horses. Did you ever see Donkey with *double* shoulder stripe? I would give some guineas to see one. — Look at any Donkey which you may meet. —//

Also, did you ever see black~~ish~~ Grey-hound <(of any sub-breed)> with tan feet, *& a tan-coloured spot over inner corner of each eye*; I want such case, & such *must* exist *because* theory tells me it ought![16]
Adios my dear old friend

Endorsement: Nov 15. 1858
Christ's College Library, Cambridge #119

Notes

1. From "William, my son, is now at Christ Coll. in the rooms above yours." to "My blood boiled with old ardour, when he caught a Licinus, — a prize unknown to me."
2. Dated by reference to Moor Park (Emma Darwin's *Diary*; Oct 24 "*Ch. went to M. Park*") and to William Darwin's rooms at Christ's College (see letter of CD to W E Darwin, CCD, Vol 7, 15th [Oct 1858]).
3. John Ralfs (1807-1890). Surgeon and botanist who lived at Penzance. A subscription was being raised. He supplied CD with specimens of *Pinguicola* for *Insectivorous plants,* 1875. A genus of the brown algae, *Ralfsia*, is named after him (*DNB*). See Letter of CD to J D Hooker, 2 Nov [1858], CCD, Vol 7.
4. J D Hooker and T H Huxley had, through the Philosophical Club of the Royal Society, organised a fund to aid the botanist John Ralfs, whose health had failed. See Note 1, Letter to J D Hooker, 2 November [1858], CCD, Vol 7.
5. CD stayed at Moor Park hydropathic establishment from 25th Oct. to 1st Nov. 1858 (CD's '*Journal*' and Emma Darwin's *Diary*).
6. William Erasmus Darwin (1839-1914). Attended Christ's College, Cambridge, from 1858-1861. The rooms of Fox at Christ's have been identified as Second Court, middle staircase, second floor (E7)[see Letter #8 (n2)], so W E Darwin's rooms would have been on the third floor.
7. See Letter #8 (n1).
8. CD's college servant or gyp at Christ's College, Cambridge (1828-1831) [see Letter #19 (n4)].
9. Francis Darwin (1848-1925). Third son of CD. Botanist and biographer of CD. (*DNB*)
10. The bombadier beetle. This beetle makes the sound of breaking wind when disturbed and fires a noxious fluid in the direction of the disturber. The action is brought about by the combination of two fluids, hydroquinone and hydrogen peroxide, from separate sacs in the body, producing *p*-benzoquinone. Ironically, this is a prime example, of those who support "Intelligent Design", of an organism that could not have evolved into its current body form.
11. Whittlesea (Whittlesey) Meer, near Peterborough; see Letter #47 (n5 & 6); the excursion to Whittlesea Meer with CD, Fox and Albert Way is recalled many times in subsequent letters [see Letters #48

(n17), #61 (n18), #90 (n17), #96 (n12), #115 (n7), #148 (n7)] and in CD's *Autobiography*. However, the insect previously recalled is *Panagæus crux-major*.

12. Licinus occurs as a genus on p 14 of Stephens' *Catalogue Br Insects*, Vol 1; however, CD's copy has no marginalia against this genus. Francis Darwin, who was ten years old at this time, relates in LLii (p 140): "*About this time my father revived his old knowledge of beetles in helping his boys in their collecting. He sent a short notice to the 'Entomologist's Weekly Intelligence', June 25th, 1859, recording the capture of Licinus silphoides, Clytus mysticus, Panagæus 4-pustulatus. The notice begins with the words. "We three very young collectors having lately taken in the parish of Down," &c, and is signed by his three boys, but is clearly not written by them. I have a vivid recollection of the pleasure of turning out my bottle of dead beetles for my father to name, and the excitement, in which he fully shared, when any of them proved to be uncommon ones.*' In his *Autobiography* CD also wrote: "*I had never seen in those old day Licinus alive, which to the uneducated eye hardly differs from many of the black Carabidous beetles: but my sons found here a specimen, and I instantly recognised that it was new to me; yet I had not looked at a British beetle for the last twenty years.*"
13. Fox's mother, Ann Fox, née Darwin (1799-1859) was suffering from skin cancer. She died on 7th April 1859 (*Darwin Pedigrees*). No diary for Fox has been found for 1859 or 1860 (see Appendix 1).
14. Fox visited CD on the 27th April 1858; see Letter #125 (n5).
15. J S Henslow (1797-1861). Teacher and friend of both CD and Fox during and after their Cambridge years. See Bibliography and Introduction to Chapter 2. Henslow is mentioned frequently in previous correspondence; see Letters #8 (n3), #25 (n8), #29 (n12), #30 (n11), #35 (n2), #36 (n1) and many subsequent letters.
16. In *The Variation of Animals and Plants under Domestication* (1868), Vol 1, p 28/29, CD stated: "*One fact, however, with respect to the colouring of domestic dogs, I at one time hoped might have thrown some light on their origin; and it is worth giving, as showing how colouring follows laws, even in so anciently and thoroughly domesticated an animal as the dog. Black dogs with tan-coloured feet, whatever breed they may belong to, almost invariably have a tan-coloured spot on the upper and inner corners of each eye, and their lips are generally thus coloured.*"

132 (CCD 2412) To: The Rev W. Darwin Fox, Delamere Rectory, Northwich[1].

Moor Park, Farnham
Surrey
Saturday [12th Feb 1859][2]

My dear Fox

It is long since we have had any communication, so I am tempted to send you a scrap to extract a scrap from you, with news about yourself. — I have been extra bad of late, with the old severe vomiting rather often & much distressing swimming of the head; I have been here a week & shall stay another & it has already done me good[3]. I am taking Pepsine, ie the chief element of the gastric juice, & I think it does me good & at first was charmed with it. My abstract is the cause, I believe of the main part of the ills to which my flesh is heir to; but I have only two more chapters & to correct all[4], & then I shall be a comparatively free man. — I have had the great satisfaction of converting Hooker & I believe Huxley & I think Lyell[5] is much staggered.

We are a very pleasant party here & are very comfortable & I am glad to say that not one of Dr Lane's patients has given him up & he gets a few fresh ones pretty regularly[6]. He is most eager to build an Establishment near the Crystal Palace; but I fear will fail for want of Funds[7]. — All my home party are with the Langtons[8] & they went there on account of Etty, who has lately fallen back grievously owing to a very slight attack of fever[9]. — William is very happy at Cambridge & he has changed into my old rooms & has taken my old engravings & with old Impey[10], it must be a sort of resurrection. Let me some time have a note telling me about yourself & belongings

My dear Fox
Ever yours most truly
C. Darwin

Christ's College Library, Cambridge #106

Notes

1. The envelope is missing so that the address is assumed to be Delamere.
2. Dated by CD's visit to Moor Park hydropathic establishment (see Note 3, below). The only Saturday available to CD to write a letter during the visit 12th February. Also by reference to William at Christ's College and Impey [see Letter #131 (n5 & 6)].
3. CD stayed at Moor Park from 5th to 19th February 1859 (CD's '*Journal*'; Emma Darwin's *Diary*).
4. In his "*Journal*" CD records for 19th March: "*Began relooking over first MS. Chs. & finished last chapter.*" This confirms the very rapid writing of the draft for *On the origin of species* begun during the visit to Sandown from 20th July–12th Aug,1858.
5. J D Hooker (1817-1911), T H Huxley (1825-1895) and Charles Lyell (1797-1875), CD's close friends and strong supporters for publishing his ideas on organic evolution.
6. CD refers to the allegation of adultery against Edward Wickstead Lane in 1858 (see Letters #126 & #127).
7. In the event, Lane took over Sudbrook Park. Petersham, near Ham, Surrey, in 1860 (Freeman, 1978; Metcalfe, 1906, p 56).
8. Charles Langton and Charlotte Langton (née Wedgwood), Emma Darwin's sister. They lived at this time at Hartfield Grove, Hartfield, Sussex, on the edge of Ashdown Forest, some twenty miles from Downe. Elizabeth Wedgwood lived nearby and offered much assistance at such times. Emma Darwin's *Diary* shows that Emma and Etty went to Hartfield on the 7th, followed by the rest of the children, except William and George, on the 8th, and stayed till the 18th Feb.
9. Etty, Henrietta Emma Darwin had been ill for much of January (Emma Darwin's *Diary*).
10. CD updates information given in the previous letter [Letter #131 (n5 & 6)].

133 (CCD 2412; LLii, part[1]). To: The Rev W. Darwin Fox, Delamere Rectory, Northwich.

Down Bromley Kent
24th [March 1859]

My dear Fox

It was very good of you to write to me in the midst of all your troubles, though you seem to have got over some of them in the recovery of your wife's & own health[2]. — I had not heard lately of your mother's health, & am sorry to hear so poor an account[3]. But as she does not suffer much,// that is the great thing; for mere life I do not think is much valued by the old. What a time you must have had of it, when you had to go backwards & forwards![4] —

We are all pretty well & our eldest daughter is improving. I can see daylight through my work & am now finally correcting my chapters for press; & I// hope in month or six weeks to have proof-sheets[5]. I am weary of my work. It is a very odd thing that I have no sensation that I overwork my brain; but facts compel me to conclude that my Brain was never formed for much thinking. — We are resolved to go for 2 or 3 months, when I have finished to Ilkley or some such// place[6], to see if I can anyhow give my health a good start, for it certainly has been wretched of late, & has incapacitated me for everything. You do me injustice when <you think> that I work for fame: I value it to a certain extent; but, if I know myself, I work from a sort of instinct to try to make out truth[7].//

How glad I should be if you could sometime come to Down; especially when I get a little better, as I still hope to be. — We have set up a Billiard Table[8], & I find it does me a deal of good, & drives the horrid species out of my head. —//

Farewell my dear old friend. —
Yours affecty.
C. Darwin

One of my Boys is turned Coleopterist[9], & how the sight of a fresh Brachinus crepitans[10] did remind me of that famous trip to Whittlesea meer[11]. —
Pray give my very kind remembrances to your sisters. —
I most sincerely hope that Mrs Fox's sufferings will not increase.

Christ's College Library, Cambridge #120

Notes

1. Life and Letters, Vol 2, p 150/151 gives the complete letter with the exception of the postscript.
2. Fox's *Diaries* for 1859 and 1860 have not been found, so the background to this statement and others in this letter unfortunately, cannot be clarified.
3. According to Fox's *Diary* for 1858 (for 1859, see note 2) Ann Fox née Darwin, his mother, had suffered from skin cancer for over a year.
4. Fox's mother and unmarried sisters lived at Kensington Park Gardens East (the address for this letter). Based on his movements for 1858, it is likely that Fox was visiting Kensington from his home in Cheshire, 200 miles away, at least every two months.
5. Proof sheets for *On the Origin of Species*, which was published on the 24th Nov 1859 (CD's *Journal*).
6. Dr Lane moved to Sudbrook Park at this time and this fact together with the impression that his visits to Moor Park were doing him less good (letter to William Darwin, 13th Feb 1859, CCD, Vol 7) induced CD to look elsewhere for hydrotherapy. Ilkley Wells House in Yorkshire was a hydropathic establishment run by Dr Edmund Smith from Sheffield (Metcalfe 1906, p 107). CD went to Ilkley in October 1859 (CD's *Journal*).
7. Evidently Fox had teased CD over his desire to publish and for fame. In a previous letter [Letter #112 (n12)] CD had stated: "*I am got most deeply interested in my subject; though I wish I could set less value on the bauble fame, either present or posthumous, than I do, but not, I think, to any extreme degree; yet, if I know myself, I would work just as hard, though with less gusto, if I knew that my Book w^d be published for ever anonymously.*"
8. In early 1859, CD had bought a billiard table (see letters to George Darwin, 23rd Feb. and William Darwin, 14th March, CCD, Vol 7).
9. Francis (Frank) Darwin, aged ten years old, at this time, and was 'mad over beetles' (see letter to W E Darwin, 22nd September 1858, CCD, Vol 7).
10. In CD's copy of Stephens' *Catalogue Br Insects* there is the annotation "*Huntingdonshire, Cam: Spring 28.*"
11. CD repeats a subject in the letter of 13th Nov 1858 [see Letter #131 (n8 & n9)].

134 (CCD 2451). To: The Rev W. Darwin Fox, 3 Kensington Park Gardens E Notting Hill London

Down Bromley Kent
Ap. 10th[1859][1]

My dear Fox

I anticipated from your former letter that your mother could not possibly survive long[2]. You must yourself have been quite prepared; but when the blow comes it is always shocking. Pray give my kindest remembrances & sincere sympathy to your sisters. What a happy time I once spent at Osmaston[3]. How all things are changed for us in this world. — I am extremely glad that you going to attend to your own health; for I am sure it must want care, after all you have gone through. Dr Lane's//

address is "Moor Park, Farnham, Surrey"[4]. He has two tarifs, 4 guineas per week for the better & larger bed-rooms & 3 guineas for the smaller, but I believe these are good. There is Bathman to add, & 5s per week if fire in Bedroom[5]. — I **think** if anyone goes as friend it is only 2 guineas per week — I do not think I have//

ever heard charge for man & wife. — Had you not better write to Dr Lane; I would under similar circumstances openly. — I am sure they would be very glad to have you: for they were much taken with you[6]. In my opinion it is a very nice place. — It will be splendid if you come to see us at Ilkley[7].

Believe me, My dear Fox

Your old & affectionate friend
Ch. Darwin

Christ's College Library, Cambridge #121

Notes

1. Dated by reference to Fox's mother (note 2).
2. Ann Fox née Darwin died on 7th April 1859 (*Darwin Pedigrees*).
3. Osmaston Hall, near Derby, had been the home of Samuel Fox, III and family from 1814 to ca1850. CD spent a week there in 1828 meeting Fox's parents and four unmarried sisters (see Letter #5).
4. Edward Wickstead Lane ran the Moor Park hydropathic establishment, but was about to move to Sudbrook Park, Petersham, near Ham, Surrey (see Letter #118, Note 4).
5. For his stay of two weeks at Moor Park in February 1859, CD recorded having paid £9. 6*s*. (*Account Books*, Down House MS).
6. Fox visited CD at Moor Park on 27th April 1858 [see Letter #125 (n2 & n3)].
7. CD intended to go to Ilkley Wells House in Yorkshire, which was a hydropathic establishment run by Dr Edmund Smith [see Letter #133 (n6)].

135 (CCD 2451; LLii, part[1]). To: The Rev W. Darwin Fox, 3 Kensington Park Gardens E Notting Hill

Down Bromley Kent
Sept 23d [1859]

My dear Fox

I was very glad to get your letter a few days ago. I was wishing to hear about you, but have been in such an absorbed, slavish, overworked state, that I had not heart without compulsion to write to anyone or do anything, beyond my daily work. —

Though your account of yourself is better, I cannot think it all satisfactory, & I wish you could soon go to Malvern again[2]. —//

My Father used to believe largely in an old saying that if a man grew thinner between 50 & 60 years of age his chance of long life was poor, & that on the contrary it was a very good sign if he grew fatter[3]; so that your stoutness, I look at as very good omen. —

My health has been as bad as it well could be all this summer; & I have kept on my legs, only// by going at short intervals to Moor Park[4]; but I have been better lately, & thank Heavens, I have at last as good as done my Book, having only the index & two or three Revises to do[5]. — It will be published in 1st week in November & a copy shall be sent you. — Remember it is only an Abstract (but has cost me above 13 months to write!!) & facts & authorities are far from given in full. I shall be curious to hear what you think of it, but I am not so silly as to expect to convert you[6]. — Lyell has read about half of the volume in clean sheets & gives me very great kudos. He is wavering so much about the immutability of species; that I expect he will come round[7]. — Hooker has come round & will publish his belief soon[8] — So much for my abominable volume, which has cost me so much labour that I almost hate it. —//

On October 3d I start for Ilkley but shall take 3 days for journey![9] It is so late that we shall not take house; but I go there alone for 3 or 4 weeks; then return home for a week & go to Moor Park for 3 or 4 weeks[10], & thus I shall get a moderate spell of Hydropathy; & I intend, if I can keep to my resolution of being idle this winter. But I fear//

ennui will be as bad as a bad stomach. — Emma has suffered much from headach of late; but I hope will soon be better. — I very much fear that my eldest girl Etty will never be strong; & Lenny has been often ailing for last 2 years with intermittent (but only symptomatically so) pulse[11]. — My poor constitution like everything else is transmitted by inheritance. —
My eldest two Boys give me//

in every way much satisfaction[12]. Farewell my dear old friend. I truly hope that you will steadily get stronger. I shd much like another note before very long. Farewell

Ever yours affecty
C. Darwin

Remember if ever you can find out whether a cross between any two coloured horses produces a dun, I shd particularly wish to hear[13]. —

Christ's College Library, Cambridge, #122

Notes

1. Most of this letter is given in Life and Letters of Charles Darwin, Vol 2, pp 167-168, from the start to "But I fear ennui will be as bad as a bad stomach"
2. Fox had gone to Dr Gully's hydrotherapy establishment in Malvern in early 1857 [see Letter #116 (n3)] and also visited his mother in Malvern, twice, later in that same year. CD had gone there

previously in the late 40s but the death of his daughter Annie there in 1851 (Letters #88 & 89), seemed to be the trigger for him to change to Dr Lane's hydropathic establishment at Moor Park [Letter #112 (n5) and subsequent letters].

3. Fox's diaries for 1859 and 1860 have not been found (see Appendix 1); his weight at this time is not recorded, but it is when he was at Christ's College (see Chapter 2, note 20). Concerning Robert Waring Darwin, his father, CD records: "*He was 6 feet 2 inches in height, with broad shoulders, and very corpulent, so that he was the largest man whom I ever saw. When he last weighed himself, he was 24 stone, but afterwards increased much in weight.*" (*Autobiography* and *LL*i, pp 11–12).
4. Both CD's *Journal* and Emma Darwin's *Diary* record that he went to Moor Park on two further occasions, after his visit in February: 21st – 28th May and 19th – 26th July.
5. CD was busy cutting down the length of his draft of *On the Origin of Species* to 155,000 words and proof reading. In the latter task, he was helped by J D Hooker, T H Huxley and Charles Lyell (see correspondence in CCD, Vol 7) and by Geogina Tollett, a novelist, who was a family friend of Emma Darwin (Healey, 2001) [see Letter #77 (n8)].
6. CD anticipates that Fox will not go along with his theory (of natural selection) and he is correct [see Letter #192 (n5 & 6)].
7. As CD suspected, and as events later would confirm, Charles Lyell still had reservations about the mutability of species (e g, see Browne, 2002).
8. Joseph Dalton Hooker incorporated an evolutionary view into his essay on the Australian flora (Hooker, 1859; *On the flora of Australia, its origins, affinities, and its distribution; being an introductory essay to the Flora of Tasmania.* Vol 1, London).
9. CD's *Journal* indicates that he started for Ilkley Wells House, Otley, Yorkshire on the 2nd Oct. 1859.
10. In the event CD did not go to Moor Park and remained at Ikley Wells House until early December. Emma joined him on the 17th October and stayed, in a rented house, until the 24th November (Emma Darwin's *Diary*).
11. Emma Darwin's *Diary* shows no indication that Etty (Henrietta) Darwin or Lenny (Leonard) Darwin had health problems at this time. Etty accompanied Emma to Ilkley.
12. William, the oldest of CD's children, was at Christ's College, Cambridge [see Letter #131 (n6)] and George, the next boy, was at Clapham Grammar School.
13. CD repeats a request previously sent [see Letter #125 (n6)].

136 (CCD2451). To: The Rev W. Darwin Fox, Delamere Rectory, Northwich

Ilkley Wells House
Otley Yorkshire
Thursday
[6th Oct 1859][1]

My dear Fox

I am in the establishment & have a sitting room & bedroom[2]. — I came only on Tuesday, & your note has just reached me. — I always hate everything new & perhaps it is only this that makes me at present detest the whole place & everybody except one kind lady//

here, whom I knew at Moor Park[3]. — It would be excessively nice if you were to come here for a time. Dr Smith[4], I think, is sensible, but he is a Homœopathist!! [5] & as far as I can judge does not personally look much after patients or anything else. — There is capital steward & the House seems well managed.//

I came here intending to stay for 3 or 4 weeks, but I very much doubt whether I shall have patience. — But this mornings post has brought me note <from Emma> telling me to look out for a House, as she is greatly inclined to come here[6]. But I have not least idea whether there is <a > House]] which would suit or how I could do my water-cure out of Establishment. — So everything is utterly uncertain. I heartily wish you would come, but I darenot not advise or press it.

Most affec^{ly} yours
C. Darwin

I grieve to hear or rather to infer that you must be considerably worse: I hope that you will come here or at least somewhere[7]

Christ's College Library, Cambridge, #123
Endorsement: "Oct 6/59"

Notes

1. Dated from the day of the week, the Endorsement and CD's visit to Ilkley Wells House.
2. CD left Down for the Ilkley Wells hydropathic establishment on 2nd October 1859. He remained there until approximately 7th December (CD's *Journal*).
3. This may have been Mary Butler, whom CD met at Moor Park, then again at Ilkley Wells, becoming a friend and visiting the Darwins at Down: see letter to Mary Butler, 11th September 1859, CCD, Vol 7.
4. Edmund Smith was the proprietor of Ilkley Wells House (Metcalfe 1906, p 107).
5. CD was staunchly opposed to Homœopathy [see Letter #85 (n9), #117 (n3 and Letter to Susan Darwin referred to therein)].
6. Emma Darwin and some of the children (including Etty), rented a house at Ilkley, where CD joined them on 17th October, staying until 24th November (Emma Darwin's *Diary*). After that CD went back to Ilkley Wells House for a fortnight, leaving on approximately 7th Dec (Note 1, above); see letter to J. D. Hooker, 15th October 1859, CCD, Vol 7.
7. Fox's diary for 1859 has not been found and his ailments and movements at this time are unknown. Fox had suffered from lumbago, in the previous year [Fox *Diary*, 1858, and Letter #121 (n2)].

137 (CCD 2533). To: The Rev W. Darwin Fox, Delamere Rectory, Northwich

Wells Terrace Ilkley, Otley Yorkshire
Wednesday
{16th Nov 1859]

My dear Fox.

I daresay you would like to hear about me, & I want to hear about you. Did you go to Malvern, & how is the place in your head[1]? — I doubt whether Dr Smith[2] would have suited you: they all say he is very careful in bad illness; but he constantly gives me impression, as if he// cared very much for the Fee & very little for the patient. — I like the place very much, & the children have enjoyed it much & it has done my wife good; it did Etty[3] good at first, but she has gone back again. — I have had a series of calamities; first a sprained ancle, & then badly swollen whole leg & face; much rash & a// frightful sucession of Boils — 4 or 5 at once. I have felt quite ill — & have little faith in this "unique crisis" as the Doctor calls it, doing me much good[4]. I cannot now walk a step from bad boil on knee. We have been here above 6 week, & I feel worse than when I came; so that I am not in cheerful frame of mind. So poor old Sir Francis[5] is gone: I never saw him// but once, on our to me memorable & pleasant visit to Sydnope[6]. The Cromptons[7] are here, & they know well all of you, & are, as they say, connected with you. — Poor Mr Crompton who has just lost his wife, is here, & the old Lady who seems very nice: I have not seen the invalid daughter[8]. I find that Mr Rhodes Darwin lives about// 10 miles off, near Arthington Stn at a very nice place[9]. I sh^d^ like to call there, but shall not have strength or spirits. We shall stay about a fortnight longer here; & possibly though not probably I may stay a week or so still longer in Establishment; but it will depend on how I feel. —

You will probably have received, or will very soon receive my weariful// book on Species. I naturally believe it mainly includes the truth, but you will not at all agree with me[10]. — Dr Hooker, whom I consider one of best judges in Europe, is complete convert, & he thinks Lyell is likewise. Certainly, judging from Lyells letters to me on subject, he is deeply staggered[11]. —

Farewell. If the spirit moves you let me have a line

Yours affectionately
C. Darwin

Endorsement: "Nov 17. 1859"
Christ's College Library, Cambridge, #124

Notes

1. Fox's *Diary* for 1859 has not been found so there is no information on whether he did go to Malvern. CD was presumably asking Fox for his current opinion of the place.
2. See Letter #136 (n4).
3. Henrietta Darwin (1843-1927).
4. CD had suffered his "*unique crisis*" at the end of October. See letters to Charles Lyell, 25th October 1859, and to J. D. Hooker, 27th October or 3rd November 1859, CCD, Vol 7.
5. Sir Francis Sacheverel Darwin, see note 6, below.
6. Francis Sacheverel Darwin (1786-1859), son of Erasmus Darwin and Elisabeth Pole née Collier, and Deputy Lieutenant of Derbyshire, died on 6th November 1859 (*Darwin Pedigrees, Alum Cantab*). CD and Fox visited him at Sydnope Hall, Derbyshire: see Letter #29 (n10) for visit, and #121 (n12) for details on Sydnope Hall.
7. Sir Samuel Crompton of Wood End was Fox's 1st cousin on his father's side. He died in 1848 leaving several children. His eldest daughter married Earl Cathcart.
8. CCD, Vol 7, suggests that "*CD may be referring to John Gilbert Crompton of the Flower Lilies, Windley, Derbyshire. His wife, Millicent Ursula Crompton, had died on 4 October 1859 (Gentleman's Magazine n.s. 7, 2 (1859): 544)*".
9. Francis Rhodes, who took the surname Darwin in 1850 under the terms of the will of his brother-in-law, Robert Alvey Darwin, lived at Creskeld Hall, Poole, Yorkshire (*Darwin Pedigrees*). Arthington is a village east of Otley. The inheritance of Elston Hall, the Darwin family seat, had been discussed in earlier correspondence, when Fox had suggested that Erasmus, CD's brother, might inherit it (see Letter #54 (n12 & 13).
10. CD seems to be fairly persuaded that Fox will not support his views.
11. Joseph Dalton Hooker and Charles Lyell. Charles Lyell never fully accepted CD's views especially on the evolution of man, although he did adopt natural selection as the driving force of evolution in his tenth edition of *The Principles of Geology*, 1866-68.

#137A (CCD 2604). To: Revd W. D. Fox, Delamere Rectory, Cheshire

Down Bromley Kent
Dec. 25th. [1859]

My dear Fox
I wish you as happy a Christmas as you deserve & I am sure that ought to be a very happy one. — Fifteen round the Fire in your room, — I never before realised

what a party you are![1] I fear you cannot say much good about yourself, as you do not say it, & seem to have gone about so little in London[2]. — We// returned home last Friday fortnight. The last 10 days at Ilkley, I was splendidly well & for the first week at home; but since then I have had as bad a week as man could well have with incessant discomfort, I may say misery. — I have necessarily been very busy during these weeks but not with work which would be any strain to any other mortal man. I was hardly able from lameness, Boils &c to give Water cure a// fair trial this time, but I think we shall go there again next early summer.

My Book has been very successful in the ordinary sense; & I have had to reprint with few corrections a new Edit of 3000 copies[3]. Kingsley has permitted me to print a good sentence on the theological bearing of my work[4]. The Book has already made a few enthusiastic & first-rate converts, viz Lyell, Hooker, Huxley, Carpenter[5] &c. — It will, I find, attract// some little attention abroad. I daresay parts may be fanciful & I look at it as certain that there will be a large amount of error, though I cannot as yet see such errors. —

I am going soon to begin my bigger book, which I shall publish as 3 separate volumes, with distinct titles, but with a general title in addition[6]. — I think this will make the work less hard labour. — Here is an egotistical note for you my dear old friend. —

Yours affect^y
C. Darwin

Endorsement: "Dec 27/59"
Christ's College Library, Cambridge, #125.

Notes

1. At this time Fox had fourteen children (five, surviving, from his first marriage). Fox's second wife, Ellen Sophia, had had, up to this time, ten children, but one daughter had died as an infant. Therefore one child must have been away: the two oldest girls would have been 23- and 24-years old at this time and were unmarried, although both married in1861.
2. CD must be referring to a recent letter from Fox, in which Fox presumably said that he had visited London but had not had the time to visit the Darwins. Fox's *Diary* for 1859 has not been found (Appendix 1).
3. This was the 2nd Edition of *On the origin of species* (5th thousand) published on 5th January 1860 (Freeman, 1978).
4. See CCD, Vol 5, letter from Charles Kingsley, 18 November 1859, and letter to Charles Kingsley, 30 November [1859].
5. CD refers to Charles Lyell, J D Hooker, T H Huxley and William Carpenter (1813-1885) (Fullerian Professor of Physiology, Royal Institution, London, and Registrar of the University of London).
6. This is first explanation of how CD meant to publish the material assembled for the book he had called, notionally, *Natural Selection* [see Letter 139 (n5); see , also, Satuffer 1975].

138 (CCD 2733). To: The Rev W. Darwin Fox, Delamere Rectory, Northwich

Down Bromley Kent
22^d [March 1860]

My dear Fox

The hybrid exhibited some years ago was examined by Owen[1] & showed in its filed hoofs clear proof of artifice. If I go to London I will look at that now shown;

but nothing less than dissection & proof that the internal organs & bones were in some degree intermediate would convince me of the possibility of so astounding a hybrid[2]. —

You ask about myself; but I cannot give a good account// of myself: I am trying under a Mr Headland[3] a course of nitro-muriatic acid, eating no sweet things & drinking some wine: but it has done nothing for me as yet & I shall go to my grave, I suppose, grumbling & groaning with daily, almost hourly, discomfort. I have begun slowly at my larger book; & this reminds me that I have found a memorandum to the effect that I *thought//* I remembered your saying that the common gander does not *always* turn white. If you have any positive knowledge will you send me a line; & if I do not hear, I will understand that you have no positive case[4]. —

Linnæus says the goslings of both dusky & white geese are similar yellow[5]; Do you know how this is? — The "Origin of species" has made quite a commotion amongst naturalists & I am well contented with its// reception. A German edition will appear immediately & *two* (!) American Editions[6]. So that what is right & what wrong in my Book will soon be sifted & known. —

Like a very bad man that you are, you do not say one word about yourself

Farewell my old friend
Yours affectly
C. Darwin

Erasmus, I grieve to say, has not yet quite lost his ague[7]. —

Postmark: Postmark: MR 22 60
Christ's College Library, Cambridge, #127.

Notes

1. Richard Owen (1804-1892). Comparative anatomist. Superintendent of the Natural History Department, British Museum and populariser of dinosaurs. Staunch opponent of evolution. (*DNB*)
2. Fox urged CD to examine an alleged hybrid deer being exhibited in London. See Letter #139.
3. Frederick William Headland (1830-1875). CD's doctor from 1860. He must have diagnosed CD as having an abnormal level of oxalic acid in the blood and an excess of urea in the urine (for further references, see CCD, Vol 8).
4. Clearly, CD was continuing his *Natural Selection* book, this part of which appeared in *Variation of animals and plants under domestication* (1868), and where he stated in Vol 1, p 288 that ganders of the species *Anser ferus* do not always become white, but he did not cite Fox.
5. Linnaeus (Carl von Linné) 1811. *Lachesis Lapponica; or a tour of Lapland*, first published from the original manuscript journal of the celebrated Linnaeus, by J E Smith, (see Vol 1, p 5). CD recorded this fact in his abstract of the work (DAR 71, 68), which he had read in 1846 (see *CCD,* Vol 4, Appendix IV, Darwin's Reading Notebooks; DAR 119: 17a).
6. There was an official version of the American Edition: see Appendix IV, New material added to the American edition of *Origin*, CCD, Vol 8.
7. Erasmus Alvey Darwin (1804-1881). CD's brother. Concerning ague: see letter of Erasmus to CD, 23rd Nov 1859, CCD, Vol 7.

139 (CCD 2809). To: The Rev W. Darwin Fox, Delamere Rectory, Northwich

Down Bromley Kent
May 18th [1860]

My dear Fox

It was very kind of you to take so much trouble about the curious dog; but unless one is on the spot it is hopelessly difficult to understand the amount & origin of difference. The dog, however, must be a most curious creature. — I am glad to hear of a naturalist taking up so difficult & neglected a subject as spiders. — If I have time & if in London moderately soon, I will// look at the deer-Hybrid[1] — But the fact is, if it appeared ever so intermediate I shd. not believe in its hybrid nature, unless after *careful dissection* <of its internal organs>; & one might easily be deceived by hoofs being filed &c &c. You will be sorry to hear that our eldest girl, Etty, has now been 3 weeks ill with odd fever, partly remittent partly typhoid[2]; but I have great hopes we see signs of// abatement. It has harassed us much, though not exactly frightened us. At one time, however, the Doctors seemed rather anxious. — But I think it must end soon. — What a household you must have had with seven with Hooping Cough!

My health has been better of late, which I am inclined to attribute to mineral acids, no sugar, & drinking wine[3]. —

I do not know whether// you ever see various Reviews, but the attacks have been falling thick & heavy on my now case-hardened hide. — Sedgwick & Clarke opened regular battery on me lately at Cambridge Phil. Socy [4]. & dear old Henslow defended me in grand style, saying that my investigations were perfectly legitimate. I have begun my bigger Book[5], but make very very slow progress. —

My dear old friend,
Yours affect
C. Darwin

Postmark: MY 19 60
Christ's College Library, Cambridge, #128

Notes

1. See Letter #138 (n2).
2. Henrietta Emma Darwin, CD's daughter, was diagnosed with "a form of Typhus fever" (CCD, Vol 8, letter to J D Hooker, 11 May [1860]).
3. See Letter #138 (n3).
4. See Letter by J S Henslow to J D Hooker, 10 May 1860, CCD, Vol 8.
5. CD seems to have decided, at this time, to publish the book, which became *The variation of animals and plants under domestication* (1868), and was referring to this. However, in Letter #137A (n6) he explained that he had decided to publish "*my bigger book, which I shall publish as 3 separate volumes, with distinct titles, but with a general title in addition*". In the course of events, CD published *The variation of animals and plants under domestication* (1868), *The descent of man* (1871) and *The expression of emotions in man and animals* (1872), which he considered the central planks of his theory. Fox was confused and enquired of CD about what had happened to his *Big Book* for the next decade and more [see Letter #201 (n7)].

140 (CCD 2836). To: The Rev W. Darwin Fox, Delamere Rectory, Northwich

Down Bromley Kent
June 18th [1860]

My dear Fox

I have been looking over old letters of yours, & I want to ask a question: but I do so on condition, that you just answer it, at your leisure as briefly as possible. — Dixon says that his Musk Ducks often took to roost & perch on high places or ridge of Barn, or perch along side Hens &c[1]. — I see you used to keep// Musk ducks — did you ever observe this or any tendency to lay eggs on high places? Case bears on retention of aboriginal habits — I see in your letter capital case of a cross between white & some other var. of Musk-duck *reverting* to black & white or aboriginal colour of Musk Duck[2]. —//

Can you give me any authortive case <with exact dates> of period of gestation of any breeds of Dog[3]? How I wish I knew for instance about Blood Hounds — I suppose no chance of Galton's knowing; indeed I do not know whether Mr H. Galton is alive[4]. —

Poor Etty has now been in bed for exactly 7 weeks with low fever; but at last, thank God,// she is beginning to gain little strength[5]. —

Yours affect.
C. Darwin

Postmark: Postmark: JU 18 60
Christ's College, Cambridge, #129

Notes

1. Dixon, E. S. 1848. *Ornamental and domestic poultry; their history and management*. London, p 66. The passage is marked in CD's copy [Darwin's Library Collection (DAR), University Library, Cambridge].
2. Fox's letter discussing this point, like most other letters from Fox, has not been found. CD cited Fox's responses in *Variation in animals and plants under domestication*, 1868, Vol. 2, p 14: "*But the Rev. W. D. Fox tells me that, by putting a white drake to a slate-coloured duck, black birds, pied with white, like the wild musk-duck, were always produced.*'
3. In *Variation in Animals and Plants under Domestication,* Vol 1, p 30, CD stated, on the period of gestation: "*Rev. W. D. Fox has given me three carefully recorded cases of retrievers, in which the bitch was put once to the dog; and not counting this day, but counting that of parturition, the periods were fifty-nine, sixty-two, and sixty-seven. The average period is sixty-three days; but Bellingeri states that this applies only to large dogs, and that for small races it is from sixty to sixty-three days*".
4. CD referred to John Howard Galton, a brother of Samuel Tertius Galton and a relative of both Fox and CD (see below), with whom Fox had corresponded about bloodhounds in 1838, see Letter 64A. John Howard Galton (1794-1862) lived at Hadzor House, Worcestershire and was the High Sheriff of Worcestershire (*Burke's Landed Gentry*, 1879). Erasmus Darwin had a daughter, Frances Anne Violetta Darwin, by his second wife, Elizabeth Collier Pole, who married Samuel Tertius Galton. Both CD and Fox were attracted to their daughter, Elizabeth (Bessy) Galton, in their younger days, and she features in earlier correspondence (see Letters #6 (n7), #7 (n10), #9 (13), #27 (n4), #45, #64A & #87) and also in Fox's *Diary* (Chapter 2). Their son, Francis Galton, became noted for his travel and scientific writings, and is known as the founder of forensic finger printing and eugenics.
5. Henrietta Emma Darwin, see Letter #139 for further details of her illness. Emma Darwin's *Diary* shows that Etty gained strength from about the 8th June.

Conclusion

The final letters of this period cover the momentous events from June 1858 to the publication of *On the Origin of Species* in late 1859 - and its early reception.

Events overtook Darwin in June 1858 (see Introduction to this chapter). Up to that time Darwin had been working steadily but slowly on his species book (notionally called *Natural Selection*), which in fact would already have been several volumes in length. By June 1858, Darwin was becoming despondent at the length and probably frustrated by a number of unresolved issues. It was clear to him that, while in general the theory of natural selection was well supported by his observations, there were difficulties with the theory. Perhaps most difficult of all was the question of what made up a species and why. His work on domestic animals indicated how plastic animals could be. Were all dogs for example one species? If not how could one decide? All breeds of dogs interbred with each other. Thus there seemed to be a continuum across this group whether it was one species or more. Yet for the most part nature seemed to be comprised of fairly different plans and animals with well defined species, each relatively different from the nearest neighbour. How could this come about? Darwin came back to his tree example. Just like a growing tree, the branch tips would produce a succession of smaller branches, a few of which would survive, but most would die. In time this would give rise to the familiar tree configuration and not a densely tangled bush. Another problem was just how this came about? How was it that only a few branches survived? Darwin had concentrated on competition between varieties up to this point. "**Survival of the fittest**", a phrase first used by Herbert Spencer, that meant that less successful varieties would die out, through a process of fierce competition. This was Tennyson's "*Nature red in tooth and claw*". There were also many other difficulties, which Darwin faced if he were to come up with a rational account of biological evolution, such as why some families of plants or animals have many members, while others have few. Furthermore, there was no adequate description of how characters were inherited.

Into this mix of deep understanding and abyssal ignorance came a letter from Alfred Russell Wallace, a collector of natural history specimens in Indonesia. Wallace had already sent one broadside on biological evolution to the Annals and Magazine of Natural History, London in 1855. That paper was entitled "*On the law which has regulated the introduction of new species*". The title alone was guaranteed to ruffle the feathers of the establishment who denied the introduction of new species, or, for more liberal thinkers, who believed that god controlled the introduction of new species. Charles Lyell warned Darwin at that time about Wallace and urged Darwin to publish rapidly a sketch of his theory; but as Darwin wrote to Fox, "*I remember you protested against Lyells advice of writing a sketch of my species doctrines; well when I begun, I found it such unsatisfactory work that I have desisted & am now drawing up my work as perfect as my materials of 19 years collecting suffice, but do not intend to stop to perfect any line of investigation, beyond current work.*" (Letter #112).

The letter, which Darwin now received, enclosed another manuscript entitled "*On the tendency of varieties to depart indefinitely from the original type*". Darwin described this paper, in passing it on to Charles Lyell for publication, as follows, "*It seems to me well worth reading. Your words have come true with a vengeance that I sh^d be forestalled. You said this when I explained to you here briefly my views on "Natural Selection" depending on the Struggle for existence. — I never saw a more striking coincidence; if Wallace had my M.S. sketch written out in 1842 he could not have written a better short abstract! Even his terms now start as Heads of my Chapters*" (18th June1858).

The response by Lyell and Joseph Dalton Hooker to this calamity was to arrange quickly a presentation of the views of both Darwin and Wallace in separate papers to the Linnean Society of London. Unfortunately, Darwin was in no position to write a cogent paper to match that of Wallace, he was far from well and there was sickness in the family, Etty with a form of cholera and Charles Waring with scarlet fever. Darwin felt that the sketch that he had written in 1842 and revised in 1844 was out of date, for it did not contain his ideas on the tree-like structure of evolution or his ideas on sexual selection. Instead they cobbled together a letter that Darwin had written recently to Asa Gray, the American botanist, together with some summary notes. Neither Darwin or Wallace was present at the meeting on the 1st July 1858, where the papers were read by one of the secretaries, with Darwin's first, amongst five others for that session. As Lyell and Hooker had hoped the papers elicited no discussion. Perhaps too the meeting was subdued by the announcement of the death of the great botanist, Robert Brown. Later, the President of the Society, Thomas Bell, reported famously, "*The year which has passed has not, indeed, been marked by any of those striking discoveries which at once revolutionize, so to speak, the department of science on which they bear*".

Now Darwin had his chance. Under the urgings of Lyell and Hooker he dropped the work on his big species book and concentrated on as short an abstract of his ideas as he could manage, without footnotes and largely without references. The result, a little over a year later, in November 1859, was "*On the Origin of Species by means of natural selection, or the preservation of favoured races in the struggle for life*". The book was an instant success and has never been out of print since. Wallace remained in the East Indies until 1862 and by the time he returned to England his role as a founder of the new discipline of biological evolution was largely overlooked by the public, and has been ever since, although he is still regarded as the father of biogeography.

In all these momentous events Darwin's contribution, apart from writing the "*Origin*", was largely a passive one. Just before the critical meeting of the Linnean Society his baby Charles Waring died of scarlet fever and the family withdrew to a safe haven on the Isle of Wight for three weeks. At the time of the publication of the "*Origin*" Darwin was taking a water cure at Ilkley Wells House in Yorkshire and in the following year with controversy raging over his ideas and critical discussion taking place, as at the meeting of the British Association in Oxford in 1860, Darwin was quietly rewriting parts of his big species book which were to be incorporated into "*The variation of animal and plants under domestication*" (2 volumes, 1868) and "*The Descent of Man*" (2 volumes, including sexual selection, 1871). In addition, during these early years following the "*Origin*", Darwin found the time to cement his long-held interest in plants by publishing two books: "*On the contrivance by which British and foreign orchids are fertilised by insects and the good effects of intercrossing*" (1862) and "*On the movement and habits of climbing plants*" (1865). It seems that Darwin relaxed from the strain of evolutionary matters by addressing these plant studies.

What is of interest here is to follow these momentous affairs and debates through the Darwin-Fox correspondence. Of course, from now on the two correspondents stand in a very different light to each other. The obscure vicar, Fox, remains the obscure clergyman - with a famous friend. The famous, and to many clergymen the infamous, Darwin has to renegotiate his relationship to his obscure cousin. Before the events of 1858/59 we get little forewarning of what is to come. And even after the publication of the "*Origin*" Darwin is diplomatic enough to avoid any confrontation. His attitude as with all his protagonists at the time is to make it appear that he is innocently trying to ascertain the truth, not trying to stir up controversy.

The first real test came when Darwin received the "bombshell" letter from Wallace in 1858. Despite the fact that several letters pass between the correspondents at this critical juncture there was only a very late mention of the Wallace letter on 21st July 1858, no mention of the Linnean Society meeting but a reference to the papers that would eventuate (Letter #130): "*After all, I am now beginning to prepare an abstract of my Species Theory. By an odd coincidence, Mr Wallace in the Malay Archipelago sent to me an Essay containing my exact theory; & asking me to show it to Lyell. The latter & Hooker have taken on themselves to publish it in Linnean Journal, together some notes of mine written very many years ago; & both of them have urged me so strongly to publish a fuller abstract, that I have resolved to do it, & shall do nothing till completed: it will be published, probably, in Journal of Linn. Socy & I shall have separate copies & will send you one. — It is impossible in abstract to do justice to subject.*" One can forgive Darwin *post eventum* for not mentioning this event as it would have been seen as indecorous to mention work in the setting of a death in the family. However, Darwin had time to mention the letter even before his son became ill and certainly could have informed Fox about the momentous Linnean Society meeting. In fact Darwin did write two letters to Fox at this time (#125A & 126), but both were concerned with the upsetting case of Dr Lane, Darwin's hydrotherapist.

One can conclude, therefore, several things from this letter. Fox, obviously, was not on the inner circle of Darwin's acquaintances: Hooker certainly was; Lyell, too; and Huxley. Secondly, Darwin obfuscates in telling Fox that what will be published are some old notes of his. Very deliberately Darwin did not publish the sketch of 1842/44, for the reason that he wanted to show that his thinking was more advanced than this, and now included the idea of the biological tree and sexual selection. Thirdly he makes no mention that his article (or rather his two articles) will be published first, before that of Wallace. Darwin was calculating that Fox did not know who Wallace was and did not understand the practice of precedence, whereby Wallace should have been given priority for the discovery of the principle of evolution, because he submitted his paper first. Today the act of submitting a ground-breaking paper to a rival in the field would be unheard of, but even in those days it was highly unusual, and even more unusual for the rival to get into print first. So Darwin glossed over the facts. In fact, Darwin and all his inner circle glossed over the facts – and urged Darwin to get his book out as quickly as possible.

Did Fox know what the Wallace/Darwin theory was? And was he sympathetic to it? Unfortunately we do not know. As stated in the Introduction to this chapter Fox probably was amongst the likes of the Rev Charles Kingsley, who felt that there was a substantial amount of truth in the theory - that nature operated mechanistically - up to a point - but that God still had a hand in it. In fact this was the very stance that A R Wallace took: nature operated freely in most directions but when it came to the evolution of the human brain, God had stepped in and directed the process to give man an intellect high above that to which natural selection, alone, was capable. This would have been the view of conservative scientists like Adam Sedgwick and Richard Owen, who thought that even this was going a bit too far and that God should be regarded as the prime cause in every step of the formation of a new species. However, Darwin would not countenance such a view. For him, either one believed that nature acted throughout by natural selection or not at all (see letter to Lyell, 19th June 1860)). For the moment, though, he guarded these views and avoided in print any arguments over contentious issues such as the evolution of man. As he wrote to Wallace, "*You ask whether I shall discuss "man"; — I think I shall avoid whole subject as so surrounded with prejudices, though I fully admit that it is the highest & most interesting problem for the naturalist*" (22nd Dec 1857).

Perhaps, it is surprising that the Darwin-Fox correspondence should survive these momentous times. However, it did, and as we see in the next Chapter, became even stronger.

Chapter 7
The Most Dangerous Man in Europe: Living in the Shadow of Fame

Letters #140A-176 (total 37); covering 18 Oct 1860 to 6 Feb 1867 (6 years)

"I cannot remember that I ever published a word directly against religion of the clergy; but if you were to read a little pamphlet which I received a couple of days ago by a clergyman, you would laugh, and admit that I had some excuse of bitterness. After abusing me for two or three pages, in language sufficiently plain and emphatic to have satisfied any reasonable man, he sums up by saying that he vainly searched the English language to find terms to express his contempt for me and all Darwinians."

"On my last visit to Down, Mr. Darwin said, at his dinner-table, 'Brodie Innes and I have been fast friends for thirty years, and we never thoroughly agreed on any subject but once, and then we stared at each other, and thought one of us must be very ill.' "

Quotations of Charles Darwin given by Rev Brodie Innes, Vicar of Downe, 1846–1869. From *The Life and Letters of Charles Darwin*, Vol 2, by Francis Darwin (1888)

Introduction

The publication of "*On the origin of Species*" in November 1859 projected Charles Darwin into such an exalted orbit that it was bound not only to affect his relationship with William Darwin Fox, but with the world in general. The first signs of this are hard to detect but probably affected both Darwin's health and those around him and his leisure time; "*I have done little of my regular work this summer; chiefly owing to incessant anxiety & movements on account of Etty. My correspondence about the "Origin" has also been gigantic, & a day hardly passes without one, two or three letters on the subject. I have amused myself with a little Natural History of other kinds...*" he wrote to Fox in 1860 (Letter #141). He took refuge in working

A. W. D. Larkum, *A Natural Calling*, DOI 10.1007/978-1-4020-9233-6_7,

on less contentious subjects than evolution, such as orchids, insectivorous plants and the physiology of plants. Why this subject area took his fancy is not clear, because he had an interest in a broad range of subjects, many of them seemingly more suitable such as animal studies. He often complained of the over-emphasis on systematics and classification, which is where many of the potential areas lay. Possibly, his evolutionary interests combined with an inclination to strike out somewhere new took him in the new direction (Allan, 1977). "*I am much pleased & somewhat surprised at your liking my orchid-book:*", Darwin later wrote to Fox (Letter #148), "*the Botanists praise it beyond its deserts, but hardly anyone, not a Botanist, except yourself, as far as I know, has cared for it. The subject interested me much, & was written almost by accident; for it was half written as a mere paper & then I found it too long, & thought I would risk publishing it separately*". This was his book, *On the various contrivances by which British and foreign orchids are fertilised by insects and on the good effects of intercrossing* (1862). This work was long-planned but immediately after the publication of the "*Origin*", Darwin was seeking solace in observing plants, as he wrote to Fox, "*& have lately worked hard at the power of the Drosera or sun-dew in catching flies; & tremendous slaughter the plant makes*" (Letter #141). This latter work would be his second book, "*Insectivorous plants*" (1865), after the *Origin*.

Whether Darwin's ill health was exacerbated by his new-found fame or whether he and Emma were more concerned to avoid visitors is a matter of much debate (see, e g, Colp, 1977). On occasions when Darwin wanted information, as with the digestive processes of insectivorous plants, he could quietly slip up to London and into a residence of an acquaintance to gain information. However, for most the door was shut; and this included Fox, despite several attempts [Letter #165 (n1)]. Nevertheless Fox did manage one last meeting, in the house of Darwin's brother, Erasmus, in 1862 (Letter #151); and although Fox had contact with Darwin's siblings and other relations at various times, and although the correspondence continued until Fox's death in 1880, the two men never met again.

Darwin, of course, found that servicing the "*Origin*" was a massive undertaking in itself. Just the American edition took up much time; in addition, translations were needed into several languages. Then there was the enormous backlash from conservative biologists and churchmen (Himmelfarb, 1959; Ellegård, 1990), from where the title to this chapter derives. Darwin tried to answer their criticisms in individual letters and in the changes to the third edition of the *Origin* and subsequent editions. "*My correspondence about the "Origin" has also been gigantic, & a day hardly passes without one, two or three letters on the subject*", he wrote to Fox in Letter #141, and then again in Letter #142, "*I am at present busy with a new & somewhat enlarged & corrected Edit. of the well abused Origin, which the Public like, whatever the Reviewers may do*".

However, as the dust settled on the "*Origin*", Darwin had already decided on his agenda for the next decade or more. This was to publish the parts of his "Big Book" not adequately covered in the "*Origin*"; most notably, "variation under domestication" and "sexual selection". Almost immediately after completing the "*Origin*" and dealing with the first round of correspondence (see Correspondence, Vol. 8), Darwin began work on *Variation* (*Journal*, 1860) and had three chapters completed by the end of the year. However, he was very reticent to admit to Fox that he had given up the idea of publishing his great opus in one book, and did not do so until the 1870s (Letter #201).

The signs of this decision, to publish *Variation* separately, as the two volume *The variation of plants and animals under domestication* (1868), is a renewed vitality in the correspondence with Fox, on all manner of subjects, and at much greater depth than previously. Interestingly, Fox was happy to go along with this. For example in Letter #145 Darwin wrote, "*I am going to bother you. Looking over some of your old notes, I see that you have kept the wild breed of Turkeys from* $L^{d.}$ *Leicester & Powis. You know that they now say that the common Turkeys have descended from a southern so-called species. Have you ever crossed intentionally or accidentally your wild & common;.....?*". And so it goes on, with 5 letters in 1860, 2 in 1861, 4 in 1862, 9 in 1863, 2 in 1864, 1 in 1865 and 2 in 1866 from Darwin (or Emma) to Fox. This does not include some letters from Darwin and most of the replies from Fox, which have not survived.

Part of the reason for the spate of letters in 1863, was the search for Annie Darwin's grave. As told in Letter #160 the grave was missed on the first attempt in 1863, when the Darwins returned to Malvern, this time to Malvern Wells, for more hydropathic treatment for Darwin and their son, George. Fox was gravely ill himself at this time but managed a clear description of the site of the gravestone and an accurate rendering of the inscription "*To a dear child*" (Fig 26). Another reason was the request by Emma for Fox to support her campaign for less cruel small-animal traps. Nevertheless, more than half of these letters were on subjects for "*Variation*".

For Fox, these years were the continuing life of a rector and family man with numerous connections to the landed gentry near his home at Delamere and to relations further afield. We are fortunate to have diaries for the years 1861 to 1865. These diaries reveal Fox as very busy, traveling frequently throughout England, attending to his son's education and overseeing a farm, as well as keeping up an interest in agricultural shows and natural history. All the while, however, his health was getting poorer and he was being advised to take longer breaks away.

For both men this was a time of consolidation, but also it was a time when their age and their illnesses were catching up on them.

140A (CCD #S06319). To: Charles Darwin [Down House, Bromley, Kent]

Deepdene
Hythe
Southampton
Aug 16. [1860][1]

My dear Darwin,

In answer to your enquiry, I believe I might truly state – not one dozen but dozens of white cats came under my observation. The first I saw was a half bred Persian at Ryde. I had a kitten of hers, from whom I had// a great number, never having less than six at a time for years. In every case, if *pure* white <& with blue eyes>, they were stone deaf. I used to shew this by making all sorts of the loudest noises close to them, which they never in the least perceived

I have remarked them in various other places, once at the large Inn at Chichester[2] (which I think was a male) – 2 Norwegian// Cats whose owner I amazed (as I have done several other) by remarking "that it was a pity such fine cats should be deaf". These were females.

Of course the greater part of those we bred & kept were females, but I also had males, as I kept my breed pure for many years. I am *certain* the females were deaf, & I have no doubt about males being so also, as I must have observed them if they had not been.//

I forget where I wrote several letters on the subject of these blue eyed cats – it is now 20 to 30 years since – but they elicited exactly similar facts.

I had several that had only one eye blue: and we were quite convinced that in those cases, they were deaf only on the blue side. All my older children remember "Lily" our first cat – as perfectly deaf & blue eyed - & she certainly brought litters of kittens, for many years[3] - So that the sex is undeniable, tho' I once had a Male Tortoiseshell (or at least so reputed) who had a family – to the great delight of my children.

Evers yours W.D. Fox

2.

After writing the above I asked my wife & cubs if they remembered the white cats. "Lily" was too far back, but "Glaucops" was well remembered for her azure eyes, and kittens innumerable.

Fanny, who is just returned from <Bardwell nr> G[t] Yarmouth[4] – added "that there is a cat there perfectly white, but with the usual green eyes, that is deaf – like the blueeyed ones." She remarked// this particularly, as opposed to the blue eyed theory.

"But at Caister[5] Rectory <close by>. Mr Stewart has a magnificent blue eyed cat which is quite deaf."

I shall probably see both these cats before long and shall very likely find a Colony about there.

The Caister Cat – being described as *very fine, may* probably be male.

Your son William was kind enough to call here// on Friday, and most favourably impressed us all.

Provenance: University Library, Cambridge; Manuscript Room: DAR164: 188

Notes

1. The year for this letter is uncertain. The major clues are, firstly, the reference to white cats with blue eyes being deaf, which echoes the earlier information in Letter #110, and, secondly, Fox's statement: "*Your son William was kind enough to call here on Friday, and most favourably impressed us all*". Since Fox's letter is written from Southampton the temptation is to assume that the visit took place in Southampton, but this is not necessarily so. The years 1853-1858 seem to be too early for William to be roaming the countryside: he would have been from 14 to 19 years old. More significantly, perhaps, Fanny Fox would have been 5 to 10 years old. The next possible date is 1859; the year William was 20, and Fanny 11, years old. In 1859, William was at Christ's College, Cambridge and made a trip to the Lake District in July/August (see CCD, Vol 7, CD letter to W E Darwin 7th July [1859]) and could have called on Fox. In 1859 Darwin was preoccupied with writing "*On the Origin of Species*". The first Edition of "*Origin*" stated "that white cats with blue eyes were invariably deaf". So CD might have written to Fox to clarify the earlier letter of 1856 (#110). The next possibility is 1860;

that year Darwin dusted off his old manuscript and notes for "*Natural Selection*" and began to write "*Variation*" (what was to become *The variation of animals and plants under domestication* published in 1868), completing three chapters by the end of the year (see CD's *Journal* for 1860). Fox *Diaries* are missing for 1859 and 1860. So either 1859 or 1860 are possibilities, albeit by default. Thereafter to 1865, the Fox *Diaries* do not offer any likely dates. In *The variation of animals and plants under domestication*, 1868, Vol 2, p. 322 there is the statement: "*The Rev. W. Darwin Fox informs me that he has seen more than a dozen instances of this correlation in English, Persian, and Danish cats; but he adds 'that, if one eye, as I have several times observed, be not blue, the cat hears. On the other hand, I have never seen a white cat with eyes of the common colour that was deaf.'* " And in footnote 23 on the same page CD added, "*The first case recorded in England by Mr. Bree related to a female, and Mr. Fox informs me that he has bred kittens from a white female with blue eyes, which was completely deaf; he has also observed other females in the same condition.*" This statement seems to be a compilation of information in Letter #110 and from this present letter, and therefore could have been written any time up to 1867.

The tentative conclusion is that this letter was written in 1860. This fits with the statement in Letter #142 (17th Dec 1860), "*Your remarks on William pleased us much: He is now at home & gives a good account of your appearance.*" The reason for the late reply may be that this letter was somehow overlooked until much later: the Darwins spent an extended period at Eastbourne (21st Sept - 10th Nov) that Autumn, following extended illness in the family. In fact this may be the letter that CD refers to in Letter #144, "*Your note has followed us here, where we arrived 4 or 5 days ago*", but the implication in that letter, that the note contained information that two of Fox's daughters were now married makes that seem less likely. There may also have been another round of correspondence that was lost. The fact that William Erasmus Darwin was also sent an inkstand, in addition to the one that Fox sent to CD (Letter #143), seems to suggest that William may have expressed interest in an inkstand during his earlier visit to Fox.

2. Fox would have visited Chichester, Hampshire, many times on his trips to the Isle of Wight, starting in 1825 (see Chapter 2); he met his first wife, Harriet, on the Isle of Wight, in 1833, and returned many times thereafter. The "20-30 years" mentioned by Fox fits with this possibility.
3. CD refers to the subject of white cats, which are blue-eyed, being invariably deaf in *On the Origin of Species* (p 9), but only in a general way. In *The variation of animals and plants under domestication* (1868), CD specifically identified Fox in a footnote, see note 1, above. This evidence fits with the 1860 date for this letter.
4. Great Yarmouth is on the East coast of Norfolk, close to the Norfolk Broads. Bardwell is inland, closer to Bury St Edmunds.
5. Caister-on-Sea lies to the north of Great Yarmouth on the Norfolk coast.

141 (CCD 2953). To: The Rev W. Darwin Fox, Delamere Rectory, Northwich

15 Marine Parade
Eastbourne
Oct 18th [1860]

My dear Fox

Many thanks to you for calling my attention to the Hybrids[1]. But I have seen them & read about them in French — I at first quite disbelieved the case; but I suppose it is true; though by chance I had a note from Lyell this morning, who reports that some French naturalists believe whole story is Cock & a Bull[2]. It is <the> most curious case that ever was if true; for so many have failed to get Hybrids from Hare & Rabbit. — We

have been here for 4 weeks & return home in a weeks time[3]. — We came here for chance of the sea doing Etty good; & it has certainly to certain extent succeeded[4]. At one time this summer I gave up all hopes of her recovery, but she does gain strength at a snail's pace, & now suffers only from indigestion & weakness[5]. We, all the others, are well; though I cannot

boast very much of myself. — I have done little of my regular work this summer; chiefly owing to incessant anxiety & movements on account of Etty. My correspondence about the "Origin" has also been gigantic, & a day hardly passes without one, two or three letters on the subject[6]. I have amused myself with a little Natural History of other kinds; & have lately worked hard at the power of the Drosera

or sun-dew in catching flies; & tremendous slaughter the plant makes. —
I shall be very glad to be at home again; for the weather is most dismal. — I hope that the world goes fairly well with you & yours. —

My dear old friend
Yours affectionately
C. Darwin

Endorsement: "Oct 18/60"[7]
Christ's College Library, Cambridge #130

Notes

1. These were hybrid offspring of hares and rabbits; see CCD, Vol 8, letters to Charles Lyell, 18th May and 5th July 1860.
2. The letter from Charles Lyell has not been found. CCD, Vol 8, suggests that the French naturalist may have been Pierre Paul Broca, since there is an undated abstract of Broca's paper on hybridity (Broca, 1858-9), which discusses the hare–rabbit cross, in DAR 205.7(2), 154. Furthermore, CD cited the work in *Variation in animals and plants under domestication*, Vol 1, p 105, n7 and said "*But from what we hear of the marvellous success in France in rearing hybrids between the hare and rabbit, it is possible, though not probable, from the great difficulty in making the first cross, that some of the larger races, which are coloured like the hare, may have been modified by crosses with this animal.*"
3. Emma Darwin's *Diary* gives the dates as 21st Sept to 10th Nov 1860. Emma was quite sick during this stay.
4. Henrietta Emma Darwin was recuperating from a long illness. In her book *Emma Darwin: a Century of Family Letters* (Litchfield, 1915), Henrietta said "*In 1860 my poor mother's thoughts and time were engrossed with care of me in a long illness (probably typhoid fever) lasting with relapses from May, 1860, till Midsummer, 1861.*"
5. For further details see letter of Emma Darwin to W E Darwin, September [1860], (DAR 210.6).
6. This signalled a continuing occupation. CCD, Vol 8, covers just one year (1860) and contains many of the letters referred to. However, correspondence in later years was similarly heavy (see further volumes of CCD).
7. The endorsement is on the accompanying envelope.

142 (CCD 2953). To: The Rev W. Darwin Fox, Delamere Rectory, Northwich

Down Bromley Kent
Dec 17th [1860]

My dear Fox

Your remarks on William pleased us much: He is now at home & gives a good account of your appearance[1]. —
I shall be proud to send you the Photograph of myself by Post. & will order a copy for you[2]. —
I have no idea whether an engraving of my Grandfather could be purchased; there are Photographs of his picture to be had at Cambridge & I will tell William to get

(if he can) one & send it you[3]. The engraving of my Father is hideous, but I presume my sisters could get it at Shrewsbury, if you wished for it[4]. —
Thank for your enquiries about poor Etty who has had a terrible succession of illnesses; she is now very slowly, but steadily improving[5].
You also ask after Caroline: she also was very ill this summer; but is now fairly well. She & her three daughters live at Leith Hill Place, 6 miles from Dorking[6].
As for myself I have been of late rather below my low mark; & so end my long Bulletin. I am at present busy with a new & somewhat enlarged & corrected Edit. of the well abused Origin, which the Public like, whatever the Reviewers may do[7]. —
I have done very little in Nat. Hist. this summer owing to Etty's illness. —
You say not a word about your innumerable progeny[8]; so that I hope that they are all well. Farewell

My dear Fox
Yours very truly
C. Darwin

I forgot. — you offer me a good inkstand. This is by far too good an offer not to be gratefully accepted[9].
You might send it to "6, Queen Anne St. — Cavendish Sq^{e}".[10] —

Postmark: Postmark: DE 18 60
Christ's College Library, Cambridge, #130A

Notes

1. It seems clear that William Erasmus Darwin had visited Fox (at Delamere?) some time before this letter. It seems too long ago for the visit to be the same as that mentioned in August (Letter 140A). According to Emma Darwin's *Diary*, "William came" on 13th December 1860. Fox's *Diary* for 1860 has not been found so the details of this meeting are unknown. It is possible that the meeting is the same one as referred to in August in Letter 140A (refer to note 1 of that letter).
2. CD's comments in the following letter suggest that he sent Fox a copy of a photograph taken by Maull and Fox. It is not known precisely when this photograph was taken. It was dated as "1854?" by Francis Darwin (*LLi*, frontispiece); Nora Barlow suggested it was taken when CD was "aged 51", i e, in 1860. An entry in CD's *Account Books* (Down House MS, February, 1858) suggests that CD may have sat for the photograph in the summer of 1857.
3. Erasmus Darwin, CD's grandfather, was the brother of Fox's grandfather, William Alvey Darwin. CD referred possibly to a photograph of the portrait by Joseph Wright of Derby, *circa* 1780. The original is in the Darwin Museum, Down House. Another portrait of Erasmus Darwin by Wright, dated 1770, is in Darwin College, Cambridge (see Fig 1).
4. CD refers to the mezzotint of Robert Waring Darwin made by Thomas Lupton in 1839 after the portrait in oil by James Pardon (see Fig 2). The engraving is at Down House and the portrait is in Darwin College, Cambridge.
5. See previous letter (#141) for details.
6. Caroline Sarah Wedgwood (1800-1888). CD's second oldest sister. She married Josiah Wedgwood III, and they lived at Leith Hill, Surrey. Their three daughters were Sophy Marianne, Margaret Susan, and Lucy Caroline Wedgwood (*Darwin Pedigrees*).
7. The second edition of *On the Origin of Species* was rapidly prepared and contained few changes apart from small corrections. It was published on 5th Jan 1860; the third edition, which contained more changes and an historical sketch, was published on the 3rd April 1861 (Freeman, 1978).
8. By 1860 Fox had 15 living children [see Letter #137A (n1)].
9. Two inkstands are referred to by CD in Letter #143 (n1).
10. The address, at the time, of CD's brother, Erasmus Alvey Darwin.

143 (CCD3046). To: The Rev W. Darwin Fox, Delamere Rectory, Northwich

Down Bromley Kent
Jan 9th [1861]

My dear Fox

I received yesterday from my Brother the two Inkstands for William & myself; & we are both very much obliged for your present. I have filled mine this morning & it seems to work capitally, but I have not yet had much writing with it, as I have been merely correcting// proofs[1] — The only fault is that I sometimes take too much ink, but with a little practice I shall get over this, & I have been eternally plagued with my Ink getting muddy, & if that is avoided I shall be for an equal eternity grateful to you, — as no doubt William will, who// writes as detestably bad a hand as I do & my good Father did before me[2]. —

We manage to keep Ettys room of a good temperature & I do not think she suffers from the extreme & dreadful cold: she has been rather better these 4 or 5 days than usual[3]. All my Boys are// now at home & very jolly[4]. Farewell my dear Fox & I hope always to keep your Inkstand, as a memorial of you by me.—

Farewell
Your affect
C. Darwin

P.S. I believe you have been a pig-Breeder. I found a strange statement in a German Book that white Sows go with young a week longer or shorter (I forget which) than piebald or black Sows[5]. — I presume it is false: but many odd peculiarities are correlated with colour[6]. — Have you kept any exact record? Or do you know any careful Breeder that does? —
Do not think of answering this unless you have materials to judge by[7]. — It must be a cock & bull story, though given rather positively by goodish man. —

Postmark: Postmark: JA 10 61
Christ College Library, Cambridge, #126

Notes

1. No details on the inkstands are available; presumably the inkstand referred to in Letter #142 was one of these. CD was correcting proofs of the third edition of *On the Origin of Species* [see Letter #142 (n7)].
2. At last CD admits that his handwriting is bad! There was a time when CD felt obliged to defend his handwriting [Letter #38 (n6)]. Interestingly, Erasmus Darwin, his grandfather, had good handwriting, so it was not altogether hereditary, in the male line!
3. For Henrietta's (Etty's) illness see Letter #141 (n4).
4. According to Emma Darwin's *Diary*, William Erasmus Darwin came home from Cambridge University on 13th December 1860. George Howard and Francis Darwin, who were attending Clapham Grammar School in South London, stayed at Down until 4th February 1861. Leonard Darwin, now nearly eleven years old, would go to Clapham Grammar School shortly after this date. Horace Darwin was nine and a half years old and still being tutored at home.
5. CD was preparing material, which would eventually go into Chapter 3 of *The variation of animals and plants under domestication* (1868). There, CD discussed variations in pigs, cattle, and horses in several places. On p 74 of Vol 1, he stated: "*The Rev. W. D. Fox has given me ten carefully recorded cases with well bred pigs, in which the period varied from 109 to 123 days. According to Nathusius*

the period is shortest in races which come early to maturity." Thus the German work may be that of Nathusius: CD's abstract, bound in the annotated copy of Nathusius, 1860 in the Darwin Manuscript Collection, University Library, Cambridge, is dated "*Jan. 14/61*". However, this particular remark has not been traced.

6. CD was interested in the correlation of colour with other characteristics and had discussed this with Fox on a number of occasions (see, e g, Letters #110 (n11) & # 124 (n7). CD used these observations in *On the origin of species* and in *The variation of animals and plants under domestication* [see Letter #110 (n11)].
7. It is clear that Fox was able to provide the information (see Note 5, above).

144 (CCD 3204). To: The Rev W. Darwin Fox, Delamere Rectory, Northwich

2. Hesketh Crescent
Torquay
July 8th [1861]

My dear Fox

Your note has followed us here, where we arrived 4 or 5 days ago[1]. — As we wished to try 6 or 8 weeks of sea-side for my eldest girl we thought we would make this awfully long (to us) stretch & come here. We are charmed with the view from this crescent & with the walks all around & we have got a very good// House. But is no joke bringing 16 souls & 3/4 tun of luggage so far. — My eldest girl is still a sad invalid; but certainly improves: she gets up twice every day now & can walk one or two hundred yards[2].

I am glad for myself to have this outing & change for I have been a poor wretch for many months. You say not one word about yourself. Are you not a pretty sort of// man[3]? It seems indeed strange to hear of your having two daughters married[4]. —

Poor dear Henslow's death has been a sad loss to many. He wrote me by dictation a most kind note from his death-bed[5]. — L. Jenyns is going to write a biographical notice of him[6]. —

I shall not go to Manchester[7]; but it would be a great temptation// to get a sight of you. — How long it is since we met! Farewell my dear old friend

Yours affect^y
C. Darwin

My eldest son is going to join as Partner in a Bank at Southampton: I had so good an offer it seemed a pity to reject it[8]. —

We hear that the Darwins of - - -[9] (Elston) are coming here[10]. —

Postmark: JY 8 61

Christ's College Library, Cambridge, # 131

Notes

1. The Darwins left for Torquay on 1st July 1861, spending the night *en route* in Reading (see CD's *Journal*. In her book *Emma Darwin; A Century of Family Letters* (Litchfield, 1915), Henrietta Darwin wrote: "*In June, 1861, we went to Torquay, and there I began to get well. It was a happy time. My father was fairly well, and the boys full of enjoyment. We had our customary visitors, Erasmus Darwin, and Hope, Hensleigh Wedgwood's youngest daughter*."
2. But see Note 1.

3. Fox's *Diary* for 1861 shows that his health had improved in this year.
4. CD was slightly ahead of events. Fox's eldest daughter Eliza Ann Fox was married, to the Rev[d.] Henry Martyn Sanders, on the 29th May 1861. However, Harriet Emma Fox was not married, to Rev[d.] Samuel Charlesworth Overton, until 24th September. Both marriages took place at Delamere (Fox *Diary* for 1861).
5. John Stevens Henslow (1796-1861), had been a teacher, mentor and friend to both CD and Fox (see Introduction to Chapter 2) and was much mentioned in their correspondence (see Letters #8 (n3), #25 (n8), #29 (n12), #30 (n11), #35 (n2), #36 (n1), #39 (n4), #40 (n3) & #41 (n8), #43/5, #44/3, #48/15, #49/26, #53/9, #58/3, #62/4, #67/3, #71/7, #72/2 and many subsequent letters). The letter mentioned by CD has not been found.
6. Darwin's memoir of John Stevens Henslow was published in *Memoir of the Rev. John Stevens Henslow*, edited by L Jenyns, London, 1862 (see CCD, vol 9, Appendix X).
7. The British Association for the Advancement of Science held its 1861 meeting in Manchester. It appears that Fox, too, did not attend this meeting, possibly because of the wedding of his second daughter in September (see Note 4).
8. CD arranged, and financially underwrote, a partnership for his son William Erasmus Darwin with the Southampton and Hampshire Bank. William was studying law at Christ's College, University of Cambridge, and this meant that he terminated his studies early. See letters to W E Darwin, 25th May 1861 and to John Lubbock, 10th July 1861, 1st August 1861 and 2nd August 1861 (CCD, Vol 9).
9. Sic. CD must have forgotten to fill in the dashes.
10. This was Francis Rhodes Darwin, husband of Charlotte Maria Cooper Darwin, the granddaughter of Erasmus Darwin's elder brother, William Alvey Darwin (Fox's great grandfather). The inheritance of Elston Hall is discussed in Letter #54, Notes 12 & 13. Fox met Rhodes Darwin at Cheltenham later this same year, on 7th November (Fox *Diary* for 1861).

145 (CCD 3544). To: The Rev W. Darwin Fox, Delamere Rectory, Northwich

Down Bromley Kent
May 12th [1862]

My dear Fox.

I am going to bother you. Looking over some of your old notes, I see that you have kept the wild breed of Turkeys from L[d.] Leicester & Powis[1]. You know that they now say that the common Turkeys have descended from a southern <so-called> species[2]. Have you ever crossed intentionally or accidentally your wild & common; & did you ever cross the hybrids inter se or//

with either pure parent & were they <quite> fertile? Have you ever given half-bred birds to other people, & did they with them become mingled with common Turkeys? Can you recognise the half-breds by their appearance? I sh[d] be grateful for any information, which I might quote on your authority[3]. — When you write tell me how *you* <&> all are[4].//

We were very glad to see your son at Torquay[5]. I am much as usual, always grumbling & complaining. We have of late had much anxiety about our youngest Boy, who has failed in same way, but worse than, four other of our children[6]. This is very shabby note, but I am tired with having written a heap of letters. — My dear old friend. —

Yours affect[ly].
C. Darwin

Do you know anything of so-called Japanned Peacocks suddenly appearing from the common Peacock?[7]

Endorsement: "/62"

Christ's College Library, Cambridge, #132

Notes

1. Fox's reply has not been found. However, in *The variation of animals and plants under domestication* (1868), CD records (Vol 1, p 293): "*Wild turkeys, believed in every instance to have been imported from the United States, have been kept in parks of Lord Powis, Leicester, Hill, and Derby. The Rev. W. D. Fox procured birds from the two first-named parks, and he informs me that they certainly differed a little in form from each other in the shape of their bodies and in the barred plumage on their wings.*"
2. In 1856, John Gould described a new species of turkey, *Meleagris mexicana*, a native of Mexico, which Gould believed to be distinct from the wild turkey of North America (Proceedings of the Zoological Soc of London, Vol 24, pp 61-63). However, in *The variation of animals and plants under domestication* (1868), Vol 1, p 292, CD stated: "*It seems fairly well established by Mr. Gould that the turkey, in accordance with the history of the first introduction, is descended from a wild Mexican form, which had been domesticated by the natives before the discovery of America, and which is now generally ranked as a local race, and not a distinct species.*"
3. In *The variation of animals and plants under domestication* (1868), Vol 1, p 292, CD states "*In England also, this same species has been kept in several parks; from two of which the Rev. W. D. Fox procured birds, and they crossed freely with the common domestic kind, and during many years afterwards, as he informs me, the turkeys in his neighbourhood clearly showed traces of crossed parentage.*"
4. From the Fox *Diary* for 1862 it appears that most of his large family was thriving. Fox still suffered from lung problems and in September was told to take a course of sea air and was advised to go to Rhyl at once [see letter #148 (n1)].
5. CD and family had spent July and August 1861 in Torquay (see Letter #144)). The son referred to may be the eldest son, Samuel William Darwin Fox, who at that time was at Wadham College, University of Oxford, reading divinity.
6. Horace Darwin had been periodically ill since the beginning of the year but was better in June (Emma Darwin's *Diary*).
7. In *The variation of animals and plants under domestication* (1868), Vol 1, Chapter 8, pp 290-292, CD described peacocks in detail and says: "*There is one strange fact with respect to the peacock, namely, the occasional appearance in England of "japanned" or "black-shouldered" kind.*" He then went on to discuss whether this type indicated descent from a distinct species or was due to natural variation. See also, letter to P L Sclater, 14th May 1862 (CCD, Vol 10).

146 (CCD 3555). To: The Rev W. Darwin Fox, Hillfields, Hampstead, London N.W.

Leith Hill Place
Dorking
Saturday [17th May 1862]

My dear Fox

Your letter has been forwarded here, where we remain till Thursday evening[1]. I am so very sorry to miss your visit[2]. Our house is utter confusion with painters & we shall, when we return have to live in a bed-room; but I think by Monday or Tuesday week one room will be ready; is there any chance of your being in London so long as that? —

Very many thanks for your letter about Turkeys, which interests me much[3]. If without very great trouble you could get me any more information about the fertility &c of the hybrids or mongrels, I shd. be very much obliged. —

In Haste
My dear Fox
Ever yours
C. Darwin

Postmark: MY 17 62
Christ's College Library, Cambridge, #133

Notes

1. From CD's *Journal* and Emma Darwin's *Diary* it is recorded that they spent from 15th to 22nd May 1862 at Leith Hill Place, the home, in Surrey, of his sister Caroline Sarah Wedgwood and her husband Josiah Wedgwood III, Emma Darwin's brother.
2. Fox and his wife visited "Hillfield", the home of his wife's parents (and whence this letter is addressed) from 5th to the 21st May 1862. Fox probably wrote to CD to suggest that he and his wife visit them at Down, but clearly the times available did not suit. In the event Fox, on the same visit, went to the Great Exhibition (Crystal Palace), his sisters at Kensington Park Gardens, Kew Gardens and the Star and Garter Hotel in Richmond (Fox *Diary*, 1862).
3. See Letter #145 (n1-3).

147 (CCD 3717). To: The Rev W. Darwin Fox, Delamere Rectory, Northwich

Cliff Cottage
Bournemouth
Sept. 12th [1862]

My dear Fox

Very sincere thanks for all the trouble which you have so kindly taken for me about the Turkeys: your information will be of use to me, whenever circumstances will permit me to finish my half completed volume[1]. — Your ancient case of the woman with a hairy face is curious[2]; for, as perhaps you know, there has of late occurred an instance in the// Malay archipelago; & the peculiarity was hereditary & accompanied by peculiarities in the teeth[3].

Thanks for all your sympathy about us: we have been most unfortunate; Horace seriously ill, in a strange manner, all the Spring[4]; & then Leonard came home from school with Scarlet Fever, had recurrent fever, with serious mischief in his kidneys & then bad erysipelas[5]. At last// we started for this place; but he suffered much from the journey & at Southampton Emma had Scarlet-fever pretty sharp[6]: we have been here for about 10 days & both my patients are going on admirably, & we have two Houses so that I trust the other children will escape. I have never passed so miserable a nine months[7]. — I hope we shall all get safe home in// about 3 weeks. I had thought of going to Cambridge[8]; but now I feel very doubtful & shall not make up my mind till the time arrives & I see how I am. If I do go it will be only for 2 or 3 days. It would indeed be most pleasant to meet you there; but I never know what I can do. All this misery has shaken me a good deal; but I am righting now. — Emma sends her kind remembrances to you.

My dear old friend
Yours affectionly
C. Darwin

If you go to Cambridge please tell me[9].
Postmark: Postmark: SP 13 62
Endorsement: "Sep 15/62"
Christ's College Library, Cambridge, #134

Notes

1. It appears that Fox supplied CD with more information on turkeys, the subject of CD's enquiry of two letters back [see Letter #145 (n1, 2 & 3)].

2. The reference is to the case of Barbara Van Beck, which Fox must have pointed out, as CD acknowledges Fox as the source in *The variation in animals and plants under domestication* (1868), Vol 2, p 4.
3. CD may have mis-cited the country. In *The variation in animals and plants under domestication* (1868), Vol 2, p 327, there is a detailed report of a Burmese family in which excessive hair growth, especially in the ears and on the face, and lack of molar teeth was observed over three generations.
4. See Letter #145 (n6).
5. Leonard Darwin was sent home from Clapham Grammar School on 12th June 1862, suffering from scarlet fever and he continued with fever into July (see Emma Darwin's *Diary*; also letter to W E Darwin, 13th June 1862 and, e g, the letter to Asa Gray, 14th July1862, CCD, Vol 10).
6. The Darwins stayed in Southampton with their son William (see note 7, below).
7. CD, Emma, and Leonard Darwin had started for Bournemouth on 13th August 1862, but were delayed at Southampton until 1st September by Leonard's relapse and then by Emma's scarlet fever [William Erasmus, CD and Emma's son, was a new partner in a bank there; see Letter #144 (n8)]. They then went to Bournemouth and stayed until 29th September, and were joined at various times by their other children (see CD's "*Journal*" and Emma Darwin's *Diary*).
8. In 1862, the British Association for the Advancement of Science held its annual meeting in Cambridge during the first week of October. Fox did not attend.
9. See note 7, above.

148 (CCD 3732). To: The Rev W. Darwin Fox, Delamere Rectory, Northwich

Cliff Cottage
Bournemouth
Saturday night 20th. [Sept. 1862]

My dear Fox

Your affectionate & pleasant letter has interested, pleased & grieved me. — I know that your lungs have often troubled you; but I greatly fear that this last attack has been more serious & has left mischief behind. If the sea-air suits you, it seems a thousand pities that you should not make everything bend to it & stay some time there.//

Your life is a precious one. Four days is nothing for the sea-air[1]. When I read of you & your dog I exclaimed to Emma, "how like Fox", but I did not know of Erasmus' saying[2]. I suppose you have given up all idea of Cambridge[3]; I have begun to turn tail & have resolved to go home on Oct 1st; but if I can screw my courage up, I may go there on Saturday or Monday; but it will depend on how my//

stomach is[4]. I shd like to see the old place once again, where I have spent so many happy hours; but I am not sure whether it will not be more melancholy than pleasant; for I know I shall feel knocked up & unable to ramble about & see the old haunts. What pleasant hours we have spent together at our alternate breakfast & teas. There were no fears & anxious looking forward in those days. And poor dear Henslow//

is gone[5]. About two years ago I stumbled at Down on a Panagæus crux major: how it brought back to my mind Cambridge days! You did me a great service in making me an entomologist[6]: I really hardly know anything in this life that I have more enjoyed that our beetle-hunting expeditions[7]; Prince Albert told Lyell, that he looked back with more pleasure to collecting insects, than he <had> ever found//

in stag-shooting[8]. I am *much* pleased & somewhat surprised at your liking my orchid-book[9]: the Botanists praise it beyond its deserts[10], but hardly anyone, not a Botanist, except yourself, as far as I know, has cared for it. The subject interested me much, & was written almost by accident; for it was half written as a mere paper & then I found it too long, & thought I would risk publishing it separately[11]. What you say about it, is//

very pleasant; for at one time I agreed with Lyell that I was an ass to publish it[12]. I have lately been making some curious observations on the "dimorphic" fertilisation of other plants[13] & likewise on their sensibility; & upon my life I am coming to the conclusion, that they must have something closely analogous to diffused nerve-matter[14]. But as you most truly say what a mystery life is; & a mystery one//

feels the more, the more one knows. — As soon as I get home, if we all can but keep well, I must return to variation under domestication[15]. I do not wonder that you found the jelly-fishes puzzles[16] to dissect; it would take weeks to get even a glimmering of their structure, — mere organised water[17]. —

I did not know that you were doubly a grandfather[18]. I will send your messages// to my sister[19]. Perhaps Erasmus, who has never stirred out of London all this summer, will come here for our last week[20]. He is very far from strong. All Darwins ought to be exterminated. Farewell my dear old friend; I do most truly hope that your health may improve & your lungs recover.

Farewell
Yours affectionately
Charles Darwin

Postmark: SP 22 62
Endorsement: "Sep 24/62"
Christ's College Library, Cambridge, #135

Notes

1. Fox wrote in his diary for 14th Sept. 1862: *Mr Morton examined my Chest well this morning and says "that in my Lungs the air passes feebly in, but fairly, but that with the Sound of the Inspiration, the out pouring of the breath is heard also, ought might not to be." He says the sound is very distant & feeble, but that the moment I speak or become excited, there is a sudden rush into the Lungs. He says the Lungs ought never to be used except as a Breathing apparatus – not for speaking especially continuously. He also said "that if it had not been for Sea Air he should say I shd have gone long ago." He wishes me to go to Rhyl at once – Sea Air & rest he hopes will quite revive me. Recommends me not to go out in damp weather. Mr John Harrison afterwards examined me and said " there was much feebleness in action of the heart, & that it did not send the blood well in the lungs."* Fox and his wife went to Rhyl, a seaside town in Flintshire, North Wales, from the 16th to the 20th September.
2. Erasmus' saying: this could refer to either Erasmus Darwin, CD's grandfather or to Erasmus Alvey Darwin, CD's brother. Sayings and epigrams of Erasmus Darwin, senior, were collected and appreciated (see, e g Letters #214 & #215). Later CD would write a book on his Grandfather, although it was officially published under the name of E Krause, 1879, *Erasmus Darwin. Translated from the German...with a preliminary essay by Charles Darwin* (see King-Hele, 2003).
3. See Letter #147 (n7).
4. CD did not go to Cambridge (see CD's *Journal*) and neither did Fox.
5. See Letter #144 (n5).
6. This is one of the few explicit statements where CD acknowledges his debt to Fox for introducing him to entomology (or, at least, beetle-hunting).

7. CD and Fox often reminisced about their beetle hunting days and especially a well-remembered excursion to Whittelsea Meer and the capture of *Panagæus crux-major*, during their undergraduate days [see *Autobiography* and Letters #47 (n5 & 6), #48 (n17), #61 (n18), #90 (n17), #96 (n12), #115 (n7), #131 (n11)].
8. CD refers to Charles Lyell and Albert Francis Charles Augustus Emmanuel, husband of Queen Victoria, and Prince-Consort of England. Prince Albert died, somewhat unexpectedly, on 14th Dec 1861, and was much mourned. As recorded at the time in Fox's Diary for 1861.
9. *On the various contrivances by which British and foreign orchids are fertilised, and on the good effects of intercrossing* was published in May 1862. CD had been interested in orchids and their mechanisms of pollination for many years [see Letter #115 (n11)], but there is little reference to this book in previous correspondence, and it seems it was a spur-of-the-moment decision (see note 11).
10. Among the botanical admirers of his book on orchids were George Bentham, Daniel Oliver, and Asa Gray (see letter to Asa Gray, 10-20th June 1862, CCD, Vol 10).
11. CD had originally thought of publishing a paper on the pollination mechanisms of various species of orchids for the Linnean Society of London, and prepared a manuscript in July and August 1861. From correspondence (see letter to John Murray, 21st September 1861 and letter from John Murray, 23rd September 1861), he proposed publishing it as a book in September 1861.
12. The exact source of this remark attributed to Charles Lyell is not known; but see letter to J D Hooker, 7th March 1862, note 4, where CD uses the description "*ass*" on himself.
13. CD read a paper, "*Dimorphic condition in Primula*", before the Linnean Society in November 1861. He then continued to work on the phenomenon of heterostyly in 1862. These studies culminated in two books: *The effects of cross and self-fertilisation in the vegetable kingdom*, Murray, 1876 and *The different forms of flowers of the same species* Murray, 1877 (see Letters #202 & #206).
14. At this time CD outlined his experiments on *Drosera* and other insectivorous plants, in letters to Daniel Oliver, 17th September 1862, and J D Hooker, 18th September 1862 and 26th September 1862. These researches culminated in two books: *Insectivorous Plants*, Murray 1875 (see Letters #202. #203 & #204) and *The power of movement in plants*, Murray, 1880 (published at the time of Fox's death).
15. As we have seen in previous letters, from 1856 onwards, CD had been preparing a "big book" for which the tentative title was *Natural Selection*, which was intended to comprise 3 volumes [see Letter #139 (n5)]. In the introduction to the *Origin*, CD described his book as an abstract of a larger work he was preparing on *Natural Selection* (p 1). The first book of the series was to be entitled "*Variation under domestication*" (see letter to T. H. Huxley, 16th December 1859, and to John Murray, 22nd December 1859). CD had been working on the manuscript intermittently since January 1860, and had reached chapter eight on "*Silk-worms Geese &c*" by the summer of 1862 (see CCD, Vol 10, Appendix II). In the event this "first" book was published in 2 volumes as The *variation in animals and plants under domestication,* by Murray in 1868. Other parts of the big book were published elsewhere, or given to George Romanes to publish or not published at all [see Stauffer, 1977; Letter 139 (n5)].
16. Francis Darwin's copy of this letter has "puzzlers" here (University Library, Cambridge, DAR 144: 135-246).
17. Jelly fish are notoriously difficult to dissect. However, T H Huxley, CD's greatest defender at this time, made his name doing just that on *HMS Rattlesnake* (Desmond, 1994).
18. Fox had two?¿ married daughters [see Letter #144 (n4)]: Eliza Anne Sanders gave birth to her first child, Charles Henry Martyn Sanders, on 21st March 1862 [Fox *Diary* 1862; *Gentleman's Magazine* n.s. 12 (1862): 638]; Harriet Emma Overton also gave birth to her first child, Frederick Arnold Overton, on 15th July 1862 (Fox *Diary*); see also *Darwin Pedigrees*, pp 15-16).
19. CD had three surviving sisters, Caroline Sarah, Emily Catherine, and Susan Elizabeth Darwin. CD may be referring to Susan Elizabeth Darwin, as she was a close friend of one of Fox's sisters, Julia.
20. Erasmus Alvey Darwin lived at 6 Queen Anne Street, Cavendish Square, London (see CDD, Vol 10). CD and family stayed with him on 29th September 1862, on their return from Bournemouth (Emma Darwin's *Diary*).

149 (CCD 3732). To: Charles Darwin, Down

Delamere Rectory, Northwich, Cheshire
Feb[y] 6 [1863][1]

My dear Darwin

It is now long since I have heard any thing of you[2]. Is there the least hope of your being in or near London from middle of next week this the following one[3]? I should so very much like to see you again. I am going to take my 3rd Boy to a school// near Maidenhead[4], and must be a few days in or about London. How our children grow up — My eldest has been now near two years at Oxford[5] — My 2[d] is in 6th Form at Repton, and promises to be clever[6] — My 3[d], as I said going to Maidenhead[7] — My 2 eldest girls each with a fine boy[8] — and the other 10 running up like willows[9].

Yours I suppose are doing rapidly — while we (as one of my little ones said to her Mother a few days since) go down as they go up.//

Are you a Believer in Free Martins[10] — I have always found them true in practise till now. A Lady near here, whom I know well, and her twin Brother also — has one fine Baby & just had a miscarriage of another. How are Caroline Susan & Catherine[11]? What ages it is since I have seen them — The former never since her Marriage. I sh[d] much like to <see> her & her two girls[12].
Susan I mean to offer to pay a visit to ere long, if she will have me.

That I may not tire you out I will conclude with our kindest regards to M[rs] Darwin — I am your old friend & cousin W D Fox

(*across the head of the letter*)
M[r] Woodds
Hillfield
Hampstead NW
after next Wednesday[13]

Provenance: University Library, Cambridge; Manuscript Room: DAR 164: 176

Notes

1. The year of this letter is established by the reference the time of Samuel William Darwin Fox, Fox's eldest son, at Oxford University (see note 5, below).
2. The last known letter from CD was on 20th Sept 1862 (see Letter #148).
3. CD arranged to meet Fox in London on 13th February 1863 (see Letters #150 and #151).
4. Robert Gerard Fox (1849-1931). In his diary Fox wrote for "*Took Gerard to Mr Prices new school at Maidenhead*".
5. Samuel William Darwin Fox entered Wadham College, Oxford in 1861: "*I was with Darwin at Oxford & he matriculated at Wadham College this day. He came out first in Exam.*" (Fox *Diary*, 5th June1861).
6. Charles Woodd Fox, went to Repton School (Fox's School), near Derby, which he entered in1859 (*Repton School register*) and then went on to Christ Church, Oxford University in 1865 (*Alum Oxon).*
7. See Note 4, above.
8. See Letter #148 (n18).
9. Fox had 15 surviving children at this time.
10. Freemartin or free martin: a "hermaphrodite or imperfect female" (*OED*). See further reference in Letter #175.

11. CD's surviving sisters [see Letter #148 (n19)].
12. Caroline Sarah Darwin married Josiah Wedgwood III and there were three surviving daughters of the marriage, Katherine Elizabeth Sophy, Margaret Susan, and Lucy Caroline Wedgwood (*Darwin Pedigrees*).
13. Fox's father-in-law, Basil George Woodd, lived at Hillfield, Hampstead, London N.W.

150 (CCD 3975). To: Rev. W. D. Fox, J. Woodds Esquire, Hillfield. Hampstead N.W.

6. Queen Anne St W.[1]
Tuesday Eveng [10th Feb. 1863]

My dear Fox

Your note has been forwarded here, where we have all been staying for a week. I have been bad enough of late & came here to see whether a change wd. do me some good & it has succeeded. — I suppose from your note that you arrive at Mr Woodd's[2] on Wednesday Evening. — We return home on Friday (*possibly*, but improbably on Saturday —) & leave this house at 2°30' for the Train. —

Now would it suit you to come here to lunch at 1° or 1°15' on Friday; or on Thursday, but on that day Emma will probably be out. — I do hope you will be// able to come for I shall be so sincerely glad to see you. — Can you come to Down; we shd be very glad; but my stomach has got to such a pitch that I can seldom stay, not even with nearest relations, for more than half-hour at a time. Let me have a line, that I may be sure to be at home, whenever// you can come[3]. —

My dear old friend
Ever yours
C. Darwin

Endorsement "Feb 11 1863"
Postmark: Postmark: FE 11 63 London 10
Christ's College Library, Cambridge, #136.

Notes

1. CD stayed at 6 Queen Anne Street, London, the home of his brother Erasmus Alvey Darwin, from 4th to 14th February 1863 (see CD's "*Journal*").
2. Presumably Basil George Woodd, the father of Fox's wife, Ellen, who lived at Hillfield, Hampstead, London NW.
3. In the event Fox met CD at his brother's house on 13th Feb. 1863 (Fox *Diary* for 1863). Fox records in his diary for that day: "*Spent 2 hours at Erasmus Darwin's – Charles and Wife – eldest Dau & youngest Son there – also Catherine Darwin*".

151 (CCD 3975). To: Charles Darwin, [6 Queen Anne St, W]

~~Dela~~ Hillfield Hampstead NW
[11th Feb 1863][1]

My dear Darwin

Many thanks for your note. I am truly sorry to hear you are so poorly[2].

I scarcely hoped to see you, and am very glad to be able to do so. I will certainly come on Friday about 1— and shall rejoice to see Er*as*mus at same time[3].

Many thanks for your *invitation*[4] to Down, but it is impossible for me to accept it—nor do I think it w[d] be anything but a teaze to you if I did.

Had I not been able to get a glimpse of you in Town, I might have tried to run down for a few hours.

Sincerely dear Darwin
Yours
W D Fox

University Library, Cambridge: Manuscript Room: DAR 164: 177
Slightly damaged

Notes

1. The probable date has been determined by the relationship between this letter and Letter #140, which was written on the evening of Tuesday, 10th Feb. 1863.
2. See Letter #150.
3. Erasmus Alvey Darwin, who lived at 6 Queen Anne Street, London W. For Fox's visit there on 13th Feb 1863, see Letter #150 (n3).
4. Obscured by damage.

152 (CCD 4033). To: Rev. W. D. Fox, Delamere Rectory, Northwich.

Down Bromley Kent
March 9th [1863][1]

My dear Fox

I have just been quoting in my M.S. on your authority from an old letter that you crossed White Muscovy Drake with slate-coloured duck & that the young were always pied black & white like the common or aboriginal breed[2]. — I have many analogous facts in common Ducks, fowls, & pigeons & the case interests me much. Now can you tell me// do the White Musk & Slate-coloured musk, (when not crossed) each breed quite or nearly true?[3] Again in same letter I have quoted that "12 white ewes of M[r] Woodd's had 23 coal-black lambs by a Ram that had a small patch of black only".[4] Can you tell me what breed these ewes were? Was Ram of same breed? Do you know how big & where the black patch on ram was? —

I should almost expect that if two very different breeds of *white* sheep were crossed there would be some tendency to dusky lambs; & so with horns, if two hornless breeds were crossed; but I know not where to enquire. I am even inclined to *suspect* that there is a tendency// to a return to primordial wildness in hybrids between two domestic species[5]. — I enjoyed much seeing you in London[6]. — My London trip did me some good; but I have since had a very bad fortnight; & Emma declares, I fear with truth, that we must all go for 6 or 8 weeks to Malvern[7].

My dear old friend
Yours ever sincerely
C. Darwin

Postmark: MR 10 63
Endorsement: Endorsement: "March 11. 1863"
Christ's College Library, Cambridge, #138

Notes

1. The date of this letter is given by the reference to the meeting with Fox on the 13th February 1863 (see Note 6).
2. The Muscovy duck originated in Mexico and the male has a unique, dark head and crest colouration. The musk duck originated in southern Australia and it seems unlikely that CD is referring to this species. Fox's reply, which has not been found, was presumably what CD used in his final word on the subject (see Note 4), but it appears from that version that he conflated musk duck and Muscovy duck.
3. See Letter #106 (CCD 1836), in which CD thanked Fox for "*particulars on the eggs and colour of the Muscovy Ducks & sheep*". The crosses conducted by Fox are briefly mentioned in *The variation of animals and plants under domestication (1868)*, Vol 2, p 40.
4. This information, presumably with updated information sent by Fox (see Letter #153), is given in *The variation of animals and plants under domestication (1868)*, Vol 2, pp 20-21. The Woodd referred to is either Basil George Woodd, of Hillfield, Hampstead, Fox's father-in-law, or Charles Henry Lardner Woodd, of Roslyn, Hampstead, Fox's brother-in-law.
5. CD was battling two major ideas here, which had been the subject of the correspondence with Fox for over a decade. Firstly, through breeding it is possible to throw up new characters, which are not found in the wild species. Secondly, there is a tendency for reversion, that is for hybrid (or bred) varieties to revert to the wild type (see CCD, Vol 12, letter to Hooker, 13th Sept. 1864, note 5). Without a knowledge of Mendelian genetics this was an impossible task, and even today it is a vexed question. Chapters 7, 8 and 9 of *The variation of animals and plants under domestication (1868)*, where these ideas were finally set down, cover this subject and are summarised at the end of Chapter 9 (*Inheritance*, continued).
6. Fox visited CD at the house of his brother, Erasmus Darwin, on 13th Feb 1863 [see Letter #150 (n1)].
7. CD was referring to the hydropathic establishments at Malvern, and especially that of Dr Gully, visited by both CD, Fox and their families in the past ten years or more [see Letters #79 (n8), #80 (n3), #82 (n1), #83 (n2), #85 (n1), #88 (n8), #114 (n2) and the Fox *Diaries*].

153 (CCD 4037). To: [Charles Darwin, Down, Bromley, Kent]

Delamere Ry Northwich
March 12 [1863][1]

My dear Darwin

I remember well what you mention about the slate coloured & White Muscovy Ducks[2]. The Slate coloured, I imagine, were the produce of some cross raised at Birmingham. I have never seen them since nor do I think I ever reared any, as I lost the Drake.

The white always breed true to colour, as far as my experience goes, & I keep a large flock of them — (I have about 20 now) —

With however this exception that they have a great tendency// to a Black Topping on head (exactly as represented in Bewicks plate)[3].

Unless some considerable pains are taken in weeding out all these black topped Birds — the *pure white* would soon cease.

I never saw however a single black feather any where but on the top of Head.

The Black Lambs I also well remember[4]. They were Southdown Ewes. The Ram I never saw — He was provided by Giblitt the Great Butcher[5] — and I was told in answer to my enquiries that he was all but white, having a small patch only of Black — I *think* about the head. The flock was the queerest I ever saw, and *every* lamb being black in the lot, looks as if there was a positive law of some kind in the matter.

I will write to ask if Mr Woodds man[6] remembers where the Black was in Ram, but I fear it will not be satisfactorily answered.

The Ram appeared cross bred with a good deal of Southdown in him.

The only *pure* cross of Sheep I know (if one can use such a term) — are between pure Leicester & Southdown. I think it was the late Earl Morton[7] who had a beautiful flock of these. They had no horns—& partook of each strain.

I have a small flock of the black & white sheep — which keep very true to shape & colour. When crossed with Leicester they produce a race// pretty nearly balanced — except that there is a great tendency to Black lambs. The only 3 born this year are black. Having once got the black strain in this way from a cross with the black & white & Leicester—it takes many generations — at least 6 or 7 — I am told by a friend of mine — to get back the white colour — by continual crossing with a Leicester. I can only speak to the *3^{d}* generation — which was as black as the first[8]. I have never seen or heard of *horned* produce from hornless crosses. I do not think they exist so in our English crosses. The Cheviot & Leicester are often crossed — & certainly have no horns — Neither have the Southdown & Leicester. There seems little tendency in crosses of *Sheep* to return to primordial wildness — as the Irish have been almost renovated by the Leicester cross: and the Shropshire Down sheep are a greatly improved cross bred sheep.//

2.
I fear I have not added much to my former notes on these points[9].
What I wrote at time, may be strictly depended upon — In this letter you will see what are surmises only.

I should be very glad to hear of your getting to Malvern for 6 or 8 weeks[10]. It is a glorious place for renovating the health in. If my time & means would allow of it, I sh^{d} go for a month every spring & be got into condition. I quite believe I owe my life to it in my last illness[11] — and it was a curious illus=tration of *the System* versus the D$^{rs.}$ as Gully[12] completely denied the existence of what I told him was my conceived idea of my wretchedness — but added "Well if it is so, the// water cure will prove its existence — and so it did. I have never met with more than one similar case, which was most carefully treated according to rule, and the young man died of it.

I cannot tell you how glad I was to see you so well in London[13]. It was a great pleasure to see so many old & unlined faces — & all looking so well[13]. I very much want to see Susan and Caroline again[14].

If you go to Malvern I shall try to come over & get a weeks bathing while you are there. I have some curious enquiries about you & your Book sometimes. You ought to occupy a cage by the side of the Gorilla in the British Museum sometimes to let the Public see you[15]. —

Remembrances to Mrs Darwin. —
Ever yours sincerely
W D Fox

Provenance: University Library, Cambridge; Manuscript Room: DAR164: 178

Notes

1. The year is established by the relationship to Letter #152, which was written in 1863.
2. See Letter #152.
3. Bewick, T. 1826. *A history of British Birds*, Vol. 2, p. 317, Newcastle. "Topping" is the crest of a bird (*OED*); see Letter #152 (n2).
4. See Letter #152, Note 4.
5. CCD, Vol 11, suggests that the reference is possibly to William Giblitt, who had a butcher's shop at 110 Bond Street.
6. The reference is either to Fox's father-in-law, Basil George Woodd, of Hillfield, Hampstead or his brother-in-law, Charles Henry Lardner Woodd, of Roslyn, Hampstead.
7. George Sholto Douglas (1789-1858), 17th Earl of Morton (*Modern English Biography*).
8. See Letter #152, Note 4.
9. Fox's original letter on these subjects has not been found; but see Letter #152 (n4).
10. See Letter #152 (n7).
11. It is not clear to what illness Fox is referring. It was possibly in the years for which Fox *Diaries* do not exist (1853-57 and 1859-60).
12. Dr James M. Gully was the proprietor of a hydropathic establishment at Malvern, which both CD and Fox went to in former years [see Letters #79 (n8), #80 (n3), #82 (n1), #83 (n2), #85 (n1), #88 (n8), #114 (n2)].
13. Fox met CD at the house of his brother, Erasmus Darwin, at 6 Queen Anne Street, London, on 13th February (see Letters #150 and #151). Also present, besides CD's brother, were Emma Darwin, Henrietta Darwin, Horace Darwin and CD's sister, Caroline Darwin [see letter #150 (n3)].
14. CD's sisters, Susan Elizabeth Darwin and Caroline Sarah Wedgwood; see also Letter #149.
15. Many cartoons at this time featured CD as a primitive man or gorilla-like creature. Note also that at this time gorillas were largely mythical beings, the first dead specimens arriving in England only in 1861.

154 (CCD 4044). To: The Rev W. Darwin Fox, Delamere Rectory, Northwich

Down [Bromley Kent]
16th [March 1863][1]

My dear Fox

Very many thanks for your to me very interesting letter[2]. — I did not think it likely that you could add much. — if you should hear what Ram was I shd. like to add it[3]. After writing I found <sentence> in your old letter, which I had overlooked saying that you believed white & slate Musks breed true, so I have// put it in *cautiously*[4]. — These facts interest me greatly[5]. I suppose we must go to Malvern; but it breaks my heart[6]. — I am tired with a lot of letters, so goodBye with many thanks

Ever yours
C. Darwin

Postmark: MR 16 63
Endorsement: "March 17/63"
Christ's College Library, Cambridge, #137

Notes

1. Dated by reference to Letter #153.
2. See Letter #153.

3. No such reply has been found, but see the letter #153, where Fox says: "*The Ram appeared cross bred with a good deal of Southdown in him.*" See also CD's description of a goat-like ram in *The variation of animals and plants under domestication (1868)*, Vol 2, p 30.
4. This letter from Fox has not been found; however, see Letter #152 (n2). The crosses conducted by Fox are briefly mentioned in *The variation of animals and plants under domestication (1868)*, Vol 2, p 40.
5. See Letter #152 (n5).
6. See Letter #152 (n7).

155 (CCD 4057). To: The Rev W. Darwin Fox, Delamere Rectory, Northwich

Down, Bromley, Kent. S.E.
March 23d [1863][1]

My dear Fox

I am particularly glad to have the authentic particulars & your own cases; both of which I shall give on your authority[2]. In the old note I, no doubt, read **2**3 for **1**3 <lambs>: at the time I supposed that the ewes had bred for two seasons. — The whole case seems a good one for showing preponderance of one colour over other[3]. —//

Many thanks. —
We doubt a little about Malvern, for I am having an attack of Eczema on my face, which does me as much good as Gout does others; & which renders cold // water to my face intolerable; so that I could not stand water-cure of any kind, as I found at Ilkley, when I had first attack of this horrid & blessed eczema[4]. —

Incomplete, second half of second page missing.
Provenance: American Philosophical Society, Philadelphia (292)

Notes

1. The year is established from the two previous letters.
2. This letter from Fox has not been found, like the majority of his letters to CD; however, see Letters #153 and #154. The information given by Fox is mentioned on his authority in *The variation of animals and plants under domestication (1868)*, Vol 2, pp 30-31 and p 40.
3. Fox's letter has not been found; however, see Letter #152 (n2).
4. CD was referring to James Manby Gully's hydropathic establishment at Malvern, [see Letter #153 (n12)], and to his visit to a hydropathic establishment at Ilkley, Yorkshire in 1859 (see Letter #123). For CD an attack of eczema was not altogether a bad thing, as he often felt better for it (see, e g, letter from J D Hooker, 5th March 1863, CCD, Vol 11).

156 (CCD 4178). To: [Charles Darwin, Down, Bromley, Kent].

[Delamere Rectory]
[Northwich]
[16-22 May 1863][1]

My dear Darwin

About two months since you wrote me you were suffering much[2] — I hope you are now quite yourself again. Do you ever see "the illustrated Times Newspaper". I was much amused by that of May 2 – 9 – 16 containing a quizz upon you, which I yesterday met with at my Neighbours[3]. If you// have not seen this additional proof of

Natural developement, I will send the papers to you. They will give you ten minutes a=musement.

I hear Dr Gullys brain has quite broken down and disabled him from work[4]. He told me when I last saw him that it must do so soon.

He will be a great loss to a vast number of people. There is no one// at Malvern fit to take up his place I fear. Do you ever see Dr Lane now? I think I saw an advertisement that he was practising near Kew — am I right[5].

With kindest regards to Mrs Darwin

Ever your aff Cousin
W D Fox

Provenance: University Library, Cambridge: Manuscript Room, DAR 164: 175

Notes

1. The possible dates are determined by reference to the issues of the *Illustrated Times* for 2, 9, and 16th May 1863 (see Note 3, below), and by the relationship to Letter #157.
2. See Letter #155.
3. Fox refers to three humorous cartoons and a poem (*Illustrated Times*, 2 May 1863, p. 317, 9 May 1863, p. 333, and 16 May 1863, p. 348). These cartoons were the first instalments in a twenty-part series (CCD, Vol 11). See Fig 27 for the cartoon on May 2nd: "*The origin of species, dedicated by natural selection to Dr. Charles Darwin.*" No. 1. — *Like a Bull at a gate* (drawn by C H Bennett).
4. James Manby Gully was the proprietor of a hydropathic establishment at Malvern where both CD and Fox had been treated in the past decade or more [see, e g, Letters #79 (n8) & #153 (n12)]. It seems that Dr Gully was under some stress at this time although the details are not known. Much later in the 1860s and early 1970s, he retired and was a suspect in the notorious poisoning death of Charles Bravo (the subject of several books and television documentaries). Dr Gully was probably innocent but it ruined his last years (Mann, 1983).
5. Edward Wickstead Lane previously the proprietor of the Moor Park hydropathic establishment, where CD had been treated many times [see Letters #112 (n4) #114 (n4), #118 (n3 & 4), #125 (n1), 125A (n 2), #134 (n4), #135 (n10)]. In 1860, CD attended Lane's new hydropathic establishment at Sudbrook Park, Petersham, near Ham, Surrey [see Letter#118 (n4)], which he operated with his brother. Kew is about three miles north-east of Petersham. CD attended Sudbrook Park from 28th June to 7th July 1860 (CD's "*Journal*"), but it was not mentioned in his correspondence to Fox.

157 (CCD 4181). To: Rev. W. D. Fox, Delamere Rectory, Northwich.

Down. Bromley. Kent. S.E.
May 23^d^ [1863][1]

My dear Fox

Many thanks for your kind note of enquiry[2]. I cannot say much for myself[3]. We have lately been staying at Hartfield & Leith Hill Place (& I gave to Caroline your kind messages) for a fortnight[4]; but the change did me no good & I have been mostly in bed for the last week from my old enemy sickness. We gave up Malvern on account of my Eczema[5]; but it is all gone & perhaps after the holidays we may go there, unless I improve. Gully will be a great loss & I hardly know whom to consult there. I must be under some experienced man, for I could not stand much hard treatment. All this everlasting illness has stopped my work much. — I am glad you told me about Gully, for I had heard only a rumour[6]. —

Thanks about Illustrated Times, but I have seen it[7]. There has been a much better squib, on Owen & Huxley about the Brain; part of which appeared in Public Opinion[8]. — // I hope the world goes pretty well with you my old friend. I cannot say it does with me. Our youngest Boy is a regular invalid with severe indigestion, clearly inherited from me[9].

Farewell
Yours ever truly
C. Darwin

Postmark: MY 24 63
Endorsement: Endorsement: "Darwin May 26 1863"
Christ's College Library, Cambridge, #139

Notes

1. The year is determined by the relationship to the previous letter (Note 2) and the endorsement.
2. See Letter #156.
3. CD's health had been poor since the end of February 1863 (see the *Diary* of Emma Darwin).
4. See Letters #149 and #153. From 27th April to 13th May 1863, according to Emma Darwin's *Diary*, the Darwin family visited Hartfield Grove, Hartfield, Sussex, and Leith Hill Place, near Dorking, Surrey, the homes, respectively, of Charles Langton and Charlotte Langton (née Charlotte Wedgwood), Emma's sister, and Josiah Wedgwood III and CD's sister, Caroline.
5. For CD's eczema and the effect on him see Letter #155 (n4).
6. According to Fox's report in his recent letter (Letter #156): "*Dr Gullys brain has quite broken down and disabled him from work*", but see note 4 of that letter. In the event, CD consulted James Smith Ayerst [see Letter #159 (n1)].
7. See Letter #156 (n3). The reference is to a series of cartoons in the *Illustrated Times*, in May 1863 – see Fig 27.
8. The references are to an anonymous publication in 1863, part of which was published in *Public Opinion,* Vol 3 pp 497-8. For the "squib", see CCD, Vol. 11, Appendix VIII. Also on the date of this letter (23rd May), in *Public Opinion*, was a letter, which continued a spate of articles on the antiquity of man (following Sir Charles Lyell's book of that name). This letter by "Anthropes" denied a close link of man with the Gorilla, due to lack of fossil finds of the gorilla in contradistinction to Man. The reference is to Richard Owen (1804-1892) and T H Huxley (1825-1895), prominent anatomists involved in the public discussion of the origins of man, and antagonistic to each other's views.
9. Horace Darwin, CD's youngest living child, had been ill since January 1862, and CD's health had been poor since the end of February 1863 (see Emma Darwin's *Diary;* and Colp, 1977).

158 (CCD 4193). To: [Charles Darwin, Down, Bromley, Kent]

Delamere Northwich
May 29 [1863][1]

My dear Darwin

I am indeed sorry to receive such a poor account of you[2]. As Gully is not to be had[3] — I should recommend you a trip to Ben Rhydding[4] near Ilkley in Yorkshire. Dr MacLeod is very careful and the place is a very nice one in many ways[5]. Why not go there for a fortnight *before* the holidays if you are so poorly — taking only Mrs Darwin with you?

I fear your work will kill you at last. You want cessation from Brain work more than any thing else.

As a proof I think so I will not add more// than our kind regards to M[rs] Darwin & yourself

Ever yours truly
W D Fox

Provenance: University Library, Cambridge; Manuscript Room: DAR 164: 179.

Notes

1. The year was determined by the relationship to Letter#157 and the reference to Dr Gully.
2. For details see Letter #157 (n4).
3. See Letter #156 (n4).
4. Ben Rhydding was at that time a small village near Ilkley where a hydropathic establishment had been started in the 1840s. Fox knew the place well because his sister-in-law, Miss Fletcher, and Admiral Mudge, his relation through his first wife, resided there, before Mudge's death on 28th October1852.
5. William MacLeod was the senior physician at the hydropathic establishment at Ben Rhydding, near Otley, Yorkshire (*Medical Directory,* 1863).

159 (CCD 4292). To: Rev. W. D. Fox, Delamere Rectory, Northwich.

Villa nuova Malvern Wells
Friday 4[th.] [Sept.1863]

My dear Fox

I have had a bad amount of sickness of late & came here yesterday; & as I could get no country in Grt. Malvern we are here & I have put myself under D[r]. Ayrehurst, though very sorry not to be under D[r]. Gully[1]. Emma came here a day or two first & took this//

house[2]. And now I come to the painful subject which makes me write at once. Emma went yesterday to the church-yard & found the gravestone of our poor child Anne gone[3]. The Sexton declared he remembered it, & searched well for it & came to the conclusion that it has disappeared. He says//

the churchyard a few years ago, was much altered & we suppose that the stone was then stolen. Now some years ago, you with your usual kindness visited the grave & sent us an account[4]. Can you tell what year this was? I was so ill at the time & Emma hourly expecting her confinement that I went home//

& did not see the grave[5]. It is not likely, but will you tell us what you can remember about the <kind of> stone & where it stood; I think you said there was a little tree planted. We want, of course, to put another stone. I know your great & true kindness will forgive this trouble.

Yours affect
C. Darwin

Postmark: SP 4 63
Endorsement: "Sep /63"
Christ's College Library, Cambridge, #140

Notes

1. Fox had said, in a letter to CD (Letter #156), of James Manby Gully, who was the proprietor of a hydropathic establishment at Malvern, "*D[r] Gullys brain has quite broken down and disabled him*

from work". This determined CD to look elsewhere. James Smith Ayerst's hydropathic establishment at Malvern Wells was reportedly opened "*in conjunction at first with Dr. Gully*" (Metcalfe 1906, p 94), although it was publicised in Ayerst's name alone (*Medical Directory*, 1858-68; *Post Office Directory of Birmingham* 1860 and 1864). Dr Gully apparently discontinued his own hydropathic establishment at Great Malvern in the late 1850s, although he continued to practise from his residence until the late 1860s or early 1870s (*Post Office Directory of Birmingham* 1854 and 1860; Mann, 1983). Dr Gully visited CD at some time during his stay at Ayerst's establishment (see Letters #160 and #162).

2. According Emma Darwin's *Diary*, Emma travelled to Malvern on 1st September 1863.
3. Anne Elizabeth (Annie), the Darwins' eldest daughter, died and was buried at Malvern in April 1851 (see Letter #89). Since, at that time, Emma Darwin was nearly at term with Horace Darwin, CD, as he explains later in this letter, left Malvern on 24th April 1851, before Anne's burial The burial was arranged by Frances Emma Elizabeth Wedgwood (CCD, Vol 5, letter to F M Wedgwood) on the 25th April 1851.
4. Fox's letter has not been found, but see Letter #112 (n3) and see also Letter #161. Fox had visited his mother in Malvern on 15th Sept 1857 (Fox *Diary*, 1857) and took the water cure at other times [see Letter #153 (n12)].
5. See note 3, above.

160 (CCD 4296). To: [Charles Darwin, Villa Nuova, Malvern Wells]

Delamere Rectory Northwich
Sep 7 [1863][1]

My dear Darwin

I have not been able to answer your letter[2] before in consequence of an attack of fever which has kept me in bed the last 5 days — & I am at present about as strong as a drowning // kitten[3] — I hope to be about again soon. I wish I was at Malvern under Gully[4].

What you mention about your dear childs grave stone is to me, almost incredible[5]. Is it possible you have overlooked it in consequence of its <being> buried up by surrounding trees & shrubs, as I can easily imagine it might be by this time.

It is very difficult to describe an exact locality — but either//my wife[6] or I could go to the spot, almost blindfold. A line drawn in line with East end of Church[7] to the road — would nearly cut it I think.

Again — entering the Yard from the lower Gate — & following the path to the main entrance of Church — it is almost 1 third of way — no great distance to right (as to go to the Church)—among se- several tombs which have shrubs & trees thickly planted round them —& even when I last saw it — a good deal covering up this stone.// I am sure it was there in June 1859[8]. It was a good strong upright stone — and I remember well "To a good & dear child".[9]

If it has been taken away it is one of the most audacious and abominable villainies I ever heard of — & without object — as the digging up — carrying off & reworking the stone would cost more than its worth.

I must not write more — Pray rest yourself & give Malvern a fair chance — Our kindest regards to Mrs Darwin

Yours ever W D Fox

Provenance: University Library, Cambridge, Manuscript Room, DAR 164:180

Notes

1. The year was determined by the relationship to Letter #159.
2. See Letter #159 (n3).
3. Fox was ill when he wrote this letter and became very ill the following day. In his diary for 8th Sept 1863, Fox wrote "*I was taken very ill this early morning. I had returned from London the week before and not been well since but this morning apparent inflammation in right lung – across the stomach and also the pleura.*" Later in the Year's Summary he wrote "*I was in Bed many days and very ill till end of month. When first taken, I quite thought I could not survive many hours – but by Gods Mercy I was spared.*"
4. Fox had previously said "*Dr Gullys brain has quite broken down and disabled him from work*" (Letter #156), and, as a result, CD was being treated in Malvern wells by Dr Ayerst [Letter 159 (n1)].
5. In Letter #159, CD told Fox that Emma Darwin had been unable to find the gravestone of their daughter, Anne Elizabeth Darwin, who died at Great Malvern in 1851, and that they feared it had been stolen. CD had left Malvern before Anne was buried and so had never seen her gravestone [see Letter #159 (n3)]. (See Fig 26).
6. Ellen Sophia Fox (1820-1887). Fox probably considered that he would be too sick to travel (see note 3).
7. The church is the Anglican Great Malvern Priory Church, Church St, Great Malvern.
8. There is no further information on this visit by Fox in June 1859. Fox's *Diary* for 1859 has not been found.
9. The gravestone of Annie in the Priory Churchyard is shown in Fig 26.

161 (CCD 4294). To: Rev. W. D. Fox, Delamere Rectory, Northwich.

Malvern Wells
Sunday [6th – 27th Sept. 1863][1].

Dear Mr Fox

I have been sending out a good many of these papers & have been very much surprized & pleased to find that a good many persons Squires Ladies & MPs are interested in the matter & anxious to find a remedy[2].
Charles recommends me to send one to you & to beg you to send one to Mr Wilmot (Sir F's son in law)[3] I send one or two more in hopes that you or Mrs Fox[4] may approve of them & use them in some way. With my very kind remembrances to her I am yours

very sincerely
E. Darwin

Several have suggested offering a prize for a humane trap[5] —

Christ's College Library, Cambridge, #142a

Notes

1. Presumably this is the letter referred to in Letter #160, which would date it after that letter. The date range is established by the relationship between this letter and Letter #162 and by the address. According to Emma Darwin's *Diary*, she arrived in Malvern Wells, on 1st September 1863; the first Sunday after her arrival was the 6th September.
2. Emma Darwin refers to *An Appeal*, a four-page circular concerning the cruelty of using steel traps for catching vermin, which Emma and CD had privately printed for distribution in August (see CCD, Vol. 11, letter from G. B. Sowerby Jr. to Emma Darwin, 22 July 1863, and "*An Appeal*", CCD, Vol. 11, Appendix IX). No replies to the circular have been found.

3. Emma Darwin refers to Sir Francis Sacheverel Darwin (1786-1859) and to his son-in-law, Edward Woollett Wilmot, who was CD's cousin-in-law and Fox's second cousin-in-law, through Edward's marriage to Sir Francis' daughter, Emma Elizabeth Wilmot née Darwin (*Darwin Pedigrees*). For information on Edward Woollett Wilmot, see Letter #105 (n1).
4. Ellen Sophia Fox (1820-1887).
5. See Letter #166 (n4), and "*An Appeal*", CCD, Vol 11, Appendix IX.

162 (CCD 4312). To: Rev. W. D. Fox, Delamere Rectory, Northwich.

Malvern Wells
Tuesday. [29th Sept 1863]

Dear Mr Fox

I am writing instead of Charles to thank you for your precise answer to his enquiry[1]. I am glad to say that by the help of your directions & the lady at whose house our poor Annie lodged we have found the tomb stone[2]. It is very much covered with trees & looks so green & old I am sure I looked at it many times thinking it quite out of the question that should be it. Also the iron palisades are gone, at least both the sexton & lady thought there had been rails round it, but that does not signify. This has been a great relief.

Charles has been quite ill last week but for the last 5 days he has decidedly improved, but I expect his recovery will be very slow[3].

We like Dr Ayerst tho' he has not the influence of Dr Gully. Dr G. it is hopeless to try to see tho' I must say he has been to see Ch. twice & he quite approves of his treatment[4]. He takes 2 or 3 wet rubbings in the day & small walks in the garden, but he is weak — Our stay here has however been of real use to our sick boy & put us on a better system with him[5]. Ch. appetite is so good I think he must get strength soon & he has struggled on for 5 days without sickness.

I sent you a paper concocted between Ch. & myself wisely directed to Delamere *Droitwich* & so I will put in another in hopes you may employ it in some way[6]. I have met with a good deal of encouragement & I see it today in the Worcester paper endorsed by the Member of Parliament[7]

Will you remember us both very kindly to Mrs Fox & believe me

Yours very sincerely
E. Darwin

Postmark: SP 29 63
Christ's College Library, Cambridge, #141.

Notes

1. Concerning the whereabouts of the gravestone of their daughter, Annie, see Letter #159.
2. Fox, who had visited Annie's grave in 1856 and 1859, sent specific directions for finding it (see Letter #160). The reference is to Eliza Partington, the house-keeper of Montreal House, where Annie had lodged (CCD, Vol 5, letter to E A Darwin, 19th April 1851; Keynes, 2001).
3. Emma Darwin recorded in her *Diary*, that CD was "*giddy and unwell*" on 19th September and then sick every day from 20th to 23rd September. CD was better over the succeeding five days, while still suffering sickness. For example on the 1st October she wrote "*Much swimming in head all day. Much giddiness and shocks in evening. Good night.*" The sickness continued periodically until their return to London on 12th October.
4. It is clear that Dr James Manby Gully, who had treated both CD and Fox in the past was still active despite Fox's description in Letter#156: "*Dr Gullys brain has quite broken down and disabled him*

from work". It is more probable that Dr. Gully had reduced his activities and was only taking on old, established patients. CD was being treated by James Smith Ayerst [see Letter #159 (n1)].

5. Horace Darwin, their youngest surviving child, had been ill intermittently during 1863, and began hydropathic treatment on 15th September 1863 (Emma Darwin's *Diary*). Henrietta Elizabeth Darwin, their eldest daughter, also accompanied them on most of this trip.
6. Emma Darwin sent Fox a letter enclosing a pamphlet warning against animal traps sometime in September (see Letter #161).
7. The newspaper has not been identified. The members of parliament for Worcester, in 1863, were Richard Padmore and Osman Ricardo (see CCD, Vol 11).

163 (CCD 4365). To: Rev. W. D. Fox, Delamere Rectory, ~~Nantwich~~ Northwich.

Down. Bromley. Kent. S.E.
Dec 8 [1863]

Dear Mr Fox

Thank you very much for your letter[1]. I must own it has discouraged me a good deal as to the necessity of steel traps[2]. I believe you are the only person in England who has energy & humanity to visit the traps at midnight & Charles// tells me you once carried a toad over the Menai Bridge for fear he should come to a bad end[3] —

I want to ask you to do me a favour. My game keeper friend[4] here who is so smooth spoken I suspect he chiefly tells me what he thinks I want to hear, tells me that the animals almost always break the leg in their first frantic efforts to escape— is this true according// to your experience?[5] But I want you to be so kind as to have a trap wound round with cloth or flannel over the iron teeth so as to see whether it will save much suffering to have the leg only pinched instead of torn with the teeth. I thought if it is true that the leg is broken the animal would suffer nearly as much in whatever way it is held fast. I am going to send out papers for subscriptions// & when I have got as many names as I can I must put it into the hands of the Soc. in Pall Mall, as the awarding the prize will be a difficult matter[6]. I send you a paper to shew what sort of a beginning we have made & I am more grateful for two 5s/ subscriptions than one at £1, for I think the number of names who take an interest in the attempt is what chiefly signifies[7]//

I can give rather a better account of Charles this week but he is never long without attacks of sickness. The water cure seemed to do him harm & Dr Gully quite owned that he was not strong enough to bear it[8] Since we came home he has consulted Dr Brinton of Guy's Hospital & I hope mineral acid is doing him some good[9]. He is wonderfully cheerful when not positively uncomfortable// He does not feel the least temptation to disobey orders about working for he feels quite incapable of doing any thing.

His good symptoms are losing *no* flesh & having a good appetite so that I fully hope that in time he will regain his usual standard of health which is not saying much for him. He desires me to remember him most kindly to you & with my very kind remembrances to Mrs Fox[10] I am dear Mr Fox yours very sincerely

E. Darwin

I will trouble you with the new paper when it is printed[11]. I wish you had told us how you are yourself.

Postmark: DE 8 63

Christ's College Library, Cambridge, #142.

Notes

1. The letter from Fox, like many others, has not been found.
2. Presumably Fox had commented on the pamphlet sent to him by Emma Darwin (see Letters #161 and #162); this pamphlet, prepared by CD and Emma, protested against the use of steel traps by gamekeepers (See CCD, Vol. 11, Appendix IX, *An Appeal*).
3. No other reference to this action of Fox has been found. The Menai Bridge spans the Menai Straits to connect the island of Anglesey to Caernarvonshire, Wales.
4. This individual has not been identified. CCD, Vol 11 gives the names of two gamekeepers in the village of Downe at the time.
5. Fox's answer, if there was one, has not been found.
6. The Royal Society for the Prevention of Cruelty to Animals, had offices at 12 Pall Mall, London. There is an entry in CD's *Account Books - Banking Accounts* (Down House MS), which records the receipt of a subscription towards funding a prize for the design of a humane trap for the destruction of vermin.
7. This enclosure has not been found.
8. For CD's activities in Malvern Wells in September 1863, see Letters #159, #160 and #162.
9. Emma Darwin mistook the hospital, since William Brinton was a physician at St Thomas's Hospital, London. The mineral acid referred to may have been hydrocyanic acid: this was recommended by George Busk, along with "*light bitters with subnitrate of bismuth*" (see letter from George Busk, CCD, Vol 11, *ca* 27 August 1863). In the same letter George Busk recommended Dr Brinton. CD had been trying nitro-muriatic acid, since 1860, under the directions of his doctor Frederick William Headland [see Letter #138 (n3)].
10. Ellen Sophia Fox (1820-1887).
11. This may be a circular from the Royal Society for the Prevention of Cruelty to Animals soliciting donations for a prize to be offered for the design of a humane vermin trap (see note 6, above).

164 (CCD4484). To: [Charles Darwin, Down, Bromley, Kent]

Hillfield Hampstead NW
Thursday 5 [May 1864][1]

My dear Darwin

I am a Man about Town for some little time to come[2], & should much like to see you and yours — if you can have me, & I can manage it. Are you at home, and if so, would you take me in & do for me for one night some day next week after Tuesday[3]. I quite forget how you are approachable.

With our kindest regards to M[rs] Darwin & yourself

Yours most sincerely
W D Fox

I can easily imagine you w[d] rather not have me — in which case, do not scruple to say so.

Provenance: University Library, Cambridge; Manuscript Room: DAR 164: 181

Notes

1. The date is established by the relationship between this letter and Letter #165.
2. The notepaper is black-edged: Fox's mother-in-law, Mrs Woodd, had died on the 13th January 1864 (Fox *Diary,* 1864). From the address it is clear that Fox was visiting his father-in-law, Basil George Woodd, and this is confirmed by his *Diary* entry ("*30th April: to Hillfield till 10th May*").
3. The visit did not take place, owing to the illness of CD (see Letter #165). Fox and CD never met again.

165 (CCD 4487). To: Rev. W. D. Fox, Delamere Rectory, Northwich.
(From Emma Darwin)

Down Bromley — Kent
Friday [6th May 1864]

Dear Mr Fox

I wish we could have accepted your kind proposal of coming here[1] but I must tell you how ill Charles has been since we were at Malvern[2]. He has had almost daily vomiting for 6 months & it was becoming a great anxiety that it should be// stopped. About 3 weeks ago Dr Jenner saw him for the 2nd time[3] & he has succeeded in stopping the vomiting & he has been 3 weeks free from it, but I am afraid of the least exertion bringing it on till his stomach has more recovered its strength. You know that the// pleasure of seeing & talking with you wd be all the worse for him the more he enjoyed it at the time.

The day we saw you in London last spring has been almost the last time of his being tolerably well[4] —

He is still in his room & goes to bed more than once in the day, but he walks out of doors & occupies himself a// good deal with experiments with his flowers, so that tho' his progress is very slow he does get on. I dare say you are enough of a Doctor to like to know what has done him so much good. It is small & very frequent doses of Chalk, Magnesia & Carbonate of Ammonia[5]. Charles desires me to give you his kind love —

I am my dear Mr Fox
Yours very sincerely
E. Darwin

(*crosswise at head of letter, in CD's (?) handwriting*)
Will you remember us both to Mrs Fox

Endorsement: Mrs C. Darwin ~~April~~ Mar 6 1864
Postmark: MY 6 64 London SE
Christ's College Library, Cambridge, #143

Notes

1. See Letter #164.
2. CD, Emma Darwin, Henrietta Darwin and Horace Darwin were in Malvern Wells in September and early October 1863, where CD and Horace (for a shorter time) were being treated by Dr Ayerst; see Letters #159, #160 and #162.
3. William Jenner first visited CD on 20th March 1864; his second visit was on 10th April (Emma Darwin's *Diary*). See also letter from William Jenner to Emma Darwin, [17 March 1864] and note 2, and CD to J D Hooker, 13th April [1864], CCD, Vol 12.
4. Fox met CD and Emma Darwin, Erasmus Darwin and some other members of the family on 13th February 1863 [see Letter #150 (n3)].

5. The treatment included "*drinking very little - enormous quantities of chalk, magnesia & Carb. of Ammonia*": see CCD, Vol. 12, letter to J. D. Hooker, 13 April [1864] and note 6, therein. As seen in previous correspondence, Fox, who had often been an invalid himself from 1834 onwards, took a keen interest in CD's health.

166 (CCD 4365). To: The Rev. W. D. Fox, Delamere Rectory, Northwich.

Down Bromley Kent.
May 16 [1864]

Dear Mr Fox

This is only a business letter upon traps[1]. I do wish you would send your invention to the exhibition of traps that the Society in Pall Mall is just going to have at St James's Hall// for the purpose of awarding a prize of 50£[2] & it would be such a good way of making it known, supposing it should answer. I think a concourse of traps must be useful in order to compare the merits & demerits. Have you heard of this one of which I send the// advertisement?[3] <I do not want it back —>

I do hope you will have no further anxiety about throats but we have found in this part of the world that Diphtheria is not nearly so alarming a thing as it used to be. Charles has escaped sickness & on the whole goes on well. He is very much obliged to your for mentioning that way of taking chalk[4] —//

The Secretary's address is J. Colam Esq[5] Soc. for the Prevention of Cruelty Pall Mall

I am dear Mr Fox with kindest remembrances to Mrs Fox

yours very sincerely
E. Darwin

I think the traps are to be on view on the from the 18th to 23rd. of May.

Endorsement: Mrs E Darwin May 17/64
Postmark: MY 16 64 London S E
Christ's College Library, Cambridge, #144.

Enclosure

R. BRAILSFORD'S PATENT WIRE VERMIN TRAPS.

Having for a number of years noticed the vast increase of running vermin, without any adequate means being taken to lessen the evil, has induced me to offer to the public my Patent Wire Vermin Traps, which are proved to be far superior to any other for the destruction of foumarts, weasels, stoats, rats, &c. Their advantage consists in being light and easily moved about, in being openly constructed, so that the vermin go in without the least fear of danger. They are easily set, and may be kept in use all the year round. They may be used amongst all kinds of game or poultry, without the possibility of their doing the least injury to them, which is not the case with steel traps; also causing less cruelty to the animals caught than any other trap yet invented. And where foxes are preserved they are invaluable. They have been used on various preserves for some time, where they have given entire satisfaction, causing great destruction of rats, stoats, &c. (testimonials of which will be given if required), and in some instances three, four, and five rats have been caught at once in a single trap. Cash price per dozen, 2*l.* 16*s.*; larger size for cats, &c., 6*s.* each; for otters or other large animals (either for home or abroad), price according to size and quality.— Address RICHARD BRAILSFORD, Knowsley, Prescot, Lancashire. R. B. has on hand his celebrated Ointment for mange in dogs, Worm Powders &c., which only require one trial to prove their efficacy. Also his Check Collars for breaking dogs, and which no sportsman ought to be without.

Notes

1. For previous correspondence about humane animal traps, see Letters #161 and #163; and also CCD, Vol 11, Appendix IX, *An Appeal*.
2. Over 200 humane traps were displayed at the annual meeting of the RSPCA held at St James's Hall, Regent Street, London, on 26 May 1864. The society subsequently mounted an exhibition of more than 100 humane traps at the Royal Horticultural Gardens, South Kensington, in June 1864 (*The Times*, 27th May 1864, p 11, and 28th May 1864, p 14). For a discussion of the outcome of the competition, see CCD, Vol. 11, Appendix IX.
3. See Enclosure.
4. Fox's letter on the subject has not been found. For the treatment recommended at this time, by Dr Jenner, see Letter #165 (n5).
5. John Colam (1827-1910). Animal welfare campaigner. (*Oxford Dict Nat Biog)*

167 (CCD468). To: [Charles Darwin, Down, Bromley Kent]

Delamere Ry Northwich
Nov 28 [1864][1]

My dear Darwin

I was much pleased to see you have the Copley Medal given you by RS[2]. I suppose it is a great honour from such a Socy—but I am sadly ignorant of such matters.

I wish I could hear that you were in better health[3], but I suppose your destiny is to let your Brain destroy your Body. If equal to writing a few lines it wd give me great pleasure to again see your well known handwriting — but do not think of it if you// are not equal to it.

I hope you are all well as, I am thankful to say we are except my poor Wife who is shortly expecting our *16*th Baby, and therefore not very comfortable[4].

We quite hoped we had fulfilled our destiny in this respect, but I have no doubt the little stranger will find a nook among *our* others. How children *do* grow — My eldes*t* *is*// just expecting her 3d child[5] — My eldest Boy has just taken his degree at Oxford[6].

It is awful to reflect upon — What old fellows we get. I do not imagine I shall leave home till the spring, when I shall probably be in or about London[7] — & if you and Mrs Darwin can have me for a night, I will try *on*ce more see you.

With our Kindest regards *My* dear Darwin
Your affec Friend & Cousin
W. D. Fox.

Provenance: University Library, Cambridge, Manuscript Room, DAR 164:182
Damaged

Notes

1. The year is determined by the award of the Copley Medal of the Royal Society to CD.
2. The Copley Medal is the premier and oldest award of the Royal Society, London. It is awarded for outstanding achievement in research in any branch of science, and alternates between the physical and biological sciences. In 1864 it was awarded to Charles Darwin for "*his important researches in geology, zoology, and botanical physiology*". See CCD, Vol 12, Appendix IV, *Darwin and the Copley Medal*, for a detailed account on the subject.
3. CD's health had improved since the spring of 1864 when Emma Darwin turned down Fox's offer to come to Down on the grounds of CD's very poor health (Letter #65); see Emma Darwin's *Diary*.
4. Gilbert Basil Fox was born on the 16th December. He was the twelfth and last child by Fox's second wife, Ellen Sophia.
5. Eliza Anne Fox had married Henry Martyn Sanders [see Letter #148 (n18)]. Their third child was Laura Katherine Sanders (*Darwin Pedigrees*).
6. Samuel William Darwin Fox, obtained his BA at Wadham College, Oxford University in 1864 (*Alum Oxon*). Fox wrote in his *Diary* for 1864 that his son passed his finals on 29th November.
7. Fox's *Diary* for 1865 shows that he was in and about London from 8th April to about the 23rd May.

168 (CCD4181). To: Rev. W. D. Fox, Delamere Rectory, Northwich.

Down Bromley Kent S.E.
Nov — 30th. [1864]

My dear old Friend. —

I was glad to see your hand-writing[1]. — The Copley being open to all sciences & all the world is reckoned a great honour[2], but excepting from several kind letters, such things make little difference to me. It shows, however, that Natural Selection is making some progress in this country & that pleases me[3]. The subject, however,// is safe in foreign lands. I am glad you are all well; but I never heard anything so awful as your sixteenth child![4] We are all fairly well, except my youngest Boy who is too invalidish for school[5], which is a great pity for he is about the cleverest of the lot. —

As for myself, I fear I have reached my sticking point — I am very weak// & continually knocked up, but able most days, to do from 2 to 3 hours work, & all my Doctors tell me this is good for me; & whether or no, it is the only thing which makes life endurable to me[6]. — I am slowly crawling on in my vol. on "Variation under Domestication"[7] occasionally recreating myself with a little Botanical work[8]. —

Whenever you come to London, do be sure, if you can spare// time, come here for a bit, for though I can see but little of anyone, I shd. very much enjoy shaking hands with you, my old & true friend.

Yours affectly
C. Darwin

Endorsement: Dec 1. 1864
Postmark: DE 1 64 London S E
Christ's College Library, Cambridge, #145.

Notes

1. See Letter #167.
2. See Letter #167 (n2) for details on this award.

3. Note that the description for CD's award of the Copley medal, "*for his important researches in geology, zoology, and botanical physiology*", made no mention of his evolutionary contributions.
4. See Letter #167 (n4).
5. Horace Darwin had been in poor health since 1862 (see previous letters, e g, #157, and Emma Darwin's *Diary*).
6. Fox has commented before that it is "*brain work*" that knocks CD up (see, e g, Letters # 158 and #167). The sense of what CD is saying is that it is only "*brain work*" that keeps him going
7. CD resumed work on *Laws of Variation* for "*Domestic Animals & Cult. Plants*" on 14th Sep. 1864 (Charles Darwin's *Journal*). He had not worked on the manuscript since July 1863 (see *Journal*). *The variation of animals and plants under domestication* was published in 1868.
8. In his *Journal*, CD records "*Began to count seed of Lythrum about April 26th. Finished Lythrum paper about May 25th. Began Tendril paper, & finished it, on Sept 13th, but afterwards had about a fortnight for additions. Hence this paper on Climbing Plants took four months!!*" See Letter #169 (n2) for details on the *Lythrum salicaria* paper.

169 (CCD 4741). To: [Charles Darwin, Down, Bromley Kent]

Delamere Ry Northwich
Jany 6 [1865][1]

My dear Darwin

I must write you a few lines to thank you much for your sending me your paper upon the 3 forms of Lythrum Salicaria[2]. I have read it with very// great pleasure, and shall look at my Garden Variety of it with much more interest than I ever did before, next year.

I forget whether I have told you that my wife brought me a very fine little Boy three weeks since to day[3].

Most of our friends *condole* with us on the affliction but we are well contented to bear it[4]. We have not// one too many in our 16[5]. 14 are now at home — but we shall soon lose my eldest Boy who has taken his degree at Oxford, & will be ordained in March[6].

I have now 5 Grand=children also[7].

But I will not bother you with more reading. May every blessing be yours and Mrs Darwins in this new year

Ever Your affece Cousin
W. D Fox

Provenance: University Library, Cambridge; Manuscript Room: DAR 164:183

Notes

1. The year was determined by the reference to CD's paper on *Lythrum salicaria*, and to the birth of a son to Fox (see notes 2 and 3, below).
2. Darwin C D (1865). *On the sexual relations of the three forms of Lythrum salicaria.* J. Linn Soc, Lond (Bot) 8, 169-196. The paper was read to the Linnean Society on the 16th June 1864. See, also, CCD, Vol 12, Appendix III, *Presentation list for "Three forms of Lythrum salicaria"* (with an introduction). Fox was on the presentation list.
3. Gilbert Basil Fox was born 16th December 1864. He was the twelfth and last child of William Darwin Fox and his second wife, Ellen Sophia Fox (*Darwin Pedigrees*).
4. On two previous occasions, CD had used the word "*condolences*" in relation to a birth in the Fox family: in 1852 he had wrote, "*I congratulate & condole with you on your tenth child; but please to observe when I have a 10th, send only condolences to me*" (Letter #88), and in 1855 he wrote, "*I*

congratulate & condole with you on your 12th child: in my own case, I shd have wished only for condolences!" (Letter #106).

5. Fox had six children by his first wife, Harriet (the first still-born). One of his children, by his second wife, Ellen Sophia, died in 1853 [see Letter #94 (n 1) and *Darwin Pedigrees*].
6. Samuel William Darwin Fox (1841-1918) took his BA at Wadham College, Oxford, in 1864. His first position was as curate of St Paul's, Manningham, Yorkshire, in 1865 (*Alum Oxon, Crockford's Clerical Directory*).
7. 1 grandson and 2 granddaughters from Eliza Ann Sanders (Charles Henry, Agnes Marian and Laura Katherine) and 1 grandson and 1 granddaughter from Harriet Emma Overton (Frederick Arnold and Annie Maud). (*Darwin Pedigrees*)

170 (CCD 4903). To: [Charles Darwin, Down, Bromley Kent]

[Delamere Rectory Northwich]
[before 26th Oct. 1865][1]

two pages missing
with you then.
I took my 2d Boy Charles to Oxford last week. I sho*uld* much h*a*ve preferred Cambridge — but a*s* he was fortunate enough to g*a*in a Studentship at Ch*rist* *Chur*ch[2] I had no *** *an half page missing**** think of you, my p*** ***, with your sad health, while I *ha*ve a perfectly enjoyabl*e* life — tho my lungs are rather *r*otten
I met a spleen*di*d man for a couple of d*ay*s l*ast* week

Provenance: University Library, Cambridge; Manuscript Room: DAR 164: 204
Damaged; very fragile

Notes

1. The date was determined by the relationship to Letter #171 and note 2, below.
2. Charles Woodd Fox matriculated at Christ Church, Oxford, on 13 October 1865 (*Alum. Oxon.*). Fox wrote in his *Diary* for 14th Oct 1865, "*to Oxford with Charles till 14th*".

171 (CCD 4181). To: Rev. W. D. Fox, Delamere Rectory, Northwich.

Down Bromley Kent. S.E.
Oct. 25[-26][1] [1865]

My dear Fox

It is a long time since we have had any communication & I should like much to hear how you are yourself as well as Mrs Fox & all your family.

I know you will wish to hear about me. I have had a bad time for the last 6 months & have been able to do no scientific work. I have// put myself under Dr Bence Jones's care & he has stopped my vomiting by a scanty diet of toast & meat; but I cannot recover my strength[2]. I know you are half a Doctor therefore I thought you wd like to hear these details.
My Sister Susan's health has lately been much failing & I fear from the last accounts that her state is becoming serious[3]. As for the rest of us// we are pretty well, at least

for Darwins, & that is not saying much. My children however are all well I am thankful to say

Oct 26. I had dictated[4] thus far yesterday, & now by an odd chance your kind & welcome letter has arrived.[5] I heartily rejoice at the good account you give of yourself, patriarch as you are with your half dozen grandchildren I congratulate you most sincerely on the brilliant success of your son & my Godson[6]. My eldest son[7] is established as a banker at Southampton & is fairly well contented with// his lot. My second son George has just passed his Little go at Trin. Coll.[8] He has a turn for mathematics, & means to try for a scholarship there this Easter[9]. As for myself I have not gone out of my grounds for the last 12 months; but on Nov. 7. we go for a week to Erasmus's in order that Dr Bence Jones may see me two or three times[10]. I hope this may coincide with part of your visit, as I sh^d^ like extremely to see you again for a few minutes, — more than that I fear I c^d^ not stand.
My wife desires to join in very kind remembrances to Mrs Fox My dear old friend
Yours affectionately
C. Darwin

This letter appears to be in Emma Darwin's handwriting – see note 3.
Postmark: OC 27 65
Christ's College Library, Cambridge, #146.

Notes

1. From second part of the letter it is clear that the letter was held up by a day by the receipt of Fox's letter (#170).
2. According to Emma Darwin's *Diary*, CD began his diet under the supervision of Henry Bence Jones on 24th July 1865. CD must have abandoned John Chapman's ice treatment, which he began on 20th May 1865 (CCD, Vol 13, letter to John Chapman 16 May 1865; Emma Darwin's *Diary*: 20th May 1865: "*No sickness great flat — nausea – put on ice at 5 p.m. Dr Chap — good night no flat*"). (See also Letter #85 (n12).
3. Susan Elizabeth Darwin (1803-1866) lived at the Mount, the Darwin family home in Shrewsbury. She died on 3rd Oct 1866 at Shrewsbury (*Darwin Pedigrees*).
4. The handwriting seems to be that of Emma Darwin, although CD did not say to whom he was dictating.
5. See Letter #170.
6. This must be Samuel William Darwin Fox (1841-1918), who was called Darwin and had just gained his BA at Oxford University [Letter #169 (n4].
7. William Erasmus Darwin, who joined the Southampton and Hampshire Bank as a partner in 1861, with CD's financial help [see Letter #144 (n8)].
8. George Howard Darwin matriculated at Trinity College in October 1864 and gained 2nd Wrangler position in the Tripos examination in 1868 (Freeman, 1978). "*Little-go*" was the popular name for the first examination for the degree of B A at Cambridge University, officially called "*The Previous Examination*" (*OED*). Had CD forgotten with what trepidation he approached his "*Little go*" (Letters #9, #10 and #13 and how thrilled he was when he passed in 1830 (see Letter #28)?
9. George Howard Darwin received a prize for achieving a first class mark in the Easter term examination, 1866 (*Cambridge University calendar* 1867, p 397).
10. Dr Bence Jones: see Note 2, above.

172 (CCD 13808). To: [Charles Darwin, Down, Bomley, Kent]

Delamere Northwich Cheshire
[before 1 Mar 1866]

(Two pages missing)
Do you know anything **about** Woking[1]. A *grand* man — a Dr Brushfield[2] — wants to become the Superintendant of a new Asylum there. —
he is a perfect man for the situation & I should much like to see him there.
(One and two thirds page severely damaged)

Provenance: University Library, Cambridge; Manuscript Room: DAR 164: 205
Very fragile

Notes

1. Woking was then a small town lying 37 km SW of central London in rolling countryside in a gap between the North Downs. It was opening up at that time as a railhead, since it was on the rail line to SW England, and also on the new private rail line to Brookwood Cemetery, operated by the London Necropolis Society, which aimed to alleviate the problem of the increasing number of burials in greater London.
2. According to research by the Darwin Correspondence team this letter must be dated before 1st March 1866, the formal date of appointment of Dr Thomas N Brushfield as superintendent of Brookwood Asylum, Woking, Surrey. From the dates of correspondence in the Surrey Record Office, his appointment was in process before December 1865.

173 (CCD 5195). To: Charles Darwin, Down, Bromley, Kent.

[Delamere Rectory, Northwich]
August 20 [1866][1]

My dear Darwin,

Though very unwilling to give you trouble I cannot quite cease writing to you and hoping for a few lines whenever you are able to write.

Erasmus[2] gave a better account of you when I saw him in the Spring — I hope this may have continued, and that your wretched health, has been less wretched than usual.

Nothing but mental rest will ever give you any comfort// and you have got into such a normal state of requiring the stimulus of mental activity, that you must have it, & the poor body go to the wall[3].

How goes on *the Book*? Requiring so many additions and corrections, that it never approaches its end I suppose[4].

I am anxiously looking for your work on domestic animals which will I suppose soon be out[5]. Possibly it may be so already, for I am sadly behind hand with knowledge of what is going on in the literary world. Except Sir P Egerton[6] I have no literary Nat Hist^y^ friends here — & he is never at home for more than a few weeks.//

I was deeply grieved a few days since, to hear of poor Susans suffering state from Frank Parker[7]. I knew she was very ill, but I hoped not in such pain as he told me she was.

Erasmus I thought looking extremely well. He gets younger, not having a Book to destroy him.

We have all been at Whitby — a famous place every point of view — I just left it before Sir C. Lyell arrived there[8]. I should have much liked to see him again. There were some grand specimens of Ichthyosaurus & Plesiosaurus there for P*ublic* Inspection[9].

Possibly you may be interested in the fact of Smerinthus Ocellatus// & Populi being now regularly bred between. They are cognate species — eyed Hawk Moth & the Poplar. I imagine there is little difficulty in procuring the mules. I have a lot of the Caterpillars of each & hope next year to try whether the mules will breed again.[10]

Possibly this may be proved by others this year.

When you write, if you ever do, tell me how you all are - and also how Caroline[11] & her belongings are.

My Brood are all well I am thankful to say — it is a good deal to say of 16.

With our kindest regards to Mrs Darwin Believe me Ever yours affec[ly]

W. D Fox

Provenance: University Library, Cambridge; Manuscript Room: DAR 164: 184

Notes

1. The year was determined by the relationship to Letter #124 (e g, hawk-moth).
2. Erasmus Alvey Darwin, CD's brother.
3. Fox astutely diagnoses CD's condition.
4. Maybe, Fox is referring to the big book that he considers that CD is writing, not realising that CD had long ago decided not to go ahead with it and to publish it in separate parts. CD was partly responsible for the misunderstanding: see Letters #139 (n5) & #137A (n6).
5. Fox refers to *The variation of animals and plants under domestication* (1868). This was not published until January 1868, owing to delays in setting out the footnotes and in proof reading [see Letter #176 (n3)].
6. Sir Phillip Malpas Egerton, Fox's neighbour at Oulton Park [see Letters #66 (n9), #73 (n6), #85 (n5), #87 (n4), #93 (n2) & #105 (n6)].
7. The references are to Susan Elizabeth Darwin, CD's sister, and Francis Parker, the son of Marianne Parker née Darwin and therefore CD's nephew. Frank Parker was a solicitor in Chester, near to Delamere. Susan was dangerously ill (see Letter #174).
8. Fox refers to Whitby, a seaside town in North Yorkshire and to Charles Lyell (1797-1875), the author of *The Principles of Geology* (1831-1833), who features occasionally in this correspondence [see Letters #56 (n10), 55 (n12), #64]. Fox and Lyell were not well known to each other.
9. In 1837, Martin Simpson was appointed Lecturer and Keeper of the Whitby Museum and continued in this post until his death in 1887. Through his efforts fossils from the nearby alum shale sea cliffs were exhibited, many collected by the Rev George Simpson. The museum today has an Ichthyosaurus and a Plesiosaurus on display, whose collection date to before the time of this letter (Keeper and Registrar, Whitby Museum).
10. Fox refers to two species in the family Sphingidae, *Smerinthus ocellatus,* the eyed hawk-moth and & *S. populi* (now *Laothoe populi*), the poplar hawk-moth. These species interbreed in captivity but only sterile hybrids are produced, as predicted by CD in his reply (Letter #174).
11. Caroline Sarah Wedgwood, née Darwin, CD's sister and Emma Darwin's sister-in-law.

174 (CCD 5197) To William Darwin Fox, Delamere

Down Bromley Kent S.E.
Aug. 24th [1866]

My dear Fox

It is always a pleasure to me to hear from you & about you, patriarch as you are with countless children and grandchildren[1] — I feel a mere dwarf by your side. My own children are all well & two of my Boys are touring Norway[2]. Poor Susan is in a terrible suffering state & I fear there// is no hope for her except the one & last hope for all[3]. We expect Caroline here with her three girls on Monday & I will give her your kind enquiries[4]. As for myself my health is very decidedly better though I am not very strong. I attribute my improvement partly to Bence Jones' diet & partly, wonderful to relate to my riding every day which I enjoy much[5]. I don't be-//

lieve in your theory of moderate mental work doing me any harm, anyhow I can't be idle[6]. I am making rapid progress with my book on domesticated animals which I fear will be a big one & has been laborious from the number of references[7]. I hope to begin printing towards the close of this year & when it is completed I will of course send you a copy, as indeed I am bound to do// as I owe much information to you. I should have begun printing before this had I not lost nearly three months by the troublesome labour of largely correcting a new edit. of the Origin. I think I have heard of hybrids between the Sphinx-moths which you mention[8]; 1 shall be surprised if the hybrids are fertile.

Believe me, My dear old Friend
Yours affectly.
Charles Darwin

Postmark: AU 25 66

Provenance: Smithsonian Institution Libraries, Dibner Library of History of Science and Technology

Notes

1. See Letter #173. Fox had sixteen children and five grandchildren at this time (see Letter #169, notes 5-7; *Darwin Pedigrees*).
2. These sons of CD were George Howard and Francis (see letter from G H Darwin to H E Darwin, June-July 1866 (DAR 251: 224000). Emma Darwin in her *Diary* recorded "*Boys went to Norway*" on 16 August 1866.
3. CD's sister, Susan Elizabeth Darwin (1803-1866), was seriously ill at this time, and died on 3rd October 1866.
4. CD refers to Caroline Sarah Wedgwood, née Darwin, his sister. Her daughters were Katherine Elizabeth Sophy, Margaret Susan and Lucy Caroline Wedgwood (*Darwin Pedigrees*).
5. CD consulted Henry Bence Jones in July 1865. CD began following his recommended diet shortly thereafter [see letter to H B Jones, 3 January 1866 (n2 and n4)]. CD started riding on 4 June 1866 (Emma Darwin's *Diary*); see also letter to John Lubbock, 2 August 1866 (n2).
6. See Letter #173.
7. CD refers to *The variation of animals and plants under domestication* (1868).
8. See Letter #173.

175 (CCD 5388). To: Charles Darwin, Down, Bromley, Kent

Delamere Rectory Northwich Cheshire
Feb 1 [1867][1]

My dear Darwin

I have so long kept my resolution of not writing to you[2], knowing how your time is taken up, and how small your powers are — that I have now determined to reward my goodness by sending off a few lines, hoping that you will some day find time and strength to rejoice my eyes with your handwriting again, and that you will tell me as much as possible about yourself — Mrs Darwin & your family.

I cannot help hoping that I shall// hear riding has done much for you[3]. If it suits you, it will be every thing to you. You get air and exercise without fatigue and must perforce, give that big brain of yours *some* rest.

How strange it seems that you and I are left alive, and poor Susan & Catherine taken away[4]. They seemed so healthy and strong and we such runtlings in comparison.

I had such a nice cheerful letter from Susan only a few months before her sad sufferings came on that I quite hoped to have again seen her cheery face[5]. What is become of the old house at Shrewsbury[6]. What heaps of genuine kindness have I met within its walls[7]. I often look back to the days of joy & sorrow I passed there.//

How is Caroline and where? I should so much like to see her and her girls[8].

I have looked anxiously for your Book on domestic animals[9]. Is it coming out this Spring? How gets on the great Book[10]. Facts from your numerous correspondents in the World must keep accumulating so much that it must seem almost to go back instead of forward.

I did fear you would never live to complete it, but I cannot help hoping now that you may live to a good old age, and become strong again.

We have just broken up our Christmas Party — & dispersed 4 of our children — to Oxford — Kings College — & London — while the rest are sitting down to their// routine habits — I to mine of teaching my 4th Boy Latin &c preparatory for school[11].

Except myself we have wintered well, & the Boys have had plenty of skating I have not had a good winter — my lungs having kept me a prisoner almost altogether. I hope however I shall now be able to get out & take air & exercise, both of which I much want. And now I release you from reading this sad tract. Do let me have a few lines from you soon.

I hope Mrs Darwin and your children are all well. I have a very pleasing recollection of the face of your eldest Girl, whom I saw in London at Erasmus's[12] .
With our kindest regards to Mrs Darwin Believe me very dear Dn

Ever yours affectly & truly W D Fox

(*written across the head of the letter*)
P.S. I have no less than *3* Free Martins[13] here who have children — 1 man 2 women but one of the latter I find from her New Man had a near escape of being barren. She has however a fine Son[14].

Provenance: University Library, Cambridge, Manuscript Room, DAR 164: 185

Notes

1. The date was determined in relation to Letters #174 and #176.
2. The previous, extant letter from Fox is Letter #173 of six months before this time.
3. In his Letter #174, CD told Fox that he thought that the improvement in his health was due partly to his riding every day (see also CCD, Vol 14, letter from H B Jones, 10th February 1866 and note 3). CD reports in his reply to this letter (#176), that riding has done him good.
4. Two of CD's sisters had died the previous year. Emily Catherine Langton (1810-1866), died at the Mount, the Darwin family home, in February, when she came to Shrewsbury to look after her ailing sister, Susan. Susan Elizabeth Darwin died later that same year at the Mount in October (*Darwin Pedigrees*). This left Caroline Wedgwood as CD's sole surviving sister (see note 8, below).
5. As related in CD's previous letter (#174), Susan fell ill several months before her death.
6. Fox refers to The Mount, the Darwin family home in Shrewsbury, where Susan had lived until her death. The selling of the house and it contents are described in Letter #176 and in note 6 of that letter.
7. Fox had seen much of the Darwin family in early years. His diaries reveal that he went to the Mount on several occasions, including a pony trip in 1834 with his new wife Harriet (see, also, CCD, Vol. 1, Letters of Catherine Darwin to CD, 29th October 1834, 28th January 1835 and 29th January 1836). Fox's *Diary* for 1847 reveals a visit "*To Dr. Darwin at Shrewsbury*" on 2nd - 5th January and a visit to Delamere by Susan Darwin on 17th May. After that the rapidly increasing family of Fox and wider responsibilities probably curtailed his visits.
8. Fox refers to CD's surviving sister, Caroline Sarah Wedgwood and to her three daughters Katherine Elizabeth Sophy, Margaret Susan and Lucy Caroline Wedgwood (*Darwin Pedigrees*).
9. Fox refers to *The variation of animals and plants under domestication* (1868).
10. Fox has in mind CD's old idea of a "Big Book" [see Letter #137A (n6)] & #139 (n5)]. By this stage CD had divided the projected contents of his "Big Book" into several books and had given up the idea of a single book that would summarise all his ideas. He finally admitted this to Fox in 1874 [Letter #201 (n7)].
11. Charles Woodd Fox, Fox's second eldest son, matriculated at Christ Church, Oxford, on 13th October 1865. Robert Gerard Fox, Fox's third son was educated at King's College, London (Hants Advertiser, 1909). The other two who left may have been his two daughters who were at school in London [see Letter #184 (n11)]. Fox's fourth son was Frederick William Fox, aged 11 at this time.
12. Fox refers to his meeting at the house of Erasmus Alvey Darwin at 6 Queen Anne Street, with CD and Emma Darwin on 13th February 1863. According to Fox's *Diary* entry for that day, CD's eldest daughter, Henrietta Emma Darwin, was also present. Catherine Darwin, CD's sister was also present. See also Letter #150 (n3).
13. Freemartin, or free martin, a 'hermaphrodite or imperfect female' (*OED*). Fox had asked CD whether he believed in freemartins in Letter #149, noting "*I have always found them true in practise till now. A Lady near here, whom I know well, and her twin Brother also—has one from Baby & just had a miscarriage of another.*"
14. CD's response to this information is not known; he did not respond to this matter in his reply to Fox (Letter#176). However, it was a particularly interesting topic and one with which CD, like others of his day, struggled. In *The Descent of Man* (1871), Vol 1, p 207, CD wrote that at a very early embryonic period, both sexes off vertebrates possessed '*true male and female glands*' and that the sexes in vertebrates evolved long before humans appeared (ibid 1, p 208).

176 (CCD 5392). To: William Darwin Fox Delamere Ry, Droitwich, Cheshire

Down Bromley Kent
Feb 6th [1867]

My dear Fox

It is always a pleasure to me to hear from you, & old & very happy days are thus recalled[1]. This is rather a joyful day to me, as I have just sent off the M.S. for two huge volumes (I grieve the Book is so big) to Printers on Domestic Animals[2]

&c &c// but my book will not appear, even if completed, before next November, as Murray has strong prejudice against publishing except during Spring & Autumn[3]. I am utterly in darkness about merit of my present book; all that I know is that it has been a most laborious undertaking. Of course a copy//will be sent to you[4]. —

It is true indeed that Death has been busy with us, & it is astonishing to me that I sh[d] have survived my two poor dear sisters[5]. The old House at Shrewsbury is on sale, but has as yet found no purchaser & I daresay will not soon. — All the furniture was sold by// Auction, having been bequeathed to the Parkers who had become like Susan's children[6].

Caroline & Erasmus are fairly well for them; but this is not saying much for them, especially for the latter, who does not often leave the House[7]. I am so sorry to hear// so poor an account of yourself with your active habits being confined must be a terrible deprivation You are quite right about riding: it does suit me admirably, & I am very much stronger, yet I never pass 12 hours without much energetic discomfort[8]. But I am// fairly well content now that I am no longer quite idle. —

Poor Bence Jones has been for months at "death's" door & was quite given up; but has rallied in surprising manner from inflammation of Lungs & heart-disease[9]. My wife is fairly well// but suffers much from repeated headaches[10] & the rest of us are well. — I hope you will get all right with returning Spring

My deal old friend
Believe me;
Yours affectionately
Ch. Darwin

Postmark: Bromley FE 7 67
Endorsement: 'C. Darwin Feb 8 1867'
Christ's College Library, Cambridge, #147

Notes

1. See Letter #175, but in a wider sense CD got great benefit from reminiscing about his days at Cambridge with Fox and others (see many reminiscences in previous letters and *Autobiography*).
2. *The variation of animals and plants under domestication*, 2 volumes, published in January, 1868.
3. Delays were caused by trying to shorten the size of the book by placing footnotes in smaller type. However, finally it was CD who held up publication because the Index was not ready (Browne, 2002).
4. Fox was on the list of recipients of presentation copies, as he was for all CD's books.
5. CD refers to his late sisters, Emily Catherine Langton and Susan Elizabeth Darwin (see Letter #173 (n3).
6. Susan Darwin had lived at The Mount, the Darwin family home in Shrewsbury, ever since the death of their father in 1848; following her death on 3rd October 1866, CD's brother, Erasmus Alvey Darwin, and surviving sister, Caroline Sarah Wedgwood, traveled to Shrewsbury to sort out some of the family belongings in the house (see CCD, Vol 14, letter from E A Darwin 11th October 1866). The house was let and eventually sold (see CCD, Vol 14). CD's eldest sister, Marianne, had married, in 1824, Henry Parker (1788-1856), physician at the Shropshire Infirmary in Shrewsbury [Letter #176 (n12)]; after their deaths the grown-up children, Robert, Henry, Francis, and Mary Susan Parker, stayed at The Mount, when in Shrewsbury, with Susan Darwin (Freeman, 1978).
7. Fox had asked about CD's surviving sister, Caroline in Letter #175. The character of CD's elder brother, Erasmus Alvey, is discussed in Wedgwood and Wedgwood, 1980, p 233. Erasmus was also

depressed following Susan's death in October 1866; Susan had been his favourite sister (Wedgwood and Wedgwood 1980, p 288, and Browne, 2002, p 267).

8. See Letter #175.
9. Henry Bence Jones, a physician, who had treated CD in 1865 and 1866 (see CCD, Vols. 13 and 14 and Letters #171 & #174), became dangerously ill in September 1866 and had been close to death in January 1867 (Kyle, 2001, p16); see also CCD, Vol 15, letter from H B Jones to Emma Darwin, 1st October 1867.
10. Emma Darwin noted, in her *Diary,* that she had "*headaches*" for 6th, 7th and 8th February 1867 (and at many other times in these years).

Conclusion

Understandably, what is missing from the correspondence between Charles Darwin and William Fox in the years following the publication of "*On the Origin of Species*" is anything about the great debate over the Darwin-Wallace theory of organic evolution and Darwin's version of it, which captivated the public imagination so much. There is no discussion of the famous Oxford debate, with the confrontation between Darwin's bulldog, T H Huxley, and the Bishop of Oxford, "Soapy" Sam Wilberforce, in 1860. Nor do we get any flavour of Darwin's behind-the-scenes marshalling of his supporters and polishing of their arguments, which is so evident in Darwin's letters to such persons as J D Hooker and T H Huxley at this time. This is because Fox, as a clergyman, was bound to stay out of the fray; despite the stance of a few clergymen such as Charles Kingsley, who were openly supportive of Darwin's thesis, the clergy as a whole either chose some middle ground or vehemently opposed it. We do not know Fox's actual opinion, since no letters exist stating his views, but even as late as 1870, and later, he was still resisting the notion that man may have evolved from the apes (Letter #192). It was a brave clergyman that did support Darwin's views at this time, when even many scientists such as Richard Owen and Adam Sedgwick were openly opposed to them. And Fox, who depended on a large social network for his activities, probably was bound to hedge his opinions, while glorifying in the fact that he was an old friend of Darwin. Another old friend of Darwin's and his adviser for many years, Charles Lyell, seemed to hesitate, in public, in his support of the full implications of the "*Origin*", while accepting many of it points (Desmond and Moore, 1991). And even the co-founder of the theory, Wallace, held back in the area of human evolution, saying that God must have intervened in the evolution of the human brain, because he, Wallace, could see no mechanism by which such a sophisticated organ as the brain could have evolved naturally (Wallace, 1871, p 555). The problem was one of semantics and philosophy. For a person such as Adam Sedgwick, the general thesis of the "*Origin*" was so mundane as to be self-obvious, but the failure to acknowledge the guiding hand of God was unacceptable (Sedgwick to CD, 24th November 1859; CCD, Vol 7). To Darwin on the other hand, one must accept every thing or accept nothing: like Newton's laws of gravity and motion, these new laws applied at all times and were not tweaked by a tinkering god.

Just what Darwin's views on Christianity were at this time are not recorded, since he did not want to offend his wife, Emma, who was deeply religious. In fact in 1861, she was so troubled that she wrote Darwin a second letter on the subject of religion, saying in part "*It is feeling & not reasoning that drives one to prayer*" (CCD, June 1861). On the bottom of this note Darwin wrote "*God bless you*". Also on a wider front, Darwin did not want to stir up adverse debate unnecessarily; he realised that there would be enough of this any way. Neither did he want to cut off correspondence with such informed naturalists as Fox, who were still essential for his two great evolutionary books to come, *The variation of animals and plants under domestication* (1868) and *Descent of Man* (1871), as well as his books on instinct, plants and earthworms. It is probable that at this time he was an agnostic in the sense that he did not believe in the personal saviour of Christianity; and while, if there was a God, then one must concede that God was theoretically capable of doing anything, these things should be put to the test: the ideas of a purely mechanistic working out of natural selection should be the subject of experimentation, not religious debate (see Moore, 1989).

The question was how to put the ideas to the test. Up to this time the only available means was inductive reasoning: from a set of qualitative observations to a general principle. Darwin was dissatisfied with this approach; as he wrote to T H Huxley in 1859, "*the inaccuracy of the blessed band (of which I am one) of compilers passes all bounds*". He realised that there was another approach and that was quantitative observation and experiment. Just how to do this required completely new disciplines, such as plant and animal physiology and genetics. Plant and animal physiology were in their infancy and genetics hardly existed at all (certainly not the name). In genetics an unknown Abbot in a monastery in Austria, Gregor Mendel, was making the first efforts to quantitatively assess "characters" or "individual differences" in plants. This work, although published in 1865, was completely ignored by the biologists of the time and was not "discovered" until 1900. It is interesting to speculate what might have happened if Mendel had been one of Darwin's correspondents. However, even if the role of genes had been known at the time it would not have helped. Long after genes were recognised in the first decades of the twentieth century, the Darwin-Wallace theory of organic evolution looked inadequate, because on a simple analysis the most common genes would predominate and new species variants would tend to disappear. It was not until the brilliant work of R A Fisher, and others, in the 1920s and 30s, showing how selection pressure could increase the frequency of rare but beneficial genes, that the modern synthesis was generally accepted and the Darwin-Wallace theory reinstated. Even then, genetics had to wait until the discovery of desoxyribosenucleic acid (DNA) by Watson and Crick in 1953, for the modern phase to begin. And as any brief study of Quantitative Trait Loci (QTLs) will show, we are still far from a full understanding of this complex area.

Nevertheless, Darwin did manage an impressive attack on this quantitative side of evolutionary theory. He began to be interested much more in quantitative facts: "*Now can you tell me do the White Musk & Slate-coloured musk, (when not crossed) each breed quite or nearly true? Again in same letter I have quoted that "12 white ewes of Mr Woodd's had 23 coal-black lambs by a Ram that had a small patch of black only". Can you tell me what breed these ewes were? Was Ram of same breed?*

Do you know how big & where the black patch on ram was? — I should almost expect that if two very different breeds of white sheep were crossed there would be some tendency to dusky lambs; & so with horns, if two hornless breeds were crossed; but I know not where to enquire. I am even inclined to suspect that there is a tendency to a return to primordial wildness in hybrids between two domestic species". He wrote this to Fox in Letter #152, and, thereafter, Fox began to reply with quantitative facts. However, most of this new approach was too late for "*The variation of plants and animals under domestication* " (1868). And there was even little room for it "*Descent of Man*", Volume 2 (1871), which was concerned with sexual selection. The books that finally saw the fruits of this new approach were "*The effects of cross and self fertilization in the vegetable kingdom*" (1876) and "*The different forms of flowers on the same species*" (1877). Nevertheless the seeds for the quantitative aspects of these works were sown at this time (pun intended); and resulted in such letters from Fox as Letters #184 and #186.

At the broadest levels of society, Darwin's *Origin* recast philosophy and Victorian morality. Or rather it polarized the debate. Over the past one hundred and fifty years, there had been a number of philosophers, who had tried to rationalize the sciences with religion, starting with Locke and going on with Hume and William Whewell, whom Darwin had first met at Cambridge. Darwin, as his reading list shows (CCD, Vol 4), tried to make sense of his own ideas on natural selection in relation to the various philosophies on offer. None appealed to him and he seems to have abandoned the attempt. This is not surprising because the world of organic evolution left little room for manœuver for the philosophers of the nineteenth century; and even for the more extreme forms of rationalism in the twentieth century (Dennett, 1999). It is interesting that John Stewart Mill was one of the few philosophers to take Darwin seriously and, in 1863, Mill published his book *Utilitarianism*. Herbert Spencer, who coined the term "*survival of the fittest*" (See *Origin*, Chapter 3) was another who took up Darwin's ideas, but Darwin found his philosophy very difficult to understand (Darwin, F, 1887; Freeman, 1978). Opposition at Cambridge was led by Adam Sedgwick; Whewell, who at this time was Master of Trinity College, banned Darwin's books from the College library (Freeman, 1978); it is also interesting that Christ's College, Darwin's old College and the College of his sons, William and Francis, does not, to this day, possess a single copy of a first edition of Darwin's works.

At a more personal level, these letters are full of family talk, since the families on both sides were growing up; and in the case of Fox still growing: his last and sixteenth child, Gilbert Basil Fox, was born in 1864. Darwin offered a saturnine comment on Fox's last child, "*I am glad you are all well; but I never heard anything so awful as your sixteenth child!*" (Letter #168). Whereas Fox replied heartily, "*Most of our friends condole with us on the affliction but we are well contented to bear it. We have not one too many in our 16*" (Letter #169).

What is also not brought out from the letters is the inner life of these two men, a life that they shielded from their closest friends and family. For Darwin we have few clues, except his ill health and what he did not say. For Fox we have the unbroken record of his diaries, from 1861 to 1865. These diaries reveal some of the inner fabric of his life and the complicated relationships of a family man at this time. Fox clearly carried on with his farm and all the worries that that entailed. His financial difficulties would not have improved greatly over the period. The death of his brother, Samuel Fox IV, in 1859, brought him some slight advantage as his brother left him all that he possessed, but this was not great, as it involved a number of properties, which were heavily mortgaged. Moreover, the death of his mother in the same year brought only further complications, for she left her rather small inheritance to her daughters, who would need every penny to sustain them at their apartment in Kensington Park Gardens, London. What seems to have been a fairly sharp rift with the strong-willed sisters Julia and Eliza, cannot have helped this situation (see Appendix 3 and below). Nevertheless, Fox kept up his visits to his sisters throughout this period, coupling them to visits to his father-in-law, Basil George Woodd, at Hillfield, Hampstead. He was still seeing the sister of his first wife, Miss Fletcher, but by this time the relationship may have become embittered. In late December 1861, there occurred an event that took up several pages in Fox's two diaries at this time. Apparently it fell to him to arrange a transfer of his mentally-deficient brother-in-law, Sir Richard Fletcher, from a place at Stamford to a new home at Pinchcombe House, near Gloucester. Having arranged this difficult move, by train and horse and carriage, with all decorum and rectitude, Fox wrote, in his diary, "*Thus ends this to me most harassing business. I have been bitterly disappointed by Miss Fletcher, Mr Br[oadrick] & my sister Eliza opposing my having care of him. I feel I could have been of the greatest use to him & that with us he would have had every comfort - & that the income w*[d] *have enabled me to help my children in many ways I now cannot & I bitterly feel that these three were not the persons who sh*[d] *have stood in my way*".

Despite his weakening lungs and heart (he was diagnosed in 1866 with angina), Fox kept up an exacting round of journeying during this period, visiting places in London, Derby, North Wales, Yorkshire and the Isle of Wight, while at the same time managing the education of his children, looking after his farm and running his parish. His children, by and large, were academically clever, went to good schools like Repton, and then on to Oxford University, where the sons read divinity and law. The exception to this rule was the third son, Robert Gerard Fox who was born in 1849 and during this period was in his teens. He seems to have been a difficult child. At an earlier time he lost a thumb in an accident with machinery on the farm. Later he was unsettled at school. Fox withdrew him from Repton School and put him at the school of Mr Price at Maidenhead (Letter #149). Later still Gerard went to King's College, London and then, no doubt on advice from Queen Victoria and her advisers, was appointed tutor to Prince Frederick William Victor Albert of Prussia (1859-1941), who became Kaiser Wilhelm, II, of Germany (Letter #213). As to his daughters, Fox seems to have had advanced ideas for the age and sent them to secondary schools. This contrasted with the Darwins' approach, where private tutors were employed; and it may have led to the upbringing of young adults with good social skills: thus six of the nine daughters who reached maturity got married. In the case of Darwin, while Henrietta, the invalid, did get married, to a much older man, at the age of 28, it was never expected that they would have children, and Henrietta's eccentricities only seemed to increase (Raverat, 1970; Healey, 2001). The other daughter, Elizabeth, never married.

For Darwin at this time, financial matters were becoming easier. He had inherited from his father in 1848 and made some good choices in railways and Liverpool Docks shares (Darwin's *Account Books*; Desmond and Moore, 1991). Then in 1859 with the publication of the *Origin*, he began to earn money for the first time from his books; an income stream that increased considerably over the years (Freeman, 1978). At this time, too, his son William, who became a partner in the bank at Southampton (Letter #144), largely took over his father's financial affairs, and made some very astute decisions, especially in Consols and other railway shares, which left Darwin with a much-enhanced fortune (Desmond and Moore, 1991, Browne, 2002).

On a broader perspective of the correspondence, it is not surprising that the world at large went wheeling by without comment. This was the time of the Civil War in the United States of America, the assassination of President Lincoln and the abolition of slavery, and, while Darwin did write expressing his long-held views against slavery to Asa Gray (CCD, Vol 8), neither Darwin nor Fox expressed any opinions to each other. Prince Albert, the Prince Consort to Queen Victoria, died in 1861. This was a time of national mourning, the like of which was probably never seen again until the death of Princess Diana in 1997. In his diary for 1861 (DAR 250), Fox wrote down how much the local shopkeepers and artisans responded spontaneously to Albert's death, by shutting up shop and openly mourning. Later still, in 1867, Disraeli, the conservative Prime Minister, pushed through a parliamentary reform act, which greatly enlarged the suffrage, a theme that the Whig supporters of 1833, Darwin and Fox amongst them, had vehemently supported [Letter #48 (n11)]. In 1867, Karl Marx published his opus "Das Kapital", but despite Marx having sent Darwin a copy of the second edition, there is no evidence that Darwin knew of him or his book. Probably both Darwin and Fox had drifted more to the right than the left, despite their whiggish, younger days.

None of this appears in the pages of their correspondence; and perhaps this is understandable because both men were ageing and drifting apart; Darwin was 58 and Fox, 63: relatively old by standards of the times, especially after a lifetime of ill health on both sides. As mentioned in the *Introduction*, they did meet once during this time (Letter #151), in February 1863, and although both talked of more meetings right up to the late 1870s, that was in fact the last time they were to meet: but the realities of their situations, with ill-health and family commitments, meant that any meeting was becoming increasingly unlikely. So at a time when the children of these two men were growing up and maturing there was little evidence of the strong friendship between them. It is easy enough to see, although hardly accurate, how Francis Darwin could write in his three volume *The life and letters of Charles Darwin* (1887, Vol 1, pp 171/2), "*Their friendship was, in fact, due to their being undergraduates together. My father's letters show clearly enough how genuine the friendship was. In after years, distance, large families, and ill-health on both sides, checked the intercourse; but a warm feeling of friendship remained*". Such a view puts greater emphasis on the 45 letters of their undergraduate days in comparison to the 110 subsequent letters. In fact, there was still a fruitful time ahead, right up to Fox's death in 1880, resulting in a further 40 letters (not including those of Charles W Fox, who wrote the last two letters to Darwin).

Chapter 8
The Final Years

Letters #177-218 (42), 1 Feb 1868 – 1880, 13 years

Very few, even among those who have taken the keenest interest in the progress of the revolution in natural knowledge set afoot by the publication of the "Origin of Species"; and who have watched, not without astonishment the rapid and complete change which has been effected both inside and outside the boundaries of the scientific world in the attitude of men's minds towards the doctrines which are expounded in that great work, can have been prepared for the extraordinary manifestation of affectionate regard for the man, and of profound reverance for the philosopher, which followed the announcement, on Thursday last, of the death of Mr Darwin.
Thomas Henry Huxley in *Obituary (to Charles Darwin);*
T. H. Huxley, *Collected Essays* (1893), Vol. 2, p. 244.

"Dull Atheist! could a giddy dance
of atoms lawless hurled,
construct so wonderful, so wise
So harmonized a World"
Erasmus Darwin, from the poem "The Folly of Atheism"

At first in this last period of the lives of Darwin and Fox, every thing continues much as before, although, in fact, we are in a very different world from the frivolous days of riding, shooting and youthful camararderie of Chapters 2 and 3. The period begins with a series of letters, concerning a wealth of subjects. Darwin had just published his two-volume book "*The variation of animals and plants under domestication*" (1868), but had realised that there was so much more to say. He had long wanted to treat that "holy of holies", the origin of man, in an evolutionary manner. When Charles Lyell did not take up this challenge in his book "*The Antiquity of Man*", in early 1863, Darwin offered his notes to Alfred Russel Wallace (LLiii), but in vain. The origin of the human brain was one thing at which Wallace baulked (Wallace,1871). So Darwin decided to take this on himself, although rather reluctantly. The attraction here was the opportunity of combining of his thoughts on human evolution, on which there was really very little evidence at that stage, with his ideas on sexual selection, which were in ferment. Of course sexual selection was pertinent to man, but it blossomed out into almost every avenue of the evolution of plants and animals. As Wallace pointed out, sexual selection was just a form

A. W. D. Larkum, *A Natural Calling*, DOI 10.1007/978-1-4020-9233-6_8,

of evolution itself. It could therefore encompass any aspect of evolution. So Darwin's book bulged, and eventually Darwin was able to persuade his publisher, John Murray, to publish "*The descent of man*" (1871) in two volumes, with the second volume - on sexual selection - being by far the larger volume and containing a wealth of information on every subject under the sun, including much from Fox.

And so we find a continuation of the same kinds of letters from Darwin to Fox, teasing Fox out on a number of subjects: "*Will you aid me? — I have references, somewhere safe, on one of a pair magpies being repeatedly shot on or near the nest; & each time a new mate was found in course of few days. ... I am working at what I call "sexual selection" & want all facts & cases of polygamy (is any bird in England besides the pheasant polygamous?) — of the females alluring the males — of victorious males getting wives — of attachment between individual birds — anything & everything — I am driven half-mad by the number of collateral points which turn up in my present enquiry. — Did your half-wild male Turkeys fight? & how, I forget whether they have spurs — Do Partridges use their rudimental spurs in fighting; I suppose no man can answer. Do you know anything about fighting of Guinea-fowls, or Peacocks — Forgive me Forgive me*" (Letter #179). And, again, a little later, "*I have vague remembrance that you once mentioned to me instances of Horses, or dogs, or cattle, or pigs, in which the male preferred some particular female, or conversely the female allowed of the advances of some particular male. .... Have you ever kept an account, or know anyone who has done so, of number of males & females born, in pigs, dogs cattle, sheep, or poultry? I much want a large collection of facts on this head*" (Letter #183). In return, Darwin got a rich haul of information. First there was the "*national magpie marriage*", an annual pairing event amongst magpies, observed by a number of gamekeepers. Fox's astute observations and his friendly relations with local gamekeepers provided a rich source of information, which Darwin distilled and placed in "*The descent of man*". This was just the sort of information that the Victorian public expected from this grand old man of science, who were brought up on Gilbert White's *Selborne* [see Letter #181 (n8)]: information on what had been under their noses all the time but just needed a new look to bring it out; nothing as racy as grandfather Erasmus' comparisons of sex in flowers with human sexual relations, but just enough of a nuance to give a frisson to the descriptions. Even Fox was induced to comment, somewhat pruriently, "*Fancy this as the state of things in general in a Forest*" (Letter #184), and forgetting conveniently the earlier affairs of the heart, which he had had.

The other request for quantitative information was more of a bug bear. Fox duly supplied figures on the ratio of sexes in sheep and cows (Letter #186) in December 1868. This was information that was much harder to massage for a good public reception and, by and large, Darwin omitted the details and put them in appendices or footnotes [see, e.g., Letter #183 n(10)]. Reading through the archives at the University Library, Cambridge on the material that Darwin collected for *The descent of man* (and other books), one finds large amounts of statistical data supplied by correspondents, which Darwin never used (CCD, Vol 16, *et seq*).

The publication of *The descent of man* was a real test for Darwin as the suggestion that Man may have evolved from "monkeys" was still a sensitive topic. "*I hear sad tales about your Book about to come forth. I suppose you are about to prove man is a descendant from Monkeys &c &c Well, Well! — I shall much enjoy reading it. I have given up that point now. The three main points of difference to my mind — were that Men drink, smoke & thrash their wives — & Beasts do not. .... There are points in your unrivaled Book "The Origin of Species" — which I do not come up to — but with these few expressions omitted, I go with it completely. I do not think even you will persuade me that my ancestors ever were Apes — but we shall see*", Fox wrote in 1870 (Letter #192). Surprisingly, however, the book caused few ripples, since the book was authoritative and much of the heat had been taken out of the subject by T H Huxley's book, *Evidence as to man's place in nature* (1863). Most people knew what to expect by now and the book was much better received than Darwin could ever have expected.

Even Fox did not seem to complain when eventually he received his copy (Letter #193), although doubts lingered on in the Victorian psyche; "*Your Books are very often in my hand, & give me great pleasure, but I cannot go with you on all points — in fact I think I diverge from some of your leading ideas more than I did, as I think more of them*", Fox wrote on receipt of *The expression of emotions in man and animals* (Letter #198; 1872).

The expression of emotions in man and animals (1872) was the only other book that Fox was to receive, apart from the plant books. He died before the publication of the *Power of movement in plants* (1880) and *The formation of vegetable mould though the action of worms, with observations on their habits* (1881); and one feels sure that Fox would have approved of these, the last of Darwin's books.

Darwin was probably well aware of the lack of quantitative information up to the time of the publication of *The descent of man*, and as he often did, he now turned to plants to assuage his feelings. For many years he had accumulated facts about the dissemination of plants, such as the viability of the seeds of all kinds of plants in seawater (see, e.g., Letter #95 (n10); and see also Letter #113 – for lizard eggs). Now with the aid of his son Frank, Darwin launched into quantitative observations on a bewildering number of aspects on the functioning of plants – flowering, pollination, propagation and out-crossing, growth, and movement. These studies eventually gave rise to three books: *Insectivorous plants* (1875), *The effects of cross and self fertilisation in the vegetable kingdom* (1876), *The different forms of flowers on plants of the same species* (1877), and two second editions, under different names and with some substantial changes: *The movements and habits of climbing plants* (1876), *The various contrivances by which orchids are fertilised by insects* (1877). These books were read by fewer people than his other books, but they were safe both in terms of evolution and the references to sex, and so sold surprisingly well: by this time it was quite acceptable to have a few books of Darwin on readers' shelves even if these had not been read. Fox quotes several episodes where he clearly gained in status by being known both as a clergyman and as a close acquaintance of Darwin (Letters #196, 200 & 213).

In passing, one should not forget that these books were substantial contributions to science. The work on the nastic movements of plants (the orientation of plants in relation to light and gravity), and the enzymic activities of the organs used by insectivorous plants to trap and digest insects, were well ahead of their time and the information was recycled in textbooks for nearly a century before any real advances were made.

For Fox, these years were one of retirement and family enjoyment. When Fox retired to the Isle of Wight in 1873 his youngest child Gilbert Basil was only 8 years old and he had 5 children still at school. This must have been a considerable concern to him. However the lack of diaries for most of the 1870s, until 1878, provide few clues as to how these concerns were dealt with. One other aspect which drew him closer to Darwin was his new garden: "*It was a wrench to leave my old parsonage where I had spent 35 years, but I felt it right every way. Here I have been busy in doing what you did at Down — not building a house — but making Gardens &c out of the most unpromising materials I ever had to do with. After buying every potato &c — I have now atchieved a good Kitchen Garden with lots of all Vegetables &c a good deal of fruit, & in another year shall look tolerably civilised*", Fox wrote in 1874, now in his 70th year (Letter #200).

Fox's letters from now on were full of reminiscences and anecdotes. Darwin managed to revive a little interest at this time over insectivorous plants and Fox was happy to give back any special knowledge that he might have, such as advice about "*Old Price*" (Letter #203). However, this was as far as he could or would go. When Darwin tried once more, "*I am in great want of a living plant of Utricularia. Have you ever seen it in clear ditches in the I. of Wight? If so, could you send me a plant (with a root if it has one) packed in damp moss in a tin box*" (Letter #202), Fox's reply was that the plants that Darwin wanted occurred at the other end of the Isle of Wight and he had no intention of going so far. In the end Darwin's son, William, who lived in Southampton, collected them. Indeed Darwin had enough help at this stage from a myriad of correspondents, many of them clergymen, that he did not need Fox's help, anymore. Perhaps significantly, this was the last time that Darwin tried to include Fox as a helper in his studies, and their correspondence ebbed.

Both men were now seeing many old friends pass away. Of special significance here was Albert Way. It was Way who had been the catalyst for their much-remembered expediton to Whittlesea Meer, during their undergraduate beetling days when they had so much fun with the bombadier beetle, *Brachinus crepitans* (Letter #131), and the rare carabid beetle, *Panagaeus crux-major* (Letters #47, #55 & #57). When Fox pointed out that Way had recently died, Darwin's equivocal reply, which could almost do justice to King Lear, was, "*And as you say one looks backwards much more than forwards & can never expect to have nearly such keen enjoyment as in old days, as when we breakfasted together at Cambridge & shot in Derbyshire. Do I remember Brachinus? Am I alive? Poor Albert Way I did not know that he was dead*".

The last years of the friendship of Charles Darwin and his second cousin William Darwin Fox reveal little that we did not know before. The friends had now grown so far apart that there was little to hold them together except reminiscences. The lives of the two men were very different. On the one side, Darwin had reached the height of his fame and, up to 1878, was still absorbed in biological studies. On the other side, Fox was already reaching the end of his active life in 1868: he did not retire until July 1873, but already in 1869 he was spending the winters at Sandown on the

Isle of Wight, where he had bought a house. Nevertheless the friendship had a little life left in it through the shared interest in observations of the countryside, where Fox was clearly still a most keen observer.

For Fox, his major occupation now was his family. He did not try to take up his old studies of fossils and birds on the Isle of Wight. Instead like Voltaire's Candide, facing the void, he adopted the principle *"il faut cultiver notre jardin"*. The result was a spluttering of the flame of friendship, rekindled briefly by reminiscences and Darwin's and his sons' interest in their ancestry, particularly in Darwin's grandfather, Erasmus Darwin.

177 (CCD 5837). To Charles Darwin, Down, Bromley, Kent

Delamere Rectory Northwich Cheshire
Feb 3 [1868][1]

My dear Darwin

It gave me almost as much pleasure to see your Son 2nd Wrangler as if he had been my own[2].

I did not at once write to congratulate you because I thought you would be burdened with// letters, and now I have another cause for writing — viz to thank you for your kind present of your new Work, which I received on Friday last[3].

I have *run* it over with much satisfaction, and shall now proceed to ruminate it over at my leisure. It will be a great treat to me.

I have not written to you of late, because I feared to give you the trouble of a reply, which I thought would, in your// health, be an annoyance.

I greatly rejoiced to hear from Erasmus[4], that you were much better than you had been. When next in Town, I shall write to ask if I can run down to you for a few hours. It would be a great pleasure to again see you.

I have not been very well of late. I fear I am threatened with Angina from the pain I often have. I have consulted 2 Medicos, who gave me exactly opposite// advice, so am going upon my own plan, till I can take a 3rd Doctors advice[5].

We are all well — when quite at liberty I should much rejoice to hear of you and yours.

With kindest regards to Mrs Darwin, and most heartfelt congratulations upon her Sons success —

Ever dear Darwin
Yours affectly W D Fox

Provenance: University Library, Cambridge; Manuscript Room: DAR 164: 186

Notes

1. The year was determined by the reference to George Howard Darwin's examination (see note 2, below).
2. George Howard Darwin (1845-1912) had achieved second place in the final examination for the mathematical Tripos Examination at Cambridge University in January (*Cambridge University Calendar*); this position was known as "second wrangler".
3. Fox refers to *The variation of animals and plants under domestication (1868)*; his name appears on CD's presentation list for the book (see CCD, Vol 16, Appendix IV).
4. Erasmus Alvey Darwin.

5. Fox did not retire until 1873; nevertheless his heart, as well as his lung problems, must have been a burden for him and he took greater periods away during his last years as a Rector (see, e g, Letter #193).

178 (CCD 5842). To: Rev. William Darwin Fox, Delamere Rectory, Northwich, Cheshire

Down Bromley Kent
Feb 6th [1868][1]

My dear Fox

I thank you very sincerely for your most cordial congratulations. George's success has been a very great pleasure to us[2]. Whenever you come to London it will give me & Emma very great pleasure to see you here, but I hope that you will stay the night with us. You give but a poor account of your own health[3]. I hope that what you fear is not the case. I have known quite a large number of men who the D[rs] declared (falsely as it afterwards appeared) to have had their hearts affected; & how many men there are who live to good old age although their hearts are certainly affected.

There is so much detail in my book that you will never be able to read it through; but some of the chapters, such as that on Reversion are I think curious. You will find several facts given on your authority[4].

Believe me my dear old friend
Yours very sincerely
Charles Darwin

No envelope: dictated to Emma Darwin
Christ's College Library, Cambridge, #148

Notes

1. The year was determined by the relationship between this letter and Letter from #177.
2. CD refers to George Howard Darwin; see Letter #177 (n2).
3. See also Letter #177.
4. CD discussed reversion in the second of his chapters on inheritance in *The variation of animals and plants under domestication (1868)*, Vol 2, pp 28-61; CD cited Fox ten times, over both volumes, for information on various domestic animals.

179 (CCD 5929). To: Rev. William Darwin Fox, Delamere Rectory, Northwich, Cheshire

Down Bromley Kent
25[th]Feb [1868][1]

My dear Fox

You have been an acute observer of Birds, & may come across sharp gamekeepers & sportsmen-naturalists in the course of next 2 - 4 months. — Will you aid me? — I have references, somewhere safe, on one of a pair magpies being repeatedly shot on or near the nest; & each time a <new> mate was found in course of few days. A gamekeeper this morning told me that he had just trapped one of a pair *with young*, & in 2//

days a *pair* <to his surprise> again frequented the nest[2]. — Can you get me other instances with magpies, carrion-crows, hawks, jays or *any other Bird*[3]. Where possible to say, which sex was killed? How can you account for this fact: the gamekeeper said he has never noticed single magpies during pairing season. It is incredible that there sh^{d} be many single birds unpaired, always ready <to pair>. Or will <one of> a pair, which have failed in rearing young, always break//

his or her troth & join a widow or widower with the attraction of a nest. The gamekeeper also said that he had *never* seen an unpaired partridge; yet surely there must sometimes be an excess of males or females. — Think, enquire & be a good man & try & illuminate me. — I am working at what I call "sexual selection"[4] & want all facts & cases of polygamy (is any bird in England ~~except~~ besides the pheasant polygamous?) — of the females alluring the//

males — of victorious males getting wives — of ~~individual~~ attachment between individual birds — anything & everything — I am driven half-mad by the number of collateral points which turn up in my present enquiry. — Did your half-wild male Turkeys fight? <& how>, I forget whether they have spurs — Do Partridges use their rudimental spurs in fighting; I suppose no man can answer. Do you know anything about fighting of Guinea-fowls, or Peacocks — Forgive me Forgive me. Your old friend & tremendous bore.

C. Darwin

P.S. Do not hurry yourself to answer — Tell me nothing except you feel very sure; for remember that there are many too glad to pitch into me. —

Annotation. *Across top of p. 4 in handwriting that is not CD's*: "C. Darwin Sexual Selection Feb 7 68 victorious male getting wives —"

No envelope
Christ's College Library, Cambridge, 148a[5].

Notes

1. The year was determined by CD's note, dated 25 February 1868, in DAR 84.2: 202 (see n. 2, below).
2. CD recorded the information on magpies from the gamekeeper William Reeves (University Library, Cambridge: DAR 84.2: 202), noting: "*We can understand magpie case only by supposing many birds do not like each other & do not pair or do not nest, & will pair with bird with nest*". Reeves was a gamekeeper for John Lubbock, the neighbour of CD [see Letter #200 (n8)]; CD reported his observations of magpies in *The descent of man* (1871), Vol 2, p 103; for CD's other references on magpies, see *Ibid*, note 5.
3. Fox complied with this request assiduously over the next ten years as is evident by later letters to CD.
4. CD was working on his book *The descent of man, and selection in relation to sex* (2 volumes, 1871). The second part of this book, as the title suggests, is concerned with CD's theory of sexual selection.
5. Francis Darwin apparently did not number this letter at first. For the origin of its present numbering as 148a, see Letter #181 (n12). See Appendix 1 for the exegesis of these letters.

180 (CCD 5762). To: [Charles Darwin, Down, Bromley, Kent]
From: [W. D. Fox]

[Delamere Rectory, Northwich, Cheshire]
[after 25th Feb 1868][1]

see any but *pairs*.

We used — a few years since, to have *Great Numbers* of Magpies in the Forest. Up to the *National* Marriage[2] in early spring — you see all numbers together — but afterwards every Bird has its mate — & I never saw an odd Magpie under any circumstances[3].

[4]The Keepers remark about partridges — of never seeing unpaired ones is similar. There may however be pairs (to speak Hibernice[5]) of males & females in all Birds — who not having met a suitable match, keep together for Company[6].

I was much surprised a few weeks since, on going into a Canary Breeders room — to see 5 Hens put to one Cock, & all apparently Breeding well. He told me — that the Cock only *pairs* with one Hen, & will only feed her — or her young ones — but that the others are Concubines, to whom he pays the necessary attentions of Love, but does not trouble himself about them or their concerns & that they have to build their nests & rear their young themselves.// the same thing occurs with wild ducks under domestication, & when only removed by one or 2 generations from perfect liberty. The Drake as a rule *pairs*, but is very salacious after other ducks besides his Wife. It is commonly said that the Wild Duck & Guinea Fowl *pair*, — But both are willing to keep Concubines, & both do much better in fact, with 2 females than with one[7].

One great difficulty often strikes me in this matter. What of pheasants Turkeys &c &c where one Male has many hens. The Sexes seem abt equal — & certainly are as much so, in these birds — as in those who strictly pair.

By the way — the extraordinary earliness of this Season was shewn by the fact of the week ending May 9 — a Pheasants nest with 16 eggs in was found in Forest.

Now what becomes of all the Cock Pheasants &c — if the sexes are about equal. Here shooting the Males of course helps the difficulty — & some few Birds kill each <other> fighting for their loves in Spring — but these are very few.

I imagine the <extra> Cock Birds in Pheasants get driven from the best haunts — & that a kind hearted hen (perhaps not herself happy//

2

in her conjugal relations) *pairs* with him, & perhaps in their native country they pair much more than here.

I have observed Birds all my life, & never saw an *odd* Bird in Spring or Summer — except when I have (to my shame be it spoken) killed one of a pair — & then you only observe the widowed Bird for a day or two, & then it disappears.

There are some Birds which you could not help seeing, if they were *lonely* Birds — viz Carrion Crows — Magpies — Wood pigeons — Fly catchers.

I never saw a *rogue* Pheasant — driven, as they say Rogue Elephants are from the herd. Stags are said to be often — or always — thus driven from the Herds by the stronger ones — but I fancy they go to the herds of fawns — or herd together as you see in Parks.

I was amused by a writer in the Field some months since — saying "that it was most fortunate there were much fewer female Wolves in Russia than males, or the Country w^{d} be eaten up by them" — he exemplified his fact by the Nos of males in pursuit of one female[8].

One w^{d} suppose any one who saw Bitches in same state — w^{d} have had sense enough to come to a truer conclusion.//

Are not Black Game[9] polygamous as well as Pheasants?

Do "females allure the Males" in Birds — It always seems to me as if an irresistible impulse drives the males to seek the female — especially in Birds.

I have sometimes thought that a strong argument might be drawn against your Theory — *when extreme* about the Origin of Species (I quite agree with you except when, as I think you get into *extremes*) when you look at the extraordinary differences in the mode & organs of procreation[10].

Each Genus — or Family — seems to have its own peculiar mode of procreation & the Organs for it — so essentially different. Look at Cats — Dogs — horses — Cows —. Did you ever see the Organ of the Male Rhinoceros? I imagine you might throw all animals into good Natural Groups merely by this one part of their structure & mode of using it, with its accompanying gestation.

And does not this also separate Man from the rest of creation? All other <female> Creatures have their *seasons* only of Love — This seems to me to separate us much more than any anatomical detail, from the Apes &c —[11].

In Birds the mode of generation is same//

Provenance: University Library, Cambridge; Manuscript Room: DAR 86: 83; 84, 89. Incomplete

Notes

1. The date was determined, tentatively, by the relationship to Letter #179. However, there are many letters covering similar subjects up to 1878.
2. In Letter #179, CD asked Fox about the mating habits of magpies. The "National Marriage" is described in greater detail in Letter #184 (n14). It is also described in *The Descent of Man,* (1871), Vol 2, pp 102 (see Letter #184 (n14) for a transcription).
3. See Letter #179.
4. CD has added blue, red or black lines to the side of each paragraph of this excised part of Fox's letter. For this paragraph and the next there are two parallel black lines. For other markings see CCD, Vol 16.
5. Hibernice: Latin for "Irish"; i e, Fox means to say "to speak Irish" or to use a word in a strange way.
6. In *The Descent of Man*, (1871), Vol 2: 106, CD stated, "*But birds of the same sex, although not of course truly paired, sometimes live in pairs or in small parties, as is known to be the case with pigeons and partridges*".
7. CD reported Fox's information on polygamy among various domesticated birds in *The Descent of Man* (1871), Vol 1, p 270 [see Letter #184 (n21) for a partial transcription].
8. *The Field*, No. 785, Vol. 31, Jan 11th 1868 had an article "*Wolf Hunting in Russia*" by "Basil" which said, in part, "*and as to every female wolf there are thirty or forty males, the ladies always at this time have a considerable following. It is well that the numbers are so proportioned by nature, else the country would soon be quite overrun with them*." See also Letter #184 (n27).
9. Fox was referring to the black grouse, *Tetrao tetrix*.

10. This is one of the few expressions we have of Fox's views on CD's theory of Natural Selection [see also Letter #201 (n7)]. It appears that, like many Victorians of his time, Fox was having trouble with accepting the full import of CD's thesis: Fox, it seems, can follow it through in given instances, but the generality, especially as applied to Man, is difficult for him to accept (for a good example of this sort of reaction, see letter from Adam Sedgwick to CD, 24th November 1859 (CCD, Vol. 7), which was said even to have upset Emma Darwin.
11. CD marked this passage with a single blue pencil line down the side of the page.

181 (CCD 6172). To: William Darwin Fox, Delamere Rectory, Northwich, Cheshire

[Down, Bromley, Kent]

14th May [1868][1]

My dear Fox

I fear from your note received this morning that you are far from well[2]; but cannot you pay us a visit[3]? It wd give me, my dear old friend, real pleasure — All days are the same to us, but we are a very small party now.// The Orpington Station on the S.E. Railway is now open, only 3 miles distant from us[4].

Your letter is an excellent one & gives me just the kind of facts I want to know[5]. If you can find the mem. about the carrion-crows pray do so.// I shall let you have no peace till you give me a full account of the "great Magpie marriage": I never heard of such a thing except with Black cocks & some foreign birds[6]. Your story about the Peacocks is so good that I must quote it[7]. Don't hunt for references about Magpies unless you know of other cases, besides those given by Macgillivray, Couch, & I think White of Selborne[8]. You ought to have// written a book like White's Selborne, for I am sure you could have done so.

I hope you will be able to pay us a visit[9] & then I shall hear all about yourself & your family.

My dear Fox

yours affectionately

Ch. Darwin

I am quite delighted that you approve of my book[10]. I do not think that many have cared for it — too many details for almost everyone — & beloved Pangenesis disagrees badly with many[11].

No envelope

In Emma Darwin's hand writing, except for signature and the postscript, which are in CD's hand writing.

Christ's College Library, Cambridge 148b[12].

Notes

1. The date was determined from the opening of Orpington Station, which occurred in March 1868 (see citation in CCD, Vol 16 to Cox, 1988, p 48).
2. No letter with information on Fox's health exists at this time. No Fox diary exists for 1868, so Fox's health is not known at this time. However, this remark reiterates a similar remark of Letter #178 (n3).
3. No visit is known to have occurred (see Appendix 2).
4. See note 1, above.
5. CD may be referring to Letter #180, although there is no information, in the parts of that letter that have survived, of a memo on carrion crows, nor on the pheasant story (see Note 7).

6. For Fox's account of the "magpie marriage", see the Letter #184 (n14). Fox had also mentioned it in his letter #180. See also *The Descent of Man*, (1871), Vol 2, pp 102–3.
7. In *The Descent of Man*, (1871), Vol 2, p 46, CD wrote, "*the Rev. W. Darwin Fox informs me that two peacocks became so excited whilst fighting at some little distance from Chester that they flew over the whole city, still fighting, until they alighted on the top of St. John's tower.*"
8. CD refers to W Macgillivray (1837-52) *History of British birds*; G White (1789) *Natural history of Selborne*; the reference to J Couch is probably to *Illustrations of instinct deduced from the habits of British birds* (1847). The first two are cited in *The descent of man* (1871), Vol 2, in the chapters on birds.
9. See Note 3.
10. This is one of the few early indications that Fox approved of any of CD's evolutionary work. However, Letters #192 (n6 & 7) and #198 (n2), indicate that Fox did not go all the way with CD.
11. The book is *The variation in animals and plants under domestication* (1868) for which Fox received a presentation copy (see Appendix IV, CCD, 1868). CD discussed his hypothesis of pangenesis in Vol 2 of that book. Wallace was one of the lone supporters of this hypothesis (letter to Wallace, 27th Feb 1868, CCD, Vol.16), which was one of the few flawed proposals made by CD (Browne, 2002, Vol. 2, Chapter 8). Soon, in 1872, Friedrich Weissman would propose that special tissue, the *germ plasm*, which produced the next generation in both plants and animals. When this was widely accepted, the hypothesis of pangenesis was quietly dropped.
12. This letter has the number 148a. However, later, Francis Darwin made the note "148a = 148b". Letter #179 has no number on it and accordingly has been assumed to be "148a" of the Christ's College letters.

182 (CCD 6187). To Charles Darwin, Down House, Beckenham, Kent

Hillfield Hampstead [London N.W.]
19 May [1868][1]

My dear Darwin

I wish very much I could have run over to Down while here, but it is quite hopeless for me to think of doing so. I return home on Saturday, and am not equal to any extra exertion I can help, tho' very comfortably well, as long as I am quite idle, as my pain tho constant, is very bearable[2].// I have not time to enter into your letter today, but will do so ere long.

I should much have enjoyed seeing you, and Mrs Darwin again — if I could.

Yours always
W D Fox

Provenance: University Library, Cambridge; Manuscript Room, DAR 164: 187.

Notes

1. The year was determined by the relationship to Letter #181.
2. See Letter #181 for CD's invitation. For Fox's recent health, see Letter #181 and #178. Fox must have been visiting his father-in-law, Basil George Woodd, at Hillfield House, Hampstead.

183 (CCD 6426). To W. D. Fox, Delamere Rectory, Northwich

Down Bromley Kent S.E.
Oct. 21[1868][1]

My dear Fox

I am rather uneasy about you. When you wrote from Hampstead, you spoke uncomfortably about your health & as you have had at various times so many illnesses I should very much like to hear how you are[2] —

If you have strength &//

inclination, but not without, tell me what you can about the great mapgie marriage that I may quote your account[3]. What I call sexual selection as applied to birds has turned out to be an everlasting subject, & I am still at work on it[4]. Some time ago I broke down & we all went for 5 weeks to Freshwater Bay & enjoyed it much[5]. I know you like//

hearing about our children so I may tell you that George has delighted us by getting his fellowship at Trinity this first year[6].

Poor old Sedgwick now 84 yrs old, with a multitude of complaints, most kindly wrote to congratulate us[7]. Leonard succeeded about half a year ago in getting in as N° 2 into Woolwich & means to be a R. Engineer if he can[8]. Henrietta has been touring in Switzerland & met there//

two of Sir F. Darwin's daughters & much liked their frank & cordial manners[9].

I have nothing more to say — for this note is written to extract an account, & I much hope a better account, of yourself —

My dear Fox
Your sincere friend
Ch. Darwin

P.S. I have vague remembrance that you once mentioned to me instances of Horses, or dogs, or cattle, or pigs, in which the male preferred some particular female, or conversely the female allowed of the advances of some particular male[10].
Have you ever kept an account, or know anyone who has done so, of number of males & females born, in pigs, dogs cattle, sheep, or *poultry*? I much want a large collection of facts on this head.

Dictated to Emma Darwin, only the postscript is in CD's handwriting.
Envelope missing.
Christ's College Library, Cambridge, 149

Notes

1. The year was determined by the relationship to Letter #182 and the "Magpie marriage" (Letter #181).
2. Refer to Letters #181 and #182.
3. See Letter #181.
4. See Note 10, below.
5. For CD's health at this time, see the letter to Asa Gray, 15 August [1868], CCD, Vol 16. The Darwins stayed at Freshwater from 17th July to 20th August 1868 (Emma Darwin's *Diary*). Henrietta Litchfield née Darwin stated in *A Century of Family Letters*, "*In 1868 we took one of Mrs Cameron's little houses at Freshwater for six weeks. It was a beautiful summer, and we had a very entertaining time. Mrs Cameron. . . . . .was sociable and most amusing, and put my father and Erasmus Darwin, who was with us, into great spirits*." The poets, Tennyson and Longfellow, also visited the Darwins there.
6. George Howard Darwin, CD's son. Was elected to a Junior Fellowship at Trinity College, Cambridge, in 1868.
7. Adam Sedgwick (1785-1873). Woodwardian Professor of Geology at the University of Cambridge, 1818-1873; Canon of Norwich, 1834-1873. (*DNB*). Sedgwick became an active opponent of "Natural Selection" after the publication of "*On the Origin of Spec*ies" in1859. His remarks in his letter to CD, 24th November 1859 (CCD, Vol. 7) were said to have offended even Emma Darwin (see Letter

#180 (n9)). The information on Sedgwick's condition at this time is contained in his letter to CD, 11th October 1868 (CCD, Vol 16).

8. In July 1868, Leonard Darwin came second in the entrance examination for the Royal Military Academy, Woolwich (see letter to Horace Darwin, 26th July 1868, n 2, CCD, Vol 16).
9. According to Emma Darwin's *Diary*, Henrietta Emma Darwin went abroad on 31st August 1868. CD refers to Francis Sacheverel Darwin, his half-uncle [see #Letters 137 (n6) and 125a (n6)]. Henrietta possibly met Violetta Harriot Darwin and Ann Eliza Thomasine Darwin, the only two daughters who were not married by 1868 (*Darwin Pedigrees*). Fox was also related to Francis Sacheverel Darwin, through his mother, and had met his family on a number of occasions [see Fox *Diaries* and Letter #125A (n6)].
10. CD was interested in the proportion of the sexes in different species; see, e.g., letter to H W Bates, 11th February 1868 (n2), CCD, Vol 16. CD was collecting material for *The descent of man and selection in relation to sex* (1871). In Part II, Chapter 8, "*Principles of Sexual Selection*", he wrote "*I have remarked that sexual selection would be a simple affair if the males were considerably more numerous than females. Hence I was led to investigate, as far as I could, the proportion between the sexes of as many animals as possible; but the materials are scanty. I will here give a brief abstract of the results, retaining the details for a supplementary discussion so as not to interfere with the course of my argument.*"

184 (CCD 6436). To: Charles Darwin, Down, Bromley, Kent Oct 29 1868

(*on a separate sheet from the opening part of the letter [next, below]: in CD's handwriting up to "near Dorchester", the rest in Fox's handwriting*)

ArchBishop Whately life, Vol II. 171

"I have sown the seeds of the White Bk Currant & the white variety of the Woody Nightshade — & all of them — as many as have flowered — have come true.

[1]On the other hand, I have sown the berries of the Florence Court Yew (which Botanical Books speak of as a distinct species) & all that have come up as yet, have been common yews."[2]

32 years since W D Fox brought berries from Dr Darwin's hedge; of Hollies of yellow berried kinds (& which Dr Darwin told W D Fox he had himself sown with yellow Berries) — and about half have grown up large enough to berry — & bear yellow fruit — The other half Red berries[3]. I cannot state exact proportions as many have not yet fruited. As far as they have about half are yellow ½ red.

Copper Beech — At Piddle Hinton Rectory — near Dorchester[4]

next day he came & found 2 other charitable hawks who had come to succour//

(*at the top of the opening part of the letter*)

the orphan. He killed these 2 & left nest. Afterwards he found 2 more charitable individuals on same errand of mercy — One of these he killed — The other also he shot, but could not find — No more came."

F. O. Morris — by the Honble & Rev W. W. Forester."[5]

Delamere Ry, Northwich
Oct 29 [1868][6]

My dear Darwin

The above extracts I copied thinking they might interest you. The ArchBishop evidently was no Botanist — & the Keeper a Brute, as most are.

I have to thank you for a very kind letter received a few days since[7]. You ask about myself so I am obliged to enter into that disagreeable subject. I have// been more or less an invalid for the last 12 month[8]. First D^{r} thought it heart & recommended Bitter Beer — 2^{d} D^{r} doubted but recommended Claret instead of Bitter Beer — Hoped it was not heart — but rather liver & Stomach — 3rd D^{r} — recommended dry Sherry thought it neuralgia of walls of stomach, &c. 4th D^{r} thought nothing was the matter, possibly Gout. Not to drink Port Wine, which I find by a side wind he does himself 5 days a week, but feels he is *y*oung. Lastly — I am assured it is not heart — or Liver — possibly not Neuralgia of Stomach — possibly is derangement of Blood Vessels — positively is not cancer or tumour of stomach <or Gout>. I have taken every vegitable poison — Strichnine — Aconite — Bella Donna &c — & most of the mineral do — — My last D^{o} was a course of Aconite. There! I hope you will understand what is the matter with me.

At times I suffer a good deal — at times I am free from pain. I am however now in very tolerably comfortable health, with care not to stoop — overdo my strength &c & get thro a good deal of work in all ways. I have very much to be thankful for <in> all ways. Of course a continual pain is annoying, but I dwell upon it as little as I can help.

I promised Dr Tanner[9] to leave home from 1 Novr to Xtmas & go Southwards but a succession of causes have kept me from being able to do so before the 9$^{\underline{th}}$ when if I can I mean to take a fortnight at either Isle of Wight or Hastings. I must then come to the Election, to vote for 2 Div^{s} of Derbyshire[10] — & shall remain at home for the Xtmas holidays. (I have now 1 Boy at Christchurch Oxford — 1 at Kings Coll — 1 Boy at School at Whitchurch & 2 girls at schools in London[11] — & I must make them a happy home at X^{ts}.//

Tanner then wants me to go to some *decided* place — Algiers — he recommends — but that is out of the question. I shall go South however if necessary.

Well I have given you — a full acc^{t} of myself, as you asked — There is no one else I w^{d} do so to.

So to work. I could not make any experiments this Summer from want of health & power. But a very clever Keeper, "Jesse England" who has all his life observed Birds — gives me the following facts.

With respect to "the Magpie Marriage". We used to have a very great Number of Magpies in the Forest till he destroyed them[12]. They well performed for us, the place of Rooks, & used to be in Numbers from 20 & upwards quite commonly together.//

He tells me "he can well remember a great Congregation of Magpies one early spring in a part of Forest called The spreading Oak Field — and that he shot 5 of them. That they were chattering & flying about amongst the trees, & sometimes fighting."

My own recollections I before told you[13]. I have seen all the Magpies apparently of the Forest — several times in the early months of year Jany or Feb — I think the Former — all congregated together with a universal chattering & Bustle.

I watched them for some time & was much amused with the evident importance of the whole Affair. In two instances certainly — I noticed that after this day — the

Magpies were all in pairs — I always called it "the Magpie Marriage" — & have frequently mentioned it as having taken place[14].//

Jesse England says that <with> Magpies & Carrion crows, he has frequently observed that when he has shot the *Cock* Birds (he says the hens are generally too wary) from the nest, that another has come & taken their place." And he gave me this case as occurring *this Summer*. "I shot the cock Bird of a Carrion Crows nest in Mr Wilbrahams Wood[15] — The hen was shy & I could not kill her" — The young hatched at least a fortnight afterwards when I shot another Cock Bird at nest — & afterwards I caught the hen Bird with a trap."
He said he had many times shot cock Birds of Magpies & Carrion Crows & found another took their place. That it is much easier to kill the Cocks, as they feed the Hens on nest & are less wary.//

3

I before told you that I had myself when a lad shot Carrion Crows from nest at Osmaston, & to my surprise observed a pair again in a day or two[16].

England tells me — (to give you an idea of the abundance of our Magpies a few years since) that he with another man — shot 19 Cock Magpies from the nest one May[17].

His experience will in some measure answer your Questions as to the sex — Probably 9 out of 10 are the cock Birds.
Partridges. He never saw an *odd* Bird. But partridges are so companionable, that there is little likelihood of ever doing so. He says he has often known 2 hens to 1 Cock bird, & that they have generally had large coveys — so as to make him sure both birds bred. On the other hand he has known several times 2 cocks to 1 hen, when//

they rarely or never have any young ones.

This year he has observed 5 old Cock Birds, which have always kept together this season — 2 of which he has killed[18].

Partridges fight very pertinaceously — he only two days since watched a Battle for a long while[19]. They scuffled at each other — pecking & scratching with their feet (they have no spurs) — & continue this for hours.

Pheasants he has known frequently kill each other (I picked up a splendid Cock in Bolton Abbey Woods[20] some years since in March, which had evidently been killed by a spur thro the head).

On one occasion — England said a Man called him into his garden to see a Battle going on, which he said he had watched above an hour.

Both cocks were so exhausted that// England picked them up & put them in his pockets. After a while one was able to run away — but the other died.
Wild-Ducks — The pairing of these may throw some light upon the matter. It is generally said that *wild* ducks must be *paired* to do any good in hatching.

I have no doubt they generally do in a wild state, but I do not believe always — At all events in a semiwild state — they do much better 2 or 3 ducks to a Mallard — & this Summer — England (who breeds a great Number on a large pond) — killed off his Mallards so as to leave 7 or 8 Ducks to each, & he never has been so successful or had larger broods[21]. His ducks are in a completely wild state except that they come to feed.

Englands experience with respect to Hawks (Kestrels & Sparrow Hawks)// is opposed to Honble & Rev Foresters acct. He says he never observed a fresh Cock come if he killed one — but that he has often found the Cock Bird will bring up the young, if he killed the hen bird. Very lately he shot a hen Kestrel or Sparrow hawk who had young — & caught the Cock Bird a fortnight afterwards on nest.

This was on a spruce tree at least 15 feet high — he left the trap set, & to his astonishment a day afterwards found a *stoat* in it.

There are cases innumerable of other Birds (not the Parents) feeding young Birds, & I can quite imagine the Parental ~~sta~~ feeling prompting this as it w^d in ourselves. This w^d account for the Magpies you mention feeding the young of those who had been killed. — With regard to there being *odd* Birds unpaired, I can easily imagine that there may an excess in particular districts of// either sex — but these would not mope singly, but w^d herd together generally, I sh^d say. Then many Birds must die during pairing season — especially females, as it is a critical time with them — & I have several times met with fine cock Birds of the smaller kinds of finches &c — dead, without any apparent cause in spring. These w^d all cause vacancies in the domestic circle requiring to be filled up.

I believe Concubinage is not uncommon among birds. I told you of the Man at Chester, who breeds Canaries — putting a *paired* cock to several other hens. He impregnates them, & *they* build & bring up the young — he, is meanwhile being an excellent husband & Father to his wife & ch^n by her[22].

I must tell you of a white Muscovy Duck, hatched here this year among some 30 others.//

To the great amusement of my children, she cocks her tail exactly like a hen.

I shewed her today to a Miss Hall with us — who immediately said "she is a Hen Duck". It is a curious variety. Mrs Fox says that it illustrates Darwin— &
finding that Ducks w^d not do any good in such hot summers, is retrograding to a hen. I am trying to take care of it — to see what her eggs do next year.

I have a list of the sexes of sheep — & cows for many years past, which I will send you shortly[23]. I shall have *satisfied* you for the present, if you read thro all I have written.

Turkeys — My wild & ½ wild Turkeys fight very much — Even at this time of year the young Cocks are constantly doing so — and the Hens also with ea: other —
They spur each vigorously — throwing the weight// of body well in the attack — They also bite & tear each other about neck & head. In spring — you cannot keep 2 Cocks in a flock — I feel sure the stronger w^d kill the weaker bird.

The Cocks go some distance from Farm yard to Farm yard to attack distant neighbours — a mile or more. They are eminently pugnacious Birds.

Both Guinea Fowl & Peacocks fight.

I think I before told you that Guinea Fowl do better — 1 cock to 2 or 3 Hens than when paired[24]. That they are not highly moral, I can give you a good proof in that this year — I had 2 *Pearl* Birds paired — all the rest — 8 or 10 — were *white*.

Out of 32 Birds reared under hens, & therefore eggs taken at random — 6 only were white — the rest pied more or less. The Pearl Cock was the master of yard & evidently made the most of his opportunities. Now if Guinea Fowl, Wild Ducks & partridges — all of whom pair as rigidly perhaps as most birds, (& whose peccadilloes we only know better, because they are larger birds & under our eyes) take these liberties, and as Canaries certainly do also — I have little doubt that a great deal of licence prevails generally among birds — even when strictly paired, and I have little doubt but that generally an odd female w^{d} be compassionately cared for by some cock bird in her neighbourhood, who w^{d} keep her in some Rosamonds Bower, where she might bring up her young unknown to the Wife[25]. Males I have no doubt generally herd together until they can meet with a lady love.

It has often puzzled me what becomes of all the cock pheasants B^{k} Game[26] — Turkey Cocks &c that are extra — as I suppose we may take 5 Hens to one cock as an average, in Breeding season. Some no doubt kill each other, but still the actual deaths must be small in number, to those beaten & driven away from their victorious Rivals' domain. What becomes of these unwilling Bachelors. — I cannot help thinking that when driven far from their Rivals haunts — some <one> of his Concubines takes compassion on his loneliness & deserts to him.

In this country Gamekeepers are a sort of providential arrangement to kill the extra cocks — but in a state of nature those kindly animals do not exist & what then?//

5

As I am sure about an equal number of the sexes are produced, upon our usual calculation of the habits of these birds — out of 20 Pheasants &c hatched — there w^{d} be 10 cocks — & only 2 being required for the 10 hens — there w^{d} be 8 odling Bachelors.

Fancy this as the state of things in general in a Forest.

And generally speaking they do not seem to herd together — as *pairing* Birds do. I cannot fancy a Community of Cock Turkeys — Cock pheasants &c in spring.

Animals do often herd together in separate sexes, as deer for instance — but I do not imagine any of these Birds do so.

If you succeed in getting the number of sexes of poultry Ducks &c. you will, I feel sure, find an almost equal Number of the sexes — I have noticed often in both ~~Poultry~~ Hens, Ducks, Geese, &c that they are about equal in the average — Not in each particular//

year. For instance this year I have 15 Mallards to 4 Ducks — & sometimes there is an excess of Turkey Cocks one year — & of hens the next — but on the average I am convinced you will find no disparity of sexes.

Well, my dear Darwin, I will dismiss you. I wonder whether you will get thro my letter.

By the "sexual selection" reminds me that this year I have 2 flocks of geese — One White Swan Geese — the other Common Irish. They always appeared to be separate, & the respective Ganders to guard each others flocks but to my surprise out of 32 Goslings there are only 4 which are common Geese tho' 2 geese sat & brought out about 20[27].

To this day the flocks keep quite distinct but the ½ bred swan Geese are undeniable & make noble birds[28].

It is common for pigs not to take to one Boar, but admit another immediately[29].

Did you see a *wise* writer in the field congratulate the World upon the excess of male Wolves over females[30]. He saw several times one female pursued by numbers of males he said, which proved his point.

Well! Well! as a neighbour says. I will conclude you will say I am grown old & garrulous. Olden times make a chat with you a pleasure. I rejoice to hear Leonards success. It almost equals your Cambridge Son[31]. Kindest regards to Mrs Darwin Ever yours W D F

Provenance: University Library, Cambridge; Manuscript Room, DAR 164: 189; DAR 193: 112; DAR 83: 187; DAR 84.1: 128-30; DAR 86: A87-9.
For annotations to this letter see note 32.

Notes

1. This passage is in CD's handwriting and is a note (up to "near Dorchester").
2. This quotation, with minor changes, is from the *Life and correspondence of Richard Whately* (1866, Vol 2, p 171). Richard Whately was the Archbishop of Dublin from 1831 to 1863.
3. In *The variation of animals and plants under domestication* (1868), Vol 2, p 216, CD stated, "*I noticed during several winters that some trees of the yellow-berried holly, which were raised from seed from a tree found wild by my father remained covered with fruit, whilst not a scarlet berry could be seen on the adjoining trees of the common kind.*"
4. In *The variation of animals and plants under domestication* (1868), Vol 2, p 19, CD had cited Alphonse de Candolle's finding that only about a third of the seedlings of the copper beech displayed purple leaves.
5. The quotation is a loose borrowing from a letter to *The Times* (6 August 1868, p 10), entitled "*The murder of British birds*"; the letter by Rev F O Morris was forwarded by the Hon and Rev Orlando Watkin Weld Forester and was an account given to him by a gamekeeper. CD quoted the account from *The Times*, not from Fox's transcription, in *The Descent of Man* (1871) Vol 2, p 107 (n8).
6. The year was determined from the reference to the success of Leonard Darwin (see note 29, below).
7. Letter #183.
8. See Letters #177, #178 (n3) and #183.
9. Not identified; no *Diaries* of Fox from 1866-1877 have been found (Appendix 1).
10. The divisions of Derby and South Derbyshire held elections on 16 November 1868 (*The Times*, 16 November 1868, p 8). The Fox family still had connections to Osmaston Hall and had property at Elvaston, both in south Derbyshire, which clearly gave Fox voting rights, which he conscientiously exercised, in favour of the Liberals. As a result of the Reform Act of 1867, new constituencies had been created, including two new MPs for Derbyshire.
11. Charles Woodd Fox, Fox's second son, was studying at Christ Church, Oxford (Letter #170 (n2); *Alum Oxon*). Robert Gerard Fox, Fox's third son, was educated at King's College, London (Boase 1894). Fox's fourth son, Frederick William Fox, born in 1855, may have been attending a nearby school in Whitchurch, Shropshire. The two daughters have not been identified; Fox had sixteen surviving children (*Darwin Pedigrees*).
12. See Letter #181. CD referred to the celebration of the "*great magpie marriage*" in The Descent of Man, Vol. 2, p 102; he cited Fox as the source of information on magpies in Delamere Forest.
13. See Letter#180. It seems from CD's reply (Letter #181) that there was more on the "great Magpie Marriage" in that letter than has survived.

14. The "*great magpie marriage*" is quoted by CD in *The Descent of Man* (1871), Vol. 2, p 102: "*The common magpie (Corvus pica Linn.), as I have been informed by Rev. W. Darwin Fox, used to assemble from all parts of the Delamere Forest, in order to celebrate the "great magpie marriage*."
15. Mr England may have worked for the Wilbrahams, of Delamere House, Cheshire, prominent local landowners (Ormerod, 1882, Vol 2, p 107; obituary in Chester Courant 4th Feb 1852). Wilbraham is mentioned frequently in Fox's *Diaries*. George, Sr, died in 1852 (obituary in Chester Courant 4th Feb 1852; *Alum. Cantab.*) and was succeeded by George Fortescue. CD noted accounts "*of either the male or the female of a pair having been shot, and quickly being replaced by another*", in *The Descent of Man* (1871), Vol 2, p 103.
16. Fox may have mentioned this in the missing part of Letter #180. CD included the case in *The Descent of Man* (1871), Vol. 2, p. 104: "*This apparently holds good in some instances, for the gamekeepers in Delamere Forest assured Mr. Fox that magpies and carrion-crows which they formerly killed in succession in large numbers near their nests were all males; and they accounted for this fact by all the males being easily killed whilst bringing food to the sitting females.*" Osmaston Hall, near Derby, had been the Fox family home (see Chapter 1 and Figure 4).
17. See *The descent of man* (1871), Vol 2, p 102: "*Some years ago these birds abounded in extraordinary numbers, so that a gamekeeper killed in one morning nineteen males, and another killed by a single shot seven birds at roost together.*"
18. CD referred to these various pairings of partridges in *The descent of man* (1871), Vol 2, p 106; Fox is acknowledged in a footnote, Vol 2, p 107 (n7): "*and to Mr Fox, of partridges*".
19. These examples are not quoted by CD in a section on the "*Law of battle*" for birds in *The Descent of Man* (1871), Vol 2, pp 40-51; but see Letter #181 (n7) for the example on peacocks.
20. Bolton Abbey is in northern Yorkshire, northwest of Leeds and Ilkley. Both Fox and Darwin had visited that region (see Letters #108A and #137).
21. Fox is cited by CD for this information in *The descent of man* (1871), Vol 2, p 270: "*The Rev. W.D. Fox informs me that with some half-tamed wild ducks, kept on a large pond in his neighbourhood, so many mallards were shot by the gamekeeper that only on was left for every seven or eight females; yet unusually large broods were reared. The guinea-fowl is strictly monogamous; but Mr Fox finds that his birds succeed best when he keeps one cock to two or three hens. Canary-birds pair in a state of nature, but the breeders in England successfully put one male to four or five females; nevertheless the first female, as Mr. Fox has assured, is alone treated as the wife, she and her young ones being fed by him; the others are treated as concubines.*"
22. See letter from E A Darwin, [after 25 February 1868], and *The descent of man* (1871), Vol 1, p 270 (see transcript under note 20).
23. See Letter #183; see also Letter #186.
24. See Letter #180 and n5.
25. Rosamund's (or Rosamond's) bower: a labyrinth that Henry II was said to have constructed for his mistress, Rosamund Clifford, at Woodstock Palace (see *EB* and *ODNB*).
26. The reference is to the black grouse, *Tetrao tetrix*.
27. CD questions this statement in the next letter (#185).
28. CD included Fox's account in *The descent of man* (1871), Vol 2, p 114, but gave alternative name of "Chinese goose" for the Irish goose: "*The Rev. W.D. Fox informs me that he possessed at the same time a pair of Chinese geese (Anser cygnoides) and a common gander with three geese. The two lots kept quite separate, until the Chinese gander seduced one of the common geese to live with him. Moreover, of the young birds hatched from the eggs of the common geese, only four were pure, the other eighteen proving hybrids; so that the Chinese gander seems to have had prepotent charms over the common gander.*" See also Letter #186.
29. This information was used in *The descent of man* (1871), Vol 2, p 273, crediting "*A clergy-man, who has bred many pigs, assures me that sows often reject one boar and immediately accept another.*"
30. The *Field*, No 785, Vol 31, Jan 11th 1868 had an article "*Wolf Hunting in Russia*" by "*Basil*" which said, in part, "*and as to every female wolf there are thirty or forty males, the ladies always at this time have a considerable following. It is well that the numbers are so proportioned by nature, else the country would soon be quite overrun with them.*" See also Letter #180 (n8).
31. See Letters #177 (n2) and #183 (n8) for the recent successes of George Howard Darwin and Leonard Darwin.

32. CD excised various parts of this letter and put them in folders for *The descent of man* (1871) and presumably for revisions to *The variation of animals and plants under domestication* (1868). This explains the multiple locations for the parts of this letter that have survived. CD placed various markings on the side of each excised portion; for information on these, see CCD, Vol 16.

185 (CCD 6447). To W. D. Fox [Delamere Rectory, Northwich, Cheshire]

Down. Bromley. Kent. S.E.
Nov. 4 1868

My dear Fox

It was very good of you to write to me fully about your health. I grieve to hear that you suffer from so much pain; but I remember that in old times you bore pain well, & that is very hard to do[1].

Although you make such fun of your D^{rs}, your case seems to be very mysterious, & I wish I c^{d} think that the one who says nothing is the matter with you was at all near the truth[2].// In y^{r} travels southward is there any chance of your being able to pay us a visit, for this w^{d} give us very great pleasure? We go on the 7th to Eras^{s}. for 8 or 9 days & after that shall be at home for ever so long[3].

I fear that it must have cost you a great deal of fatigue writing me so long a letter. I have read it twice over & have been much interested by yr various on inheritance and especially on the Courtship of birds. I shall use some of your facts, & have been very glad// to consider all of them; for I am in a great muddle on many points. I wish I cd believe that when one of a pair is shot & is soon replaced, that it was always male, but from some trustworthy accounts this likewise occurs with the female. I wonder how yr keeper recognised the male Magpies and crows. I have not yet got together all my notes (which are but scanty) on the numbered relations of the sexes; so that if yr strength permits I shall be extremely glad to receive the account of the births of yr sheep &c. With race horses alone I have moderately large table, viz above 20,000 & the// sexes turn out almost exactly equal[4]. It is a shame to trouble you but I do not quite understand yr goose case. Is the Swan-goose the same as the China-goose?[5] You say "out of 32 goslings there are only 4 which are common geese, tho' 2 geese sat and brought out about 20" But how many geese altogether sat? ~~& did the~~ & which kinds were they? As all the goslings, except 4, showed signs of a cross, how many of the geese played false with the Swan-gander[6]. But do not answer this, if you are bothered with things to do. —

Many thanks for all your kind help. I do most deeply hope, my dear old friend, that your health may improve & that you may suffer less pain.

Yours affectionately
Ch. Darwin

(*Written across the head of the letter*)

Do you know anyone, to whom I c$^{d.}$ write, who keeps Merino sheep, of which the ewes are Hornless, for I want to know whether the Horns are developed earlier or later in life, < in rams,> than with sheep, in which *both* sexes are Horned[7].

No envelope
In Emma Darwin's handwriting, except for the last page (half way from "as all goslings. ....."), the signature and Postscript
Christ's College Library, Cambridge, #150.

Notes

1. See Letter #184. There is no specific reference to Fox bearing pain well in previous correspondence, but it is all too evident from his life of illness, mainly from lung disease, and from his *Diary* for 1824-26 (see Chapter 2).
2. See Letter #184.
3. CD stayed in London at the home of his brother Erasmus Alvey Darwin from 7th to 16th November 1868 (see "*Journal*").
4. William Bernhard Tegetmeier had placed a notice in the 22 February 1868 issue of the *Field* asking for information on the proportion of the sexes born to various domestic animals (see CCD, Vol 16, letter to W. B. Tegetmeier, 11 February [1868] and note 2). For racehorses, see CCD, Vol 16, letter from W. B. Tegetmeier, [before 15 February 1868] and note 4. [See also Letter #107 (n12)].
5. See Letter #184 (n27), and Letters # 96 & #108). The swan goose is *Anser cygnoides*. The domestic swan goose is also known as the Chinese goose [Madge, S. and Burn, H. (1988). Wildfowl: an identification guide to ducks, geese and swans of the world. C Helm, London.]
6. See Letter #185 (n27).
7. In *The descent of man* (1871), Vol 1, p. 84, CD states, "*On the other hand, the differences between the sexes may be increased under domestication, as with merino sheep, in which the ewes have lost their horns.*" However CD did not get any information from Fox, see Letter #186.

186 (CCD 6455). Charles Darwin, Down, Bromley, Kent

Delamere Ry Northwich
Dec ~~Nov~~ 9 [1868][1]

My dear Darwin

Very many thanks for your letter[2]. You ask several questions in it, which I will at once get off my mind.

I did not mean to assert that when one of a pair of Crows — Magpies &c is shot & replaced, "it is *always* the male". In all the cases I have met with, or heard of, it has been the case, and for this reason possibly, that the Male Bird in Crows & Magpies, seems, until the young are hatched, continually engaged in collecting food for// the female & conveying it to her on the nest, where she receives it with much the same demonstration of satisfaction, as young Rooks do.

There is no difficulty whatever in recognizing the Male Birds of either Crows or Magpies, and any keeper w^{d} do so easily; but they also know them from the fact I have mentioned of their continually going to & from the nest to feed the incubating female, — and in doing this they are generally shot by the waiting Keeper, who knows of the nest.

The best way to ascertain about the sex of the shot & replaced Bird, would be to try the experiment with *Rooks.//* Each sex might easily be shot in any Rookery where the trees are young — and the difficult fact as to where the single Birds come from to replace the shot ones, might be more easily traced in a Rookery than elsewhere.

Rooks so much resemble Carrion Crows, that I have no doubt similar phenomena would shew themselves in both[3].

I have little doubt that you will find the numerical relations of the sexes in birds & beasts, as nearly as possible coincide, if your tables are large enough. In single cases they differ largely — but in the aggregate, will, I think be found to be nearly equal. In domestic poultry some years in a flock of Turkeys the Males preponderate often largely — but I have noticed// that the next year the Hens are in the ascendant as much.

The Swan Goose is the China Goose. L Jenyns gives it that name from Bewick, and it so well describes the Bird, which is really quite an intermediate ~~one~~ species, that I always call it so[4].

I have today examined exactly into the facts <respecting the hybrid Geese>, which are not quite as I stated them in last letter. The old ones — were 1 common Gander with 3 Common Geese. 1 <white> China Gander with 1 white do Goose. The two latter were paired, and the other four kept together, until the China seduced one common Goose to live with him, and occasionally made inroads into his Neighbours Harem besides. 1 common Goose produced no young ones The white China brought 3 pure bred China only — The other two common brought out 4 common// Geese and 18 half bred Common & white China. There is no possibility of doubting about the cross, as they are *far* more Chinese than common — with long swan necks, & beautiful birds. Both the Common Geese which hatched, must have played false — the one which hatched none, of course I cannot accuse of incontinency.

The cross is so superior in size & beauty, that I am killing off the common Gander & mean to have no others <but ½ breeds>.

I told you I have a white China Gander most *affectionately* paired with a Canada Goose. Now for the sexes of sheep — Cows &c as far as I can give them. I shall go backwards for convenience.//

	Leicester Sheep					Alderney Cows		
	Males		females			Males		females
1868	23		12	—	1868	1	—	3
1867	23		23	—	1867	3		4
1866	24		18	—	1866			
1865	15	—	11	—	1865	5	—	3
1864	26	—	24		1864	7	—	3
1863					1863			
1862					1862	3		6
1861	15	—	11					
1860	14	—	9					
1859	16	—	9					
1858	5	—	9					
1857	7		12					
1849	18	—	16		1849	3		6
1850	12	—	11		1850	3.	—	7
1851								
1852	11.	—	5		1852	2.		5
	209		170			27		37

This is a very beggarly Table — but I find I cd not give *correct* lists of other years.//

Merino sheep I know nothing whatever about. I am not sure that I ever saw one.

Did I tell you of a Herd of Cows at Trusley[5] in Derbyshire — where last year 33 cow calves came to 1 Bull. (I feel sure this was the proportion, but will write & ask, if worth while to you. <(I have since found my mema made *at time* from E Fosters lips — the Farmer they belonged to.)>[6] I have been delayed horribly in forwarding this by a multitude of causes — Terrible illness in Parish — The horrible Election which took me nearly a week to vote twice for liberals[7] — & a host of other things — which made it impossible to see another Keeper I wanted to.

I missed him 2ce — but caught him on Saturday. He has never observed that either crows or mag//

pies ever are replaced when one is shot[8] — but he has known *Cock* birds go on sitting the eggs after he has killed the *hen* bird & hatch the young. He has also seen the Magpie gatherings early in Spring — or rather in Winter in the Forest, & that they were all paired afterwards.

To give you an idea of the number we used to have here, he once killed 7 Magpies at a shot out of a fir tree — while roosting.

Very many thanks for your kind expressions of regard &c about my health. I have been better during the last month, & have done a good deal. I am tomorrow going to Mr Woodds at Hillfield with my wife to stay a few days — see my Sisters[9] — Dr &c[10]. Were the time of year rather more favourable I wd have tried to accept your kind invitation[11]. I shd much like to have a day with you once more, and again see Mrs Darwin. I hear you are having a Bust made — at which I rejoice[12].

Kindest regards to Mrs & Miss Darwin[13]

Ever yours, W D Fox

Endorsement: "Decr 9."
Provenance: University Library, Cambridge; Manuscript Room, DAR 84.1: 126-7; DAR 85: B36-7.
For annotations: see CCD, Vol. 16.

Notes

1. The year of this letter is determined by its relationship to Letter #185.
2. Letter #185.
3. See Letter #184.
4. See Letter #185 and n 6. The references are to Leonard Jenyns (mentioned often in previous correspondence - Letters #10 (n7), #19 (n9), #25 (n7), #27 (n13), #28 (n2) #25, #32 (n4), #34 (n5), #42 (n10) and Thomas Bewick [see Letter 153 (n3)]. Bewick gave the principal name as "swan goose" and the alternative names: Chinese, Spanish, Guinea, or Cape goose.
5. The Foxes had owned land at Trusley.
6. Philip Foster was a farmer in Trusley, Derbyshire. (CCD, Vol 16). CD cited Fox for this information in *The descent of man* (1871), Vol 1, p 305.
7. The reference is to the recent parliamentary election, see Letter #184 (n10). As the holder of a Cambridge degree, Fox would also have been entitled to vote for the parliamentary representative for the University of Cambridge, whose election was held on 16 November 1868 (*The Times*, 17 November 1868, p 5). However, he may have been able to vote both in mid-Cheshire (21st November

1868) and in South Derbyshire (16th November). In the 1857 elections Fox travelled to Derby to vote liberal (Fox *Diary*, 2nd April 1857).
8. The keeper has not been identified. For the views of Fox's other keeper, see Letter #184.
9. Fox had three unmarried sisters still living and they resided at Kensingon Park Gardens (Eliza, Emma, and Julia; *Darwin Pedigrees*)).
10. Basil George Woodd (1881-1872), the father of Ellen Sophia Fox, lived in Hillfield, Hampstead, London. Ellen Sophia Fox had two sisters still living, Maria Jane Nevinson and Louisa Gertrude Walker. The doctor Fox refers to has not been identified, and Fox's *Diaries* for 1867-1878 have not been found (see Appendix 1); Fox's possible health problems are given in the letter #185 [for more reliable information, see Letter #148 (n1)].
11. In Letter #185, CD had invited Fox to visit him at Down.
12. CD sat for the sculptor Thomas Woolner in November 1868 (see CCD, Vol 16, letter to J D Hooker, 26th November 1868 and note 6).
13. Fox is referring to Emma and Henrietta Darwin.

187 (CCD 6500). To W. D. Fox, [Delamere Rectory, Norhwich, Cheshire]

Down. Bromley. Kent. S.E.
Dec. 12th [1868][1]

My dear Fox

I write a line to say that I received safely your letter this morning, & to thank you very much for it[2]. — You have told me all that I wished to know most clearly & fully.

If I receive the other returns promised to me I shall find yours on the sheep & cattle very useful to add to the others[3]. —

My dear Fox
Yours affectionately
C. Darwin

My work will have to stop a bit for I must prepare a new edit. of that <everlasting> origin, & I am sick of correcting[4]. —

Provenance: American Philosophical Society (357)
No envelope.

Notes

1. The year was determined by the relationship to Letters #184 - #186.
2. See Letter #186.
3. CD was interested in the proportion of the sexes in domestic animals (see Letter #186) and had asked for information on this in letters published in *Gardeners' Chronicle*, 15th February 1868, and in the *Field*, 22nd February 1868 (see CCD, Vol 16, letter to W B Tegetmeier, 11th February 1868). William Bernhard Tegetmeier had undertaken to tabulate the results and forward them to CD (CCD, Vol 16, letter from W. B. Tegetmeier, 9th March 1868). For details on Tegetmeier, see Letter#108 (n8). CD's tabulations of results for sheep and cattle, including those sent by Fox in Letter #186 are in DAR 85: B25, B40; they were mentioned in a supplementary discussion, but CD clearly thought that they were too scanty and said in Vol 2 of *The descent of man* (1871), "*I have remarked that sexual selection would be a simple affair if the males were considerably more numerous than females. Hence I was led to investigate, as far as I could, the proportion between the sexes of as many animals as possible; but the material are scanty. I will here give a brief abstract of the results, retaining the details for a supplementary discussion so as not to interfere with the course of my argument.*" [see Letter #183 (n10)].

4. CD recorded beginning work on the fifth edition of *On the origin of species* on 26th December 1868 (see Charles Darwin's *Journal*).

188 (CCD7107). To Charles Darwin, Down, Bromley, Kent

Broadlands Sandown I. Wight
Feb 15 [1870][1]

My dear Darwin

It is so long since I have heard anything of you and yours, that I trouble you with a few lines, which perhaps Mrs Darwin will answer for you — to ask how you are this very trying Winter. We have all been here since October and are become quite settled here, tho' at present *** *** *** ***// Delamere up, and some of us, possibly all — may go there in Summer.

We have all been laid up with this extraordinary cold weat*her*. Sandown is a complete *Tem*ple of the Winds[2], and this East one has been very biting.

What was the accident from riding, which Mrs Darwin alluded to, with you[3]. I hope you have quite got over it, and that it has not prevented your riding as usual, as that seemed to do you *so* much good.

I hope George got quite// strong and well again[4].

I must try to make the acquaintance of your Banker Son[5] at Southampton when days get a little longer & brighter & brig*hter*. I never thought Sou*thampt*on a very healthy *p*lac*e* *** *** *** *** I used to live in *those* parts[6].

What old fellows we get — you however are the younger of the two.

I should very much enjoy seeing you again, are you likely to be in Town this Spring? I suppose I shall be there for a short time next month, if able to go. How is Mrs Wedgwood — are her daughters[7] *** ***// But I will not go on bothering your poor Head with questions. Give this to Mrs Darwin & I shall feel immensely grateful if she will a*nswer* it, when she has time, *&* will tell me all ab*out* *you* and your belongi*ngs*.

In the meanwhile, with our kindest regards

Believe me
Ever your attached Friend
W. D Fox

Provenance: University Library, Cambridge; Manuscript Room: DAR 164: 190
Damaged, fragile

Notes

1. The year of this letter was determined by the reference to CD's fall from his horse, Tommy, which occurred in April 1869 (see note 3).
2. This phrase is obscured by damage, but CD quotes it in next letter (#189).
3. No letter from Emma Darwin has been found. CD's horse, Tommy, rolled on 9th April 1869 and injured CD, but not badly (Emma Darwin's *Diary*; Healey, 2002).
4. George Howard Darwin (1845-1912). Emma Darwin recorded in her *Diary* for 1869 that George came to stay for a week and on 24th October 1869 wrote "*G poorly*".
5. CD's eldest child, William Erasmus Darwin (1839-1914) who was a partner of the Southampton and Hampshire Bank in Southampton, see Letter #171 (n6).
6. There is no record of Fox having lived in the area of Southampton, although he did live near it on the Isle of Wight during his convalescence in 1833-34, and spent time at Inns in the region *en route*

to the Isle of Wight. At the end of his 1824-26 *Diary* (Chapter 2 and Appendix 4), Fox was near Southampton. Fox and family also spent some time near Southampton, probably 16th Aug 1860, as indicated by the undated Letter #140A.

7. Probably Caroline Wedgwood, CD's sister, who married Josiah Wedgwood, III. Their daughters were Sophy, Margaret (mother of Ralph Vaughan Williams, the composer) and Lucy.

189 (CCD 7113). To: William Darwin Fox, Delamere Rectory, Northwich, Cheshire.

Down. Beckenham Kent. S.E.
Feb. 18th [1870][1]

My dear Fox

I was very pleased to receive your letter with some news of yourself. We hope (but it is not quite certain) to be at 6. Q. Anne St on March 4th & stay a week there[2], & I shall be very glad to see you there once again. — But if not, or indeed in any case, will you not come down here & pay us a visit? — We shall all be heartily glad to see you. — From what I remember// of Sandown, I can quite believe it the "temple of the winds"[3]. This horrid cold has made me miserable: nothing will keep me decently warm. I have resumed my rides[4], & am able to do a good deal of writing every day, but I never pass 6 hours without a fit of extreme discomfort, & so I shall go <on> to the last of my uncomfortable days. Yet with all my discomfort I am very happy, thanks//

to my dear wife & good children. They are all pretty well, but George is not strong, & I greatly fear he will find the Bar too severe work[5]. I regret this much for he is such an indomitable worker & has so clear a head that I think he wd be very successful. He dined lately at Mr Bristowe's, who has been very kind to him & advised him[6]. — I hope you will get to know William at Southampton[7], who flourishes in his business & has got a little House at Basset near the common. — Frank[8], (nor 3) is going to be educated as a Doctor, & is going// next year to try for Honours in the Science tripos. Leonard[9] is head man, ie. Divisional officer, at Woolwich, & will be sure of his commission as Engineer this time next year. — Horace[10] is rather an invalid, at a tutor in Suffolk & has great mathematical aptitude, but is <I fear> too weak to do much

Henrietta[11] is staying in S. of France, partly for health & partly for pleasure, & Bessy[12] alone is at home —

By Heavens I have told you enough about us all! One of Caroline's[13] daughters is married to a clergyman, son of Judge Williams.

I hope we shall meet. Farewell my dear old friend.

Yours affectionately
Ch. Darwin

Provenance: Museum of Isle of Wight Geology, Sandown Library

Notes

1. The year was determined by the relationship to Letter #188.
2. This was the home, at this time, of Erasmus Darwin, CD's brother [see Letter #89 (n9)].
3. Quoted from Fox's previous letter [#188 (n2)].
4. CD refers to the fall from his horse, Tommy, referred to in the previous letter [#188 (n3)].
5. George Howard Darwin (1845-1912), 5th child and 2nd son of CD. He did give up his law career and return to Cambridge on a fellowship at Trinity College [see Letter #183(n6)]. For his recent health, see Letter #188 (n4).

6. This is likely to have been Samuel Boteler Bristowe (1822-1897), Barrister at the Inner Temple, who CD had probably met earlier [see Letter #108 (n11)]. His brother, Henry Fox Bristowe (1824-1893), was also a Barrister at Lincoln's Inn. Fox was their uncle, through the marriage of his eldest sister Mary Ann to Samuel Ellis Bristowe (1800-1855) of Beesthorpe Hall. Nottinghamshire. Fox was a great friend of the Samuel Ellis Bristowe (see Chapter 2) and of his nephews, who both had successful careers both in the law and the political sphere.
7. William Erasmus Darwin (1831-1914), a partner of the Southampton and Hampshire Bank in Southampton [see Letter #144 (n8)].
8. Francis Darwin (1848-1925), 7th child and 3rd son of CD, who gave up medicine to pursue botanical studies at the School of Botany, University of Cambridge.
9. Leonard Darwin (1850-1943), 8th child and 4th son of CD, who became a Major in the Royal Engineers, MP for Lichfield (briefly) and author.
10. Horace Darwin (1851-1928), 9th child and 5th son of CD, who became an engineer and founder of the Cambridge Instrument Company.
11. Henrietta Litchfield née Darwin (1843-1930). Edited the Letters of Emma Darwin (Litchfield, 1915).
12. Elizabeth Darwin (1847-1928), 6th child and 4th daughter of CD. Remained unmarried.
13. Caroline Wedgwood, CD's 2nd sister, married to Josiah Wedgwood, III. Her 2nd daughter, Margaret Susan, married the Rev Arthur Charles Vaughan Williams, and their third child was Ralph Vaughan Williams, the composer.

190 (CCD 7353). To: Charles Darwin, Down, Bromley, Kent.

Delamere Rectory Northwich Cheshire
Oct 28 [1870][1]

My dear Darwin

What an age it is since we have met, or even corresponded. I have often thought of you and wished to write, but knowing how much you are engaged, I have not done so. Dr Hooke*r* told me a few weeks since, that you are going to astonish us all very shortly by a new// work[2]. I shall anxiously look for it. There is such a freshness about your Books that I never tire of them, & when tired & worn out, take down a Volume, with the certainty that I shall soon be relieved from present troubles.

I now write in consequence of *our* having lost the direction of a Governess, which we believe Mrs Darwin applied to us to help[3]. If we are right in Mrs Darwin being the Applicant — She will fill up the enclosed for 3 Votes at the proper place.// We applied to several others I believe, & I hope they have kept the name & properly applied. Should we be wrong in imagining Mrs Darwin to be the applicant — you will kindly return the paper.

We go next week to our Winter Quarters at Sandown in Isle of Wight — staying some days with Mrs Fox's Father at Hampstead[4].

Should you, by any accident be in London about that time I should greatly enjoy five minutes intercourse with you.

Tell Mrs Darwin, with our kindest regards, to// reply to this letter, and Believe my old & much valued Friend to remain Yours always

W. D Fox

Provenance: University Library, Cambridge; Manuscript Room: DAR 164: 191

Notes

1. The year was determined by CD's reply (Letter #191).
2. *The Descent of Man* (1871). The reference to J D Hooker indicates, probably, that Fox had attended a scientific meeting or soirée in London: his brother-in-law, Charles Henry Lardner Woodd, was a

fellow of the Geological Society of London. There is no indication from any records that Fox was personally acquainted with Hooker.

3. The Fox household like the Darwin household employed governesses to look after their children. This would have been some kind of printed sheet providing space for recommendations and references.
4. Basil George Woodd (1781-1872), Fox's father-in-law. Fox had very nearly retired, from ill-health, at this time, but did not retire officially until 1873. His pastoral duties would have been performed by a curate.

191 (CCD 7370). To: W. D. Fox

Down. Beckenham ~~Bromley~~ Kent. S.E.
Nov. 15 [1870]

My dear Fox

I suppose that you are now settled for the winter (which has come with a vengeance today) in the Isle of Wight. — My wife wrote about Governess[1] when I got your letter of Oct. 28th & asked you whether you could come down here, but I suppose that was impossible. — You are a good-for-nothing man, not to have said something about your health & strength.// I am a good deal used up with the excessive labour to me of correcting proofs of my present book in 2 Vols; & now ~~at~~ it will not be finished till end of year. I owe many facts to you in the larger part viz on sexual differences of Birds[2]. I shall be delighted to send you a copy, whenever it is published, though I have sometimes had misgivings that I ought not//

to do so, as I fear you will disapprove of my main conclusion on the origin of man[3]; but I can most ~~truy~~ <truly> say that I have written nothing without deliberate consideration — & acquiring all the knowledge which I possibly could. — I will send it, as I know well that you are a charitable man & do not without good evidence believe in bad motives in others. — It is very delightful to me to // hear that you, my very old friend, like my other books, & you were one of my earliest masters in Nat. History.

I wish I had got a little more strength.

I feel that each job as finished must be my last. —

Farewell
Your's affectionately
Charles Darwin

Provenance: American Philosophical Society (385)

Notes

1. See Letter #190.
2. This identifies the book as *The descent of man and selection in relation to sex* (1871), which was published in 2 volumes. Fox had contributed many examples, especially on birds, that were used by CD; see Letters #179-#186.
3. "*origin of man*": many Victorians were willing to accept the theory of evolution, but it was difficult for them to go all the way and accept that the evolution of humans followed inanimate, evolutionary principles (and Fox was one of them, see Letters #180 (n10), #192 & #210 (n7). In fact, Wallace, the co-founder of the theory of evolution with Darwin, always maintained that man's brain could not have

evolved without the direction of a guiding force (see Wallace, A R, 1871, *Natural selection*, p 355). When CD offered Wallace his notes on "Man" and suggested that Wallace write the book instead of CD, Wallace refused (Darwin, F, 1892).

192 (CCD 7376) To: Charles Darwin, [Down, Beckenham, Kent]

Broadlands Sandown I. Wight
18 [November 1870][1]

My dear Darwin

The sight of your hand writing did me much good. It would have greatly rejoiced me to have been able to run down to Down for a day, as M[rs] Darwin kindly asked me to do, but I am obliged to be careful at this time of year, for fear of being laid up, and I felt I ought not to delay getting to Winter Quarters[2]. As you ask me how I am, I am glad to say much better than I have been for some months, & quite hope//

to be able to get about thro the winter. I was very ill a few months ago, and hardly thought I should get over it.

I get very stiff and old in my feelings of body, and childish in my mind I think. I do not think I was ever more *young* in mind — in fact I have a very enjoyable ex=istence, & I know few I w[d] exchange with. There now.

I wish you could give yourself a little rest, but I know you cannot. In cœlo quies, in terra nulla[3]. I hear *sad* tales about your Book about to come forth. I suppose you are about to prove man is a descendant from Monkeys &c &c[4] // Well, Well! — I shall much enjoy reading it. I have given up that point now. The three main points of difference to my mind — were that Men drink, smoke & thrash their wives — & Beasts do not.

But alas my faith is overthrown entirely. The <Lady> Monkey from the Andamans — drinks & smokes like a Christian; & evidently the Gentleman w[d] thrash, if not kill the Lady, if he had an opportunity[5].

I always look at Books as I do Newspapers. I am not bound to tye my mind to that of the writer. There are points in your unrivaled Book "The Origin of Species" — which I do <not> *come up* to[6] — but with// these few expressions omitted, I go with it completely. I do not think even *you* will persuade me that my ancestors ever were Apes[7] — but we shall see.

I have no *religious* scruples about any of these matters. I see my own way clearly thro them — — but I see many points I cannot get over, which prevent my going "the whole Hog" with you.

In a few years — if not sooner — we shall know a great deal more than we do now. — We are sadly cribbed *here*, and ones mind feels the impossibility of grasping what one longs to do. Well, well! (as a friend of mine always says in a difficulty) let us do *our* best, & hope for better things. I must run over & see you some day.

Why not you & M[rs] Darwin run over here, when you have finished your//

(*across the head of the letter*)

Book — you can study my little Apes & Apesses — Kindest regards to M[rs] Darwin & thanks for her note —Always yours Affec[ly]

W D Fox

Provenance: University Library, Cambridge; Manuscript Room: DAR 164: 192

Notes

1. The date was determined by the relation to letters #191 and #192 and the reference to "*The descent of man*" published in 1871.
2. No letter from Emma Darwin exists for this time. As explained in Letter #190, W D Fox and family removed to the Isle of Wight from Delamere Forest, Cheshire, in winter. This was to provide warmer conditions and sea air which were thought to help Fox's lung condition (see Letter #145 (n4).
3. Latin phrase, meaning "Peace in Heaven, but not on Earth"
4. It is clear that these remarks refer to "*The descent of man*", published in 1871. CD did indeed discuss the origins of man, and suggested that man's forerunners, extinct apes, came from Africa.
5. The Illustrated London News of 18th September 1869 carried an Article entitled "*Andaman Monkey with an engraving of a female monkey smoking a pipe*", with the title "*Andaman Monkey at the Zoological Society Gardens*" (see Fig 28).
6. See Letter #191 (n3). CD was right to suspect that Fox would not "go the whole Hog" with him.
7. See Letter #191 (n3).

193 (CCD 7505). To: Charles Darwin, Down, Beckenham, Kent

Broadlands Sandown I. Wight
Feb 21 [1871]

My dear Darwin

I have to thank you much for "the Descent of Man". I have often wondered when it would come out, and need not tell you how anxious I was to see it. That its perusal will give me *intense* pleasure I have no doubt — tho' very probably I may not agree with you wholly. I have been too much engaged since// Sunday, when I received your Book, to be able to throw myself into it, and luxuriate in it; I have only roamed a little thro' it, somewhat with the feeling of a Miser gloating over his treasure without enjoying it. I hope very shortly to be able to give it my full attention, and anticipate no little pleasure from doing so.

How I wish we were within a moderate distance of each other, so that we might sometimes meet.

If I am in Town, as is// probable in April, if not sooner, I shall certainly try to run down to Down if only for an hour, if you are at home at the time.

How old we get! — and yet it seems but the other day when we spent those glorious times together at Cambridge.

I have always hoped that as you grew older, you might possibly much improve in health, but I fear you cannot confirm my hopes, tho' by this time I trust you have quite got rid of the miseries attending your fall from horseback, and I hope that misfortune did not make you give up the exercise.//

I was much shocked to see Frank Parkers death in the paper[1]. He seemed as likely to live as most men, unless he met with an accident on horseback, when his great weight would tell against him.

We have been here since Nov^{r}, and all had excellent health. I have been very busy in making a large garden out of the most wretched field I ever had to deal with. By means of sea Sand and a Profusion of Sewage Manure I have succeeded beyond my expectation, & we have already an abundance of most things.

I suppose we shall migrate Northwards in April, unless we c^{d} let our house for six months, when *w*e should leave at once; but//

(*across the head of the letter*)
this is not likely. My Wife desires her kindest regards to Mrs Darwin. I wish she would write me a line saying how you are and also how Mrs Wedgewood[2] is.

Ever Dear Darwin
Yours affec^{ly} W. D Fox

Provenance: University Library, Cambridge; Manuscript Room: DAR 164: 193

Notes

1. Francis Parker (1829-1871) was the third child of Henry Parker and Marianne Parker née Darwin (CD's sister). He lived in Chester. See Letter #176 (n6).
2. Caroline Wedgwood née Darwin, CD's only surviving sister.

194 (CCD 8408) To: Charles Darwin, Down, Beckenham, Kent.

Broadlands Sandown
July 13 [1872][1]

My dear Darwin

I have thirsted very much to see you again for some time past, and quite hoped to have been able to offer myself for a day, if you could have me next week. But our plans are now altered and we d*o* not go thro London*.* I supposed we should *on* our way to Cheshire.// What an age it is since we have met. I should be so glad if you can find ten minutes spare time, in which to tell me about yourself and Mrs Darwin and your family. Are you a Grandfather yet?[2]

We have much enjoyed *ou*r stay here since last November[3], tho' owing to my lingering too long in the North, I got a nip in the Lu*ngs* which invalided me for some months[4]. Now I wish we could induce// you and Mrs Darwin to come and inhale our air here, when we return in the early Autumn. We have let our house here from next Tuesday for six weeks — and when that is over, we hope to return here. Do write me a few lines to Delamere Rectory

Northwich
Cheshire.

Mrs Fox unites with me in kindest regards to Mrs Darwin (if not forgotten by her) & Believe me Always yours affec^{ly}

W. Darwin Fox

Tell me how Mrs Wedgewood[5] is & se*nd* my *remembrances*[6] to her.

Provenance: University Library, Cambridge; Manuscript Room, DAR 164: 195
Slightly damaged

Notes

1. The year was determined from the relationship of this letter to Letter #193.
2. Darwin's children were rather slow producing grandchildren. The first was Bernard Darwin in 1876 and the next, just before Darwin's death, was Erasmus Darwin in late 1881. By the time of this letter, Fox had at least six grandchildren by his first two daughters; the other, surviving, 14 of his children produced numerous grandchildren, but much later than this.
3. Fox was spending the winters on the Isle of Wight at Broadlands, Sandown (see Letters #190 and #193). He finally retired there in 1873.
4. Fox suffered from chronic lung impairment and angina [see Letters #48 (n3), #49 (n)4, #66, #148 (n1)].
5. Caroline Wedgwood, née Darwin, CD's sister and Emma Darwin's sister-in-law. She was CD's only surviving sister at this time.
6. As a result of damage, the word is obscure.

195 (CCD 8413) To: W Darwin Fox, Delamere, Northwich, Cheshire.

Down, Beckenham, Kent.
July 16th. 1872

My dear Fox

I am very sorry to hear that your lungs have been a little affected[1]. It is indeed a long time since we have met, & I much wish you could have paid us a visit; though, as I grow older I am able to talk with anyone or stand any excitement less & less. It is very odd, for I can work slowly for 3 or 4 hours// daily, if nothing disturbs me. — I am now correcting proofs of a small book on the Expression of the Emotions in Man & animals[2]. — My wife is fairly well; but I cannot say much for my children; two of my sons being now in Germany for health sake[3]. Henrietta has no// child[4], & I hope never may; for she is extremely delicate. Erasmus[5] is moderately well, & so is Caroline[6]. Here is a full bulletin of health!

Many thanks for your kind invitation; but I never feel up to visit anywhere. A Cambridge man, a friend of one of my boys, was visiting here lately, & he is, an ardent collector of Beetles; &// old days with you were so vividly recalled by hearing about Panagæus![7] I wonder whether Albert Way is alive[8]. —

Farewell my dear old Friend
Yours most sincerely
Ch. Darwin

Pray give my & Mrs. Darwin's very kind remembrances to Mrs. Fox. —

Provenance: American Philosophical Society (B/D25.269)

Notes

1. See Letter #194 (n3).
2. *The expression of emotions in man and animals*. John Murray, London, 1972.
3. George Howard Darwin was certainly on the continent at this time. On 6th August Emma Drain wrote in her diary, "*G. came from Switzerland*". On the 2nd August she wrote, "*Horace from Ireland*". There is no indication that any other of the sons was absent at this time.
4. Henrietta (Etty) Darwin married Richard Buckley Litchfield (1831-1903), in the Church at Downe on 31st August 1871; he was a scholar and philanthropist working for the Ecclesiastical Commission.

He was 40 years old at the time. Henrietta was 28 years old; her health had always been delicate and she had often been sick (see, e g, Letters #118, #127, #135, #137, #139, #140; Healey, 2001).

5. Erasmus Alvey Darwin (1804-1881), CD's brother.
6. Caroline Wedgwood née Darwin, CD's last surviving sister. Fox had enquired after her in Letter 194 (n4).
7. *Panagæus crux-major,* a rare large carabid beetle, which is mentioned frequently over the years in association with an expedition to Whittlesea Meer by CD, Fox and Albert Way (note 8, below) in their Cambridge days [see Letter #47 (n6 & 7], also Letters #48 n17), #55, #57, #62 (n7), #90 (n17), #96 (n10), #7 (n8), #115(n6), #148 (n6); mentioned in CD's *Autobiography.*
8. Albert Way (1805-1874). Antiquary and traveler. B A, Trinity College, Cambridge, 1829. Cambridge friend of CD; see Fig 15; (see Letters #2 (n24); #13 (n159); #19 (n8), #48 (n18), etc.). He is mentioned in CD's *Autobiography.* (*Alum Cantab, DNB)*

196 (CCD 8577). To: Charles Darwin, Down, Beckenham, Kent

Broadlands Sandown I. Wight
Oct 25 [1872]

My dear Darwin

I am always glad of an excuse to ask how you and yours are, but the present one is one that will amuse you.

A very nice old Lady here will have it that you// and Mrs Darwin are at the present time in an unknown part of America somewhere by the Yellow River. That Mrs Darwin was the only lady except a Mrs Blackmore (I think) who unfortunately died just when the Expedition had got too far for her husband to return without having his throat cut, so he wisely buried his poor wife & went on his way.

Can you unravel this *ve*ry cock & Bull story.//

It is so long since I have heard of you, that I cannot deny the truth of the above acc^t — and can only say, that I believe you are safe at Down.

We are in our Winter Quarters[1], much enjoying the delicious healthy air of this place — I believe the very healthiest in England — and I have every reason at present to speak well of it, as I came here three weeks since a regular Invalid[2] — & tho' not yet strong, am in a most comfortable state of health whereas I was the reverse.// I shall rejoice if you can give a good account of your own health & that of Mrs Darwin & your family.

What an age it is since we have met. I should so much enjoy seeing you again. It is above a *Quarter of a Century* since I & my wife were at Down[3].

I wish you & Mrs Darwin w^d now repay the visit by coming to us. With our united kindest regards to Mrs Darwin Believe me my dear old Darwin Ever yours

W. Darwin Fox

Provenance: University Library, Cambridge; Manuscript Room, DAR 164: 196.
First page with black surround.

Notes

1. For the past two years Fox and his family had spent the winters at their house, Broadlands, at Sandown on the Isle of Wight (see Letters#190, #193 and #194).
2. See Letter #194 (n3) for Fox's health.

3. Fox visited Down alone on 2nd November 1849 [see Letter #83 (n3)]. Fox visited Down with his young bride, Ellen Sophia, from 29th May to 1st June, 1846 [see Letter #77 (n1)].

197 (CCD 8585). To: W. Darwin Fox, Broadlands, Sandown, Isle of Wight

Down, Beckenham, Kent.
Oct. 29th [1872][1]

My dear Fox[2]

Your old Lady might just as well have said we were gone visiting to the moon[3]. We have, however, been away from home & returned only on Saturday from a villa which we took <for 3 weeks> at Sevenoaks, close to the magnificent Knole Park[4]; as I had been working at the microscope & my head had failed. Three weeks of extreme dullness have done me some good; but my// strength fails more & more, & I find I require a rest every 6 weeks or so. — I really have no news to tell about myself, except that I finished some two months ago a small book on Expression[5], & which will be published at Murray's sale early in November. I shall soon receive copies & will send you one. Whether you or anyone will care// much about it, I know not. — How I wish I could accept your invitation & pay you a visit at Sandown; but I have long found it impossible to visit anywhere; the novelty & excitement would annihilate me. —

I am heartily glad to hear that you are better. You mentioned when you wrote last, that before leaving to Isle of Wight you had been bad. — I cannot// give much <of> an account of my children. Three of my sons are ailing more or less[6], & have inherited my poor constitution. It is a fearful misfortune.

Farewell
My dear old Friend
Ch. Darwin

Postmark: BECKENHAM c OC 29 72
Christ's College Library, Cambridge, #151.

Notes

1. The year is given by the reference to Letter #196 and to the publication of *The expression of emotions in man and animals*, John Murray, London, 1872.
2. The handwriting is that of CD, but it is much more shaky than hitherto.
3. This CD's reply to Fox's anecdote from the previous letter #196.
4. The stay was from 5th to 26th October 1872 (Emma Darwin's *Diary*); it was at Sevenoaks Common (CD's *Journal*), near to where Horace Darwin had lodgings (Freeman, 1978). Knole House and gardens was owned by the Sackville family at this time, and the house and the magnificent 1000 acre Elizabethan deer park were open, largely, to the public.
5. See note 1, above.
6. CD refers to George, Howard, Francis and Horace Darwin.

198 (CCD 9023). To: Charles Darwin, Down, Beckenham, Kent

Delamere Rectory Northwich Cheshire
August 22 [1873]

My dear Darwin

What an age it is since I have heard any thing of you. It has been my fault, as I do not think I have ever written even to thank you for sen*ding* me a copy of your Expression of Emotions[1] &c. I do *very* greatly *appreciat*e your so kin*dly*//

thinking of me when you publish. Your Books are very often in my hand, & give me *great pleasure*, but I cannot go with you on all points — in fact I think I diverge from some of your leading ideas more than I did, as I think more of them[2].

Your works have the very great advantage over most of this day, that they make you think and medi=tate — and the delicious freshness of y*our* *conc*clusions *in* *ear*lier ones//

~~are~~ is always a source of intense pleasure. I do not think I have heard from you, since a letter last year recalled the glorious days of Panagæus Crux Major[3]. I never now see an Entomologist. My own Boys hunt Lepidoptera but that is all. If your ~~Brother~~ Sons friend the Beetle hunter ever comes to the I*sle* of Wight, send him to me[4].

It would be quite refreshing to see a general Entomologist.

I take the same pleasure in all branches of Nat: History as I did 50 years ago — but Eheu labuntur fugaces anni[5]// and I have not much time or strength to follow up the pursuit.

I have a lovely little Natterjack[6] staring at me from under some moss at this moment which I brought from Island. What pretty little creatures they are.

You gave me but a poor description of the health of your children last year[7]. I hope they are better & stronger, and that you are as well as usual, which is not I fear saying much. We are all here for some weeks *to* come, when I mean to resign my little living here, where I have *li*ved some 35 years in much *en*joyment[8]. It is a severe wrench, but as I cannot do the work of a Parson, it seems right *to* resign to one who can. If I ever *h*ave a day in London — I shall run *d*own & have a look at the Old Philo*sopher* at Down. My heart warms//

(*across the head of the letter*)
when I think of the happy days we spent together. A line telling me how you and Mrs Darwin are, will be always a treat.

Ever yours affec Friend
W. D. Fox

P.S.
I must send you an extract from a letter received a day or two since from a Mr Gover residing at present in my house at Sandown.

First he says "I see you have a complete set of Darwins works[9]. Many of his facts are valuable in connexion with the hereditary principle in which I am deeply interested."

He then adds — to show how much he has profited by reading your works

*"*I have been much interested with the Birds flying to and from their nests in the Verandah. One has a *black patch like a collar* on *under part of neck, & a*

transverse black line to *the Bill.* I do not know what Birds they are, and whether the marked one is Male or Female"
One can scarcely imagine a man so well describing a Cock Sparrow, & not recognising the birds, especially as he *was* a Livery man of London & has a house at Blackheath*.*
This tickles me immensely.//

I once told you of a connexion of mine who regularly shed her toe nails[10]. She is now an old woman & though she still sheds them, they are not such good specimens as they were.

I heard from her a few weeks since & she tells me that a Nephew (a sisters Son) sheds his

Farewell my dear old friend. I cannot tell you how many of the pleasantest memories of my life are connected with your good Father[11], Caroline[12], Susan[11], & yourself.

Kindest regards to Mrs Darwin
Ever yours W D Fox

Provenance: University Library, Cambridge; Manuscript Room: DAR 164: 199
Damaged

Notes

1. *The expression of emotions in man and animals*. John Murray, London, 1872.
2. Fox does not elaborate on his views. He expressed doubts in Letters # 180 (n10), #192 (n6 & 7) and #195 (n4).
3. See Letter #195 (n7), where CD remembered *Panagæus crux-major*. This letter was on the16th July 1872 and CD wrote another later letter on 29th October 1872 (Letter #197).
4. See Letter #195.
5. Latin quote; usually "Eheu! fugaces labuntur anni": "Alas! Our fleeting years pass away".
6. Natterjack toad, *Epidalea calamita*; formerly *Bufo calamita*.
7. See Letters #195 and #197.
8. Fox retired in October, 1873 to the Isle of Wight (*Crockford's Clerical Directory, 1873*). The Church and Rectory at Delamere are shown in Figs 23 & 24. Fox was much missed by his parishioners and the local elementary church school (see Fig 25), which, in the 1980s, was still referred to as Fox's School (Marion Williamson, Tarporley, pers comm).
9. CD sent Fox a copy of every book that he published. These were in the possession of Col Richard Crombie in 1990, but have since disappeared.
10. See Letter #86 (n1).
11. Robert Darwin (1766-1848).
12. Caroline Wedgwood née Darwin (1800-1888) and Susan Darwin (1803-1866), two of CD's four sisters.

199 (CCD 9040). To: [W. Darwin Fox, Delamere Rectory, Northwich, Cheshire]

Down, Beckenham, Kent.
Sept 1. [1873][1]

My dear old friend.

I sh^d^. have written before, but I have been in bed for some days with ugly Head symptoms, but the Doctors thinks <that these> are only secondary[2]. They tell me, however, to do as little as possible for some time. I will therefore only write a line to say how glad I sh[d]. be to see you here ~~if I could~~ whenever I am decently well. — Your pleasant letter has pleased me *much* in many ways[3]. Ah the old days! This is the first of September[4]. How well I remember some of our shooting excursions[4].

My head swims, I must write no more[5].

Farewell
My dear old friend
Ch. Darwin

My son George[6] is better, but he has been compelled to give up his profession. — Emma is fairly well. — We are a poor lot. —

Christ's College Library, Cambridge, #152.
The envelope is missing

Notes

1. The year was determined by the reference to George (Darwin) giving up his career in the law.
2. In her *Diary* Emma Darwin recorded for 26th Aug 1873: "*Ch. v. unwell partial loss of mem for 12 h and numerous bad sinking fits*". And again on the 28th, "*v bad in early mg — shock in head*". On the advice of T H Huxley, Andrew Clark was summoned and as related by Browne (2002), "*The verdict was reassuring: 'there was a great deal of work in him yet'. Clark put Darwin on a strict diet, which seemed to work well enough for a while, and prescribed special pills packed with strychnine and iodine.*"
3. See Letter #198.
4. The opening of the shooting season, see Letter #4 (n58) where CD writes, "*Talking of that, upon my soul it is only about a fortnight to the first. And then if there is bliss on earth that is it.—*" For Fox's love of shooting in his youth, see Chapter 2.
5. It is surprising, perhaps, to learn that CD published another 7 books after this date.
6. George Howard Darwin (1845-1912), see note 1, above.

200 (CCD 9446). To: Charles Darwin, Down, Beckenham, Kent.

Broadlands Sandown I. Wight
May 8 [1874][1]

My dear Darwin

You will be surprised by the package which accompanies this — the account of which I will give you at once. One of my daughters was staying some months since at Sydney Dobells in Gloucestershire — who showed her a collection of Butterflies & moths arranged in a Book// presented as you see the enclosed. They are done by a M[r] Merrion an Entomologist at Gloucester, who hopes to introduce them into Schools &c.

M[r] Merrion, hearing that I was a connexion of yours, begged that I would be the means of sending these preserved B.s to you. — The great desire of his heart being that he should have the honour and glory of presenting them for your approbation.

I told him I would take care you had them, & here they are. The plan is ingenious and answers sufficiently for practical purposes, but alas// the poor Butterflies lose a sad deal of their lustre of beauty in the process. I know nothing whatever of the man, but hear he has published some small works on Entomology.

You will have made him a happy man by accepting them, & they will not burden you by their bulk.

I felt glad of the occurrence as I had been wishful to write to you for some times past but had no particular excuse for doing so.

It is now long since I have heard of you, & I sh^{d} much like a few lines from Mrs Darwin or yourself, telling// me (as I hope) that you are in better health, and that Mrs Darwin & your family are well. Tell me also how Caroline[2] is. I should so like to see her & you again. If I go to London this Summer, I shall certainly try to do so.

I am now a regular resident here, having given up Delamere Altogether. I hoped to have met with some Naturalist in the Island, but have not done so at present.

I have outlived almost all my old Isle of Wight friends — & entirely my Naturalist ones[3].

It was a *wrench* to leave my old parsonage where I had spent *35*[4] years, but I felt it right every way.//

2

Here I have been busy in doing what you did at Down — not building a house — but making Gardens &c out of the most unpromising materials I ever had to do with. After buying every potato &c — I have now atchieved a good Kitchen Garden with lots of all Vegetables &c a good deal of fruit, & in another year shall look tolerably civilised. Your good Father[5] & Susan & Catherine[6] so often come into my mind when in my Garden.

I have got all my old furniture — & remembrances about me, and am very busy and happy. I find old age en=creeping upon me, however, more rapidly than I expected.//

I had looked for its approaches more gradually, but it comes by leaps — and I believe if I live another couple of years I shall be quite an old man.

I am in my 70th year[7] — does it not seem strange how old we get (you are however younger than me) — but it seems but a few years since we were lads together. You have worked so hard & gone thro' so much bad health that it probably does not seem so to you —the worst of it is with me that while I lack the power of youth — I feel in myself as young & foolish as ever I was. It seems to me// as if at my age — one dwells upon the far past — & the deep future — far more than the present.

But I shall weary you out with my scribbling. I have been delighted with Sir John Lubbocks Thysanura in Ray S^{y}.[8] It seems to me a most wonderful Book to come from a man known as he is — as a Philosopher, Politician & Banker. The world in general cannot understand how such a mind can care about *silver fishes*, as they are commonly called —about the most contemptible & least cared for of the infinitely small. He is somewhat like what has been said of the Elephant proboscis — able to root up trees & pick up a needle.//

Is your son still a Banker at Southampton & if so what is his Firm there. I sometimes meditate going over there, & if I did, should much like to see him[9].

Do you intend publishing a work containing the Sum total of your gatherings — as I think you once did, or are you content to leave them scattered thro' your various Vols?[10]

I imagine you have got so into the habit of real work that it is as necessary to you as Idleness seems to me[11].

That you may have increased health & many years to work in is the sincere wish of your old & attached Friend

W. D Fox

Kindest regards to M[rs] Darwin from my Wife & self.

P.S.

Do you remember Brachinus crepitans[12], and our delight on finding it. I have never seen it since those glorious days, until my Boy brought in some a week since, as much pleased as we were with it & as much surprised — They fired away most wonderfully. I dont think they have practised any improvement in gunnery[13].

This reminds me of poor little Albert Way. I had no idea when my friend ~~La then~~ Lady Natterton told me she was going to the South of France to help to nurse a friends Husband — that Way was the man — until I saw *his* death lately[14].//

You probably have often met him — I never have since we three *gloried* over Panagæus Crux Major[15] — Those were days — & I always rejoice over them when I see the old specimen — & think of Henslow[16] &c &c.

Provenance: University Library, Cambridge; Manuscript Room, DAR 164: 197 3 parts, damaged

Notes

1. The year was determined by reference to CD's reply in Letter #201.
2. Caroline Wedgwood née Darwin (1800-1888)
3. It is interesting that Fox makes no reference to the other Rev William Fox of Brighstone, Isle of Wight, who was a famous collector of dinosaurs on the Isle of Wight.
4. Damaged – probably "35".
5. Robert Darwin (1766-1848).
6. Susan Darwin (1803-1866) and Catherine Darwin (1810-1866), two of CD's four sisters, who both died in1866.
7. Fox was born on the 23rd April 1805 (*Darwin Pedigrees*), so he was 3 years and ten months older than CD.
8. John Lubbock, 4th Baronet and 1st Baron Avebury (1834-1913). FRS, banker, politician and biologist, studying entomology and anthropology. He was the son of John William Lubbock, 3rd Baronet, and both were neighbours of CD at Down until John, Jr, moved to Chislehurst, Kent, in 1865. Lubbock, eventually, sold to CD the land on which the famous "Sand Walk" was constructed; he was also a key figure in arranging CD's burial in Westminster Abbey. Emma Darwin's *Diary* records many visits between the families before and after that date (see *DNB*). Fox was a member of the Ray Society, London, and would have received a copy of the volume containing Lubbock's *Monograph of the Collembola and Thysanura*, 1873.
9. William Erasmus Darwin (1839-1914); see Letters #144 (n8), #171 (n6) and #189 (n7) concerning his bank in Southampton and CD's role in establishing William there. In Letter # 140A, Fox writes that he had met William, but that seems to be before William's move to Southampton. William also received an ink stand from Fox at that time [Letter #143(n1)].
10. CD's answer, in the negative, is in the next letter (#201).
11. CD was to publish another 7 books after this date.
12. This is the bombadier beetle (see letter #131 (N8), hence the reference to gunnery. Fox has forgotten that CD related a similar story with his sons [Letters #131 (n8) and #133 (n11)]. This insect brought

back memories of a famous trip to Whittlesea Meer in their college days [see Letter #131 (n9)] and of Albert Way (see note 14, below).

13. This letter is reminiscent of CD's own account, in these letters [Letters #131 (n8) and #133 (n11)] and in his Autobiography (p 51), of his children's encounters with beetles (including the bombadier beetle).
14. Albert Way (1805-1874). His biographical details are to be found in Letters #2 (n2), #48 (n18), #195 (n8). See also notes 12, above, and 15, below. See also Fig 15.
15. CD, Fox and Way were members of the famous expedition to Whittlesea Meer in their college days when they collected *Panagæous crux-major* and *Brachinus crepitans* (see note 12, above).
16. John Stevens Henslow (1796-1861), Darwin's mentor and Fox's friend, who is mentioned often in letters between CD and Fox [see Bibliographic Register and Letters 8 (n3), #25 (n8), #29 (n12), #30 (n11), #35 (n2), #36 (n1), #39 (n4), #40 (n3), #41 (n8) and many subsequent letters].

201 (CCD 9454). To: Revd. W. Darwin Fox, Broadlands, Sandown, I. of Wight

Down, Beckenham, Kent.
May 11th [1874]

My dear Fox

I was very glad to get your interesting letter[1] which was not a word too long, & did not tell me a thing which I did not wish to hear. — I opened the parcel before reading your letter & speculated much how the quasi-drawings had been made, for I saw hairs & scales on them[1]. They are very curious & if they can be produced cheaply, they would really be very valuable & would tell anyone the names of the British Lepidoptera very quickly. I cannot conceive how they can be done, for the// same butterfly or moth must I suppose serve for several copies. I do not know Mr Merrion's address else I would write & thank him very sincerely. Will you do this kind act for me, as I generally have such a lot of letters to write every day. — I will show the plates to any one interested in Nat. Hist. who may be here. —

I am glad to have so good an account of yourself, & as for one's body growing old there is no help for it, & I feel as old as// Methuselah; but not much in mind, except that I think one takes everything more quietly, as not signifying so much. And as you say one looks backwards much more than forwards & can never expect to have nearly such keen enjoyment as in old days, as when we breakfasted together at Cambridge & shot in Derbyshire. Do I remember Brachinus?[2] Am I alive? Poor Albert Way I did not know that he was dead: it is a good job, for I have heard that his life & mind have been wretched for some years[3]. —//

Like you I always associate my Father with his gardens, & *last* evening was speaking to Emma[4] about his sitting so long in the garden listening to the Birds singing. You ask about Caroline[5]: she is aged in her body & infirm, but very brisk in mind: a few days ago, another daughter was married, & now the old pair will be left with only one nestling, & this will be dull for the House. — You ask also about myself; I have been rather better of late, but I never pass// six hours without much discomfort; & I forget myself only when I am at work. I have just finished correcting new Editions of my Descent of Man & Coral Reefs; & this is very tedious work, & <I> am delighted to be at new investigations. I am preparing a book almost wholly on Drosera or the Sun-Dew, which is a wonderful plant under a physiological point of view, & I think I have made some curious discoveries[6]. One of the chief new points is that it// secretes a fluid analogous to gastric juice, for it contains a ferment, closely analogous to pepsine, with an acid, & can thus in a few hours dissolve the hardest cartilage, bone & meat &c. &c. —

I shall never have strength & life to complete more of the series of books in relation to the Origin, of which I have the M.S. half// completed[7]; but I have started the subject & that must be enough for me. —

It will be job for you to decipher this scrawl. Farewell my dear old friend — My wife sends her very kind remembrances & do pay us a visit whenever you can.

Farewell Yours affectionately
Ch. Darwin

My son's Bank is in High St "Hawkinson, Atherley & Darwin" it is a very flourishing concern[8]. —

(*written on the side of the last page and then crosswise on the top of the last page*)
My son George[9], I fear, is a confirmed invalid & has given up the law — My third is going to be married & will be my Secretary[9]. —
Leonard R.E. is going to N. Zealand on the Venus Expedition[9].
Horace is working in an engineering workshop[10], but is often showing Hereditary ill-health.

Postmark: Becke*nham* MY 11 74
Christ's College Library, Cambridge, #153.

Notes

1. See Letter #200.
2. See Letter #200 (n9).
3. Fox had mentioned that Albert Way had died in the previous letter; for past associations see the previous letter (n9 & n10) and Fig 15.
4. Emma Darwin, CD's wife.
5. Caroline Wedgwood née Darwin. Fox had enquired after her in Letter #200. Her 4th daughter, Lucy Caroline, married Matthew James Harrison, R N, on the 29th March 1874.
6. *Insectivorous plants*, John Murray, London, 1875.
7. This was the admission of a *de facto* situation which arose as the *On the origin of Species* was published in November 1859. CD had already decided to publish a substantial part of his "Big Book" or *Natural selection* as, what became, *The variation of animals and plants under domestication* (1868) and most of his work on sexual selection went into *The descent of man, and selection in relation to sex* (1871). This left very little to publish separately, anyway. Some of the latter, he offered to George Romanes (Stauffer, 1975) and the rest was published posthumously as *Natural Selection* by Stauffer in 1975.
8. This refers to CD's first son, William Erasmus Darwin (1839-1914). It is clear from the address that the name of the bank had changed since CD established his son in the Southampton and Hampshire bank in 1861 [see Letter #144 (n8)].
9. George Howard Darwin (1843-1912) gave up a career as a Barrister and went back to Cambridge University as a Fellow of Trinity College (see Letter #). The 3rd son is Francis Darwin (1848-1925), who married Amy Ruck on the 23rd July 1874 (*Darwin Pedigrees*). Leonard Darwin (1850-1943) was attached to the U.S Transit of Venus Expedition; this expedition actually went to Tasmania, and, as part of the activities, the British Army contingent erected, at the Anglesea Barracks, Hobart, a

memorial to the soldiers who had fallen in New Zealand in 1845. Horace Darwin (1851-1928) was training to be an engineer at Sevenoaks [see Letter #197 (n4)].

202 (CCD 9499). To: W. D. Fox, Broadlands, Sandown, Isle of Wight

Down, Beckenham, Kent.
June 18. 1874

My dear Fox[1]

I know that you observe all natural objects. I am in great want of a living plant of Utricularia[2]. Have you ever seen it in clear ditches in the I. of Wight? If so, could you send me a plant (with a root if it has one) packed in damp moss in a tin box. I enclose habitats from Broomfield[3]; but know not whether they are within a drive from you — I do not care about the flowers. I once saw the plant near Eastbourne.//

Have you ever met with Pinguicola Lusitanica[4] in the I. of Wight? If you could find it, I w^{d} ask you to observe a point about it whilst in a state of nature; & indeed I sh^{d} like to see a specimen. I have been working at some very interesting points in P. vulgaris[5] —

Yours very sincerely
Ch. Darwin

Christ's College Library, Cambridge, #154

[Notes on a separate sheet (by Francis Darwin?)]

Pinguicula lusitania

In spongy bogs & moist heathy places in W. Medina; rare; probably attaining here its eastern limit. Plentiful in piece of boggy ground Called Little Moor, just below Cockleton Farm near W. Cowes July 1839!!! on Colwell heath, Sparingly.

Utricularia vulgaris.

W. Medina — in several drains & ditches in Marsh at Freshwater Gate, never seen by me in Flower. — Ditches in marsh at Easton, plentifull, FL: Vect: !!!

U. minor — very rare

E. Medina — abundantly in the meadows immediately below the farm at Langbridge by Newchurch, but flowering very sparingly. In Jacobs 1842

Notes

1. This letter is probably in Emma Darwin's or Francis Darwin's hand.
2. CD was preparing his next book *Insectivorous plants* (1875, John Murray, London); see Letter #201. *Utricularia* is a genus of insectivorous plants, of which the most common species is *U. vulgaris.* It is found in wet and boggy habitats. It has small bladders on its leaves, which trap insects. CD was able to obtain specimens from Rev H M Wilkinson, Bistern, New Forest and Mr Ralfs of Penzance, Cornwall (*Insectivorous plants*, p 395).
3. Bromfield, William Arnold (1801–1851), botanist but trained in medicine at Glasgow, where he came under the influence of the Professor of Botany of Glasgow, Sir William Hooker. Upon gaining his inheritance, Bromfield gained further experience in France, Germany and Italy before settling in Ryde, Isle of Wight, in 1836 (and must surely have been known to Fox from that time). A preliminary version of Bromfield's *Flora Hantoniensis* was published in the *New Physiologist* between 1848 and

1850. His flora for the Isle of Wight, the *Flora Vectensis,* which he never considered to be ready for publication, was published posthumously. He died from typhus in Damascus in 1851.This was the Dr Bromfield of Letter #203. Bromfield was quoted by CD in *The different forms of flowers on the same species* (1877), but not in *Insectivorous plants* (1875). The notes, which were copied out by Francis Darwin (? – see Notes, above), were partly from the *Flora Vectensis*. Fox had relations in London, whose name was Bromfield (see Appendix 4), but whether there is any connection to Dr Bromfield is not known. (*ODNB; Archives, Royal Botanic Gardens, Kew*)

4. *Pinguicula* is another genus of insectivorous plants, closely allied to *Utricularia*. It has sticky glands on its leaves, which trap insects. CD was able to obtain material from a friend in North Wales and from Mr Ralfs in Cornwall (see *Insectivorous plants*, p 391 and note 5, below).
5. In *Insectivorous Plants* (see note 2, above), CD presented many experiments on the ability of the leaves of *Pinguicula vulgaris* to respond to, and digest, organic materials. The plant material was supplied by a friend in North Wales. A smaller number of experiments were performed on *P. lusitanica,* supplied by Mr Ralfs from Cornwall, and the results were generally similar.

203 (CCD 9507). To: Charles Darwin, Down, Beckenham Kent

Broadlands Sandown I. Wight
June 22 [1874][1]

My dear Darwin

I fear I have not the least probability of getting you either of the plants you want. Had I been at Delamere — I might possibly have got Utricularia[2], but even then it was a mere chance. I have found both species in some *abundance* once or twice, & never *was*// able to again find them in the same place. It seems singular also to me that Dr Bromfield[3] seldom saw Utricularia in flower, even when abundant, as when I have found it, it has been a mass of flowers — both species. I have written to "Old Price" asking him to try to find it — & if he can to send it to you[4]. *Pinguicola* Lusitanica I never *saw* but on my return// from Hampstead — where I am going to Charles Woodds[5] this week — I will try to find it about Freshwater[6] — The locality is as far as possible from me, we & P Lusitanica — occupying the extreme West & East points of the Island.

I will also have a hunt for Utricularia — but scarcely expect to find it.

My children are much delighted at you taking up Drosera[7], as they used// to keep plants and feed them with Beef at Delamere. All the 3 species (if such they are) were plentiful in a pet Bog of mine there — very small in extent but abounding with them — Cranberry &c with a good sprinkling of vipers[8] & Lizards.

Vipers are abundant here in their haunts — but few people know it. Singularly I never heard of any one being bit here. A noble snake 3 feet 4 inches long was killed last week. —They abound & are always killed except by my family. *A* London Barrister in full practise, who//

(*written across the head of the letter*)
often comes here — will stand & scream like an Infant, if he sees a slow worm[9], un-till the Gardiner comes & valiantly slays it.

Provenance: University Library, Cambridge; Manuscript Room: DAR 164: 198
Fragile

Notes

1. The year was determined in relation to Letter #202.
2. Fox replied to CD asking for specimens of two insectivorous plants (Letter #102). For details of the plants, *Utricularia* and *Pinguicola*, see Letter #202 (n2) and annotations to that letter.
3. See Letter 202 (n3).
4. John Price (1803-1887). St John's College, B A, 1826. Botanist, Welsh scholar and school teacher; known to CD and Fox from their college days (see Letter #11 (n3), Freeman, 1978). In *Insectivorous plants* (1875) CD states "*Utricularia minor. This rare species was sent to me in a living condition from Cheshire through the kindness of Mr. John Price.*"
5. Charles Henry Lardner Woodd, FGS (-1893) of Roslyn, Hampstead, London. He was a brother of Fox's wife, Ellen Sophia. During his visits to London, Fox had previously stayed at the house of his father-in-law, Basil George Woodd, who died in 1872. See, also Letter #121 for his projected election to the Athenæum Club of London.
6. Freshwater on the Isle of Wight.
7. *Drosera* or sundew plant; see Letter #201 (n6).
8. The viper snake or "Adder" (*Viper berus*) is the only venomous snake of Great Britain (and does not occur in Ireland).
9. The slow worm (*Anguis fragilis*) is often mistaken for a snake, but is in fact a legless lizard.

204 (CCD 10073). To: Charles Darwin, Down, Beckenham, Kent.

Bedwyn Lodge Sandown I. W
July 16 [1875][1]

My dear Darwin

I have to thank you, which I do very *heartily* — for "Insectivous Plants". It is very kind of you remembering me, and I very much value you doing so. I have only peeped into it as yet — partly because I have been more than usually engaged since the Volume arrived, in hunting for a house for self and fami*ly*

We hoped to have got one//

in the New Forest, but finding this impossible, we have been obliged to content ourselves with one on its verge.
Mr Halliwell[2] — (now Phillipps) wrote some weeks since to take our house, and as all the woman kind voted for a three months change, I was obliged to succumb, and we are houseless till August when we enter upon one exactly opposite Southampton on a pretty rising ground, out of reach of the mud of River. As there is a steamer hourly to Southampton we can get into Forest by that route, or drive on our own side of water a few miles into it on all sides.

We shall be there this Aug & Sep^t.//

What are you going to do this Autumn. You generally, I think get an outing[3].

Next Monday I go to London till Saturday. I have many things to see about there. Is there any probability of your being in Town. If you are at home, I would try hard to run down to see you once again, for an hour or two[4].

I get old, and older still in bodily strength, and may not again be able to get to Down. The last winter hit me pretty hard & kept me a prisoner for some months, but I am now well for me, and able to do as usual.

I should so very much like to see you once again[4] —//

There is no one so much in my thoughts as you are, and it is so long since we have met, that we shall scarcely know each other in the next world, if we do not in this.

Do not however let me come to you unless you are pretty well, as I should be indeed sorry to be a Bore.

I shall trust to you or Mrs Darwin writing me a line telling me the exact truth.

I shall be at my Sisters[5]

54 Kensington Park Road

Ladbroke Square W

How former times crowd upon ones mind — and former loved ones —
Dear Susan — & Caroline — & your excellent Father[6] — Well well!
I must not run on — Kindest regards *to* Mrs Darwin <from my wife & self> &
Believe me

Always yours W. D. Fox.

Provenance: University Library, Cambridge; Manuscript Room: DAR 164: 200

Notes

1. The year was determined from the receipt of *Insectivorous plants* (published 2nd July 1875: see CD's *Journal*).
2. Not identified; clearly Fox was renting his house out for the Summer and trying to get a place in the New Forest, which lies on the west side of the Southampton Water.
3. While the Darwins did visit their son William at his house at Bassett, Southampton (see note 4, below), there is no evidence that they tried to see Fox at Sandown on the Isle of Wight or arrange for Fox to see the Darwins at Bassett.
4. This projected meeting did not take place as far as we know. CD's *Journal* and Emma Darwin's *Diary* record that she and CD went to Bassett, Southampton to visit their son William Erasmus, from 29th August to 11th September 1875. However, there is no record of a meeting with Fox. No diaries for Fox have been found for the years 1866-1877 (Appendix 1).
5. Fox had four surviving sisters at this time, Eliza, Emma, Frances Jane and Julia (*Darwin Pedigrees*). The three single sisters lived together at Kensington, but it is not known what Frances Jane did after the death her husband, Rev John Hughes, in 1873.
6. Fox refers to two of CD's sisters and father, Robert Darwin. See Letter #198 (n11) for further details on the sisters: Fox had a special attachment to all three; see Letter #198.

204A (S 13810). To: Charles Darwin, Down House, Bromley, Kent

[Broadlands Sandown Isle of Wight]

[1876][1]

Ann Fox died
April 7. 1859

My dear Darwin

My Sisters tell me you wish for the date of my dear Mothers death[2]. I have placed it above.

I returned to Town on Friday night and leave again on Monday. I was very sorry (both for cause & effect) that you could not come here[3].

Owing to what you said — I went to Maskelines performance at Egyptian Hall, and never was more mystified as to cause and effect there[4]. The Spirits are Fools to it

I wish your Doctor wd order you to the Isle of Wight for a month or two. If he should do such a truly sensible thing, come to us at Sandown for a week, while you chuse your lodgings.

Yours very truly
W. D. Fox

Provenance: University Library, Cambridge; Manuscript Room: DAR 210.14: 36

Notes

1. The year for this letter is difficult to ascertain. 1878/79/80 are unlikely because there are letters from that time that do not mesh well with the comments in this letter. It is also possible that Fox was replying to a request from George Darwin, who was interested in the Darwin genealogy from a young age [see Letter #112 (n7)]: "*Our second Boy, George (an enthusiastic Herald! & Entomologist)"*. George went to see "Maskelyin" at the end of 1874 (Emma to Leonard Darwin, University Library, Cambridge, DAR 239 (LD/23/1/26) and the same letter also mentions George's scepticism over spiritualism.
2. The sisters of Fox, still alive at this time, were Eliza Fox (1801-86), Emma Fox (1803-85) and Julia Fox (1809-1894), who lived at Kensington Gardens, London and Frances Jane Hughes (1806-1884), whose whereabouts at this time are unknown [see Letter #205 (n4)] (*Darwin Pedigrees*).
3. There is no letter concerning a visit by CD to Fox's home at Broadlands, Sandown. However, there were many invitations in previous years and the proximity of Darwin's son, William, and his wife, at Southampton made a visit quite feasible. Clearly the offer of a visit had been turned down on the grounds of ill health.
4. John Nevil Maskelyne (1839-1917), magician, performed at the Egyptian Hall, London, from May 1873 to 1904. He was a descendant of the Astronomer Royal, Nevil Maskelyne (*DNB*). The Egyptian Hall was on the site of what is now 170-173 Piccadilly. CD always doubted spiritualism and one of the prime motives of Maskelyne was to demystify his conjuring tricks – and disprove any connection to spirits. Fox may have been more attracted to spiritualism as were many at this time, including Wallace (see Wallace, A R, 1875. *Miracles and modern spiritualism. Three essays*. London).

205 (CCD 10515). To: the Revd W. Darwin Fox, Woodlands Hampstead London N.W.

At H. Wedgwoods Esqre Hopedene Dorking
May 26th [1876]

My dear Fox

You will see that we are not at home & shall not return for 12 days, as I stood much in need of a rest & change; so that I am very very sorry that I shall not be able to see you this time[1]. —

You enquire about poor Caroline[2]: she is in a piteous state & has now been confined to her//
bed for 20 months — She is a little better & we have been scheming whether she cd. return to Leith Hill in a special train & bed carriage, but it is decided she is yet too weak, & God knows whether she will ever return.

My son William <of Southampton> has had// a serious accident; his horse fell heavily ~~with~~ <& caused> concussion of brain; but he is going on very well & is here[3]. I hope next week he will be able to go to see Paget, & then we shall learn how long precautions must be taken, but I fear it will be months before he will be quite himself// again. —

I sincerely hope Mrs Hughes[4], whom I remember very well, may go on well. — What a deal of illness & misery there is in the world.

Remember me very kindly to Mrs Fox. — I had forgotten it was 30 years since her visit![5]

Your old friend
C. Darwin

Woodward Library, University of British Columbia

Notes

1. Fox must have been visiting the home of his brother-in-law, Robert Ballard Woodd, FRBS, FSA, to whose address this letter was sent. No letter from Fox suggesting a meeting at this time is extant. It may be noted that the last letter from Fox nearly a year before (Letter #204), had suggested a meeting, but there is no indication that it ever took place.
2. Caroline Wedgwood, née Darwin, (1800-1888), CD's surviving sister and Emma Darwin's sister-in-law. She had had a nervous breakdown in the early 60s (Healey, 2002), but after 12 year's of seclusion she emerged in the 70s, at her home of Leith Hill Place, near Dorking. CD visited her there often. This letter was written from Hensleigh and Fannie Wedgwood's home at Dorking. It is possible that they were caring for Caroline Darwin. Clearly Caroline recovered since she lived on until 1888. It may also be noted that living nearby, in Dorking, were Sir Thomas Henry Farrer and his 2nd wife, Effie (Euphemia), the daughter of Hensleigh and Fanny. It was their daughter, Ida, whom Horace Darwin married in 1880.
3. William Erasmus Darwin (1839-1914) fell from his horse while riding in 1877 and was seriously affected for sometime, but this did not stop his marriage to Sara Sedgwick a year later [see Letter 207 (n2)]. Later in life he lost a leg during a hunting accident (Litchfield, 1915, p 229).
4. Mrs Hughes was Fox's sister who married the Rev John Hughes Vicar of Penally, near Tenby, Wales. Hughes died in 1873, but it is not known where Mrs Hughes resided thereafter. There seems to be a letter from Fox missing here (see note 5, below).
5. CD is clearly replying to a recent letter (see Note 1, above). Fox and his wife visited Down House on their honeymoon in late May 1846, which is thirty years ago [Letter #78 (n7)]. In Letter #196 (n3), Fox points out that it is about 25 years since he and his wife visited Down.

206 (CCD 10923). To Charles Darwin, Down, Beckenham, Kent

Broadlands, Sandown, I.W.
Ap 3 [1877][1]

My dear Darwin

It is much easier to ask questions than answer them. "Pulmonaria" does not exactly answer either of your propositions. It does not "look up to the sky nor down to the ground".

After looking at it several times I brought Mrs Fox & held a council or Inquest upon the body. It is agreed that P: angustifolia looks ~~mo~~ up, but not *very much up*. "P. grandiflora" (as Sir P Egertons Gardener gave it me[2]) but which I never can see any difference between it & the broader leaved varieties (this has a bit of tape tied round it) does not look quite so much up.

But decidedly neither look "down to Earth".

These lines are about the angle, upwards decidedly —

In haste yours very truly
W D Fox

Provenance: University Library, Cambridge; Manuscript Room: DAR 110.2: 62

Annotation
In CD's hand: "Say flowers projecting obliquely upwards"

Notes

1. The year was determined on the basis of available evidence.
2. From this letter, Fox had clearly consulted with Sir P Egerton's gardener at Oulton Park near Delamere [see Letter #115 (n12)]. *Pulmonaria* has dimorphic flowers, that is, in the case of *Pulmonaria*, with two types of styles. CD was clearly collecting information for his book *The different forms of flowers on plants of the same species*, 1877. CD must have contacted Fox because *Pulmonaria angustifolia* (= *P. longifolia*) is found only in Dorset, Hampshire and on the Isle of Wight. In *The different forms of flowers* (1877) CD acknowledged the collection of *P. angustifolia* by his son William on the Isle of Wight (pp 105-107).

207 (CCD 11261). To: Charles Darwin, Down, Beckenham, Kent.

Broadlands Sandown I.W.
Nov 29 [1877][1]

My dear Darwin

I have often wished to write to you to know how you and yours were getting on but did not like to bother you with a letter. As I have however today heard that your Son William is about to be married[2], I write to most heartily congratulate you and M[rs] Darwin upon an event which I know you wished to take place. I am sure she will get a good husband, and I hope he wil*l*// a good wife[3]. I have not stirred from home this Summer, and therefore not heard much about you. I was sorry to hear that your Son in law, (M[r] Litchfield I think) had undergone a severe illness in Switzerland. I trust he is now well and at home. Was it the Typhoid Fever which is now become so prevalent in the Continental Cities & Hotels?[4]

I also heard one of your Sons had severely sprained his knee at Lawn tennis[5] — that is I hope likewise well.

It is some months since I heard that you had at last overcome the antipathy between // your system and tobacco. I remember the fearful headaches you went thro' at Cam=bridge, and I think you told me, only a few years since, that it continued to always overcome your nervous system[6].

I am very glad you are become a smoker, as I hope you will find it a great comfort, and do you much good. Up to middle age I think it does more harm than good, but after, it is often most useful.

I am almost afraid to ask after your sister M[rs] Wedgwood — (Caroline of old and a dear good creature she was — My heart always *bumps* when I think of her// and dear Susan centuries ago) I should be delighted to hear she was free from suffering[7].

Do not be in any hurry to answer this, but when you have ten minutes to spare I should like vastly to see your well known handwriting again. Has your Son given up his ideas upon consanguinity in marriage being unobjectionable[8]. I think upon more reflection and enquiry he would do. Forgive this long yarn — M[rs] Fox desires me to give her kindest regards and hearty congratulations to you *a*nd M[rs] Darwin —

Ever yours affec[ly]
W. D. Fox.

(*across the head of the letter*)
P.S. I was so pleased to hear of the marriage, that it drove out of my head your Honours at Cambridge, which I think do Honour to the University[9]. How I sh[d] have liked to see you Doctored.

Provenance: University Library, Cambridge; Manuscript Room: DAR 164: 201
Slightly damaged

Notes

1. The year was determined by references to William's marriage (note 2) and CD's Cambridge doctorate (note 9).
2. William Erasmus Darwin (1839-1914) married Sara Price Ashburner Sedgwick, daughter of Theodore Sedgwick, Counsellor-at-Law, of Stockbridge, Mass. and New York, USA, on 29th November 1877 (*Darwin Pedigrees*; *Litchfield*, 1915; *Healey*, 2001).
3. CD wrote to his prospective daughter-in-law, "*You will believe me when I say that for many years I have not seen any woman whom I have liked and esteemed so much as you*" (Litchfield, 1915, p 229). See also his remarks in Letter #208.
4. Richard B. Litchfield (1832-1903), Henrietta's husband, nearly died of appendicitis at Engelberg, Switzerland in 1877 (*Litchfield*, 1915, p 227). See also CD's remarks in Letter #208.
5. Leonard Darwin (1850-1943) had fallen at tennis and injured his knee (see Letter of CD to Henrietta Litchfield, 4th Oct 1877, *Litchfield*, 1915, and footnote 1, p 228). See also remarks in Letter #208.
6. Francis Darwin in *Reminiscences of my father's everyday life* (Cambridge University Library, DAR140.3.1-159) related, "*Smoking he only took to of late years — he only smoked little Turkish cigarettes — one at 3 o'clock and another (at) 6 [o'clock]. On both occasions in his bedroom. The consequence was that the bedroom got a good strong smell of tobacco. His cigarettes were not of very good tobacco & he used to enjoy & praise in his pleasant way a few good Turkish cigars which I gave him but which he only smoked on occasions as a great treat.*" For other observations on his fathers smoking and snuff-taking, see LLi, pp 121-122. George Darwin in his recollections of his father (University Library, Cambridge, DAR 112.B9-B23) said: "*He always worked on a slab in the window of the room that in later years was the smoking room & sat on a curious swiveled stool with these large castors, which had been made at Shrewsbury for his father.*"
7. Fox refers to Caroline Wedgwood, née Darwin, and Susan Darwin, two of CD's sisters, to whom he had also referred in the previous letter [#204 (n5)]. Caroline Wedgwood lived with her daughters at Leith Hill Place for another 11 years.
8. The son was George Howard Darwin (1845-1912). He produced a paper for the *Journal of the Statistical Society* (1875). "*Marriages between first cousins in England and their effects*" (Vol 18, part 2, p 34). The paper, which was a statistical analysis, showed that there was evidence for slightly lower fitness in the offspring of first cousins, but no evidence for an high death rate of these offspring in infancy (cf Browne, 2002). It had been a fear of CD for a long time that marrying cousins was not

good for the children of such unions and perhaps had contributed to lack of fitness in his own offspring [see Letter #90 (15)]. He may have been worrying more about this lately with his observations on outcrossing in plants [see *The effects of cross and self fertilisation in the vegetable kingdom* (1876)]. In his book *On the various contrivances by which British and foreign orchids are fertilised by insects, and on the good effects of intercrossing* (1862) Darwin ended with the words "*Nature thus tells us, in the most emphatic manner, that she abhors perpetual self-fertilisation... May we not further infer as probable, in accordance with the belief of the vast majority of breeders of domesticated productions, that marriage between near relatives is likewise in some way injurious*". These words were taken out in the second edition in 1877. In 1871 Darwin lobbied, but failed, to have a question on cousin marriages inserted in the National Census.

9. CD was awarded an honorary doctorate (LL D) by the University of Cambridge at Senate House on the 17th November 1877. This was the occasion on which some undergraduates staged a diversion with a monkey-marionette sent dangling over the proceedings (Desmond and Moore, 1991).

208 (CCD 11266). To: W. Darwin Fox, Sandown, Isle of Wight

Down, Beckenham, Kent. Railway Station Orpington, S.E.R.
Dec 2. 1877

My dear Fox[1],

Your sympathy is very warm to make you wish to hear all about us. Litchfield's illness has been a very serious one, namely inflammation of the caecum[2]; but he is now able to sit up for a short time every day. Leonard's knee has been a bad job//

& it would have been better if it had been broken[3]. But he now goes on one crutch instead of two. On the favourable side of the balance, Williams marriage has pleased us greatly, for we considered him an inveterate bachelor. He was married two days ago to a charming american[4]. As for myself I am better than usual & am working away very hard on the physiology of plants[5].//

We had a grand time of it in Cambridge[6] & I ~~went~~ saw my old rooms in Christ's where we spent so many happy days. I see that you ask two other questions: Caroline is better in general health & comes down stairs every day but I fear will never leave Leith Hill Place[7]. Secondly George has seen no reason to change his conclusions about the marriage of//

cousins[8]. He is very hard at work in Cambridge on Astronomical problems[9], which are so deep I cannot understand what they are about.

Farewell my dear old friend
Yours affectionately
Charles Darwin

Christ's College Library, Cambridge, #155.
Dictated (to Francis Darwin?)

Notes

1. This is the last extant letter of CD to Fox.
2. "*inflammation of the caecum*": see letter #207 (n4).
3. For "*Leonard's knee*": see letter #207 (n5).
4. "*charming american*" –Sara Sedgwick - see letter #207 (n3).
5. CD was working with his son, Francis, on *The power of movement in plants* (1880).
6. For CD's award of an honorary doctorate, see Letter #207 (n9).
7. Caroline Wedgwood née Darwin, CD's sole surviving sister [see Letter #207 (n70].

8. See letter #207 (n8).
9. George Howard Darwin's calculations on the orbiting of the moon around the Earth were still used to calculate tidal cycles around the world until recently.

209 (CCD 11355). To Charles Darwin, Down, Beckenham, Kent.

Broadlands Sandown I.W.
Feb 12 [1878][1]

My dear old Friend

For surely I may call you so, as you and one other, are all left of our friendships at a time when life was glorious. It is not bad now tho' I am nearly 73[1] — and few have been happier than I have. I rejoiced to hear from your Son at Southampton that you had all a most happy family party at Christmas[2] — (How I should have liked a peep at you all from behind a curtain).//

May you have many happy Christmas's — and therefore many happy returns of your Birthday[3].

I saw one day, that you were born in 1809[3] — I am glad to hear you are four years younger than myself — I always thought you were only two.

We were so glad to hear of W Es marriage[4]. He is just the man to make a splendid Husband — and seems as happy as possible. You and I have been so happy in our marriages that we can only wonder all do *no*t marry. But it is not//

always that a man can meet with his double.

We have an anxious house this winter in nursing one of my daughters who I believe slept in a damp bed last summer, and has never been well since[5]. We have feared (& still do) consumption — but all the D^{rs} say there is no really consumptive symptoms, but that there is inflammatory action of the air cells, com=plicated with Asthma.

A fortnight since, my only comfort in watching her, was that Asthma was the then agent of evil, and so it proved — and ever since she has been im=proving, but I have more fear than hope of the issu*e.*// I have, like all the world, (except I hope you and yours) been laid up with Bronchitis, and have not been out of doors this year.

However I am much better, and we are longing for sun and warm air for our dear Invalid.

But why bother you with these troubles — When I sat down I only meant to con=gratulate you and Mrs Darwin on your Birthday, and wish you all blessings thro the remainder of your life.

Mrs Fox joins most heartily with me in these wishes

Ever Dear old Darwin
Yours affectly
W. D. Fox

Provenance: University Library, Cambridge; Manuscript Room: DAR 164: 202. Slightly damaged.

Notes

1. The year was determined from Fox's age (73); he was born in 1805.
2. This was at Down, probably. Emma Darwin's *Diary* finishes the year on 22nd Dec at Down and begins the year at Down. Thus William and his new wife, probably, were visiting Down.
3. CD was born on 12th February 1809. In 1877 he was therefore 68 years old.
4. The marriage of William Erasmus Darwin and Sara Sedgwick: details in Letter #207 (n2).
5. This was Theodora (Dora) Fox (1853-1878); see Letter #211.

210 (CCD 11358). To: W. Darwin Fox, Broadlands, Sandown, Isle of Man.

Down Beckenham Kent
Feb 14th 1878

My dear Friend

I thank you warmly for your affectionate & most kind letter. If we may not call each other friends, I do not know who have a right; though from my unfortunate health & from our mutual distance, we have seen little of each other since those happy days in Cambridge & at Osmaston & elsewhere[1]. — I am//

very sorry to hear about your daughter[2]. It is the greatest misery possible. We have had <much> & now have some <anxiety> about my son-in-law Litchfield, who was so ill in Switzerland & <who> had a relapse & last night had another accession of pain. I dread organic mischief following from so much & such//

repeated inflammations[3]. George[4] is in Algiers & is enjoying himself there; but his health is not at all better. I grudge his want of health especially, as he has indomitable energy & a constant craving to work.

Frank & I are hard at work on physiological points with respect to plants, & I find it adds greatly to my//

interest in ~~his~~ being able to discuss all subjects with him. — William is extremely happy with his wife.

Farewell my dear old Friend

Yours affectionately
Charles Darwin

University of British Columbia, Woodward Library.

Notes

1. The number of documented meetings between Fox and CD since the voyage of the Beagle is seven (Appendix 2).
2. See Letter #209 concerning Fox's daughter Theodora (Dora).
3. For the illness of Richard Litchfield in Switzerland, see Letters #207 (n4) and #208 (n2).
4. George Howard Darwin (1845-1912).

211 (CCD 11598). To. W. Darwin Fox, Broadlands, Sandown, Isle of Wight.

Down Beckenham Kent
July 10th 1878

My dear Fox

I was indeed grieved to see yesterday in the Times the great loss which you have suffered[1]. I know well what misery this loss will cause to your affectionate disposition & to poor M^{rs}. Fox. Emma desires me to say how much she sympathises with you both.

It is no use whatever my writing, but I could not endure// not to do so. — I hope she did not suffer much.

Do not write to me, unless doing something is a little relief to you.

God bless you my dear old friend. I hope you & Mrs Fox are fairly well in bodily health.

Yours affectionately
Charles Darwin

Woodward Library, University of British Columbia.

Notes

1. Fox's daughter, Theodora (Dora), died on 5th July 1878. Fox's *Diary* for 1878 documents the pathos of the last days of his daughter and his grief at her death. It may well have been a breaking point in his life, since he died in 1881 (see *Conclusion* to this chapter).

212 (CCD 11625). To: [Charles Darwin, Down House, Bromley, Kent]

Broadlands Sandown I. Wight
July 22 [1878]

My dear Darwin

I am *deeply* obliged to you for so kindly writing to me upon our very sad loss[1]. She was a very dear and good child, and very much beloved by us.

But all this ought to do away our sorrow — I cannot say it does so// Your handwriting is so exactly as of old, that I cannot help hoping you are in tolerable health.

I have been looking over many of your old, old letters of late. They have recalled days, seldom long out of memory — I always love to dwell upon Shrewsbury in those times. Your good Father[2] hangs up in my little study — reminding me of his great kindnesses to me; and at the same time of *Ca*roline who made the likeness[3] — of Susan full of goodness &c. Those were gold letter days indeed[4].

We have both had many happy days since — but they want the joyousness — the overflowingness of youth, and youths friendship.

Some six weeks since I attended the Funeral of my oldest Naturalist Friend Hewitson[5]. I knew him when we were both very young, & afterwards we were intimate friends thro a very chequered life on his part. He spent much of his time, and the Income of a large fortune, upon his favorite pursuit — Lepidopte*ra.*// Having lost his wife & without family he could afford to keep Naturalists in Africa and South America — and for very many years has published a beautiful work which he drew *a*nd coloured all the plates for at a loss of 30£ per month.

He has left a splendid col=lection to the British Museum. I shall miss him very much.

Owing to our childs long illness, we have not yet <made> the acquaintance of your daughter at Basset[6]. I hope we shall now do so, and get them to pay us a visit. He kindly wrote to me lately[7]. M[rs] Fox sends her kindest regards to M[rs] Darwin —

Excuse this stupid **** *Yours*[8] always

W. D. Fox

Provenance: University Library, Cambridge; Manuscript Room, DAR 164: 203

Notes

1. Fox refers to the loss of his daughter Theodora, who died on 5th July 1878, and to CD's letter offering condolences (see Letter #211).
2. Robert Waring Darwin (1766-1848).
3. This refers to Caroline Wedgwood, née Darwin, CD's only surviving sister at this time. The drawing of Robert by Caroline has not survived.
4. For more other reminiscences, see Letters # 200 (n4 & n5), #204 (n5) & #207 (n7).
5. William Chapman Hewitson (1806-1878). Naturalist and specialist in diurnal lepidoptera as well as birds and birds eggs. He was born at Newcastle upon Tyne and educated at York. He trained as a surveyor for the developing rail system in Great Britain, working for George Stevenson on the London to Birmingham Railway. In 1848 he inherited a fortune and "retired" to Oatlands Park, Surrey, where he published the majority of his books. He was an accomplished artist. Hewitson published nine books - on eggs of British birds and on insects and butterflies from many countries. His book *British Oology* (1831-1844); was borrowed by CD from Fox [see Letter #51 (n14) and following letters]; in that book Fox was thanked in the preface for supplying information and specimens of nests and eggs and was cited frequently in the text. Fox corresponded with Hewitson personally (see 11 letters between W D Fox and Hewitson, 1830 – 1835; The Fox/Pearce (Darwin) Collection, Woodward Library, University of British Columbia).
6. Fox was referring to CD's new daughter-in-law, Sara, wife of William Erasmus Darwin, at Bassett, Southampton [see Letters #207 (n2) and #208 (n3)].
7. This is the first indication of correspondence between Fox and William Erasmus Darwin, despite the evidence that they had previously met [Letters #140A and #142 (n1)].
8. Two words are obscured here. The second word is probably "yours".

213 (CCD 11913). To: [Charles Darwin, Down, Beckenham, Kent]

Broadlands Sandown. I. W.
March 3 [1879][1]

"Long years have pass'd, old friend, since <we>
First met in lifes young day;
And friends long lov'd by thee and me,
Since then have dropp'd away; —
But enough remain to cheer us on" &c[2]

Dear old Darwin

How you will laugh at the above, and think it a proof of my dotage — to send them. They however so exactly express my thoughts, when I enter my little study, and see your dear Fathers happy face, as Caroline coloured the lithograph for me — years, years ago, when she and dear Susan & Catherine filled//

your old Shrewsbury home, and when you and I, were really in "lifes young day" — as happy as creatures could be. By your Fathers lithograph, (which is as like as a daughters loving hand could make it) — I have your Sons Photograph of you[3], as you now are — no longer in lifes young day, but looking very grave and sedate — widely different from a fellow photo — of you which I always call you "in your rollicking days — sitting on your chairs, just beginning to feel your strength[4].

Well! Well! We are both old fellows now, but somehow I fancy that last Photo: must have made you older and graver than you really are[5]. I shall try to get a half hour with you this Summer, if I can manage it and you will have me. I am just emerging from my Winter Chrysalis having kept the house all this winter, and feel rather like the Imago creeping out of it with my wings undried and rather shivery.

But as *my* Narcissus obvallaris[6] (our earliest daffodil) began to shew colour yesterday — I quite hope for 8 months flutter (with other Butterflies this Summer.
But, what a fool you will think me — and a nuisance to boot — if I go on in this way.

My reason for writing is to ask for a few lines from you telling me how you have past thro' this trying winter, and how your excellent little wife is. I hope also your children are better than in the last account you gave me of some of them.

I hope we shall induce W^m^ and his wife[7] to come to us for a few days soon, when the spring opens a little. I liked so much the little I have seen of him, that I should much like to see more. I fear they have no children, as I have never heard of any[8] — but for all that they may have some.

What are you about now? For I feel sure that you are busy about something. I have all your Books within a yard of my Study Chairs always, & dip into them with much satisfaction.

We get smaller and smaller in our numbers. I have only three daughters left now[9], — and one of those leaves us this Summer.

Our seven boys join us in the holiday Season[10] — but they are only ephemerals. I am just reminded (by a *scratch*) that I have a most lively family that I am now watching — A lot of Harvest Mice (M:messorius[11]) They are *many* pretty little creatures — quite distinct from any group of English Mice — Their tails prehensile and remind one of a Monkey in a small way, they are always *feeling* with the extremity & catch hold of a wire or straw beautifully. These with a flock of Larus redibundus black headed Gull — and a lot of Mole crickets[12] — form my Menageries at present. I wish I had a wall Garden for my Gulls, as I feel sure they would breed if they had space enough.

I am watching for the black heads to come on now.

I see a "Revd Richard Lubbock" mentioned as observing their habits by Yarrell[13] — I conclude he is some relation of your *wonderful* neighbours — whom I should be delighted to know.

Well, it is time I set you at liberty — that is supposing you have not long since set yourself so by throwing my letter into the fire, as it deserves. — Commend my wife and self to Mrs Darwin — we often talk of her — and wish we could see her sometimes and Believe me, old friend of some 56 years standing, Your always
W D Fox//

If you were without a subject you might write a Book upon the little German Badger hound.

The way in which every part of the frame — is adopted from their underground life — is extraordinary — The skeleton must be very curious. One would imagine that these dogs must have been very much more used for subterraneous work than they now are — as it must have taken long to form their skeleton. I sometimes almost wish my sow would die that I might examine the frame work. Of course you must have a well bred one to shew their peculiarities — Ours is from the Emperor of Germanys particular breed — The Princess Royal having given hers to my 3^{d} son Gerard[14]. I wish we could have a ten minutes examination & talk over him —

farewell.

Provenance: University Library, Cambridge; Manuscript Room: DAR 99: 172-74

Notes

1. The year was determined by the reference to William Darwin and to Fox having only three daughters still at home.
2. This set of couplets comes from The Poetical Works of Thomas Moore (1841). Moore's poems were very popular in Ireland and he was sometimes regarded as Ireland's national bard. He was also the literary executor of Lord Byron.
3. This must be the photograph by Leonard Darwin, ca 1874, with CD sitting in a chair.
4. Is Fox referring to the drawing of CD and Catherine, sitting on a chaise long when CD was about 5 years old? Or could it be the daguerrotype of CD with his son William in 1852?
5. Fox makes these remarks about Darwin who is 70. As one can observe easily in photographs at this time, CD did look older than his years (see, e g, Fig 29). This was partly due to the beard which CD grew after the age of 55, owing to problems with shaving. However, clearly the stress of illness and overwork also had an effect.
6. The Welsh daffodil.
7. William Erasmus Darwin (1839-1914) and his young wife Sara.
8. William Darwin and his wife (note 7, above) had no issue.
9. These were Agnes Jane, Julia Mary Anne and Edith Darwin. Three daughters died unmarried and five had married (one of whom had died). Dora Fox died in July 1878 (see Letter #211).
10. Of the 7 boys, 4 were either at school or university and 3 in professions [2 as clergymen and 1, Gerard, as a tutor to a future Kaiser of Germany, (see note 14, below)].
11. *Mus messorius*, the harvest mouse, the smallest European mouse.
12. Mole crickets (family Gryllotalpidi) are thick bodied insects with a burrowing habit, including shovel-like forelimbs, but unlike moles they have large beady eyes (see Fig 16).
13. William Yarrell (1784-1856). Naturalist and bookseller. Wrote standard works on British birds and fishes (*DNB*). Rev. Richard Lubbock published a book in 1845 (new edition in 1879) on *Observations on the fauna of Norfolk and more particularly on the fauna of the Broads*. Jarrold and Sons, Norwich, in which included the black-headed gull. For Yarrell, see Letters #31(n4), #95 (n4), #96 (n5), #98 (n6) & #107.
14. Robert Gerard Fox (1849-1909). Tutor to Prince Frederick William Victor Albert of Prussia (1859-1941), a grandson of Queen Victoria, who became Kaiser Wilhelm II of Germany.

214 (CCD 11995). To George Howard Darwin

Broadlands Sandown I. W.
April 15 [1879][1]

My dear Darwin

I fear I cannot give any information of Dr Darwin. I am a generation too late. My Father and Mother abounded with remembrances of him. He was most kind and useful to// my dear Mother when a girl and thro life.

When in her teens the Med: Man at Stamford feared she was going into a decline, and recommended great care as to diet &c — The D^{r} heard of her illness and invited her to pay him a visit. To her great surprise, he encouraged her to take all sorts of forbidden food —giving her after dinner a large Bowl of rich cream and strawberries, & repeating the same treatment at breakfast.// A most liberal diet completely set her up and she returned to her Mother a strong healthy lassie.

He was always most kind to her. I will try to recollect some of the many anecdotes I have heard from my Father and Mother, but fear I shall remember none worth ****. I fear we have next to no letters of his.

My Mother never kept letters on principle.

Of course the Wedgwoods must have many letters of his[2].//

I have always thought him a very great man — & compared him in my mind with Dr Johnson[3]. In those days Men of Mind did not exist by the million as they do now.

I hope your Father will take time in his work, & probably things may turn up when it is known that he is looking for letters &c.

The D^{r} once got me into a great mess. I was travelling as a Boy & rather a shy one, when a lady on the Coach with me, my only fellow passenger — found out my name was Darwin. She attacked me with great vehemence asking if I was related to that Brute D^{r}. D^{n}. On my acknowledging the crime, she told me with much impetuosity & anger — "that she was a young lady with beautiful teeth, when that Brute had them all taken out, to cure some nervous pains." I remember I was// much alarmed & feared she would attack me. I rejoice to hear your Father is strong —

Ever yours
W D Fox

Provenance: University Library, Cambridge; Manuscript Room, DAR 99: 175-6

Notes

1. The year of the letter was determined by the collection by the Darwins, in this instance, George Darwin, of letters and memories of Erasmus Darwin (1731-1802), for the biography of Erasmus Darwin, which first appeared under the authorship of E Krause (1879), and which was said to be "translated from the German with a preliminary essay by Charles Darwin". The book was republished as *The life of Erasmus Darwin... being an introduction to an essay on his scientific work,* posthumously, under CD's name in 1887.
2. For a biographical account of the Wedgwoods, based on their letters, see Wedgwood and Wedgwood, 1980. Also Litchfield (1915) in *Emma Darwin: A century of letters* concentrates on many letters to and from Emma Darwin, CD's wife, but in abridged form. Healey (2002) also gives previously unpublished information. However, there are still many unpublished letters and other documents, on the Wedgwoods, in the Keele Univerity Library, Keele, Staffordshire.
3. Dr Johnson: this is a well supported view (King-Hele, 1999, 2007).

215 (CCD 12006). To: George Howard Darwin

Broadlands Sandown I. Wight

Ap 21 [1879][1]

My dear Darwin

As I feared was the case, I have no letters or paper of your G^r^ Grandfather of any interest. They have been kept merely for the handwriting — and one for Hooping Cough.

I think there is an interesting paper upon D^r^ D^n^ in the Encyclopædia Britannica which yr Father might like to look over[2]. D^r^ E Darwin has so freely been called an Atheist &c that some of his Poems may be cited in answer[3]. Do you know this ode entitled

"The Folly of Atheism"

"Dull Atheist! could a giddy dance
of atoms lawless hurled,
construct so wonderful, so wise So
harmonized a World"

It is not one that would quite suit the *German* Mind.
It is said that Coleridge, after an interview with him, said,
"he was a wonderful man — &c any thing but a Christian[4] —// or some such expression.

I wish much I could be of any use to your Father, but I am a generation too late.

I have often heard my Father & Mother talk of him.

There is one very remarkable story which I have often heard from my Father, as shewing the D^r^'s great sagacity & Daring. I believe my Father was of the party At Mr Joseph Strutts[5] a dinner party was assembled at which the D^r^ was present — when Mr Strutt asked him to see his Butler who was said to be dying. He found him on the point of suffocation from Lung disease — & fast dying. He sent for a Kettle// of Boiling water — bared the mans breast & poured it over Lungs. The effect was magical at the time — & the man got well.

I dont know what our Modern D^rs^ would say to this *wild* practise.

I hope your Father will give himself time to really look into the D^rs^ life. There are many works which touch upon it more or less — many very unfairly.

In much haste
Yours very truly
W. D. Fox

We are just off into Warwickshire for a fortnight.

(Across the top of the first page) The only thing in the Encyclopædia Britt: is a statement that Dr D published several papers in the Philosoph: trans. in 1758. And they are worth referring to.

Provenance: University Library, Cambridge; Manuscript Room: DAR 99: 177-8

Notes

1. The year was determined in relation to Letter #214. The top of the letter has "1879" in pencil.
2. See the note, above, that was written across the top of the first page.

3. For further information on this subject see King-Hele, 1999 and 2007. Erasmus Darwin was labelled an atheist by his opponents (see Chapter 1), but he was probably a liberal thinker of his day and a skeptic of the Church as it then existed.
4. See King-Hele (1999 and 2007). The poet, Coleridge, much admired Erasmus Darwin on his first meeting, but later was fairly cynical in his observations. This may have been related to the fact that Erasmus prescribed opium to both Coleridge and Thomas Wedgwood, both of whom subsequently suffered from addiction.
5. There were several Joseph Strutts, in and around Derby, in the Eighteenth Century. This one (1765-1844) was the youngest son of Jedediah Strutt, who became active in local affairs and became a mayor of Derby, like Samuel Fox II - and also like Samuel Fox III, a silk merchant. He was a philanthropist and social reformer, and at 75, in 1840, he made a benefaction of an Arboretum to the town of Derby.

216 (CCD 12554). To: Charles W. Fox Esq.[1], Broadlands, Sandown, Isle of Wight

Down Beckenham Kent
March 29th [1880]

My dear M^{r}. Fox

I had heard that your Father was out of health, but had not the least idea that he was seriously ill. Your letter grieves me much & you all have my deep sympathy. It has troubled me & gratified me much that your Father should have thoughts of me at such a time, but he was always full of sympathy.// I saw a great deal of him in old days at Cambridge, & we used to breakfast together daily. In the course of my life, now a long one, I can truly say that I have never known a kinder or better man. — I can therefore feel what a loss// he will be to you all. I gather from your letter that he does not now suffer much, & this is some comfort. Believe me my dear Cousin, for we are cousins though in a remote degree, that I am grateful to you for having written & I remain

Yours very truly
Charles Darwin

Postmark: Beckenham MR 29 80
Provenance: Woodward Library, University of British Columbia, Vancouver, Canada.

Notes

1. Charles Wood Fox (1847-1908). 2nd s and 1st child of Fox and 2nd wife, Ellen Sophia. Repton School. Christ Church, Oxford University, B A (1870). Barrister; Lincoln's Inn. Lived at Hampstead, London. (*Darwin Pedigrees*)

217 CCD16886. From Charles W. Fox[1] to Charles Darwin, Down, Beckenham, Kent.

Broadlands Sandown [Isle of Wight]
Thursday April 8. 1880

Dear Mr Darwin

You will hardly be surprised to hear that my dear Fathers' sufferings are ended. He died this morning, apparently in no pain, and conscious almost to the very last. I had only returned home last night: and during the night he suffered twice from a

sort of spasm of the heart. then the breathing became somewhat laboured, and after a while it simply//
ceased[2]. Those who were present scarcely knew when the last moment came — It is a great comfort for us all to think that he suffered less in the latter part of his illness and that his death was so mercifully painless and free from conscious discomfort. I can frame no better wish for myself, or for any one that loved him, than that they and I may be like him in life and in death. ~~I write~~ My poor Mother still bears up wonderfully — and was with my dear Father to the last. Almost his last words were expressions of affection and gratitude to her. I believe the very last words he uttered were "I am sorry to give you//

all so much trouble" — words which are so characteristic of him that I do not hesitate to quote them to you.

I must thank you sincerely for the kind letter which you sent in reply to my former letter. Your name has always been an honoured one in my Father's house: And we shall always associate you with his memory – You will pardon a some what incoherent letter, I trust; the pressure of details today is very great.

Believe me to remain
Yours most sincerely
Charles W. Fox

Provenance: University Library, Cambridge; Manuscript Room: DAR 164
Black surround on first page.

Notes

1. Written by Charles Woodd Fox; for details see Letter #216 (n1).
2. William Darwin Fox died on 8th April 1880. His death certificate stated that he died of general atrophy over 9 months and exhaustion, and stated that R Gerard Fox, son, was present at the death.

218 CCD12572. To: Charles W. Fox Esq, Broadlands, Sandown, Isle of Wight.

Down, Beckenham, Kent.
Mar[1] 10th [1880]

My dear Mr. Fox

I write only one word to thank you for your second most kind letter. I am glad that your poor dear Father's sufferings are over & that his ending was so tranquil[2]. I have now before my eyes his bright face as a young man, so full of intelligence &// I hear his voice as clearly as if he was present. Your mother must have gone through terrible suffering during his long illness. My Wife joins me in saying how deeply we sympathise with her. Pray believe me

Yours very sincerely
Charles Darwin

Postmark: Guildford, E, AP 10 80
Provenance: Woodward Library, University of British Columbia, Vancouver, Canada.

Notes

1. Wrongly dated March instead of April by CD.
2. See Letter #217 (and note 2) for details Fox's death.

Conclusion

The lives of the two correspondents in their last years were so very different that there was little left except for reminiscences. On the one side, Darwin had reached the height of his fame and influence and, up to 1879, he was still absorbed in biological studies. On the other side, Fox was retired and doing very little, surrounded by his large family, including his sisters, at Sandown on the Isle of Wight. Both men enjoyed, for them, relatively good health in the middle 1870s. Darwin had had a bad spell in August 1873. Emma Darwin recorded in her *Diary*, "*Very unwell partial memory loss 12 hours sinking fits*". However, Dr Andrew Clark from London took an optimistic view saying that "*there was a great deal of work in him yet*", and put him on a diet with additional pills (of strychnine and iodine). This worked for a while (Browne, 2002) until Darwin overworked himself again. "*As for myself I am better than usual & am working away very hard on the physiology of plants.*" Darwin wrote in December 1877, "*We had a grand time of it in Cambridge & I saw my old rooms in Christ's where we spent so many happy days.*" (Letter #208).

For Fox, the prognosis was good provided he avoided cold and wet weather. And this too worked well for a while.

Both were family men and both families were doing well, especially on the male side, where it counted. Darwin's sons were all doing well: William the banker; George now becoming a recognised mathematician and statistician and elected to the Royal Society; Leonard in the Royal Engineers and on an expedition to Tasmania [Letter #201 (n9)]; and Horace making his way in Cambridge. Frank had always been Emma's favourite, perhaps because of his musical ability and he was happy to help his father over botanical research. It was Frank, however, who brought tragedy into the house. With his marriage to Amy Ruck in 1874 and their acceptance by the Darwin family all seemed set for a happy future. Then Amy died two days after giving birth to Bernard Richard Meirion in September 1876 and Francis was devastated. However, Charles and Emma Darwin had seen tragedy before and managed to contain their grief. They doted on this, their first grandchild, and helped to pull Frank through the crisis in his life, encouraging him to do more botanical work for his father.

It is interesting that neither the marriage of Francis Darwin nor the death of Amy Ruck are mentioned in the correspondence between the two men. There are several letters from Darwin to Fox in the first part of 1874, but there is no mention of the impending wedding. In 1877 there are two letters from Fox to Darwin, one after Amy's death; but again there is no mention of the death and neither does Darwin mention it in December of that year, although he does refer to William's marriage, which occurred on the 29th November. Clearly, Fox's network of correspondents, such as Darwin's sisters, had diminished to the point that important information was not getting through and he was not keeping up with notices in the newspapers. Also clearly Darwin no longer felt the imperative to share this misfortune with his old and valued friend, as he did with J D Hooker (Browne, 2002).

On the Fox side, the good period was up to 1878. His sons were coming along well. Most, so far, had followed their father into the Church of England, to become clergymen. His third son, Robert Gerard, had always been a difficult child, as indicated by Fox's diaries. Now even Gerard was settling down. Having started a degree at King's College, London, and given it away, he finished a BA at Exeter College, Oxford, and would soon become the tutor to Prince Frederick William Victor Albert of Prussia, a grandson of Queen Victoria, and later to become Kaiser Wilhelm II of Germany ([Letter 213 (n14)]. Still, one son was at college in Oxford and three sons were still at school, which must have been a trying time for a man in his 70s.

Then, for Fox, came the terrible year of 1878. In contrast to Darwin, Fox tells the story himself in a letter wishing Darwin "Many Happy Returns" on his birthday, "*We have an anxious house this winter in nursing one of my daughters who I believe slept in a damp bed last summer, and has never been well since. We have feared (& still do) consumption — but all the* D^{rs} *say there is no really consumptive symptoms, but that there is inflammatory action of the air cells, complicated with Asthma. A fortnight since, my only comfort in watching her, was that Asthma was the then agent of evil, and so it proved — and ever since she has been improving, but I have more fear than hope of the issue*". Despite Fox's hope against hope the situation did not improve and his daughter, Theodora, now 25 years old, deteriorated in health. Clearly from his diaries, Dora was Fox's favourite daughter and he watched the decline throughout the first half of 1878 with increasing concern. She died on the 5th July, probably of tuberculosis, and Fox was devastated. As was his custom, he devoted a whole page to this event in his special diary. A letter of condolence was received from Darwin, who, typically, had spotted the event from the "*Births, marriages and deaths*" column in "The *Times*" (Letter #211).

Fox replied to Darwin in a letter of reminiscences, recording also the death of their old friend, the naturalist William Hewitson, but which did not reveal the true grief that Fox had suffered. No matter what Fox did after that, he could not shake off his melancholia over the death of his daughter. In August he was diagnosed with a hernia and had to wear a truss. On the wedding day of his daughter, Fanny, on the 17th September 1878 he wrote "*& here I am writing very desolate and forlorn – (which is very wrong) but I cannot get rid of a constant thinking of dear Dora, & now Fanny will be gone & soon we shall be left almost alone*". This is so like something out of a Charles Dickens novel that it is difficult to accept the reality of the situation; however, more like "Little Nell" of the *Old Curiosity Shop* than "Dora" of *David Copperfield*. For Fox, however, as his diary makes utterly clear, the decline was real. Christmas that year was "*One of the most melancholy* X^{ts} *days I ever remember - very cold raw & dismal, & my own feelings much in unison*".

In the midst of all this in March 1879, Fox did manage one last letter to Darwin (Letter #213). This is the letter beginning with a quote from Thomas Moore,

> "*Long years have pass'd, old friend, since we*
> *First met in lifes young day;*"

and went on in a long rambling letter of reminiscences and comments on natural history. He did not mention his underlying distress and no reply seems to have been forthcoming from Darwin. One can imagine Fox sitting in his study at this time, faced with Darwin's books – and wondering if Darwin had a point about a self-ordering world, in which God played a minor role. If he did he did not put his thoughts on paper.

Even one year after Dora's death, Fox wrote in his diary. "*This is the anniversary of dear Doras being taken from us. I often wish I could get her loved memory somewhat more out of my thoughts, as it seems to pull me to pieces – & yet I would not lose the memory of that dear, sweet smile – so deeply eloquent – in our room, for anything that could be offered me.*" By the time of the wedding of his daughter, Gertrude, to the Rev^{d}. Fred Bosanquet on the 10th Sept 1879, Fox could not find the energy to leave his bedroom. Throughout November he had many visits from the doctor and the last entry in his diary, for 1st December 1879 said, in part, " *I had a very bad night of it – not getting to sleep till after 5 & little of it then. Felt very weak & heartsick all day.*"

He died just 15 days short of his 75th birthday on the 8th April 1880. The death certificate signed by his son Gerard states that the cause of death was exhaustion.

Charles Darwin was notified of the rapid decline by Charles Fox in early 1880, but that letter has not been found. Darwin replied with all the decorum of the situation. "*I had heard that your Father was out of health, but had not the least idea that he was seriously ill. .. ...It has troubled me & gratified me much that your Father should have thoughts of me at such a time, but he was always full of sympathy. I saw a great deal of him in old days at Cambridge, & we used to breakfast together daily. In the course of my life, now a long one, I can truly say that I have never known a kinder or better man*" (Letter #216).

On the day following his father's death, Charles Fox wrote to Charles Darwin stating "*I believe the very last words he uttered were "I am sorry to give you all so much trouble" — words which are so characteristic of him that I do not hesitate to quote them to you. .. ...Your name has always been an honoured one in my Father's house: And we shall always associate you with his memory*" (Letter #217). Darwin replied with a short letter of condolence, saying amongst other things, "*I have now before my eyes his bright face as a young man, so full of intelligence & I hear his voice as clearly as if he was present*" (Letter #218).

William Darwin Fox was buried in the graveyard of St John's, at Sandown, surrounded by his large family including his wife of 34 years, his 13 surviving children, many of them with partners, and six grandchildren, who had been called from all over England.

By the time that Darwin learned of the death of his friend for 60 years he was within two years of his own death. Whether he had intimations of this imminent event is not recorded. He was just under 4 years younger than Fox and with luck should survive at least as long as Fox had done. Nevertheless, all accounts seem to suggest that Darwin slowed down considerably in these last two years. The reason for this is not clear. He had more or less completed his agenda but his mental system, now totally geared to churning out facts, could not adapt to a more relaxed, or even different, regime. He was also in retreat to some extent.

In 1873, he had met a young and ardent student. Young, handsome and talented, George Romanes was a new inspiration for Darwin, not only because he worked in the same laboratory as John Sanderson, the neurobiologist, but because he was so interested in Darwin's theories (Romanes, 1896). Darwin in the 1870s was under intense attack from his opposition. St George Mivart, a vocal opponent, had scored some blows but more importantly Fleeming Jenkin with his review in the North British Review (1867) had joined in too, pointing out that the hypothesis of natural selection could not work because of blending inheritance. And Darwin did not have a good counter argument. With Romanes, Darwin entered into a new round of observation: did blacksmith's pass on their muscular arms to their offspring? Did race horses or cart horses pass on their physique to their offspring? These were the sort of questions that the two now tried to answer. And the answers seemed to suggest that some factor was transmitted to the offspring. Of course, this was tantamount to a Lamarckian view of inheritance and not good news for someone who had gained points by heavily criticising Lamarck. Darwin did not go into print with Romanes, but he was certainly shaken by the young man's work and dedication and in his revisions of the *Origin*, he backed further and further away from natural selection, as the sole mechanism of evolution, adding a whole new chapter on the subject in the sixth edition. This was also not good news for his staunch adherents, like Wallace and Huxley. And until recently this back-sliding has always been held up to ridicule. Of course, it is today a very viable subject given the new name of epigenesis – or the multiple effects on genes of epigenetic factors passed on from one generation to the next. And no one scoffs any more.

For Darwin, the possibility that he might have been wrong and that a young man would replace his theories with others, just as he had done, must have been a dispiriting thought. Nevertheless he encouraged the young man and gave him notes left over from his Big Book for possible publication. But as it turned out his ideas were not replaced and Romanes coined the name that would carry the day, "neo-darwinism".

Thus in his last years, there was an air of lassitude, spurred on by the death of his closest friends and relations. Erasmus Darwin, his brother died in 1881, at the age of 77. Erasmus had always been a close friend of his brother, but in ways that were a foil to the serious-minded Darwin. He was the clown in the court, in a very Shakespearian way, and knew how to take his brother down a peg or two and get away with it. He was loved by Darwin's children and by the Darwin relatives in a way that was totally different to their respect for Charles Darwin. In the event, however, it was only Fanny "Hensleigh" (Fanny Wedgwood), his true friend, and, to a lesser extent, her daughter, Effie, and Darwin's daughter, Hetty, that formed a vigil to comfort him to the end.

Darwin inherited a reasonable sum of money from his brother. He also did well by his investments, as we have seen in Chapter 7. He left a personal estate valued at sum £146,911, which would have been enough to adequately provide for his wife and children. In addition to this, there was a larger sum that was bequeathed to Darwin by Anthony Rich, an admirer, who thought that Darwin should be recognised for his contribution to science (Darwin negotiated that this sum should be inherited by his children). As a result Darwin, unlike Fox, was able to leave all his children a very comfortable inheritance. With this in mind Darwin now made a codicil to his will leaving £1000 each to T H Huxley and J D Hooker. Notably missing was any sum for A R Wallace. On the other hand, Darwin was the moving force, at this time, in organising a state pension for Wallace, which came through as a rather meagre £200 a year in mid 1881 (amounting to the very reasonable sum of £3200 by the time that the long-lived Wallace died in 1913). Darwin also left sufficient money to the Royal Botanic Gardens, Kew, to launch what has become the magnificent "Index Kewensis".

During that last year, Darwin's heart was showing signs of wear and tear. He had a frightening spasm on the doorstep of George Romanes' house in London in late 1881 and Emma was warned that his heart was deteriorating (Browne, 2002). In early 1882 it was clear that all was not well and Darwin relapsed slowly with good and bad periods. The end came on the 19th April. Emma recorded in her diary two days before, "*good day a little work – out in orch. twice*", but she knew that the end was approaching. Then on the day before she wrote, "*Ditto fatal attack 12*". Probably Emma got the second insertion wrong, as she often did, and had meant "*12 noon* " on the day he died. Darwin knew that he was dying, too, and said his farewells to Emma. Most of his sons, except Francis, were absent and arrived after his death. One of the last things Darwin said, as reported by Emma, was: "*I am not least afraid to die*". He died on the afternoon of the 19th and the cause of death was named as "*Angina Pectoris Syncope*".

The family plan had been to bury Darwin in the local churchyard at Downe, alongside his brother. His sons all arrived on 20th April. However, the news had travelled fast, via Francis to Hooker, Huxley and Francis Dalton, and from Dalton to the President of the Royal Society, William Spottiswood, with the request that a fitting resting place would be Westminster Abbey. Darwin's long-time friend and politician, Sir John Lubbock, took the request from there to the Prime Minister and the Dean of Westminster, along with the signatures of twenty members of Parliament. Soon it was all settled – except that the family had to be persuaded and they were initially reluctant. "*It gave us a pang not to have him rest quietly by Eras*", wrote Emma to Fanny "Hensleigh". In the end after deliberation, it was decided to go with the offer, despite some family feeling to the contrary.

The funeral was held a week later on the 26th April. Westminster Abbey was packed with family, friends, politicians and the general public, all eager to share a piece of history. Amongst the pall bearers were the four influential men in Darwin's life, Wallace, Huxley Hooker and Lubbock as well as the Duke of Argyll, the Duke of Devonshire, the Earl of Derby and the President of the Royal Society.

Darwin's body was laid to rest not far from Sir Isaac Newton. This was a political statement of the greatest kind. The Victorian nation had finally taken a simple step to honour one of its highest achievers in a way that marked the maturing of a nation, and without apology for being the greatest centre of materialism in the world. Were there some on both sides who decried this salute to science and the over-riding of religious sanctity? Possibly, and possibly Emma was amongst them. She did not attend the funeral, for reasons, which were not fully explained. Emma had always been a woman of simple faith and simple manners. She decried pomp and elaborate expressions of feeling. Had Darwin's last words, that he was not afraid to die, affronted her. He seemed to have given up so much of his later years to his science and now it had left him dead, and without apparently a single thought of God. And the final irony was to bury him in Westminster Abbey, the spiritual symbol of the nation and its aspirations. Emma stayed away.

No one gave a thought at this time to Darwin's life long friend, William Darwin Fox. Not too far away, however, on the Isle of Wight, the family of Fox, still grieving over his death, sorted out his papers and made arrangements to show these to the family of his old friend, who had joined him in the grave.

Epilogue

This has been the story of two "Victorian" gentlemen. Gentlemen that seem, in many ways, to characterise the age and its ambiguities. Both aspired to be country clergymen, following up at the same time, their bent to be naturalists. And if it was not for the offer of passage on the second voyage on the *Beagle*, Darwin might have ended up there. Yet the offer of the *Beagle* voyage did materialise and changed Darwin for the rest of his life. He went away on the voyage as a young man, accepting, like Fox, the norms of his upbringing and the societal pressures to fit into one of the avenues of professional life available in early nineteenth century England. The long and exacting voyage somehow acted as a trigger, which set Darwin on another course.

Yet this reasoning is not sufficient to explain the man. Many other men were recruited into similar trips by the British Navy, at this time, to go out and map the world, both in geographical and natural history ways: T H Huxley and J D Hooker, amongst them. And of these, only Darwin saw the path ahead and stuck to it. And, of course, A R Wallace, who came from a very different stratum of society with a different background of experience, also came to very similar conclusions about organic evolution.

William Darwin Fox on the other hand, Darwin's second cousin and from very similar bloodstock, saw his role in life as the country parson and stuck to that through thick and thin.

We do not know why Fox chose the Church as his profession. His great great grandfather, Timothy Fox, had been a clergyman; though after he was dispossessed of his living in the Restoration, he had a hard time, and his descendents chose to be active merchants in Derby. From Fox's diaries at Repton School one would not have seen a path in the Church as the vocation of such a child. Fox loved outdoor sports, especially game shooting, he had drunken parties and he had girl friends. Nevertheless he entered Christ's College, Cambridge, in 1824, intending, as most other undergraduates of the time, to enter the Church of England. His *Diary* for the next two years shows that he continued in all these normal activities of students of the time. And in the case of the opposite sex, it seems that his interest intensified, with Fox having two girlfriends for much of the time – until the family priest intervened and the family governess was sent away. It is also possible that there was an illegitimate offspring from the girl friend that he referred to as φαννι (Fanny) and the illness that Fox suffered in 1825/26 had many of the symptoms of a venereal disease (Chapter 2 and Appendix 4).

Then, in 1828, Fox met Charles Darwin and was transported into a new world of natural history. Beetles were the catalyst that seem to have fired both their imaginations and their life long friendship.

Soon, however, Fox had to leave this cosy Cambridge circle and take up a curacy in the counties. We know very little about this period of Fox's life since there are no diaries of the time. It came about a year before Darwin went away on the *Beagle*. We do know for certain that it all came to a halt in early 1832, with a serious lung infection, for which Fox had to recuperate, first at his father's country house at Osmaston Hall, Derby, and then on the Isle of Wight. It was there that Fox met and fell in love with his wife-to-be, Harriet Fletcher. That marriage lasted a happy 8 years until Harriet died in 1842, probably from tuberculosis after having had 6 children. Now, Fox was distraught and turned to his sister-in-law for consolation. Here we have some of the most quintessentially torn feelings ever written down. Fox was not allowed by the rules of the Church of England to marry his sister-in-law, however much he was attracted to her. From the letters in Appendix 5, this agonising affair went on for some ten years, until discovered by his second wife, Ellen. Yet the second marriage survived, with Ellen producing 12 children.

Fox, from all his writings, was a Christian, believing in Christ and the second coming, when all believers would be resurrected and reunited with their lost partners. Yet at the same time it is clear that hot blood ran in his veins and, in his profession of country clergyman, he risked offending Victorian propriety. One must ask therefore why he acted as he did? And even more perhaps, one must ask why he left a clear trail in his diaries and personal letters of his personal "lapses"? Of course, this makes him so much more an interesting human being, in retrospect. But why write it all down? He could so easily have removed the few offending passages in his diaries or thrown away the incriminating letters, as he obviously did with many others. Perhaps like Samuel Pepys, without a soul mate to whom he might unburden his innermost thoughts, he wrote in his diaries to get something "off his chest" about the human condition for a later audience.

Fox's contribution for us now is, therefore, a magnificent diary of a student at Christ's College, Cambridge, in the early nineteenth century, as the times moved from the freer, Georgian times of the Enlightenment to the more religious and sanctimonious Victorian era. This is the incomplete, but for all that sufficient tale of a man of these times, torn between spiritual, human, family and societal pressures. If he had no one to turn to except his diaries and a possible audience beyond the grave to whom to unburden his soul, so much for that! By some strange series of events it did emerge in a later age.

Is there a whisper here, perhaps, from beyond the grave, concerning Fox's lifelong friend, Charles Darwin? If there is it is a very muted one, for Fox wrote no incriminating things about Darwin in his extant *Diaries*.

That still leaves the putative diary of Fox, in 1828, covering the time that he and Darwin were together at Christ's College. Did this diary ever exist at all? The answer is unclear. It is very probable that Fox did keep a diary for that year [and a fragment does exist; see Appendix 5]. If it did exist it would now be a very valuable object - and the fact that it has not turned up suggests that it was either lost or destroyed. Either way we can conclude, by comparison with the Fox *Dairy*, from 1824-26, that it would show a life of the student Darwin very similar to that lived by Fox; not exactly saintly but typical for a student of the times, with a life of drinking, gambling, breaking out of college at night, etc, etc.

And what of the flesh and blood on the bones of Charles Darwin? Are we to believe that up to the age of nearly thirty, when he married, he had no girl friends or sexual encounters? It is possible, but the Fox record suggests otherwise. And we know from the Darwin-Fox letters, and elsewhere, that there was Fanny Owen (Letters #5 & #7) and that he admired a number of other women!

Darwin was so very good in laying down a paper trail of scientific correspondence, so voluminous as to snuff out the scent for anything more interesting. His family after his death, also engaged in redressing his image. It was not until the 1920s and 1930s, and in some cases later, that this sanitization was stripped away and the public could read about what Darwin had actually written, for instance, on religion. If material was thrown away concerning a rich inner life, then we are all the losers. Our knowledge and respect for Fox is not lessened to know that he was a full and complicated human being. The same would be true for Charles Darwin. In the absence of a shread of evidence either way on Darwin, we should do well to remember that, as we elevate him to an iconic figure of the nineteenth century, he was, also, human.

Appendix 1
The Fox Materials and Their History

There are four known sets of Fox papers that belonged either to Charles Darwin or William Darwin Fox:

I. The letters of Charles and Emma Darwin to William Darwin Fox in the College Library, Christ's College, Cambridge (155) and associated letters (7), not in the College Collection (a total of 162 letters).
II. Those materials in the Crombie family (currently under the guardianship of Gerard Crombie, Norwich) and largely deposited in the University Library, Cambridge.
III. Those materials in the possession of The Woodward Library, University of British Columbia [Fox/Pearce (Darwin) Collection].
IV. Letters from Fox to Darwin in the Manuscript Room, University Library, Cambridge (DAR 77, 83, 84, 85, 86, 99, 110, 164).

The letters from Charles and Emma to William Darwin Fox in the College Library, Christ's College, Cambridge and associated letters (a total of 162 letters)

Of these letters, 155 are held in the College Library, Christ's College, Cambridge. Of the remainder, 5 are held in the Library of the American Philosophical Society, Philadelphia, 1 is held in the Smithsonian Museum and 1 in the Ryde Museum, Isle of Wight.

The history of these letters is quite straight-forward. They were the letters sent to W D Fox by Charles Darwin (150) and Emma Darwin (5), from the time Fox was at Cambridge, with Darwin, until 2nd Dec 1877 (see List of letters, at the front of this book). After the deaths of Fox and Darwin (in 1880 and 1882, respectively), Francis and George Darwin solicited and advertised for letters to and from Darwin. We do not have any letter to show what correspondence arose in the case of the Fox family. It is likely that Charles Woodd Fox was involved, as he was the last to correspond with Darwin (Letters #215 & 217). It seems that it was at this stage, from 1883-1885, that the letters were sent to Francis Darwin. They were transcribed at the time and a typed copy exists in the Manuscript Room, University Library, Cambridge (DAR 144:135-296). At some stage after this and before the main body of the letters were donated to Christ's College, Cambridge, in 1909, 7 of these letters were separated off; of these letters, 5 now exist in the American Philosophical Society Library, Philadelphia, USA (Letters #109 (including #109A), #155, #187, #191 and #195), 1 in the Smithsonian Museum, Washington, DC, USA (Letter # #174) and 1 in the Sandown Museum, Isle of Wight ((Letter #189). It is not known how any of this distribution came about. Francis Darwin may well have been responsible for the donation of the letters to the USA. Gerard Fox, a son of W D Fox, who was active both politically and in the community of the Isle of Wight, may well have been responsible for the letter that came to the Sandown Museum, and on Letter #129 has been inscribed, on the side of the first page, vertically: "*Part of a letter of C. Darwins Came with Parcel from Isle of Wight Gerard Fox* -". However, in the absence of other documented evidence most of this is conjecture. In addition to these letters are the five letters of Charles Darwin to W D Fox or Charles Woodd Fox, held in the Woodward Library, University of British Columbia (see Section III, below). How these letters were separated off is not known. Since two of them were sent to Charles Woodd Fox, it seems likely that he was the conduit and we know from his will that he left his books, etc, to his sister Frances Maria Pearce and it seems clear that these descended to the "*Fox/Pearce (Darwin Collection)*" (Section III) – but at least one of these letters (#129) was amongst the parcel sent by Gerard Fox (see above).

In 1909, at the Darwin Centenary celebrations at Christ's College, the 155 letters now in the College's possession were given to the College on a 50 year basis, under a legally-binding document, and on behalf of the Fox family, by Victor William Darwin Fox, on condition of that the letters should be in the care of the College, with restrictions on access for this period. After this period it was agreed that the letters would become the possession of the College.

The materials in the Crombie Collection, which are held in the Manuscript Room, University Library, Cambridge (DAR 250)

These materials are as follows:

- Diaries 1820, 1821, 1822, 1823, 1824-1826, 1833, 1846 (fragment), 1847 (summary), 1848, 1849 (fragment), 1849 (first page), 1850 (fragment), 1851, 1852, 1857, 1858, 1861, 1862, 1863, 1864, 1865, 1878, 1879.
- A book of verses in Latin by W D Fox from the time that he was at Repton School (DAR 250: 35).
- A journal of a tour from Antwerp to Germany to Switzerland, 1842 (DAR 250: 31).
- A journal of a tour of Italy (no date) (DAR 250: 33).
- A collection of letters written in the 1840s and 1850s (see Appendix B) (DAR 250: 45–58).

A memoir by Rev S C Wilks (1831) on Rev Basil Woodd MA, late rector of Drayton Beauchamp & Minister of Bentinck Chapel, Mary-Le-Bone (DAR 250: 38).
An announcement of the marriage of W D Fox to Harriet, 2nd dau of Sir Richard Fletcher, Bart. (DAR 250: 39).
Fragmentary Notes, apparently for Fox's Diary, 1828 (DAR 250: 6).
Photographs of W D Fox (3, including a Daguerrotype) and Aunt Julia (2).
A drawing of W D Fox

There are other records in the possession of Gerard Crombie (Norwich). These include an illuminated family tree tracing the family to a descendent of John of Gaunt and 14 colour drawings of natural history specimens. The artist responsible for the drawings is not known, but family history indicates that it was Julia Fox, the younger sister of W D Fox. One or more of the older sisters my also have been responsible for some of the drawings. A photograph of Julia Fox is reproduced, with permission, in Fig 22.

Not present in the extant collection, but which are known to have been present until the late 1980s, and sighted by myself, were a complete set of first editions of Darwin's books up to 1880 and some 3-4 oil paintings of the Fox family from the eighteenth and nineteenth century. As noted in various letters of the Darwin-Fox collection (see also appropriate volumes of CCD), Darwin sent copies of his first editions to a select list of friends and colleagues, and Fox was on this list from as early as 1850 [see for example the list of recipients of the first edition of the *Origin of Species* (CCD, Vol 7)]. Unfortunately, none of these books was signed by Darwin. They were probably sold by the late Col Richard Crombie. The paintings were in the possession of Ann Darwin Fox of Blyth House, Southwold, and were stolen during a burglary in 1989. They have not been found.

The materials in the possession of the Woodward Library, University of British Columbia, "*The Fox/Pearce (Darwin Collection)*"

These are listed on the Library Web Site (www.library.ubc.ca/woodward/memoroom/collection/inventory_docs/fox-pearce-darwin.pdf).

The materials fall into 5 categories:

i. Five letters of Charles Darwin to W D Fox or Charles Woodd Fox.
ii. Fragments of diaries from ca 1830-1832.
iii. Letters from Fox to naturalists and friends.
iv. Letters to Fox from family and friends.
v. Photographs of W D Fox and family.

The possible history of these materials is given below.

Materials in the Manuscript Room, University Library, Cambridge (DAR)

There are letters and scraps of letters in DAR 77, 83, 84, 85, 86, 99, 110 164. This was material that was found at Down House after the death of CD. It is known that Darwin would often cut out important information and place the material in a scrap-book under the heading of one of the books that he was working on, such as *The variation of animals and plants under domestication*, or under a topic such as *Illigitimate progeny of dimorphic and trimorphic plants*.

It has been possible to resurrect large parts of Fox letters from these scrap-books (see, e g, Letters #181 &184). In addition there are many complete letters that were placed in other folders and grouped in alphabetical order of the correspondent. For example, the majority of the Fox letters to Darwin are to be found in DAR 164. Many of these letters have suffered through damp and microbial decay prior to their accession to the University Library.

It is clear that Darwin did not have a policy of keeping all the Fox letters. Letters prior to 1840 are very rare. Letters from 1840 to 1860 are rare and letters from 1860 to 1880 are much more common, but it can be established from a close reading of the correspondence that many other letters were not kept.

A tentative history of the Fox materials

W D Fox died in 1880, surrounded by his family, including his wife, Ellen Sophia who lived on in the same house until she died in1887. The wills of W D Fox, his wife and his children give clues as to the early transmission of the Fox's diaries, letters and other papers. Fox left the bulk of his effects to his wife, Ellen Sophia. However, it seems likely that she allowed his sons access to his papers and must have given permission to Charles Woodd Fox and Gerard Fox to loan the 155 Darwin/Fox letters to Francis Darwin, when Francis advertised in the Athenæum and other media for Darwin letters in 1883. Francis Darwin made a typescript of the letters of Charles Darwin to W D Fox; the typescript of the 155 letters is in the University Library, Cambridge (DAR 144.135-296); the 155 letters themselves are in the College Library, Christ's College, Cambridge. In addition, there are 5 letters in the Woodward Library, University of British Columbia, 5 letters in the American Philosophical Society, Philadelphia, USA, 1 letter in the Smithsonian Museum, Washington, DC, USA, and 1 at the Sandown Museum, Isle of Wight (as documented in sections I, II & III). Many of the 155 letters appeared, in part or in full, in *The Life and Letters of Charles Darwin* (F Darwin, 1887). The absence of any mention of the other letters (apart from Letter #129), seems to imply that Francis Darwin was not aware of them, although it must be said that they are not on subjects of great importance.

Whether the family was aware of the diaries of Fox, as well as the Darwin/Fox letters, and whether Francis Darwin was alerted to the existence of the diaries is not known. They certainly would have been of interest, especially if one ever existed for 1828, when Fox and Darwin were together at Christ's College, Cambridge. In fact, no diary of Fox for 1828 has been found, although a small fragment of a record for June, 1828, does exist (Appendix 5). However this in no way negates the existence of a diary for that year as Fox often made notes, while away from home, for example, and then wrote up his diary later. He even kept two diaries for some years, entering up daily details in one and significant events and opinions in another.

W D Fox left all his effects to his wife, Ellen Sophia and when she died on the Isle of Wight at the relatively young age of 67 years in 1887, her will asked her executors, Charles Woodd Fox (son) and Alexander Pearce (son-in-law), to dispose of "plate, jewels, pictures, books, furniture, etc", according to lists for distribution of special objects. It is likely that there were no specific directions for the disposal of the Fox letters and diaries. It seems likely that these were distributed to any interested parties. From the fact that two separate sets of these materials descended i) to the Crombie family and ii) to the Pearce family makes it certain that Charles Woodd Fox or Gerard Fox (who was an inveterate collector) obtained one set and Alexander Pearce obtained another set. Since whole series of the diaries of Fox are missing it seems possible also that someone else took these and other papers (see later). From the wills of the Fox children, we can track the descent of some materials. We will deal first with the Crombie Collection (I). Charles Woodd Fox left a portrait of Ann Fox, née Darwin, to Shrewsbury Town Council "(or if refused to be sold)" and his books and papers to his sister, Frances Maria Pearce. Samuel William Fox, the son of Fox's first wife, Harriet, did not indicate any special effects in his will and probably had lived too long away from home to have inherited any. This left Robert Gerard Fox, the inveterate collector, and a long-time resident and political figure on the Isle of Wight. He died in 1913 and left his effects to his son, Victor William Darwin Fox. It was the latter who signed off on the paper work that bequeathed the Darwin/Fox letters to Christ's College in 1909. So it seems that the Christ's College letters came via Gerard Fox. Victor Fox was killed in action in 1915 in the First World War and his estate went to his surviving sisters, Ellen Mary Daly and Agnes Letitia Darwin Fox. The daughter of Ellen Daly, Susan Frances Daly inherited from both her mother and from her Aunt, Agnes, and married Colonel Richard Crombie. It is likely that most of the papers descended directly through this line. Many of the other sons of W D Fox were clergymen and like Samuel William Fox were probably not interested in Fox's letters and diaries. Amongst the youngest children of W D Fox were Gilbert Basil, an anglican clergyman, and Frederick William Fox, a school teacher. Their possessions were left mainly to two children of Frederick Fox: Ann Darwin Fox and Anthony Basil Fox. Anthony Fox was a stockbroker and set up Blyth House in Southwold, Suffolk, as an old peoples' home. When he died, the Fox materials went to Ann Fox and from there to her niece Susan Crombie, and hence to the Crombie collection. It is clear from other wills that a strong line of inheritance went to the community of Fox descendants that congregated in Southwold from the 1930s onwards. It is not difficult to see how Set I finally consolidated in the hands of Col Richard Crombie and his wife, Susan Frances, and formed Collection I (the Crombie Collection and other materials, above). This left, apart from Charles Woodd and Gerard, only the daughters. We have seen that some papers went to Frances Maria and Alexander Pearce, probably because he was an executor of the estate. One or more of the other daughters may have also expressed an interest in the papers, but at present there is no evidence.

The line of descent of collection II is not hard to trace. Charles Woodd Fox left his books and papers to his sister, Frances Maria Pearce. She may have been interested in the life of her father, W D Fox, and it is possible that other brothers and sisters either gave or bequeathed to her further materials. Frances Maria and Alexander Pearce had two children, Nathaniel and Gertrude Marjory. The latter would have inherited papers from their parents, but also the youngest sons of W D Fox, Gilbert Basil, Reginald Henry and Erasmus Pullein left books and papers from their estates to this nephew and niece. Upon the predeceasement of Gertrude Marjory, these effects would have gone to Nathaniel and from there to his sons Charles and Christopher, both of whom entered the British Royal Airforce after the Second World War. Group Captain Christopher Pearce then gained possession of some or all of the materials. An interesting story is told in a letter held by the Woodward Medical Library, University of British Columbia, and written by the historian R D French. The letter says, "*In August 1968 I was working at the Wellcome Institute of History of Medicine in London, when I ran into a personal friend from Toronto, Philip Beudoin. He invited me down to the home of his brother-in-law & sister in Canterbury, Kent, Where Capt Pearce was stationed. At that time, quite fortuitously (under the bed in which I slept), I found a collection of nineteenth century letters and autographs. The core of the collection is the Pearce Collection, which Capt. Pearce inherited from his ancestors in the Fox family At that time, I had a microfilm made of the material, and some Xerox copies. The genuineness & importance of the material was verified for me by Dr. S. Smith of St Catherine's College, Cambridge and by Dr. Robert Stauffer of U. of Wisconsin. Microfilms are in the Wellcome Institute of the History of Medicine, London,*". The content of the material was described in a typewritten "*Addenda to Darwin and Henslow*" written by R D French (*Welcome Institution Archives*). These were clearly the same set of materials that were offered later to the University of British Columbia. The next we know is that Christopher Pearce, who had recently been discharged from the Royal Airforce, emigrated to British Columbia and offered substantially the same materials to the Librarian of the Woodward Library in 1970. The value of the documents, particularly 5 letters written by Charles Darwin was recognised and the materials were bought soon afterwards. A report made at that time notes "***The Fox/Pearce (Darwin) Collection***". *This collection of approximately eighty items was acquired in 1970 from Captain Christopher Pearce, a descendent of the Fox family. Richard French has documented his role in bringing this material to public notice. (He also drew up a preliminary list of the collection, to which Brenda Sutton provided an addendum).*" (Woodward Biomedical Library, Rev/77).

The possible existence of further material

This completes the history of the known materials. The question arises as to whether there were other materials and whether these have been lost by another branch of the family. This does seem to be a real possibility. From the will of Edith Darwin Fox (1857–1892) it seems that books and papers were left to her sisters Ellen Elisabeth Webster (1852–?) and Gertrude Mary Bosanquet (1854–?), as well as to Frances Maria Pearce ("according to written instructions"). These materials could have included some of the missing Fox diaries particularly those from 1844 to 1850 and those from 1866 to 1877. Julia Fox, the longest lived of Fox's sisters, who died in 1896, left books, etc, to Charles Woodd Fox, as did her niece, Agnes Jane Fox, who died in 1906 and was amongst the last of the Sandown Foxes. As we have seen, the Charles Woodd Fox papers descended to the "*The Fox/Pearce (Darwin Collection)*". It is unlikely that the diary for 1828, if it existed, would have been amongst any of these papers as its value would have been appreciated by the executors of Fox's will.

Appendix 2
Calendar of the Major Events in the Life of W D Fox and Documented Meetings between Darwin and Fox

Calendar of the major events in the life of W D Fox

1805 William Darwin Fox born 23rd April at Thurlson Grange, near Derby

1809 Charles Darwin born on 12th February at The Mount, Shrewsbury.

1816–1823 Fox attended Repton School, near Derby.

1824–1828 Fox read for B A degree at Christ's College, Cambridge, from Michaelmas term 1824 to Easter term 1828 with periods of absence due to illness. Darwin matriculated for Christ's College on 15th October 1827 but did not start there until early January 1828. Fox overlapped with Darwin at Christ's for Lent Term and Easter Terms 1828. Darwin lodged above Bacon's, the tobacconist on Sidney St and developed an enthusiasm for collecting insects stimulated by Fox. Darwin visited Osmaston Hall, the Fox family home, for about a week, for the Derby Music Meeting in late September. On this visit they may have called on Sir Francis Sacheverel Darwin at Sydnope Hall (see Letters # 29 & 121).

1829–1830 Fox read for divinity at Cambridge under the direction of Prof Henslow and a tutor (Thompson), with long periods at home at Osmaston Hall, near Derby. Fox and Darwin may have met only twice during this period, since their periods at Cambridge did not overlap.

1831–1832 Fox was a curate at Epperstone, near Nottingham. He did not meet Darwin during the period leading up to the Voyage of the Beagle.

1832–1837 Fox became ill through lung infection; recuperated at Osmaston Hall and at Ryde on the Isle of Wight; there he met his first wife, Harriet Fletcher, daughter of Sir Richard Fletcher, and married her on the 11th March 1834 [they had 6 children (see Appendix 7)]; took extended tours with his wife in Yorkshire and Wales in 1834 and 1835 in a pony and trap, and visited Dr Darwin and his daughters in Shrewsbury for two weeks from 28th May-11th June 1835; first child still-born in early 1835; second child born at Binstead, near Ryde, on the Isle of Wight, 10th Jan 1836; third child born at Ryde in 1837 (the other 3 children of this marriage – see Appendix 7 – were born at Osmaston Hall, near Derby). Darwin visited Fox on the Isle of Wight on the 21st November 1937.

1838 Fox obtained Crown living as the Rector of Delamere, near Chester, Cheshire, and lived there until retirement in 1873.

1842 Death of Harriet Fox, the first wife of W D Fox, on 19th March; of disease of the lungs.

1844 Susan Darwin visited the Foxes at Delamere, 17th-22nd March.

1846 Fox married his 2nd wife, Ellen Sophia Woodd, daughter of Basil George Woodd, of Hillfield, Hampstead, London, on the 20th May. Visited the Darwin's at Down House at the end of their honeymoon.

1847 Fox visited Dr Robert Darwin in Shrewsbury, 2nd-5th January.
Susan Darwin visited the Foxes at Delamere, 17th May.

1846–1873 Fox and his second wife continued to live at the Rectory, Delamere. They had 12 children: 6 boys and 6 girls (see Appendix 7); Fox developed a farm which Darwin called a "Noah's Ark".

1851 Death of Samuel Fox III, Fox's father, on 20th March.
Fox attended the opening of the Grand Exhibition in London on 1st May.

1853 Death of Louisa Mary, age 2 1/2, Fox's 2nd daughter by his second wife, Ellen Sophia on the 29th August; from scarlet fever. Also, at the same time there occurred the deaths of Frederick and Zachary Mudge, from scarlet fever.

1859 Death of Ann Fox, née Darwin, Fox's mother, on 7th April 1759.
Death of Samuel Fox IV, Fox's step-brother, on 1st June.

1861 Marriage of Fox's 2nd child, Eliza Ann, to Rev Henry Martyn Sanders on 29th May, Marriage of Fox's 3rd child, Harriet Emma, to Samuel Charleswoth Overton on 24th September.

1864 The last child, Gilbert Basil born to Ellen Sophia Fox; the 12th child of Fox and his 2nd wife and Fox's 18th child.

1868 Fox bought Broadlands at Sandown, Isle of Wight, and spent the winters there.

1873 Fox retired to the Isle of Wight.

1874 Death of Eliza Ann Sanders, Fox's 2nd child (by his 1st wife).

1875 Marriage of Ellen Elizabeth, Fox's 3rd daughter by his 2nd wife, to Baron Dickinson Webster II.

1876 Marriage of Fox's 1st son, by his first wife, to Euphemia Rebecca Bonar on 10th February.

1978 Death of Dora Fox, aged 25, 4th daughter of Fox and his second wife, Ellen Sophia; of consumption (?).
Marriage of Frances Maria, Fox's 1st daughter by his 2nd wife, to Alexander Pearce on 17th September.

1879 Marriage of Gertrude Mary, Fox's 4th daughter by his 2nd wife, to Rev Frederick Bosanquet.

1880 Death of Fox, aged nearly 75, on 7th April, at Sandown; of exhaustion.

1882 Death of Charles Darwin, aged 73, on 26th April, at Down House.

1887 Death of Ellen Sophia, Fox's 2nd wife, on the Isle of Wight, on 5th October; of diarrhoea for 4 days.

Documented meetings between Darwin and Fox

Before the Voyage of the Beagle

1. Cambridge, Lent Term and Easter Terms (approximately 7th Jan- 30th May), 1828.
2. A visit of 1–2 weeks in September 1828 to Fox at Osmaston Hall, including the Music Meeting, Derby, 9th-12th September 1828 [Letters #4 (n2) and #5 (n6)]. It was probably on this visit that CD and Fox visited Sydnope Hall, where Sir Francis Sacheverel Darwin lived, at this time [see Letters#29 (n10), #121 (n12) and #125A (n6)].
3. Tues 8th Sept1829, for several days [see Letter #23 (n2)].
4. Meeting in Shrewsbury in July or August 1830 for a few days – and may have also included a visit to the Wedgwoods at Maer Hall [see letters #31 (n1), #32 and #35 (n13)].

After the Voyage of the Beagle

1. 21st Nov 1837. CD visited Fox on the Isle of Wight (Ryde?). (CD's *Journal*).
2. 19th Dec 1842. Fox visited Darwin. "*you will hardly believe yourself how bad you looked when I last saw you in Gower St.*" (Letter #72).
3. Before 20th Nov. 1843. "We continue in the same profoundly tranquil state as when you were here" (Letter #74).
4. Fri 29th May – Mon. 1st June 1846. Fox recorded in his diary that he visited Down House on his honeymoon with his second wife Ellen Sophia. (Fox *Diary*, University Library, Cambridge: DAR 250).
5. 2nd Nov 1849. Both Darwin, in his *Journal* and Emma, in her *Diary*, recorded a visit by Fox.
6. 27th April 1858. Fox recorded in his diary "*Went to Moor Park to see Charles Darwin, who is at Dr Lanes Hydropathic Establishment – spent day there*" (Fox I, University Library, Cambridge: DAR 250). CD refers to this visit in Letter #125.
7. 13th Feb. 1863. Fox visited CD at the house of Erasmus Darwin. "*Spent 2 hours at Erasmus Darwin's – Charles & Wife – eldest Daughter & youngest Son there – also Catherine Darwin*". (Fox *Diary*, University Library, Cambridge: DAR 250). Letter #150 (n3).

Appendix 3
Family Letters of William Darwin Fox in the University Library, Cambridge: DAR 250: 50, 53, 54, 55, 56, & 58

DAR 250: 54

There is no date on this letter but it seems clear from the contents that it was written soon after the death of Fox's first wife, Harriet (19th March 1842) and that it was written to the sister of Harriet, whom Fox always referred to as Miss Fletcher. Harriet Fox was buried in the churchyard of Delamere Church, where the gravestone can still be found. The place cited at the head of the letter, "66 Marina", was probably on the Isle of Wight, where the Fletcher family lived and where the Fox family had had connections for a long time. The first page of the letter has a black surround. Miss Fletcher was clearly looking after two of the four daughters of William Darwin and Harriet Fox, probably at Delamere.

66 Marina Sunday after^{n}.

I hoped this mornings Post might bring me some tidings from you, my dear Sister, as I have not heard from you since Thursday last — and tomorrow we have no Post here. I dare say however I shall have 2 or 3 letters on Tuesday morning, and that it is owing to some mistakes. I came here yes=terday as I wrote you word I should do, and had a most beautiful day to travel in. I find all here very well except my Father, and I hope he is better than he has been tho' still rather too excitable. — My Mother seems very well & my sisters all as well as they ever are. Agnes[1] is in great preservation MA[2] rather bilious & baby[3] tiresome certainly much improved. He has now got 8 teeth & another is coming, so I hope his troubles will soon be over poor little fellow. The Woodds[4] were as kind as possible & I could have gone on staying there very well for some time longer, as I was quite one of the Family. I never met more truly kind//

People than they all are. Tell Miss Hutchinson that as there were no tidings of Col Hutchinson before I left, Mr Richards had no doubt he had become pretty well again. I do not know of what nature his attack was, but it must have much alarmed Mrs Hutchinson at the time.

With regard to myself I have really nothing to tell you in any way. I have not had any conversation with any one about what will be best to do hereafter. I have however little doubt that I shall return to Delamere as fixed before we parted & I must then look out for a Curate[5] if I can get one. I am very weak but have not coughed much to day. I think my stay at Hampstead has been of use to me altogether, as though my

lungs were far from comfortable there I got more sleep than have had & eat a good deal more — in fact quite enough.
I saw Sir Wm Newton[6] twice — having called upon him to replace dear Harriets[7] picture in the frame, as it was carelessly done before. I asked him if he kept any memorandum of his pictures — he told me he did & in referring to it, he found he finished her picture in March 1833 the very month you first called upon//
my sisters. How singular it is that it should prove to be that very month in which I first saw & loved dear Harriet. I sat some time with Lady Newton & was pleased with her.

While I was waiting at the Coach Office to leave London — a very great wish came over me to run & find out the Inn we were last at together in Spring Gardens — I had just time — the Balcony of our room looked (of course) just as when I remembered her in it, & I saw the Sopha where I laid her dear form thro the window. I cannot tell you how every thing brings her to my remembrance. In London every street brought her to my mind & when I am as I am now, where she was not — the worst of things to recall her, seems still more vividly to make me desolate & widowed. I do not encourage the feeling, but many times a day, the tear starts into my eyes — & broken hearted feeling into my chest, and I feel that I am alone & long to be where she is. Dr Brights was within 3 doors of Sir B Brodie when within 20 months dear Harriet was told by him she could not long survive. A perfect thrill came thro' me as I saw his horrid man servant at the door just as when we were there together. Oh how my heart bled for poor dear Harriet//

on that occasion — and indeed all that visit to London. I ought to feel thankful that Dr B. did not give me the same sentence. My feelings on this head are a strange mixture. For myself I sh^{d} like to stay — & feel that Duty bids me do all I can to recover my lost health, and be useful. How to do this is however the question. I am looking forward with much pleasure to returning home & beginning my Church Duty again; and tho' I sometimes felt rather alarmed at my F^{r} M^{r} &c coming for the Summer, & would rather have been by ourselves. Still I hope that attending to my people & children might have soothed me if not comforted me. Dr B. however is most positive in this part — and says I must totally rest from all parish concerns for months if I mean to recover my health. But this is so disagreeable a subject to dwell on I will say no more — as I have nothing else to write about.

If I hear I will add it before this goes to Post. If not I shall send it as it is, as it may be better than none — Pray give my Kindest love to Mrs B. — Kind regards to Miss H^{n} & Miss Ward & a kiss for the 5 children[8] & believe me

Your affec Brother W D Fox

DAR 250: 50 These two letters were contained in an envelope with the postmark: Chester 27th March 1852. This date is difficult to fit with the content of the letters, which appears to be from about the time of Fox's marriage to his second wife. Both appear to to have been written to Miss Fletcher, Fox's sister-in-law. The two letters appear to be separate. Without any dates it is difficult to put the two letters in sequence. They were both in an envelope marked Miss C. U. U. Fletcher, at

Mrs Howarths, Ipswich with a post mark of 25th March 1852. This date seems too early for the letters enclosed and are probably there by mistake. The present order was determined form the contents of these two letters and the family members, especially the children, referred to. Note that W D Fox wrote these letters in a very small script with a fine pen and his handwriting is often very difficult to decipher.

DAR 250: 58
(Letter 1 DAR 250: 58)
(*The first sheet is missing*)

in any way an impure passion on any part. I also feel that a *young person* is much more likely to make me happy & be so herself under my circumstances.

There are few who could have bent themselves to this circumstance as did dear Harriet. I feel that even I could make her ten years older – I would not do it – feeling she is much more pliable & elastic so to speak much more likely to make a happy & a useful life. Women generally if simple – so much fixed in their habits & ways of life after 30 that I always feel it a risk to themselves & thus their managing.
Had I been closer – of course this would be different but strange as it may seem to you – I always felt & do still feel – that a young woman is much more likely to make a good mother to my children. This has been always my first great care & I am certain that you will find E:[9] as fond & careful & assiduous for them (not as yourself) but as you could wish

But my time tells me I must draw to an end. You may now if you please – tell dear Jane[10] from me anything you can on the subject – with however this proviso. – *that it is still a strict secret* – as after all – there are some such difficulties as may not be overcome – tho I hope & feel they will. Pray upon her & yourself that no new connexion I form will ever in the least alienate my affection for her or & you. The possession of a Wife I can love & cherish will not make me forget dearest Harriet's love – tho' I hope it will take away the pining wish I have felt to be where she is. I know my feelings about her & have no wish to supplant her memory from my heart. There is just one more thing I wish to impress upon you and Jane – I hope *nothing* will ever make either of you allude to what passed// at Chester with C. R. I shall I doubt not tell E., as I hope never to have a secret from her – but on my account I do not wish it to be mentioned, & you are its sole depositories. After the first trial was over (& it was a deeply severe one – the more I reflected on her character & what passed – the more I felt I had a *most narrow escape*. I know this is not your feeling - & I think most highly of her in every way that in point of personal attraction I scarcely see her equal excepting Harriet (whose equal in any way I never have or ever expect to see). But now when I was most struck with her, I could not disguise <from> myself that there was a great want of stability. I think (I may say to you) that nothing could excuse her conduct (I do not mean of the Cost for there I think her right). but if we never had reason to at all counts – have some feeling shown towards him – I had from her. Before I in the least thought of EW. I felt the escape I had had I now feel that while E's attractions in a personal point of view are not to be thought of with her – that she has what the other has not – & if such a thing were possible as I had made my choice – I should feel deeply truly thankful – as I do - that I have been led in the way I have. C. is child & will always remain one – E is a child – but will become a woman when her duty calls for it. Do not let this ever go beyond you & Jane[11] if you please.

xI hope todays post will bring a letter from you to say when you are coming here – as they are getting very anxious to know. What with the post here & the Post where you are – it is enough to drive a person mad.x

I have crossed that part of my letter you are at once to answer. & now my dear sister – May God bless you in our further connexions with each other, May we be led by his Providence - & may be as much blessed as I feel the past has been – Love & kisses to the Ch.

Your affect[ly] attached Brother WD Fox

DAR 250: 58

(Letter 2, DAR 250: 58
Miss CUU Fletcher
At Mrs Howarths
Ipswich

I am writing this, my dear sister, while dinner is going in, for I fear I shall be so much engaged in the morning that I may not be able to answer your kind & much valued note received today. I was getting *very anxious* to hear from you, and any little thing now makes me more anxious than usual. You are right in imagining that I am not very happy – though in one way much more so than when you last heard from me. I received such a very sweet letter from dear E[12] on Friday that it has given me much comfort. You all do her much injustice in imagining she does not *most deeply* feel this responsibility she incurs in marrying me. No words can express it more fully than she *has al=ways* done to me. After being at Kendalls, she was so overwhelmed with the sense of it again – for I had managed to allay it somewhat before – that I much found she would have promised her Father to give me up. I wish I could show you her letters – I think I never saw more character – originality & powers shown in one who has had her disadvantages & is so *unused* to the world. I feel quite sure she will make a very valuable & superior // person, as I have always felt I have no doubt she is much attached to me as I am to her – which is saying a great deal – for I do not love by *Nature* – I wish sometimes I could put a little sense of the *fray* into my constitution.

I have now therefore no longer any doubt about E. I never have had much, but I wrote to her so strongly that - in conjunction with other things – I felt very wretched till I heard from her. I am however very miserable about the family generally. I must seem to behave so very ill, in writing nothing, nor saying anything. But my hands are completely tied. I do not hear any thing from my Mother or FJ[13]. They must think that I have no longer any feeling left, & that the whole thing is more a matter of convenience.
FJ – wrote about a week since to say that they all agreed a year at least must elapse before I could think of marrying. I do not see any reason, unless it be to punish me

which I own I richly deserve – but if there are reasons – at all events – I should know them & at once tell E.
I feel I am now losing very fast the esteem of the whole family except E – who knows me too well to blame me in the matter.
I cannot go on very long in this matter without // being regularly ill – I was getting very much out of sorts – not being able to sleep & feeling feverish till I get E[14] better – which has given me new life again – but it is most painful to feel that I am losing the esteem of all the Ws[15] – without any reason that I can see (except as I said that I should be severely punished). They cannot see the effect it will lastingly have on the minds of all the party, if it goes on much longer. If you had been here, I should have been in London before this but I cannot *leave* Eliza[16]. I should have told Mr Woodd[17] denying things - & if I had received my papers – at all events I should have felt it was doing right. If I had not full confidence in E's affection and constancy, I should feel my little hopes after all. – They must think I am acting just as Mr Martin[18] did – I am writing to E weekly & keeping up her feelings of affection - & steering perfectly clear of anything like a promise, or coming to the point with the proper persons.
I suppose before long, I shall be having some stiff letter from Mr W or one of his brothers, to know what my intentions are. I am sure I should be furious if I was one of them.

I am so very sorry I have made your visit so uncomfortable to you. *Many thanks* for trying // to educate my cousin[19] so much. I think you seem to have done so *most kindly* & judiciously. I shall much rejoice to see you again. It does indeed seem a perfect age since we had a talk together, and I long *very much* to have one. I have written to engage that Parlour Maid I spoke to you, but have not heard from her yet. She seems very nice one but I should have liked to have seen her first if I could. I quite feel we must part with Margret. Poor girl what can we do with her. I wish you could talk over with them what place she is fit for.
I have no Delamere[20] news for you at all. I was amused yesterday at a letter from Jane[21]. I wrote back her when I could send him a brace of Turkeys - & she answered my letter without in any way *alluding to them* & in her postscript said "you must not be angry if I have not answered your questions – because I, as you told me, burnt your note." I have again written to ask the question. Will you ask my Mother – if she would like her 4 turkeys to be sent off from here next Saturday or Monday direction Watford Station – so as I have ready for the carriage that brings you to Watford or to wait till Eliza comes some weeks hence, I have bought her a charm tell her. Could you bring My F^{r} and M^{r} & sisters subscription to the Clothing Club. Write to me when you can get the time to do so without annoy=ance & with a good kiss to all the children likewise – your affectionate Brother WDF

Tell Fanny[22] we have got a very nice little white kitten for her with a brown nose. Aunt Eliza[23] & she sit & purr at each other by the fire.//
Across head of letter
I have a piece indirectly for you from Miss Nutley. She is better. Mrs Robert has a very bad cold.

(Postmark on the envelope but not necessarily of the enclosed letter: Chester Mr 25 1852)

DAR 250: 55

To: Miss Fletcher
Rev. W.D.Fox
Delamere Rectory
Delamere
Chester

Mrs Broadricks
Hamphall Staffs
nr Doncaster

Postmark: Zurich 17 July
Suisse 18 Juil 42

Hotel DAngleterre
Frankfurt am Main
Friday night July 10

I sent off a long, but I fear a very stupid letter to you, my dear sister, yesterday; and am going to begin this before going to bed to night in order to go on with it en route. I am writing this at the very Inn where you & dear Harriet were at when here. You may possibly recollect it as large Inn round a courtyard - with a nice square in front. I cannot help thinking it possible it may be your room we occupy, & I sat up last night till one oclock with dear Harriets picture, at the open window and feeling sure that at all events she enjoyed the air & scene from some one of these front windows. Frankfurt is an exceedingly nice place – It is a remarkably clean comfortable town – with very nice gardens round it – a railway to Maintz – is very quiet – has excellent shops - & in fact al=together is quite the place that an English Family would like to reside at. There were fully a hundred at the English service this morning at the Consuls – and among these Mr Meales who now has Wilberforces living in Isle of Wight. I spoke to him on going in & we were to have met coming out but as we had a carriage, we did not. It is quite curious how many acquaintances we keep making as we progress on.

Tomorrow we go by Railway at ½ past 6 to Maintz – there take the Steamer to Manheim & then hope to get to Heidelburg to sleep, but this depends upon our time. I am sadly home sick today, & would give no little to see you sitting with Mr and Mrs Harrison at Delamere where everything I love is, & there always long to live with you & hear all that has passed since I left you.

My health is however much im=proved — I eat voraciously — & when tired out, sleep: -at other times my brain is too busy. I exceedingly long for your letters at Geneva, but I suppose it will be 8 or 10 days before we get there. I almost dread to see your handwriting while I long to do it above every thing. I see from your Journal, you did not go when here to see Dannikers Statue of Ariadne. It is the most splendid effect of I ever saw. A perfect triumph of art — the marble really all but lives – In general I am not fond of statuary. but really is so very exquisite that no one can help *feeling* its excellence. The Museum of Natural History is likewise very splendid — and the Cemetery — new since you were here a very pretty one, with some of the chastest Monuments I have yet seen.

Thursday night Sackinggen
on way between Basle & Shoffhauen

I have been so regularly knocked up my dear sister since I began this on Sunday night that I really could not go on with my letter. I have however thought of you very much and exceedingly wished you had been of our party, tho' I fear we should have knocked you up. We left Frankfurt early on Monday morning by Railway to Mayeneve & then again took Railway to Heidelberg where we arrived late at night. What a lovely spot it is. I rambled about the ruins that night and thought how very much dear Harriet must have enjoyed it- and still more so the day after when I got to that most exquisite Castle & its grounds, where you will remember rambling about with her till 10 at night. I never have felt more widowed than I did while there, but yet the feeling (tho' dreadfully trying to me) is at the same time most pleasing. I do so love to dwell upon dear Harriet and our most blessed intercourse together & look forward to the time when we shall again meet. I had fixed with Henry to spend the night at the ruins — He was to smoke and I to *dream* — but my sisters wished to go on that night - & perhaps it was better it was so. We were very fortunate in hearing Haydns Creation by above 200 Musical & vocal performers in the great Square of the Castle which was thronged with well dressed people (tickets 2 ½ each). The Castle &c was gaily dressed up with flags &c –altogether it was very pretty & this Music which would have been good any where, have had a very striking effect. We left this lovely place in the Evening- got to Mannheim by Railway about 10 and passed the night in the Steamer – I as I always do, on deck. There is nothing to see in this part of the river so that we missed nothing by travelling thro' the night We reached Strasburg yesterday about 6 and much enjoyed this noble Cathedral — I think the finest to my mind I have yet seen. This evening we were off again early and came to Basle by Railroad. The way pretty & the agriculture *to me* very interesting but scenery between the beautiful port of the Rhine to look back to & Switzerland forward- must be very good to be *much admired*. All however has its charms for me. I see so much to //

admire when others see little – just as I eat all German dishes bad & good as they come, & enjoy all in their way. We dined at Basle & then at once hired a W*itarior* to bring us on there for tonight & we keep him as long as we like by the day. The next place to this Walshut where you all passed such a miserable night when here — Our Inn *at present* seems pretty fair. Frances Jane[24] has hitherto borne her travels very well. — Julia is a little knocked up with the change of foods, but I hope will soon be right again.

Saturday night Zurich

I have not hitherto had time my dear Sister to find my letter to you indeed it is such a scrawl that I scarcely like to send it to you at all, but know you will like to receive a poorly written —letter than none at all. I have so longed for you & dear little EA[25] & Harry[26] the last two days. You would have so much enjoyed the Falls of the Rhine with us & so would your two little charges tho' I think they would have rather clung to us when near the magnificent scenes. We arrived at the Inn close to the Fall in

time to pay them a visit by Moonlight ~~las~~ last night — & very lively they were. I sat by them a long while thinking of my dear Wife — at one time calling to my mind her transport of delight when she was here with you in such noble scenery — at another calling her to mind as she was on that night 4 months – when I was first summoned to her on that awful Friday night. I sat there musing with myself & dreaming about dear Harriet, till I found myself looking by my sides, as I cannot help doing at times. If there ever was a place & time when one might hope to see again the form of one who was separated from us it might be in such a scene as this.
I had fully intended to have returned to spend some time by the Falls at mid=night, which I should have particularly enjoyed, for my enjoyment is not now to be with living forms – My only pleasure is to be alone & commune with my own spirit & think over other days. I however could not do this – as I was taken so unwell that I was very glad to lie down in my bed for some time. The Bread &c had disagreed with FJ Julia & myself so as to quite invalid us & make us but very poor creatures — & as we can get no Bread or Biscuits made with yeast, we are rather fast to it. I hope tomorrow will find us all improved. We spent only time at the Falls this morning – on both sides of the River – the Iris was beautiful across the spray & altogether it was I thought more lovely than by moonlight. I cannot tell you the loneliness that comes over me in these beautiful scenes. I feel as if I had been quite alone & must remain so till again with dear Harriet when the weary are at peace. The dream which you know of, & which H[t] herself gave the ground for, for unless she had put it into my mind, it certainly would never have entered when it did - & which always comes connected with her — I feel to be indeed a dream — I wish it had never entered my thoughts. I was desolated //

enough before, without having another blow. I cannot however now help it, & must try to *bow before* it. In my desolate times dear Harriet seems to whisper to me to hope that I may have peace during this troubled state of being — & I listen to the delusion feeling it to be a delusion. I long so very much to receive your letters which are I know lying at Geneva for me. It seems ages since I heard anything of you & my dear little ones at Cheshire, where all my thoughts are now concentrated. I feel as if I would give any thing for an hour this night in our Harriets room, 4 months since – we were left desolated by her dear spirit having fled. I was so benumbed that night by the 24 hours of misery I had gone thro', that I scarcely knew what passed. I felt that she was my hope, my joy, my love, my all, was gone from me, but I was too much deadened to think upon my loss. but Oh how I have felt it since. X

X I have her sweet picture now before me & the dear little pillow to rest my head on but alas my Partner in Joy & Woe is gone. The scene from my little balcony here is just what dear Harriet would have enjoyed – the beautiful lake close under me with a bright Moon gleaming across its waters — & the lightning harmlessly playing over the distant mountains. It seems to my mind to allegorise ourselves. Here am I still on Earth, with this bond still between me & my dear Wife whose pure spirit may be likened to the playful electric streams which dance in the far mountains. I must tread my way thither now alone & uncheered by her who was my comfort at all times — the only blessed comfort is – that we shall meet again – perhaps very soon.

I shall now tear myself away from my converses with you – Oh that I was with you, if you are still where dear Harriets Spirit took its flight. I long very much to know what your plans are — where you are, that you are, for when we parted you were far from strong. Pray for my sake & dear little EA & Harrys sake you can of yourself you can. When you write to Jane give my kindest love to her — May God in all things bless you & be with you

Ever your aff Brother W D Fox

There are 2 other letters to Miss Fletcher from W D Fox on his European tour at this time

1. DAR 250: 56

26 JULY

Miss Fletcher
~~Revd W.D.Fox~~
~~Delamere Forest Rectory Chester~~
Mrs Broadrick
Hamphall Stables
Nr Doncaster

Not transcribed.

2. DAR 250: 57 2

Miss Fletcher
Revd R Mallock
Cockington Court

Geneva
14 AUG 42

Postmark: Torquay Devon Au 20 1842

Not transcribed

DAR 250: 53
From: Samuel Fox IV
To: To Rev W.D. Fox, Delamere Rectory Northwick, Cheshire

S. Fox.
Clinton Army
Newark
Nov 19th 1857

My dear William

I rec^{d} your letter just as I was setting off for Newark, so had no time to answer it from Derby. Before I reply to the business part of it, I may as well mention that when I was in London 3 weeks ago, I saw Mr. Robt Woodd[27], & arranged every thing with him quite satisfactorily on the way I proposed viz. I left with him a Mortgage for £500 from In.^{o} Bateman, a Lincolnshire Farmer & also a *Declaration* signed by Lord Belper[28] & myself that £500 (part of a mortgage of £1500 from

Boteler Bristowe[29]) belongs to the Trustees of your 2nd marriage & I have Mr R Woodd's acknowledgement to that effect. Then there the other part of the mortgage from Boteler (£1000) belongs to the Trustees of y^{r} 1st marr^{ge} & the Deeds are in Lord Belpers custody – Now I think you had better write out all these particulars & put them in some safe place, so as to be referred to at any future time if necessary. With regard to your wanting to borrow the £300, I have an insuperable objection to asking Lord Belper to lend it. At the time he lent me the £3000 on Trusley, he very kindly //

offered to me without my asking & at very moderate rate of Int, 4 per Cent, & I don't think he will ever raise it to me, whatever the Int. of money maybe. This however, does not preclude you making application to ~~himself~~ him yourself if you like to do it, but if he was to refuse, it might cause some little uncomfortableness to both of you. — I think on consideration the best way might be to obtain it from some other quarter within way of business & I daresay I could manage to borrow it for you in the course of 5 or 6 weeks, but at the present moment it would be quite impracticable. No one would now part with money under 8 or 9 per Cent. Would you have the kindness to write & tell me the latest time you could do without it & I will then write to you again on the subject. I think now we are on the subject of money matters I may as well mention that when I was at K.P.G.E.[30] last winter or early into the spring, I had a long conversation with my mother Eliza & Julia about your affairs & also about your farm. They were all most strenuous about your giving it up & asked my opinion on the subject, but I said I was not competent to form one without knowing all the circumstances, but that so far //

as I wd know I was rather in favour of your keeping on. My mother then went into a number of particulars about how many men you had, what their wages came to &c &c which I told her was quite a onesided view & unless she knew what came in as well as what went out, neither she nor I could judge of what was best. They were very much excited & talked of you as if you were on the verge of ruin & Julia[31] was very violent & said if you were ruined, nobody could pity you. Altogether it was a very disagreeable conversation & I was very glad when it was ended. I don't know whether I am right in mentioning all this, but it *is* not with the idea of making any mischief, but that you may know what their opinions are. I think if you were to apply to my mother for assistance, it would not be answered, & further, if she was disposed to do anything Julia would stop her. I therefore think you are quite right not to let my mother know what you are going to do, as it might have a serious effect on her health. I have written so long a letter I have scarcely room to enter into any other topic except to say that Bonnie[32] is as well in health & as happy as it is possible to be. He had something like the mange soon after I brought//

him home, but Hough the Blacksmith soon cured him. He plays very nicely with Rough, who is a most good tempered Creature, but Bonnie is quite master of him.

With much love to all y^{r} circle & hoping you are all well. Believe me

Your affec. Brother

S. Fox

Postmark No 19 1857 Newark E
Provenance: Manuscript Room, University Library, Cambridge: DAR 250: 53

Notes

1. Agnes Jane Fox (1839-1906).
2. Julia Mary Ann Fox (1840-?).
3. Samuel William Fox (1841-1918).
4. Basil George Woodd and family. He was a successful merchant living at Hillfield, Hampstead, London. Fox would eventually marry one of the daughters in this family, Ellen Sophia, in 1846.
5. Fox eventually obtained a curate whose stipend was paid for by Basil George Woodd (Fox *Diary*).
6. Sir William Newton (1785-1869). A painter of miniature portraits, several of which are in the National Portrait Gallery and elsewhere; Newton was an early supporter of photography (Zillman, L. G., Sir William Newton: Miniature painter and photographer (1785-1869). History of photography series. School of Art, Arizona State University, 1986).
7. Harriet Fox, née Fletcher, Fox's first wife; she died on the 19th March 1842.
8. Five children. Since it appears that 3 of Fox's children are with him, and there were five children in his family at this time, he must have been referring to his two elder daughters and to three other children with Miss Fletcher.
9. E refers to Ellen Sophia Woodd, who became Fox's second wife. It is not clear from this letter if he was married to Ellen at the time of the letter. They were married on 20th May 1846.
10. Jane is not identified. She may have been a close friend of Miss Fletcher.
11. Not identified.
12. Ellen Sophia Fox, née Woodd.
13. Frances Jane, the sister of Fox.
14. This "E" seems to refer to Eliza, Fox's daughter (see note 5, below).
15. This refers to the Woodds, the family of Ellen Sophia Fox, née Woodd, Fox's wife.
16. Eliza Ann (1836-1874); the first daughter of Fox by his first wife Harriet,
17. Basil George Woodd (1781-1872). The father of Ellen Sophia, Fox's second wife,
18. Not identified.
19. Not identified. Fox may have been using this term in a loose sense. At this time he was probably putting up Sir Richard Fletcher, brother of his first wife. This identification seems the most likely as later in the 1860s Miss Fletcher was looking after Sir Richard Fletcher in Ben Rhydding, Yorkshire.
20. The village of Delamere, which was Fox's Parish.
21. Not identified: It is clear that this is another woman besides Frances Jane, Fox's sister. This Jane may be Miss Fletcher's sister, who may have been married to one of the Mudge family (see Biographical Register).
22. Possibly Frances Maria (1848-?), Fox's daughter by his 2nd wife, Ellen Sophia.
23. Eliza Fox (1801-1886), one of Fox's elder sisters.
24. Frances Jane, Fox's elder sister. Referred to below as FJ.
25. Eliza Ann, Fox's 1st daughter, by Harriet, his 1st wife.
26. Harriet Emma, Fox's 2nd daughter, by Harriet his 1st wife. She became Harriet Emma Overton and died in 1922.
27. Robert Ballard Woodd, brother of Ellen Sophia Fox. He was a solicitor.
28. Edward Strutt, 1st Baron Belper.
29. Boteler Bristowe, Fox's nephew.
30. Kensington Park Gardens East, where Fox's mother and unmarried sisters lived.
31. Julia Fox, Fox's youngest sister.
32. Bonnie was probably a dog.

Appendix 4
The Diary of William Darwin Fox: 1824 to 1826: Part 2. 2nd June 1825 to 13th (continued from Chapter 2)

Thursday.
I was fully occupied this morning with rambling about the grounds, seeing the dogs &c. My Garden looked beautiful, and indeed every thing here looks like Paradise after Cambridge.

p 56 June 3. 1825
I this morning wrote a letter to φαννι[1] and took it to Derby.
Sunday 5.
It was evening service to day. Mr Bligh[2] preached as usual, and dined here.
Monday 6.
This afternoon I accompanied my mother and sisters to see Adam's Circus, now at Derby. We were all much pleased with it, particularly my sisters, who had never seen any thing of the sort before.
Tuesday 7
Miss Woodcock[3] came to Osmaston[4] to day to stay for some time.
Wednesday 8
I was at home all today. Miss F. Strutt[5] drank tea here and spent the evening. A very hot day.
Friday 10.
I this morning walked over to Derby and brought a silk handkerchief and spoon for φαννι j .. and sent them. I had a long letter from her likewise.

p 57 June 10. 1825
I also saw Frank Hall[6] in Derby, as he was returning from calling on me, at Osmaston, and walked with him all morning. I returned home, to Dinner, very much tired with the excessive heat of the day. very hot.
Saturday 11.
I was at home all this morning, and dined in the evening at Mr Hadens[7], and spent a very pleasant evening there upon the whole. very hot.
Sunday 12.
It was a beautiful summer's day again almost to hot for any thing.

Monday 13
Frank and Arthur Wilmot[8] rode over to Osmaston this morning, but did not stay long. Lakne[9] went with them to Derby, and I suppose lost himself as he did not return.

Tuesday 14
The weather continued very hot. I went with Clarke[10] and Mr Haden[6] to dine at Thurlston[11]. My sisters and Miss Woodcock[12] came there to tea. We all returned home in the evening.

p 58 June 15th 1825
I rode over to Derby this morning to see if I could hear any thing bout Lakne, but could not, and began to fear I shall lose him. I saw the Wilmots and Mr Mundy[13] in Derby.

Thursday 16.
I was at home all this morning, and in the evening accompanied my Sisters to Mr Hadens to tea, whence we all went to the Church Concert, and were a good deal amused. We go home about 11 oclock.

Friday 17.
I went to Derby this morning and took a letter for φαννι. I also received a letter from her in the afternoon. I came home to Dinner and stopped there this Evening.

Saturday 18.
This morning about 12, I left Osmaston in the car, and went with my Grandmother to Derby, from whence I started at ½ past one in the //

p 59 June 18. 1825
Regulator[14] for Loughbro'. I was very much surprised to see Lakne walking after a gentlemans carriage. As I could not stop the Coach I called on the gentleman (A Dr Hardy) when we got to Lough=bro' and he said he had had him, ever since I had lost him. I sent him home by the Coach. It was a delightful day for travelling, and I enjoyed my journey very much. About 10 oclock I got inside as the night was rather cold and damp.

Sunday 19
I got safe into London this morning, rather tired with sitting up all night. I went directly to Sam's inn (Andertons Coffee House) where I found him in bed of course. I lay down a couple of hours, and after breakfasting, we walked all over the West End, and returning to Dinner, met Ed Strutt[15], who joined us, and we afterwards took coffee with him.

p 60 June 20th. 1825
I did not get up till after 9 this morning, and as Sam had to attend the Silk Sale, I set out to amuse myself. I spent an hour in the Exeter Change[16], and was much pleased with the excellent collection of animals there. More particularly with the Camelions and snakes of several kinds I had not before seen. I afterwards walked about and saw all I could see till 1 oclock, when I went to see the exhibition of pictures in Somerset House with Ed Strutt[15], and was much gratified with it. I then dined with Sam[17], and

afterwards went to Drury Lane to see Kean[18] in Richard the third. Robertson met me there, and we walked afterwards to several Coffee Houses and ended by supping at Offleys. I got to Andertons about one, a good deal tired with being on my legs all day.

p 61. June 21. 1825

I went this morning after and early breakfast to the West End and walked about till about one, when I returned towards the City. I then waited till past three at the Adelphi[19], for my Father & Sisters, when at length they arrived. We all dined, and my Father and Mother with Emma[20] set off to Mr Bromfields[21]. I went with FJ[22] and Julia[23] to the Exeter Change[15], and then to Burlington Arcade[24]. My Father returned about 10, when we all went to bed.

Wednesday 22.

This morning I joined them at the Adelphi[19] after breakfast, and we all went to see the Diorama of Holyrood Chapel and the Cathedral of Chartres[25], <the 1st of Whitly> which was beautiful. We then went to Bullocks Columbian Museum[26], and afterwards to the Bazaars &c, and then came back to a late dinner. Sam returned to Derby to night. I went to bed early.

p 62 June 23. 1825.

This morning after an early breakfast at the Adephi, we all set off in Hackney Coaches to the Tower Stairs, and from thence in the Venus Steam Packet[27] for Margate, where we arrived, after a very pleasant voyage at about 5 in the afternoon. None of us were at all unwell. We then got dinner, and after walking about the Town, all went to bed.

Friday 24

This morning after breakfast, My Mother and myself, with John and Lydia, set off to Ramsgate[28], to take lodgings, which we did very easily indeed, as there were plenty at liberty. We took a pleasant House, next to Sir Wm Curtis's[29] Garden, looking at the Harbour (No 2 Prospect Hill). My Father and Sisters joined us about 2 oclock, and towards evening we got quite comfortable, and domestic.

p 63 June 25.1825.

We were very busy this morning walking about the place, buying in, &c, and were very well satisfied with the whole arrangement.

Sunday 26.

We were walking about, and enjoying ourselves all day, as we did not know when we would go to Church.

Monday 27

Julia[23] began to bath this morning in the sea. It rained a good deal here. My father quite as he used to be.

Tuesday 28.

I began sea bathing this morning. We subscribed to day to a library, which will add very much to our comfort here.

Thursday 30
It was a roughish day, and rained several times. I bathed again this morning. The wether has been very uncertain for the last day or two. As it was near Full Moon we had a Spring Tide.

p 64 July 1st 1825
This morning a large West Indiaman was brought into the harbour to be repaired, as she had been injured by another vessel going against he in the Downs, and lost her fore Mast head &c. It was a very stormy day. We had a newspaper from home. All were well there. My father's health increases daily.

Saturday 2.
We this morning to our great surprise received a letter from Mary Ann[30], to inform us that she and Bristowe[31] were going in a short time to Geneva, where they intended to stop the Winter and that they wished Frances Jane[22] to accompany them. It made My Father very unhappy, and indeed in some degree all of us, as we looked upon it, as owing to his embarrassment, and we thought it might end in his staying there even years. My Mother answered it in rather a strong (but not too much so) way, and mentioned her suspicion.

p 65 July 3. 1825
It was a beautiful morning as could possibly be. I bathed before breakfast. Afterwards my Father and self attended the New Chapel here. We all afterwards walked out and en=joyed this beautiful day. We heard from home. All well.

Tuesday 5
I walked again today and Julia did so every day as well as my father, and both seemed better for it. Emma[20] bathed in the warm baths likewise. I was occupied most of the day in reading Mr Scotts[32] new novel of the Crusaders, and was rather disappointed in it.

Thursday 9.
We had a very rainy day here, & we feared from its general gloom=iness, that it must be so at Osmaston which (as the Hay is there making) rather fidgitted My Father. The Sea certainly does us all good, as my Father and Julia grow fat and we (lean kind) all feel stronger and better for it I think.

July 8th 1825
This morning my Mother received from Bristowe[31] a most ungentlemanly and passionate reply to the letter she had written a few days back upon the subject of MAs and FJs going abroad. Very luckily my father was out at this time and it has not been shewn him as I think he would never forget it. My Mother read it with much more temper than I could have imagined, and immediately answered it as coolly, and in a proper manner. A vessel of 100 tons struck this morning upon the Goodwin Sands and it is with difficulty that the Sailors and Passengers were saved. There was also an accident in the evening occasioned by a boat oversetting in the Harbour, by which two boys were nearly drowned. One was got out without much hurt; the other was not recovered without great difficulty, and the

p 67 July 8th 1825
the instant application of a warm bath and even then imperfectly as it is feared he will yet die, owing to some internal injury from a blow of the boat. My Mother and Julia saw the whole of it, and the rest of us the greater part.

Saturday 9th
The greater part of the cargo of the vessel wrecked on the Sands was this morning recovered by boats. The Vessel was broken to pieces. We received a long letter from Eliza, saying how well every thing goes at Osmaston. M. A. and her children were there, the Former pretty well considering her late illness.

Sunday 10th
It was a very fine hot day, indeed too much so to be pleasant till night. The Carolina West Indiaman which had been for some time refitting left this Harbour to day, and set out on her voyage.

July 11 1825
We were all quiet this morning till 2 when we got an early dinner, and set off in an open carriage to see some of the villages and towns in the Isle of Thanet. We first went to Broadstairs, from thence to Kingsgate where King Charles II landed at the restoration, & where there was formerly a handsome Archway which has within the last few years been washed away. There are a few very large Houses here, particularly one built by Lord Holland, (after a Roman Villa) which is now falling to decay very fast, and inherited by the poorest people. There are several orna=mental towns about which give the village, upon the whole a pretty effect. The Lighthouse at the North Foreshore is also near. We then came back by St Peters where we stopped about an hour at the Gardens, and were a good deal entertained there

p 59 July 11. 1825.
upon the whole; We got home all very much pleased to tea, and all the better for our excursion

Wednesday 13th.
The weather which had been very fine for some days became dreadfully hot.

Thursday 14.
It was one of the hottest days I ever felt, altho' there was a little breeze. The Thermometer in our sitting room from 78 to 80 degrees quite in shade, in the country it was much more.

Friday 15
Quite as hot. It was impossible to stir out till night

Thursday 19
The weather continued quite as hot and clear. The Newspapers were filled with accounts of animals killed while working from heat. As it was the anniversary of the Coronation of George IV, the Cannon here, at Dover and the 74 Grand Ship in the Downs

p 70 July 19. 1825
each fired a royal salute at noon. I had never seen cannon fired so near, of any size before.

Thursday 21
It was a much pleasanter day than any we had experienced some time, as it was cloudy, and a very pleasant wind sprang up

Saturday 23
The Weather continued to be very pleasant as the wind remained at N. E. My Father in bathing this morning got a violent knock on the Head from the machine (a wave drove him against.) I made a resolution to day, which I hope to keep, that I will not read a novel for a year, when I hope to still continue it, as I am sure they are a great waste of time, and have frequently made me very uncomfortable and idle owing to my giving way to them. Also

$_x$Remind me memorise assorted$_x$

p 71 July 24. 1825.
My fathers bruise was so much better that he was able to put his hat on this morning. We received a long letter from Eliza[33]; all were well, and going on well at Osmaston.

Monday 25.
This morning F.J. and Julia went with an old school acquaintance (a Miss Morris) who they asked to come this evening. We took a row before tea and were all soon tired of it, as it made us unwell. Miss Morris stayed to tea and seemed a nice girl I thought.

Wednesday 27.
This morning I was very busy rather boring set of acct &c. In the afternoon we all went to St Lawrence[34] (My Mother and sisters in a pony Carriage) and we walked. We saw the village which was not worth the trouble, and then all walked back to tea, after a pleasant stroll, as it was a delightful Evening. The Weather for the last four days has been pleasant as it has been cool and yet fine. Wind N. E.

p 72 July 28. 1825
It was a very cold windy day, but fine. Sir William Curtis arrived from Russia during the night in his yacht, which is a very handsome one. We were all busy preparing our departure, which is fixed for Monday. The Dungerness Steam packet put in for the night, as it was too stormy to pass the channel safely.

Friday 29th.
A most beautiful day. We took a walk all round Ramsgate.

W. x

Saturday 30.
Was a very fine day. We heard from home, all well. At night we were very glad to see the Sovereign Steam Vessel come in, as it was the one we all wished to go by on Monday. We heard from M. A.

Sunday 31st
A very fine hot day, indeed. My Mother had a headache, in the morning, but was better in the evening. We got all ready to leave Ramsgate.

p 73 *August 1st. 1825*
We all got up very early this morning, breakfasted at six, and at seven went on board the Sovereign Steam Packet in high spirits. We stayed about an hour and there

appeared but little sign of her starting owing to some pipes having come unsoldered. After a little more delay, we perceived a bustle among the sailors, and we then found out that the vessel was on fire somewhere but they could not find where. They immediately hacked up the deck in 5 places and at length found out where the fire lay, and we left. Of course it created a great bustle and every body was anxious to leave the vessel, we amongst the rest. We all agreed we had better not go in it, even if it start, which seemed very uncertain, as the vessel was a good deal injured. We therefore got a coach and set off straight to Canterbury, and from thence came on to London, where we arrived about ½ past nine after one of

p 74 August 1. 1825
the most broiling day, I ever experienced. The country thro' which we passed was beautiful, had I been able to enjoy it for the heat which was intense. The Thermometer in the shade was 89½ degrees. We went for the night to the Adelphi Hotel and hoped we had all got thro' our fatigues pretty well, but found ourselves disappointed for
Tuesday August 2. 1825
this morning when we all got up, poor Julia who had been so dreadfully bitten by Bugs that she was quite covered with bumps all over, and had one eye mostly swelled up, the pain was of course excessive. After breakfast Mr Hicks came up to us, an we found he had taken us lodgings at No 16, Dover Street, Piccadilly, whither we all moved directly. It was owing to a mistake we made that we did not get into there the night before as they were quite ready for us, had we known of it. We found there very comfortable, and soon got settled in

p 75 August 2. 1825
there. We got some medicines also for poor Julia, who was made quite ill by the pain she was in. We were kept rather at home in the evening by a shower of rain, which was very welcome, as it cooled the air. We all got to bed early.
Wednesday 3rd.
We all on rising this morning, found ourselves hearty. Julia was very much better, indeed and nearly well As my Mother and sisters were going to be employed in millinery &c my Father and self left them after breakfast and called on Mr Theobold[35], Mr Hicks[36] and Mr. Harry Bromfield[37] who I had never before seen and was on the whole rather pleased with. We all eat a very good dinner together at five, and after taking a short walk, and having tea, all got to bed.
Thursday 4th.
I this morning walked about the West End with Emma and F. Jane, while the rest were shopping. It began to rain very hard about 3, which drove us in till dinner time.

p 76 August 4. 1825
I afterwards went to see Corbett's Museum[38] in Piccadilly, which is one of the best for birds. I ever saw, except perhaps Bullocks old one. It was parti=cularly rich in English specimens. Among others I remarked a perfectly White rook (Full grown) with red eyes. There was also a fine specimen of the larger Bustard from Norfolk.

I afterwards went with my Father and Mother to Mr Harry Bromfields[37]. I had never before seen my Aunt; she was very unwell indeed. We drank tea there, and returned about ½ past 9 when we all went to bed.

*Friday 5th*

This night was a very windy (almost a hurricane) and stormy one; more than I ever experienced in London. After breakfast, we all except my Father, walked to see Westminster Abbey; but saw (on our way first) the two houses of Parliament, and Westminster Hall, which was extremely new to my sisters and self. We then went over Westmin=ster Abbey, and were most pleased.

p 77 August 5. 1925

with it. It was nearly new to me, as I had forgotten many of the monuments. We went from thence to the Soho Bazaar[39] where we stayed till near dinner time when we returned home.

Saturday. 6th.

After breakfast this morning I went with F. J. and Julia shopping &c, until 2 oclock when we dined. And all the party (myself excepted) set off with poor Julia to Bromley to stay with Mrs Chalklen. After they were all gone, I sallied forth to Skinner Street, where I found Wood[40] & John Theobold, and as they were just sitting down to dinner I stayed and took a glass of wine there. I left them (after a very pleasant hour) at 7 oclock, and as I happened to be passing by Exeter Change, at the time the beasts are fed, I walked in and was very much gratified indeed. I got home just as they were at tea, and found they had left Julia in better spirits than they had expected. F. J. stopped with her till Monday.

p 78 August 7th 1825

We (i.e. My Father, Mother, Emma and self) joined Sam, who had come from Derby this morning, went to see service performed in the Foundling Chapel, and were much pleased with the Establishment on the whole. On our return, we met Jasper Dr St Croix[41]; who accompanied us back and dined with us and rather annoyed us by staying till nine, ready for our departure, as we intended to leave London the next day.

*Monday Aug 8th*

This morning I went to meet F. J. at the Coach office, and we were quite in time to see the Horse Guard Change guard. It had rained very much in the night, which made the streets so dirty we could scarcely walk in them. About 3 oclock we left London, after an early dinner; my Father and self on the box and the rest inside. We got to Wooburn about nine oclock after a very pleasant tho' rather

p 79 Aug 8th 1825

cold ride. Every body on the road were very busy at the corn, which was all down nearly, and a good deal gathered in. We got some tea on our arrival and then all went to bed much tired.

Tuesday 9.

We breakfasted at ½ past seven and then proceeded on our journey home, where w arrived without a single accident, or unpleasant occurrence about 8 oclock, and found my grandmother, Eliza, Miss Woodcock and Mrs Mayor[42] (who was staying at Osmaston) all well, and expecting us. We sat down to a hearty meal soon after our arrival and then retired to rest early all pretty well fatigued.

Wednesday 10.

We were truly glad (I think) to get home again and were busy this morning, looking about us, and unpacking. Osmaston looked delightful after London and Ramsgate. My Father was teased with an inflammation in the eyes rather from the dust in travelling.

p80 August 12. 1825.

I this morning went to Derby to see Shaw and Foster, and to do several other little commissions. I saw Frank Wilmot. I had received a note from Wood on my first arrival at Osmaston to go with him a grouse shooting to day, but had declined it. I got home to dinner, where, I found Mr Needham[43], who had ridden over to see my Father. My Mother, Mrs Mayor[42] &c drank tea at Mr Strutts[44]. Mr Needham went home in the evening. My Fathers eyes continued to plague him.

Saturday 13.

I was home all to day. It rained a good deal in the morning. Uncle and Aunt were here in the morning.

Sunday 14.

We had Evening service, and Mr Bligh dined here. After dinner I rode over to Thurlston to bid Uncle and Aunt[45] goodbye, as they were going a journey next morning and drank tea there.

p 81 August 16. 1825.

This morning Mrs Mayor left us by the Regulator for London. I was upon the whole, glad when she was gone, tho' she certainly made herself very pleasant, while here. In the evening Eliza, Cleark and Mr Haden came over, and we shot till tea. ~~? ?~~

Thursday 15

I went with Miss Woodcock and my sisters in the car to Kedleston[46] and Quarndon[47], and enjoyed the ride much as the road was new to me, and there were several very pretty villas on it. As we were retuning thro' Derby I met Mr J. B. Crompton[48] who asked me to dine with him, which I accordingly did, and I spent a rather pleasant evening there, tho' most of the party were strangers to me, or nearly so. We had an eel for dinner weighing about 5 pounds. I got home in pretty good time, as there was nothing to be done after tea.

p 82 August 19. 1825.

I was at home all this morning, a part of it, shooting with F. J. We dined (my Father, Mother, Emma & self) at Dr Bents[49], and had but a dullish evening on the whole. I was not sorry it was over. ※

Saturday 20

Eliza, F. J. Miss Woodcock and myself went this afternoon to Mr Hadens, where we shot till 2 oclock, after which we drank tea and returned home by car.

Cleark[50] informed me, of what I had before feared I should hear all about my *lamentable misfortune*[51]. I fear it is getting abroad every where. What I shall do, I know not. I wish the suspence were over, at all events, for that is the worst of all.

Monday 22.

I rode over this morning to Trusley[52] and spent most of the day there. I returned home to tea and shot in the evening with F. J.

p 83 August 23.1825.

I was at home all today. Frank Wilmot[53] called to see me and sat a good while. I wrote to Wood to day asking him to come here for a few days. I feel so ashamed of myself wherever I go, that I am quite wretched and miserable except when by myself. When it will end I know not, and scarce dare guess[51].

Wednesday 24

I this morning rode over to Thurlston[11] directly after breakfast to see Foster[54] about a pointer I wanted. I happened to see good carnations at Wards at Elvaston[55], and I ordered some of him. I rode by Lord Harrin=gtons[42] and admired the new Gates very much indeed.

Thursday.

I had fixed to go to Swanswick[56] with F. Wilmot and went to Derby to meet him, but when there I received a message of him to say he could not come. I went to Mr Farnsworth[57] to day for the first time.

p 84 Aug 26. 1825

I walked over to Derby this morning and took a letter for φαννι, which I sent by J Thorpe[58].

Wednesday 31.

This morning Miss Woodcock went by the early coach to London, to live there with her sisters[59]. Would she had never come to Osmaston, I should be much lighter. Hugh Wood[60] came to dinner fully prepared for slaughter. We went to bed early, but neither of us slept, tho' from different causes.

Thursday Sep^{t} 1st

Wood and I got up about ½ past four, and when we were at breakfast F. Wilmot joined us. We then set out with high glee to shoot. We were out all day; I shot very ill indeed, and only killed 1 bird or two; Wood and Wilmot killed every thing that got up. We came home in the evening with 11½ brace of birds and a hare, rather tired. Jim Mugleston[61] joined us shooting and I had a good deal of conversation with him about money &c.

p 85 September 2. 1925

Wood and myself went out for half an hour after breakfast and killed one bird. We then returned and dressed ourselves for the 3$^{\underline{rd}}$ and last Archery meeting for this season. It was the first meeting I had attended since I was made a member of it. We all went except my Father who preferred staying at home. We spent (on the whole) the day very pleasantly, especially as I saw many old friends, I had not seen some time. J. Holden[62] par=ticularly. I shot uncommonly well for me, as I was second best at the 60 yard target. We had a pleasant dance in the evening and all came home much pleased at night.

Saturday 3.

F. Wilmot came to breakfast, and we afterwards all three set out to shoot in Osmaston. I only killed two birds and a land rail. They killed 11 brace and two land rails. H. Wood left us to night, and I proposed to join him on Monday.

p 86 September 4. 1825

We had morning service this day. James Holden[49] came over here before dinner and stayed the day with me. We walked over Derby in the afternoon.

Monday 5.

This morning J. Holden and I set off in a Gig with J Hs horse for Swanswick[43]. We got there to breakfast and found Hugh Wood very busy. We shot all day. & I as usual very badly and only killed one bird. We altogether killed 14 brace and 3 hares. We slept at Swanswick.

Tuesday 6.

We went out shooting a short time this morning. I killed 1 hare. Hugh and Jas H. killed 6 brace and a hare. At one oclock JH and I got into our Gig and returned home. I went with the rest of our people, to Darley to dine, and spent a pleasant evening, as F. and Arthur Wilmot[63] were there. I was very glad to get to bed when we got home.

p 87 September 7. 1825

This morning J. H. and myself set off in a gig to Trusley[53], where we breakfasted and then shot all day. We killed 2 ½ brace of birds and a hare. I killed 2 birds. We came back to Derby together at night, and I walked from thence home.

Thursday 8.

I shot this morning a few hours in Osmaston, and killed 1 bird and a hare. I was not quite well at night.

Friday 9.

As I was not quite well, I did not shoot to day, but stayed at home. I received a long letter from φαννι, all was well as I could expect. I heard also from Sam[64].

Saturday 10.

I had written yesterday to ask Simon Bristowe[65] (who was at Barrow) to meet me on Simkin Moor this morning, which he did and we shot all morning, but killed nothing. He came on to Osmaston to stay till Tuesday. We had a dinner party. The Hopes, Jewels, Mr Murphy, Mrs Nossel, Colonel Newton &c[66] ——

p 88 September 11. 1825

I had not been well some days: so took a little calomel and henna, which confined me to House to day. It was evening service, so Mr Bligh dined here s usual.

Monday. 12.

Simon and I were out shooting this morning for a few hours, but were neither of us up to much walking, and therefore killed nothing. I was much better this evening than I had been for some days.

Tuesday. 13.
Simon Bristowe left us this morning; I accompanied him with my Gun half way; he killed one bird on the way. I returned home with nothing but a pigeon. Upon the whole I like Simon B. better than before, as he is a good straight fellow, but rather opinionated I think. We dined to day at St Helens[67], where we met the Locketts &c &c. Not a pleasant party on the whole.

p 89 September 14. 1825
I enjoyed a days rest to day, as both myself and dogs rather required it. I was rather busy taking the layers from my carnations in the evening. Mr Farns=worth brought me eight new sorts at night, for which I paid him 7 shillings[68].
A Miniature Painter (Mr Cruikshank[69]) called this afternoon by appointment, and took a rough sketch of Frances Jane and myself.
Thursday 15
I was out shooting this morning, but it began to rain so hard, that I was glad to get home again with a wet skin. Mr Cruikshank called again today and got on a good deal. I dined at St Helens, where I met the Evans's Hopes &c.
Friday 16
I was at home all day, and sat for Mr C. in the morning. He dined here. Sam came in the Evening (he had returned from Burlington[70] the day before) and we played a rubber.

p 90 September 17. 1825
I was at home all this morning, and sat a good while for Mr C. who dined and drank tea here. This minature of myself began to be very like. F. J.s not so much so at present. In the evening I rode over to Elvaston. A good deal of rain.
Sunday 18.
Morning Service. A very wet day. I wrote a long letter to Miss Woodcock to day.
Monday 19.
It was a very wet day again. My sister sat to Mr Cruikshank in the evening and her miniature improved very much.
Tuesday 20.
I was out shooting today, but was very unsuccessful, as I only killed one bird. I rode over to Darley at night. Mr C. was here in afternoon. I found a man shooting at pigeons today and took his gun from him.
Wednesday 21
I sat all day with Mr Cruikshank. F. and A. Wilmot called this morning, and said they would shoot here tomorrow.

p 91 September 22. 1825
Frank and Arthur Wilmot breakfasted here, and we afterwards all went out shooting. We killed altogether 4 brace of birds a land rail and a hare and a brace of birds. They dined here, and returned home directly after tea.
Friday 23.
I was at home all this morning. We (i.e.) My Father, Mother, Eliza and self dined at Dr Simpsons[71], where I met John Every[72], whom I had not seen sometime.

Saturday 24
I went out shooting this morning, on the Moor this morning and found a bevy of Quails to my great surprise; I was not lucky enough to kill any of them. I shot three very fine birds, which I thought very good shooting for me, as I had only a single barrel. We dined at Mr Hadens tonight and met a very large party there indeed, among the rest Edward Strutt[15], who was just returned from Paris.

p 92 September 26. 1825
This morning I rode over early to Trusley to shoot. Arthur Wilmot met me at Shaws and we shot till four oclock in the afternoon. We killed 2 hares and 8 birds, of which I killed 1 hare and 5 birds. We went to Radburne[73] for dinner and I slept at Mr Shaws.

Tuesday 27.
I set out a shooting again today and out of nine or ten shots at birds, I killed eight. A. Wilmot shot with me about half an hour during which time he also killed one bird. I returned home to tea, and found all much as I left it.

Wednesday 28.
I received a letter from A. Power[74]. I sent two brace of birds to Uncle[75], and a brace to Mrs Darwin[76]. I was at home all day as both my=self my and dogs required rest. Sam came home to day and dined with us. I sat two hours to Mr Cruikshank in the evening. My Father went to Mr Needhams[44] to day.

p 93 September 29. 1825
I was out shooting at Osmaston all to day and killed to brace of partridges. Miss Fanny Strutt[5] spent the evening here. My Father did not return from Mr Needhams to day.

Friday. 30th.
I was at home all day expecting Mr Cruikshanks, but he did not come. F. Wilmot came over and stayed the evening. Mr Whelan and Sam drank tea here. It rained at night.

Saturday. October 1st.
I stayed at home this morning till eleven, when F. Wilmot came and we went out shooting a short time but killed nothing. We went to Darley[77] to dinner. I heard from Power to say he could not come to Osmaston for Music Meeting.

Sunday. 1.
Morning Service. Sam dined here. I recd a letter from Simon Bristowe to tell me he had bought a dog for me, which I was very sorry indeed to hear, as I had plenty, but fear I cannot help keeping it, as it is so.

Monday 2.
I was at home all to day, and sat the quarter part of the morning to Mr Cruikshank.

p 94 October 4. 1825
This morning was the first day of the Derby Music Meeting[78] and We most of us went to it (viz.) my Mother 3 sisters and self. We were all much pleased at the Music, tho' it was not in full band, as it was the day on which Service was performed. In

the evening 2 sisters and self went to the theatre to the Concert with which I was quite delighted as the singing by Madam Caradori[79], Misses Stevens, Wilkinson and Travis were beautiful, par=ticularly that of the first. We saw every body there almost, & among the rest The Tertius Galtons[80] who were staying at the Priory[81]. Bessy[82] seemed very glad to see us and looked quite as pretty and interest=ing as ever, without any more affectation or nonsense. We did not come home until near one oclock.

Wednesday. 5.

I did not go to the Church, as there was no ticket for me. I accompanied my sisters to

p 95 October 5. 1825

The Evening performance, and were quite as much pleased as at the prev=ious one.

Thursday 6.

I went with my sisters this morning to the Church to hear the Messiah, and certainly never was so delighted with any thing in my life. I went to the concert at night with Frances Jane, when we were equally pleased with the music.

Friday 7.

The Music Meeting closed today, with a ball, which was very well attended, as there were 450 present. We all went except Emma, and enjoyed it much as it was a very pleasant one. I danced all night, and was a good deal with Bessy Galton[83], of whom the more I see, the more I like. The Miss Darwins of Shrewsbury were also there with their brother Erasmus[83], and were very pleasant. We did not leave the Ball till past 3 oclock.

p 96 Oct 8. 1825

I forgot to mention yesterday, that Mrs Bristowe with her two little boys[84] had arrived here this night before. They seem the nicest children I ever saw, so perfectly under control, and yet so playful and natural. We were all much less tired this morning, than we expected to have been from the late hours we had lately kept. My Aunt Crompton[85] and two cousins called at Osmaston and they stayed a good while. They were more pleasant than usual I thought. We had all the party from the Priory with us to day and we had a very pleasant party. Soon after dinner Bristowe arrived. I never spent a pleasanter evening than tonight, and I much regretted my not being likely to see the party which made it so, for some time to come. Pon the whole this week has been one of the most delightful I ~~ever~~ < have> spent for some years.

p 97 October 10. 1825.

This morning I went over to Trusley to shoot. I was not very well there, so did not exert myself much, but as I did not miss any shots I killed 1 hare and 5 birds, which for the time f year was very fair.

Tuesday 11.

I was at home all to day, very busy gardening.

Wednesday 12.

I went with Bristowe today to Twyford[86] to shoot, and altho' we did not find a great deal of game, we had very pretty sport. We killed a hare, five pheasant and

3 birds. Of which I killed the hare and a brace of the birds. When we got home to dinner I found a M^{r} and M^{rs} ~~Sheppards~~ Bridgeman[87], very <much> old friends of my Father there, come to stay some days. I was rather pleased with them on first appear=ances, as they seemed very good natured cheerful people.

p 98 October 13. 1825
I was at home all to day, while Mr and Mrs Bridgeman were occu=pied in seeing Kedleston &c &c.

Octr 14. Friday
I was at Derby to day on a new mare of Bristowes. We had all the Strutts[88] of Derby and Belper to dinner with us.

Saturday 15.
This morning Mr Bridgeman and myself set off pretty early to shoot at Bristowes Osier hide near Repton, and had fairish sport there: I killed 2 pheasants, a cock and a hen, and D W George. We returned home to dinner. Sam Evans[89] and M^{r} and Mrs Jessopp[90] dined with us.

Sunday 16
Mr Bligh preached in the morning and stayed dinner as he had given up going to *Nerthon* any more.

Monday 17.
I was at home all today. The Miss Holdens[91] came to dine here and spend two or three days.

p 99 Octr 18. 1825
It was a very miserable wet day and we all stayed in the house all day. We had a dinner party in the evening. The Fosbrookes[92] and Holdens &c

Wednesday 19
The Miss Holdens left us this morning. It was a very uncomfortable cold day.

Thursday 20
It was a fine cold windy day and Mr Bridgman and myself went a shooting in Osmaston and about. The birds lay very ill and we only killed one which I killed and a crow as well. We all dined with Mr Joseph Strutt[93], where we met the Strutts of St Helens[94], and spent a pleasant evening.

Friday 21.
I walked over to Derby this morning with Mr Bridgeman and Bristowe. I wrote a long letter to φαννι . It was an excessively cold day. I had a bad cold

Saturday 22
Mr B. and I went in the car to Trusley to shoot but had very bad sport. He killed nothing, and I a pheasant hen & a pigeon.

p 100 October 22. 1825
We returned home to dinner. Mr Mill came to tea and stayed to a rubber rather latish. My cold was worse. M A left us

Sunday 23.
We had Evening Service. Mr Bligh dined here as usual and Sam who had returned the day from London. My cold was much better to day.

Monday 24.
I was at home all day gardening. They all went to Belper.
Tuesday 25.
This morning they all went to Elvaston. I stayed at home.
Wednesday 26
This morning I set off on Toffy to Beesthorpe, where I arrived after a very cold ride to a five oclock dinner and found all well.
Thursday 27
I was out shooting this morning but did not kill any thing. I went to dine at Mr Huttons with Bristowe, where we had a very good dinner and some delightful music by the Miss Huttons who sing delightfully.

p 101 Oct 28. 1825
This morning I set out shooting again and killed a brace of birds with hard work.
October 29.
As the hounds met at Mather Wood, we went there to see if we could shoot any deer. The keeper killed a buck, and I knocked a doe down but she got away. I also killed four hens. The Beak was so much out of Season as to be good for very little.
Sunday 30.
It was Evening Service and we all attended.
Monday 31.
This morning I left Beesthorpe in com=pany with Bristowe. We got to Osmas=ton about 6 oclock after a very plea=sant ride, tho' I was rather tired at last.
Tuesday Nov 1.
I walked over to Derby this morning. At night we received a letter from Cheltenham to give the melancholy intelligence that Mrs Bromfield[95] was so dangerously ill, that there was but little chance of her recovery.

p 102 Novr 1. 1825
tho' we had anticipated that she should not live long, we were much shocked to hear it so suddenly, as from the Medical Mans account it seemed all but impossible for her to live many hours. My Father took the intelligence better than I should have thought he would.
Wednesday 2
This mornings Post confirmed the melancholy tidings of yesterday. Mrs Bromfield had died on Monday morning. Few people had I should think lived a more virtuous or even life, and her death corres=ponded with her life. May my end be like hers, and accompanied by the same cheering reflections she must have experienced. Alas I dread much it will not be so. My Father had so fully expected it from yesterdays letter, that it was almost a relief to him to be relieved from his anxiety.

p 103 Novr 3. 1825
I finished packing up my things for Cambridge this morning in good time & came to day as far as Leicester by the Regulator. A couple of pictures which I had been entrusted to have at Leicester I forgot and left in the inside, and did not find out my

mistake till the coach had proceeded to London. I immediately wrote to London by the Guard of the Mail, but it put me in such a fidget, that I was quite unhappy about it. I left them all in great bustle at Osmaston, as my Mother and Eliza were going in a day or two to Cheltenham to attend my Aunts burial. I was truly sorry to be obliged to leave my Father as he would be left necessarily so much alone, and so his spirits were very low indeed. – I got some dinner at Leicester and went to bed early, but could get no sleep.

p 104 Novr. 4. 1825
I got up very early this morning and after breakfasting, proceeded on to Cambridge, by a new Coach, which went by Harbro' Kettring[96] &c, and arrived there at a little after four after a very pleasant days journey, as I had some very entertaining com=panions. I found all my old Friends at Christ Coll but Charlton[97] whom I was extremely sorry to find was gone into the army. I was quite happy to see all the others again. I got to bed pretty early.

Saturday 5th
I did not get up till late and then set out to find different men. Then were several freshmen arrived, whom I used to know well, and among them Lingard[98] and W. Walker[8], the former I found, but not the latter. I had one or two men to wine with me in the Evening, and then I went to Breynton's[99] rooms at Magdalen, & played a rubber there till about twelve when I got to bed.

p 105 Novr. 6th 1825
I got up very early for Chapel this morning, and afterwards took a longish walk with Brewin. Brewin, Wilmot and Wood wined with me. I got to bed early.

Novr. 7. Monday
I was very busy this morning paying bills &c. I wined with Lingard[8], and played a Rubber there till about 11.

Tuesday 8.
This morning I set out shooting by myself after breakfasting with F. Wilmot. I killed a hare and got home about 6, when Wilmot, Wood and Bond had a stake with me.

Wednesday 9
I was shooting this morning at Hornsey then after 4 shots I blew out both my touch holes and was obliged to return with only one Partridge. I dined with Wilmot.

Thursday 10.
It was a very rainy day. I attended the Divinity Lectures for the first time. I dined with Wood.

p 106 Novr. 11. 1825
I was at home all to day. W Walker wined with me.

Saturday 12
I went after lectures to day shooting with Evans in a Gig. We killed 3 birds of which I killed 1 and we both shot at another which we killed. I sent this morning a large packet of cakes &c to Julia. I also sent Mr Bridgeman a *brace* of birds. I dined with Brewin and I am ashamed to mention, got most dreadfully and awfully malt. I was obliged to be carried home by my legs and arms and put to bed.

Sunday 13.

I breakfasted with St John. Was very seedy all day. To get rid of which I got a horse and took a smart ride. I wined with Comb, but drank nothing as I was too much disgusted with wine.

Monday 14

I attended the Divinity lectures again to day. I did not go out much, as I was very stiff in my joints from my drunkenness.

p 107 Novr 14. 1825

I had six men to take a quick dinner with me at night.

Tuesday 15.

I again attended Divinity Lectures. I afterwards went to a Picture Auction, where I was foolish enough to lay out 2£ in oil Paintings. Packer and Robert=son wined with me and we had a quiet Rubber in the Evening,

I heard to day from Frances Jane; all at home were well. I also received a note from Mr Bridgeman, informing me that he had safely received the two pictures I had so carelessly lost coming here.

Wednesday 16th

I went out shooting this morning with *Coombes*, but it rained so *heavily* that we did not get a shot. I dined with Parker, and afterwards supped with Walton from whence I did not get till 2 oclock in the morning.

p 108 Novr 17. 1825

I attended lectures this morning, and as at home the whole day. I dined with Robertson in a great way.

Novr 15. 1825

I had a large wine party to day, chiefly of freshmen.

Novr 20. 1825

This morning after lectures I went shooting with Evans. He killed a brace of birds, I nothing. We had a mutton chop in my room and I afterwards wined with him.

Sunday 21.

I overslept myself this morning, & was to late for Chapel. I received a long letter from Julia, thanking me for the things I sent her. I wined to day with Parker.

Monday 22,

I attended Divinity Lectures to day for the 5th time. I wrote a long letter home. Dined with Hugh Wood, and supped with Will Walker. I began to day with a bad cold and cough, so that I could eat little & drink nothing.

p 109 Novr. 23. 1825

I attended Divinity Lectures this morning, and afterwards read Herodotus for some hours but did not make much progress. I dined with Robertson. We had a very flash party and drank immense=ly. 12 of us drank 35 bottles of wine and were none of us very drunk. I came over the wall with Evans at ½ past 4 oclock. I sent Miss Woodcock a Cottenham[100] cheese.

Wednesday 24.
I did not get up till 12 when I felt rather seedy and spent most of the day in walking about. I supped with Thorpe and got to bed in good time. I sent home some snipes.
Thursday 25
I was in my rooms most of to day after attending Classical and Divinity Lectures. I supped and played a rubber with Lingard.
Friday 26
I was walking about most of to day. I received a letter from Mr Bridgeman. I supped with Combe and got to bed in pretty good time.

p 110 Nov^r: 27. 1825
It was a miserable rainy day, and I sat a good Deal of it, with Frank Wilmot who was laid up. I wined with Brewin and afterwards got quietly to bed, as I had a badish cough.
Novr. 28 Sunday
I heard this morning from home and φαννι. All was well with the latter. The former contained the melancholy news of the death of Mrs Maling. She died of dysentery I attended Chapel twice to day. I had four or five men to sup with me on oysters.
Monday 28
I attended Divinity Lectures this morning for the 8th time.
Tuesday 29
I went to Divinity Lectures in the morning and in the afternoon supped with Stirrard of X^ts who had a large party.
Wednesday 30
I was in my room most of to day, and

p 111 Novr 30
was beginning a bad cold and cough. I supped with Sharp and a large party.
Thursday Decr. 1.
I attended Divinity Lectures again to day. I dined with Coombe who had a large party but my cold was so bad that I neither eat nor drank any thing. I got to bed early.
Saturday 3.
My cold got to day into my head, and was so bad that I resolved to keep to my rooms. I heard to day from home.
Sunday 4.
I did not get up till late, and then never stirred out my rooms all day. I heard from Miss Woodcock.
Monday 5.
I was in my rooms all day again. I wrote home and to Miss Woodcock.
Tuesday 6.
I was still in my rooms all day. I had Wilmot, St John, Wood Junr and Williams to dine with me on broth &c, all being invalids.
Wednesday 7
I was much better today, and got out a little. I dined with Arkwright, a very stupid party.

p 112 Decr. 8. 1825

Arthur Wilmot came over to Cambridge to see Frank. I breakfasted with them. I after=wards went to my 11th Divinity Lecture. I wrote to Julia and Jas Holden to day. I dined with Frank Wilmot, where I spent a very pleasant Evening, and stayed so late that I was obliged to get over College walls.

Friday 9.

I received a latter from home this morning, containing a 60£ note, which I began to be in great want of. I paid Mr Shaw his bill. I had a wine party to day; several men got very drunk; I drank no wine on account of my cough.

Saturday 10.

I was at home all to day. I wrote to Mr Bridgeman to day. I should come to him on Monday.

Sunday 11.

I attended Chapel and lectures to day. I wined with Power, and afterwards came to my rooms and was packing up till near eleven when I went to bed rather tired.

p 113 Decr. 12. 1825

This morning Bond and Welby break=fasted with me, and after that I set off from Cambridge in the Telegraph, for Cheshunt. It was a most miserable foggy morning, and threatened for rain. After a very uninter<est>ing ride, and a very cold one, I got to Cheshunt about ½ past 3, when I found Mr Bridgemans man waiting for my luggage, and I walked with him to M^{r} B's. I found neither he nor Mrs B. quite well, but very glad to see me. At dinner there was a Mr Mayo and two ladies. I liked the former very much; he seemed a very pleasant unassuming young man. After rather a dull Evening I got to bed and slept very well indeed.

Tuesday 13.

After breakfast this morning, I looked about the place, as much as the weather would permit me. After a good lunch, I left this hospitable place, Mr B. accompanying me as far as Waltham Cross, where I got into a coach and set off for town.

p 114 December 13. 1825

I got to London just at dusk; I immediately took a coach and went to the Tavistock Hotel in Covent Garden. I found Henry Metcalfe in the Coffee room there. I got some Coffee with him, and then went to Covent Garden Theatre with him. The Play was the Marriage of Figaro, in which Madam Vestris[101], Miss Paton and Miss Love performed. The first and last, I had not before seen. I was delighted with Madam Vestris, her acting seemed so natu=ral, and ladylike. The celebrated Monsieur Mazurier[102] performed the part of a monkey afterwards in a most amazing manner. I never saw a man so active, or one that had such perfect command of his limbs. This performance was on the whole, tho' very extraordinary, one which led one to disagreable reflexions, on our similitude to the monkey tribe.

p 115 Decr. 13. 1825

This morning after a very hearty breakfast which I took with Metcalfe, I took leave of him, and went to call on Miss Woodcock[103]. It took me nearly 2 hours to find her out, as she lived on the very extreme of London. When I once got there I could not get away, but was obliged to stay to dinner, and tea there. I then left there and after getting a cup of coffee at the Bedford Divan and a segar[104], I went into the Covent Garden Theatre at ½ price, & saw the end of the three strangers, and afterwards 2 after pieces performed in which Mr Mazurier performed again, & if possible, twitched himself about more than before. Madam Vestris sang a good deal, and delightfully. After the Theatre, I immediately returned to Tavistock and got to bed, very tired with walking about so much.

p 116 Dec^{r}: 15. 1925
I this morning immediately after breakfasting, walked to Charing Cross, to meet Julia. I had not been there very long before she arrived, looking very well, and very glad to see me. I then walked down to Miss Woodcocks with her, where I left her till Evening, while I went to get several purchases &c done. I got a chop at Affley's and then proceeded to Miss W^{s} to tea, and stay the Evening. I left there about 10 oclock, walked to the Hotel and got to bed much tired.

Friday 16th
This morning after breakfast, I went again to Miss W^{s} and walked about with Julia and Miss W. till 3 oclock when we dined, and about 5 Julia and I took leave of them, and walked as far as the Haymarket, fom whence we took a Hackney Coach, with our baggage down to the Bell and Mouth, and after having tea got into the Edinbro' and York

p 117 Dec^{r}: 16. 1825
Mail, and left London for Beesthorpe. We had as pleasant journey, as could be expected thro' the night, except that I was rather alarmed upon finding, that I had not money enough to carry us thro', I however managed till we got to Newark, where luckily I found James waiting for us, and he supplied us with money. Frances Jane was waiting in the carriage, tho' she had been up till 6 at the Newark Ball. She looked very well. We got to ~~Newark~~ <Beesthorpe> about 12 and found all well and very glad to see us. Boteler and Henry[105] grown immensely and as much improved in their man=ner and looks, as in size. Of course I did not go out much to day, as I was rather tired, and contented myself in merely looking round the House. All London and indeed all England were now in a great bustle, owing to many of the banks in London and the country having stopped payment, and the panic was so great, that nobody thought them=selves safe. Banks kept buckling daily every where[106].

p 118 Decr. 18. 1825
It was an unpleasant wet day. We all went to Church at Causton, in the afternoon, except Emma, & Frances Jane.

Monday 19.
I went out shooting today, with the Keeper in the Mather Wood, but had very bad luck, as he got all the shots. We killed 3 hares 2 pheasant 3 rabbits and a Woodcock. Of which I only killed A Rabbit and Pheasant. Bristowe was out this evening, so we

had quite a family party. It seemed quite curious for all my sisters and my self to be together at Beesthorpe, when we were so lately completely di=vided; M. A and Emma at Beesthorpe, Eliza at Cheltenham, F. J. at Osmaston and Julia at Bromley and I at Cambridge.

Tuesday 26.

As I was going this evening to a dance at Mrs Der Sœux's, my sisters would <make> me to go to Newark to have my hair cut, which I accordingly did on one of Bristowes horses; called

p 119 Dec$^{\underline{r}}$ 20. 1825

on Simon Bristowe[107], but he was not at home. I came back for dinner, about 2 oclock; M. A. and Bristowe, with F. J. and myself set off to M$^{\underline{rs}}$ Des Bœuxs. There was a very large party there and after the first stiffness was over I enjoyed myself very much, as I think F. J. did also. Sir Oswald Moseley with his son and 2 daughters were there, and I was introduced to him. His daurs were the most delightful girls, I almost ever met, so pretty, modest and unassuming. I was quite in love with them both, tho' I thought the youngest the Prettiest. We danced till near one, when we sat down to a very handsome supper, after which we commenced dancing again and did not leave off till about four oclock. F. J. danced the whole night. We all got home rather tired, about 6 oclock, and immediately got to bed.

Wednesday 21

We got up about 11 and after breakfast, I rode over to Newark with M$^{\underline{r}}$ Bristowe. We stayed there till very late, and did

p 120 Decr. 21. 1825.

but just save our dinner, after which we soon went to bed. I, this morning saw all the family of Bristowes at Newark for the first time[108].

Thursday 22.

This morning Bristowe and myself were to have ridden over to Nottingham Cricket Ball, but because it was rather a wet morning, he would not go, and tho' he offered me a horse and servant, I did not like to go, as I knew he would not like my doing so. I was much disappointed as I would have given any thing to have been there, that I might have seen the Miss Moseleys again. However I was forced to stay where I was, and amuse myself here instead. My sisters were very busy packing up, as all except Eliza left Beesthorpe for Osmaston in the morning.

Friday 23.

We, i e. Emma, F. J. Julia and myself, set off this morning as soon as it was light in the carriage, and I on

p 121 Dec$^{\underline{r}}$ 23. 1825.

box to Osmaston. We were all day in doing it, as the Roads were very busy, and got home, all safe, about ½ past 6 in the evening. I found my Grandmother, my Father, and Mother very well, and all looking, I think, better than I left them. After a good snack we all went to bed; I had the mortification to hear to day on the Road what an excellent Ball it had been at Nottingham &c &c.

Saturday 24.
This was the first Christmas Eve, I ever remember, that we were not all assembled together happy again and I hope will be the last for the present. We had nobody at night, and tho' we made an attempt to be merry at a game of Pope, it was not as it has used to be and I felt that Auld Lang Syne was gone never to return and that we were all growing old: and altered. I am sure I have altered more in mind in the last year, than if I was 20 years older. Oh, that I was as I had been, innocent and happy[109].

p 122 Decr. 25th. 1825
We had morning service. It was a very damp miserable day, and not at all like Christmas Day. Sam came to dinner.
Monday 26.
We were all at home all to day. The wet damp weather still continued.
Tuesday 27th
This was a very fine frosty morning. I went out shooting for a short time and killed a partridge.
Wednesday 28
We had a smallish fall of snow today and the frost seemed like
ly to break up.
Thursday 29
It was a cold frosty morning. Sarah Crompton[110] came to Osmaston to dinner and stay, till Saturday morning.
Friday 30
I walked over to Derby this morning, it was very dirty indeed as the snow was thawing. As New Years Eve w$^{\underline{d}}$ be on Saturday, the Derby Ball took place to night, instead of the 31.

p 123 January 30th. 1825.
My Mother, F. J. Sarah Crompton and myself went to it. It was not quite so good a Ball as usual at this time of year, but was still a very fair one as to numbers. I never enjoyed a Ball so little since I can remember going to them. I danced very little, Edward Coke[111] was there, in regimentals, having just gone into the 42nd. I saw Jas Holden. We returned home about 4 oclock in the morning.
Dec$^{\underline{r}}$ 31st. 1825.
It was a fine sharp frost. We were a good deal tired with our nights debauchery. Sarah Crompton left us.

p 124 blank

p 125 January 1st 1826
As it was evening service to day, Mr B. and Sam dined here. I asked M$^{\underline{r}}$ Bligh if he would read Herodotus with me during this vacation and he said he would be very happy. It was a very sharp frost.
Monday 2$^{\underline{nd}}$
We all drank tea this evening at S$^{\underline{t}}$ Helens, and while there, were very much shocked at hearing that M$^{\underline{rs}}$ Edward Hunt (who was staying with her father M$^{\underline{r}}$ J. Strutt) had

been attacked during dinner time by a very violent Paralytic attack, and was then laying in a very dangerous state. On account of this we came home sooner than we otherwise should have done. The Strutts were very much shocked by it indeed.

Tuesday 3rd

We heard a good account of Mrs Hunt this morning, tho' not so much as we had hoped. My Father, Mother, F. J. and myself dined at Dr Bents, where we passed rather a pleasant evening. Mr Jed. Strutt and the Fieldings were there.

p 126 January 4th 1826

It was a very cold frosty day. We dined at Colonel Newtons. Mrs Hunt was slightly better, tho' but very little so at present.

Thursday 5.

I went this morning to Mr Bligh, as he had appointed, when he told me, that upon receiving the thing, he found he could not find time, but referred me to a Mr Saunderson[112], who I went to, & liked very much. I took a lesson from him this morning. Bristowe came unexpectedly to tea, having had a very great escape; as the coach by which he came from Nottingham was totally overturned and many of the outside passengers were very much injured indeed. He being in the inside escaped with out any injury at all. Mrs Hunt was rather better again to day.

Friday 6

I was at home all to-day, reading & gardening. I wrote a long letter to Miss Woodcock. Bristowe was at Derby most of the day. Poor Mrs Malings corpse arrived at the Priory today from Valparaiso.

p 127 January 7. 1826.

I rode over this morning to Trusley where I shot two or three hours, but had very bad sport indeed, as I did not get a single thing, tho' I had 2 or 3 shots. I got home to tea, and went to bed early as I did not feel very well. Sam went to London.

Sunday 8th

Morning Service. I was reading all day. I wrote a letter to Simon Bristowe, to try if I could get rid of the dog I had bought of him in the beginning of the season, and sent it by Bristowe, who was going home next morning. Mrs Hunt much better. There was a good deal favour.

Monday 9th

I got up this morning at 6 oclock to breakfast with Bristowe and set him off on his return to Beesthorpe. At 11 oclock I went to Derby and took my second lesson from Mr Saunderson, whom I liked quite as much as before.

Tuesday 10.

This morning I set off from Osmaston about 12 with James to go to Mr Jed Strutts at Belper, for a day or two to shoot.

p 128 January 10. 1826

As it was a very cold frosty day, I only rode half way. I found Mrs J. Strutt very ill in bed. Mr. J. S. was very glad to see me and I spent a very pleasant sociable evening.

Wednesday 11.
After breakfast this morning I went out shooting (tho' there was a good deal of snow on the ground) with Freeman Strutt, and had very pretty sport con=sidering the weather. We killed 2 hen pheasants and 6 rabbits. I shot very badly indeed and only killed 2 rabbits. I was back to dinner and was very glad to go to bed, as I did not feel at all well, having a good deal of inflammation flying about me. Mrs H had spoke to Day and was much better.
Thursday 12th
As Mr Jed Strutt was coming this morning to Derby in the carriage, I took that opportunity of getting home. I walked from Derby in time for dinner, and found all, as I had left it on Tuesday.

p 129 January 13th. 1826
After breakfast this morning I rode over to Derby to receive some rents for my Father and to send a letter to poor φαννι I had written her enclosing a 2£ note. I got back to Dinner. The frost was now very severe indeed; more so than any we have had for the last to years I think. Fanny and Catherine Strutt came to tea.
Saturday 14th
I stayed at home all to day, indeed I was not at-all well, and had a good deal of pain about me.
Monday 16th
I walked over to Derby this morning and read several hours with Mr Saunderson.
Tuesday 17
I went this morning to Derby in the carriage with my Mother and Sisters, intending to see Henry Haden[113] but he was not at home, I therefore left a note for him, to say I should call on him the next morning. I was not at all well all day.

p 130 January 18th.
This morning after breakfast I rode to Derby, and was very glad to find Mr Haden at home. I consulted him abut two swellings (that I had twice before asked him about) in my groin, and he had told me, they were merely local inflamma=tion and would soon go of themselves. He now however found they were getting serious, and ordered me to take great care and keep quiet, to lay down, and to apply leaches to them as soon as I got home and to take some medicines he gave me. I read an hour or two with Mr Saunderson, and then, after getting some leaches and medicine went home, and immediately after dinner applied them, but they soon gave me very little relief. I told my Father Mother &c and took regularly to my sopha, and kept the inflammation down as well as I could with cold poultices[114].
Jany. 19.
I passed a very bad night, and was fully occupied all day with my medicines &c. &c.

p 131 January 20. 1826
M. Haden came over to see me and found me no better. He ordered me to apply 14 leaches. I heard from φαννι all were well there. I was in great pain all day, and worse at night.

Saturday 21st
M. Haden came again to day. I got rather worse, if any thing. <He solemnly assured me it was merely *inflammation*>
Wednesday 25th.
I was now regularly laid up, and never stirred from my bed room. This evening a grand Fancy dress Ball was given at Derby by the Yeomanry Officers. I was to have gone to it but of course could not, at which I was rather vexed, as I much wished to have seen so novel and beautiful a sight. Eliza came to day from Beesthorpe with Mr and Mrs Trebeck who came here for dinner to stay to day and tomorrow. The George Strutts[115] were also all here. My Mother, Eliza and Frances Jane in very becoming Spanish dresses, and all the visitors set off at 9 in spirits to it, and left poor I to myself.

p 132 January 26. 1826
The Ball had been most splendid. Betweeen four and five hundred people there, and the dresses, supper and general effect magnificent. Indeed every body seemed delighted with it. We had a dinner party to day, with guests in their dresses. Fanny Strutt was kind enough to come up and show me hers, which was one of the Merry Wives of Windsor, and was a very handsome one. Mr Haden was one of the party, and came up after dinner to see me, and he undertook me for the future instead of Henry, which I was rather sorry for.
February 10. Friday
The void between this and the last mentioned day was filled up on my part, with the usual accompaniments of sickness; Medicine, lots of leeches, pain in abundance, and plenty of time for reflection; the latter I hope will do me good. Poor Julia set off this morning early with Catharine Strutt for school.

p 133 Feb 10. 1826
I was to have accompanied them, to London and then gone to Cambridge[116], instead of which I was literally roaring with pain on my couch, as one of the swellings had now been trying to burst some days, but was prevented by constant bleeding with leeches &c. I had not been in bed some days, but my couch was made as like one as could well be done. My Mother sent my bills amount to Mr Shaw[117] and a certi=ficate of my illness. I was very much hurt to day by Mr Hadens telling me, that Bristowe (who was here) had told my Uncle in Derby that my illness was altogether venereal, and that he knew I had frequently had such before. The falsehood of it made me very angry, and I mentioned it ~~at Osmaston~~ to my Mother, which afterwards caused a great hubbub, which I was very sorry for, as it forestalled me in my intention of mentioning it to him myself, and got others embroiled in my business.

p 134 Feby. 13th. 1826
I had been by no means better the two preceding days, and continued much the same, and in a god deal of pain. Bristowe arrived here to night very unexpectedly.
Tuesday 14
I did not know until early this morning the reason of Bristowes sudden appearance here. I t was in consequence of what I had said to my Mother, on the subject which he

had mentioned to Uncle on the 10th and which I find she had mentioned to Eliza, and she again had written word of it to M. A. at Beesthorpe, who had shewn Bristowe the letter, and he had come over, very much hurt about my expressing such an opinion of it. I talked it all over with him, and he acknowledged having said as much to my Uncle, but said it was from a good motive, and he came over under the idea that it was my Mother who had made one of the expressions which I myself had used. We however agreed to think no more about it; I because

p 135 Feby. 14th. 1826

he said he had done it for my good, and he, because ~~he said~~ it had been much exag=gerated to me by both Mr Haden and my Uncle to him. After breakfast he returned to Beesthorpe with Eliza and Frances Jane in the Carriage; the former was only going to stay a week, and bring the children back with her to Osmaston, while Bristowe and Mary Ann went to London, and the latter was going to Town with them. This fracas put us all out for the day, as Bristowe's account had raised suspicion in both my Fathers and Mothers Minds that I really was unwell in the way in which he maintained, and I had great difficulty in persuading them to the contrary. On the whole perhaps it may prove a good thing to all parties as it made us all known to each other, and it certainly relieved me from some embarrassments, and I was very near disclosing the great secret which binds me down so, but could not muster resolution: would I had!

p 136 February 16. 1826.

I had had a very bad night, with a great deal of pain, when I was surprised early this morning to find myself much relieved, and I got some good sleep, which I afterwards found was occasioned by my right tumour having broken, which made me feel quite comfortable in comparison, tho I had of course a good deal of pain as yet.

February 18. Saturday

I kept rather mending. Eliza returned to night from Beesthorpe, with Boteler, and Henry[105], to stay here till M. A's return from Town.

Feb 25th. Saturday

I had been getting much better all the week, and was now able to sit up a little once or twice a day, which strengthened me very much, as I was uncommonly weak with laying so long, and losing so much blood. I was still obliged to apply leeches every now and then. I got out a little to day.

p 137 February 27th. 1826

I was out of doors about an hour to day and was much refreshed by it. I was also down satirs some time.

Tuesday 28.

I was out a long time to day, tho' it was not near so fine a day as the two or three last had been. I kept gradually getting stronger, and now left off leaches, as the inflam=mation was gone; I had now also plaister to my right side, instead of a poultice.

Wednesday March 1.
I was much better, and indeed began to feel myself get much stronger. It was a fine day on the whole, and I was out a good deal. I weighed my three hedgehogs to-day, and was surpris=ed to find they had lost so little weight during their residence in their hyber=naculum[118].
bid & Nat Cal.[119]

Thursday 2.
Splendid day. I felt much stronger & being out a good deal, did me much good I went to Evans.

p 138 March 3rd. 1826
A most beautiful day: quite summer. The birds and insects all on the alert. I was out a good while and enjoyed it much. Haden as usual.

Saturday 4th.
Rained a good deal in the night and in the morning likewise, but got fine towards twelve oclock, when I got out for an hour. I was much the same. Haden as usual.

Sunday 5.
A very fine day. Morning service. Sam came as usual. I got out for some time. I was much as before, but I hope, getting stronger.

Monday 6.
Very fine indeed. I was out a good deal. I wrote to F. Wilmot.

Tuesday 7.
A good deal of rain at night: day fine. I was busy out of doors a good while. Haden as usual. I am now certainly much more comfortable in every way. As I not only enjoy myself out of doors, but

p 139 March 7. 1826
eat and sleep well, at lest well for an Invalid, for all which blessings. I have every cause to be thankful, and enjoy them much more from having been so long deprived of them; but I still fear I am far from well, and I fear shall be long before I am perfectly so.

Wednesday 8.
It was cold in the morning, and rained a little, but got finer after=wards. Aunt and Uncle were here this morning. Haden as usual.

Thursday 9.
It was by far the most beautiful day, I think I ever saw for this time of year; it was more like June, than March. The bees covered every plant, and all the insect tribe were abroad, even butterflies. I was out a good while. I was not quite so well again to day as there was a slight return of inflamma=tion, two glands on the left side, which had ceased to be affected. I used a leach in consequence. My Father, Mother &c dined at Mr Hadens.

p 140 (smudged) Monday 10th 1826
Today was still more beautiful than the preceding, not a cloud, and except the want of leaves to the trees quite summer. I was out a good made deal. I was much the

same as yesterday and used two leaches in the evening, which made the number I had used altogether this illness 126, a pretty fair quantum. Mr Juncks of Repton came here this morning, and I was very glad to find had been keeping a pointer I had bought in Nottinghamshire, and thought I had lost, supposing it belonged to Bristowe. He agreed to keep it on for me, and do as well for it as he could. M$^{\underline{r}}$ H as usual.

Saturday 11.

This morning (which was very fine) I went to Derby (for the 1st time the last 8 weeks) in the carriage with Eliza &c my Mother. I sent to Mr Haden & then prevented his comig to Osmaston for the final time since he had attended me. I also saw Ed Strutt who had just

p 141 March 11. 1826

come from London. I got home about 1/2 past 12 and was rather tired with my exertions. It was very fine but nothing like the two preceding days as there was a coldish wind.

Sunday 12th.

It was a sharpish frost in the night and thus was a cold air all day, but the sun shone beautifully, and the south side of the house was quite warm. Mr Bligh dined here, as it was Evening Service, and Sam as usual. I was much the same. Mr Haden as usual but he had the modesty at last to find out that he need not come every day so is to miss tomorrow.

Monday 13.

It was a frost, but beautifully fine. I was out about an hour, & should have been longer, but had to write several letters which were to go to London by Pascal, viz a tre=mendously long letter to Liles in India, a letter to Julia, and notes to Bristowe M. A. and Frances, all which I accomplish by dinner time.

p 142 March 13$^{\underline{th}}$ 1826

We received a very poor account to day of M$^{\underline{rs}}$ Walter Evans, who is, I should fear dangerously ill at Dawlish. This would have been the first day of My Little Go, had I been at Cambridge this term. I was pretty much as usual to day. M$^{\underline{r}}$ Haden missed coming for the first time of his own will.

Tuesday 14.

It was a fine day but rather cold. I was out a good deal gardening &c. M$^{\underline{r}}$ Haden came to see me, and or=dered me three leaches, which I put on left side after dinner. Will Needham rode over to day to dine and slept here. M$^{\underline{r}}$ Haden and Mary also dined here. The Silk Mill at the bottom of M$^{\underline{r}}$ Hadens garden took fire this morning at 5 oclock, and was very much injured; it was supposed to be done malici=ously. It was worked by M$^{\underline{r}}$ Taylor but he having lately failed, this property, was in fact his creditors[120]. Sam dined with us.

p 143 March 15. 1826

Fine frosty morning. I walked out half an hour before breakfast. Mr Needham agreed to stay to day. I was at Derby all morning, but returned to dinner. I hope I am daily mending, tho' not very fast. Sam dined here

Thursday 16
Mr. Needham left us this morning before breakfast. Mr Haden came to day; I was much the same. It was a fine cold day. Mrs Evans much better
Tuesday 17.
Frost but a very fine day. I was out a good deal. Doll came from Mr Fosters today, with but a bad character.
Saturday 18.
Fine morning. Mr Haden came to day, and made an incision on the left side, which caused me much pain. At night we were all rather thrown into confusion by a packet of letters from Bristowe, to say that Eliza must be in London by Tuesday morn=ing to see to M As Marriage Set=tlement to which she was witness.

p 144 March 19. 1826.
It was morning service. I did not go, as Mr H said he should come, which he did. Eliza was very busy packing up. It was a very cold day. I was much as usual.
Monday 20.
Eliza left us this morning for London by the early Derby Coach. F. Wilmot called to day, and we talked over Cambridge affairs. I was out all morning. Mr Haden came. I was very busy Entomo=logising and found several sorts of Beetles in cow dung[121].
Tuesday 21.
It was a fine but cold day. I was out gardening for most of the morning. I was out gardening for most of the morning. I received letter from Julia today all were well.
Wednesday 22.
Fine but cold and threatening rain. I was out all morning. Mr H. came and gave me leave to take a glass of wine every day.

p 145 March 22. 1826
My Mother and Emma went to Derby this morning. We heard of Elizas safe arrival.
Thursday 23.
I was not quite so well to day, as I had rather more pain, owing I suppose to some little thing or other. It was a very cold day indeed, almost as if there was snow in the air. I was out a good deal. My Mother and Emma went to Thurlston. I finished potting most of my Carnations; tho' the weather is too cold for them, they look well on the whole.
Good Friday 24.
It was a very cold day, but as I felt well I went to church in the afternoon, but found it colder than I had expected. Sam and Mr Bligh dined here. Mr. H. came
Saturday 25.
I arose this morning with a very bad headache, and very luckily Mr Haden happen-ing to pass Osmaston called on his road, and told me that I had taken a bad cold, and must be quiet. I was very full of fever and unwell at night.

p 146 March 26. 1826
I was in bed all day and very little if any better, as I was very full of fever, headache, and was unable to bear the slightest light. Mr Haden came of course.

Monday 27
I was still unwell, and kept my bed entirely, untill
Thursday 30.
When I was so much better as to get up, about 12 oclock and stay up till evening in my room. Hugh Wood called this morning and stayed with me about an hour. I was left very weak again by this fresh illness. My old affair is much as before.
Friday 31.
I continued better and got up as usual, and was down stairs all morning. Mr H as usual
Saturday April 1.
I was pretty well this morning, and tho' it was <not> a very fine day got out for an hour. Heard from London; all well.

p 147 April 2nd. 1826
I continued improving. Sam and Mr Bligh dined here. Mr Haden came. I did not get out to day.
Tuesday 4
I still continued rather feverish, which hindered me from getting up my strength as I could wish. Mr Haden came to day. My Father Mother and Emma went in the Evening to the Mint Ball; it was a very full one, but rather a stupid one.
Wednesday 5.
My Mother was in bed all day with a bad headache. The Weather very fine indeed. I was out a good while to day.
Thursday 6
Weather beautiful. I was very busy gardening, and was out all morning. My Mother was pretty well again. My Dr. Mr H. came this morn=ing and finding me still rather feverish, ordered me to take Fever Draughts for a day or two.

p 148 April 7. 1826
A very fine day. I go on (I hope) very well indeed. I was out rather too much this morning I think, but I was so very busy finishing carnation's potting, and getting my Garden in order with Wheeldon. We had a long letter letter from Eliza to day; and rather a poor account of them all, as they had colds &c. I sent Bristowe his two pedigrees this afternoon, as he wished for them. I finished to day reading Wiffens translation of Tassos Gerusalem liberata[122], and was so much pleased with it, that I feel almost determined to attempt learning Italian, when I have ano opportunity.
Saturday 8.
Mr and the Miss Hadens dined here to day in a quiet way. I was much better.
Sunday 9
Morning service. Sam dined here. I continue very prosperous. I wrote to Eliza a long letter. My Mother

p 149 April 9th 1826
was not very well; she had had a headache, which left a good deal of fever hanging about her.

Monday 10
We had a little rain in the night, which begins to be very much wanted. My Mother was rather better. Mr Haden came to day, and found me not quite so well as I had been lately.
Tuesday 11
I was not at-all right this morn=ing when I got up, but felt better afterwards. I went with my Mother and Emma to Derby in the carriage and did not feel at-all tired, which I rather wonder at. Indeed I hope now I shall soon be strong enough to get to Cambridge again.
Wednesday 12
Mr H. came this morning, and found me much better.
Thursday 13.
This morning I went to Derby in the carriage and felt much better for it I think.

p 150 April 14. 1826
I this morning rode over to Thurlston[11] in the Gig, and stayed there for some time. My Aunt was in excellent spirits tho not quite well. I got back to dinner after a pleasant ride.
Saturday 15.
I stayed at home this morning, on account of Mr Hadens coming. I hope to day will be his last visit, as I have agreed to go to breakfast with him on Monday.
Sunday 16.
It being Morning service, Mr B. did not dine here. Sam did. A beautiful day.
Monday 17
I rode over in my little Gig to breakfast with Mr Haden. He pronounced me going on swim=mingly. I returned home about 12 very little tired.
Wednesday 19
I breakfasted again with, Mr H. Went on very well. My Aunt Darwin[123] was this morning deliver=ed of a very fine boy and both

p 151 April 19. 1826
did very well.
Thursday 20
I this day was at home, preparing my Garden &c &c for my departure to Cambridge, as I hope to be able to go there next week.
Friday 21.
This morning I again breakfasted with Mr Hadens, who found me still going on perfectly well. I returned home early, and took Boteler and my Mother a ride. I feel my strength returning almost hourly. Mr Mills came and drank tea here, and stayed the Evening.
Saturday 22
I went over this morning to Thurlston[11] with Emma in the Gig, and had a very pleasant ride. I saw my Aunt who was surprisingly well, and also the baby which seemed a very sprightly one[124]. I was very busy in the afternoon preparing my cloths &c, as I am in hopes to go to Cambridge on Monday.

p 152 April 23. 1826
I this morning I am ashamed to say, completed my 21st year. There is abundant cause for much reflexion and that humiliating & painful in it. However I so busy in packing up &c that it entered much less into my thoughts than it otherwise would. Sam dined here as usual. I wrote to φαννι and sent her 2£ to day.
Monday 24
I finished packing up this morn=ing and after dining at 2 oclock, with rather a heavy mind set off by the Regulator to Leicester inside, where I arrived very safe about 7 and after eating a mutton Chop, went to bed, not near so much tired, as I expected I should have been with my journey this far.
Tuesday 25.
I got up this morning at five oclock, and after breakfasting &c set off at seven by the Cambridge Coach to Cambridge;

p 153 April 25. 1826
I called upon Mr MacGuffay, at Stamford, and got some dinner with them. It was a fine tho cold day and I got into Cambridge about 6 oclock, feeling pretty well. I found all my friends well, and apparently very glad to see me again. My room was full of callers all evening. I got to bed soon after ten.
Wednesday 26.
I had all my old friends about me this morning, and felt pretty well as well. After dinner I thought it would be safe to put 2 leaches, on my right side, as I perceived a hardness there. Just as I took them off, who should knock at my door to my great surprise but Bristowe and Frances Jane, who had just arrived in Cambridge, on their way to Beesthorpe from London. I spent the Evening with them and M A at the Inn. They all then looked uncommonly well, and we spent the Evening

p 154 April 26
very pleasantly together, tho' I did not feel quite in spirits.
Thursday 27
This morning I breakfasted with Bristowe and my sisters at the Inn, and afterwards took F. J. about Cambridge to see the Lions. M. A. was too prudent to go with us. They afterwards lunched in my rooms on some Lobster and then set off again for Beesthorpe in a very disagreeable rain. I kept still all Evening as I felt I had already done rather more than I ought.
Friday 28
I did not feel at-all right this morning and my right side hurt me a good deal, so I kept very quiet. It was a most wretchedly cold day. I went to Chapel in the Evening for the first time.
Saturday 29.
I this morning, as I was no better got leaches, intending to apply

p 155 April 29. 1826
them but could only get one to bite. In the afternoon I got two more which did very well indeed, and I hope will do me much good. I went to Chapel.

Sunday 30.
I found on getting up this morning that a hardness I had felt in my right side had suppurated and discharged which relived me considerably. I went to Chapel again to day morning and Evening. It was still very cold indeed.
Monday May 1
This morning I wrote 2 letters, one to Mary Ann, and the other to my Mother and enclosed them in a basket of Cottenham Cheese[100] I was to send Mary Ann. I paid several small bills also this morning. I was quite as well to day, tho' there was more matter from each side than I liked.
Tuesday 2.
It was a very fine sunny morn=ing, and as I knew several men

p 156 May 2. 1826
who were going to see Ely Cathedral and I felt pretty well, I took a Gig with T. Bond and accompanied them. I was very pleased with the Cathedral indeed. The im=mmense piles of building, and its great variety of architecture, from the earliest Norman downwards, make it very interesting. I was however rather too tired with my journey to perfectly enjoy it as I should, if I been well. We dined at Ely, and got home very comfortably about 8 oclock. The Party Drank tea with me, and I got to bed about 11 very tired (tho I hope no worse) for my days journey.
Wednesday 3.
I was much pleased to find myself no worse for my yesterdays jaunt. It was so very cold to day (a N. E. Wind) that I never left my rooms all day.

p 157 May 4. 1826
It was so cold again to day, that I still kept at home, and never left my rooms. I had a letter from my Father to day. All were well. I was not myself quite so well to day, as I felt rather bilious.
Friday. 5th.
It was not quite so cold to day, so I ventured forth, in hopes it would bring me about a little, as I felt rather queer. I wined in a quiet way with Thos Bond, where I met Power.
Saturday 6$^{\underline{th}}$
I heard this morning from φαννι, where all was well. It was still cold.
Sunday 7th
I went to Chapel twice to day. Wined with Chapman, who I like much, as a new acquaintance.
Monday 8.
A beautiful day. I was out all morning, enjoying it. At night I wrote a long letter to Julia, another to Emma and a third to Miss Woodcock.

p 158 May 9 1826
I got up early this morning, and dispatched a whole box of good things off to Julia's. As it was a beautiful day I was very busy walking the whole of it with Wood till dinner, and Evans, Bond &c afterwards. I certainly am now gaining strength fast,

but do not find my glandular swellings get any better which discourages me much; I heard from my Father this morning who is in London. Bristowes Act of Parliament is nearly concluded[125].

Wednesday 10

I felt pretty well this morning and as it was a beautiful day walked out a good deal. I wined with M. Bond.

Thursday 11

I was not at all well to day and did not stir out much. It was a very nice day. I got Donovans Insects[126] from University Library.

p 159 May 12 1826

As I did not feel quite so well this morning, and my right side was rather more inflamed, I went and consulted Dr Haviland. He gave me some drafts to take and ordered me to be quite quiet and to poultice my sores.

Saturday 13

I was perfectly still all day. Went out in a Gig (as Dr H. had recommended me) with Jas Holden. Was very full of pain indeed. The weather was very hot. Dr H. saw me today.

Sunday 14

I was on my sopha all day, and very full of pain indeed. The weather was very hot. Dr H. saw me to day.

Monday 15

This morning the gland on my right side burst, which relieved me very much indeed. I took a drive out with St Johns, which tho' it tired me, did me good I hope. I wrote a long letter home, about going to the sea &c &c.

p 160 May 16. 1826

I was much better this morning. Had a long letter from Julia and Miss Woodcock. Took a drive with Jas Holden. A very warm day indeed.

Wednesday 17

I continued getting better. Took a drive this morning.

Thursday 18

Better. Was out a good deal in Gig to day. Had along letter from home, all well.

Friday 18

I continued improving. I drove out with St John this morning. Dined in Hall. Took a Gig for the coming week for 2" 10"0. Paid my subscription to Union up to this time.

Saturday 20

Continued much better. Dr. Haviland saw me to day. I was out in Gig some time with Jas Holden and took several insects, I had not before seen.

p 161 May 31. 1826

I attended Chapel twice today. It was not quite so hot as it had been lately. Took a drive in Gig as usual. Wrote a long letter to Eliza, and sent it.

Monday 22nd.
I was uncommonly well to day. I went up River in a boat as far as Trumpington with Bond, Evans, St John &c and did rather too much as I felt a good deal tired in the Evening.
Tuesday 23rd.
I was not quite so well this morning for my labours. The two Bonds, and Williams left Cambridge to day. I did not go out in my Gig to day.
Wednesday 24
It was a very cold day, and I accordingly staid in my Room.
Thursday 25
It rained most of today. I was much as usual. Cambridge began to grow very thin indeed of gowned men. I fear I shall soon be severely alone.

p 162 May 26th. 1826
I did not feel quite so well to day. I received a long and kind letter from home telling me that they perfectly agreed with Dr. H. about my going to the Sea[127] &c. I answered it by return of Post. St John took me on drive in the evening. The weather was delightfully fresh after the rain which had fallen.
Saturday 27
This morning Evans left Cambridge for this term and St John for 10 days. I took a drive in the Evening with Harper to see a well at Madingley, which we were told had a putrifying power. We were a good deal disappointed when brought to it, as it was merely a small bricked well, tho' I had no doubt from the sediment of the water that it had the power ascribed to it. There were very few left now at this College.

p 163 May 28. 1826
I got up in good time this morning; attended Chapel. A miserable day, and I likewise rather so. Cambridge very dull indeed. I was not so well. Wrote to Miss Woodhouse.
Monday 29
A miserable, cold, raining, morning. I was far from well, at least, to my own feelings, but they may deceive me.
Tuesday 30 Wednesday 31
Miserable raining day. I was in my rooms all day, and as I did not feel near so well, sent for Dr H. who said it was only a little indisposition.
Thursday June 1.
Received and answered a very long letter from home. Also one from Miss Woodcock. A much finer day, and I better in proportion. Every man has gone, I know, except Thorpe and Chapman.
Friday 2
A finish day. Chapman took me out in his Gig. St John came back to stay till he takes his degree.

p 164 June 2. 1826
I went into the University Library this morning and saw there a splendid Edition of Buffons Histoire Naturelle des Oiseaux in 14 vols folio, with most splendid coloured engravings. It has just been bought for 70 Guineas[128]. F. Wilmot left Cambridge.

Saturday 3.
It was a very pleasant day. I only went out in the Town a little. Dr Haviland called on me to day, and I think I shall not want him any more, at all events, hope so. I was not very well to day.

Monday 5
It was a splendid day. I took a drive with $S^{\underline{t}}$ John, and was very busy Entomologising. I caught a large hairy caterpillar on the Phalona Potatonia[129], which while I held it in my hand, tho I had gloves on, either from some liquid it exuded, or from its fine hairs working in, hurt my hand a good deal – much the same feeling as if I

p 165 June 5. 1826
had been stung by Ants

Tuesday $6^{\underline{th}}$.
It was a fine day. I was pretty well, but did not go out much. My hand itched and swelled good deal.

Wednesday 7.
This morning St John took his degree and I had him, Chapman and 3 others to dine with me.

Thursday 8.
It was a beautiful day. I was not quite well. I dined with Chapman in a quiet way. I received a letter from home enclosing 50£, and also a large parcel containing things for my journey to the sea.

Friday 9th.
I had a good deal of headache all to day, and stirred out very little. I wrote a letter to my Mother. a fine day.

Saturday 10th
A beautiful day. I took a drive with Chapman in the Evening. I was better to day than I had been some time –.

p 166 June 11. 1826
It was a beautiful day. After Chapel I was fully engaged in packing up things I meant to send home and take with me to the coach. I mean to start tomorrow if fine.

June 12 Monday.
I left Cambridge this morning at 9 oclock by the Union and proceeded to London. It was most intensely hot day, and I could scarcely bear it on the outside with an Umbrella. We had a great traveller on the Coach. Who amused us much. We got to London about 5 oclock. I immediately went to Hatchetts, Piccadilly, where cleansing myself from the dust and dirt, I got a stake, which I enjoyed much. I then went to Bond Street and ~~ordered~~ <bought> a hat, and ordered a coat from Stutly. I next got into a H. Coach and set off to my Uncles[130] Lodging, in Norfolk Street, Strand. He was not at home, but after I had spent an hour in

p 167. June 12. 1826
Exeter Change (where I saw nothing new, and regretted the melancholy fate of its late Elephant) I again called and found him at home. He was very much surprised to see me, and wanted me much to sup with him and spent the next day in

London with him, which however I determined not to do, tho I should have liked it, as I thought London in the present very hot weather, would not suit me in my present weak state, and my nights restlessness confirmed me in my resolution, as I spent a very feverish night owing to the heat and closeness, and slight cold I had caught.

Tuesday June 13.

I got up this morning in a great bustle at 8, as the Coach was to start at ½ past, got a hasty breakfast and then mounted it, and had a much pleasanter ride, than I expected as there was a pretty brize, which rather took away the excessive heat.

p 168. June 13. 1826

We passed thro Kensington, Hammersmith, and close to end of Windsor Park (when we caught a slight glimpse of Virginia Water) thro Bagshot, Basingstoke, (where we leave the Salisbury Road and Winchester, which I regretted much we did not go near the Cathedral, or stop a minute to allow of running down to it, and at length at 6 in the Evening arrived safely at Southampton. The road is rather stupid and flat for the first 30 miles, but very pretty afterwards, as the cultivation, in some parts is very rich, and the heaths and Wood Scenery, beautiful. We pass close to Sandhurst Military College, which is a handsome building. Near Basingstoke are several Roman remains, and many ruins made during the Revolution. I put up at Southampton, where the Coach stopped, as it was very near the Quay (the Royal George)

p 169 June 13 1826

where tho small, I was pretty comfort=able. I wrote a long letter to my Father from here this evening, and afterwards took a walk in the town which seems a nice clean one in the main streets. As I intend coming back here from the Isle of Wight[131] and was a good deal tired I did not see anything this evening, indeed in the town I believe there is nothing worth seeing, but near are several beautiful walks in the New Forest and more particu=larly Netley Abbey[132], which is a splendid Ruin. I got to bed very early, but being rather feverish, after travelling and the night being very hot, and my room small, I could not get a wink of sleep ~~all night~~ untill it was quite morning.

p 170

My weight	1826
Apr 28	9 · 9
May 7	9 · 11
May 27	9 · 12
June 6	9 · 13
1827	
March 2	10 · 6 with Gt Conbon[133]
June 8	10 · 4

Notes

1. Fanny – see Chapter 2, Note 63.
2. Bligh, the clergyman at St Mary's Chapel, Osmaston (not to be confused with St Martin's Parish Church at Osmaston, near Ashbourne).
3. Miss Woodcock is mentioned in earlier entries of the diary in 1824 (e.g. Tuesday 9th November; see also Chapter 2, note 62)
4. Osmaston Hall, the family residence, just outside Derby, of Samuel Fox III (see Chapter 1, note 20).
5. Likely to have been Frances Strutt, the 3rd daughter of William Strutt (see note 15) and granddaughter of Jedediah Strutt I.
6. Frank Hall, an undergraduate at St John's College; see note 38, Chapter 2.
7. Mr Haden, physician to the Fox family.
8. Francis Sacheverell Wilmot, the son of the Rev Edward Wilmot; Arthur Wilmot: not identified – one of the many Wilmots in the Derbyshire area, a major branch of the family having its seat at Chaddesden Hall, another at Osmaston Hall (until 1815) and another at Trusley; see Chapter 2, note 59).
9. Fox's dog.
10. Not identified
11. Thurlston Grange was in the grounds of Elvaston Castle and had previously been listed as the residence of Samuel Fox III, see Chapter 1 and note 17. At this time it seems that Fox's grandmother, Jane Darwin née Brown, resided there with other dependents of the Foxes (see note 123). Both were in the parish of Elvaston, which is now on the boundary of the city of Derby.
12. Miss Woodcock, probably a governess to the Fox sisters; see Chapter 2, note 62.
13. Not identified.
14. The stage coach to London.
15. Edward Strutt (1801-1888), son of William Strutt, the inventor, and grandson of Jedediah Strutt. Second cousin of Fox and at this time a young barrister in London. Became 1st Baron Belper. Liberal MP for Derby. (*Burke's Peerage, Biographical Register*)
16. Exeter Exchange, popularly known as the Exeter Change, was a building on the north side of the Strand in London. It was demolished in 1829 and the site is currently occupied by the Strand Palace Hotel. It was most famous for the "Menagerie" on the upper floors, which competed with the Royal Menagerie at the Tower of London, before the London Zoological Gardens were opened in 1828, by the Zoological Society of London.
17. Fox's brother, Samuel Fox IV; see Biographical Register.
18. Edmund Kean (1789-1833), considered to be the best Shakespearean actor of his time.
19. Adelphi Hotel or Theatre. On the 1st August (below) the Fox family booked, impromptu, into the Adelphi Hotel, but Julia was bitten by bed bugs and the family hastily removed to another abode. The Adelphi Hotel is likely to have been in the Adelphi Buildings (a Robert Adams building, now demolished), opposite the Adelphi Theatre (near The Strand, London).
20. Emma Fox (1803-1885), an elder sister of Fox.
21. Mr Bromfield, a friend of the Fox family. Possibly William Arnold Bromfield who is referred to in Letter #202 (n3).
22. Frances Jane Fox (1806-1852), a younger sister of Fox.
23. Julia Fox (1809-1896), Fox's youngest sister.
24. Burlington Arcade; built in 1819 by Samuel Ware for Lord George Cavendish as an arcase of small shops for the sale of "jewelery and fancy articles of fashionable demand". It was situated between Piccadilly and Burlington Gardens, behind Bond Street. Samuel Fox III operated one of the shops to display silk stockings and other silk apparel. The shop was under the charge of Samuel Fox IV: see note 61, below.
25. The London Diorama was based on the "Diorama", Paris, of J M Daguerre. It was situated along the east side of Park Square, Regents Park, London. The aim was to produce naturalistic illusion for the public employing huge pictures up to 25 × 15 m in size. The custom was to display 2 dioramas at a time, one military and one landscape. The "Holyrood Chapel" referred to here is likely to have been "The Ruins of Holyrood Chapel" situated near the monastery of Sainte Croix, France: a diorama of "Holyrood House Chapel, Edinburgh" was not displayed until 1827.

26. Bullocks Columbian Museum. William Bullock made collections in Mexico of Pre-Columbian artefacts and casts of Aztec treasures in 1823. These were displayed in the "Egyptian Hall", Piccadilly in an exhibition entitled "Ancient and Modern Mexico". It has been hailed as the first exhibition of Pre-Columbian antiquities anywhere in the world. Later in 1825, the collection passed to the British Museum.
27. A steam-powered boat.
28. Ramsgate, a seaside resort on the North Sea side of eastern Kent.
29. Sir William Curtis, Lord Mayor of London. He built Cliff House at Ramsgate from where he operated his luxury yacht (see entry for July 28th, below).
30. Mary Ann (1800-1829); Fox's eldest sister, married to Samuel Bristowe (see next note).
31. Samuel Ellis Bristowe of Beesthorpe Hall. Married to the eldest of the Fox sisters (see preceding note).
32. Sir Walter Scott, novelist.
33. Eliza Fox (1801-1886), an elder sister of Fox.
34. St Laurence church (Church of England) is the oldest church in Ramsgate. It has a number of old gravestones, but the brasses within the church seem to have disappeared before 1825 (Kent Archaeological Society).
35. A business associate of Samuel Fox III.
36. Not identified.
37. Harry Bromfield. Possibly a son of the Bromfield mention above (see note 21).
38. Corbetts Museum.
39. The Soho Bazaar (London), the first of its kind, was formed in 1816 to help relatives of soldiers killed or injured in the Napoleonic wars.
40. Possibly Frank Wood, see note 4, above.
41. Not identified; possibly related to Elizabeth de St Croix, who was Fox's Aunt. Charles St Croix was mentioned as visiting Fox in Cambridge (Chapter 2, note 100).
42. Not identified, but the Mayors are mentioned several times in the Fox diaries (1844-1864).
43. Possibly Francis Needham who became the 2nd Earl of Kilmorey in 1832. Mr Needham was a business friend of Samuel Fox III and Samuel Ellis Bristowe. The latter married his sister Lady Alicia Mary Needham, as his 2nd wife, in 1836 (*Darwin Pedigrees*).
44. Likely to have been Jedediah Strutt, Jr (see Chapter 2, note 85), but there were many Strutts living in Derby at this time, the sons and grandsons of Jedediah Strutt I (see note).
45. Not identified. There were many close relatives of the Foxes in and around Derby at this time.
46. Kedleston Hall, the home of the Curzon family and built from 1759-1765 with a major input from Robert Adams.
47. Quarndon is a village 4 miles due north of Derby, next to Kedleston Hall. It was noted for a chalybeate well, which is now inside the grounds of Kedleston Hall.
48. John Bell Crompton (1785-1860). Banker of Irongate, Derby and uncle of W D Fox. s John Crompton (1753-1834; Mayor of Derby,1792,1800, 1810, 1817 & 1827) and Elizabeth Fox, dau Samuel Fox I. Cousin of Sir Samuel Crompton, who was member of Parliament for Derby from 1830-1834 (and then for Thirsk). m Jane Sacheverell Sitwell. (Note that Samuel Crompton I, the grandfather of J B Crompton married Eliza Fox, dau of Samuel Fox I. A son of Samuel Crompton I, Samuel Crompton II, and therefore an uncle of J B Crompton, married, Sarah Fox, a sister of Samuel Fox III.)
49. Not identified.
50. Cleark; possibly a family functionary. This person is mentioned on several occasions.
51. The reference seems to be to the affair with Miss Woodcock, which abruptly came to an end a few days later on 31st August (see note 59, below).
52. Trusley is a village on the outskirts of Derby. It was the ancient seat of the Coke family who owned much of the land there, but the Rev Francis Wilmot was the owner of the Manor and the Rector of the village. Samuel Fox III owned property in the village.
53. Frank Sacheverell Wilmot, son of Rev Edward Wilmot – see Chapter 2, note 59.
54. Not identified; presumably a dependent of the Foxes.

55. Elvaston Castle, the residence of Lord Harrington. There were a number of houses on the estate including Thurlston Grange, where Samuel Fox II had his residence before moving to Osmaston Hall in 1814 (see note 11, above).
56. Swanswick Hall, the residence of Rev John Wood, the father of Hugh Wood, a college friend of Fox (see Biographical Register).
57. Not identified.
58. Not identified.
59. Miss Woodcock was probably a governess at Osmaston Hall. Fox appears to have been having an affair with her and she apparently returned to London in some disfavour. Fox and his sisters visited her, in London, at least once later on (see Appendix 5).
60. Hugh Wood: see note 43, above, and Chapter 2, note 41.
61. Not identified.
62. James Holden; a friend of Fox from Repton School and Christ's College (see Chapter 2, note 52).
63. Arthur Wilmot may have been a brother of Frank Wilmot, see note 39, above.
64. Samuel Fox IV, Fox's brother.
65. Simon Bristowe (1806-1868), a brother of Samuel Ellis Bristowe. The father , mother and siblings of Samuel Ellis Bristowe had come back from New York at this stage. They lived for some time at Twyford Hall, Derbyshire (see note), but then moved on. On 20th December 1825 (see note 107, below), Fox visited Simon Bristowe in Newark, which suggests that the family may have moved there.
66. The people at the dinner party have not been identified.
67. St Helens House; a large mansion in King Street, Derby, built in the Palladian style in 1726. In the early 19th century it was the home of George Benson Strutt (see note 94, below).
68. Carnations; see entry for 24th August 1825.
69. Mr Cruikshank. Clearly an artist, who made a pencil drawing of Fox over the next ten days. This drawing does not seem to have survived.
70. Burlington: likely to refer to Burlington Arcade in London, where Samuel Fox had a shop for his silk products (see note 24, above).
71. Not identified.
72. Not identified.
73. Radburne Hall, near Derby; an elegant Georgian house built in 1735 and extended by Colonel Edward Sacheverell Pole. The latter married Elizabeth Collier, and their son Edward Sacheverell Chandos-Pole, a 1st Foot Guards officer during the Napoleonic wars, resided there in 1825. Elizabeth Collier, became the second wife of Erasmus Darwin, who at this time resided at the Priory at Breadsall, some 10 miles north of Derby.
74. Power; possibly related to J. Power, who was private tutor to Fox in Cambridge.
75. Not identified. Fox had several uncles in the Strutt, Darwin and Crompton families.
76. Possibly Jane Darwin, the widow of William Alvey Darwin of Elston, who probably resided at Thurlston Grange after ca 1817 (see note 123) or Elizabeth Darwin, née Collier (1747-1832) of Breadsall Priory, about 5 miles north of Derby; she was the widow of Erasmus Darwin.
77. Darley Abbey and Darley Hall, a fine Georgian mansion, lay on the River Derwent on the outskirts of Derby; they belonged, at this time, to the mill-owning Evans family, but it is likely that these were shared with the Strutts, since two of the offspring of Jedediah Strutt Sr (William and Elizabeth) married offspring of Thomas Evans.
78. The annual music meeting in Derby, which Darwin attended in 1828 [see Letter #5 (n6)].
79. Maria Catarina Rosalbina Caradori-Allan (1800-1865). Operatic soprano from Alsace.
80. Samuel Tertius Galton (1783-1844) and family (*Darwin Pedigrees*).
81. Breadsall Priory, see note 77, above.
82. Elizabeth (Bessy) Ann Galton, the daughter of Samuel Tertius Galton (note 80, above). Fox and Darwin were very attracted to Bessy during the years 1827-1831 (see Letters #6 (n7), #7 (n10), #9 (13), #27 (n4), #45, #64A & #87).
83. It is clear that in 1825, Darwin's unmarried sisters (Caroline, Susan and Catherine) and brother attended the Derby music meeting.

84. Fox's sister, Mary Ann Bristowe with her two sons, Samuel Boteler and Henry Fox Bristowe.
85. Likely to have been Jane Sacheverell Sitwell; see Note 49, above.
86. Twyford Hall, on the banks of the Trent River, Derbyshire. This was one of the seats of one branch of the Bristowe family. The two seats were joined by Samuel Bristowe I when he purchased Beesthorpe Hall near Nottingham from a relative. On his death in 1818 he left all his estates not to his nephew Samuel Bristowe II, but to his great nephew Samuel Ellis Bristowe. On learning of the outcome of the inheritance the whole family returned to England from New York and Bristowe Sr and the family spent some years at Twyford Hall. In 1825 they were probably elsewhere.
87. Mr and Mrs Bridgeman; not identified apart from the information given in Fox's diary. Fox visited them at Cheshunt, Hertfordshire (14 miles north of London), on the 12th December 1825.
88. This was likely to include the three sons of Jedediah Strutt, Sr, William (1756-1830), George Benson (1761-1841) and Joseph (1765-1844), and their children including Edward Strutt.
89. Sam Evans: possibly a member of the mill-owning family (see note 78).
90. Not identified.
91. Sisters of Jem Holden (see note 62)?
92. Fosbrookes; an old established family in Derbyshire. Leonard Fosbrooke was a Cambridge contemporary of Fox, at Pembroke College, and knew Fox well (see Chapter 2, note 32).
93. Joseph Strutt (1765-1844). Youngest son of Jedediah Strutt, Sr. Actively involved in the commercial business of the family he was also active in the town affairs of Derby, serving twice as mayor. Lived in Thorntree House, St Peter's street. Founded the Mechanics Institution and bequeathed an Arboretum to the town. m Isabella Archibold Douglas and had three children.
94. George Benson Strutt (1761-1841); since William Strutt and his son Edward resided mainly in Belper and Joseph Strutt in Thorntree House in St Peter's street, Derby (note 94), the conclusion is that George Benson Strutt and family lived in St Helens House, when not in Belper.
95. Mrs Bromfield; probably one of the Bromfields, including Harry Bromfield, who were visited by Fox and family in London in August 1825 (see, e g, entry for 4th August 1825). There Fox refers to her as his aunt.
96. Fox refers to Market Harborough and Kettering, two old coaching towns on the route to Cambridge. The route from there was likely through Huntingdon.
97. Henry Wilmot Charleton, see note 55, Chapter 2.
98. John Lingard, adm Christ's College,1825; B A 1829. (*Peile, 1913*); W. Walker: not found in *Peile* (1913) or *Alum. Cantab.*
99. William Breynton. Attended Repton School. Magdalene College, Cambridge, B A, 1827. Barrister, InnerTemple. (*Alum Cantab*)
100. Cottenham is a small village on the outskirts of Cambridge.
101. Lucia Elisabeth Vestris, née Bartolozzi (1797-1856). She probably played Cherubino, a favourite part of hers.
102. Charles Francois Mazurier, who danced at Covent Garden and other theatres in London from ca 1806, often as a clown – see, e g, Gentleman's Magazine, 1825.
103. Clearly Fox continued to see Miss Woodcock, but under restricted circumstances. He continued the contact up to a visit in June 1828 (Appendix 5), but beyond that we have no information.
104. Segar: cigar.
105. Samuel Boteler Bristowe ((1822-1897) and Henry Fox Bristowe (1824-1893), the two young boys of Samuel Ellis Bristowe and Mary Ann Bristowe (Fox's sister).
106. Bank panic of 1825. This was a stock market crash partly over stock in South America. The panic started in the Bank of England. Six London banks and 60 county banks closed their doors temporarily. The situation was alleviated by a loan of gold from the Banque de France (Frank W Fetter. 1967. *A Historical Confusion in Bagehot's Lombard Street.* Economica, New Series, 34, 80-83).
107. Simon Bristowe was living evidently in Newark with his family (see entry for 21st December 1825 and notes 66, above).
108. This entry confirms that Samuel Ellis Bristowe's father, mother and siblings, including Simon Bristowe, were living at Newark at this time (see note 107, above).

109. Fox reiterates his sombre feelings of the year before (see 31st Dec 1824, Chapter 2).
110. Sarah Crompton; Sarah Fox (17??-1842) who married Samuel Crompton, and was Fox's aunt.
111. Probably Edward Ralph Coke (1795-??), the son of Edward Coke, the former M P for Derby.
112. Mr Saunderson, of Derby, coached Fox in classics (Herodotus). No other information has been found.
113. Henry Haden, a physician in Derby who attended the Fox family.
114. Fox's illness. This is the first explicit reference we have to a medical problem that disables Fox over the next months. However, it may have been a recurrent problem since in 1824 when Fox wrote of medical problems that he was having. The symptoms are suggestive of a venereal disease and this is what Samuel Ellis Bristowe and possibly his doctors believed (see events for 10th February 1825), but were strongly denied by Fox (see events for same day an subsequently days).
115. George Benson Strutt, see notes 89 & 95.
116. Fox was due to start Lent term in Cambridge, but due to his illness he was absent for the whole of this term.
117. Fox's tutor at Christ's College, Cambridge.
118. Hybernaculum (hibernaculum); a green house or orangerie for over wintering plants; or the winter quartes for overwintering animals (*O E D*). This entry demonstrates that Fox was carrying out experiments with animals on hibernation.
119. Not clear what this refers to.
120. Silk mill fire; this would have been of great interest to the Foxes, since Samuel Fox III & IV sold silk stockings and other silk apparel.
121. Entomologising; this attests to Fox's interest in insects long before he met CD at Cambridge in 1828.
122. Gerusalemma liberata (Jerusalem delivered) by Torquato Tasso (1544-1596). This masterpiece of sixteenth century epic poetry was never published by the author, but was published without the author's consent in 1580 (and later)..
123. Robert Alvey Darwin (1826-1847). Fox's aunt Darwin was Elizabeth, née de St Croix, married to William Brown Darwin of Elston. The baby was born at Thurlston Grange, a property in Elvaston Castle, near Derby, owned by Samuel Fox III (see notes 11 & 55, above). All the children of this marriage were born at Thurlston Grange. The reason for this is not known since it is far from the country seat of the Darwin's at Elston. However, William Alvey Darwin was Fox's maternal grandfather, and his wife, Jane (née Brown), Fox's maternal grandmother, died at Osmaston in 1835. Therefore, some time after her husband's death in 1783, she may have moved to Thurlston Grange, possibly soon after 1814 when Samuel Fox III moved from Thurlston Grange to Osmaston Hall. Conveniently, this would have removed her from Elston, where her son and new wife would have been installed after their marriage in 1817. Thurlston Grange would then have been a good place to give birth, having a mother-in-law and a sister in law, Ann Fox, to oversee the births. Ann Fox may have had a special capacity in this respect, and at least three of Fox's children were born at Osmaston Hall: Agnes Jane (1839), Julia Mary Anne (1840) and Samuel William Darwin (1841). There is a discrepancy between the date given by Fox for the birth of 19th April and that given in Darwin pedigrees of 17th April.
124. Elisabeth Darwin and her new son Robert Alvey. See entry for 19th April and note 123, above
125. See entry for 18th March.
126. Edward Donovan (1791). The natural history of British insects. London.
127. Seaside recuperation; the advice of Dr Haviland seems to have been the reason that Fox travelled to Southampton at the end of the diary. From there he presumably went to the Isle of Wight, as indicated in his entry for 13th June, and later visited many times during his convalescence in 1833/34 and met his first wife, Harriet.
128. This fine set was first published in 1775 in Paris; it currently seems to be missing from the University Library, Cambridge.
129. Probably Fox meant the grass, *Phalona patagonia.*
130. Not identified. This may have been someone of the Strutt family, who represented the family business in London.

131. Isle of Wight; this entry confirms that the Isle of Wight was Fox's destination. The Fox family had a long-standing connection with the Isle of Wight. Fox's mother Ann Fox née Darwin, had clearly stayed there often and there are drawings in the family made by Ann Darwin at Binstead, Isle of Wight. [Bingham, P and Crombie, G. *William Darwin Fox, Charles Darwin, the Isle of wight and WDF's sons.* Proceedings Isle of Wight Natural History and Archaeological Society, Volume 24 (in press, 2009)].
132. Netley Abbey, near Southampton, Hampshire: a Cistercian monastery, abandoned in the dissolution of the monasteries in the sixteenth century; it was allowed to go to ruin in the eighteenth century, when it became a national attraction, remaining so ever since.
133. Perhaps this refers to Fox's sickness at this time.

Appendix 5
Fragmentary Notes, apparently for Fox's Diary, 1828

University Library, Cambrige, DAR 250: 6
Written on blank sheets of octavo paper.

p 1 June 1828
24 Tuesday — came from Mr Bridgemans, & found Evans & Bond in Town — was with them all day
25 Went with Bond & Evans to Lens Nursery in morning, & at night to Vauxhall[1] with Evans where we were both much tired. — —
26 — Saw Bond off – Walked with Evans all day
27 — Saw Evans off today —
28 — Called this morning on Mr Yarrell[2] in Bury Street, & saw his collections of Eggs & Birds. — Was much gratified with them. —
29 — Went to Breakfast with Ed. Strutt[3], from there to Temple Church, where saw Mr Jessop in Evening walked to Kensington Gardens with Julia — We dined with Ed. Hunt — & I was quite miserable for rest of Evening, & got in warm bath by way of amusement. —
30. Called on Ed Hunt in the morning, dined with Ed. Strutt at Gardens, drank tea with him
July — Walked about London till 4 when I set off to Bromley, when after waiting about an hour & half my Mother &c joined me — We were all drinking tea very quietly when a most awful flame burst out not far off in village, which proved afterwards to be a Chandeliers shop on fire — a Mr Tyzard, it was caused from his Boilers going over — Blazed with much fury &c &c.
2 — I was so tired & *crouped* with my yesterday nights work, that I was good for nothing —
3. — Was still very tired — we did nothing partic*ularly* but had a tremendous storm of Thunder and Lightn*ing* — at night I almost ever heard —

p 2
4. We this morning set off with Miss Woodcock to Tonbridge Wells —We stayed at Knowle Park on our road which is close to Seven Oaks to see the House — it belongs to the Duke of Dorset, and is well worth the seeing, both on account of the ancientness of the building & its splendid collection of pictures. By far the finest collection of Portraits I ever saw, & by no means a mean one of others — — All the old Poets Ben Johnson — Chaucer &c &c and a beautiful Milton among these — Also a portrait of the fair Quaker by Sir J. Reynolds, of whom I was not aware, any was in existence — There was a very fine picture by Sr J. Reynolds, of John Court

and his four Sons who were stormed in prison which was very striking indeed — The furniture is also well worth seeing — the grounds seemed beautiful, but are not shewn. —
We then went on thro' a beautiful country to Tonbridge Town & then to the Wells which was 6 miles beyond — the ride is splendid but there is nothing to see at the Wells there about worth going 5 miles to see, except the celebrated Tonbridge ware[4], which is very pretty and exceedingly cheap — My sisters bought some. We got home about 9. —
5 — My Mother & I came to London by Coach, took Lodgings and returned in Evening —
6. Sunday — Church twice — Nothing positive
7 Monday — Left Bromley at 10 and set off for London by ~~the~~ way of Rotherhithe (which is 2 miles round) when we all went into the

p 3
the Thames Tunnel, & with which we were much pleased — I had seen it only last week but was still much pleased to see it again We got comfortably into our Lodgings about 2 & did not do much this evening, as my sisters were pretty well tired with the morning. —
3 Tuesday — I went before breakfast to Ed Strutt, but found him gone out — I after breakfast went to consult Mr John Scott about my eyes in New Broad Street City — I also called on Mr Theobold. — In the evening I went with Julia to Astleys where we were much pleased with a re=presentation of Waterloo[5]. The Napoleon was most splendid — quite to the life. One might fancy Nap: Buonaparte stood before you — In force, expression, figure, every thing alike him — We got home about ½ past 10. — —
9 Wednesday — We this morning went out in a Glass Coach to shop in City, rather unfortunately, as it happened as it rained in torrents all morning — We went on to Hackney & called on Mrs Grace *Aiken*. After dinner we took another drive to see Belgrave Square[6], with the magnifi=cence of which we were much pleased.
10 — Thursday — My Mother Julia & I drove about all morning shopping — At evening we all went to Opera — Semiramide, Posta splendid, Branbilla very sweet, Zuchelli Catriona &c good — The 2nd Opera was

p 4
part of the Barber of Seville in which *Frantery* introduced a most splendidly executed song in her style — really magnificent in execution. Emma got thro' it all astonishingly —
Friday 11 —
Miss Woodcock came to Town this morning to stay *~~till t~~* a day or two — I went and called on Ed= Strutt in morning — Evening was spent at home.
Saturday 12.
Went to Somerset House and Panorama of the Battle of *Nav***, with which we were much pleased — also Exeter Change — Horse Bazaar, & many shops. Miss W left us in Evening — We took a drive around Regents Park ——————

Sunday 13
Went to St James Piccadilly & Foundling Hospital —
Monday 14

****** (crossed out) Mrs Scott came to Emma and I went to Mr John Scott — In Evening saw Zoological Gardens <& *** *** *** >

15

We went to the Picture Gallery the British Institution and British Gallery At eight my Mother Julia & self went to Opera & were highly delighted

16

Set off about ¼ past 1 for Oxford whence we arrived after a very pretty drive about ½ past 8 —

Notes

1. Vauxhall Gardens, in what is now Kennington, south of the River Thames in Greater London. The Gardens were a favourite site where Londoners could relax and see entertainments; from the Restoration to about 1840.
2. Mr Yarrell; stationer, naturalist and publisher of natural history books. See Bibliographic Register and Letters #31 (n5) & 96 (n6).
3. Edward Strutt (1801-1888), son of William Strutt, the inventor, and grandson of Jedediah Strutt (see Appendix 4, note 15). Became 1st Baron Belper. Liberal MP for Derby. (*Burke's Peerage*)
4. Tonbridge ware or Tunbridge ware consisted of finely inlaid decorative wooden cabinets, boxes, etc; a sort of marquetry.
5. This was a play by J H Amsherst (1776-1851) entitled "The Battle of Waterloo: a grand military melodrama in three acts", which was first performed in 1824. It was probably performed at Astley's Royal Amphitheatre on the Surrey side of Westminster Bridge, where circus performances were also held. The other possible venue would have been the Olympic Theatre, Drury Lane, which was built by Astley.
6. Belgrave Square is one of the grandest 19th century squares in London and gave its name to Belgravia. It was begun in the 1820s and was not fully occupied until 1840.

Appendix 6
Transcriptional Variations

The following is a list of transcriptional variations between the version given here and that in the Correspondence of Charles Darwin (Cambridge University Press). The published volumes are Vols 1-16. The Darwin Correspondence team released to me a transcript of all remaining letters, but later work may alter some of these variations.

The version given here is placed first and that in the CCD version following.

Letter #1

p 1 line 5	yours	your's
p 1 line 8	the bank of sail	the bark of rail
p 3 line 8	pectinated	pectinate

Letter #4

p 2 line 7	Nebria	Nebris

Letter #6

p 3 line 5	an expedition	on expedition
Valedictory	Yours affectionate	Yours affectionat

Letter #12

p 3 line 4	Poecilus	Pæcilus

Letter #23

p 1 last line.	for the present & take rooms & live	for the present. & take rooms & live

Letter #30

p 2 last line	& then from there// Wonder	& then per the // Wonder

Letter #32

Valedictory	Yours sinc,	Yours, sinc,

Letter #35
p4 1st line with every wish for your happiness with every wish your happiness

Letter #52
3rd para, line 1 & 6 crustacea crustaceæ
3rd para, line 14 bonâ fide bona fide
P S I have watched the manners of the whole set: I have patched the manners of the whole set:

Letter #66
Para 1 Line 6 conception concepcion

Letter #68
Para 1 Line 2 fearful aweful

Letter #69
Para 2 line 4 Mss. W. D. Fox Mr. S. W. D. Fox

Letter #75
Para 1 Line 3 whenever wherever

Letter #91
3rd para line 3 2 Gower St Queen Anne Street

Letter #106
1st Para Line 4 Call Drake Call Duck

Letter #114
2nd page, line 3 noticed note6
2nd page, line 3 by in

Letter #120
2nd para, line 3 domed, doomed

Letter #124
2nd para, line 2 (and three times thereafter) chestnut chesnut
(N B, in *On the Origin of Species* it is chestnut)

Letter #131
Para 4, line 2: immortal, immotal

Letter #137
Side 2, line 4 Rhodes Rhoades

Letter #141

Para 3, line 2	movements	movement

Letter #153

Page 2, line 1	Topping	Topper
Page 4, line 5	unlined	valued
Valedictory	Remembrances	Kindest regards

Letter #169

Valedictory	Ever your	I am your

Letter #179

Para 2 2nd line	where	when

Letter #184

Para 7 line 1.	respect	regard
Page 3 4th line	may	morning
page 4 (Wild Ducks) line 2	these	them
line 10	kestrels	kestril

Letter #189

Line 1	pleased	glad

Letter #198

Para 2 line 2	freshness	alertness
extra page. 1st line	I once told you of a	I am told of a

Letter #200

P.S. line 4	gunnery	Germany

Letter #201

Postscript	often showing hereditary ill-health	often unwell hereditary ill-health

Letter 213

Para 2 line 2	like as a daughters	like as light as a daughters
last line	this	the
Para 3 line 3	Narcissus	Harupus
line 5	I	you

Para 4 line 1	any	him
Para 6 (P S) line 5	sow	sons

Letter #214

Para 1 line 3	thro	this
Para 6 line 1	mess	mass

Appendix 7
The Family Trees

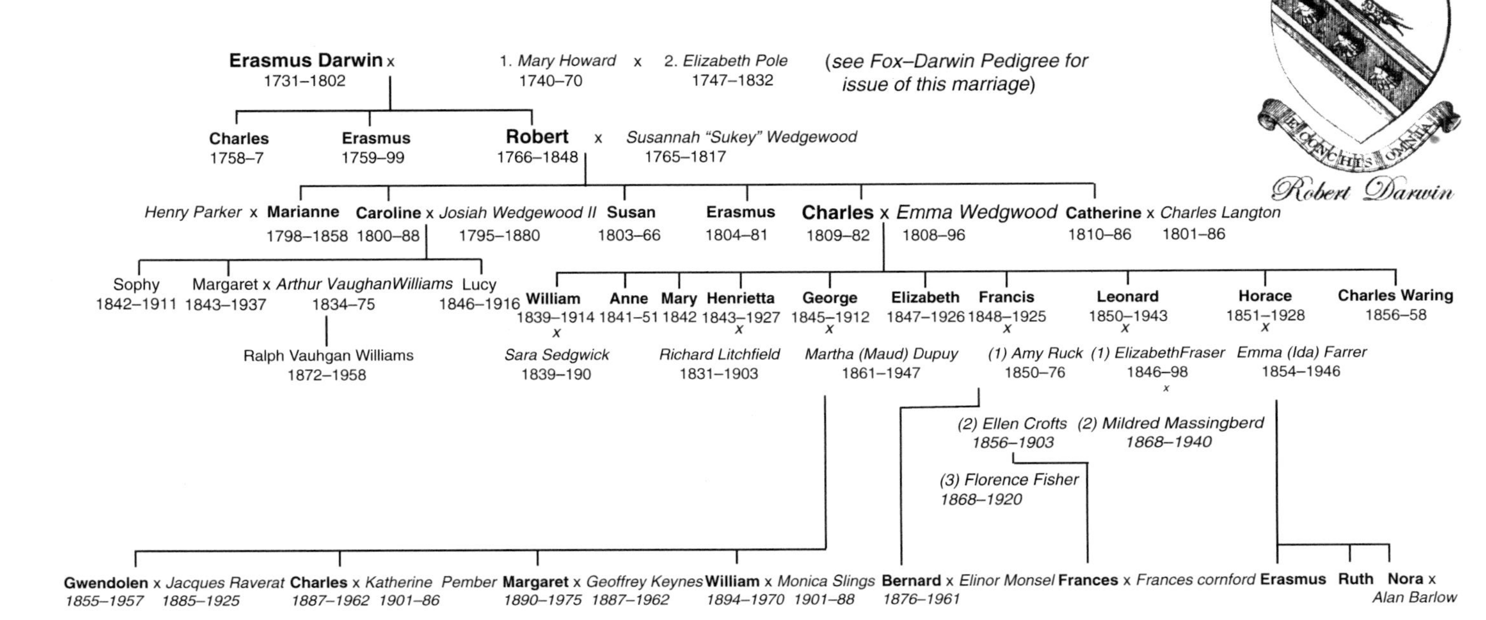

The Darwin Pedigree
Erasmus Darwin x
1731–1802
1. Mary Howard x 2. Elizabeth Pole
1740–70
1747–1832
(see Fox–Darwin Pedigree for issue of this marriage)
Charles
1758–7
Erasmus
1759–99
Robert x Susannah "Sukey" Wedgewood
1766–1848
1765–1817
Henry Parker x Marianne
1798–1858
Caroline x Josiah Wedgewood II
1800–88
1795–1880
Susan
1803–66
Erasmus
1804–81
Charles x Emma Wedgwood
1809–82
1808–96
Catherine x Charles Langton
1810–86
1801–86
Sophy
1842–1911
Margaret x Arthur VaughanWilliams
1843–1937
1834–75
Lucy
1846–1916
Ralph Vauhgan Williams
1872–1958
William
1839–1914
x
Sara Sedgwick
1839–190
Anne
1841–51
Mary
1842
Henrietta
1843–1927
x
Richard Litchfield
1831–1903
George
1845–1912
x
Martha (Maud) Dupuy
1861–1947
Elizabeth
1847–1926
Francis
1848–1925
x
(1) Amy Ruck
1850–76
(2) Ellen Crofts
1856–1903
(3) Florence Fisher
1868–1920
Leonard
1850–1943
x
(1) ElizabethFraser
1846–98
x
(2) Mildred Massingberd
1868–1940
Horace
1851–1928
x
Emma (Ida) Farrer
1854–1946
Charles Waring
1856–58
Gwendolen x Jacques Raverat
1855–1957
1885–1925
Charles x Katherine Pember
1887–1962
1901–86
Margaret x Geoffrey Keynes
1890–1975
1887–1962
William x Monica Slings
1894–1970
1901–88
Bernard x Elinor Monsel
1876–1961
Frances x Frances cornford
Erasmus
Ruth
Nora x
Alan Barlow
Robert Darwin

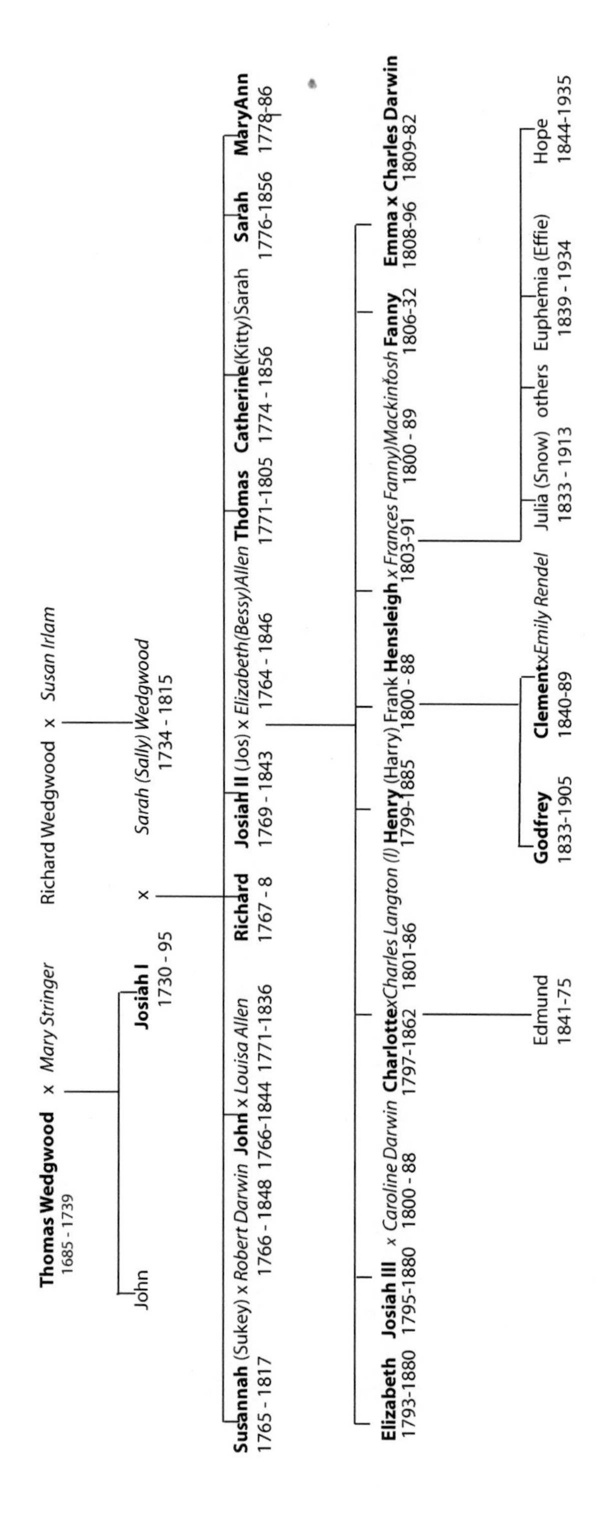
The Wedgwood Pedigree
Thomas Wedgwood
1685 - 1739
x Mary Stringer
Richard Wedgwood x Susan Irlam
John
Josiah I
1730 - 95
x
Sarah (Sally) Wedgwood
1734 - 1815
Susannah (Sukey) x Robert Darwin
1765 - 1817 1766 - 1848
John x Louisa Allen
1766-1844 1771-1836
Richard
1767 - 8
Josiah II (Jos) x Elizabeth (Bessy) Allen
1769 - 1843 1764 - 1846
Thomas
1771-1805
Catherine (Kitty) Sarah
1774 - 1856
Sarah
1776-1856
MaryAnn
1778-86
Elizabeth
1793-1880
Josiah III x Caroline Darwin
1795-1880 1800 - 88
Charlotte x Charles Langton (I)
1797-1862 1801-86
Henry (Harry)
1799-1885
Frank
1800 - 88
Hensleigh x Frances (Fanny) Mackintosh
1803-91 1800 - 89
Fanny
1806-32
Emma x Charles Darwin
1808-96 1809-82
Edmund
1841-75
Godfrey
1833-1905
Clement x Emily Rendel
1840-89
Julia (Snow)
1833 - 1913
others
Euphemia (Effie)
1839 - 1934
Hope
1844-1935

The Fox - Darwin Pedigree

Robert Darwin 1682 - 1754 x *Elizabeth Hill (Sleaford)* 1702 - 1797

William Alvey Darwin 1726-1783 x *Jane Brown* 1746-1835

Erasmus Darwin 1731-1802 x *1. Mary Howard m. 1757* 1740-1770 x *2. Elisabeth Collier (Pole) m. 1781* 1747-1832

Samuel Fox

William Brown 1774-1841 | Elisabeth Hill 1775-1776 | **Ann Darwin** 1777-1859 | John Hill 1780-1781 | John Alvey 1784-1784

Charles 1758-1778 | Erasmus 1759-1799 | Robert Waring 1766-1848 | William Alvey 1767-1767 | Elizabeth 1763-1764

Edward Darwin 1782-1829 | Frances Anne Violetta 1783-1874 | Francis Sacheverel 1786-1859 | John 1787-1898 | Henry 1789-1790

Emma Elizabeth Georgina 1784-1818 | Harriot 1790-1825

Martha Strutt 1763-1798 m 1791 x **Samuel Fox III** 1765-1851 ← **Samuel Fox II** ← **Samuel Fox** ← **Timothy Fox**

2nd marriage - 1799

Samuel Fox IV 1793-1859 | Daughter

MaryAnn 1800-1829 | Eliza 1801-1886 | Emma 1803-1885 | **William Darwin Fox** 1805-1880 | Frances Jane 1806-1891 | Julia 1809-1894

Charles Darwin (See Darwin Pedigree)

1. Harriet Fletcher m. 1834 1799-1842 x *2. Ellen Sophia Wood m. 1846* 1820-1887

Eliza Ann 1836-1874 | Harriet Emma 1837-1922 | Agnes Jane 1839-1906 | Julia Mary Ann 1840 - | **Samuel William** 1841-1918

Charles Woodd 1847-1908 | Frances Maria 1848- | **Robert Gerard** 1849-1909 | Lousia Mary 1850-1853 | Ellen Elizabeth 1852 - | Theodora 1853-1878 | Gertrude Mary 1854-1900 | **Frederick William** 1855-1931 | **Erasmus Pullein** 1858-1939 | **Reginald henry** 1860-1933 | **Gilbert Basil** 1864-1941

Biographical Register

Bristowe, Anna Maria (1826-1862). Dau of Samuel Boteler Bristowe and Mary Ann Fox; niece of William Darwin Fox; m. (as 2nd wife) Baron Dickinson Webster of Penns; had issue: first son, Baron Dickinson Webster of Newland Court, Great Malvern, who m. Ellen Elizabeth Fox, 7th dau of William Darwin Fox, at Dibden, Hants in 1875. (*Darwin Pedigrees*)

Bristowe, Charles John (1862-1911). 2nd s Samuel Boteler Bristowe. Repton School and Trinity Hall, Cambridge (BA, 1884). Clergyman and school master (St Paul's, London), First Director of Education, Nottinghamshire Co. Council. Rowing 'Blue' and coach. (*Alum Cantab)*

Bristowe, Henry Fox, 1st Bart Q C (1824-1893). Trinity Hall, Cambridge (adm 1845). Barrister Lincoln's Inn. Vice Chancellor of County Palatine, Lancaster (1881-1893). m Selina Bridgman (1850); 2 s, Henry Orlando and Leonard Hugh, and 2 dau, Georgina Emily and Alice Mary. (*Alum Cantab*)

Bristowe, Samuel, I (1694-1761). Of Twyford Hall, Derbs. m. Mary Savage (1730). Father –in-law of Samuel Fox, II; great grandfather of William Darwin Fox. Portrait in Notts Archives. (*Burke's Landed Gentry*)

Bristowe, Samuel, II (1736-1818) 1^{st}s of Samuel Bristowe, I. High Sheriff for Notts (1799). Bought Beesthorpe Hall from relative, William Bristowe (1695-1770). Unmarried. Portrait in Notts Archives. (*Burke's Landed Gentry*)

Bristowe, Samuel, III (1766-1846). 1^{st}s Thomas Bristowe; merchant; emigrated to New York. m Eliza Ann Fox née Banks (1797); children: Eliza Welby, **Samuel Ellis Bristowe** (1800-1855; inherited Beesthorpe Hall from great uncle, Samuel Bristowe, II) and 9 other children. Subsequently returned to England (1809) and lived at various places including Twyford Hall and Beesthorpe Hall. (*Bristowe Family Archives*)

Bristowe, Samuel Boteler (1822-1897). 1st s of Samuel Ellis Bristowe. Trinity Hall, Cambridge (M A, 1848, Fellow, 1849-56). Barrister and QC, Inner Temple, Judge of Co Courts; MP for Newark on Trent 1870-80. m Albertine Lavit (1856); children: 3 s, Albert William (1858-1893, emigrated to USA), Charles J (1862-1911) and Frederick Edward (1866-1930) and 3 dau; no grandchildren in this line. (*Alum Cantab, DNB*)

Bristowe, Samuel Ellis (1800-1855). Born in USA and inherited Beesthorpe Hall from his great uncle, Samuel Bristowe, High Sheriff of Nottingham (1818). Repton School; Emmanuel College, Cambridge (1818); JP and DL, Beesthorpe Hall, Nottingham and Twyford Hall, Derbs. m 1. Mary Anne Fox (1821); children: Samuel Boteler, Henry Fox and Anna Maria (1826-1862); m 2. Lady Alicia Mary Needham (1836); no children. (*Alum Cantab; Burke's Landed Gentry*)

Bristowe, Thomas (1739-1814). 2^{nd}s Samuel Bristowe (1694-1761). m Mary Fosbrooke of Shardlow (1765); children: Samuel Bristowe, III (1766-1846) and Mary. (*Bristowe Family Archives*)

Collier, Elizabeth (1747-1832). Natural dau of Charles Colyer, Earl of Portmore. m 1. Edward Sacheverel Chandos-Pole; children: 2 s & 3 dau; m 2. Erasmus Darwin (1781); children 4 s and 3 dau; including Sir Francis Sacheverel Darwin, Knight, of Breadsall Priory (1786-1859). (*Darwin Pedigrees*)

Darwin, Ann (1777-1859) m. **Samuel Fox, III** (1779); children: Eliza, Emma, William Darwin, Frances Jane and Julia (see Family Tree, Appendix 7). (*Darwin Pedigrees*)

Darwin, Anne Elizabeth (1841-1851). 2nd dau of **Charles Darwin** and Emma. Died at Malvern over complications with tuberculosis (?). CD's favourite child. CD's tribute (CCD, Vol. 5, Appendix 2). Annie's gravestone in Great St Mary's, Malvern (Fig. 26). (*Darwin Pedigrees*)

Darwin, Caroline Sarah (1800-1888). 2nd child of **Robert Waring Darwin** and Susannah, née Wedgwood. m. Josiah Wedgwood, III; children; 3 girls: Sophia Marianne, Katherine Elizabeth, Margaret Susan, & Lucy Caroline (grandson was Ralph Vaughn Williams, the composer). (*Darwin Pedigrees*)

Darwin, Charles Robert (12 Feb 1809 – 19 April 1882). Shrewsbury School; Christ's College Cambridge, BA (1833). Biologist and geologist; founder with A R Wallace of theory of organic evolution. Lived at Down House, Bromley, Kent (1843-1882). FRS, Royal Medal of RS (1853), Copley Medal of RS (1864). Author of 19 books including *On the origin of species* (1859). Buried in Westminster Abbey. m **Emma Wedgwood** (2 May 1808 – 2 Oct 1896)(9th child and 4th dau of Josiah Wedgwood, II); 7 s & 3 dau: William Erasmus, Anne Elizabeth, Mary Eleanor, Henrietta Emma, George Howard, Elizabeth, Francis, Leonard, Horace & Charles Waring; Anne Elizabeth, Mary Eleanor and Charles Waring died when young, the rest lived long lives - see Family tree, Appendix 7. (*Alum Cantab, Darwin Pedigrees, DNB, Desmond & Moore, 1991, Browne, 1995, 2002*)

Darwin, Charles Waring (1856-1858). 10th and last child and 6th s of **Charles Darwin** and Emma. Died of scarlet fever (28 June 1858). (*Darwin Pedigrees*)

Darwin, Elizabeth (1847-1928). 6th child of **Charles Darwin**. Known as Bessy. Stout, nervous and impractical; unmarried (see *Period Piece*). (*Darwin Pedigrees*)

Darwin, Emily Catherine (1810-1866). **Charles Darwin**'s younger sister. Late marriage to Charles Langton (at the age of 43) in 1863, as his 2nd wife, after the death of Landgton's first wife Charlotte Wedgwood. (*Darwin Pedigrees*)

Darwin, Erasmus (1731-1802). St John's College, Cambridge and Edinburgh University. Physician, Litchfield and Derby. Inventor; helped to found the Lunar Society. Botanist. Poet: major works Botanic Garden (1790/1), Zoonomia (1794/6), Phytologia (1800) & Temple of Nature (1803). m 1. Mary Howard (1757); children: Charles, Erasmus, Elizabeth, **Robert Waring Darwin**, & William Alvey; m 2. Elizabeth Chandos-Pole née Collier (1781); children: Edward, Frances Anne Violett, Emma Georgina Elizabeth, Francis Sacheveral, John, Henry & Harriot. (*Darwin Pedigrees, DNB, Darwin, 1887, King-Hele, 1999, 2007.*)

Darwin, Erasmus Alvey (1804-1881). 2nd child of Robert Waring Darwin and brother of **Charles Darwin**. Christ's College, Cambridge and Edinburgh University, MD, but never practised. Considered an invalid and given a pension by his father. Residences in London, 1) 24 Regent St, 2) 43 Gt Marlborough St, 3) 5 Queen Anne St. (*Darwin Pedigrees, Freeman, 1978*)

Darwin, Francis, Knight (1848-1925). 7th child of **Charles Darwin**. Clapham School. Trinity College, Cambridge, BA (1870). Trained as a physician but never practised. Helped CD with botanical books from 1865 and co-authored *The power of movement in plants* (1880). Fellow of Christ's College and Reader, School of Botany, Cambridge. FRS (1882) Edited *Life and Letters of Charles Darwin, including an autobiographical chapter*, 3 Vols. (1888), and *More letters of Charles Darwin* with AC Seward, 2 Vols. 1903. Major work: Elements of Botany (1895). m 1. Amy Ruck (1874); child: Bernard Richard Meirion (CD's first grandchild); m 2. Ellen Crofts (1883); child: Frances Crofts Cornford née Darwin; m 3 Florence Henrietta Fisher (1913). Lived at 80 Huntingdon Rd, Cambridge. (*Alum Cantab, Darwin Pedigrees, DNB*)

Darwin, Francis Sacheveral, Knight (1786-1859). Emmanuel College, Cambridge. Physician and traveller. Of Breadsall Priory nr Derby. Deputy Lt, Derbyshire. 4th child of Erasmus Darwin and Elizabeth Pole née Collier. m Jane Harriett Ryle (1815); children: Reginald Darwin, Edward, John Darwin, Mary Jane, Georgina Elizabeth, Violetta Harriot, Ann Eliza Thomasine & Millicent Susan. (*Burke's Peerage, Darwin Pedigrees*)

Darwin, George Howard, Knight (1845-1912). 5th child of **Charles Darwin**. St John's and Trinity College, Cambridge (BA, 2nd Wrangler 1868). Barrister, 1872, but never practised.

Fellow, Trinity College (1868). Plumian Professor of Astronomy & Experimental Philosophy, Cambridge, 1883. FRS (1879) Major work: The tides (1898). m Maud du Puy (1884): children: 2s & 2 dau: Gwendolen Mary (Raverat), Charles Galton, Margaret Elizabeth & William Robert. Lived at Newnham Grange (now part of Darwin College). (*Alum Cantab, Darwin Pedigrees, DNB*).

Darwin, Henrietta Emma (1843-1930). 4th child of **Charles Darwin**. Sickly child and valetudinarian; she never got up for breakfast (Period Piece). m (1871) RB Litchfield (1831-1903); no children. Helped CD with *Descent of Man* (1871). Edited *Emma Darwin: A century of letters* (1915). Known as "Etty". (*Darwin Pedigrees*)

Darwin, Horace, Knight (1851-1928). 9th child of **Charles Darwin**. Trinity College, Cambridge (BA, 187x), later a Fellow. Engineer; founder of the Cambridge Instrument Co. m Emma (Ida Cecilia Farrer (1880); children: Erasmus, Ruth, Emma Nora (Nora Barlow). (*Alum. Cantab, Darwin Pedigrees, DNB*)

Darwin, Leonard, Major and Knight (1850-1943). 8th child of **Charles Darwin**. Clapham School. Royal Engineers. FRS (1903) President Brit Eugenics Soc and Royal Geographical Soc. MP for Litchfield 1892-95. Friend and supporter of RA Fisher, the statistician of organic evolution. m 1. Elizabeth Frances Fraser (1882); m 2. Charlotte Mildred Massingberd (1900). (*Darwin Pedigrees, DNB*)

Darwin, Marianne (1798-1856). **Charles Darwin**'s sister. m Henry Parker (1824); 4 s & 1 dau; after the deaths of father and mother their grown-up children were adopted by Emily Elizabeth at the Mount. (*Darwin Pedigrees*)

Darwin, Mary Eleanor (1842-1842). 3rd child and 2nd dau of **Charles Darwin**. Died at 3 weeks at Down House. (*Darwin Pedigrees*)

Darwin, Susan Elisabeth (1803-1866). **Charles Darwin's** elder sister. Lived on at the Mount Shrewsbury until her death. She adopted the grown up children of her sister, Marianne Parker. (*Darwin Pedigrees*)

Darwin, Susannah (1765-1817) – see Susannah Wedgwood.

Darwin, Robert of Elston (1682-1754). Barrister. m. Elizabeth Hill (1723-24); great grandfather of **Charles Robert Darwin** and **William Darwin Fox**. (*Darwin Pedigrees*)

Darwin, Robert Waring (1766-1848). 4th child of Erasmus Darwin and Mary Howard. Edinburgh University, medicine. Physician, The Mount, Shrewsbury. M Susannah Wedgwood (1796); children: Marianne, Charlotte Sarah, Susan Elizabeth, Erasmus Alvey, **Charles Robert** & Emily Catherine. When lasted weighed, was 24 stone. (*Darwin Pedigrees*)

Darwin, William Erasmus (1839-1914). 1st child of **Charles Darwin**. Rugby School. Christ's College, Cambridge, BA 1862. Partner in Grant & Maddison, Bankers of Southampton (also called Southampton and Hampshire Bank) with share supplied by CD. Lived at Ridgmount, North Stoneham, Basset, Southampton. m Sarah Sedgwick (1877). Looked after CD's finances after 1862. After the death of his wife in 1902, lived next door to brother Leonard at 12 Egerton St, London. (*Alum Cantab, Darwin Pedigrees, Christ's Coll Mag, 1914*)

Fletcher, Charles Orlando. 2nd s of Sir Richard Fletcher.

Fletcher, Elizabeth Mallock. 1st dau of Sir Richard Fletcher.

Fletcher, Jane. 3rd dau of Sir Richard Fletcher. Possibly married one of the Mudge family.

Fletcher, Harriet (1799-1842). 2nd dau of Sir Richard Fletcher and his wife Jane née Mudge. 1st wife of **William Darwin Fox**; m 1834, children 1 son and 4 dau: Eliza Ann, Harriet Emma, Agnes Jane Samuel William. Lived at Ryde, Isle of Wight, until marriage, then Sandown and Delamere, Cheshire. The first 2 children were born at Ryde, Isle of Wight, and her last 3 children were born at Osmaston Hall, the family home of Samuel Fox III. (*Darwin Pedigrees*)

Fletcher, Miss (C M M = Elisabeth Mallock Fletcher?). 1st dau of Sir Richard Fletcher? Looked after children of Fox, at Delamere, after her sister, Harriet, died in 1842 and later Sir Richard Mudge at Ben Rhydding, Yorks and lived, also, at Ipswich, Norfolk (see Appendix 5).

Fletcher, Richard, 1st Bart (1768-1813). Royal Engineers. Died heroically in the Peninsular War fighting under the Duke of Wellington. m Elizabeth Mudge (1771-1808); children: 2 s and

3 dau: Richard, Elizabeth Mallock, **Harriet** (1799-1842, wife of **William Darwin Fox**), Jane, Charles Orlando.

Fletcher, Richard John, 2nd Bart. Probably mentally handicapped and cared for all his life. Unmarried. Last known residence at Pinchcombe House, near Stamford, d 1876.

FitzRoy, Robert (1805-1865). s of General Lord Charles FitzRoy. RN. Hydrographer and metereologist. Took over the command of HMS Beagle after the death of Commander Stokes on 1st voyage. In command of 2nd voyage (with **Charles Darwin**), 1st as commander and then as Captain (1835); promoted to Vice Admiral in 1863. MP for Durham (1841). Governor of New Zealand (1843-1847). Founder of what was to become The Meteorological Office and, in 1860, used the new telegraph system to produce the first synoptic charts. Committed suicide, leaving a widow and dau. (*DNB, Nichols, 2003; Gribbin and Gribbin, 2003; Thompson, 2004*)

Fosbrooke, Leonard (1804-1892) s of Leonard Fosbrooke of Shardlow Hall, Derbs. Rugby School. Pembroke College, Cambridge, B A, 1827. Barrister, Lincoln's Inn, Inner Temple. D L of Leicesters. (*Alum Cantab*)

Fox, Agnes Jane (1839-1906). 3rd child of **William Darwin Fox** and 1st wife, Harriet, née Fletcher. Unmarried. (*Darwin Pedigrees*)

Fox, Charles Woodd (1847-1908). 2nd s and 1st child of **William Darwin Fox** and 2nd wife, Ellen Sophia, née Woodd. Repton School. Christ Church, Oxford University, BA (1870). Barrister; Lincoln's Inn. Lived at Hampstead, London. Wrote Letter #217 to CD informing of the death of W D Fox. (*Darwin Pedigrees*)

Fox, Edith Darwin (1857-1892). 9th child of **William Darwin Fox** and 2nd wife, Ellen Sophia, née Woodd. Umarried and buried at Sandown, Isle of Wight. (*Darwin Pedigrees*)

Fox, Eliza (1801-1886). 1st child of **Samuel Fox III** and his second wife, Ann, née Darwin. Buried at Sandown, Isle of Wight. (*Darwin Pedigrees*)

Fox, Eliza Ann (1836-1874). 1st child of **William Darwin Fox** and 1st wife, Harriet, née Fletcher. m Rev Henry Martyn Sanders (1861); 9 children. (*Darwin Pedigrees*)

Fox, Ellen Elizabeth (1852-). 5th child of **William Darwin Fox** and 2nd wife, Ellen Sophia, née Woodd. m 2nd Baron Dickinson Webster, solicitor of Newton Abbot, Devon, in 1875 (her 1st cousin once removed) and had issue. The father, the 1st Baron Dickinson Webster, married Fox's niece (dau of Fox's sister Mary Ann and Samuel Ellis Bristowe) and it was their son that Ellen Elisabeth married (see Letter #61). (*Darwin Pedigrees*)

Fox, Ellen Sophia (1820-1887), wife of **William Darwin Fox**, see Ellen Sophia Woodd.

Fox, Emma (1803-1885). 2nd dau of **Samuel Fox III** and his 2nd wife Ann, née Darwin. Unmarried; buried at Whitchurch, Hants. (*Darwin Pedigrees*)

Fox, Erasmus Pullein (1858-1939). 10th child of **William Darwin Fox** and 2nd wife, Ellen Sophia, née Woodd. Lived at St Leonard's on Sea, Sussex. (*Darwin Pedigrees*)

Fox, Frances Jane (1806-1884). 4th child of **Samuel Fox III** and his 2nd wife Ann, née Darwin. m. Rev. John Hughes of Penally, nr Tenby (1852); no children. (*Darwin Pedigrees*)

Fox, Frances Maria (1848-). 2nd child of **William Darwin Fox** and 2nd wife, Ellen Sophia; m Alexander Pearce (1878); 1s Nathaniel (1870), civil engineer; & 1 dau Marjory (1881). (*Darwin Pedigrees*)

Fox, Frederick William (1855-1931). 8th child of **William Darwin Fox** and Ellen Sophia. Brasenose College, Oxford; BA 1879. Clergyman/school teacher. Lived at 20 North parade, Southwold, Suffolk. m Rebecca, 3 s, Samuel Frederick, Francis Darwin, Anthony Basil, and 2 dau, Joan Darwin and Ann Darwin. (*Alum Oxon, Darwin Pedigrees, National Archives*)

Fox, Gertrude Mary (1854-1900). 7th child of **William Darwin Fox** and 2nd wife, Ellen Sophia, née Woodd. m Rev Frederick Charles Tindal Bosanquet (1879); children (incl Theodora, 1880). (*Darwin Pedigrees*)

Fox, Gilbert Basil (1864-1941). 12th and last child of **William Darwin Fox** and 2nd wife, Ellen Sophia, née Woodd. Repton School. St John's College, Oxford, BA (1887). Clergyman. Lived at "Nuthatch", Queen's Road, Shanklin, Isle of Wight. (*Darwin Pedigrees*)

Fox, Harriet Emma (1837-1922). 2nd child of **William Darwin Fox** and Harriet, née Fletcher. m Samuel Charlesworth Overton (1861); 7 children; died in Zurich, Switzerland. (*Darwin Pedigrees*)

Fox, Julia (1809-1894). 5th and last child of Samuel Fox, III and 2nd wife, Ann, née Darwin. Unmarried. (*Darwin Pedigrees*)

Fox, Julia Mary Anne (1840-). 4th child of **William Darwin Fox** and 1st wife, Harriet, née Fletcher. Broke off affair with Basil Kilvington Woodd, in 1863. m Samuel Everard Woods (1883); no children. (*Darwin Pedigrees*)

Fox, Louisa Mary (1850-1853). 4th child of **William Darwin Fox** and 2nd wife, Ellen Sophia. Died in childhood. (*Darwin Pedigrees*)

Fox, Mary Anne (1800-1829). 6th child of Samuel Fox III and 2nd wife, Ann. m Samuel Ellis Bristowe (1821); children: Henry Fox and Anna Maria. (*Darwin Pedigrees*)

Fox, Martha, 1st wife of **Samuel Fox III**; see Martha Strutt.

Fox, Reginald Henry (1860-1933). 11th child of **William Darwin Fox** and his 2nd wife, Ellen Sophia, née Woodd. Lived at "Nuthatch", Shanklin, Isle of Wight. (*Darwin Pedigrees*)

Fox, Robert Gerard (1849-1909). 3rd child of **William Darwin Fox** and his 2nd wife, Ellen Sophia, née Woodd. King's College, London (1869); Exeter College, Oxford (BA, 1872). Tutor to Prince Frederick William Victor Albert of Prussia (1859-1941), grandson of Queen Victoria, who became Kaiser Wilhelm II of Germany. Lived at Sandown, Isle of Wight and at Almathiea, Hythe; JP, Hants. Took a great interest in local affairs including chair of the board of visiting magistrates at Parkhurst prison. m Emily Mary Chance (1880) and had children. (*Darwin Pedigrees*)

Fox, Samuel, I (1676-1755), Draper. Mayor of Derby, 1741/2. Married and had children, including **Samuel Fox II**. (*Local Studies Centre Archives, Derby*)

Fox, Samuel, II (1715-1767). s of Samuel Fox, I. Soap Boiler. m (1755) Mary Bristowe (1733-1812); children: Elizabeth, Mary, Sarah, Letitia Anne & **Samuel Fox III** (1765-1851). (*Local Studies Centre Archives, Derby*)

Fox, Samuel, III (1765-1851). s of Samuel Fox, II. Silk merchant. Lived at 1) Thurlston Hall, Derby & 2) Osmaston Hall, Derby. m 1. Martha Strutt (1791); children: Samuel Fox, IV and a dau, who died in infancy; m 2. Ann Darwin (1799); children: Eliza, Emma, Frances Jane, **William Darwin Fox** & Julia. (*Darwin Pedigree*)

Fox, Samuel, IV (1795-1859). 1st child of **Samuel Fox III** and his 1st wife, Martha, née Strutt. Repton School. Silk merchant in the business of his father, **Samuel Fox III**, in Derby. After his father retired to the environs of London in ca 1845 he lived at Osmaston Hall, where he was visited many times by W D Fox. (*Darwin Pedigrees*)

Fox, Samuel William Darwin (1841-1918). 4th child of **William Darwin Fox** and Harriet, née Fletcher. Wadham College, Oxford, BA (1864). Clergyman (St Peters, Lymm, Cheshire; St Paul, Maidstone, Kent; Christchurch). m Euphemia Rebecca Bonar (1876); no children (*Darwin Pedigrees*).

Fox, Theodora (1853-1878). 6th child of **William Darwin Fox** and 2nd wife, Ellen Sophia, née Woodd. Unmarried. Died of lung disease at Sandown. Favourite dau of W D Fox. (*Darwin Pedigrees*)

Fox(e), Timothy (ca1628-1710). Aston School, Birmingham. Christ's College, Cambridge, adm 1646. Rector, Drayton Bassett, Staffs, 1651-62; ejected, 1662. Imprisoned at Derby Jail, 1684. (*Peile, 1910, Alum Cantab, DNB*)

Gully, James Manby (1808-1883). Physician. Proprietor of an hydropathic establishment at The Lodge, Malvern. Treated Darwin and W D Fox [Letters #79 (4), 82 (1), 83 (2), #85]. Annie Darwin died in Malvern in 1851. In1872 he had an affair with a young woman, Florence Ricardo. Ricardo later married Charles Bravo, who died of poisoning in 1876. Gully testified at the trial but appears to have been innocent and no-one was found guilty. (*DNB, Mann, 1983*)

Hall, Frederick (Frank). s of J Cressy Hall of Alfreton, Derbyshire. Repton School. St John's College Cambridge, B A 1828. Frederick Hall. Became a conveyancer and a JP. (*Alum Cantab*)

Henslow, John Stevens (1796-1861), St John's College, Cambridge, BA, 16th Wrangler (1818), Fellow (1819); Clergyman, botanist & mineralogist. Professor of Botany at the University of Cambridge (1827-1861), co-founded the Cambridge Philosophical Society, with Adam Sedgwick, and founded the second Cambridge University Botanic Garden (1831), from 1837 spent much of his time in his parish at Hitcham, Suffolk, and was especially active in education; teacher and mentor to CD and Fox; m Harriet Jenyns (dau, of George Leonard Jenyns and sister of Leonard Jenyns); child: Frances Harriet, who married Joseph Dalton Hooker. (*DNB, Walters and Stow, 2001*).

Herschel, John Frederick William, Bart. (1792-1871). s of Frederick William Herschel, the discover of the planet, Uranus. St John's College, Cambridge, B A (Senior Wrangler) 1813. Astronomer and chemist. Master of the Mint (1850-1855). Spent several years at the Cape of Good Hope, from 1834, observing the Southern stars; it was here that Darwin met him during the Beagle visit, 8-15th July 1836. Wrote, inter alia, *Preliminary discourse on the study of natural philosophy*, which impressed Darwin [Letter #39 (n2)]. (*DNB*)

Hewitson, William Chapman (1806-78) Naturalist and specialist in diurnal lepidoptera as well as birds and birds eggs. *British Oology* (1831–44), Newcastle upon Tyne. *Illustrations of New Species of Exotic Butterflies*. Vols 1-5 (1851, 1862-1871, 1878) (Hewitson paid for the last volume when the Trustees of the British Museum refused). Fox corresponded with Hewitson personally (see 11 letters between W D Fox and Hewitson, 1830 – 1835; The Fox/Pearce (Darwin) Collection, Woodward Library, University of British Columbia). See Letter #121. Fox was thanked in the preface to Hewitson (1831- 44) for supplying information and specimens of nests and eggs and was cited frequently in the text. (*CCD*)

Hill, Elisabeth (1702-1797). m. Robert Darwin of Elston (1723-24) and had issue. Great grandmother of **Charles Robert Darwin** and **William Darwin Fox**. (*Darwin Pedigrees*)

Hooker, Joseph Dalton, Knight, OM (1817-1911). 2nd son of William Jackson Hooker. Glasgow High School. Glasgow University, MD (1839). Assistant surgeon on Ross's Antarctic expedition (1839-1843), when he made botanical collections. In 1847 he mounted a private expedition to India and the Himalayas to collect plants for The Royal Botanic Gardens, Kew, where his father was, later, the director. Books, especially *Flora Antarctica*, made his reputation. Was invited by Darwin to treat the Galapagos plant specimens from the Beagle, in 1848, and a strong friendship began. Hooker with Charles Lyell, arranged the joint presentation of Darwin's and A R Wallace's ideas on organic evolution at the Linnean Soc, London, in 1858. Hooker was the first biologist to support, in print, the Darwin's theory of natural selection (*Introductory Essay to the Flora Tasmanieae*, 1859). Director of the Royal Botanic Gardens, Kew, and President of the Royal Society. m 1. Frances Henslow: 4 s & 3 dau. m 2. Hyacinth Jardine: 2 s. (*Desmond, 1999, ODNB*)

Hope, Frederick William (1797-1865). Entomologist and clergyman. Bequeathed his insect collection to, and founded the chair of Zoology at, Oxford University. (*DNB*)

Huxley, Thomas Henry (1825-1895). s of George Huxley a mathematics teacher who fell on hard times. Self taught, surviving as a medical assistant. MB, Charing Cross Hospital. Assistant Surgeon on HMS Rattlesnake (1846-1850). Work on *Hydrozoa*, a name he coined, established his reputation. Worked on invertebrates and later vertebrates. Professor of Natural History at the Royal School of Mines, South Kensington (1854) & Hunterian Professor, Royal College of Surgeons (1863-1869). President of the Royal Society. Darwin's "bulldog"; close friend and early supporter, never wavering in his support of organic evolution but questioning of Natural Selection; not on such intimate terms as J D Hooker. (*Desmond, 1994, ODNB*)

Jenyns, Leonard (1800-1893)(later changed his name to Blomefield). Youngest son of George Leonard Jenyns, of Bottisham Hall, Bottisham and Vicar of Swaffham Prior, Cambs. St John's College, Cambridge, BA (1822), where he came under the influence of J S Henslow. Clergyman; Vicar of Swaffham Bulbeck, Cambs. Early member of the Cambridge Philosophical Society; founding member of what became the Royal Entomological Society; many papers on natural history subjects. Turned down offer to accompany Capt Robert FitzRoy on the 2nd voyage of the Beagle. Edited "Fishes" for "Zoology of the Voyage of the Beagle" (1840-1842), Vol 3. Wrote Memoir of John Stephens Henslow (1862), with recollections by Charles Darwin. His sister, Harriet, married J S Henslow. (*CCD, DNB*)

Lane, Edward Wickstead (1822-1891). Proprietor of Moor Park, a hyropathic establishment near Farnham and later Sudbrook Park, Petersham, nr Han, Surrey [Letter #156 (n5)]. (*CCD*)

Lyell, Charles (1797-1875) 1st Bart. Father, Charles, was a lawyer. Grew up in Scotland and the New Forest, Hants. Exeter College, Oxford (1820); studied under William Buckland. 1st career in law, then turned to geology from 1827. Wrote the *Principles of Geology*, 3 Vols (1830-1833) in which he propounded a theory of Earth history based on Hutton, and sometimes called unifomitarianism; this made his reputation and had a profound influence on Charles Darwin [Darwin probably received the last volume of *Principles* on the Falkland Islands in 1834; see Letter #56 (n10)]. Professor of Geology, Kings College, London (1830s). *Elements of Geology* (1835) was a field guide for students. Lyell and Darwin became firm friends after the return of the Beagle in 1836 ["*Amongst the great scientific men, no one has been nearly so friendly & kind, as Lyell*"; Letter 58 (n13), 1836] and from 1842 onwards Darwin shared his evolutionary views with Lyell, but the first 9 editions of *Principles* did not mention Natural Selection; the 10th did but in luke-warm terms. Lyell, with J D Hooker, arranged the joint presentation of Darwin's and A R Wallace's ideas on organic evolution at the Linnean Soc, London, in 1858. *Geological Evidences of the Antiquity of Man (1863)* was also weak on the possible evolution of humans. Buried in Westminster Abbey. m Mary Horner (1832); no children. (*DSB, ODNB, Stafford, 1989*)

Mudge, Elizabeth (1771-1808) m Sir Richard Fletcher 1st Bart had 1 son and 4 daughters, including Harriet Fletcher (1st wife of **William Darwin Fox**).

Mudge, Frederick William (<1844-1853) s of Zachary Mudge II and Jane. Died of scarlet fever in the care of W D Fox at Delamere in 1853 [Letter #93 (n19)].

Mudge, Jane, see Jane Fletcher

Mudge, John (1721-1793), s of the Rev Zachary Mudge. Physician of Plymouth. FRS (1777); Copley Medal for his '*Directions for making the best Composition for the Metals for reflecting Telescopes* ...'. Married 3 times and had 20 children, including Elizabeth, who married Sir Richard Fletcher (the mother of Harriet Fletcher), and Admiral Zachary Mudge (1770-1852), the grandfather of Zachary and Frederick Mudge, who died at Delamere in 1852 (see below and Letter #93 (n19).

Mudge, Zachary, I (1770-1852). Admiral. s of John Mudge, physican of Plymouth. Brother of Lt. Genl. William Mudge, Trigonometrical Survey of Great Britain and Ireland and Lt. Governor of Woolwich and Addiscombe; brother-in-law of Sir Richard Fletcher. Most famous for Vancouver Expedition. m. Jane Granger (1807) of Souton near Exeter, and had 1 son, Zachary Mudge of Sydney. Spent last years at Ben Rhydding, Yorkshire. (*Flint,1883; Burke's Landed Gentry; DNB*).

Mudge, Zachary II (1813-1867). Oriel College, Oxford; BA (1834). Barrister. m Jane E Dickson (1844); children: 3 s including Zachary Granger Mudge and Frederick William Mudge, who both died of scarlet fever in the care of W D Fox at Delamere in 1853 [Letter #93 (n18)]. Of Sydney, Plympton, Devon. (*Alum. Oxon.; Burke's Landed gentry; Gent Mag 1868 i, 120*).

Mudge, Zachary Granger (<1844-1853) s of Zachary Mudge II and Jane. Died of scarlet fever in the care of W D Fox at Delamere in 1853 [Letter #93 (n18)].

Needham, Alicia Mary (1798-1885). m Samuel Ellis Bristowe (1836).

Nevinson, Maria Jane, née Woodd. 4th child and 1st dau of Basil George Woodd; sister of **Ellen Sophia Fox**, née Woodd. m George Henry Nevinson, Esq of Leicester and had issue. Lived at Oughtershaw on the Yorkshire Moors.

Power, Joseph (1798-1868). s of John Power M.D. of Market Bosworth, Leics. Clare College, Cambridge, 10th Wrangler in 1821. Fellow of three colleges. Junior Proctor, 1830-31. Clergyman of several parishes. Cambridge University Librarian (1845-64). (*Alum Cantab, DNB*)

Romanes, George John (1848-1894). s of Rev George Romanes. Schooling in Canada. Caius College, Cambridge, B A,1871. University College, London, 1874-76. Professor, Edinburgh University, 1886-1890. FRS. Friend of CD from early 1870s to 1882. Coined the term "neo-darwinisism". Major works: *Mental evolution in animals* (1883) (including Darwin's essay on *Instinct*), *Darwin and after Darwin* (3 vols,1892-1897), *An examination of Weissmannism* (1893), etc. m Ethel Duncan (1879); 6s and 1 dau. (*Alumni Cantab, Romanes, 1896*)

Sedgwick, Adam (1785-1873). Trinity College, Cambridge, B A, 5th Wrangler, 1808). Geologist and clergyman. Woodwardian Professor of Geology at Cambridge University (1818-1873).

Senior Proctor, 1827-28. Promoted the early geological interests of CD and went on tour of N Wales with him in Aug 1831 [Letter #44 (n2)]; but was a strong opponent of "*On the Origin of Species*". (*Alum Cantab, DNB*)

Strutt, Edward, 1st Lord Belper (1801-1880). s of William Strutt and grandson of Jedediah Strutt (1726-1797). Trinity College, Cambridge, B A 1823. Barrister, Lincoln's Inn and Temple. Liberal MP (for Derby, 1830-1848, Arundel 1851-52, Nottingham 1852-56). Chancellor of the Duchy of Lancaster 1852-54; President of University College, London, 1871-79. A second cousin of Fox. (*Burke's Peerage, DNB, Fitton and Wadsworth, 1958*)

Strutt, George Benson (1761-1841). s Jedediah Strutt Sr. Carried on family business in Belper but also lived in St Helens House, Derby. m Catherine Radford and had children, including Jedediah Strutt Jr. (*Fitton and Wadsworth, 1958, Strutt Archives, Derbyshire Record Office*)

Strutt, Martha (1763-1798) dau Jedediah Strutt. 1st wife of **Samuel Fox, III**. (m 1791) ; 1 s Samuel Fox, IV (1793-1859) and 1 dau (died in infancy). (*Darwin Pedigrees, Fitton and Wadsworth, 1958*)

Strutt, Jedediah (1726-1797). Hosier and Cotton spinner of Belper, near Derby. With brother-in-law, William Woolatt, developed a stocking frame attachment that allowed the production of ribbed stockings. m Elizabeth Woolatt (1755); children: Willliam (1st Baron Belper), Elizabeth, Martha, George Benson and Joseph Strutt. Martha (1763-1798) was the 1st wife of Samuel Fox, III and mother of Samuel Fox, IV. (*DNB, Fitton and Wadsworth, 1958*)

Strutt, Jedediah Jr (1785-1854). s George Benson Strutt and grandson of Jedediah Strutt (1726-1797). Carried on family textile business in Belper. Married and had a son, George Henry Strutt (1826-1895) who continued the family business. (*Fitton and Wadsworth, 1958, Strutt Archives, Derbyshire Record Office*)

Strutt, Joseph (1765-1844). Youngest son of Jedediah Strutt, Sr. Actively involved in the commercial business of the family, he was also active in the town affairs of Derby, serving twice as mayor. Lived in Thorntree House, St Peter's street. Founded the Mechanics Institution and bequeathed an Arboretum to Derby. m Isabella Archibold Douglas and had three children. (*Fitton and Wadsworth, 1958, Strutt Archives, Derbyshire Record Office*)

Strutt, William (1756-1830. s Jedediah Strutt Sr. Inventor and engineer and cotton industrialist. FRS (1817). Lived both in Belper and Derby. Built many industrial and municipal structures, including many mills and bridges. Founder of the Derby Philosophical Society with Erasmus Darwin. m Barbara Evans, dau of Thomas Evans, weaving industrialist of Darley Abbey; 1 s, Edward, who became 1st Lord Belper, and 3 dau. (*Fitton and Wadsworth, 1958, Strutt Archives, Derbyshire Record Office*)

Tegetmeier, William Bernhard (1816-1912). Author, journalist and naturalist, specialising in domestic animals and birds (pigeons and fowls). Editor of the Field (1864-1907) and Secretary of the Apiarian Society of London. (*DNB*)

Walker, Louisa Gertrude. 5th child and 2nd dau of Basil George Woodd. Sister of **Ellen Sophia Fox**, née Woodd. m Rev John Walker of Old Malton, Yorks. (*Fox family archives*)

Way, Albert (1805-1874), B.A., Trinity College, 1829. Antiquary and traveler. A founder of the Society of Antiquaries. He and Fox introduced CD to insect collecting. See also Fig. 15. (*Alum Cantab, DNB*)

Wedgwood, Josiah I (1730-1795). Potter and 13th child of a potter. Founder of Josiah Wedgwood & Sons and established Etruria Hall, Staffs. Close friend of Erasmus Darwin and member of The Lunar Society. m Sarah Wedgwood, a cousin (1764); 4 s & 3 dau: **Susannah** (who became the wife of Robert Waring Darwin and mother of **Charles Darwin**), John, Richard, Josiah II, Thomas, Catherine & Sarah Elizabeth. Known as Josiah. (*Meteyard, 1865; King-Hele, 1999; Healey, 2001; Dolan, 2004*).

Wedgwood, Josiah II (1769-1843). 4th child of Josiah Wedgwood I. Reluctantly took over the pottery works from his father. Moved from Etruria to Maer Hall. Friend of Robert Waring Darwin, his son-in-law. m Sarah Elizabeth Allen (1792); 4 s & 5 dau: Sarah Elizabeth, Josiah III, Mary Anne, Charlotte, Henry Allen, Francis, Hensleigh, Frances, **Emma** (who became the wife of **Charles Darwin**). Known as Jos. (*Meteyard, 1865; Litchfield, 1915; Healey, 2001*)

Wedgwood, Josiah III (1795-1880). 2nd child of Josiah Wedgwood II. Country gentleman and senior partner of Josiah Wedgwood & Sons. m Caroline Sarah Darwin, his cousin (1837); 4 dau: Sophia Marianne, Katherine Elizabeth, Margaret Susan, & Lucy Caroline. Known as Joe. Lived at Leith Hill Place, Surrey. (*Wedgwood and Wedgwood, 1980, Healey, 2001*)

Wedgwood, Susannah (1765-1817). 1st child of Josiah Wedgwood I. Helped father with pottery business and gardens at Etruria Hall. m Robert Waring Darwin (1796); 2 s & 4 dau: Marianne, Caroline Sarah, Susan Elizabeth, Erasmus Alvey, **Charles Robert Darwin** & Emily Catherine. Known as Sukey. Mother of Charles Darwin and aunt of Emma Darwin née Wedgwood. Died when Charles Darwin was 8; he had few memories of his mother. (*Darwin Pedigrees, Healey, 2001*)

Wedgwood, Emma (2 May 1808 – 2 Oct 1896). 9th and last child of Josiah Wedgwood II and Sarah Elizabeth née Allen. m **Charles Robert Darwin** (1839), her 1st cousin; 7 s & 3 dau: William Erasmus, Anne Elizabeth, Mary Eleanor, Henrietta Emma, George Howard, Elizabeth, Francis, Leonard, Horace & Charles Waring; Anne Elizabeth, Mary Eleanor and Charles Waring died when young, the rest lived long lives. Lived at Maer Hall until married at age of 30, then 2 years at 12 Upper Gower St, W London, then 40 years at Down House, Orpington, Kent, with Charles Darwin. After the death of Charles Darwin in 1882, she spent the winters at The Grove, Huntingdon Rd, Cambridge and the summers at Down House. She died at The Grove. (*Darwin Pedigrees, Litchfield, 1915, Healey, 2001*)

Whewell, William (1794-1866). Trinity College, Cambridge, B A, 2nd Wrangler (1816). Master of Trinity College, Cambridge (1841-1866). Astronomer and philosopher. Professor of Mineralogy (1828-1832) and Knightsbridge Professor of Philosophy (1838-1855), Cambridge University. FRS. Of his many works, perhaps *The history of the inductive sciences* (2 Vols. (1857), is the best known. CD knew Whewell at Cambridge and after the *Beagle* voyage but preferred Herschel's philosophical approach to science [see Letter #39 (n2)]. (*Alumni Cantab, DNB, Ruse, 1975*)

Wilmot, Francis Sacheverel (1804). s of Rev Edward Wilmot, Kirk Langley, Derbs. Repton School. St John's College, Cambridge, B A, 1829. Barrister, Lincoln's Inn. (*Burke's Landed Gentry, Alum Cantab*)

Wilmot-Horton, Robert, Bart. (1784-1841). s of Sir Robert Wilmot of Osmaston Hall. Christ Church, Oxford, B A 1806. MP for Newcastle under Lyme, 1818-30. Under Secretary to the Colonies, 1825-28; where he made several reforms and influenced, particularly, the colonial development of New South Wales. Governor of Ceylon 1831-37. (*Alum Cantab, DNB*)

Woodd, Basil George (1781-1872). Merchant, m Mary Mitton dau of Rev Robert Mitton of Harrogate (1814); 3 s and 4 dau. Fox knew him from as early as 1835 (see letter to Woodd from Fox held in Fox-Pearce (Darwin) Collection, Woodward Library, University of British Columbia). (*National Archives*)

Woodd, Basil Kilvington, (1842-1886). Eldest s of Basil Thomas Woodd. Harrow School, Trinity College, Cambridge, LL B, 1865. C of E clergyman, Ely (deacon); All Saints, Huntingdon. Ended affair with Julia Mary Anne Fox, 4th child of W D Fox and his 1st wife, Harriet, in 1863; unmarried. (*Alum Cantab*)

Woodd, Basil Thomas (1815- 1895). Trinity College, Cambridge, B A, 1837. Barrister, Lincoln's Inn. Latterly lived at Conygham Hall, Knaresborough, Yorks. M P for Knaresborough 1852-1868, 1874-1880. m Charlotte Dampier of Colinhays, Somerset. Issue of 3 s and 4 dau; inherited Hillfield from father in 1872. (*Alum Cantab*)

Woodd, Charles Henry Lardner (1821-1893). FGS. Merchant. m 1. Lydia Sole with issue of 2 dau; m 2. Jane Harris (1864) with issue 2 dau. Lived at Rosslyn, Hampstead, London and Oughtershaw Hall, Yorks. (*National Archives*)

Woodd, Ellen Sophia (1820-1887). 3rd dau of Basil George Woodd, m (1846) **Rev William Darwin Fox**: 5 s and 6 dau: Charles Woodd, Frances Maria, Robert Gerard, Louisa Mary, Ellen Elizabeth, Theodora, Gertrude Mary, Frederick William, Edith Darwin, Erasmus Pullein, Reginald Henry, Gilbert Basil. (*Darwin Pedigrees*)

Woodd, John Henry Townsend (1866-1939). s of Robert Ballard Woodd. King's College, Cambridge, B A, 1875. Architect and Surveyor, Guy's Hospital, London. Married with 2 dau. (*Alum Cantab*)

Woodd, Maria Jane, 1st dau of Basil George Woodd, m George Henry Nevinson, Esq. of Leicester (married by W D Fox in 1851) and had issue. Known as Minnie.

Woodd, Mary Emma, 2nd dau of Basil George Woodd. m (1845) Thomas Henry Roper, St John's College, Oxford and Barrister, Lincoln's Inn Court. (*Alum Oxon*)

Woodd, Robert Ballard (1816-1901). FRBS, FSA; Accountant. m Barbara Bathume with issue 2 s and 2 dau. Lived at Woodlands, Hampstead, London. (*National Archives*)

Yarrell, William (1784-1856). Stationer, bookshop owner, publisher and naturalist. Major works: *The History of British Fishes*, 2 Vols (1836) and *The History of British Birds*, 2 Vols (1843), which became a standard work. Described Bewick's swan, distinguishing it from the Whooper swan [see Letter #31 (n4)]. Fellow and Treasurer of the Linnean Society; founding member of the Zoological Society; founder of what became the Royal Entomological Society. (*DNB*)

References

Arbuckle, E.S. (1983). Harriet Martineau's letters to Fanny Wedgwood. Stanford Univ. Press.

Barber, Lynn. (1980). The Heyday of Natural History. Jonathan Cape, London. ISBN 0-224-01448-x

Barrett, P. H. (Ed)(1977). The collected papers of Charles Darwin. 2 Vols. University of Chacago Press. Charle's darwins notebooks

Barlow, N. (1958). The Autobiography of Charles Darwin, 1809-1882; with original omissions restored; edited with Appendix and Notes by his grand-daughter. Collins London.

Barrett, P. H., Gautry, P. J., Herbert, S, Kohn, D. & Smith, S. (Eds)(1987). Charles Darwin's notebooks, 1836-1844: geology, transmutation of species. British Museum (Natural History) and Cornell University Press, USA.

Bateson, William (1902). Mendel's Principles of Heredity, a Defense, First Edition, Cambridge University Press, Cambridge, UK.

Blackburn, Julia (1989). Charles Waterton (1782-1865): Traveller and Conservationist. Bodley Head, London.

Bowler, P. J. (1983) The eclipse of Darwinism: anti Darwinian evolution theories in the decades around 1900. Johns Hopkins Univ. Press, Baltimore.

Brooks, J. L. (1984). Just before the origin. Alfred Russel Wallace's Theory of Evolution. Columbia University Press, New York. ISBN 0-231-05676-1.

Brooke, C.N.L. (Ed.) 1997. A History of the University of Cambridge. Vol III. 1750-1870. Cambridge University Press, Cambridge, UK. ISBN 0521 35060 3.

Brown, F. B. (1986). The evolution of Darwin's religious views. J. Hist. Biol. 19, 1-45

Browne, Janet (1996). Charles Darwin. Vol. I. Voyaging. Princeton University Press.

Browne, Janet (2002). Charles Darwin. Vol. II. The Power of Place. Princeton University Press.

Burckhardt, F. & Smith, S. (1985, 1994). A calendar of the correspondence of Charles Darwin, 1821–1882: revsd edn, 2002. Garland, New York.

Burckhardt, F. & Smith, S., and others (Eds)(1985-2008). The correspondence of Charles Darwin. Vols 1 – 16. Cambridge University Press, Cambridge, UK.

Evans, E. J.(1983). The Great Reform Act of 1832. Methuen, London.

Chadarevian, S. de (1996). Laboratory science versus country-house experiments: the controversy between Julian Sachs and Charles Darwin. British J. Hist. Sci. 29, 17–41.

Chalkin, C. W. (1975). Provincial towns of Georgian England. E. Arnold, London. ISBN 978-0713157611.

Colp, R. Jr. (1977). To Be an Invalid: The Illness of Charles Darwin. University of Chicago Press.

Curtis, John, (1823–1840). British Entomology - being illustrations and descriptions of the genera of insects found in Great Britain and Ireland. Printed for the author; London.

Darwin, C. R. (Ed)(1838–1842). The zoology of the voyage of HMS Beagle under the command of Captain Fitzroy, R. N., during the years 1832 to 1836. 5 parts. Smith Elder & Co., London

Darwin, C. R. (1839). Journal of researches into the geology and natural history of the various countries visited by HMS Beagle. Smith Elder & Co., London

Darwin, C. R. (1842). The structure and distribution of coral reefs. Smith Elder & Co, London.
Darwin, C. R. (1845). Journal of researches into the natural history and geology of the various countries visited by HMS Beagle round the world. John Murray, London.
Darwin, C. R. (1846). Geological observations on South America. Smith Elder & Co, London.
Darwin, C. R. (1851). Geological observations on coral reefs, volcanic islands, and on South America. Smith Elder & Co, London.
Darwin, C. R. (1851/1854). A monograph of the sub-class Cirripedia, with figures of all the species. 2 Vols. Ray Society, London.
Darwin, C. R. (1851/1854)(1854=1855). A monograph of the fossil Lepadidae, or pedunculated cirrepedes, of Great Britain. A monograph of the fossil Balanidae and Verrucidae of Great Britain. 2 Vols. Ray Society, London.
Darwin, C. R. (1859, 1872). On the origin of species by means of natural selection, or the preservation of favoured races in the struggle for life. John Murray, London. (Sixth, standard edition, 1872.)
Darwin, C. R. (1862, 1877). On the various contrivances by which British and foreign orchids are fertilized by insects, and on the good effects of intercrossing. John Murray, London. (2nd Edition, 1877 with modified title and changes to text).
Darwin, C. R. (1865, 1880). On the movements and habits of climbing plants. John Murray, London. (1880 publication with title "The power of movements in plants" with substantial changes and additions and contributions from Francis Darwin.)
Darwin, C. R. (1868). The variation of animals and plants under domestication. 2 Vols. John Murray, London.
Darwin, C. R. (1871). The descent of man, and selection in relation to sex. 2 Vols. John Murray, London.
Darwin, C. R. (1872). The expression of emotions in man and animals. John Murray, London.
Darwin, C. R. (1875). Insectivorous plants. John Murray, London.
Darwin, C. R. (1876). The effects of cross and self fertilization in the vegetable kingdom. John Murray, London.
Darwin, C. R. (1877). The different forms of flowers on plants of the same species. John Murray, London.
Darwin, C. R. (1881). The formation of vegetable mould through the action of worms. John Murray, London.
Darwin, C. R. (1887). The life of Erasmus Darwin.... Being an introduction to an essay on his scientific work. John Murray, London. (2nd Edition of Krause, E. 1879.)
Darwin, C. R. (1887). The Life of Erasmus Darwin, with an essay on his scientific works by E. Krause, James Murray, London, 2nd Ed.
Darwin, C. R. (1888). Autobiography. Murray, London.
Darwin, E. (1803). The Loves of Nature: or, the Origins of Society: a poem with philosophical notes. J. Johnson, London.
Darwin, F. (1887). The Life and Letters of Charles Darwin (including an Autobiography by Charles Darwin). 3 volumes. James Murray, London. (For the Autobiography, see the unexpurgated text in Barlow, 1958, and King-Hele, 2005)
Darwin, F. (1892). The autobiography of Charles Darwin and selected letters. James Murray. London. (See the unexpurgated text of the Autobiography in Barlow, 1958, and King-Hele, 2005.)
Darwin, F. (1909). The foundations of 'The Origin of Species'; two essays written in 1842 and 1844. Cambridge University Press, Cambridge, UK.
Darwin, F. & Seward, A. C. (1903). More letters of Charles Darwin. 2 Vols. Cambridge University Press, Cambridge, UK.
Dennett, D. C. (1995). Darwin's dangerous idea: evolution and the meanings of life. Simon & Schuster, New York.
Desmond, A. (1989). The Politics of Evolution. Univ of Chicago Press. ISBN 0-226-14346-5.
Desmond, A. (1994). Huxley: Vol. 1 The Devil's Disciple, London: Michael Joseph, ISBN 0-7181-3641-1

Desmond, A. & Moore, J. (1991). *Darwin*. Michael Joseph, Penguin Group, London, ISBN 0-7181-3430-3.
Desmond, R. (1999). Sir Joseph Dalton Hooker: Traveller and Plant Collector. Antique Collectors' Club and The Royal Botanic Gardens, Kew. ISBN 1-85149-305-0
Dolan, B. (2004). Wedgwood: The First Tycoon, Viking Adult, ISBN 0-670-03346-4.
Ellegård A. (1990). Darwin and the general reader. New edition of the original publication of 1958. University of Chicago Press, Chicago.
Fitton, R.S. and Wadsworth, A.P. 1958. The Strutts and Arkwrights, 1758-1830: A study of the early factory system. Manchester Univ Press.
Flint, A. S. (1883). Memoirs of the Mudge family. Netherton and Worth (Plymouth Library Archives).
Foster, J. (1888). Alumni Oxonienses, 1715-1886. Parker and Co., Oxford, UK.
Freeman, R. B. (1977). The works of Charles Darwin. An annotated bibibliographical handlist. Dawson Publishing, Folkestone, UK
Freeman, R. B. (1978). Charles Darwin. A Companion. Dawson Publishing, Folkestone, UK
Freeman, R. B. (1984). Darwin Pedigrees. (Reprinted by R B Freeman from a privately printed edition of 60 copies of "*Pedigree of the family of Darwin*" by H Farnham Burke, 1888.)
Ghiselin, M. T. (1969). The triumph of the Darwinian method. University of Chicago Press, Chaicago. ISBN 0-226-29024-7.
Glover, S. (1831 & 1833). A History and Gazeteer of Derbyshire. 2nd Ed. Vol. 1 & 2. Derby.
Glass, B., Temkin, O. and Strauss, W. L. (1959). Forerunners of Darwin. The Johns Hopkins Press, Baltimore.
Gribbin, J. and Gribbin, M. (2003). FitzRoy: the remarkable story of Darwin's captain and the invention of the weather forecast. Headline Book Publishing, London ISBN 0-7553-1182-5.
Healey, E. (2001). Emma Darwin. The inspirational wife of a genius. Headline Book Publishing, London. ISBN 0 7472 6248 9.
Herbert, S. (1980). The Red Notebooks of Charles Darwin. British Museum (Natural History), London.
Herbert, S. (2005). Charles Darwin, geologist. Cornell University Press.
Hopkins, M. A. (1952). Dr. Johnson's Lichfield. Peter Owen Ltd.
Hutton, William (1836). The History of Birmingham. J Guest.
Himmelfarb, G. (1959). Darwin and the darwinian revolution. W W Norton & Co, New York.
Keynes, Randal (2001). Annies's box. Charles Darwin, his daughter and human evolution. Fourth estate, London. ISBN 1-84115-660-4
Keynes, Richard (Ed)(1988). Charles Darwin's Beagle Diary. Cambridge University press, Cambridge, UK.
Kohn, D. (1985). Darwin's principle of divergence as an internal dialogue. Kohn, D., Ed., *The darwinian heritage*, pp 245–257. Princeton University Press.
Kottler, M. (1985). Charles Darwin and Alfred Russel Wallace: two decades of debate over natural selection. Kohn, D., Ed., *The darwinian heritage*, pp. 367-432. Princeton University Press.
King-Hele, D., (1999). Erasmus Darwin. A life of unequalled achievement. Giles de la Mare, London. ISBN I 900 3 57-08-9.
King-Hele, D. (Ed.) (2003). Charles Darwin's the Life of Erasmus Darwin, Cambridge University Press, Cambridge. ISBN 0-521-81526-6.
King-Hele, D. (Ed.) (2005). Autobiography of Charles Darwin. CUP.
King-Hele, D. (Ed.) (2007). The Collected Letters of Erasmus Darwin. Cambridge University Press, Cambridge. ISBN 0-521-82156-8.
Krause, E. (1879). Erasmus Darwin. Translated from the German . . . with a preliminary essay by Charles Darwin. John Murray, London. (see also King-Hele, 2003.)
Lamarck, Jean Baptiste Pierre Antoine de Monet de (1815–22). Histoire naturelle des animaux sans vertèbres. 7 Vols. Paris.
Litchfield, H. (1915). Emma Darwin: A century of family letters. Cambridge University Press, Cambridge, UK.

Logan, D and Sanders, V. (Eds.) (2007) The collected letters of Harriet Martineau. Pickering and Chatto Pub., London.

Manier, E. (1978). The young Darwin and his cultural circle. D Reidel Pub Co, Dordrecht. ISBN 90-277-0856-8.

Mann, P. G. (1983). James Manby Gully, M. D. privately printed. ISBN 0 9508938 0 3.

Meteyard, E. (1871). A group of Englishmen (1795-1815) being records of the younger Wedgwoods and their friends. London.

Metcalf, Richard. (1906). The rise and progress of hydropathy in England and Scotland. Simpkin, Marshall, Hamilton, Kent & Co., London.

Moore, J. R. (1977). On the education of Darwin's sons: The correspondence between Charles Darwin and the Reverend G. V. Reed, 1857-1864. Notes and Records of the Royal Society of London, 32, 41-70.

Moore, J. R. (1989). Of love and death: why Darwin "gave up Christianity". J. Moore, Ed. *History, humanity and evolution*, 195-229. Cambridge Universtiy Press.

Nichols, Peter (2003). Evolution's Captain: The Dark Fate of the Man Who Sailed Charles Darwin Around the World. HarperCollins. ISBN 0-06-008877-X.

Ormerod, G. (1882) The history of the county palatine and city of Chaster. 2nd edition, revised and enlarged by Thomas Helsby. 3 Vols. G. Routledge & Sons, London.

Ospovat, D. (1981). The development of Darwin's theory: natural history, natural theology, and natural selection, 1839-1859. Cambridge University Press.

Page, W. (Ed) (1970). Derbyshire, Vol. 2 in the series "The Victoria History of the Counties of England. Published for the University of London, Institute of Historical Research. Ed. William Page, FSA. Reprinted from the original edition of 1905 by Dawsons of Pall Mall, London.

Peile, J. (1913). Bibliographical Register of Christ's College, 1505-1905. Vol. II 1666-1905. Cambridge University Press, Cambridge, UK.

Pickering, G. (1974). Creative malady: illness in the lives and minds of Charles Darwin, Florence Nightingale, Mary Baker Eddy, Sigmund Freud, Marcel Proust and Elizabeth Barrett Browning. Allen & Unwin, London.

Pinney, T. (1974). The letters of Thomas Babington Macaulay. 7 volumes. Cambridge University Press. See Vol. 1, pp 122-3 & p. 169.

Rackham, H. (1927). Statutes of Christ's College. Farr and Tyler, Cambridge.

Raverat, G. (1952). Period piece: a Cambridge childhood. Faber and Faber, London. ISBN 1-904555-12-8

Reynolds, D. (Ed.) (2004). Christ's; A Cambridge college over five centuries. Macmillan, Basingstoke. ISBN 0 333 98988 0

Romanes, Ethel (1896). Life and letters of George John Romanes. Longmans, Green, London.

Rudwick, M. (1982). Charles Darwin in London: the integration of public and private science. Isis 75, 186-206.

Ruse, M. (1975). Darwin's debt to philosophy: an examination of the influence of the philosophical ideas of John F. W. Herschel and William Whewell on the development of Charles Darwin's theory of evolution. Studies in history and philosophy of science 6, 159-181.

Schofield, Robert E. (2004). The Enlightened Joseph Priestley: A Study of His Life and Work from 1773 to 1804. University Park: Pennsylvania State University Press. ISBN 0271024593.

Secord, J. A. (2000). Victorian sensation. The University of Chicago press, Chicago. ISBN 0-226-74410-8

Shipley, A. C. (1924). Cambridge Cameos. Jonathan Cape, London.

Slotten, R. A. (2004). The Heretic in Darwin's court. The life of Alfred Russel Wallace. Columbia University Press, N.Y. ISBN 0-231-13010-4.

Stafford, Robert A. (1989). Scientist of Empire. Cambridge University Press, Cambridge, UK.

Stauffer, R. C., Editor (1975). Charles Darwin's Natural selection: being the second part of his big species book written from 1856 to 1858. Cambridge University Press, Cambridge, UK.

Stephen J. F. (1828). A systematic catalogue of British insects: being an attempt to arrange all the hitherto discovered indigenous insects in accordance with their natural affinities. Containing

also the references to every English writer on entomology, and to the principal foreign authors. With all the published British genera to the present time. London.
Sturges, R P, (1979). The membership of the Derby Philosophical Society, 1783-1802. Midland History 4, 212–229.
Thompson, Harry (2005). This Thing of Darkness. Headline Review, London. ISBN 0-7553-0281-8.
Wallace, A. R. (1871). Natural selection. MacMillan & Co, London.
Walters, S.M. and Stow E.A. (2001). Darwin's Mentor. Cambridge University Press, Cambridge, UK. ISBN 0-521-59146-5
Webb, R. K. (1980). Modern England: from the eighteenth century to the present. George Allen & Unwin, London
Wedgwood B and Wedgwood H, (1980). The Wedgwood Circle: four generations of a family and their friends, 1730–1897. Collier, London. ISBN 0029902304.
Woodham-Smith, C. (1991). *The Great Hunger, 1845-49.* Penguin, London.
Venn, J. A. 1940–1954. Alumni Cantabrigienses. Part II, 1752-1900. Cambridge University Press, Cambridge.

Index

N

Quotations from the Darwin/Fox correspondence

Charles Darwin

William Darwin Fox